COMPREHENSIVE SURGICAL IN-SERVICE AND BOARD REVIEW

COMPREHENSIVE SURGICAL IN-SERVICE AND BOARD REVIEW

A GUIDE TO THE 99TH PERCENTILE

Ravi Morchi, MD

OXFORD
UNIVERSITY PRESS

OXFORD
UNIVERSITY PRESS

Oxford University Press is a department of the University of Oxford. It furthers the University's objective of excellence in research, scholarship, and education by publishing worldwide. Oxford is a registered trade mark of Oxford University Press in the UK and certain other countries.

Published in the United States of America by Oxford University Press
198 Madison Avenue, New York, NY 10016, United States of America.

Library of Congress Cataloging-in-Publication Data
Names: Morchi, Ravi, author.
Title: Comprehensive surgical in-service and board review :
a guide to the 99th percentile / Ravi Morchi.
Description: New York, NY : Oxford University Press, [2024] |
Includes bibliographical references and index.
Identifiers: LCCN 2023020439 (print) | LCCN 2023020440 (ebook) |
ISBN 9780197630525 (paperback) | ISBN 9780197630549 (epub) |
ISBN 9780197630556 (online)
Subjects: MESH: Surgical Procedures, Operative—methods |
Biological Phenomena | Study Guide
Classification: LCC RD37.2 (print) | LCC RD37.2 (ebook) | NLM WO 18.2 |
DDC 617.0076—dc23/eng/20230706
LC record available at https://lccn.loc.gov/2023020439
LC ebook record available at https://lccn.loc.gov/2023020440

DOI: 10.1093/med/9780197630525.001.0001

Printed by Integrated Books International, United States of America

FSC
MIX
Paper
FSC® C183721

CONTENTS

LIST OF FIGURES

ABBREVIATIONS

AA	Arachidonic acid
AAA	Abdominal aortic aneurysm
AAs	Amino acids
AB	Antibody
ABI	Ankle-brachial index
ACE	Angiotensin converting enzyme
ACh	Acetylcholine
ACL	Anterior cruciate ligament
ACT	Activated clotting time
ADH	Antidiuretic hormone
ADH	Atypical ductal hyperplasia
ADLs	Activities of daily living
ADP	Adenosine diphosphate
AFB	Acid fast bacilli
AFLP	Acute fatty liver of pregnancy
AFP	Alpha fetoprotein
AG	Aminoglycoside
AG	Angiotensin
AG	Anion gap
AKI	Acute kidney injury
AKs	Actinic keratoses
ALH	Atypical lobular hyperplasia
AMP	Adenosine monophosphate
ANOVA	Analysis of variance
ANP	Atrial natriuretic peptide
APC	Antigen presenting cell
APC	Argon plasma coagulation
APCT	abdominal pelvic CT scan
APR	Abdominoperineal resection
APRV	Airway pressure release ventilation
ARDS	Acute respiratory distress syndrome
ASA	Aspirin
ASD	Atrial septal defect
ASIA	American spinal injury impairment
ASIS	Anterior superior iliac spine
ATG	Antithymocyte globulin
ATN	Acute tubular necrosis
AVM	Arteriovenous malformation
AVN	Avascular necrosis
AxLND	Axillary lymph node dissection
AXR	Abdominal x-ray
BBI	Brachio-brachial index
BCA	Brachiocephalic artery
BCC	Basal cell carcinoma
BCS	Budd Chiari syndrome
BCV	Brachiocephalic vein
BD	Bowen's disease
B2	Billroth-II
BLQ	Bilateral lower quadrant
BMI	Body mass index
BMS	Bare metal stent
BP	Biliopancreatic
BP	Blood pressure
BPH	Prostatic hypertrophy
BRTO	Balloon retrograde transvenous obliteration
BT	Blalock-Taussig
BTM	Biodegradable temporizing matrix
CABG	Coronary artery bypass graft surgery
CAD	Coronary artery disease
CAH	Congenital adrenal hyperplasia
CAM	Confusion assessment method
cAMP	Cyclic adenosine monophosphate
cAMP PDE	Cyclic AMP phosphodiesterase
CaO_2	Arterial oxygen content
CBD	Collecting bile duct
CCA	Cholangiocarcinoma
CCK	Cholecystokinin
CCY	Cholecystectomy
CD	Collecting ducts
CDD	Choledochoduodenostomy
CDH	Congenital diaphragmatic hernia
CEA	Carcinoembryonic antigen
CEA	Carotid endarterectomy
CF	Cystic fibrosis
CFU	Colony forming unit
CGD	Chronic granulomatous disease
cGy	Centigray
CHA	common hepatic artery
CHD	Common hepatic duct
CHF	Congestive heart failure
CHX	Chlorhexidine
CI	Confidence interval
CIA	Common iliac artery
CIS	Carcinoma in situ
CKD	Chronic kidney disease
CLISA	Central line insertion site assessment
CMV	Cytomegalovirus
CN	Cranial nerve

CNBx	Core needle biopsy
CND	Central neck dissection
CNS	Central nervous system
CO	Cardiac output
COHb	Carboxyhemoglobin
COMT	Catechol-O-methyl-transferase
COPD	Chronic obstructive pulmonary disease
COX	Cyclooxygenase
CP	Cricopharyngeus
CPAP	Continuous positive airway pressure
CPB	Cardiopulmonary bypass
CPP	Cerebral perfusion pressure
CRH	Corticotropin releasing hormone
CRRT	Continuous renal replacement therapy
CRS	Cytoreductive surgery
CSF	Cerebrospinal fluid
CSL	Complex sclerosing lesion
CST	corticospinal tract
CT	Computed tomography
CTA	CT angiogram
CTAP	CT abdomen and pelvis
CTD	Connective tissue disorder
CTZ	Chemoreceptor trigger zone
CVA	Cerebrovascular accident
CVP	Central venous pressure
CXR	Chest x-ray
D1	Duodenal segment 1
D5W	D5 water
DA	Duodenal atresia
DAI	Diffuse axonal injury
DBP	Diastolic blood pressure
DCD	Donation after cardiac death
DCIS	Ductal carcinoma in situ
DCT	Distal convoluted tubule
DD	Duodenoduodenostomy
DDAVP	Desmopressin acetate
DES	Drug eluting stent
DFSP	Dermatofibrosarcoma protuberans
DHEA	Dehydroepiandrostenedione
DHFR	Dihydrofolate reductase
DI	Diabetes insipidus
DIC	Disseminated intravascular coagulation
DIP	Distal interphalangeal joint
DIP	Distal interphalangeal joint
DIT	Di-iodotyrosine
DJ	Duodenojejunostomy
DLCO	Diffusion capacity for carbon monoxide
DO_2	Oxygen delivery
DOAC	Direct oral anticoagulant
DPL	Diagnostic peritoneal lavage
DRE	Digital rectal exam
DSA	Donor specific antibody
DS-BPD	Duodenal switch-biliopancreatic diversion
DTI	Direct thrombin inhibitor
DAPT	Dual antiplatelet therapy
DVT	Deep vein thrombosis
EAC	External auditory canal
EBSLN	External branch of the superior laryngeal nerve
EBUS	Endobronchial ultrasound
EBV	Epstein Barr Virus
EC	Enterochromaffin cells
ECF	enterocutaneous fistula
ECL	Enterochromaffin-like
ECMO	Extracorporeal membrane oxygenation
ECT	Ecarin clotting time
EDH	Epidural hematoma
EDV	End diastolic velocity
EEG	Electroencephalogram
EF	Ejection fraction
EGD	Esophagogastroduodenoscopy
EGJOO	Esophagogastric junction outflow obstruction
EIA	External iliac artery
EIV	External iliac vein
EMG	Electromyography
EMR	Endoscopic mucosal resection
ePTFE	Expanded polytetrafluoroethylene
ER	Endoplasmic reticulum
ERCP	Endoscopic retrograde cholangiopancreatography
ERV	Expiratory reserve volume
ESD	Endoscopic submucosal dissection
ESR	Erythrocyte sedimentation rate
ESRD	End stage renal disease
$ETCO_2$	End-tidal CO_2
ETT	Endotracheal tube
EUA	Exam under anesthesia
EUS	Endoscopic ultrasound
EXBx	Excisional biopsy
FAP	Familial adenomatous polyposis
FAST	Focused abdominal sonogram for trauma
FB	Foreign body
FENa	Fractional excretion of sodium
FEUrea	Fractional excretion of urea
FEV1	Forced expiratory volume in 1 second
FFA	Free fatty acid
FFP	Fresh frozen plasma
FHF	Fulminant hepatic failure
FHH	Familial hypocalciuric hypercalcemia
FIT	Fecal immunohistochemical testing
FLR	Functional liver remnant
FMD	Fibromuscular dysplasia
FNA	Fine needle aspiration
FNH	Focal nodular hyperplasia
FO	Fossa ovalis
FOBT	Fecal occult blood test
FPL	Flexor pollicis longus
FRC	Functional residual capacity
FSP	Fibrin split products

FTSG	Full thickness skin graft
FVC	Forced vital capacity
GABS	Group A beta hemolytic streptococci
GAN	Greater auricular nerve
GB	Gallbladder
GCA	Giant condylomata accuminata
GCS	Glasgow Coma Scale
GDA	Gastroduodenal artery
GEJ	Gastroesophageal junction
GER	Gastroesophageal reflux
GERD	Gastroesophageal reflux disease
GFN	Genitofemoral nerve
GFR	Glomerular filtration rate
GH	Growth hormone
GI	Gastrointestinal
GIB	Gastrointestinal bleed
GIST	Gastrointestinal stromal tumor
GJ	Gastrojejunal / gastrojejunostomy
GNR	Gram-negative rods
GOO	Gastric outlet obstruction
GPC	Gram-positive cocci
GPR	Gram-positive rods
GSV	Greater saphenous vein
GSW	Gunshot wounds
GVHD	Graft versus host disease
H2B	Histamine blocker
HA	Hepatic arteries
HAP	Hospital associated pneumonia
HBcAb	Hepatitis B core antibody
HBIG	Hepatitis B immune globulin
HBO	Hyperbaric oxygen
HBsAb	Hepatitis B surface antibody
HBsAg	Hepatitis B surface antigen
HBV	Hepatitis B virus
HCAP	Healthcare associated pneumonia
HCC	Hepatocellular carcinoma
HCl	Hydrochloric acid
HCV	Hepatitis C virus
HD	Hemodialysis
HFOV	High-frequency oscillating ventilation
HGB	Hemoglobin / Hgb hemoglobin
HGD	High grade dysplasia
HIPEC	Hyperthermic intraperitoneal chemotherapy
HIT	Heparin-induced thrombocytopenia
HJ	Hepaticojejunostomy
HLA	Human leukocyte antigen
HM	Heineke-Mikulicz
HNPCC	Hereditary non-polyposis colon cancer
HPO	Hypertrophic pulmonary osteoarthropathy
HPV	Human papilloma virus
HR	Heart rate
HRA	High-resolution anoscopy
HS	Hidradenitis suppurativa
HSV	Herpes simplex virus
HSV	Highly selective vagotomy
HTX	Hemothorax
HU	Hounsfield units
HV	Hepatic vein
I-131	Radioactive iodine
IABP	Intra-aortic balloon pump
IBC	Invasive breast cancer
IBD	Inflammatory bowel disease
IBSLN	Internal branch of the superior laryngeal nerve
IBW	Ideal body weight
IC	Ileocecal
IC	Inspiratory capacity
ICAM	Intracellular adhesion molecule
ICG	Indocyanine green
ICP	Intracranial pressure
IDC	Invasive ductal carcinoma
IDDM	Insulin-dependent diabetes mellitus
IEA	Inferior epigastric artery
IF	Intrinsic factor
IFN	Interferon
IFN-a	Interferon alpha
IHC	Immunohistochemistry
ILC	Invasive Lobular Carcinoma
IM	Intramedullary
IMA	Inferior mesenteric artery
IMRT	Intensity modulated radiation therapy
IMV	Inferior mesenteric vein
INH	Isoniazid
iNO	Inhaled nitric oxide
IOC	Intraoperative cholangiogram
IOP	Intraocular pressure
IP	Interphalangeal
IPAA	Ileal pouch anal anastomosis
IPC	Intermittent pneumatic compression
IPH	Intraparenchymal hematoma
IPH	Intraparenchymal hematoma
IPMN-B	Intraductal papillary mucinous neoplasms of biliary origin
IPOM	Intraperitoneal onlay
IRA	Ileorectal anastomosis
ITC	Isolated tumor cells
ITP	Idiopathic thrombocytopenic purpura
IUGR	Intrauterine growth retardation
IVC	Inferior vena cava
IVDU	Intravenous drug user
IVH	Intraventricular hemorrhage
IVIG	Intravenous immune globulin
JGA	Juxtaglomerular apparatus
JJ	Jejunojejunostomy
LA	Left atrial/atrium
LAD	Left anterior descending
LAR	Low anterior resection
LCIS	Lobular carcinoma in situ
LDH	Lactate dehydrogenase

LES	Lower esophageal sphincter
LGD	Low grade dysplasia
LGI	Lower gastrointestinal
LGIB	Lower gastrointestinal bleed
LHA	Left hepatic artery
LIFT	Ligation of intersphincteric tract
LIMA	Left internal mammary
LLQ	Left lower quadrant
LN	Lymph node
LOS	Length of stay
LMWH	Low molecular weight heparin
LPR	Laryngopharyngeal reflux
LPV	Left portal vein
LR	Lactated ringers
LR-LPJ	Limited resection and longitudinal pancreatojejunostomy
LST	Lateral spinothalamic tract
LT	Ligament of Treitz
LUQ	Left upper quadrant
LV	Left ventricle
LV	Left ventricular
LVAD	Percutaneous left ventricular assist device
LVAD	Left ventricular assist device
LVEDV	Left ventricular end diastolic volume
MA	Mafenide acetate
MAC	Membrane attack complex
MAC	Minimum alveolar concentration
MACE	Major adverse cardiac events
MALT	Mucosa-associated lymphoid tissue
MAP	Mean arterial pressure
MCHC	Mean corpuscular hemoglobin concentration
MCT	Medium chain triglyceride
MCV	Mean corpuscular volume
MDR	Multidrug resistant
MELD	Model for end stage liver disease
MEN	Multiple endocrine neoplasia
MFH	Malignant fibrous histiocytoma
MG	Myasthenia gravis
MI	Myocardial infarction
MIE	Minimally invasive esophagectomy
MIS	Müllerian inhibitory substance
MIT	Mono-iodotyrosine
mIVF	Maintenance intravenous fluid
MMC	Migrating motility complex
MMF	Maxillomandibular fixation
MMF	Mycophenolate mofetil
MMSE	Mini-mental status exam
MOF	Multi-organ failure
MPNST	Massive transfusion protocol (MTP)
MRCP	Magnetic resonance cholangiopancreatography
MRI	Magnetic resonance imaging
MRI	Magnetic resonance image/imaging
MRM	Modified radical mastectomy
MRND	Modified radical neck dissection
MRSA	Methicillin-resistant staph aureus
MSI	Micro satellite instability
MSSA	Methicillin-sensitive *S. aureus*
MV	Minute ventilation
MVO_2	Mixed venous oxygen saturation
NAC	N-acetylcysteine
NAC	Nipple areolar complex
NAFLD	Non-alcoholic fatty liver disease
NASH	Non-alcoholic steatohepatitis
NCS	Nerve conduction studies
nDI	Nephrogenic diabetes insipidus
NEC	Necrotizing enterocolitis
NF	Neurofibromatosis
NGT	Nasogastric tube
NH	Nursing home
NHL	Non-Hodgkin's Lymphoma
NIF	Negative inspiratory force
NJ	Nasojejunal
NK	Natural killer (cells)
NMJ	Neuromuscular junction
NMJ	Neuromuscular junction
NOTES	Natural orifice transluminal endoscopic surgery
NPO	Nil per os (nothing by mouth)
NPPV	noninvasive positive pressure ventilation
NPWT	Negative pressure wound therapy
NS	Normal saline
NSAIDs	Nonsteroidal anti-inflammatory medications
NSE	Neuron-specific enolase
NSM	Nipple sparing mastectomy
OCP	Oral contraceptive
OM	Otitis media
OPSI	Overwhelming post-splenectomy sepsis
ORIF	Open reduction and internal fixation
OSA	Obstructive sleep apnea
PA	Pulmonary artery
PAF	Platelet activating factor
PAIR	Percutaneous transhepatic aspiration and alcohol or hypertonic saline injection with re-aspiration
PALS	periarteriolar lymphatic sheaths
pAO_2	Alveolar pressure of oxygen
PaO_2	Arterial pressure of oxygen
PAP	Pulmonary artery pressure
PBD	Percutaneous biliary drain
PCC	Prothrombin concentrate complex
PCL	Posterior cruciate ligament injuries
PCR	Polymerase chain reaction
PCT	Proximal convoluted tubule
PCWP	Pulmonary capillary wedge pressure
PD	Pancreatic duct
PD	Parkinson's disease
PDA	Patent ductus arteriosus

PDGF	Platelet derived growth factor
PE	Pulmonary emboli
PEEP	Positive end expiratory pressure
PEG	Percutaneous endoscopic gastrostomy
PEG	Polyethylene glycol
PEG-IFN	Pegylated interferon
PEG-J	Percutaneous endoscopic gastrojejunostomy
PEP	Post-exposure prophylaxis
PET	Positron emission tomography
PFD	Pelvic floor dysfunction
PFO	Patent foramen oval
PFT	Pulmonary function test
PGI	Prostaglandin
PHA	Proper hepatic artery
PICC	Peripherally inserted central catheter
PIP	Proximal interphalangeal joint
PJ	Pancreatojejunostomy
PMNs	Polymorphonuclear cells
PNA	Pneumonia
pNET	Pancreatic neuroendocrine tumor
PNMT	Phenylethanolamine N-methyltransferase
POD	Postoperative day
POEM	Per oral endoscopic myotomy
PONV	Postoperative nausea and vomiting
PP	Pancreatic polypeptide
PPCT	Pancreatic protocol CT
PPD	Purified protein derivative
PPFC	Peri-pancreatic fluid collection
PPHN	Persistent pulmonary hypertension
PPI	Proton pump inhibition
PPV	Patent processus vaginalis
PRA	Panel reactive antibody
pRBCs	Packed red blood cells
PSA	Prostate specific antigen
PSARP	Posterior sagittal anorectoplasty
PSC	Sclerosis cholangitis
PSC	Primary sclerosing cholangitis
PSDSS	Peritoneal surface disease severity score
PSV	Peak systolic velocity
PT	Patellar tendon
PT	Prothrombin time
PTC	Percutaneous transhepatic cholangiography
PTFE	Polytetrafluoroethylene
PTH	Parathyroid hormone
PTHrP	Parathyroid hormone related protein
PTHrP	PTH related peptide
PTLD	Post-transplant lymphoproliferative disorder
PTU	Propylthiouracil
PTX	Pneumothorax
PUD	Peptic ulcer disease
PV	Portal vein
PV	Pulmonary vein
PVE	portal vein embolization
PVR	Pulmonary vascular resistance
Q	Perfusion
QALY	Quality-adjusted life year
qSOFA	Quick SOFA
QT	Quadriceps tendon
RA	Right atrium
RCA	Right coronary artery
RCRI	Revised cardiac risk index
RCT	Randomized controlled trial
RDS	Respiratory distress syndrome
R/Q	Respiratory quotient
RDW	Red cell distribution width
RER	Rough endoplasmic reticulum
RES	Reticuloendothelial system
RFA	Radiofrequency ablation
RHA	Right hepatic artery
RIF	Rifampin
RLD	Right lateral decubitus
RLN	Recurrent laryngeal nerve
RLQ	Right lower quadrant
RPF	Retroperitoneal fibrosis
RPV	Right portal vein
RS	Radial scar
RSBI	Rapid shallow breathing index
RSI	Rapid sequence induction
RUQ	Right upper quadrant
RV	Residual volume
RV	Right ventricle
RVF	Rectovaginal fistula
RxY	Roux-en-Y
RxY GB	Roux-en-Y gastric bypass
RxY HJ	Roux-en-Y hepaticojejunostomy
SA	Sino-atrial
SA	Splenic artery
SA	Staphylococcus aureus
SAH	Subarachnoid hemorrhage
SaO_2	Arterial saturation of hemoglobin
SB	Sengstaken-Blakemore
SB	Small bowel
SBFT	Small bowel follow-through
SBP	Spontaneous bacterial peritonitis
SCA	Subclavian artery
SCC	Squamous cell carcinoma
SCFA	Short chain fatty acids
SCIP	Surgical care improvement program
SCJ	Squamocolumnar junction
SCM	Sternocleidomastoid
$ScVO_2$	Central venous oxygen saturation
SDH	Subdural hematoma
SE	*Staphylococcus epidermidis*
SEMS	Self-expanding metal stent
SEPS	Self-expanding plastic stent
SGLT	Sodium glucose transporter
SIADH	Syndrome of inappropriate antidiuretic hormone
SILS	Single incision laparoscopic surgery
SIRS	Systemic inflammatory response syndrome

SLNB	Sentinel lymph node biopsy
SLND	Selective lymph node dissection
SMA	Superior mesenteric artery
SMV	Superior mesenteric vein
SN	Silver nitrate
SOFA	Sequential organ failure assessment
SQ	Subcutaneous
SR	Sarcoplasmic reticulum
SRS	Stereotactic radio surgery
SS	Silver sulfadiazine
SSI	Surgical site infection
SSM	Skin sparing mastectomy
STC	Slow transit constipation
STSG	Split thickness skin graft
SUV	Standardized uptake values
SV	Stroke volume
SV	Splenic vein
SVC	Superior vena cava
SVG	Saphenous vein grafts
SVR	Systemic vascular resistance
T&S	Type and screen
TACE	Transarterial chemoembolization
TACE	Transarterial chemoembolization
TAE	Transanal excision
TAH BSO	Total abdominal hysterectomy and bilateral salpingo-oophorectomy
TAMIS	Transanal minimally invasive surgery
TAP	Transversus abdominis plane
TAPP	Transabdominal preperitoneal repair
TAPVR	Total anomalous pulmonary venous return
TARE	Transarterial radioembolization
TB	Total bilirubin
TB	Tuberculous mycobacteria
TBG	Thyroid binding globulin
TBI	Toe-brachial [blood pressure] index
TBI	Traumatic brain injury
TBSA	Total body surface area
TBW	Total body weight
TCD	Transcranial Doppler
TE	Tracheoesophageal
TEF	Tracheoesophageal fistula
TEG	Thromboelastography
TEM	Transanal endoscopic microsurgery
TEN	Toxic epidermal necrolysis
TEP	Total extraperitoneal repair
TF	Tissue factor
TFT	Thyroid function test
TH	Helper T cells
TG	Thyroglobulin
TGD	Thyroglossal duct cyst
TI	Terminal ileum
TIPS	Transjugular intrahepatic portosystemic shunt
TLC	Total lung capacity
TM	Tympanic membrane
TME	Total mesorectal excision
TMJ	Temporal-mandibular joint
TPA	Tissue plasminogen activator
TPN	Total parenteral nutrition
TPO	Thyroid peroxidase
TRALI	Transfusion related acute lung injury
TRAM	Transverse rectus abdominis muscle
TRH	Thyrotropin releasing hormone
TSH	Thyroid stimulating hormone
TTP	Thrombotic thrombocytopenic purpura
TV	Tidal volume
TVUS	Transvaginal ultrasound
TXA	Tranexamic acid
TXAn	Thromboxane
UC	Ulcerative colitis
UES	Upper esophageal sphincter
UGI	Upper gastrointestinal
UGIB	Upper gastrointestinal bleed
UPPP	Uvulopalatopharyngoplasty
URI	Upper respiratory tract infection
US	Ultrasound
UTI	Urinary tract infection
UV	Ultraviolet
VA	Venoarterial
VAP	Ventilator associated pneumonia
VARD	Video assisted retroperitoneal debridement
VATS	Video assisted thoracic surgery
VHL	Von Hippel-Lindau
VMA	Vanillylmandelic acid
VO_2	Oxygen consumption
VPI	Velopharyngeal insufficiency
VRE	Vancomycin-resistant enterococci
VSD	Ventricular septal defect
VTE	Venous thromboembolism
VWD	von Willebrand disease
VWF	von Willebrand factor
VZV	Varicella zoster virus
WBAT	Weight bearing as tolerated
WBCs	White blood cells
WBRT	Whole breast radiation
WLE	Wide local excision
XDR	Extensive drug resistance
XI	Accessory nerve
XRT	Radiation therapy
ZMC	Zygomaticomaxillary complex

FUNDAMENTALS

1.

CELLULAR, TISSUE, AND BODY FLUID FUNDAMENTALS

CELL BIOLOGY

- Cells have a lipid bilayer membrane with cholesterol for fluidity. The interior of a cell is negative compared to the extracellular environment. The cytosol is high in K and HPO_4.
- Cell wall antigens include **glycolipids** comprising the ABO blood types, and **glycoproteins** comprising human leukocyte antigen (HLA) types A, B, C, and DP, DQ, and DR.
- **Desmosomes** are cell membrane proteins that provide cell–cell adherence via cytoskeletal elements, and **hemidesmosomes** provide cell adherence to the extracellular matrix and basement membrane.
- **Tight junctions** are formed by proteins that allow neighboring cell membrane fusion. They prevent movement of ions, thereby producing a polarity with distinct apical, lateral, and basal sides.
- **Gap junctions** are composed of connexin subunits functioning as aqueous channels that allow communication between the cytoplasm of neighboring cells.
- A variety of **cell membrane receptors** exist, including ligand activated ion channels, ligands triggering enzyme activity, and ligands triggering intra-membrane G proteins that transduce an intracellular signaling cascade involving phosphatidylinositol or cAMP.
- **Intracellular receptors** may be in the cytosol or nucleus, and are bound by ligands that have entered the cell's interior.
  - Steroid hormones are bound to carrier globulins that cross the cell membrane and bind to a cytosolic receptor. The complex then moves into the nucleus to effect nucleic acid transcription factors.
  - Thyroid hormones cross both the cell and nucleus membranes and directly bind to receptors within the nucleus, regulating transcription factors.
- The **cytoskeleton** is a complex latticework of microfilaments, microtubules, and intermediate filaments. It runs from the nucleus throughout the cytoplasm to provide the cell with structure and deformability. It interconnects with neighboring cells through cell membrane proteins to provide tissue and organ turgor.
  - **Microtubules** provide pathways for intracellular transport, form the mitotic spindle and anchoring chromosomes, and contribute to cellular motion.
  - Intermediate filaments include desmin in muscle, keratin in hair and nails, and vimentin in fibroblasts.
- The **sarcomere** is a specialized construct of myosin thick filaments sliding along actin thin filaments. It is responsible for contraction of skeletal and cardiac muscle.
- The **cell cycle** is composed of four phases:
  - **G1 phase** is the most variable in duration, and determines cycle length. It is affected by growth factors and may go to a quiescent G0 phase in some tissue.
  - **S phase** is a time of protein synthesis and chromosomal duplication, pairing, and matching.
  - **G2 phase** is a pause between chromosome duplication and mitosis.
  - **M phase** encompasses mitosis and consists of multiple stages.
    - **Prophase** is when chromosomes undergo **centromere attachment** and spindle formation. **Metaphase** is when **chromosomes align. Anaphase** is when they are pulled apart along microtubules. **Telophase** is when two separate nuclei reform.
- **Apoptosis** refers to programmed cell death. It is a normal part of the cellular lifespan and serves to control cell populations during development and aging, keeping tissues and organs in balance. Apoptosis occurs via two main pathways.
  - The extrinsic pathway involves CD8 T cell death receptor binding at the cell surface, followed by the activation of intracellular caspases resulting in nuclear DNA breakdown.
  - The intrinsic pathway is via intracellular Bcl proteins, increased mitochondrial permeability leaking cytochrome C, and intracellular caspase activation dismantling nuclear DNA.
  - Once DNA is fragmented, the cell is phagocytosed by macrophages with no extracellular inflammatory response.
  - Apoptosis is facilitated by p53 and inhibited by Bcl-2.
- The **nucleus** is the largest organelle, delineated by its own membrane and nucleoplasm. It houses DNA tightly complexed with protein (chromatin).

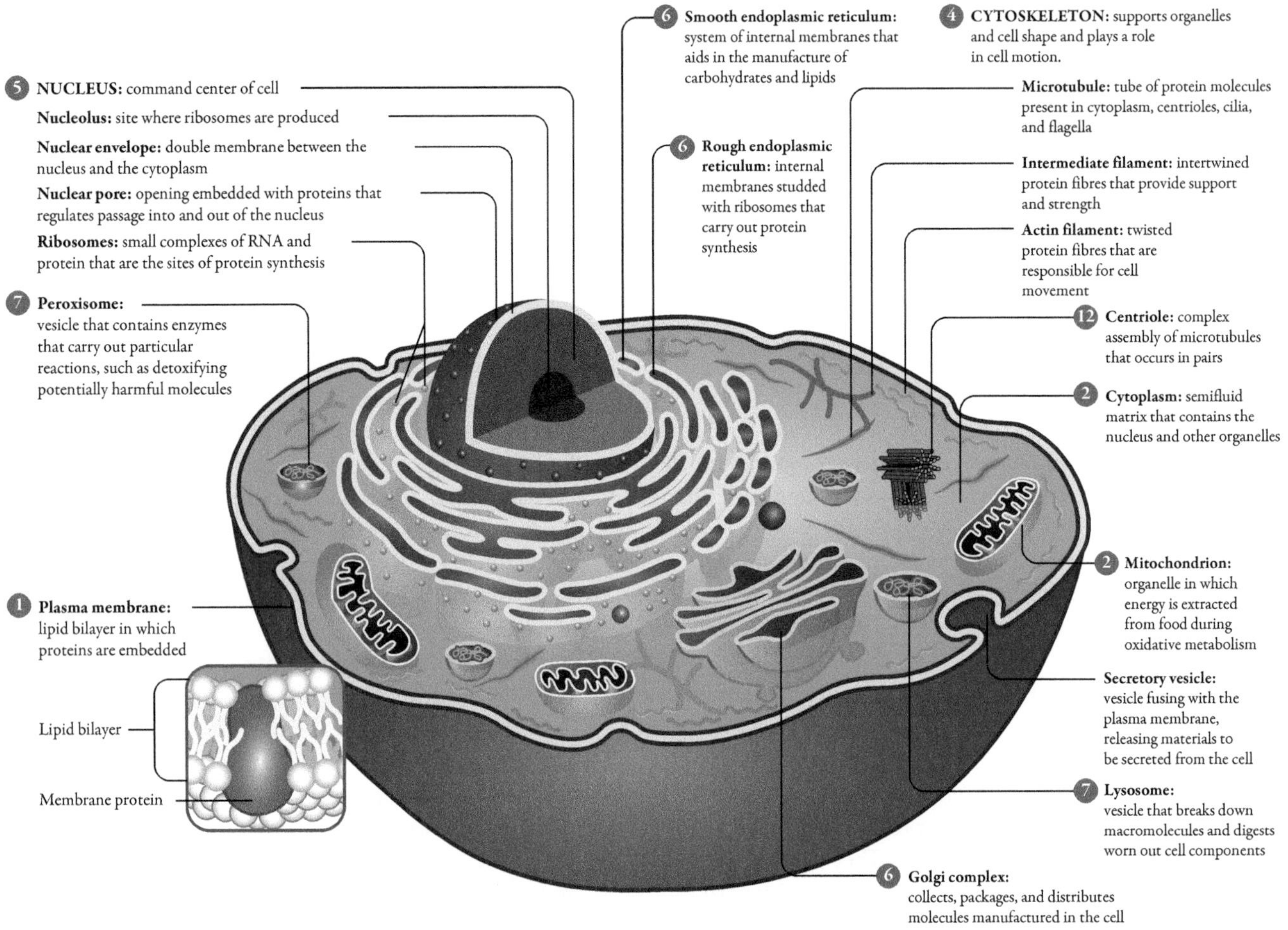

Figure 1.1 Cellular architecture.

- Its outer membrane is contiguous with the cytoplasmic **smooth endoplasmic reticulum,** which functions in lipid and steroid synthesis, gluconeogenesis, and detoxification of endocytosed compounds.
- DNA and RNA consist of a ribose and phosphate backbone with purine (adenine or guanine) and pyrimidine (cytosine, thymidine, or uracil) interfacing with one another.
  - **Transcription of DNA to mRNA** is coordinated by topoisomerase, initiation factors, RNA polymerase, and transcription factors. DNA polymerase carries out DNA duplication.
  - Inaccuracies in nucleic acid transcription and duplication can result in single base substitution, addition, or subtraction. Such errors can lead to missense or nonsense mutations. The former result in an amino acid substitution at the time of protein translation. The latter result in a stop codon that ends protein translation.
- The **nucleolus** resides within the nucleus and is the site of ribosome production.
- **Translation of mRNA to amino acids (AAs)** requires ribosomal rRNA and tRNA complexed to the corresponding AA.
  - Cytosolic **free ribosomes** translate proteins for intracellular use.
  - The **rough endoplasmic reticulum (RER)** contains ribosomes responsible for translating proteins destined for exportation.
- The **Golgi apparatus** modifies proteins with carbohydrates and shuttles them to lysosomes or the cell surface for exportation.
- **Lysosomes** degrade proteins received from the Golgi apparatus, from fusion with ingested phagosomes, or from fusion with pinocytosed endosomes. Degradation can be enzyme-mediated or from O_2-dependent radicals. Proteins not degraded by endolysosomes may be recycled on MHC II molecules to the surface of an antigen presenting cell (APC).
- **Glycolysis** takes place in the cytosol, and 2ATP are generated. Pyruvate and NADH produced enter the mitochondria.

- **Mitochondria** take products of glycolysis from the cytosol and run the **Krebs cycle (tricarboxylic acid cycle)** within its inner membrane. The NADH and FADH2 produced deliver electrons to the inner membrane transport chain. Oxygen uptake at the end of the transport chain allows for ATP production.
- **Gluconeogenesis** takes place in the cytosol. Amino acids and lactic acid are converted to glucose. AcetylCoA, however, cannot be used for gluconeogenesis.

INFLAMMATION

- The inflammatory response is composed of a number of local factors and cytokines produced in response to tissue damage of any kind (mechanical trauma, thermal, infectious). This is followed by recruitment of neutrophils, macrophages, and lymphocytes which augment the inflammation, increase vascular permeability, and increase coagulability.
- Tissue injury exposes collagen, releases platelet activating factor (**PAF) from endothelium**, and liberates **tissue factor,** which in turn activates the extrinsic coagulation pathway.
- **Growth factors** are released from platelets, fibroblasts, and epidermal cells. They are chemotactic for neutrophils, macrophages, and fibroblasts, and promote epithelialization and angiogenesis.
- **Nitric oxide (NO)** is produced from arginine during the inflammatory process. NO activates guanylate cyclase leading to increased cGMP and local vasodilation, increasing blood flow.
- **Kininogen and bradykinin** result in vasodilation and increased capillary permeability. They are inactivated by angiotensin converting enzyme (ACE). Of note, patients using ACE-inhibitors can have increased bradykinin levels, vascular permeability, and angioedema.
- **Phospholipids** in cell walls are broken down to **arachidonic acid (AA)** by phospholipase. *See Figure 1.2.* **Glucocorticoids** inhibit phospholipase activity on phospholipid, reducing arachidonic acid production.
  - AA can then be broken down to **prostaglandins (PG) and thromboxane (txa)** by cyclooxygenase (COX). PG and TXAn cause **vasodilation, bronchodilation,** increased capillary permeability, and fever. COX is inhibited reversibly by NSAIDs and irreversibly by aspirin.
  - Alternative to COX, lipoxygenase activity on AA produces **leukotrienes.** These are chemotactic and slow anaphylatoxins that cause **vasoconstriction, bronchoconstriction,** and increased capillary permeability.

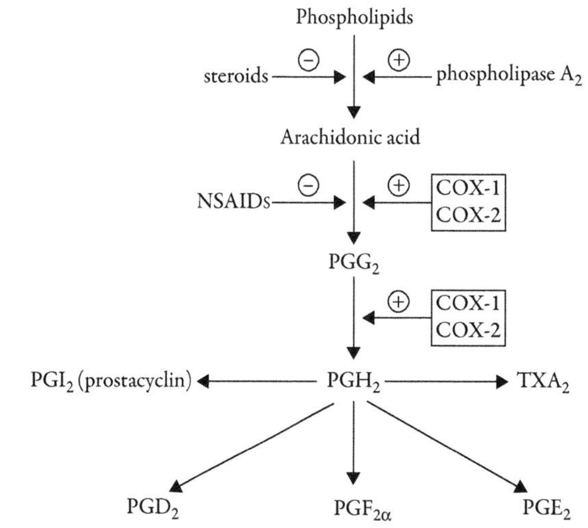

Figure 1.2 Arachidonic acid pathway.

- **Norepinephrine and dopamine** are released in response to afferent nerves from the site of injury, contributing to the sympathetic response to injury.
- **Endothelin** is a potent vasoconstrictor originating from endothelial cells. It is also pro-inflammatory and possibly assists the production of oxygen radicals (below).
- **Cytokines** are compounds produced for cell-to-cell communication during the inflammatory response. They act in autocrine, paracrine, and endocrine fashion at the level of cell surface receptors. Any given cytokine can be produced by a number of different types of cells.
  - **TNF-alpha** primarily comes from macrophages. Other cells can produce it to a lesser extent. It activates PMNs, lymphocytes, and other macrophages. TNF-alpha is a pro-coagulant, mediates cachexia, increases NO and PAF, and can cause multisystem organ failure. TNF-alpha inhibitors include **infliximab or adalimumab** which are used for inflammatory bowel disease (IBD) treatment.
  - **IL-1** is from macrophages and synergizes with TNF-alpha. It is responsible for PGE2 mediated hypothalamic fever, which can be inhibited by NSAIDs. IL-1 Increases acute phase reactants and activates fibroblasts.
  - **IL-2** is produced by T lymphocytes and activates other lymphocytes. It results in increased Ig production, improved gut integrity, and has anti-tumor effects.
  - **IL-4** is from B cells and is anti-inflammatory.
  - **IL-6** increases **hepatic acute phase proteins** (CRP, fibrinogen, haptoglobin, ceruloplasmin, alpha-1 antitrypsin, ferritin, C3) and decreases albumin, prealbumin, and transferrin.
  - **IL-8** activates **intrapulmonary inflammatory cells** and is implicated in acute respiratory distress

syndrome (ARDS). IL-8 levels can be predictive of multi-organ failure (MOF).
  - **IL-10** is produced by lymphocytes, basophils, and mast cells. It is **anti-inflammatory and decreases IL-1 and IFN-alpha** as well as decreasing MHC II. The latter leads to macrophage and monocyte hyporesponsiveness for up to 7 days.
  - **Interferon (IFN)** activates macrophages, natural killer (NK) cells, and CD8 cytotoxic T cells. The result is that **viral replication, viral protein translation, and its associated intracellular signaling, are inhibited.** IFN also enhances Th1 differentiation to help fight neoplasia. There are several types of IFN (alpha from macrophages, beta from epithelial cells, fibroblasts, and macrophages, gamma from T lymphocytes and NK cells).
- **Complement (C1–C9)** are intravascular proteins that function to augment the antibody response and assist phagocytes in clearing micro-organisms.
  - The classic pathway starts with IgG or IgM binding to an antigen. This activates C1, C2, C4, then C3 and assembly of C5–C9 as a membrane attack complex (MAC) resulting in organism cell membrane destabilization.
  - The alternate pathway starts with factors B, D, and P activating C3 and finally the C5–C9 MAC.
  - C3a, C4a, C5a produced during this cascade function as anaphylatoxins that mediate vasoconstriction, bronchoconstriction, and increased capillary permeability. It also promotes mast cell and basophil histamine release, as well as TNF-alpha, IL-1, and eicosanoid production. C3a and C5a are chemotactic, and C3b and C4b function in opsonization.
- **Oxygen radicals** include superoxide anion and hydrogen peroxide. They are generated in an inflammatory state and can help fend against microbes, but can also damage endogenous cells. Reducing agents like superoxide dismutase, catalase, glutathione peroxidase, and vitamin A and E can help control oxygen radical levels in tissue. Patients with NADPH oxidase defects have trouble making oxygen free radicals.
- **Platelets** arrive early in the inflammatory response and adhere to exposed collagen via GP 1b receptors. They contain dense granules with adenosine and serotonin and release platelet derived growth factor (PDGF) for PMN, macrophage, fibroblast, and smooth muscle recruitment as well as angiogenesis and epithelialization. PAF4 and TXAn contribute to aggregation with other platelets via GP2b3a receptors. Platelets have a lifespan of 7–10 days.
- **Polymorphonuclear cells (PMNs)** include neutrophils, basophils, and eosinophils. They have a lifespan of 7 days in blood and 1–2 days once they are recruited into tissue.
  - Diapedesis is the process by which cells migrate from the vascular lumen, through endothelium, and into target tissue. L-selectin (leukocytes) or P-selectin (platelets) bind to E-selectin (endothelium) for rolling adhesion as the cell begins the process of diapedesis. Beta2-integrin on cells then binds intracellular adhesion molecule (ICAM) on the endothelium for anchoring adhesion and trans-endothelial migration of the cell, allowing it to enter the sub-endothelial tissue space.
- **Macrophages** play a dominant role in coordinating inflammation and wound healing by production of the cytokines TNF-alpha and IL-1. Macrophage and monocyte activity can be decreased for up to one week after the initial inflammatory response.
- **Postoperative/Post-traumatic Phases**
  - The **ebb phase** is characterized by hypothalamic sympathetic nervous system activation leading to vasoconstriction, sodium retention, and antidiuretic hormone (ADH) release. This lasts for 24–36 hrs postoperatively or post-injury.
  - The **early flow phase** is from catabolic endocrine and adrenal medullary release of catecholamines and can last for days.
  - **Late flow phase** is marked by anabolism and diuresis and can last for days or weeks.

WOUND HEALING

- Tissue Injury first results in immediate **vasoconstriction** of injured vessels; this is followed by vasodilation of intact surrounding vasculature.
- **Inflammatory phase** is the first week after injury. **Platelets** arrive in minutes and release PDGF and other chemoattractants. **PMNs** arrive in 24–48 hours unless inhibited (uncontrolled diabetics). **Macrophages** arrive in 48–72 hours. Fibronectin acts as a chemoattractant for them. **Lymphocytes** arrive in 5–7 days. Mesenchymal cells start making **type 3 collagen** from intracellular procollagen. Chronic wounds are halted in the inflammatory phase and do not progress due to infection, malnutrition, foreign bodies, glucocorticoids, or low tissue oxygen tension.
- **Proliferative phase** is 1–3 weeks after injury. **Fibroblasts** arrive and produce type 1 collagen, which replaces type 3. Neovascularization proceeds, and **granulation tissue** is laid down to allow for further epithelialization over the wound and eventual secondary intent closure. Epithelialization begins at hair follicles, glands, and wound edges, and proceeds at 1–2 mm/day.
- **Remodeling (maturation phase)** is from 3 weeks to 1 year. Collagen content has been maximized, and **collagen deposition and breakdown occur at equal pace.** Vitamin C–dependent **cross-linking of type 1 collagen** augments scar strength up to 80% of the original tissue strength. This peaks by 8 weeks and contributes to tensile strength for primary intent

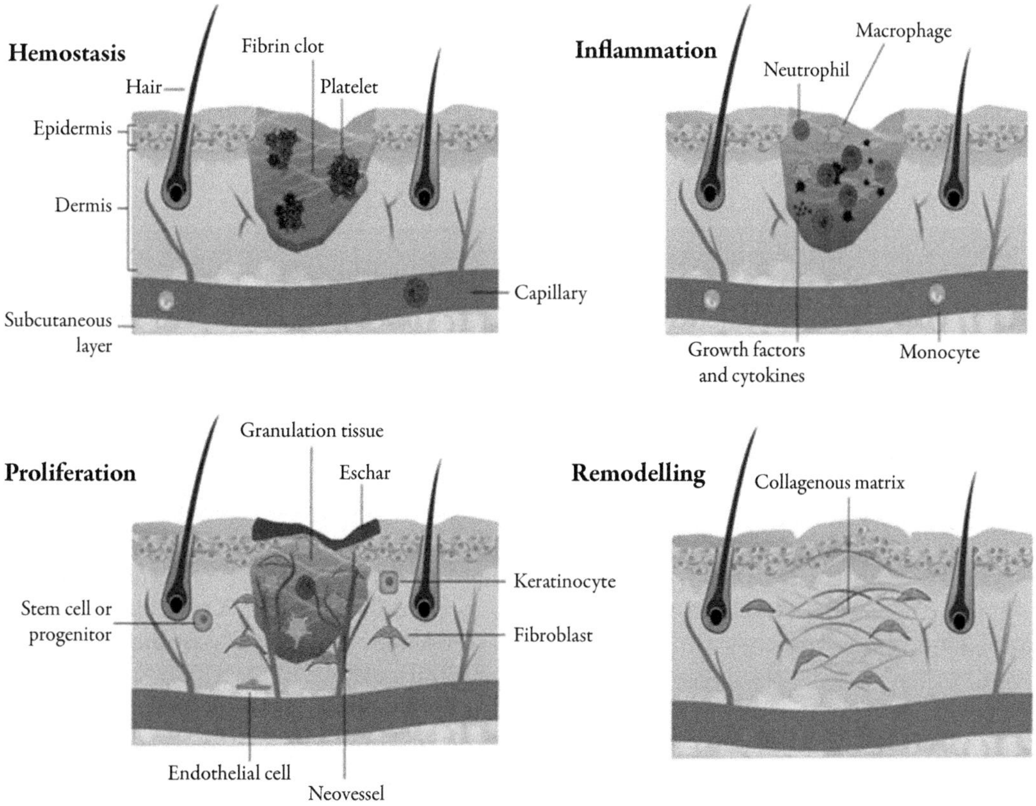

Figure 1.3 Stages of wound healing.

closure. There is **decreased vascularity** to the wound, and **myofibroblasts** contribute to **wound contraction**. This occurs to a greater extent in the perineum compared with other injured sites. Nerve regeneration proceeds at 1 mm/day.

- **Collagen.** Alpha-ketoglutarate, oxygen, and vitamin C for hydroxylation of proline are necessary components to collagen production. Decreased deposition is associated with low tissue oxygen (tcO_2 <25 mmHg), low lactate, and low growth factor.
  - Type 1 is present in skin, bone, tendon, and healed wounds. Osteogenesis imperfecta is a deficiency of type 1 collagen.
  - Type 2 is in cartilage, and nourished via diffusion.
  - Type 3 is present early in wound healing and comprises vasculature.
  - Type 4 is in the basement membrane.
  - Type 5 is in the cornea.
  - Ehlers-Danlos has multiple subtypes that cover all variants of collagen insufficiency.
- **Vitamin A** promotes epithelialization and reverses the effects of glucocorticoids.
- **Zinc** contributes to granulation tissue and RNA/DNA polymerization.
- Iron and Vitamin E are not essential. Anemia does not affect wound healing.
- Denervation and chemotherapy initiated more than two weeks after tissue injury or surgery have no effect on healing.
- Radiation given 4 weeks prior to, or more than 4 weeks following, surgery will not affect wound healing.
- **Impediments to healing** include bacterial infection, necrotic tissue, diabetes and glucocorticoids inhibiting the inflammatory phase of healing, albumin <3, immune suppression (impaired DNA/RNA synthesis), or any form of ischemia producing low tissue oxygen content.
- Scar tissue is made up of proteoglycans, hyaluronic acid, and type 1 collagen.
  - **Hypertrophic scars** are characterized by an excess of type 1 collagen within the confines of the scar. It is not genetically determined but rather due to wound infection, burns, or impaired healing. Treatment is with lidocaine and steroid injection early on, silicone sheeting, pressure garments, or re-excision of the scar.
  - **Keloid scars** are characterized by excess type 1 collagen that extends beyond the confines of scar. There is a **genetic component to keloids**. Treatment is with lidocaine and steroid injections early on, intralesional chemotherapeutics, silicone sheeting, pressure garments, radiation, or excision and local tissue rearrangement for closure.

- Primary closure is generally with suture for tensile strength. Secondary closure is via granulation tissue formation and epithelialization.
  - **Plain gut** is from sheep submucosa and cow serosa. It has the shortest absorption time.
  - **Vicryl (polyglactin)** is a synthetic multifilament suture with a **70-day** absorption time.
  - **Chromic** is chromium infused gut suture with a **100-day** absorption time.
  - **Monocryl (poliglecaprone)** is a synthetic monofilament with **100-day** absorption time.
  - **PDS (polydiaxanone)** is a synthetic monofilament that slowly absorbs over **180 days.** It is used for abdominal fascial closure in a continuous format with bites taken 5 mm off the edge and every 5 mm along the cranial-caudal length of the wound. Bites should be en masse. The overall suture to wound length ratio should be 4:1.
  - **Nylon, polypropylene, and polyester** sutures are synthetic non-absorbable sutures.

NUTRITION

- **Ideal body weight (IBW)** is calculated as 106 lbs. + 6 lbs. per inch over 5 feet, in men. For women, it is 100 lbs. + 5 lbs. per inch over 5 feet.
- Nutritional parameters include the following:
  - **Albumin:** half-life of 18 days and so provides a long-term assessment of nutrition.
  - **Transferrin:** half-life of 10 days.
  - **Prealbumin:** half-life of 2 days, so it is a good indicator of acute nutritional status.
- Poor preoperative nutritional status should be suspected in the setting of the following:
  - Albumin <3.
  - 10%–15% weight loss over the preceding 6 months.
  - Body weight <85% of IBW.
  - In such cases, patients may need **preoperative supplementation** of megesterol, tube feeds, or total parenteral nutrition (TPN).
- Caloric intake should be 50% carbohydrates, 30% lipids, and 20% protein.15% of all protein intake should be composed of essential amino acids.
  - **Normal caloric needs** are 20–25 kcal/kg/day and 0.8–1 gm protein/kg/day of actual body weight for non-obese patients.
  - **Trauma patients,** including those that are sedated and intubated, need 25–30 kcal/kg/day and 1–1.5 gm of protein/kg/day actual body weight.
  - **Septic patients** need 30–35 kcal/kg/day and 1.5–2.0 gm protein/kg/day actual body weight.
  - **Burn patients** have the highest needs at 35–40 kcal/kg/day and 2.0–2.5 gm protein/kg/day.
    - Alternatively, burn patient caloric needs can be calculated based on their normal, baseline daily requirement plus 30 kcal/kg/day for every 1% surface area burned, and **3** gm protein/kg/day for every 1% of surface area burned
  - For **obese patients** with a BMI >30, one should use **adjusted body weight** instead of actual body weight, since the excess adipose is not as metabolically active. Adjusted body weight is calculated as IBW + 1/4 (actual weight − IBW). Using this adjusted weight, caloric needs are 10–15 kcal/kg/day and 2 gm protein/kg/day. This is a **hypocaloric high protein diet.**
- **Resting energy expenditure** (or resting metabolic rate) is the number of calories burned at rest over a 24 hour, non-active period and corresponds to the organism's basal metabolic rate. It can be calculated using the Harris Benedict equation or measured indirectly via a metabolic cart that analyzes respired gases. Respiratory quotient (R/Q) is the ratio of CO_2 production to O_2 consumption, and is accurate provided there is no pleural air leak.
  - **R/Q >1** indicates excess intake of carbohydrate calories and will yield lipogenesis.
  - **R/Q of 0.85** represents mixed protein, lipid, and carbohydrate utilization.
  - **R/Q of 0.8** is primarily protein metabolism, and **0.7** is primarily fat metabolism.
- Basal metabolism is increased by 10% for every degree body temperature above 38C.
- **Enteral feeding** is the ideal means of providing calories in the peri-operative period. Enteral nutrients prevent mucosal atrophy, lymphoid atrophy, and villi shortening. It maintains brush border enzyme activity and mucosal IgA levels. It also decreases bacterial growth, systemic infection from bacterial translocation, prolonged ventilation, and ICU stay. One may start feeding early postoperatively, as long as cardiac preload is adequate and there are limited vasopressors in use.
  - **Oral feeding** is preferred, unless the patient cannot swallow safely, cannot take adequate quantities orally, or does not have an intact proximal enteric tract. Artificial enteric feeding is then employed to meet caloric needs (below).
  - **Nasogastric** feeding is sufficient if mentation and laryngeal reflexes are intact. Otherwise **post-pyloric** feeding into the duodenum or jejunum can decrease aspiration pneumonia in these patients.
  - Gastric tube residuals should be <200 ml every 4–6 hours. High residuals indicate that the foregut is not peristalsing sufficiently, and the metabolic demands of feeding could outweigh its energy supply.
  - If the patient is likely to need feeding for >30 days, then a **percutaneous endoscopic or radiologic gastrostomy or gastrojejunostomy** can be placed.

- Percutaneous endoscopic gastrojejunostomy (PEG-J) tubes enter the stomach and have a lumen within the stomach as well as a second lumen distally along the tube within the jejunum. This allows feeding distally into the jejunum, and venting of the stomach to prevent distension. Such tubes suffer from malposition and occlusion more frequently than simple gastric tubes.
- **Bolus** feeds can be given at 25–500 ml each time via a gastric route. Jejunal feeding should not be bolused, since there is little reservoir capacity to the small bowel.
- **Intermittent** feeding can be given over a half-hour each time.
- **Continuous** feeding is at 10–50 ml/hour via a gastric or jejunal route.
- Elemental formulas can be used for those patients with poor gastrointestinal function or prolonged malnutrition. A protein-rich formula may assist with muscle preservation.

- **Total parenteral nutrition** should be started for adult patients who are **NPO >7–10 days** or have **baseline malnutrition** or **loss of >15% body weight** in the recent past, and inability to receive any form of enteral feeding due to intestinal inactivity, discontinuity, or suspect viability.
  - **Glucose based** parenteral formulas have 50% glucose and 10% amino acids plus Na, Cl, K, Ca, Mg, PO_4, acetate.
  - **Lipids** are given separately at 10% (1 kcal/cc) or 20%. Excessive lipids may saturate the reticuloendothelial system (RES) and decrease immune function.
  - TPN can result in cholestasis producing cholelithiasis, biliary cirrhosis, portal hypertension, and hepatic dysfunction. These changes are generally worse in infants. Carnitine can help treat hepatic steatosis.
  - Metabolic bone disease, nephrolithiasis, catheter-related thromboses, and bloodstream infections are other long-term complications.
- Essential minerals and vitamins and symptoms of their deficiencies are noted below:
  - **Chromium:** hyperglycemia, encephalopathy, neuropathy.
  - **Copper:** pancytopenia.
  - **Selenium:** cardiomyopathy and skeletal muscle weakness.
  - **Phosphate**: skeletal muscle weakness and RBC hemolysis.
  - **Zinc**: poor wound healing, and an intertriginous and perioral eczematous rash.
  - **Niacin:** dermatitis, diarrhea, dementia (known as pellagra).
  - **Iron**: microcytic anemia.
  - **Fatty acids: scaly dermatitis, alopecia,** and thrombocytopenia.
  - **Vitamin A** normally promotes epithelialization, prevents night blindness, and negates the effects of glucocorticoids. Deficiency results in failure of maturation of epithelial surfaces.
  - **Vitamin B_1**: Wernicke's encephalopathy, cardiomyopathy, and neuropathy.
  - **Vitamin B_6:** sideroblastic anemia, glossitis, and neuropathy.
  - **Vitamin B_{12} and folate**: megaloblastic anemia, glossitis, and neuropathy.
  - **Vitamin C** is an antioxidant. It decreases crystalloid requirements in critically ill burn patients, increases lymphocyte and PMN activity, and assists in collagen production. Deficiency results in collagen defects.
  - **Vitamin D** deficiency results in rickets and osteomalacia. The D2/3 parent compound is taken either orally or liberated from the skin by UV light. This is converted to 25(OH)D by the liver and measured in blood. The kidney will convert 25(OH) D to 1,25(OH)2D and then to 1,24,25 (OH)3D (calcitriol). Calcitriol increases calcium reabsorption from the intestine.
  - **Vitamin E**: neuropathy.
  - **Vitamin K**: coagulopathy.

DIGESTION

- The gastrointestinal tract produces a large amount of endogenous secretions, independent of any oral intake. Water intake averages 2 L–2.5 L in a day.
  - Saliva accounts for 1.5 L, gastric 2.5 L, bile 0.5 L, pancreas about 0.7 L–1.5 L, and small bowel 1.5 L–3 L.
  - Small bowel reabsorbs 9 L of fluid and the colon 1 L–2 L. Stool losses account for about 0.5 L.
- Insensible losses include sweat and evaporation from respiration and can range from 0.5 L–4 L.
- **Starvation** results in lipid breakdown and an R/Q of 0.7. Brain, myocardium, renal, and skeletal muscles shift to ketone use during starvation.The peripheral nervous system (PNS), adrenal medulla, RBC, and WBC remain obligate glucose users.
  - **Refeeding syndrome** is characterized by decreased K, Mg, and Phos secondary to insulin-mediated intracellular shift of these ions once carbohydrate feeding is initiated and the patient is transitioning to anabolism. The result of this intravascular ion depletion is cardiac, neural, and muscle deficiencies that can lead to death.
  - Results of long-term malnutrition are **kwashiorkor** (protein deficiency) and **marasmus** (general calorie deficiency). The latter are irritable, weak, and emaciated.
  - Cachexia is mediated by TNF-alpha production in chronic inflammation or malignancy.

- **Physiologic stress, physical injury, or surgery** all result in a mixed lipid, protein, and carbohydrate breakdown with an R/Q of 0.85. This is due to cortisol and systemic catecholamines. As a result, a glucose infusion does not necessarily spare proteolysis or halt gluconeogenesis in these stressed states.
- **Carbohydrates** should comprise 50%–60% of daily caloric intake.
  - Dietary fiber consists of **fructooligosaccharide** complex carbohydrates that remain in the enteric lumen for bacterial degradation to **short chain fatty acids.** These can be used by colonocytes.
  - Digestion of carbohydrates is with salivary **amylase**, pancreatic amylase, and brush border **disaccharidases** which help produce monosaccharides. In this manner, lactose is converted to glucose and galactose, maltose to two glucose molecules, and sucrose to fructose and glucose.
  - Secondary active transport via **sodium glucose transporter (SGLT)** in intestinal mucosa and renal tubule cells is insulin independent. This allows for the movement of glucose and galactose into the bloodstream.
  - Various other **glucose transporters** exist. GLUT1 is in RBC, neurons, and endothelium, GLUT2 in hepatocytes and intestine, and GLUT3 in neurons. **GLUT4 is insulin dependent** and present in adipose, skeletal, and cardiac muscle. GLUT5 receptor is responsible for facilitated diffusion of fructose across the intestinal mucosa.
  - Intracellular glucose metabolism can follow a few pathways:
    - Hexose monophosphate shunt producing nucleotides.
    - Glycolysis-producing pyruvate, and lactate, and a small amount of ATP.
    - Or, ultimately, the Krebs cycle producing acetyl cod and NADH/FADH2, which go on to deposit electrons on the inner mitochondrial membrane and generate larger quantities of ATP.
  - **Glycogen** stores are two-thirds in muscle and one-third in liver. The latter has glucose-6-phosphatase, so the glucose produced can actually be exported from the cell for use elsewhere.
  - **Gluconeogenesis** is the process by which lactate, alanine, or other amino acids are converted to glucose. Deamination of glutamine or glutamate allow the kidneys to contribute to gluconeogenesis in late starvation. The **Cori cycle** is when hepatocytes utilize lactate generated from glycolysis anaerobically in muscle, and convert it back to glucose to be transported back to muscle for use.
- **Protein** should comprise 10%–20% of caloric intake.
  - Digestion begins with **pepsin** from gastric chief cells, then **trypsin** (generated from pancreatic trypsinogen by enterokinase). Trypsin activates **chymotrypsin and carboxypeptidases** for further protein breakdown.
  - Intraluminally, proteins are digested to polypeptides and then oligopeptides and tri- and dipeptides or single amino acids. Intracellularly, peptides are broken down to urea or single AAs, or may be reorganized into new protein.
  - Glutamine is important for **small bowel enterocyte and immune cell nutrition.** It also functions to transport NH_4 to kidneys for nitrogen excretion. This produces glutamate.
  - AAs like alanine contribute to gluconeogenesis after **deamination in late starvation.**
  - Nitrogen balance is calculated as protein intake/6.25 gm – (24 hrs. urine nitrogen + 4 g).
    - Normal endogenous protein synthesis is about 250 gm/day.
    - Urinary N_2 is **7–10 gm/day, but can be as high as 30 gm/day in starvation or stress** due to muscle protein breakdown. Urinary N_2 decreases to 20 gm/day once ketones are being used by the brain, myocardium, kidney, and skeletal muscle.
  - Branched chain AA include leucine, isoleucine, and valine. They are metabolized in muscle and hence do not contribute to hepatic encephalopathy.
  - Essential AAs include phenylalanine, Valine, tryptophan, threonine, isoleucine, methionine, histidine, arginine, leucine, lysine.
- **Lipids** should comprise 30%–40% of total calories.
  - **Triacylglycerols (TAG)** are composed of a glycerol backbone and fatty acid side chains.
  - **Cholesterol** is the other component to fat, and used by the body to make hormones and bile salts.
  - Lipid digestion starts with pancreatic lipase, cholesterol esterase, and phospholipase. This produces **free fatty acids (FFA)** and glycerol.
    - Unsaturated fatty acids are a structural component to the cell membrane, so fatty acid metabolism does affect cell membrane structure and immunocyte function. Essential fatty acids in this regard are linoleic (omega-6) and linolenic (omega-3).
    - Saturated fatty acids are used for energy.
  - **Micelles** are aggregates of bile salts (cholic and chenodeoxycholic acid), long chain FFA, and monoacylglycerol. Micelles fuse with the enterocyte membrane and help absorb lipids as well as fat-soluble vitamins ADEK. *See Figure 1.4.*
  - **Chylomicrons** are formed within enterocytes and are composed of 90% TAG-long chain FFA plus 10% phospholipids, cholesterol, and protein. Chylomicrons enter lymphatics on the basal side of the enterocyte.
  - Medium and short chain FFA are complexed with albumin within the enterocyte and enter the portal system complexed with albumin, heading directly to the liver.

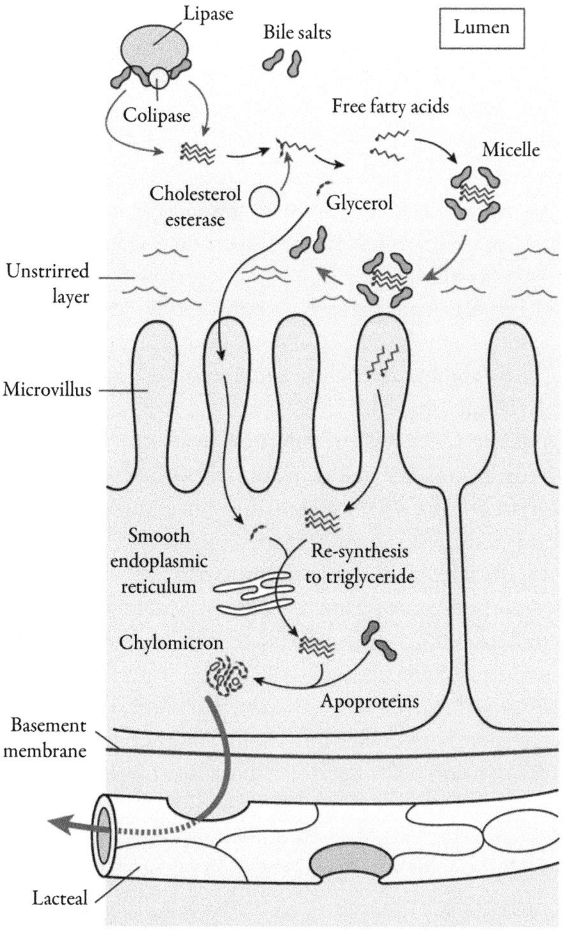

Figure 1.4 Intestinal fat absorption.

- **Lipoprotein lipase** resides on the endothelium of liver and adipose tissue and breaks down circulating TAG in the bloodstream to FFA and glycerol for absorption into the cell. Its activity is increased by insulin.
  - **FFA binding protein** on endothelium of liver and adipose binds medium and short chained FFA. Its production is inhibited by trauma, sepsis, and TNF.
- The **carnitine shuttle** moves cytoplasmic FFA into mitochondria for breakdown to acetyl-CoA (AcCoA). This enters the tricarboxylic acid cycle (TCA). If no 3-carbon skeletons are available to combine with AcCoA, then it is transported as acetoacetyl-CoA to the liver, where deacylase produces ketones. Ketones can be used by the brain, myocardium, kidney, and skeletal muscle.
- **Hormone sensitive lipase** is activated by growth hormone (GH), cortisol, and catecholamines, and inhibited by insulin. It breaks down intracellular TAGs to FFA and glycerol, which are released into the bloodstream.

FLUIDS & ELECTROLYTES

- Na and K determine the intracellular to extracellular osmotic gradient.
- Proteins determine intravascular to interstitial gradient across the venular side of the capillary membrane. The intracellular environment has increased cations and negatively charged proteins, compared to the external environment.
- Normal plasma osmolarity is 275–295 per liter of H_2O and can be calculated as 2(Na+k) + BUN/2.8 m + glc/18. Osmolality, on the other hand, is measured per kg H_2O.
- Urine osmolality is from 100–1200 mOsm/kg. If **>500 then serum osmolality is probably also high.**
- Water comprises 60% of total body weight (TBW) in men and infants, but only 50% of TBW in women.
  - **Dehydration** is defined as low TBW, while **hypovolemia** refers to a low intravascular volume.
  - Two-thirds of TBW is intracellular, while one-third is extracellular. Two-thirds of extracellular fluid is interstitial and one-third is intravascular.
  - Average daily **insensible loss** is 600 ml, but can be up to 1.5–4 L lost in perspiration and respiration. Loss increases by 10% for every 1C increase in temperature. 500–1000 ml/hour evaporate during a laparotomy. Insensible losses can be replaced with D5W.
  - Rehydration should be with isotonic crystalloid (NS, LR, Plasma-Lyte).
  - For maintenance IV fluid rate after rehydration is complete, use 4 ml/kg/hour for the first 10 kg weight, then add 2 ml/kg/hour for the next 10 kg weight. Thereafter, above 20 kg, you can add 1 ml/kg/hour. All resuscitation and maintenance fluids should be **isotonic crystalloid,** except for **neonates who can receive 1/4 NS for maintenance.**
- GI losses include:
  - **Saliva** totaling 1.5 L and high in K.
  - **Gastric juices** totaling 2.5 L and high in HCl but low in Na. Known gastric losses should be replaced with **D51/2NS + 20 KCL.**
  - **Pancreatic secretions** totaling 0.5–1 L and **bile** totaling 0.5–1 L. Both are high in Na and HCO_3 and can be replaced with lactated ringers (LR).
  - **Small bowel** secretions totaling 0.5–1 L excreted and up to 9 L reabsorbed. Lost small bowel fluid should be replaced with LR.
  - Stool totaling about 250 ml.
  - **Ileostomy output or colonic diarrhea** can total 1 L–3 L, and are low in Na but high in K. Such losses can be replaced with LR + 20 KCL.
- **Potassium (K)** requirement is **0.5–1 mEq/kg/day.** It is primarily an intracellular ion, with toxicity determined by excess or deficient *serum* concentrations. K shifts out

of cells during acidemia, and into cells (for excretion by the renal tubular epithelium) in the setting of alkalemia.

- Serum hyperkalemia produces characteristic peaked T waves followed by a widened QRS and then a slow wide complex rhythm, finally resulting in asystole. Treatment includes calcium chloride or calcium gluconate to stabilize cardiac myocytes and 10 units of insulin with an ampule of D50 as well as an ampule of $NaHCO_3$, both of which help shift serum K into cells.
- Hypokalemia at 3–3.5 is asymptomatic. Arrhythmia, **T wave flattening and enlarged U waves can be seen when serum levels are <3.** Patients have generalized **weakness at levels <2.**

- Sodium (Na) requirements in the adult are **1–2 mEq/kg/day.** Na derangements are compensated for by brain osmolar equilibration within 48 hours. As a result, if patients are asymptomatic, correction of Na values should proceed cautiously at 0.5 mEq/hour.
  - If a patient is **seizing** due to hyponatremia, then immediately give **100 ml of 3% hypertonic saline (HS) over 10 minutes.** This can be repeated until the seizures cease. Sedatives and other anti-epileptics should be used concurrently.
  - If a patient is **encephalopathic** without seizures then **3% HS can be given at 1 ml/kg/hour.**
  - Na deficit or excess is calculated using 140 as the normal Na value and noting the difference from the patient's Na level. This difference is multiplied by 0.6 L/kg.
  - **Hyponatremia** is managed based on the patient's volume status.
    - Hypovolemic hyponatremia is most common and is due to a loss of sodium in excess of free water losses. It is treated by volume repletion with isotonic crystalloid.
    - Hypervolemic hyponatremia is due to excess sodium and a relatively greater amount of free water. It is treated by diuresis and free water restriction.
    - Euvolemic hyponatremia is an excess of free water with a normal total body sodium. It is treated with thyroid and adrenal replacement or free water restriction. Demeclocycline can be used in the chronic setting for SIADH (syndrome of inappropriate secretion of antidiuretic hormone).
  - **Hypernatremia** should also first be approached by determining the patient's volume status.
    - Most patients are hypovolemic and should receive crystalloid.
    - D5 water (D5W) can be given to those who are euvolemic. They are losing free water, and may have diabetes insipidus (DI) or nephrogenic diabetes insipidus (nDI). Alternatively they may just have restricted access to free water.
      - **Water deprivation** will normally result in an increased urine Osmolarity (OSM) as water is reabsorbed along the nephron to temper the serum hypernatremia. However, if a patient has DI or nDI, this will not happen and the serum OSM will increase without an increase in urine OSM.
      - **Desmopressin** is similar to vasopressin in its ADH and platelet effects, but without vasoconstriction. When given in the above scenario, it will normalize the response to water deprivation if the patient has DI, but will have no effect if the patient has nDI.
    - Hypervolemic patients probably received too much sodium in the setting of renal insufficiency. They should have sodium restriction and diuresis or another form of volume removal.
- **Calcium.** Over **95% of the total body calcium resides within bone.** The rest is in serum, 50% of which is protein bound, 45% ionized, and 5% complexed to Phosphates, HCO_3, or acetate.
  - Hypercalcemia is from malignancy or hyperparathyroidism. Treatment is with volume repletion, mithramycin, calcitonin, bisphosphonates, and loop diuretics.
  - Hypocalcemia can occur after parathyroidectomy, transiently after thyroidectomy, or after a rapid blood transfusion. Chvostek and Trousseau signs may be present on exam. The QT interval on an ECG is prolonged. Treatment is with repletion of magnesium and calcium IV or oral calcium chloride, calcium carbonate, or calcium gluconate. Vitamin D_3 can be added, too.
- **Magnesium** is primarily in **bone and intracellular.** Hypermagnesemia produces neural and respiratory depression and is from renal failure or iatrogenic administration. Treatment is with **calcium**, volume repletion, and diuresis. Hypomagnesemia is from diuresis, chronic TPN administration, or alcohol abuse. Treatment is with repletion over 1–2 weeks. Since Mg is a cofactor for parathyroid hormone (PTH), patients may need hypocalcemia treatment as well.
- **Phosphate is 80% within bone, 19% intracellular,** and about 1% intravascular.
  - Hyperphosphatemia treatment is with dialysis in an acutely ill end stage renal disease (ESRD) patient, the enteric phosphate binder aluminum hydroxide for non-emergent ESRD patients, and volume expansion if the patient has preserved renal function.
  - Hypophosphatemia is from an intracellular **shift** (secondary to insulin, beta agonists, alkalemia, bone hunger, refeeding syndrome), **Vitamin D deficiency** (which increases PTH levels and hence renal phosphate loss), or other causes of increased GI or renal losses. Symptoms include cardiac myocyte and diaphragm muscle insufficiency. Treatment is with IV phosphate if the level is <1, otherwise oral replacement can suffice.

- **Acute Kidney Injury (AKI)**
  - A fractional excretion of sodium (FENa) should be calculated for patients with oliguria and azotemia to distinguish between pre-renal and intra-renal etiologies. Fractional excretion of urea (FEUrea) can also be used.
  - Pre-renal causes include hypovolemia, hemorrhage, cardiac dysfunction, or bilateral severe flow-limiting renal arterial and arteriolar lesions.
  - Intra-renal causes are usually from acute tubular necrosis (ATN) or inflammation, and may have an active urinary sediment.
  - Post-renal causes are from bilateral ureteral obstruction, bladder outlet obstruction, or obstruction of a Foley catheter.
  - **Contrast nephropathy** has a combination of **pre-renal and ATN features.** Risks are highest for those with preexisting renal disease and elderly diabetic cardiovasculopaths with hypotension who are receiving an endovascular procedure.
    - Patients have a non-oliguric increase in creatinine by 25% and epithelial casts in urine sediment with a FENa <1%.
    - Creatinine peaks in 3–5 days, but if the pre-contrast renal function was adequate, then likely there will be no permanent renal damage.
    - Water soluble iodinated non-ionic, low osmolarity contrast has lower rates of nephropathy and anaphylactoid reactions.
    - Prehydration can help prevent nephropathy from intravenous contrast, rhabdomyolysis (myoglobin toxicity to tubules), or tumor lysis (uric acid and phosphate toxicity to tubules).
- **Acid–Base disorders** can be deciphered with an arterial blood gas and chemistry panel in a series of steps.
  - First determine the overall pH as alkalemic or acidemic.
  - If the PCO_2 is elevated in the setting of acidemia, then there is a **primary respiratory acidosis.**
    - The pH should drop 0.08 for every 10 mmHg rise in the PCO_2 if the respiratory acidosis is acute. In the chronic state, the pH should drop by 0.03 for every 10 mmHg rise in the PCO_2.
    - The calculated **compensatory HCO_3 should be 2-4 mEq higher for every 10 mmHg rise in the** PCO_2. If the actual HCO_3 is outside of this range, there is a primary metabolic acidosis or alkalosis.
  - If the PCO_2 is decreased in the setting of alkalemia, then there is a **primary respiratory alkalosis.**
    - The pH should rise by 0.08 for every 10 mmHg drop in the PCO_2 if the respiratory alkalosis is acute. In the chronic state, the pH should rise by 0.03 for every 10 mmHg drop in the PCO_2.
    - The **calculated compensatory HCO_3 should be 2-4 mEq lower for every 10 mmHg drop in the** PCO_2. If the actual HCO_3 is outside of this range, there is a primary metabolic acidosis or alkalosis.
  - If no primary respiratory acidosis or alkalosis is apparent (i.e., the pH and PCO_2 are both either elevated or depressed) then there is a primary metabolic disorder.
  - If the HCO_3 on the chemistry panel is below 24 and the pH demonstrates acidemia, then there is a **primary metabolic acidosis.** The buffer system of this acidosis includes **extracellular HCO_3** and **intracellular phosphates and proteins (including hemoglobin)**.
    - Calculate the anion gap (AG) with the equation Na – (HCO_3 + Cl). If the AG is >12, then the patient's metabolic acidosis is termed an **anion gap acidosis** and may be from salicylates, methanol, uremia, diabetic ketoacidosis, paraldehyde, isoniazid, iron, lactic acidosis, ethanol, or ethylene glycol.
    - Determine the delta gap by subtracting 12 from the AG and then add this number to the measured HCO_3. The result is an estimate of the original bicarbonate concentration prior to the development of the AG acidosis.
      - If this original value is below 24, there was a concomitant non-AG acidosis to begin with.
      - If this original value is above 24, then there was a concomitant metabolic alkalosis to begin with.
      - If the calculated AG is 12 or below, then the metabolic acidosis is termed **non- anion gap acidosis.** This is most often due to bicarbonate loss in diarrhea or pancreatic output, saline administration, parathyroid hormone excess, renal tubular acidosis, or hypoaldosteronism.
      - If after volume repletion the HCO_3 level drops, then the patient probably had an initial volume contraction metabolic alkalosis as well.
      - In either case, the expected compensatory PCO_2 level in response to the bicarbonate can be calculated using Winter's formula: 1.5 (HCO_3) + 8. The PCO_2 will range from 2 above to 2 below this value.
        - If the actual PCO_2 is above the range calculated by Winter's formula, then this represents a concomitant primary respiratory acidosis (not merely compensation).
        - If the actual PCO_2 is below the range calculated by Winter's formula, then this represents a concomitant primary respiratory alkalosis (not merely compensation).
  - If the HCO_3 on the chemistry panel is above 24 and the pH is alkalemic, then there is a **primary metabolic alkalosis.**
    - The common causes are volume depletion, loss of gastric volume, hypokalemia, and hyperaldosteronism.

- In a state of volume depletion, proximal convoluted tubule (PCT) reabsorption of Na is accompanied by reabsorption of HCO_3, which produces the alkalosis of a volume contracted state.
- In the state of gastric losses, the volume lost happens to be acidic (hydrochloric acid) so this further accentuates the alkalosis generated at the PCT via Na and HCO_3 reabsorption. K and H are also lost in the urine.
- If after volume repletion the HCO_3 drops, then this is uncovering an underlying non-anion gap acidosis that was likely originally present.
- The expected **compensatory** PCO_2 **for any metabolic alkalosis can be calculated as follows: 0.67(HCO_3 – 24) + 45.** If the actual PCO_2 is above or below this value, then the patient has a primary respiratory acidosis or alkalosis as well.

2.

IMMUNOLOGY, ONCOLOGY, AND STATISTICAL CONCEPTS

- The cells involved in inflammation (PMN, macrophages, basophils, mast cells, etc.) are part of the **innate immune arm** and do not specify their response to particular antigens. *See Figure 2.1.*
  - **Natural killer (NK) cells** are also part of the innate system and recognize an activation signal from a target cell, as well as a lack of an inhibitory signal (absence of self major histocompatibility complex [MHC]) on that same cell. If confirmed as non-self, the NK cell will commit to attack. It is involved in cancer surveillance, and its function is enhanced by IL-12.
- The **adaptive immune arm** involves T and B cells that have targeted a learned response to specific antigens. *See* Figure 2.1. The liver, thymus, and bone marrow are primary lymphoid organs. Lymph nodes and spleen are secondary lymphoid organs.
  - **T lymphocytes** mature in the **thymus** and then travel to the spleen, lymph nodes, and mucosal surfaces that interface with the external environment—oropharynx, enteric, colonic, respiratory, and genitourinary. T cells have CD3 protein complex on their cell surface.
    - **Helper T cells (TH)** have CD4 receptors and are activated by external endocytosed bacterial antigens recycled and presented on **MHC2 molecules (DR, DP, DQ)** in antigen-presenting cells (APCs), including macrophages, dendritic cells, and B cells. Helper T cells are involved in delayed hypersensitivity type 4 reactions
      - **TH1** cells release IFN-y and IL-2 for cytotoxic T cell activation. They also stimulate lymphokine-activated killer cells and tumor infiltrating cells for defense against malignancy.
      - **TH2** cells release IL-4 for B cell activation to plasma cells and memory cells.
    - **Cytotoxic T cells** have CD8 receptors and are activated by viral and endogenous antigens presented on **MHC1 molecules (A, B, C)** situated in the membrane of all nucleated cells.
    - **Suppressor T cells** also have CD8 and regulate helper T cells and cytotoxic T cells.
  - **B cells** mature in the **fetal liver and bone marrow** and become plasma cells after IL-4 stimulation from TH2 cells. They produce antibodies (ABs) contributing to humoral immunity. Polyclonal ABs recognize multiple epitopes on one antigen. Monoclonal antibodies recognize one epitope. Types of antibodies are listed below.
    - **IgM** is a pentamer and the initial antibody produced. It functions in opsonization and complement activation. Surface IgM may interact with MHC2 molecules on APCs.
    - **IgG** is a dimer and the most abundant antibody in the serum. It functions in opsonization and complement activation. Either IgM or IgG antibodies can be directed against a cellular antigen or circulating antigen in type 2 or 3 hypersensitivity reactions respectively.
    - **IgA** is the most abundant antibody overall, and serves to provide mucosal protection in secretions, Peyer's patches, and breast milk. It adheres to luminal antigens, preventing microorganism movement across the epithelial lining into the internal environment. IgA also prevents bacteria from adhering to epithelium. It does not participate in opsonization or complement fixation.
    - **IgD** is a membrane-bound antigen receptor.
    - **IgE** is involved in allergic reactions and parasitic infection. The antibody resides **on the surface of eosinophils** and binds antigens there. The eosinophil releases major basic protein which activates mast cells and basophils, promoting their release of histamine, serotonin, and bradykinin. The result is peripheral vasodilation, capillary permeability, and pulmonary vasoconstriction in a **type 1 hypersensitivity response.**

TRANSPLANT IMMUNOLOGY

- **MHC1** molecules correspond to human leukocyte antigen (HLA) A, B, and C. They provide antigen

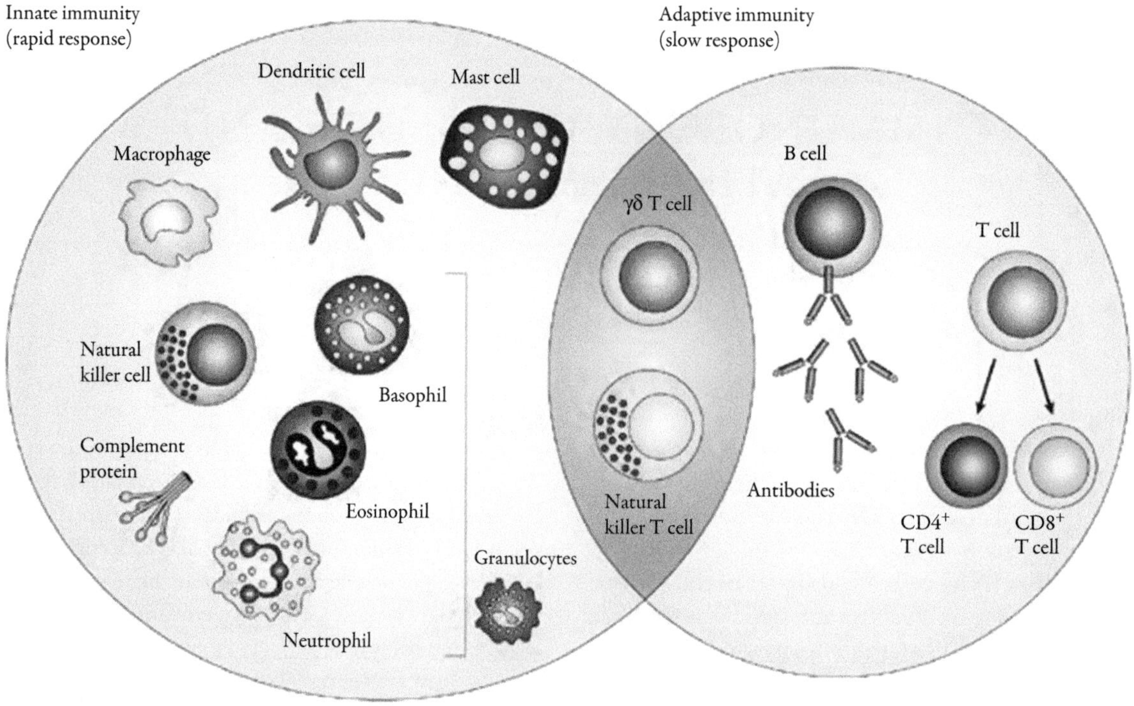

Figure 2.1 Innate and adaptive immune cells.

presentation on all cell surfaces. The antigens are generally endogenous or viral.

- **MHC2** molecules correspond to HLA DP, DQ, DR, and help with antigen presentation on specialized antigen-presenting cells (APCs), including macrophages, dendritic cells, and B cells. The antigens presented are generally exogenous or from non-viral microorganisms.
- **Panel reactive antibody (PRA)** is when the recipient serum is mixed with a panel of antigens to determine the percentage of cross-reactivity of the potential recipient with the general population. A recipient's antibody profile (and hence their PRA) will change with transfusions, pregnancy, autoimmune conditions, and other exposures.
- **Donor specific antibodies (DSA)** are measured by mixing the recipient's serum with a panel of antigens that are representative of a particular donor to ensure that it is worth considering the organ. This is also known as a virtual crossmatch.
- An actual **crossmatch** involves mixing the recipient serum with actual donor lymphocytes (containing HLA, A, B, C, DR, DP, DQ) prior to the transplant. If positive, then the recipient has antibodies against donor HLA and so will have hyperacute rejection in response to the organ. In some cases of living donor transplants, the recipient may be treated with **preoperative intravenous immunoglobulin (IVIG) and plasmapheresis** to manage the antibodies and allow for transplantation.
- **Primary graft dysfunction** can occur in a variety of grafts in the early postoperative period, and is a diagnosis of exclusion after a workup for rejection or vascular insufficiency. It is thought to be secondary to prolonged warm or cold ischemia times, reperfusion injury after implantation, or a compromised status of the donor organ prior to implantation.
- **Humoral rejection** is from preformed or post-transplant antibodies against either donor HLA or another donor antigen.
  - **Hyperacute rejection** is mediated by preformed IgG against donor HLA or incompatible donor ABO. *See* Figure 2.2. This occurs within hours after transplantation, activates complement, and causes vessel thrombosis. It does not happen in liver transplants. Treatment is with organ removal and re-transplantation.
  - **Acute humoral rejection** can be treated with glucocorticoids, IVIG, plasmapheresis, and rituximab. Chronic rejection can also be antibody mediated.
- **Cellular rejection** is when T cells attack donor HLA or another donor antigen.
  - Accelerated rejection occurs within one week post-op, and **acute cellular rejection** is after 1 week and generally up to 3 months post-op. It

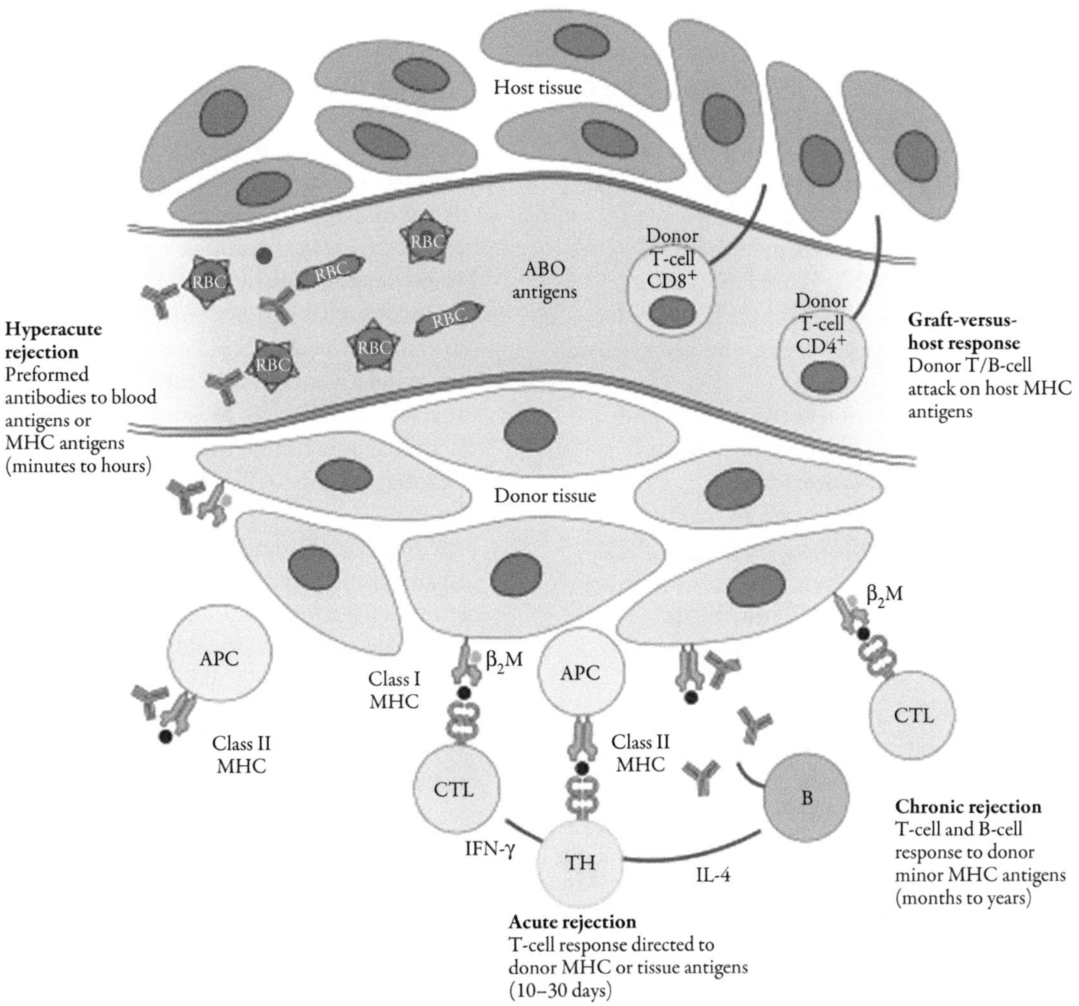

Figure 2.2 Interactions between recipient immune cells and donor tissue.

can occur years later, however. Treatment is with increased immunosuppression, glucocorticoids, and antithymocyte globulin.
- **Chronic rejection** occurs months to years after transplantation and is a combination of a cellular and humoral attack causing graft fibrosis. *See Figure 2.2.* Bronchiolar and coronary occlusion develops in lung or heart grafts respectively. Treatment is supportive and can include increased immunosuppression. Chronic rejection is the main cause of graft failure over the long term.

- Antibody therapy is the cornerstone of immunosuppressive **induction** after transplantation or an episode of acute rejection
  - **Antithymocyte globulin** is used for preoperative induction of immunosuppression as well as in the management of acute rejection. Antibodies are against T and B cells, integrin, and Intracellullar adhesion molecule (ICAM). The results is a depletion of complement and increased cytokine release. Patients can be pretreated with acetaminophen and diphenhydramine
  - **Basiliximab** is an antibody against T cell activation and can also be used for induction or the treatment of rejection.
  - **Alemtuzumab** targets mature lymphocytes via CD52.
  - **Rituximab** is an antibody against B cell CD20 and is used for the treatment of humoral rejection or desensitization in prospective recipients with a positive crossmatch.
  - Other adjuncts for treating rejection include glucocorticoids, IVIG, and plasmapheresis.
- After induction, patients will be on a variety of medications for **maintenance** of immune suppression:
  - **Cyclosporine** binds cyclophilin, inhibiting calcineurin, and IL-2 and IL-4 production. Side effects include hyperglycemia, nephrotoxicity, central nervous system symptoms, hepatotoxicity, hirsutism, gingival hyperplasia, thrombocytopenia, and hepatotoxicity.

Cyclosporine undergoes hepatic metabolism and bilio-enteric recirculation and reabsorption.

  - **Tacrolimus** binds FK protein preventing IL-2 synthesis. It is more potent than cyclosporine and generally used instead. Complication profile includes hyperglycemia, nephrotoxicity, and central nervous system symptoms.
  - **Sirolimus** binds the FK protein and inhibits T and B cell response to IL-2. It is used for patients intolerant of tacrolimus. Side effect profile includes pancytopenia, hypercholesterolemia, and poor wound healing.
  - **Belatacept** blocks interaction between APC and T cells and can be used in place of tacrolimus or cyclosporine with less nephrotoxicity.
  - **Glucocorticoids** are also an important part of maintenance therapy, and usually tapered to 10-15 mg/daily at some point.
  - **Mycophenolate mofetil (MMF)** inhibits de novo purine synthesis and causes leukopenia and gastrointestinal (GI) symptoms. Enteric coated versions are available that mitigate the GI effects.
  - **Azathioprine** inhibits de novo and salvage purine synthesis. It causes bone marrow suppression, but is the preferred alternative to MMF if the patient has GI symptoms or is pregnant.
- **Post-transplant malignancies** are most commonly skin cancer. Post-transplant lymphoproliferative disorder (PTLD) from Epstein Barr virus (EBV) is the second most common malignancy in recipients.
- **Post-transplant Infection**
  - Early infections occur at <1 month and are usually bacterial.
  - Late infections occur at >1 month and are more often viral, protozoal, or fungal.
    - **Cytomegalovirus (CMV)** presents with malaise and leukopenia, with tissue invasion resulting in pneumonitis, hepatitis, colitis with GI ulceration, or retinitis. It may also produce graft failure. Prophylaxis is with acyclovir, and treatment is with ganciclovir and decreasing the immunosuppressive regimen.
    - Other **herpes virus family** infections include HSV, VZV and EBV. The latter can produce hepatitis or PTLD; treatment includes a decrease in immunosuppression and use of rituximab.
    - Progressive multifocal leukoencephalopathy (PML) is a devastating CNS infection caused by JC virus (JCV), a polyomavirus that commonly establishes persistent, asymptomatic infection in the general population.
    - **Fungal infections.** Aspergillus is a mold and the most common invasive lung infection post-transplant. Yeast infections include candida and cryptococcus. Dimorphic fungal infections produce pulmonary and systemic symptoms; organisms include *histoplasmosis*, *blastomycosis*, and *coccidiomycosis*.
    - **Pneumocystis jiroveci pneumonia (PJP)** is a common fungal infection, and produces an alveolitis. Prophylaxis and treatment is with trimethoprim-sulfamethoxazole.
- Authority for deceased donor permission to harvest starts with a spouse and, if not available, moves on to an adult son or daughter, parent, adult brother or sister, or official guardian in that order.
  - **Donation after brain death** is the most common circumstance and allows organs to be perfused naturally, immediately prior to procurement.
  - **Donation after cardiac death** is an option if a patient cannot be pronounced brain dead. This increases warm ischemia time, since one must wait for cessation of cardiac activity of 2–5 minutes before procurement of organs.
  - Procurement of infradiaphragmatic organs is with infrarenal aorta and inferior mesenteric vein (IMV) cannulas for administration of cold solution, biliary and duodenal flushes of cold solution, and an inferior vena cava (IVC) vent. Cold perfusion would mark the end of warm ischemia time.
  - For specifics of each transplantation procedure, see the respective organ chapter.

ONCOLOGY

- Neoplasia can result from an inherited or acquired mutation affecting an oncogene or tumor suppressor gene, an exogenous carcinogen, viral infection, or immune deficiency resulting in poor surveillance of cell replication and maturation. Neoplasia may, but need not always, develop from a spectrum of metaplasia and dysplasia.
- **Tumor suppressor genes** address errors in replication by coordinating DNA repair, cellular arrest, or cellular apoptosis. Mutations in tumor suppressor genes can result in neoplasia. Examples include the following:
  - RB1 mutation and retinoblastoma.
  - p53 mutation disrupting regulation of G1 and apoptosis. The result can be Li-Fraumeni syndrome with brain, breast, sarcoma, neuroblastoma, nephroblastoma, adrenal tumors, and leukemia.
  - APC mutation seen in familial adenomatous polyposis and sporadic colon cancer.
  - DCC and CDH mutations affecting cell adhesion. The latter is associated with hereditary gastric cancer.
  - BRCA mismatch repair gene mutation causing breast cancer.
- **Oncogenes** control movement through the cell cycle. Mutations can dysregulate the cell cycle and result in neoplasia. Examples include Ras and K-Ras (transmembrane protein controlling intracellular signaling), Src and RET (tyrosine kinase), sis (PDGF),

ErbB (EGFr receptor), MYC (transcription factors), and VEGF gene (angiogenesis).

- **Tumor markers** are a means of detecting and following the progression or regression of local or distant tumors via serum sampling.
  - **Carcinoembryonic antigen (CEA)** correlates with mucin production in pancreatic neoplasms. Serum levels can correlate with colon cancer persistence or recurrence after treatment.
  - **Alpha fetoprotein (AFP)** is a specific marker for hepatocellular carcinoma. It can also be elevated in non-seminomatous germ cell tumors, and yolk sac tumors
  - **Prostate specific antigen (PSA)** is sensitive for prostate cancer and prostatitis, but has a low specificity for cancer.
  - **CA 19-9** is elevated in pancreatic cancer or biliary obstruction.
  - **CA 125** correlates with ovarian epithelial cancers. **Inhibin** is for epithelial ovarian and sex cord stromal tumors. Beta **hCG** is for nonseminomatous germ cell tumors, choriocarcinoma, and embryonal cancer
  - **Neuron specific enolase (NSE)** is for small cell lung cancer and neuroblastoma
  - **Chromogranin A** is elevated in carcinoid tumors.
- **Chemotherapy** targets rapidly dividing neoplastic cells as well as native cells that replicate regularly.
  - Drugs that affect all phases of the cell cycle include the following:
    - **Alkylating agents** cisplatin, carboplatin, cyclophosphamide, and busulfan that disrupt DNA replication.
    - **Doxorubicin** intercalates in DNA and produces O_2 radicals disrupting nucleic acid bonds. It is cardiotoxic.
    - **Bleomycin** generates oxygen radicals that cause DNA strand breaks. It causes pulmonary toxicity.
  - Drugs affecting the M phase include:
    - **Taxanes** which stabilize microtubules resulting in apoptosis.
    - **Vinca alkaloids** which destabilize microtubules.
  - Drugs affecting the S phase of DNA replication include:
    - **Methotrexate (MTX)** which blocks purine production by dihydrofolate reductase (DHFR). Leucovorin enters the metabolic pathway distal to DHFR and so can help salvage certain cells from MTX toxicity.
    - **5-fluoruracil (5-FU)** inhibits purine production by blocking thymidylate synthetase Leucovorin enters proximal along the metabolic pathway, so it will enhance the toxicity of 5-FU.
    - **Etoposide** inhibits topoisomerase DNA unwinding during attempted replication.
  - **Levamisole** was originally an antihelminthic that was found to stimulate the immune system against cancer. Side effects include neutropenia, encephalopathy, disulfiram reaction.
  - **Imatinib** is a competitive inhibitor of tyrosine kinase in GISTs associated with cKIT, CD 117, and PDGF receptors.
  - Checkpoint inhibitors include ipilimumab (CTLA-4) and pembrolizumab (PD-1).
- **Radiation (XRT)** targets the M phase of the cell cycle by directly damaging DNA and constructing oxygen radicals which also induce DNA injury.
  - XRT dosing is fractionated to allow repair of normal cells, redistribution of tumor cells in their cycle, and re-oxygenation of tumor cells to enhance O_2 radical production.
  - Higher energy XRT has skin preserving effect.
  - Brachytherapy is the term used when the radiation source is placed within the body, adjacent to the target tissue.
  - The most radiosensitive tumors include seminoma and lymphoma. Radio-resistant tumors include dense solid masses, sarcomas, and some epithelial carcinomas.

STATISTICAL AND EVIDENCE-BASED CONCEPTS

- **Incidence** is the number of new cases diagnosed over a time period.
- **Prevalence** is the number of cases at any given point in time. The prevalence of a disease has an impact on the positive and negative predictive values of a test.
- In screening programs for cancer, **lead time bias** refers to early identification of asymptomatic tumors with screening, and hence a perceived survival benefit compared to those who were diagnosed later after symptoms developed.
- A **length time bias** refers to the fact that screening detects indolent tumors earlier than aggressive ones, and hence these patients, and their associated prolonged lifespan, make it appear the screening test has improved survival.
- **Sensitivity** is the number of patients who truly have the disease and tested positive (true positive), divided by all patients who truly have the disease (whether they had a true positive or false negative test result). Tests with high sensitivity can rule out disease if the result is negative.
- **Specificity** is the number of patients who truly do not have the disease and tested negative (true negative), divided by all patients who truly do not have the disease (whether they had a true negative or false positive test result). Tests with high specificity can definitively diagnose disease if the result is positive.

- **Accuracy** is all true values (negative and positive) divided by the entire group (true and false negatives and positives).
- **Quantitative variables** are measurements that exist along a continuum.
  - **Ratio variables** are absolute values with a definite zero point. Examples include vitals and weight.
  - **Interval variables** are ordered measurements, but there is an arbitrary zero point. Examples include dates of the month.
  - **Student's T test** (also known as the **unpaired T test**) compares quantitative variables, often the mean values, between two groups.
  - **Analysis of variance (ANOVA)** compares quantitative variables across three or more groups.
  - **Paired T test** compares quantitative variables within a group across time such that one population is both the control group (before an intervention) and experimental group (after an intervention).
  - **Mean** is the average of all values.
  - **Mode** is the most common single value among all the results. It is best for an **even distribution** of data points.
  - **Median** is the middle value of all results aligned in order. It is best for a **skewed distribution** of results.
- **Qualitative variables** are categories. They can be binary, nominal, or ordinal. In the latter case, the groups have a definite hierarchy.
  - **Chi squared test** compares **nominal** variables between two groups.
  - **Mann-Whitney U test** compares **ordinal** variables between two groups.
- A **Kaplan-Meier curve** is used to estimate survival after a discrete event
- **Absolute risk reduction** is the incidence of a disease or condition in exposed minus the incidence in non-exposed groups.
- **Number needed to treat** is 1/absolute risk reduction.
- **Relative risk** is the incidence of an outcome in the exposed group/incidence of an outcome in the non-exposed group. It is computed in a prospective manner in a cohort or randomized control trial.
- **Odds ratio** is the chance of being part of the case (or control) group in a retrospective evaluation, based on some initial factor or variable. In such situations, the number in the "case" and "control" groups were determined at the outset of the study, so incidence and relative risk cannot be calculated.
- A **type 1 error** is assuming there was a difference between groups when it was actually only due to chance. In these instances, the null hypothesis was rejected incorrectly.
- A **type 2 error** is assuming there was no difference between two groups, usually because the study population was underpowered. In these instances, a difference actually existed, but the null hypothesis was accepted incorrectly.
- The **p value** measures the power of a test. If <0.05, then there is a $>95\%$ chance that the difference in results between the two groups is true. And there is a $<5\%$ chance that the difference was due to chance alone.
- **Standard deviation** is the range of deviation from the mean value of a variable in a group. It is the square root of variance, determined by the difference between each individual data point and the mean.
- The **95% confidence interval (CI)** is within 2 standard deviations from the mean. It represents the range around the mean within which the true value will exist 95% of the time. The narrower the CI, the more precise the result. A CI of likelihood ratios should not cross 1.0.
- **Quality of care** refers to differences in patient morbidity and mortality across institutions that are not attributable to chance (type 1 or 2 errors) or variations in case mix
  - **Case mix** includes patient comorbidities and severity of illness. It will vary greatly between hospitals and result in a variety of different procedures performed at these hospitals. In such instances, risk adjustment must be done before comparing outcomes across institutions. When comparing any given procedure, case mix and patient factors seem to be a smaller contributor to variations in mortality between institutions.
  - High hospital volume for esophageal, pancreatic, and aortic surgery are associated with measurably better results since they are relatively common, high-risk procedures where such differences can be measured. Other case outcomes are less closely tied to overall hospital volume.
  - Surgeon-specific volume and surgeon subspecialty training are closely related to outcomes across all procedures.
  - Although morbidity among multiple institutions may be similar after risk adjustment, higher mortality rates are associated with those hospitals that had **"failure to rescue"** in any given post-surgical morbidity or complication.
- **Systematic analyses and meta-analyses** search and review all relevant data to a particular question, attempting to draw a cause and effect relationship. If the data set are randomized controlled trials (RCTs), then this is the highest level of evidence on a topic. Bundles are 3–5 evidence-based, level 1 practices tied together, often based on the results of a systematic or meta-analysis.
- **Randomized controlled trial (RCT)** is the second highest level of evidence on a topic. It measures efficacy and attempts to demonstrate a cause and effect relationship. RCTs avoid selection and treatment

bias. If observers are blind to results, then it avoids observational bias as well.
  - RCTs may be pragmatized by relaxing inclusion and exclusion criteria, performing intention-to-treat analyses, performing randomization by clusters of patients (not on an individual basis), removing blinded result interpretation, and providing adaptive modifications to the methodology as results return and are reviewed throughout the study period.
  - A **patient centered outcomes trial** allows patient and family to choose their treatment arm.
  - **Pragmatization** mimics effectiveness in the real world, but can weaken the results obtained.
  - Patients may be analyzed by their original randomization group (intention-to-treat analysis) rather than a group they may have crossed over into during the study period.
- **Observational studies** draw broad conclusions, without actual causal relationships. They are useful in settings where randomization is not possible or ethical. They can generate hypotheses upon which RCTs are later conducted
  - A **cohort study** is the third highest level of evidence. One or more at-risk populations are followed over time to see if an outcome of interest occurs. Interventions, if undertaken, are not randomized. Once outcomes are available, the factors particular to each group are reviewed to identify any correlations. Relative risk may be calculated.
  - A **case-control study** is the fourth highest level of evidence. Two or more populations that differ by a single trait or exposure are followed over time to see if the outcome of interest occurs. This can be done **retrospectively** by comparing a group with a particular outcome to another group without that outcome, and then looking back at initial distinguishing features. Odds ratio can then be calculated.
- **Case reports** are the fifth level of evidence, and **expert consensus opinion** is the sixth.
- Cost effectiveness is measured in costs (dollars, pain or suffering, productive years lost) and benefits (quality-adjusted life years, QALYs)
- **Beneficence** refers to the healthcare provider's duty to protect the patient.
- **Autonomy** refers to the patient's right to informed consent and making their own decisions.
- **Justice** means that each patient should have an equal share of medical care.

3.

MICROORGANISMS AND ANTIBIOTICS

INFECTION CONTROL

- **Universal precautions** include wearing a gown, gloves, and an eye shield as well as regular hand cleansing. This is also called *contact precautions*.
- **Droplet precautions** include the addition of a **mask**. This helps protect against particles larger than 5 micrometers that travel short distances in the air. Examples include *H. influenza* and *N. meningitidis*.
- **Airborne precautions** include a **negative pressure room and N95 mask**. These additional measures help protect against particles less than 5 micrometers that travel longer distances and are suspended in air. Examples include varicella, tuberculosis, and measles.
- **Methicillin resistant staph aureus (MRSA) decolonization** can be done universally or used selectively for patients with a positive nasal screen. The process includes isolation, intranasal mupirocin, and chlorhexidine bathing. Decolonization decreases the total number of MRSA infections but does not specifically alter the rate of surgical site infections (SSIs).
- **Antiseptics** kill organisms topically on the body. **Iodophors** cross into bacteria and replace intracellular ions in Gram-positive and -negative organisms. **Chlorhexidine (CHX)** destabilizes cell membranes. It is not used near the eye or ear. CHX is better than iodophors for **central lines**, and equivalent to iodophors for a surgical preparation in the OR.
- **Disinfectants** kill organisms on inanimate objects.

ORGANISMS

- Microflora vary based on location in the gastrointestinal (GI) tract.
  - Gastric and small bowel flora are mostly Gram-positive cocci (GPC).
  - The distal small bowel contains GPC, Gram-positive rods (GPR), and Gram-negative rods (GNR).
  - The colon can contain approximately a trillion organisms, mostly GNR, anaerobes, and some GPC.
- **Gram-negative rods (GNRs)** have lipopolysaccharides in the outer cell membrane that contribute to endotoxemia. Most GNR are from the enteric or colonic tract. *Escherichia coli* and *Klebsiella* are common examples and are found in urinary tract infections (UTI) or enteric origin infections. Salmonella can be particular to atherosclerotic plaques and mycotic aneurysms.
- **Gram-positive cocci (GPC)** take up Gram stain in their cell wall because they have no outer cell membrane.
  - ***Staphylococcus aureus*** **(SA)** is coagulase positive and produces collagenase. It is the main cause of all SSI. If it is resistant to methicillin and related compounds, it is termed methicillin resistant SA (MRSA).
  - ***Staphylococcus epidermidis*** **(SE)** is coagulase-negative and produces an exoslime matrix. It is the most common cause of a central line infection.
  - ***Streptococci*** produce pharyngeal and skin infections. *Streptococcus pyogenes* occasionally causes a necrotizing soft tissue infection. *Streptococcus pneumoniae* is common in lung infections, *viridans* in endocarditis, and *bovis* bacteremia can be seen in colon cancer. *Enterococcus* is a type of *Streptococcus* seen in the enteric tract.
- **Gram-positive rods (GPR)** include *Corynebacterium* (diphtheria), *Listeria*, *Bacillus* (*cereus* or *anthracis*), and Clostridial species. The latter is the most clinically relevant for surgical specialties.
  - ***C. difficile*** grows in the colon when other organisms are suppressed by antibiotics. It makes toxins A and B, resulting in mucosal injury and diarrhea Treatment is with PO vancomycin, IV Flagyl, PR vancomycin, or PO fidaxomicin. Up to 50% of cases recur. Fecal transplant by nasogastric tube (NGT) or colonoscopic tube is 90% effective at preventing multiple recurrences. Surgical options include a total colectomy with Hartmann's pouch for progressive septic shock despite all medical measures. A less commonly employed technique is laparoscopic ileostomy with antegrade vancomycin and polyethylene glycol (PEG) enemas.
  - ***C. tetani*** is inoculated from the environment into soft tissue wounds. Tetanospasmin exotoxin results in activation of motor neuron endplates and spasm of skeletal muscle. Treatment is by cleansing and debriding the infected wound and providing

antibiotics active against this anaerobe. **Tetanus toxoid** is used as prevention for the following:
    - A clean wound in a patient whose last booster was more than 10 years ago.
    - A dirty wound in a patient with an unknown vaccine history or whose last booster was more than 5 years ago or who never completed the tetanus series. **Tetanus immunoglobulin** is added in the case of an unknown or incomplete tetanus immunization history.
  - ***C. perfringens*** inhabiting contaminated wounds can result in a necrotizing soft tissue infections in muscle, fascia, or subcutaneous tissue.
  - ***C. septicum*** bacteremia can be seen in colon or hematologic malignancies and also inoculate soft tissue, producing a necrotizing infection in the absence of any trauma.
  - ***Actinomyces*** are Gram-positive rods that produce oral abscesses as well as localized pulmonary or gastrointestinal involvement. Cervicofacial and thoracic infections may result in sinus tracts. Microscopically they can form hyphae that appear fungal-like. Treatment is with drainage of any abscess and a penicillin-based antibiotic.
  - ***Nocardia*** are Gram-positive rods that produce chronic pulmonary and central nervous system (CNS) infections in immunocompromised hosts. Microscopically they can appear as branching filaments resembling fungi. Treatment is with debridement and trimethoprim-sulfamethoxazole.
- **Gram-negative cocci (GNC)** are less likely to be encountered in surgical patients. Organisms include *Neisseria gonorrhoeae*, *N. meningitidis*, and the coccobacillus *H. influenza*.
- **Tuberculous mycobacteria (TB)** are intracellular obligate aerobe bacilli that stain with acid fast. They may be multidrug resistant (MDR) to isoniazid (INH) and rifampin (RIF) or have extensive drug resistance (XDR) to four anti-tuberculous medications.
  - Primary infection may be symptomatic or become latent and contained within granulomas. Reactivation at a later date is due to a lapse in the cellular immune arm.
  - **Diagnosis** is by 3 sputum samples for smear, culture, and polymerase chain reaction (PCR). Skin testing with purified protein derivative (PPD) or serum testing of interferon gamma in response to PPD are other methods of testing latent TB.
  - **Treatment** is with a four-drug regimen (often RIF, INH, pyrazinamide, ethambutol) for 2–12 months depending on the extent of pulmonary disease, presence of extrapulmonary infection, CNS involvement, and immune status of the patient. A two-drug regimen of RIF and INH can also be used in certain instances.
    - For latent TB in a patient <35 years old, INH alone can be given for 9 months. However, acid fast sputum testing and a chest X-ray should be negative for any active disease.
- **Fungal** growth may be as yeast (candida, Cryptococcus), mold (aspergillus, mucormycosis), or dimorphic (histoplasmosis, blastomycosis, coccidiomycosis). Treatment is required if any of the following are present:
  - Fungemia.
  - Single-site fungal growth with corresponding severe symptoms.
  - Any two separate sites without symptoms.
  - Single mucocutaneous sites with no symptoms may represent colonization by a fungal organism and so may not need treatment.
  - Any indwelling catheters associated with the infection should be removed.
  - Most fungal infections are due to immune suppression (HIV, diabetes mellitus, glucocorticoid use, or neutrophil dysfunction), surgery or a breach in the gastrointestinal barrier, intravascular or intraperitoneal catheters, obesity, and the use of broad spectrum antibacterial drugs that disrupt the microorganism community allowing fungal overgrowth.
  - **Candida** is located in the foregut and can cause respiratory, ocular, or bone infection.
  - **Treatment** is with the following:
    - Nystatin, clotrimazole, or miconazole for skin, oral, or vulvovaginal infection.
    - Fluconazole for esophageal disease.
    - Voriconazole or caspofungin for invasive cases.
    - Amphotericin as a last-line therapy.
    - **Candiduria** may be colonization of a long-standing urinary catheter or a new infection.
      - Treatment with **fluconazole and/or amphotericin bladder washes** is recommended if the patient is symptomatic, neutropenic, or is pending an intervention. The urinary catheter should be exchanged.
    - **Candidemia** should always be treated. Fluconazole is a first-line option unless the organism is *C. glabrata* or the patient is immune compromised or neutropenic. In that case, use **caspofungin.** Continue treatment for 2 weeks after blood cultures become negative.
    - Fungal intra-abdominal abscess, deep space SSI, or burn wound infection should receive treatment for at least 4 days after source control.
  - ***Cryptococcus*** produces CNS pathology in AIDS patients and is treated with amphotericin or fluconazole.
  - ***Aspergillus*** presents in three main forms:
    - A **hypersensitivity reaction** to its antigens resulting in reactive bronchoscopasm in distal airways.

    - An **aspergilloma** passively filling a parenchymal lung cavity. This needs surgical removal only if bleeding or symptomatic and not responsive to medical therapy.
    - **Invasive disease** actively destroying pulmonary tissue. In this case, treatment includes voriconazole or amphotericin and early surgical debridement.
  - **Mucormycosis** invades nasal sinuses or pulmonary parenchyma in an immunocompromised patient. Treatment is with amphotericin and early surgical debridement.
  - **Histoplasmosis** is from the Mississippi River valley and Ohio. It is inhaled and produces pulmonary pathology, mediastinal lymphadenopathy which may calcify, mediastinal fibrosis, superior vena cava (SVC) syndrome, and dysphagia. It can result in broncholithiasis from a granuloma eroding into airways. Treatment for severe disease is with amphotericin. Surgery is reserved for airway or vascular complications unresponsive to medical therapy.
  - **Coccidiomycosis** is from the southwest United States and produces mediastinal and hilar lymphadenopathy. 60% of patients are asymptomatic. Treatment for symptomatic patients is usually fluconazole.
  - **Blastomycosis** is from the central United States and produces similar pulmonary pathology.
- **Select viral infections pertinent to surgical care.**
  - **Human immunodeficiency virus (HIV)** can be sexually transmitted or from blood exposure
    - There is a **0.3% needlestick transmission rate.** This is increased if the needle is visibly contaminated with the patient's blood, the patient had a high viral load, or there was deep tissue injury in the exposed individual.
    - Transmission can be decreased by 80% with **post-exposure prophylaxis (PEP),** which includes Truvada daily and raltegravir BID given within 1–2 hours after exposure and continued for 4 weeks thereafter.
    - There is a **0.1% transmission rate from mucous membrane exposure** to HIV-positive blood.
    - There is a **30% transmission rate from an HIV-positive mother to infant.**
    - Screen the source patient with **bedside HIV antibody or HIV p24 antigen testing, and confirm with polymerase chain reaction (PCR).** Check the exposed individual at day 0 and again at 6 weeks, 12 weeks, and 6 months.
    - A CD4 count of 800–1200 is normal; 300–400 correlates with symptomatic disease, and <200 opportunistic infections.
    - A viral load less than 500 correlates with low infectivity, but >30,000 indicates increased infectivity and progression to AIDS.
    - Specific surgically relevant presentations in AIDS include the following:
      - Gastrointestinal bleeding from cytomegalovirus (CMV) colitis induced ulcerations, Kaposi's sarcoma, or gastric or rectal lymphoma.
      - Secondary sclerosing cholangitis from *Cryptococcus*, CMV, microsporidia, or *Cyclospora*.
      - Focal neurologic symptoms from toxoplasmosis. Empiric treatment is with pyrimethamine/sulfamethoxazole or clindamycin with or without glucocorticoids for increased intracranial pressure (ICP). CNS lymphoma can also produce focal neurologic findings in AIDS.
  - **Hepatitis C virus (HCV)** is most often from IV drug use or sexual transmission.
    - There is a **3%** needlestick transmission rate. Source patient and the exposed can be screened with anti-HCV antibody and confirmed with HCV PCR.
    - 30% of cases of HCV spontaneously clear. 60% of patients develop chronic infection. 15% of these have cirrhosis, and up to 5% develop hepatocellular carcinoma (HCC). HCV and non-alcoholic fatty liver disease are the two main causes for liver transplantation in the United States.
    - There is no passive immunity (immunoglobulin) or PEP for HCV.
    - Treatment is with antiretrovirals if the exposed person tests positive and remains so after 12 weeks of watchful waiting, implying they are developing a chronic infection.
  - **Hepatitis B virus (HBV)** is a double-stranded DNA virus transmitted sexually or from blood exposure.
    - There is a **30%** needlestick transmission. The source should be checked for surface antigen (HBsAg), which indicates active infection.
    - The exposed individual can be checked for **hepatitis B surface antibody (HBsAb) titers.** These indicate either the person received the vaccine or they were previously infected and recovered from HBV. For those with inadequate titers, **hepatitis B immunoglobulin (HBIG)** can be given as passive immunity within one week of exposure and again at 4 weeks after exposure. An **HBV booster vaccine** can be given at the same time.
    - **Hepatitis B core antibody (HBcAb)** is specific for recovery from a previous infection, and does not rise with immunization alone.
    - Most acute infections of HBV resolve. Only a minority develop chronic infection, cirrhosis or HCC. However, given how common HBV is, it remains the primary cause for cirrhosis and HCC worldwide.
  - **Cytomegalovirus (CMV)** may be symptomatic in immunocompromised patients. Intranuclear inclusions are seen on histology. CMV is present in up to 50% of transplant recipients within 6 months. Primary infection can occur in a CMV-negative recipient who received a CMV-positive organ. Infection may occur after treatment of acute rejection with OKT3 or other instances of enhanced immunosuppression. CMV can cause accelerated coronary arteriosclerosis in heart

transplant recipients and gastrointestinal bleeding from penetrating colonic ulcerations.

ANTIMICROBIALS

- **Antivirals**
  - Acyclovir inhibits DNA polymerase and is used for herpes simplex, varicella, and Epstein Barr virus.
  - Ganciclovir and valganciclovir are for CMV.
  - Oseltamivir is for influenza.
  - Interferon works as an antiviral via RNA ribosomal translation inhibition.
- **Antibacterials** *See Figure 3.1.*
  - Resistance can be mediated by transfer of plasmids making B-lactamase, a cell wall binding protein mutation (in MRSA, vancomycin resistant entercocci), or decreased active transport of the antibacterial into the cell (as in aminoglycoside resistance).
  - **Polymyxin** is an outer cell *membrane* inhibitor active against Gram-negative organisms.
  - **Cell Wall Inhibitors**
    - **Penicillins** treat GPR and GPC except *S. aureus*. Oxacillin and nafcillin target *S. aureus* that is methicillin-sensitive (MSSA). **Aminopenicillins** have added activity against enterococcus and GNR. Ticarcillin and piperacillin cover a broader array of GNR as well. B-lactamase inhibtors are often added to help target methicillin-sensitive staph aureus (MSSA).
    - **Cephalosporins** target GPC, GPR, and varying degrees of GNR depending on the generation of the drug. They have no activity against MRSA or enterococcus.
      - First-generation cephalosporins do not penetrate the CNS and are active against GPC.
      - Second-generation cephalosporins have activity against some GNR as well.
      - Third- and fourth-generation drugs target GPC and GNR including pseudomonas, *Acinetobacter*, and *Serratia*. They also treat anaerobes.
    - **Vancomycin** is used for **MRSA and enterococcus.** It can also be given orally or rectally as a topical agent for ***C. difficile.***
    - **Carbapenems** treat a broad array of organisms including GPC, many GNR, and anaerobes. They do not target MRSA, enterococcus, or *Proteus*. Cilastatin can be added to increase the half-life.
    - **Aztreonam** is a monobactam that targets the **cell wall of GNR**, including pseudomonas, *Acinetobacter*, and *Serratia*.
    - **Fosfomycin** affects GN and GP cell walls.
  - 30s ribosome inhibitors include the following:
    - **Aminoglycosides (AGs)** need oxygen to cross the cell wall and reach the ribosome. They are for aerobic GNR, including pseudomonas, *Acinetobacter*, and *Serratia*. AG may synergize with aminopenicillins. Side effects include **reversible nephrotoxicity, irreversible ototoxicity, and neuromuscular junction blockade.**
    - **Linezolid** is for MRSA and VRE.
    - **Tetracycline** treats GPC and GNR. Side effects include **tooth discoloration.**
    - **Tigecycline** is a less commonly used ribosome inhibitor, and is not active against pseudomonas, *Proteus*, or vancomycin-resistant enterococci (VRE).

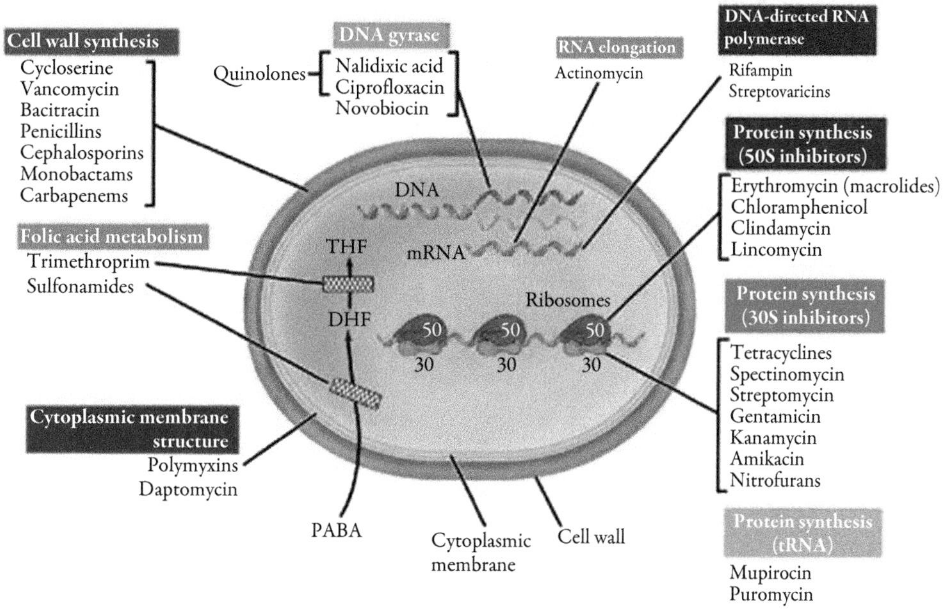

Figure 3.1 Site of action of common antibiotics.

- 50s ribosome inhibitors.
  - **Erythromycin** is used for GPC in community acquired pneumonia. Side effects include nausea and cholestasis. Erythromycin binds motilin prokinetic receptors and can enhance bowel motility.
  - **Synercid** targets MRSA and VRE.
  - **Chloramphenicol (CAF)** is not commonly used in the United States.
  - **Clindamycin** covers most **anaerobes and also decreases toxin production** in group A beta hemolytic streptococci (GABS) and *C. perfringens*. This drug has been known as a common precipitant of *C. difficile*.
- **Trimethoprim/Sulfamethoxazole** inhibits folate metabolism by blocking dihydrofolate reductase (DHFR) and **dihydropteroate synthase**. It treats both GNR and GPC.
- **Quinolones** inhibit DNA gyrase and are used to treat most GPCs, other than enterococcus. GNRs, other than pseudomonas, *Acinetobacter*, and *Serratia*, are also covered. Side effects include **weakening of tendon and cartilage, seizures, and a prolonged QT.**
- **Metronidazole** increases O_2 radical production and is active against anaerobes and parasites. Side effects include **neuropathy.**
- **Anti-tuberculous** medication includes the following:
  - Isoniazid, which targets mycolic acid in the cell wall. It is hepatotoxic and can exacerbate vitamin B6 deficiency.
  - Rifampin inhibits RNA polymerase and RNA transcription.
  - Pyrazinamide and ethambutol.
- **Antifungals** have three general groups:
  - **Polyene (amphotericin and nystatin)** disrupt ergosterols in the fungal cell membrane leading to depolarization and lysis. There is a liposomal version of amphotericin that is less nephrotoxic
  - **Voriconazole, itraconazole, and fluconazole** also disrupt ergosterols in the fungal cell membrane and are used for aspergillosis. All three are nephrotoxic.
  - **Echinocandins (anidulafungin/caspofungin/micafungin)** disrupt fungal cell wall glucan. These agents are used for candida.

4.

HEMATOLOGY

PHYSIOLOGY

- Erythropoietin is produced by the renal interstitium adjacent to the proximal convoluted tubule (PCT). It stimulates bone marrow production of red cells.
- **Tissue injury** leads to initial vasoconstriction and then vasodilation to bring in cellular mediators of inflammation and hemostasis (see the inflammation and wound healing sections in Chapter 1).
- Platelet **adhesion** to exposed collagen is by **GP1b receptors** with **assistance from von Willebrand factor (VWF)**
- Adhesion is followed by platelet **aggregation via GP2b3a receptors.**
- Lastly, there is platelet **activation** with **fibrin cross-linking of GP2b3a receptors** on adjacent platelets. This forms the platelet plug.
  - Activation is encouraged by **thromboxane (TXA_n), serotonin, adenosine diphosphate (ADP), and bradykinin**
  - Activation is inhibited by **aspirin (ASA), clopidogrel, nonsteroidal anti-inflammatory medications (NSAIDs), prostaglandins (PGI_2), cyclic adenosine monophosphate (cAMP), and nitric oxide (NO).**
- Fibrin is generated by either the intrinsic or extrinsic coagulation pathways. *See Figure 4.1.*
  - **Tissue factor (TF)** released after trauma or surgery activates the extrinsic pathway via factor 7. This in turn activates factor 10.
  - Prothrombin complex (10, 5, and prothrombin) catalyzes the formation of thrombin and fibrin.
  - Prothrombin time (PT) measures the extrinsic pathway and Partial thromboplastin time (PTT) measures the intrinsic pathway. When anticoagulation is with heparin, the PTT goal for most thrombotic disease processes is 60–90.
  - **Anti-Xa levels** should run between 0.3–0.7 for therapeutic heparin.
  - **Activated clotting time (ACT)** is measured as a point of care test intra-operatively and provides rapid results determining the adequacy of anticoagulation with heparin. The ideal range is >200 seconds for vascular cases and >400 seconds for cardiopulmonary bypass.
  - **Bleeding time** is the time it takes for a platelet plug to form after a calibrated dermal cut is created in a patient. It is a measure of platelet adhesion, aggregation, and activation, more so than the coagulation cascade.
  - **Ecarin clotting time (ECT)** tests only the terminal portion of the clotting cascade by activating prothrombin to thrombin and then fibrin. ECT is specific for monitoring the effectiveness of direct thrombin inhibitors (DTIs).
  - **Thromboelastography (TEG)** is a dynamic point of care test that can graphically depict clot formation and destruction. *See Figure 4.2.*
    - Insufficient factor results in an increased R time.
    - Insufficient fibrinogen results in a decreased alpha angle.
    - Decreased MA (clot strength) is secondary to severely depressed fibrin levels or decreased or poorly functional platelets. The latter is treated with desmopressin acetate (DDAVP) or a platelet transfusion.
    - Increased LY30 >3% represents enhanced fibrinolysis. Treatment is with tranexamic acid.
- The coagulation cascade is regulated by a number of auxiliary anticoagulant proteins
  - Antithrombin III (AT III) inhibits thrombin, 9, 10, and 11.
  - Protein C and S degrade factors 5 and 8, which are synthesized in the endothelium.
  - Plasmin degrades factors 5, 8, and fibrin.
  - Both 5 and 8 degrade in stored plasma unless frozen when freshly obtained.

BLOOD PRODUCTS

- **Packed red blood cells (pRBCs)** are stored at 4°C for up to 42 days. They may accumulate K, H, lactate, and NH_3 in storage and eventually deplete ATP and 2,3 DPG levels. This leads to a **left shift** of the oxyhemoglobin curve, impairing oxygen delivery. Red cells have a 24 hours lifespan once thawed. Complications of pRBC transfusion are listed below.
  - **Acute hemolytic transfusion reactions** are due to **recipient pre-formed circulating IgM** in the setting

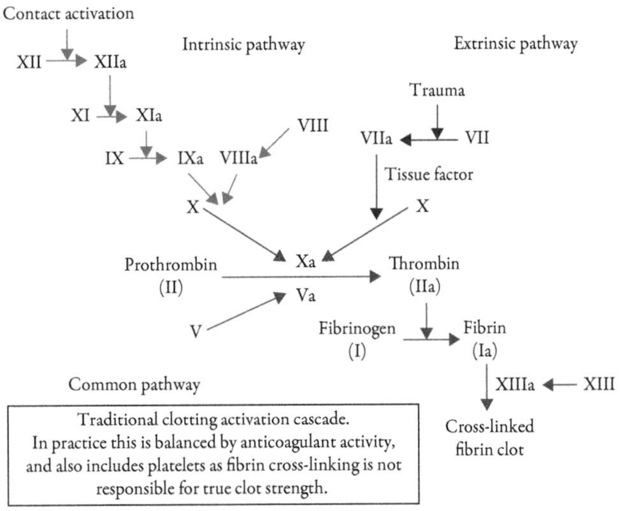

Figure 4.1 Coagulation cascade.

of ABO incompatibility. This leads to intravascular hemolysis, low haptoglobin, high unconjugated bilirubin, hemoglobinuria, distributive shock, and diffuse bleeding. This is now the second most common cause of post-transfusion reaction death.

- **Subacute, delayed hemolysis** is due to **IgG production** after a prior transfusion with minor antigen incompatibility. The transfusion was generally 3–10 days earlier. To prevent this, one needs to repeat the type and screen (T&S) every few days to ensure minor antibodies in the recipient have not changed.
- **Febrile reactions** result from **recipient antibodies to donor white blood cells (WBC)** with IL-6 and cytokine production by those donor WBCs. Treatment is to stop the transfusion and use leukocyte filters in the future.

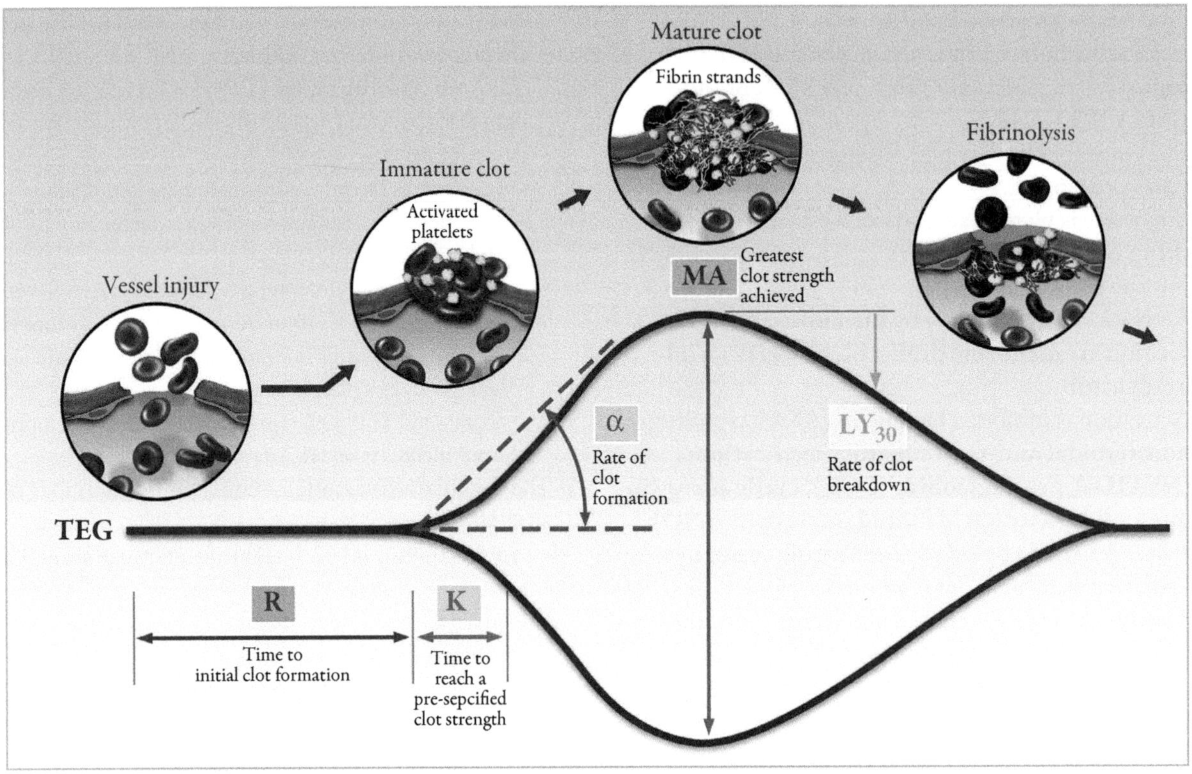

	R-time (R)	K-time (K)	Alpha angle (α)	Maximum Amplitude (MA)	Lysis % at 30 minis. (LY30)
Definition	Time to first deviation from baseline	Time for tracing to reach 20 min amplitude	Angle between baseline and tangent line that intersects initial deviation	Maximum deviation of tracing from baseline	Decrease in curve amplitude (relative to MA) at 30 minutes
Controlling Pathways	Coagulation cascade	Fibrinogen cleavage Fibrin polymerization	Fibrinogen cleavage Fibrin polymerization	Fibrinogen activity Platelet count/quality	Fibrinolysis
Interpretation	⬆ = hypocoagulable ⇩ = hypercoagulable		⇩ = hypocoagulable ⬆ = hypercoagulable		
Therapeutic Implications	⬆ = administer fresh frozen plasma	If K time is ⬆ or alpha angle is ⇩, then: administer cryoprecipitable or fibrinogen concentrate		⇩ = administer platelets	⇩ = administer tranexamic acid

Figure 4.2 Thromboelastography.

- At times, these recipient antibodies can lead to **post-transfusion thrombocytopenia** and purpura at a future date. Patients may bleed 5–6 days after that transfusion. Treatment is with intravenous immunoglobulin (IVIG).
- **Anaphylaxis** occurs after recipient **antibodies interact with donor proteins and IgA.** The treatment is to stop the transfusion and use washed pRBCs in the future. If there is only local tissue reaction, like urticaria, one may continue the transfusion at a slow rate and institute H1-receptor blockade.
- **Transfusion related acute lung injury (TRALI)** is the most common cause of transfusion reaction related death. **Donor antibodies to recipient WBCs** result in their sequestration in the pulmonary circulation within 2–6 hours of initiation of a transfusion. Alveolar inflammation sets in, and the serum WBC count may be decreased.
- **Graft versus host disease (GVHD)** is when donor leukocytes attack recipient tissue in the setting of recipient immune deficiency. This can be prevented in the future by using irradiated pRBCs.
- **Hypocalcemia** results from citrate in the packed cells binding free calcium in the setting of large volume product transfusions. This can lead to coagulopathy and cardiogenic shock. The process reverses once the liver metabolizes citrate.
- **Bacteremia** can be from gram negative rods (GNR) that infected pRBCs or gram positive cocci (GPC) that grew in a platelet transfusate.

- **Cryoprecipitate** contains factors **13, 1, Von Willebrand factor (VWF)- VIII.** This is the portion of plasma that freezes last or thaws first. It can be used for any type of von Willebrand disease (VWD), for hemophilia A, or in a setting of low fibrinogen.
- **Fresh frozen plasma (FFP)** contains all factors including protein C and S. The freezing process helps preserve factors. It may be used for international normalized ratio (INR) reversal in urgent situations for patients who can tolerate the volume.
- **Platelets** may be transfused for levels <10,000 to prevent spontaneous bleeding, <20,000 for central line placement, <50,000 for general surgery or lumbar puncture, and <100,000 for neurosurgery or epidural catheter placement.
- **DDAVP** releases VWF-VIII from the endothelium. It may be dosed every 12–24 hours x 2–3 doses for types 1 and 2a VWD.
- **Vitamin K** takes 6 hours for effect on carboxylation of factors 2, 7, 9, and 10, increasing their production. IV vitamin K is for urgent reversal of warfarin. PO vitamin K is adequate if the patient is just supra-therapeutic but has no active bleeding.
- **4 factor prothrombin concentrate complex (PCC)** includes 2, 7, 9, 10, and protein C and S. It is for reversal of INR in an emergency situation or a volume intolerant patient.
- **Topical application** of exogenous thrombin and fibrin can augment hemostasis. Oxidized cellulose (Surgicel) increases RBC lysis and TF release. Topical collagen matrix or glutaraldehyde cross-linked peptides and albumin can provide a scaffold for platelet plug and fibrin formation

ANTIPLATELETS

- **NSAIDs** are reversible inhibitors of cyclooxygenase (COX). *See Figure* 4.3.
- **Aspirin** is an irreversible inhibitor of COX. Platelet prostaglandin (PGI) and thromboxane (TXAn) production decreases. The endothelium can re-synthesize PGs, but the relative TXAn deficit leads to impaired aggregation and activation of platelets.
  - Aggrenox is aspirin combined with dipyridamole. The latter inhibits adenosine reuptake and cAMP phosphodiesterase, preventing platelet activation.
- **Ticagrelor** is a reversible inhibitor of the platelet P2Y12 receptor at a site separate from ADP binding.
- **Clopidogrel** is an irreversible inhibitor of platelet P2Y12 receptor at the ADP binding site. Gp2b3a is also blocked.
- Current recommendation is to hold antiplatelets for 5–7 days prior to surgery unless the patient is within **1 month of bare metal stent (BMS) placement or within 6 months of drug eluting stent (DES) placement within a coronary artery.** In that case there are a few options:
  - 1. If the surgery is not urgent, delay the procedure until after the 1-month or 6-month window.
  - 2. If the surgery is urgent and the procedure is a high bleeding risk, proceed to hold the clopidogrel for 5–7 days before performing surgery. One may consider an IV 2b3a blocking agent as a bridge to surgery (while the clopidogrel is being held) if the patient is of high thrombotic risk, in addition to their high operative bleeding risk.
  - 3. If the surgery is emergent or if it is urgent and the procedure is not a high bleeding risk, one can just operate with clopidogrel onboard.
  - Antiplatelets may be continued through all endoscopic procedures that do not include a major biopsy or resection.

ANTICOAGULANTS

- **Heparin** potentiates antithrombin III (AT III). It is given as a continuous infusion with an 80 u/kg bolus and 18 u/kg/hour for a goal **PTT 60–90.** Heparin

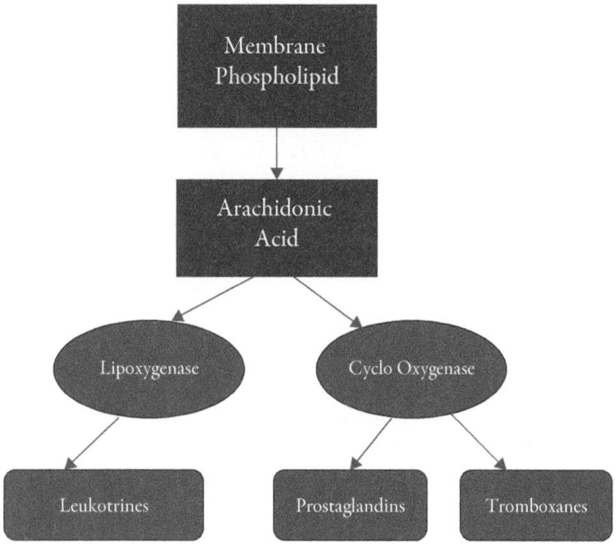

Figure 4.3 Cyclooxygenase.

has a T1/2 of **45–90 minutes**. Its efficacy can also be measured by **Anti-Xa levels,** which should run between 0.3–0.7 for therapeutic uses.
  - Heparin vasodilates peripheral arterioles via release of NO.
  - Metabolization is by the reticuloendothelial system (RES) in the liver.
  - There is a 5% bleeding risk in general. Long-term use can cause osteoporosis and alopecia.
  - Protamine 1mg per 100 u heparin is used for reversal. Beware of a bradycardic, hypotensive reaction with protamine administration.
- **Low molecular weight heparin (LMWH)** has increased bioavailability and activates AT III only against factor Xa and thrombin preferentially. It is administered as twice-daily subcutaneous injections with a **T1/2 of 4–6 hours**. Its efficacy can also be measured by **Anti-Xa levels,** which should run between 0.3–0.7 for therapeutic uses.
  - It relies on renal excretion, so dosage may need to be adjusted, or the drug avoided altogether in the setting of renal insufficiency.
  - LMWH be continued up to 4 weeks post-op for VTE prevention after abdominal or pelvic tumor resection, inflammatory bowel disease (IBD), or bariatric surgery.
  - LMWH has **decreased thrombosis, bleeding, and mortality** compared to heparin when used for the treatment of venous thromboembolism (VTE).
- **VTE prophylaxis** recommendations include
  - **Intermittent pneumatic compression** (IPC) **for low VTE risk patients,** those undergoing laparoscopy, or with a history of smoking or obesity only.
  - **LMWH if medium risk.**
  - **Both LMWH and IPC if high risk.**
  - Pre-op VTE prevention dosing should be at least 2 hours prior to induction for medium- or high-risk patients undergoing major surgery.
- **Fondaparinux** is an indirect Xa inhibitor administered subcutaneously once daily with a **17 hours half-life.** It is renally excreted.
- **Rivaroxaban and apixaban** are oral direct Xa inhibitors taken twice daily with T1/2 of **12 hours**. May follow Anti-Xa level for efficacy.
  - Both drugs prolong the PT and PTT, have hepatic and renal excretion, and can be partially reversed with PCC.
  - Hold either drug for 1 day prior to minor surgery, and 2 days prior to major surgery.
  - They have lower bleeding complications than heparin.
  - Both can be used for proximal iliofemoral deep vein thromboses (DVTs) in patients who do not have a history of malignancy.
- **Argatroban** is a reversible IV DTI given as a continuous infusion with a T1/2 of **50 minutes.** It is monitored with PTT. Argatroban, like heparin, relies on **hepatic metabolism** and is commonly used in the setting of heparin induced thrombocytopenia (HIT) with thrombosis (HITT).
- Bivalirudin is a reversible IV DTI with T1/2 of 25 minutes.
- Hirudin is an irreversible DTI with T1/2 of 60–90 minutes.
- **Dabigatran** is an **oral DTI** given twice daily with a T1/2 of about **12 hours.** It is renally excreted and reversed completely with **idarucizumab**, or partially with PCC or hemodialysis (HD). Dabigatran is used for treatment of VTE and CVA.
- **Warfarin** inhibits vitamin K dependent carboxylation of factors 2, 7, 9, and 10. Its T1/2 is **24 hours.** Efficacy is measured by INR. It is contraindicated in pregnancy, but tolerated during breastfeeding.
  - Either withhold dosing or use vitamin K PO or IM for reversal if a patient is supra-therapeutic but not bleeding.
  - Use FFP and vitamin K IV for active bleeding. Otherwise PCC can be used for emergent reversal in a volume intolerant patient with a supra-therapeutic INR and critical intracranial bleeding.
  - Hold warfarin for 5 days prior to surgery. Only bridge in a hypercoagulable state, mechanical valve replacement, or a patient with a high CHA_2DS_2-VASc score (particularly the elderly with a TIA or CVA history).
  - One may continue warfarin in all endoscopic procedures that do not entail a major biopsy or resection.
  - VTE regimens include **3 months treatment for triggered DVT and 6 months treatment for unprovoked DVT**. Some patients will need therapy

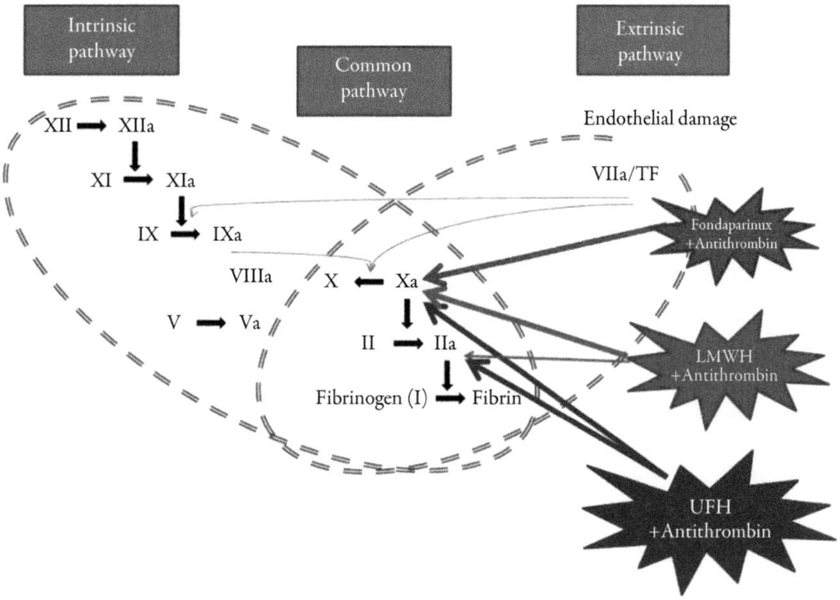

Figure 4.4 Anticoagulants.

for their **lifetime** if they've had serial DVTs, a large PE, or a hypercoagulable state.

- **Thrombolytics** encompass streptokinase, urokinase, or tissue plasminogen activators (TPAs) including alteplase, reteplase, or tenecteplase. TPA converts plasminogen to plasmin systemically. Thrombolysis is used for the following:
  - 1. Arterial or venous thromboses with ischemia that is non–limb threatening, or
  - 2. Large pulmonary thromboembolism (PE) with right ventricular strain or hypotension.
  - TPA is contraindicated in the setting of active bleeding, intracranial mass, CNS trauma, recent CVA, or recent neurosurgery.
- **Retrievable inferior vena cava (IVC) filters** can be placed for patients with lower extremity DVT or PE, and:
  - 1. A contraindication to anticoagulation.
  - 2. Ineffectiveness of anticoagulation, or
  - 3. Intolerance to anticoagulation.

HYPERCOAGULABLE STATES

- **Factor 5 Leiden** is the #1 congenital hypercoagulable disorder. Inheritance is autosomal dominant. Factor 5 is resistant to the effects of protein C and S.
- **Prothrombin defect** rendering factor 2 hyperfunctional is the #2 congenital hypercoagulable disorder.
- **Tobacco** is the #1 acquired hypercoagulable state.
- **Heparin induced thrombocytopenia with thrombosis (HITT)** is the #2 acquired condition.
- **Antiphospholipid** (anti-cardiolipin or lupus anticoagulant) results in an elevated PTT but a hypercoagulable state.
- **Hyperhomocysteinemia** can be seen in infants who also have skeletal abnormalities, mental deficiencies, or seizures. It characteristically increases **arterial** thromboses. Treatment is with folate and B_{12}.
- **Antithrombin III (AT III) deficiency** can be treated with recombinant AT III or FFP in addition to heparin. Heparin alone will not work since AT III must be present.
- **Polycythemia vera** is a hyperviscous state with elevated cell lines that can lead to thrombosis. Treatment is with phlebotomy and antiplatelets.
- **Protein C or S deficiency** results in dermal or subcutaneous vessel thromboses. This can be seen in patients started on warfarin without another immediate-acting anticoagulant bridge.

COMMON DISORDERS

- **Von Willebrand disease (vWD)** is the most common congenital bleeding disorder. There are multiple subtypes
  - Type 1: reduced VWF quantity.
  - 2a: decreased VWF function.
  - 2b: increased VWF function.
  - 2n: no function of VWF leading to an increased PTT.
  - 3: absence of vWF also leading to an increased PTT.
  - Treatment is with DDAVP for 2–3 doses if type 1, or 2a. For other subtypes, the best therapy is

recombinant factor VIII-vWF complex or cryoprecipitate. One may monitor factor VIII levels post-op.

- **Hemophilia A** is factor VIII deficiency.
  - Severe forms have <1% function and there is **spontaneous** bleeding.
  - Moderate forms have 1%–5% function with bleeding after minor trauma.
  - Mild forms have 6%–25% function and bleed excessively only after major trauma or surgery.
  - Neonates generally do not bleed because mother's factor VIII can cross the placenta.
  - Treatment is with **recombinant factor VIII** at 50 u/kg to restore 100% activity before any surgical procedure, and keep at >80% for the following 2 weeks.
    - May also use factor VIII to treat spontaneous hemarthrosis or after any traumatic injury, even before signs of bleeding develop.
    - **Cryoprecipitate** is another option.
    - **Recombinant factor VII** is for any patient that has factor VIII or IX inhibitors.
- **Hemophilia B** is factor IX deficiency, and can be treated with **recombinant factor IX or FFP**, but not cryoprecipitate.
- **Factor VII deficiency** can be treated with recombinant factor VII or ffp, but not cryoprecipitate
- **Bernard Soulier** describes absent Gp1b, and **Glanzmann thrombocytopathy** describes absent GP2b3a. Treatment for both is with platelet transfusion.
- **Uremic thrombocytopathy** is treated with hemodialysis or DDAVP.
- **Heparin Induced Thrombocytopenia (HIT)** is when heparin-PAF4 complexes form after the initiation of heparin. These complexes then act as a hapten for antibody production 5–7 days later. The antibodies result in clumping of platelets and their subsequent destruction. This same process can occur with LMWH.
  - **Diagnosis.** Platelet counts can drop by >50% within 5 days or more after heparin initiation. Venous or arterial thromboses from the clumped platelets, also known as white clot, can occur in addition to simple thrombocytopenia.
    - ELISA testing for the antibodies can suggest the diagnosis; however, many patients will have circulating antibodies for months after heparin exposure, and not develop platelet clumping and consumption. So this result must be taken in the appropriate clinical context, and a serotonin release assay is considered confirmatory.
    - Rapid diminution of platelets immediately after initiation of heparin is unlikely to represent antibody mediated HIT as it is too early for antibodies to have formed. The one exception is if the patient had previously received heparin and made antibodies at that time.
  - **Treatment** is by stopping heparin and using a DTI. **Argatroban IV** is the preferred agent, especially if there is renal insufficiency.
- **Disseminated intravascular coagulation (DIC)** is initiated by systemic TF released into circulation. This results in consumption of platelets and all factors with elevated D-dimer, fibrin split products (FSP), and eventually prolonged PT and PTT. Thrombosis and bleeding can occur simultaneously, in different locales.
- **Primary fibrinolysis** involves the release of plasminogen activators leading to plasmin functioning outside of the thrombus site. The result is bleeding with a prolonged PT and PTT, but **normal platelets and D-dimer**, since thrombi are dismantled prior to fibrin cross-linking.
  - Plasmin is inhibited by amicar or tranexamic acid (TXA). The latter is best delivered as a bolus within 3 hours of the initiation of hemorrhage from major trauma, followed by an 8 hour infusion. TXA decreases hemorrhage, MI, and mortality, and is most effective if given <1 hour after the traumatic event.
- **Carbon monoxide** poisoning creates carboxyhemoglobin (COHb) and left shifts the oxyhemoglobin curve. This impairs offloading O_2 to tissue and results in "100%" arterial saturation with high venous saturations as well. COHb levels of >30% result in confusion, >40% lethargy or coma, and >60% death. Treatment is with 100% FiO_2, which decreases the half-life of COHb to <60 minutes. Positive pressure ventilation can be used if necessary.
- **Cyanide** toxicity poisons mitochondria and impairs oxygen utilization leading to high venous oxygen levels and distributive and cardiogenic shock. Treatment is with hydroxycobalamin to create cyanocobalamin, or sodium thiosulfate to create thiocyanate. Both siphon off cyanide. Nitrites are another option; they create methemoglobin, which can bind cyanide and keep it from mitochondria.
- **Methemoglobinemia** results from nitrites converting Fe_2+ to Fe_3+ in heme entities. O_2 saturation by pulse oximetry often reads 85%, and blood appears chocolate brown. Treatment is with methylene blue.

5.

ANESTHESIA AND PERIOPERATIVE CARE

PHARMACOKINETICS AND DYNAMICS

- **Oral medication** undergoes first-pass hepatic metabolism. Inhalational, intravenous, sublingual, dermal, and rectal administration bypass hepatic metabolism and have higher bioavailability.
  - **ED50** is the blood level of a drug at which the desired end organ effect is reached in 50% of patients. **LD50** is the blood level of a drug at which it is lethal in 50% of patients. The therapeutic window and safety profile correlates with the difference between ED50 and LD50.
  - Zero order kinetics describes a constant volume of drug being eliminated regardless of the drugs' concentration. **First order kinetics** is when elimination is proportional to the drug concentration.
  - The **half-life** is the time it takes for the serum concentration of an agent to decrease by 50%. At five half-lives, a drug is essentially eliminated from the serum, with only 3% left. Administration of an agent serially over five half-lives results in a steady state serum level.
  - Anesthetics, narcotics, and sedatives with an increased **volume of distribution** have a high extravascular or adipose reserve. This may delay emergence.
  - Metabolism occurs in hepatocyte smooth endoplasmic reticulum via the p450 system. **Phase 1** of metabolism is demethylation, oxidation, reduction, and hydrolysis. **Phase 2** of metabolism is glucuronidation or sulfation of a water soluble metabolite in order to facilitate excretion.
- **Inhalational anesthetics** are the mainstay of achieving all components of general anesthesia: analgesia, immobility, loss of consciousness, amnesia, and blunting of autonomic and airway reflexes.
  - Inhaled agents with high blood solubility will stay dissolved within blood, and hence take a longer time to cross into the central nervous system. Less blood soluble agents therefore have a faster onset of action (**nitrous oxide**). Nitrous oxide should be avoided in patients with a bowel obstruction.
  - Minimum alveolar concentration (MAC) is the alveolar concentration of an inhaled agent at which 50% of patients do not react to a surgical stimulus. Lipid-soluble agents cross into the CNS easily and have enhanced potency and a lower MAC. Age and any other sedatives in circulation will decrease the MAC of an inhaled agent.
  - **Sevoflurane** has a fast onset, and **isoflurane** improves hepatic perfusion and may be cardioprotective. Enflurane, desflurane, and halothane are used less commonly.
- **Intravenous agents** are often used to supplement inhaled anesthetics.
  - **Barbiturates** act at the GABA receptor, but are no longer commonly used.
  - **Propofol** alone can work to diminish consciousness, memory, motor response, and reflexes. It does not provide analgesia. It is metabolized by the liver. Infusions over 48 hours can lead to an infusion syndrome with metabolic acidosis, bradycardia, hypotension, rhabdomyolysis, acute kidney injury, and hyperlipidemia. Treatment is by discontinuing the infusion and starting hemodialysis.
  - **Ketamine** is an NMDA receptor antagonist dissociative agent. It has sympathomimetic properties and increases secretions. It does not provide analgesia.
  - **Dexmedetomidine** is a central alpha-2 agonist used to assist in sedation while preserving cognition.
  - **Etomidate** is a rapidly acting sedative used for rapid sequence induction (RSI) and does not result in hypotension. Repeated use can result in adrenocortical suppression.
  - **Opioids** activate mu receptors, undergo liver metabolism, and are renally excreted. They can be delivered orally, transdermally, or intravenously as a push or patient-controlled analgesia. The latter may or may not include a basal rate depending on the patient's opioid tolerance.
    - **Morphine** is a derivative of codeine. It has multiple active metabolites that accumulate in renal failure. Morphine accentuates histamine release.
    - **Methadone** is similar to morphine but has less euphoria and prolongs the QTc.

    - **Hydromorphone** undergoes liver metabolism but has inactive metabolites, so it is better for patients with renal failure.
    - **Fentanyl** has 80 times the strength of morphine. Sufentanil is longer acting and more potent. Remifentanil is more potent and also very fast acting, and is metabolized by plasma esterases.
    - **Meperidine** has neurotoxic normeperidine metabolites that build up in renal failure. It is no longer commonly used in the United States.
    - **Opioid bowel disorder** is a dose-related dysmotility common to the majority of recipients.
    - **Narcotic bowel syndrome** is paradoxical diarrhea and increased pain with increased opioid dosage.
    - **Naloxone** is the universal reversal agent, competitive antagonist, for opioid overdose.
  - **Benzodiazepines** are sedatives that act at the GABA receptor and undergo hepatic metabolism.
    - **Midazolam** is short-acting but has lipid-soluble active metabolites that build up in prolonged infusion.
    - **Diazepam** is intermediate in duration of action.
    - **Lorazepam** is the longest acting and has no active metabolites, so is better if liver disease.
    - **Flumazenil** can be used as an antidote to counteract iatrogenic benzodiazepine overdose.
  - **Monitored anesthesia care** utilizes intravenous agents to obtain some degree of analgesia, sedation, decreased mobility, and impairment of motor function. Autonomic and airway reflexes are preserved. Although the main components of general anesthesia are not fulfilled, the patient can undergo short procedures without the need for airway intervention.
  - **Total intravenous anesthesia** is the use of IV agents to achieve all components of general anesthesia, without inhalational gas. However, electroencephalogram monitoring (EEG) would be needed to ensure that the CNS is quiescent.
- **Paralytics** block the neuromuscular junction (NMJ) and produce immobility at the face and neck followed by diaphragm paralysis. Recovery starts with the diaphragm, after which the face and neck regain functionality.
  - **Succinylcholine** is a depolarizing NMJ blocker with a rapid onset and produces fasciculations with increased intracranial pressure (ICP) and intraocular pressure (IOP). It is metabolized by plasma pseudocholinesterases. If these are atypical or insufficient, then the patient may suffer prolonged paralysis.
    - **Malignant hyperthermia** is inherited in an autosomal dominant fashion. An abnormal ryanodine receptor results in increased cytosolic calcium from sarcoplasmic reticulum (SR) release. This leads to muscle contraction and prolonged masseter rigidity. Diagnosis is clinical with an elevated temperature and CO_2 noted intra-operatively after receiving a causative agent. Treatment is by stopping all inhaled anesthetics, cooling, and giving **dantrolene 2.5 mg/kg every 5 minutes.** Definitive diagnosis can be made later by a muscle biopsy contraction test on exposure to halothane.
    - **Hyperkalemia** can result from upregulated nicotinic receptor activation if succinylcholine is given to patients having suffered a recent soft tissue injury or denervation more than 1–3 days prior.
    - **Pupil dilation** can result in closed angle glaucoma.
  - **Non-depolarizing NMJ blockers** also work at autonomic ganglia and can have some effect at muscarinic receptors, leading to tachycardia. This class of drug is the main cause for intraoperative anaphylaxis (with latex and antibiotics being the other etiologies).
    - **Cisatracurium** is short-acting and metabolized by Hoffman degradation. Therefore it is preferred in patients with renal failure.
    - **Rocuronium and vecuronium** are intermediate-acting, and **pancuronium** is slow-acting and undergoes renal metabolism.
    - NMJ blockade is potentiated by the concurrent use of antiribosomal antibiotics.
    - Reversal of NMJ blockade is with acetylcholinesterase inhibitors. Neostigmine acts peripherally, and edrophonium or pyridostigmine act centrally.
- **Local Analgesia**
  - **Amides** include lidocaine and bupivacaine.
    - 5 mg/kg **lidocaine** is the safe dose (up to 7 mg/kg is safe if you add epinephrine). CNS symptoms of dizziness, tinnitus, seizures, and confusion precede cardiac toxicity in the setting of an overdose.
    - 2.5 mg/kg **bupivacaine** is the safe dose (3.5 mg/kg is safe if epinephrine is added). It is longer acting than lidocaine and demonstrates cardiac before neurotoxicity in an overdose.
    - 10%–20% intralipids are the antidote for amide toxicity.
  - **Esters** include tetracaine, procaine, and cocaine. They have a shorter duration of action, increased allergic reactions, and are metabolized by plasma pseudocholinesterase.
- **Epidural analgesia** is via local bathing of nerves running extradurally. *See Figure 5.1*. This results in sympathetic denervation and vasodilation followed by a sensory blockade. Motor blockade only occurs at high doses. There is improved analgesia with opioids added into the epidural space.
  - Epidurals are associated with **faster return of enteric function and decreased pulmonary complications.** Pain control and length of stay are similar to other modalities.
  - Hold anticoagulation 12–24 hours prior to placement and 4–6 hours after removal. May remove the Foley while epidural is still in place.

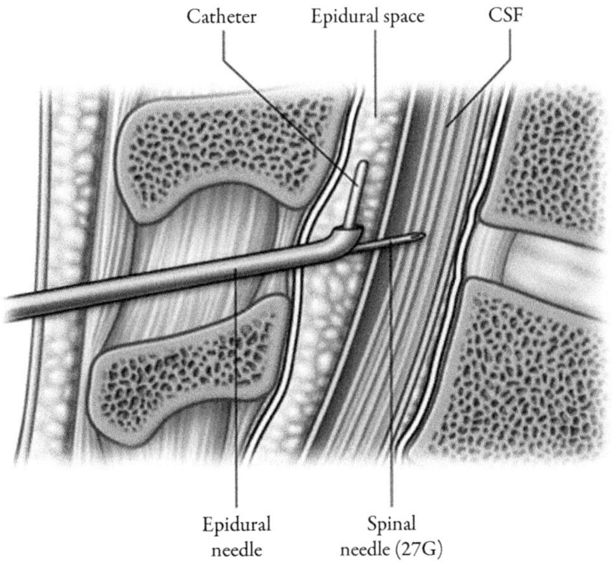

Figure 5.1 Accessing the epidural and subarachnoid (spinal) spaces.

- **Spinal analgesia** is injected into the subarachnoid space with spread determined by baricity. *See Figure 5.1.* Patients have a sensory, motor, and sympathetic block that can be higher than the point of needle entry.
  - Patients with persistent leakage from the thecal space after spinal puncture can have a spinal headache that is worse with standing. Management is with intravenous fluid, caffeine, and a 10–20 ml autologous blood patch over the site of dural breach.
- Nausea and vomiting emanate from the **medullary vomiting center**. *See Figure 5.2.* Stimuli input into the vomiting center is via the following:
  - Cerebral cortex: sight, sound, smell, anxiety, high ICP.
  - Vestibular nerve.
  - Distension of a hollow viscus within the proximal gastrointestinal tract.
  - Serum compounds interacting at the **chemoreceptor trigger zone (CTZ)**, including dopamine, serotonin, and opioids. Drugs that can modulate the vomiting reflex and sensation of nausea act at the CTZ and include the following:
    - **Prochlorperazine, chlorpromazine,** and **metoclopramide**, which block central D2 dopamine receptors.
    - **Promethazine** blocks D2 dopamine, H1 histamine, and muscarinic receptors.

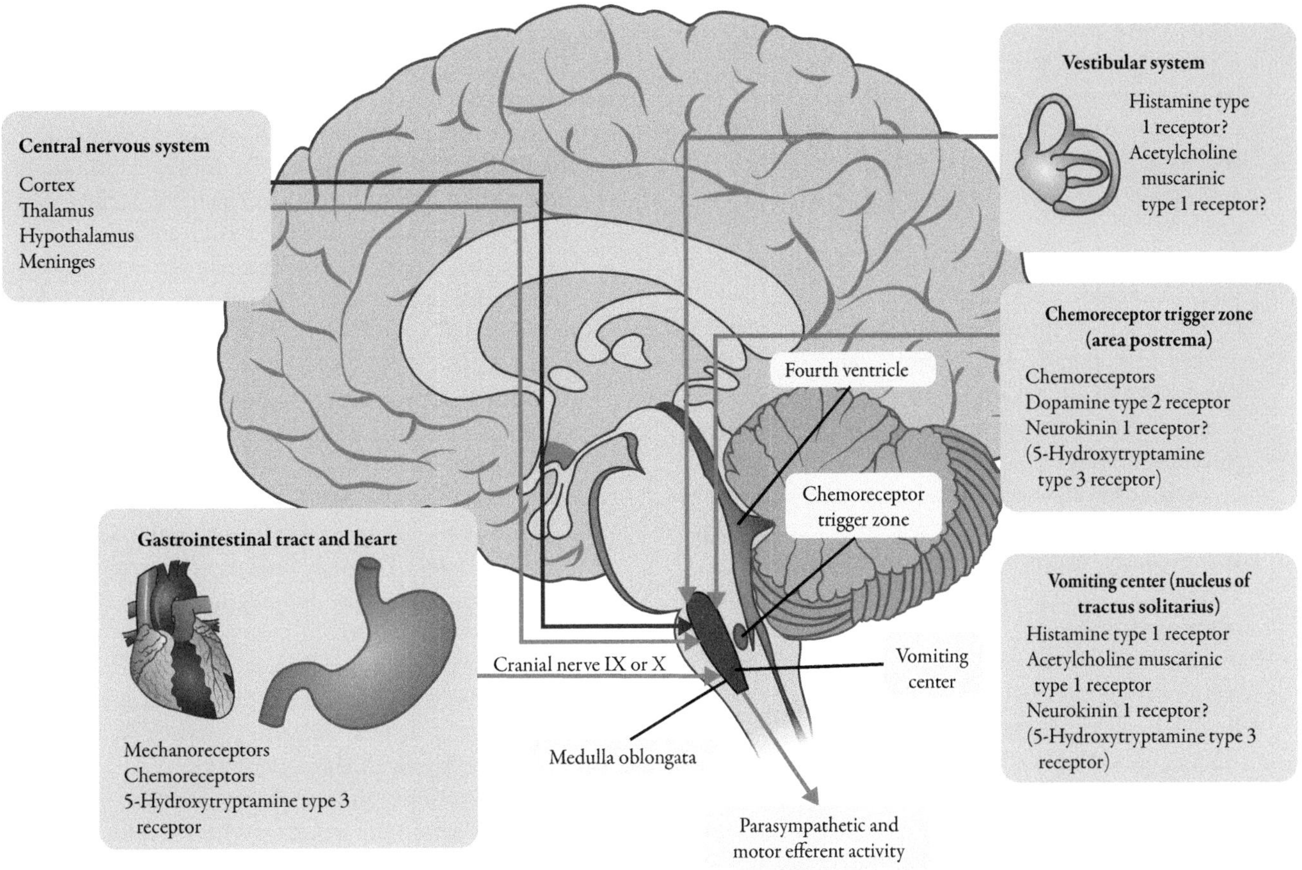

Figure 5.2 Medullary vomiting center and inputs that generate nausea.

- **Haloperidol** blocks central dopamine receptors and can result in extrapyramidal symptoms, so is less commonly used for nausea or vomiting.
- **Ondansetron** is an antagonist for 5HT receptors on enterochromaffin (EC) cells and the nucleus tractus solitarius.
- **Scopolamine** blocks M1-5 muscarinic receptors and can be used perioperatively via transdermal route to mitigate postoperative nausea and vomiting (PONV).
- Preoperative **oral sodium citrate** may minimize the deleterious effects of aspiration.

PERIOPERATIVE CONSIDERATIONS

- **Case classification** is used to estimate surgical site infection (SSI) risk.
  - **Clean cases** have a 1%–5% wound infection rate. These involve soft tissue only, for example an elective hernia repair. Technically, **no preoperative antibiotics** are needed for Class 1 cases, unless a prosthesis or mesh is being inserted. In that case, **cefazolin** may be given.
  - **Clean-contaminated cases** have a 3%–11% wound infection rate. These involve a controlled bowel or biliary transection. An example would be a simple, uncomplicated appendectomy. Patients generally need a preoperative dose of a **2nd-generation cephalosporin**. One exception to this would be an elective biliary procedure with no prior instrumentation of the biliary tract. Even though the cystic duct will be transected and controlled, these patients do not necessarily need preoperative antibiotics.
  - **Contaminated cases** have a 10%–17% infection rate and include instances of an uncontrolled breech of non-prepped bowel or the biliary tract, either intra-operatively or immediately prior to the operation (e.g., in a trauma situation). Antibiotics should be given pre-incision as well as for **24 hours postoperatively.**
  - **Grossly contaminated cases** have a >20% infection rate. Patients have preexisting peritonitis and soilage that predates surgery by hours or days. Pre-incision antibiotics should be continued for up to **4 days after source control**.
- **Patient classification** is by the American Society of Anesthesiologists (ASA)
  - ASA 1: healthy patient.
  - ASA 2: patients with a single, controlled comorbidity.
  - ASA 3: patients with multiple comorbidities that are sub-optimally controlled. Examples include end stage renal disease (ESRD) on hemodialysis (HD) and stable angina.
  - ASA 4: patients with comorbidities that are a constant threat to life. Examples include ESRD not on HD and unstable angina.
  - ASA 5: are moribund patients.
  - ASA 6: are organ donor patients.
  - The suffix "E" is added for any emergency case.
- **Frailty.** Patients designated as frail have abnormalities in the components below:
  - **Cognition** evaluated by a mini-mental status exam (MMSE) or mini-cog exam.
  - **Depressed mood** associated with exhaustion or social isolation.
  - **Comorbid diseases** as measured by the Charlson index.
  - **Weight loss and malnutrition**, including a low albumin.
  - **Disability** and limited mobility as measured by the capacity for activities of daily living (ADLs) and a timed up-and-go test of > 12 seconds to stand, walk 3 meters, and sit back down. Diminished psoas and quadriceps muscle volume may correlate with frailty.
  - Frailty is associated with longer hospital length of stay (LOS), readmission, need for discharge to a nursing home, and postoperative morbidity, delirium, and mortality.
- **Revised cardiac risk index (RCRI)** includes age >70; coronary artery disease (CAD) or previous myocardial infarction (MI), especially if within the prior 6 months; valvular disease, congestive heart failure (CHF) or impaired exercise tolerance <4 METs (inability to climb one flight of stairs or perform anything more than basic physical tasks); cerebrovascular disease; renovascular disease (creatinine >2); and insulin-dependent diabetes mellitus (IDDM).
  - **Low-risk procedures** include ocular, endoscopic, or soft tissue interventions. These do not need any preoperative testing unless the patient has MET <4 or multiple RCRI. An ECG and laboratory evaluation may be performed in those cases.
  - **Medium-risk procedures** include carotid endarterectomy (CEA) and most general and thoracic surgery cases. These patients require only an ECG and labs, unless they have **MET <4 or multiple RCRI**, in which case a **preoperative functional workup should be done.** This may include echocardiography, stress myocardial perfusion imaging, and possible cardiac catheterization.
  - **High-risk procedures** include major vascular and cardiac surgery. These patients generally all receive preoperative functional workup given their RCRI status.
- **Preoperative Medical Management**
  - It is best if patients **quit smoking** >4 weeks prior to surgery for maximum benefit. Otherwise, there is still a decrease in SSI rate for patients who stop smoking within this timeframe.

- **Beta-blockers should be continued** perioperatively for patients who are taking them normally. ACE inhibitors may be held due to preoperative hypotension.
- **Statins should be continued** to prevent sudden plaque rupture.
- **Stop coumadin 5 days prior to surgery**, and there is no need to bridge with heparin or enoxaparin unless the patient has a mechanical valve or is hypercoagulable or has a high **CHA_2DS_2-VASc score**,
- **CHA_2DS_2-VASc** helps predict stroke risk in patients with non-valvular atrial fibrillation. Its components include: **C**HF, **H**TN, **A**ge >5 (2 points), **D**iabetes, **S**troke (2 points), **V**ascular disease, **A**ge >65, **S**ex category (1 point for female).
- Patients are normally anticoagulated at baseline if their score is >1–2 for men or >2–3 for women.
- **Continue aspirin** right up to, and through, surgery if the patient has coronary stents. Avoid elective surgery within one month of placement of a bare metal stent or within 6 months of placement of a drug-eluting stent (DES).
  - If surgery is absolutely necessary within these time periods, **stop clopidogrel 5–7 days prior** to surgery if there is a high bleeding risk.
  - If the surgery has a low bleeding risk and the patient has a very high thromboembolic risk, it may be best to just continue clopidogrel throughout the perioperative period.
  - For a patient with both a high bleeding and high thrombosis risk, one may consider a **bridge with IV 2b/3a inhibitors** while holding clopidogrel.
- Stop oral hypoglycemics and short-acting insulin the morning of surgery. For intermediate or long-acting insulin, patients should take 3/4 dose the prior evening and 1/2 dose on the morning of surgery
- For patients taking daily glucocorticoids equivalent to a prednisone dose of 10 mg or greater, they should receive a **preoperative stress dose of hydrocortisone** 100 mg IV. It is recommended to continue their oral regimen after surgery. If they cannot take oral meds postoperatively, then IV supplementation with hydrocortisone is recommended.
- **Surgical care improvement program (SCIP)** includes a set of immediate perioperative considerations to improve outcomes
  - Smoking cessation preoperatively.
  - Continuing perioperative **beta-blockade** for patients previously taking them.
  - Chemical or mechanical **venous thromboembolism (VTE) prevention.**
  - Hair removal in the areas of a planned incision.
  - **Surgical site infection (SSI)** prevention with **intravenous antibiotics within 1 hour** of incision and intraoperative re-dosing based on the drug's half-life and the patient's blood loss. Approved preoperative antibiotics include cefazolin, cefuroxime, cefoxitin, cefotetan, ertapenem, and vancomycin as an alternative agent in patients with methicillin resistant staph aureus (MRSA) or a beta-lactam allergy. Prophylactic antibiotics should be discontinued within 24 hours of the procedure.
  - Perioperative temperature and glucose control.
  - Enhanced perioperative oxygenation in colorectal surgery.
  - Removal of urinary catheter within two days postoperatively.
- **Enhanced Recovery after Surgery (ERAS)**
  - These are a set of evidence based, multidisciplinary interventions to enhance postoperative recovery.
  - Principles are applied to elective resection in an ambulatory, well nourished, ASA class 1 or 2 patient who has not had prior surgery. They result in a **decrease in LOS and in *nonsurgical* complications** like urinary tract infection (UTI), pneumonia (PNA), and myocardial infarction (MI). However, they **do not alter the rate of *surgical* complications or mortality.**
  - Preoperative aspects include the following:
    - Prehabilitation exercises.
    - Nil per OS for **solids 6 hours prior** and for **liquids 2 hours prior,** so the patient is not in a fasting, catabolic state heading into surgery.
    - Placing a **thoracic epidural** for laparotomy.
    - Performing a **transversus abdominis plane (TAP) block** for laparoscopy.
  - Intraoperative aspects include:
    - Dexamethasone at induction and ondansetron at the end of surgery for **PONV prophylaxis.**
    - Maintain **normothermia** during the case.
    - Provide **local anesthetic** into areas prior to incisions being made.
    - Perioperative **supplemental O_2** to decrease SSIs in colorectal surgery.
    - Use a **separate instrument tray for closure.**
  - Postoperative aspects include the following:
    - **Scheduled Tylenol and nonsteroidal anti-inflammatories (NSAIDs)** and minimize narcotics.
    - **Oral intake** initiated immediately after surgery.
    - Sugar-free gum chewing.
    - The peripheral mu receptor antagonist, **Alvimopan,** to enhance recovery of motility.
    - **Avoiding the use of surgical drains and nasogastric tube (NGT)** postoperatively. NGT should be removed in OR, except for the case of esophagogastric surgery with a proximal anastomosis. In this case the NGT can be removed postoperative day 2.
    - Removing the urinary catheter early, regardless of the presence of epidural analgesia.

POSTOPERATIVE INFECTIONS

- **UTI** is the most common infection and can be prevented by avoiding urinary catheters, using sterile technique when inserting, and removing a urinary catheter within 2 days postoperatively.
- **Pneumonia (PNA)** is the most common cause of infectious related death
  - **Healthcare associated pneumonia (HCAP)** can occur in nursing home or long-term care residents, anyone with hemodialysis (HD) within 30 days, patients on home infusions, home wound care, or any person in the hospital for more than 48 hours. Anaerobes are common causative organisms.
  - **Ventilator associated pneumonia (VAP)** can occur in any patient intubated for more than 48 hours. Staphylococcus or Gram-negative rods (GNRs) are causative. Preventative measures include head of bed elevation, chlorhexidine oral swabs, and clearance of secretions above the endotracheal tube balloon.
- **Surgical Site Infection (SSI)**
  - Superficial SSI are in the dermis or subcutaneous layer
  - Deep incisional SSI are the level of fascia or muscle
  - Organ space SSI occur at the visceral surgical site
  - Risk factors include: older age, chronic comorbidities (higher ASA), obesity, malnutrition, immune suppression, alcohol, drugs, tobacco, long operations, hypothermia, hyperglycemia, hematoma or seroma, and excessive blood loss (with loss of intraoperative circulating antibiotics)
  - **Preventative measures** for SSI include:
    - Preoperative smoking cessation, weight loss for obese patients, enteral or parenteral nutrition for malnourished patients.
    - Enteric **topical antibiotics with a mechanical bowel prep.**
    - Whole body **chlorhexidine showering** to decrease MRSA complications but not necessarily SSI.
    - **Clipping hair** only when it interferes with the surgery site.
    - Surgeon **hand hygiene** (alcohol or traditional scrub).
    - Surgical prep with **chlorhexidine or an iodophor with alcohol.**
    - SCIP measures (see above) including **the appropriate timing and re-dosing of antibiotics,** perioperative temperature, and glucose control <180.
  - **Treatment** of most SSI is by opening the wound, irrigation, debridement if needed, and wet to dry dressings. Negative pressure wound therapy (NPWT) can be applied to larger wounds when they are clean. Antibiotics can be given for areas of cellulitis that extend beyond the wound edges or areas of debridement.
    - If SSI occurs within 48 hours post-op, consider bowel injury or necrotizing soft tissue infection (NSTI) by *Clostridium perfringens*, group A beta hemolytic streptococcus (GABS), or multiple organisms including bacteroides. *C. perfringens* produces an alpha exotoxin, and there are usually minimal WBC in the wound. GABS produces an exotoxin. Treatment of NSTI is with resuscitation, antibiotics, and immediate broad debridement until reaching well-anchored tissue. The wound may require serial debridements, and the antibiotics can be tailored once a causative organism is identified.
- Preoperative antibiotics given for abdominal sepsis should be continued for no more than 4 days after the source is controlled intra-operatively. A localized infected fluid collection can be considered controlled by an interventional drain of at least 7F size, and need not require operative control.
  - Patients should improve within 72 hours of true source management (whether surgical or drain).
  - Antifungals can be added if there is an upper gastrointestinal perforation, immune suppression, uncontrolled diabetes, or the patient is on parenteral nutrition.
  - May remove postoperative or IR drains if the output is persistently <20 ml per day.
- **Central line and peripherally inserted central catheter (PICC) infection** rates are similar. **Central line insertion site assessment (CLISA)** is a scoring system from 0 to 3. A score of 2 corresponds to 3 mm–6 mm of erythema around the catheter entry site without purulence and should prompt consideration of removal.
  - Diagnosis is confirmed by noting **>25 colony-forming units (CFUs) from catheter tip or 8 CFU greater than peripheral cultures.** If the line culture is positive but peripheral culture negative, consider line colonization. Can be prevented by **chlorhexidine scrub.**
  - Coagulase negative staph epidermidis is the primary organism; coagulation-positive staph aureus is #2, and fungi are #3.
  - Fibrin sheaths develop around catheters by 5–7 days.
  - PICC have increased thrombosis rates, especially if longer duration, patient is bacteremic, or tip is distal to the superior vena cava (SVC) rather than at the cavoatrial junction.

OTHER POSTOPERATIVE COMPLICATIONS

- **Acute kidney injury (AKI) or CHF** generally occur in older diabetic hypertensive patients with high ASA class. Of all complications, they have the highest increase in postoperative **mortality** and cardiovascular complications.

- Postoperative **hyperglycemia** doubles mortality, quadruples major adverse cardiac events (MACE), and increases AKI, infection, and pulmonary embolic risk.
- Postoperative **atrial fibrillation** lengthens ICU stay (no change in total LOS), increases cerebrovascular accident (CVA) risk, and possibly increases mortality.
- Postoperative **MIs** occur at 1–2 days due to thrombosis atop in situ plaque in elderly, diabetic patients with a history of CHF or symptomatic CAD.
- **Pulmonary complications** are more common if the patient has chronic obstructive pulmonary disease (COPD), pulmonary hypertension, CHF, an ASA score greater than 2, obstructive sleep apnea (OSA), smoking, age >50 years, or albumin level <3.5.
  - OSA is scored by the **STOP BANG criteria**. **S**noring, **T**ired, **O**bserved apnea, **P**ressure elevation, **B**MI>35, **A**ge>50, **N**eck circumference >40, **G**ender-male. A score >5 is moderate or severe OSA.
  - **Preoperative inspiratory training** for 2 weeks or more can decrease postoperative complications and ICU stay.
  - Baseline pulmonary function tests (PFTs) are not prognostic for general, non-thoracic surgery cases.
- **Postoperative nausea and vomiting (PONV)** is secondary to pain and opioids. Young female patients undergoing laparoscopic surgery are at higher risk. Prophylaxis is with antiemetics and glucocorticoids.
- **Delirium** occurs in elderly patients ASA class 3 or 4 designated as frail (see above) who are receiving opioids or anticholinergics. It increases LOS, postoperative nonsurgical morbidity and mortality, and the need for nursing home transfer. Initial non-pharmacologic treatment is preferable to medication.
- **Ileus** can be preemptively managed with gum chewing, early oral feeding, and epidural catheters. See Chapter 9, Small Bowel, for details.
- **Parotitis** occurs 4–12 days after surgery and is secondary to salivary stasis and duct obstruction. Prophylaxis is with sialogogues and treatment with anti-staphylococcal antibiotics and promotion of drainage.

6.

LAPAROSCOPIC PRINCIPLES

- **Positioning, Anesthesia, and Insufflation**
  - Most cases are done under general anesthetic; however, local or epidural anesthesia alone has been used in select cases like diagnostic laparoscopy and extraperitoneal hernia repair.
  - Arms are usually abducted for upper abdominal surgery. Arms are tucked on one or both sides for lower abdominal or pelvic surgery.
  - Allen stirrups are preferable to candy cane stirrups for lithotomy or modified lithotomy positions.
  - A modified decubitus position (between supine and lateral decubitus) with table rotation can give a practitioner the benefits of both positions.
  - Lateral or semilateral position is for surgery on retroperitoneal structures above the aortic bifurcation.
  - Trendelenburg position helps with working on retroperitoneal structures below the aortic bifurcation.
  - There is an increased likelihood of **postoperative nausea and vomiting (PONV)** in young, female, nonsmoking patients with a history of motion sickness who undergo laparoscopic procedures and receive opioids. It can be prevented by prophylactic antiemetics, glucocorticoids, decreased opioid usage, and minimization of nitrous oxide and volatile anesthetics.
- **Insufflation**. The insufflator can go up to 10 L/min. The practitioner sets the maximum pressure and flow rate, and the insufflator displays the amount of gas instilled thus far. Warm, humidified CO_2 gas may be better than regular CO_2 in longer operations.
  - Veress needle insertion can be directly through the umbilicus, through a periumbilical incision, or at **Palmer's point** (mid-clavicular line in a subcostal position).
  - One may **aspirate and perform a saline drop test** to confirm intraperitoneal positioning. Alternatively, one may just attach tubing at low insufflation flow to confirm correct position, before transitioning to higher flow. The abdominal wall may be lifted to facilitate saline drop or insufflation. Diffuse tympany should be noted as the peritoneal cavity insufflates.
  - Do not place a Veress needle through mesh or blindly in the vicinity of a prior surgical scar.
  - In a hostile abdomen, an open (Hasson) entry technique can be utilized for direct port placement (see Hasson technique below).
  - **1–3 L of CO_2 is usually needed to reach 10–15 mmHg** pressure within the abdominal cavity, provided the patient is relaxed. One may lower insufflation pressures in pediatric and pregnant patients.
  - CO_2 has a high diffusion quotient into blood and then across the alveolar membrane to be exhaled. This increases pCO_2 and end-tidal CO_2 ($ETCO_2$). Less than **1%** of patients have a CO_2 pulmonary gas embolism.
  - Pneumoperitoneum **displaces the diaphragm cephalad** and decreases lung functional residual capacity (FRC), **increases afterload**, transiently increases preload followed by a sustained **decrease in venous return** from the lower extremities due to intraperitoneal pressure, and decreases renal perfusion, resulting in elevated antidiuretic hormone (ADH), renin, aldosterone, and a diminished urine output. *See Figure 6.1.*
  - Vagally mediated bradycardia can also occur with insufflation. Bradycardia plus the impaired venous return and high afterload result in **decreased cardiac output.**
  - Lower flow rates initially allow for hemodynamic monitoring during insufflation, and less abdominal and shoulder pain postoperatively.
  - One may lose working space if the insufflator is insufficient, if there is aggressive suctioning in the field, or if there is an increase in abdominal muscular tone.
  - **High pressures** may indicate tubing obstruction or occlusion, or inadequate anesthesia.
  - **Low pressure** but high flow indicates a leak into the external environment, abdominal wall, or hollow viscous. Low pressure/low flow indicates the insufflator is not on, or the gas tank is low.
  - Most of the CO_2 pneumoperitoneum resolves in less than 1 hour. Post-op shoulder pain can last for 1–3 days, however.
  - Nitrous oxide as an inhaled anesthetic should be avoided in the setting of a small bowel obstruction.

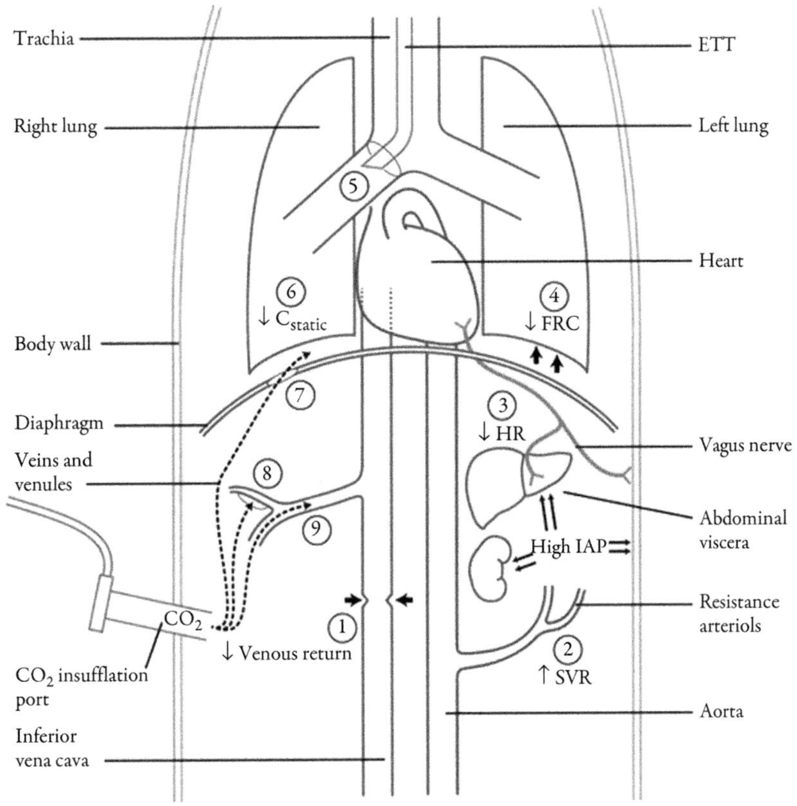

Figure 6.1 Physiologic effects of CO_2 pneumoperitoneum for laparoscopy.

Nitrous oxide should not be used intraperitoneally because of fire risk from an unprepped colon perforation releasing methane.

- **Cameras, Trocars and Ports**
  - Zero degree scopes are better in crevices, deep in the pelvis, or in the mediastinum. Diameters range from 2 mm–10mm. Larger diameter decreases susceptibility to damage and improves light and image quality. Lengths range from 30 cm–45 cm. Longer scopes have increased susceptibility to damage.
  - Trocars can have a knife blade tip, plastic tip, protective shield, or optical tip.
  - The **Hasson technique** uses a blunt tip trocar via a periumbilical incision. The umbilical stalk is grasped, fascia incised longitudinally up the stalk, the pre-peritoneal space bluntly opened, peritoneum grasped and opened, and blunt tip trocar inserted.
  - Trocars specific for obese patients can be 10 cm or greater in length.
  - Prior to exiting the abdomen, look for incidental injuries from trocar placement, and decrease pneumoperitoneum for evaluation of venous hemostasis.
  - **Port positioning** will be dependent on the operation being performed, and is reviewed in each pertinent chapter. For a diagnostic laparoscopy or ventral hernia repair, all ports can be placed on one far lateral side of the abdomen. In diagnostic laparoscopy, **biopsies** may be obtained via the following:
    - 20g–22g **fine needle aspiration (FNA) for** cell types without architecture. The plunger should be pulled back as multiple passes are taken through the specimen. Suction should be released before removing the needle.
    - **Core needle biopsy (CNBx)** is via a 14-18g needle and provides architecture.
    - **Incisional or excisional biopsies** both provide more tissue than core needle samples. The latter is used for lymph nodes or small peritoneal lesions. Incisional or excisional wedge resections are useful for peripheral liver lesions.
    - **Peritoneal wash** is done by collecting any ascites or infusing 100 ml or more of saline and retrieving it after agitation for 3–5 minutes. This can be used for staging malignancies.
- **Instrumentation**
  - **Cautery.** The current density transmitted with electrocautery is increased with decreased tissue resistance and a small surface area of contact.
    - **Cutting mode** is a **continuous waveform with low voltage** and rapid heating of deeper tissue. Vaporization and cell lysis occur quickly, and there is minimal

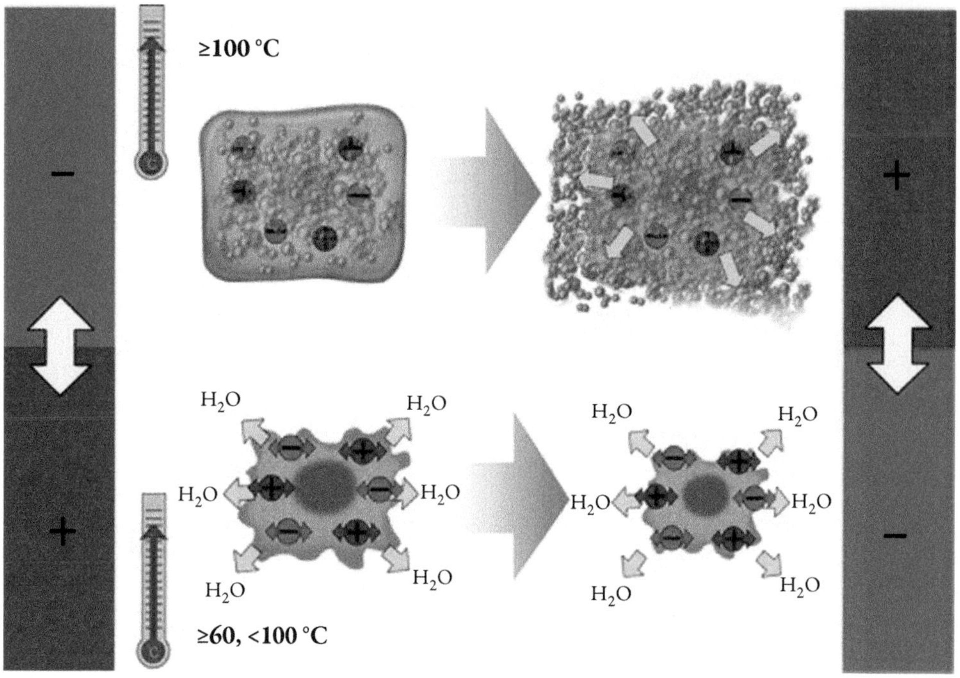

Figure 6.2 Cellular effect of cautery temperature.

charring and coagulation. Temperatures reach 100°C or greater. *See Figure 6.2.*

- **Coagulation mode** is an intermittent waveform with **high voltage and superficial coagulation.** Temperatures are generally below 100°C.
- Blended modes are a combination of the two.
- **Monopolar current** traverses from the application device, through the body, and exits by way of a pad. It may be diverted unintentionally due to proximal contact between the cautery device and an unseen structure (**direct coupling**) or may be unintentionally transferred to another instrument that stores the energy and dissipates it to tissue upon subsequent intermittent contact (**capacitive coupling**). If a ligated structure with no vascular inflow or outflow is cauterized, heating can be excessive and result in tissue damage. This is termed the **narrow return circuit phenomenon**.
- **Bipolar current** transmits only between the two jaws of the instrument. It can work as a sealing device for vessels up to 7 mm in diameter. Bipolar current cannot cut through metal within its jaws. There is less lateral spread of current, and it works better in a "wet" environment.

- **Argon** provides superficial coagulation at a depth of 3 mm without direct instrument to tissue contact.
- **Cryoablation** is to −40°C with 1 cm margins. This is not used near vessels given the warmth of blood.
- **Ultrasonic coagulation** is via a rapidly oscillating blade at 50,000 HZ and a static second jaw. The device can transmit energy to adjacent structures unintentionally, since there is a longer cooling time after each activation. Lower power settings are for hemostasis, and high power settings for cutting. Temperatures can range from **50°C–300°C.**
- **Laparoscopic suturing** is with braided, colored suture on a tapered needle through two ports at least 10 cm apart ~~and a central~~ camera port. The needle can be inserted through a 10mm or 12mm port. Suture should be 15 cm length for **intracorporeal knot tying. Extracorporeal knot tying** puts greater strain on tissue due to the angle and pull of thread. Suture should be about 75 cm length for extracorporeal knots.
- **Laparoscopic staplers** are 30 mm–60 mm in length and have varying staple heights. Staple height of 2.5mm has better hemostasis and is useful for thinner tissue and vasculature. Staple height of 4 mm is for thicker tissue like the stomach.
- **Single incision laparoscopic surgery (SILS)** can afford improved cosmesis and possibly improved pain, but longer operative time.
- **Natural orifice transluminal endoscopic surgery (NOTES)** includes a variety of procedures like percutaneous endoscopic gastrostomy (PEG), endoscopic retrograde cholangiopancreatography (ERCP), transgastric or transvaginal cholecystectomy, per oral endoscopic myotomy (POEM), endoscopic submucosal dissection (ESD), transanal endoscopic microsurgery (TEM), and transanal minimally invasive surgery (TAMIS).

ALIMENTARY TRACT AND ADJUNCT SOLID ORGANS

7.

ESOPHAGUS

ANATOMY AND PHYSIOLOGY

- The esophagus has no serosa. It has longitudinal and circular muscle layers, within which lies the submucosa, muscularis mucosa, lamina propria, and squamous mucosal lining.
- Striated muscle is present in the cephalad one-third, and a transition to smooth muscle occurs in the caudal two-thirds.
- There are three main types of peristaltic waves mediated by esophageal muscle.
  - **Primary waves** start at the initiation of swallowing.
  - **Secondary waves** start from esophageal distension with a food bolus.
  - **Tertiary waves** are nonperistaltic and nonprogressive.
- The **upper esophageal sphincter (UES)** is the cricopharyngeus muscle and resides at 15 cm from the oral aperture on endoscopy.
  - The UES has a baseline of 60 mmHg intraluminal pressure at rest and relaxes to 15 mmHg to allow passage of a food bolus.
- The **lower esophageal sphincter (LES)** is a 2 cm–4 cm long high pressure zone of the distal esophagus residing at about 40 cm from the oral aperture. *See Figure* 7.1.
  - It is a composite of the phrenoesophageal ligament, diaphragm crura, circular and longitudinal esophageal muscle fibers, and gastric sling fibers.
  - It has a 6–15 mmHg intraluminal pressure at rest, relaxing to 0 mmHg with a food bolus.
  - Relaxation is mediated by the vagal nerve. If relaxation is unguarded by upstream incoming peristaltic waves, the patient may have physiologic reflux. If this ascends to the hypopharynx and larynx, it is known as laryngopharyngeal reflux (LPR).
- The **diaphragmatic hiatus** is marked endoscopically by noting the area of pinch on inspiration or sniff.
- The **gastroesophageal junction (GEJ)** is marked endoscopically by the cephalad **end of gastric rugal folds**. If it is above the diaphragmatic hiatus then the patient has a hiatal hernia.
- Squamous epithelium lines the entirety of the esophagus. The **squamocolumnar junction (SCJ)** is where the squamous cells transition to gastric mucosal cells. It should be located at the same level as the GEJ. If the SCJ is cephalad to the GEJ, then the patient has Barrett's columnar metaplasia within the esophagus (see below).
- **Arterial perfusion** originates from separate sites along the length of the esophagus. *See Figure* 7.2.
  - Cervical segment perfusion comes from inferior thyroid arteries.
  - Thoracic segment perfusion is through branches directly from the aorta.
  - Abdominal esophagus perfusion is via the left gastric artery and inferior phrenic arteries.
- **Venous drainage** heads into the azygous and hemizygous systems.
- **Lymphatic drainage** of the upper two-thirds of the esophagus runs cephalad, and the lower one-third runs caudal to the celiac nodes.
- The **right vagus nerve** runs posteriorly along the length of the esophagus and bifurcates into a celiac plexus branch and the posterior gastric branch.
  - There may be a high takeoff of a posterior gastric branch. This is known as the criminal nerve of Grassi.
- The **left vagus nerve** courses anteriorly and may interdigitate with esophageal muscle. It gives off a hepatic branch that runs in the hepatogastric ligament, with the main body of the nerve continuing as the anterior gastric branch.
- The esophagus is extrinsically compressed at the following sites, where intraluminal contents may consequently lodge:
  - Cricoid cartilage—14 mm esophageal diameter at this site.
  - Aortic knob and left bronchus—16 mm esophageal diameter in this area.
  - Left atrium.
  - Diaphragmatic hiatus—19 mm esophageal diameter here.

MOTILITY DISORDERS

- **PATHOPHYSIOLOGY**
  - **Achalasia** is defined by a failure of the LES to relax and an absence of upstream esophageal body peristalsis. It is secondary to a lymphocytic myenteric

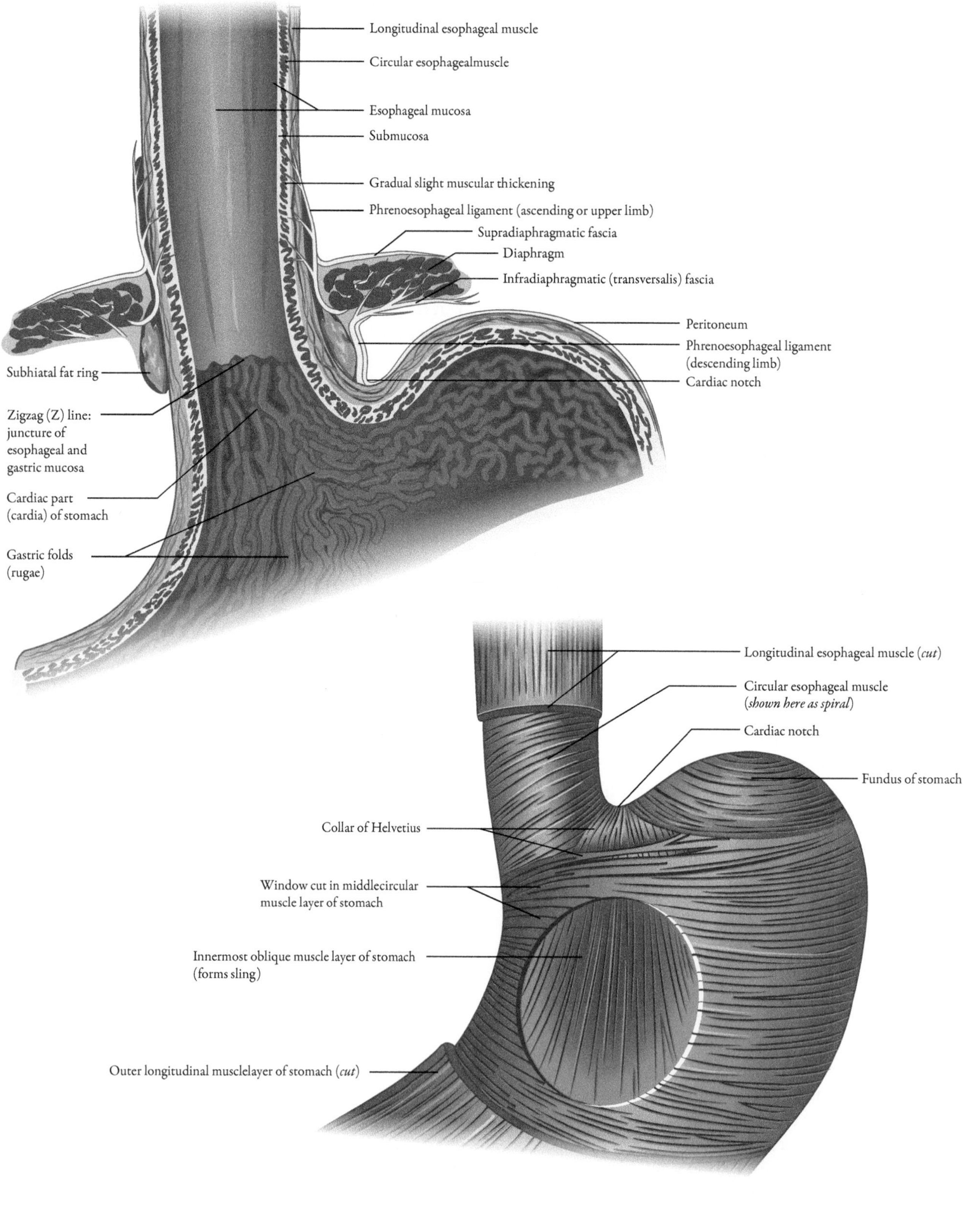

Figure 7.1 Components of the lower esophageal sphincter.

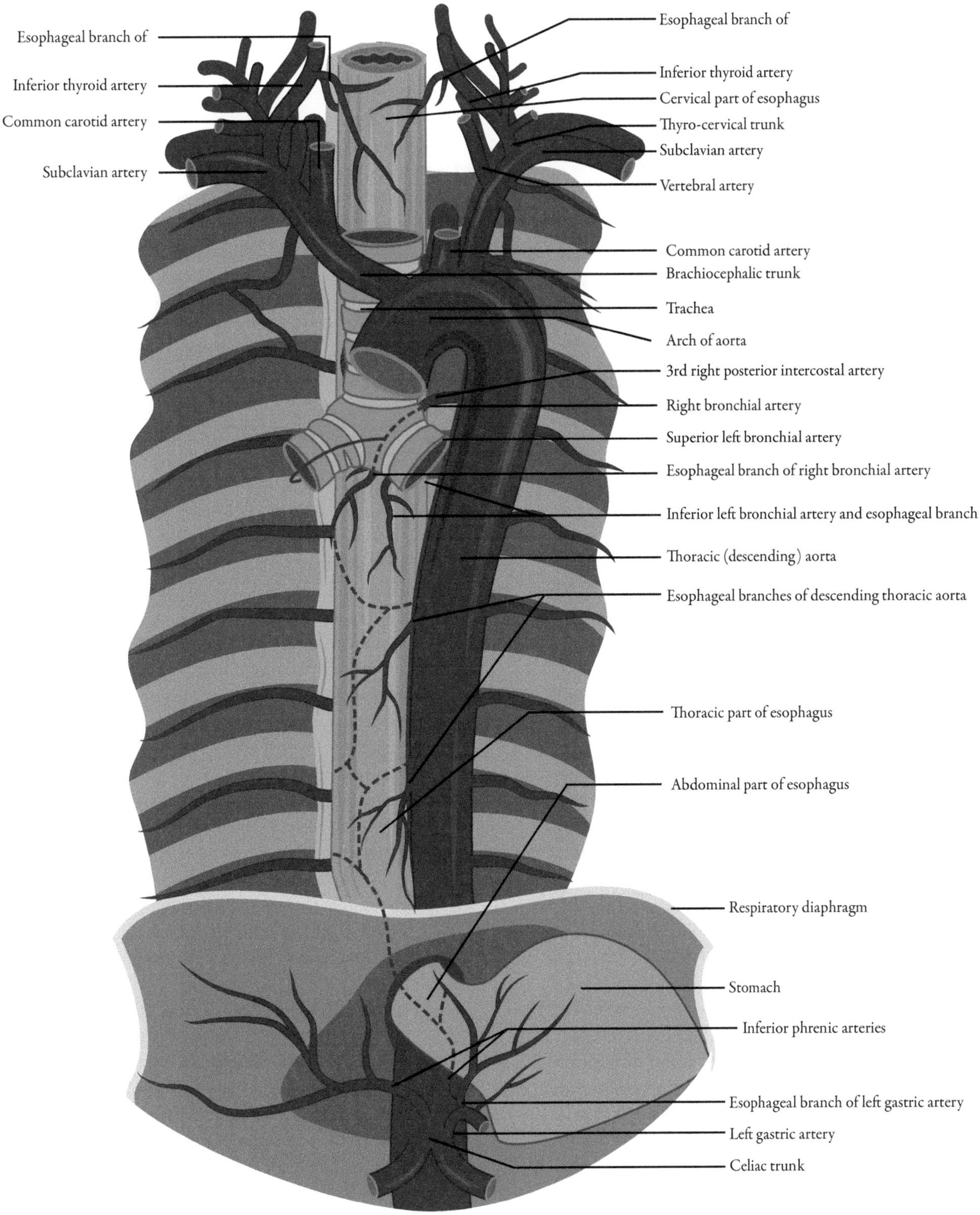

Figure 7.2 Arterial perfusion of the esophagus.

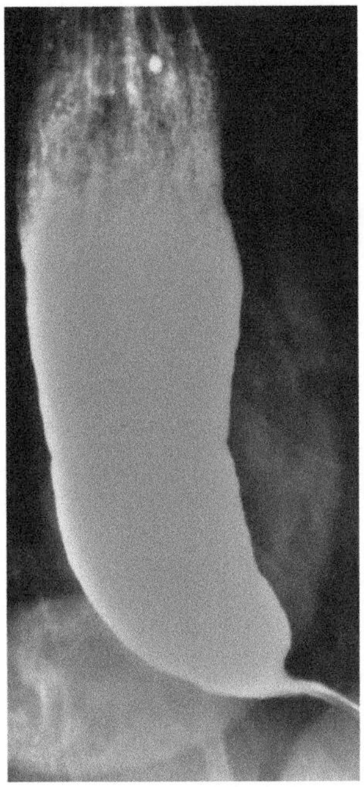

Figure 7.3 Achalasia with a dilated proximal esophagus on contrast examination.

inflammation, resulting in fibrosis and neuronal degeneration at the LES. *See Figure* 7.3.

- **LES pressure remains >5 mmHg while swallowing.** This and the proximal aperistalsis produce the classic symptoms of dysphagia to liquids as well as solids.
- Patients may also complain of regurgitation of ingested liquids or particulate matter, weight loss, and respiratory insufficiency
- Three types of achalasia are defined by their manometric findings:
  - Type 1: aperistalsis and a non-relaxing LES.
  - Type 2: aperistalsis, non-relaxing LES, but diffuse pressure columns in the esophagus.
  - Type 3: impaired LES relaxation with multiple uncoordinated low-amplitude esophageal body contractions. These patients may have chest pain.
- Over years, the esophagus may become dilated and tortuous, termed a sigmoid or megaesophagus. Patients may develop epiphrenic pulsion diverticula as well.

- **Esophagogastric junction outflow obstruction (EGJOO)** is the term used if a patient still has some meaningful peristalsis in the setting of a non-relaxing LES. Technically this is not true achalasia.
- **Hypertensive LES** can occur without impaired peristalsis.
- **Nutcracker esophagus** is the term used for **high-amplitude *peristaltic* contractions** with normal LES tone.
- **Diffuse esophageal spasm** is characterized by exaggerated, ***unorganized, nonperistaltic* waves** with normal LES tone.
- **Scleroderma** is an autoimmune mediated fibrous replacement of esophageal smooth muscle resulting in loss of peristalsis and a hypotonic LES. This promotes gastroesophageal reflux which can lead to secondary distal esophageal stricture formation and eventual dysphagia.

- **DIAGNOSIS**
  - Motility disorders are diagnosed by **manometry after an upper gastrointestinal (UGI) contrast study and esophagogastroduodenoscopy (EGD)** have ruled out obstruction, high-grade dysplasia, or malignancy.
  - Achalasia may have a bird's beak narrowing of the distal esophagus on UGI contrast study.
  - A **pH probe** may be needed to diagnose gastroesophageal reflux if the above studies do not yield a diagnosis.
- **TREATMENT AND OUTCOME**
  - For achalasia one can perform a laparoscopic myotomy, pneumatic balloon dilation, per oral endoscopic myotomy (POEM), chemical myotomy with botulinum toxin, or medical management.
    - 1. **Laparoscopic Myotomy**
      - Ports should be placed at the umbilicus, B midclavicular subcostal spaces, and B anterior axillary spaces.
      - Dissect hepatogastric ligament to right crus, as well as greater curvature short gastrics past angle of His to left crus, and have the two planes meet. Extend dissection on anterior esophageal surface cephalad into mediastinum and protect the anterior vagus nerve.
      - Incise longitudinal and circular layers at 11 o'clock or 2–3 o'clock from **6 cm above GEJ to 3 cm below,** onto stomach cardia. Bluntly separate esophageal muscle from submucosa for 50% of the enteric circumference to prevent postoperative scar from bridging the myotomy. *See Figure* 7.4.
      - Add a ***partial* wrap** and gastropexy the wrap to crus. The myotomy treats dysphagia, and the wrap prevents reflux. 360-degree wraps are not used, because the esophagus generally has impaired motility.
        - **Anterior Dohr wrap** anchored to myotomy edges is used most often, since it does not require posterior dissection around the esophagus.
        - If a retroesophageal dissection had to be performed, then a **posterior partial Toupe wrap** can be done.
        - Another option is the **intrathoracic Belsey 270-degree wrap** around the distal 4 cm of esophagus via a left thoracotomy. This is used in the setting of a hostile abdomen with a short intra-abdominal esophagus.
      - If the patient has a very dilated sigmoid esophagus, myotomy still provides relief and should be done before resorting to an esophagectomy. The dilated esophagus will likely have poor tone; hence no wrap need be performed, since this will result in

Figure 7.4 Laparoscopic esophagomyotomy. Should extend from 5–7 cm on esophagus down through 3 cm over the cardia of the stomach.

postoperative dysphagia. The patient will need postoperative proton pump inhibition (PPI).
- Laparoscopic myotomy and a partial wrap is complicated by perforation in 2% and postoperative gastroesophageal reflux (GERD) in <10%.

- 2. **Pneumatic Dilation**
  - Performed with a 3 cm–4 cm diameter balloon. This dilates the LES to a goal resting pressure of <10 mmHg. Reflux after the procedure is treated with **PPI**. Patients may require serial dilations.
- Both pneumatic dilation and myotomy may result in **relapse after 5 years in 10%–20%** of cases. This may have been due to an inadequate myotomy, a tight wrap, malignancy, progressive fibrosis, or postoperative GERD causing a stricture.
- Recurrence can be treated with the other modality, POEM, or an esophagectomy if megaesophagus has set in despite an adequate Heller myotomy.
- 3. **Per Oral Endoscopic Myotomy (POEM)**
  - Upper endoscopy allows the practitioner to make a mucosal incision well above GEJ, permitting the scope to traverse inferiorly in a submucosal tunnel. A **circular muscle myotomy** from 7 cm above to 2 cm below GEJ is performed. The longitudinal muscle is left alone. *See Figure 7.5.*
    - This relieves dysphagia well, but may result in a higher rate of **postoperative reflux** at up to 50%, in comparison to a wrap.
- 4. **Chemical Myotomy**
  - Botulinum toxin is injected at four quadrants above the squamocolumnar junction. This relaxes the LES musculature for 1–6 months.
  - It is reserved for patients who are not candidates for surgery or pneumatic dilation. About 50% of patients need repeat treatment.
- 5. **Medical Management**
  - Nitrates, calcium channel blockers, or cyclic-AMP phosphodiesterase inhibitors may be used for those who are not candidates for surgery or dilation, and have failed Botox injection.
- **Hypertensive LES** may be treated with medical management followed by myotomy if that fails.
- **Nutcracker esophagus** is managed with calcium channel blockers (CCBs) and nitrates rather than surgery.
- **Diffuse esophageal spasm** is treated first with CCB, nitrates, and cAMP phosphodiesterase inhibitors. A **myotomy of the esophagus from the level of the aortic arch down to 1 cm above the GEJ** is performed if medical management fails. Since the LES is not hypertonic, there is no mandate for the myotomy to extend through it.
- **Scleroderma** management is medical and aimed at the underlying process. However, end stage patients may need an esophagectomy.
- Isolated impaired peristalsis can be managed with metoclopramide.

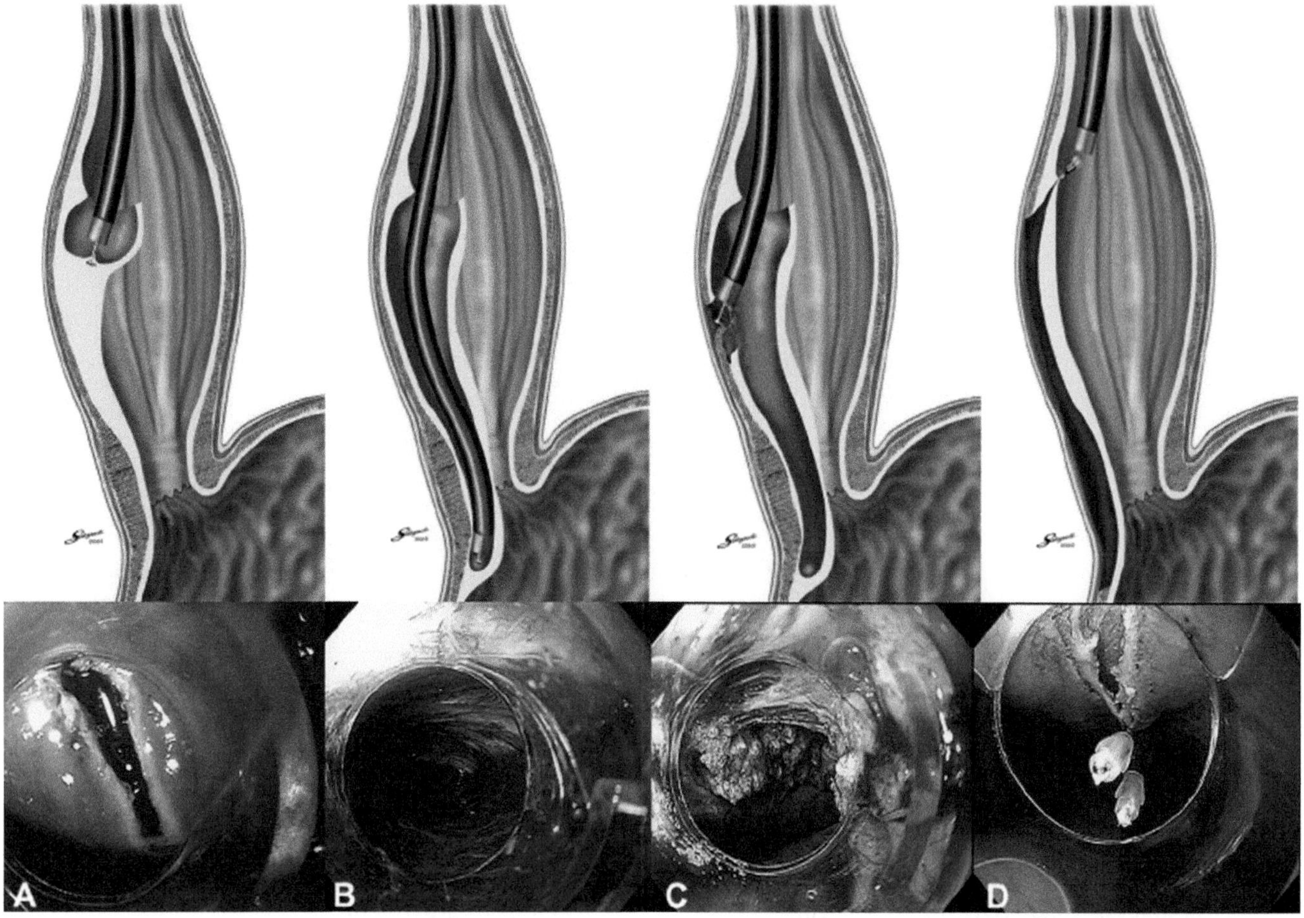

Figure 7.5 Per oral endoscopic myotomy. The circular muscle layer is cut after a submucosal tunnel is traversed by the endoscope. The longitudinal muscle is left intact.

GASTROESOPHAGEAL REFLUX (GER)

- **PATHOPHYSIOLOGY**
  - The LES guards against retrograde flow of gastric acid.
  - If the muscular apparatus becomes hypotonic (resting pressure diminishes below 6 mmHg) and the length of the LES is shortened (<2 cm total length or <1 cm intra-abdominal length) then the SCJ is exposed to gastric acid.
  - Also, gradual unrolling of the GEJ and loss of the angle of His end up exposing the SCJ to gastric acid.
  - Gastric acid bathing esophageal mucosa results in postprandial or supine heartburn symptoms and eventually metaplasia of squamous to columnar mucosa.
  - Acid may ascend to the hypopharynx if propagating esophageal waves are inadequate to clear the refluxate. This is termed laryngopharyngeal reflux (LPR).
  - Any associated respiratory symptoms can be due to (1) a vagal reflex in response to gastric acid in the esophagus, or (2) actual refluxate entering the hypopharynx with passage of contents past the laryngeal inlet and into the tracheobronchial tree.
- **DIAGNOSIS**
  - The diagnosis of GERD is clinical and based on the characteristic episodic retrosternal discomfort relatable to the postprandial state or recumbency.
    - If there are no *concerning associated* symptoms, one may empirically start a PPI and escalate dosage if needed. Most cases respond to this.
    - **Failure of PPIs, or the presence of concerning associated symptoms listed below, require further workup.**
      - Dysphagia or odynophagia can be suggestive of a mechanical obstruction or mucosal pathology that needs to be ruled out with either UGI contrast or endoscopy.
      - Bloating can be from impaired gastric emptying that can be diagnosed with a nuclear medicine gastric emptying study.
      - Epigastric or lower chest pain is suggestive of esophagitis or a peptic ulcer in the distal esophagus or stomach.

- Early satiety could indicate a paraesophageal hiatal hernia residing within the mediastinum and decreasing available stomach volume.
- Dyspnea occurs when the herniated stomach occupies a large portion of the mediastinum.
- Regurgitation of particulate matter, aspiration, weight loss, or anemia from a peptic or Cameron ulceration are other concerning features.

- **Additional Workup**
  - **EGD** may demonstrate reflux esophagitis, stricture, esophageal ulcer, or Barrett's metaplasia with or without dysplasia.
    - **Esophagitis** is the term given to mucosal inflammation, most commonly after long-standing exposure to gastric acid.
      - Grade A: hyperemia, <5 mm mucosal break.
      - Grade B: superficial erosions or ulcers, >5 mm mucosa break, limited by mucosal folds.
      - Grade C: extensive ulceration or narrowing. The mucosal break extends across mucosal folds, but <75% circumference.
      - Grade D: fibrotic stricture or mucosal break >75% circumference.
    - A **stricture** at the squamocolumnar junction is generally the fibrotic sequelae of recurrent inflammation and becomes symptomatic once the esophageal lumen is narrowed to <12 mm. It requires a **biopsy** to exclude dysplasia or neoplasia.
    - **Barrett's metaplasia** is a mucosal transformation from squamous to columnar as a protective strategy against acid exposure.
      - It appears as red velvet mucosa with **goblet cells** effectively extending the squamocolumnar junction of variable lengths cephalad from the GEJ in an asymmetric or circumferential manner.
      - It carries an adenocarcinoma risk of 0.2%–0.5% per year, so **four quadrant biopsies** should be taken at **every 1–2 cm intervals** to evaluate for dysplasia or neoplasia.
      - **Surveillance endoscopy** with biopsies should continue every year if there is a >3 cm length of Barrett's, and every two years if <3 cm length of metaplasia, provided there is no dysplasia.
      - Some Barrett's changes may reside deep into the squamous mucosa, termed buried subsquamous Barrett's. This may not be detected by endoscopy. Such metaplasia would persist after superficial mucosal ablation.
    - **Ulcers** are focal areas of erosion through the squamous mucosa or within a metaplastic columnar segment. They can extend into the submucosa and muscularis propria.
    - **Eosinophilic esophagitis** appears as ringed "feline" mucosa with eosinophils on biopsy. Patients have odynophagia and symptoms of heartburn. However, they are refractory to PPI.
  - **pH probe is used if an EGD is not diagnostic for reflux esophagitis.**
    - Patient should hold their PPI for 7–10 days, and the practitioner can place a continuous naso-esophageal probe or a Bravo probe clipped 5 cm above the GEJ.
    - A **DeMeester score** is calculated based on the number of episodes of pH<4, duration of episodes, and position. A score of >14.72 is abnormal.
  - Endoscopic impedance testing can diagnose reflux regardless of pH, for example in the setting of bile reflux. However, this is used less often.
  - **Manometry** can be used to diagnose esophageal dysmotility or a short and hypotonic LES.

- **TREATMENT**
  - The mainstay of treatment for GERD is **lifestyle and diet modifications, and PPI**. Surgery or other invasive therapy is reserved for patients who are refractory to, or cannot tolerate, medical management.
  - **Fundoplication with or without cruroplasty.**
    - In addition to failure of medical management, this procedure is utilized for individuals with persistent Barrett's metaplasia (without high-grade dysplasia or carcinoma on biopsy), nocturnal reflux on pH probe, or high-grade reflux esophagitis on EGD.
    - Young patients with severe symptoms who do not want to take lifelong PPI may also be candidates for surgery.
    - **Severely obese patients with a body mass index (BMI) >35 should be treated with weight loss surgery, not fundoplication.**
    - The procedure is most commonly performed laparoscopically and begins with division of short gastric vessels, takedown of the phrenoesophageal ligament, exposure of left crus, and opening of the hepatogastric ligament to expose the right crus.
    - **Cruroplasty** can then be performed if the hiatus is indeed enlarged. A biologic mesh may be placed in a horseshoe fashion.
    - Some patients with preoperatively diagnosed gastroparesis may receive a pyloromyotomy at this time to minimize postoperative gas-bloat syndrome.
    - A fundal wrap is usually 1 cm–3 cm in length, with sutures incorporating part of the anterior longitudinal esophageal muscle fibers.
      - A **Toupe or partial** wrap is 270 degrees, and **Nissen** wraps are 360 degrees. *See Figure* 7.6.
      - Having a 50–60F bougie in place is an option during wrap construction.
      - **Belsey Mark IV** anterior 270-degree wrap through a left posterolateral thoracotomy may be employed for a hostile abdomen, failed laparoscopy, very large diaphragmatic hernia, or a predicted short abdominal esophagus. The intrathoracic dissection allows the esophagus to be freed cephalad, and therefore a

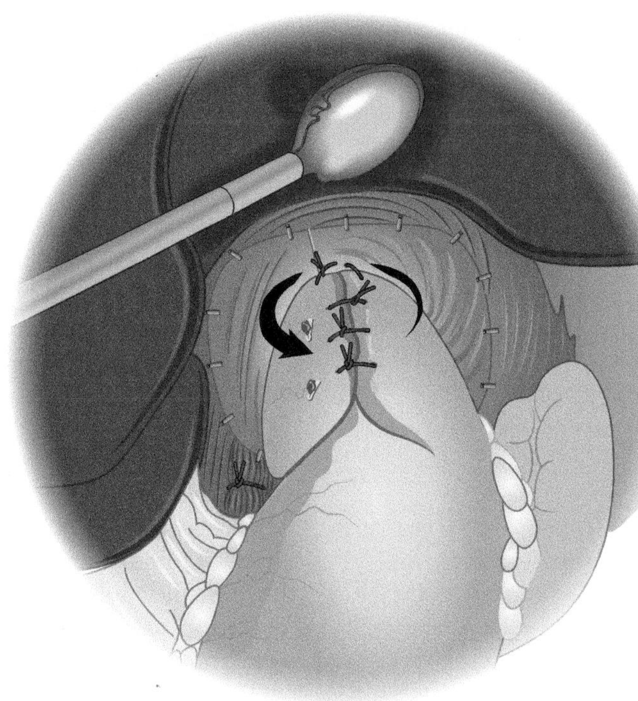

Figure 7.6 Laparoscopic 360-degree fundoplication with cruroplasty.

greater length of intra-abdominal esophagus can be achieved.

- **Collis gastroplasty** involves stapling at the angle of His inferiorly into the cardia of the stomach at an angle parallel to the left lateral esophageal wall.
  - This can be performed if the length of abdominal esophagus is inadequate. One must ensure the fundal wrap lies above the native stomach cardia after the gastroplasty.
- **Magnetic sphincter augmentation** is a device placed laparoscopically between the posterior vagus and esophageal wall with minimal dissection of the phrenoesophageal ligament. It can be used in patients with medically refractory reflux, adequate intra-abdominal esophagus length, and a hiatal hernia <3 cm. In this population it has been proven non-inferior to standard anti-reflux surgery at 5 years follow-up. Erosion of the device into the enteric lumen has been reported and requires device removal.
- **Transoral incisionless fundoplication** is an endoscopic route to constructing a wrap in a patient with the appropriate anatomy. Patients may still need a laparoscopic cruroplasty.
- **Treatment for other reflux related conditions.**
  - **Barrett's metaplasia with dysplasia**
    - If **low grade dysplasia (LGD)** is noted, surveillance continues every 6 months and may warrant anti-reflux surgery, after which dysplasia and Barrett's changes may regress.
  - For **high-grade dysplasia (HGD) or carcinoma in situ (CIS)** on biopsy
    - **Radiofrequency ablation (RFA)** treats the superficial 0.5 mm of mucosa and is superior to cryotherapy, argon plasma coagulation, or photodynamic therapy. Since it is superficial, long lengths or circumferential areas of dysplasia can be treated, provided there is no nodularity.
    - **Endoscopic mucosal resection (EMR)** can be attempted if the area of HGD or CIS is nodular, provided that it is <2 cm and endoscopic ultrasound (EUS) or fine needle aspiration (FNA), or any other imaging employed, is not concerning for lymphatic or metastatic involvement. EMR involves a submucosal saline lift injection and snare polypectomy, sometimes with the assistance of a suction cup.
- **Strictures** are managed with dilation via a bougie or balloon after a biopsy excludes dysplasia and neoplasia.
  - The target diameter for dilation is 14 mm. Most patients improve after 1–3 dilation attempts, but generally still need a fundoplication to address the reflux that caused the stricture. A Collis gastroplasty may need to be performed at time of surgery, since the presence of a stricture is often predictive of a shortened intra-abdominal esophagus.

    - If any HGD or CIS are discovered, management would be similar to these entities in the setting of Barrett's metaplasia above.
  - **Esophageal ulcers** are managed similar to gastric ulcers, after biopsies exclude dysplasia or neoplasia.
  - **Eosinophilic esophagitis** is treated with **oral corticosteroids and allergen avoidance.** Dilation is not employed.
  - Note that if any **invasive carcinoma** is found in the above workups for metaplasia, strictures, or ulceration, the patient must be staged and the treatment algorithm should be dictated by the staging. Details in the carcinoma section below.
- **OUTCOMES**
  - If Barrett's metaplasia (with or without low grade dysplasia) was present initially, then endoscopic surveillance should continue after anti-reflux surgery. However, **if HGD or carcinoma were seen, this must be treated before any anti-reflux surgery is performed.**
  - An **iatrogenic capnothorax** from breach of the pleural lining during the mediastinal dissection usually self-resolves in the early postoperative period.
  - A barium esophagram can be done on postoperative day (POD) 1 to establish a baseline appearance and ensure adequate gastric emptying. This will exclude an iatrogenic vagal nerve injury causing high pyloric tone.
  - A soft diet is instituted for 6 weeks thereafter.
  - Over **75% of patients are free of PPI therapy within 10 years of surgery.**
  - Up to 6 weeks of **dysphagia or bloating** with an inability to belch (known as the gas-bloat syndrome) can be managed conservatively, as it is due to edema at the wrap site. It is more common with a Nissen 360-degree fundoplication.
  - A **wrap that is too tight** is the main reason for dysphagia or the gas-bloat syndrome that persists for greater than 6 weeks post-op. This is treated with **serial dilations** after the 6 weeks postoperative period has passed.
  - The **wrap may herniate superiorly into thorax,** and is the most common cause for recurrent reflux. This is not necessarily decreased by pexy of the wrap to the diaphragm.
  - Alternatively, the **wrap may slip inferiorly** and trap the cardia of the stomach above it, exposing the esophagus to acid-producing parietal cells in the cardia and resulting in recurrent GERD symptoms and a postoperative positive pH study. This would need surgical revision.
  - A less common failure would be if the **wrap entirely breaks down.** This may be managed expectantly.
  - An **injury to the vagus nerve** impairs LES, fundal wrap, and pyloric relaxation, and can result in post-op dysphagia or gas-bloat syndrome from retention of contents in the stomach.
  - **Recurrent reflux** as time progresses is due to gradual loosening of the wrap and can be treated medically or with repeat surgery.
  - Most long-term wrap failures are actually **asymptomatic** and only seen radiographically, so they can be managed expectantly.

DIVERTICULA

- **PATHOPHYSIOLOGY**
  - Diverticula are outpouchings of the esophageal wall.
  - The underlying cause is either high intraluminal pressure causing pulsion diverticula or external traction. *See Figure 7.7.*
  - **Pulsion diverticula** generally are herniations of the mucosa in between the muscular layers of the esophageal wall. In this regard, they are considered pseudodiverticula, since they do not contain all layers of the esophageal wall.
    - Zenker's pulsion diverticulum lies between the inferior constrictor and cricopharyngeus (CP) and is secondary to poor cricopharyngeus relaxation with pulsion of the esophageal mucosa posteriorly.
    - Epiphrenic pulsion diverticula in the distal 10 cm of esophagus are secondary to LES hypertonicity. Many are asymptomatic.
  - **Traction diverticula** occur secondary to an adjacent mediastinal or peribronchial fibrotic process pulling at the mid-thoracic esophageal wall. They contain all layers of the esophageal wall and hence are true diverticula. They are much less common than pulsion diverticula.
- **DIAGNOSIS**
  - Symptoms include a "wet voice," dysphagia, regurgitation of undigested food, and halitosis.
  - **UGI contrast study** will demonstrate a Zenker's or traction diverticulum.
  - For epiphrenic pulsion diverticula workup should include a **manometry in addition to UGI contrast study.** This will evaluate achalasia or any underlying discordance in esophageal muscular function that produced the diverticulum in the first place.
- **TREATMENT**
  - For Zenker's pulsion diverticula:
    - No treatment is needed if the outpouching is small, asymptomatic, and incidentally found.
    - Small (<1 cm) *symptomatic* diverticula are treated with a **3 cm–5 cm longitudinal cricopharyngeal myotomy** through the inferior constrictor and extending inferiorly past CP and onto the esophagus. The diverticulum itself need not be intervened upon since it will involute if small.

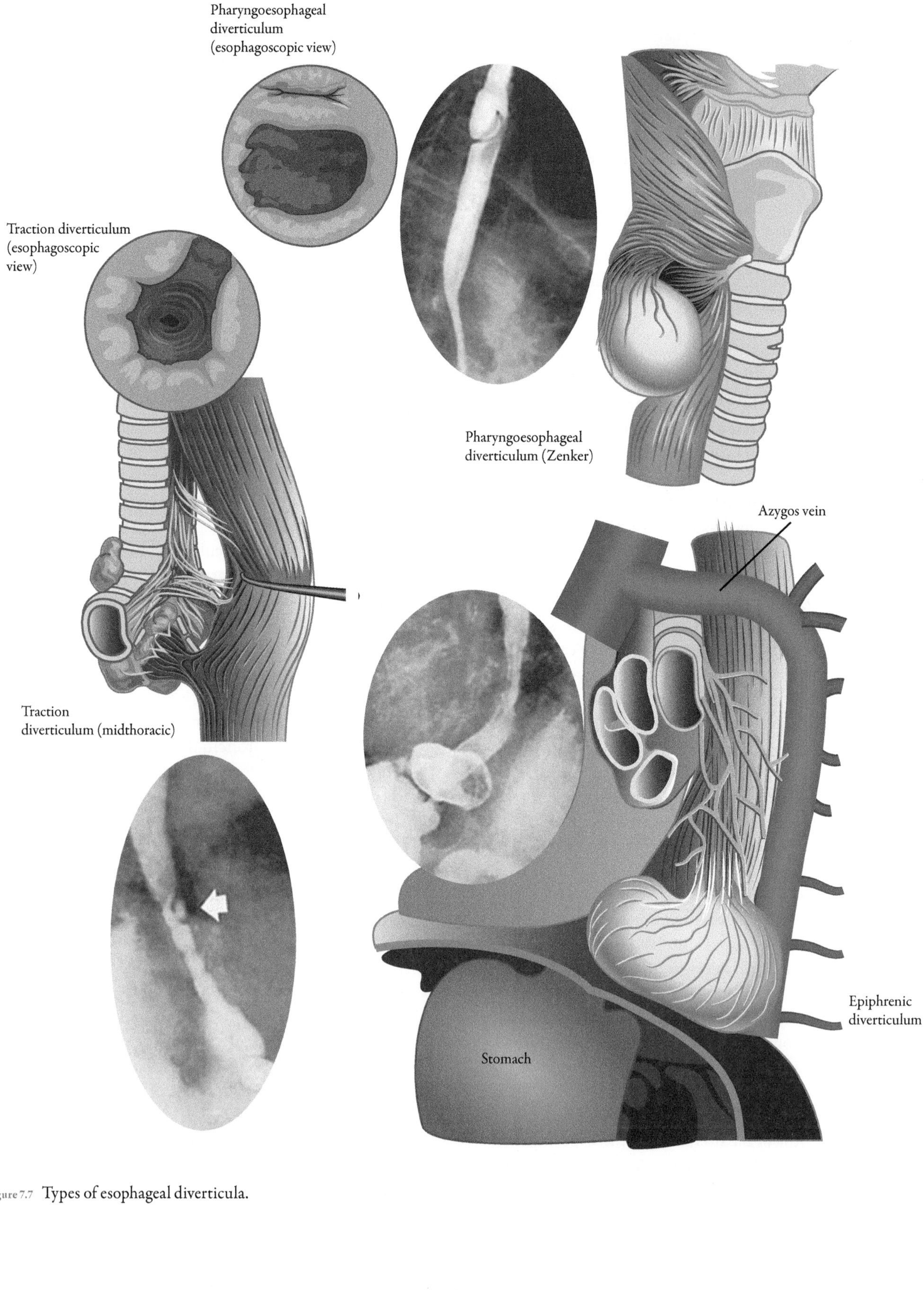

Figure 7.7 Types of esophageal diverticula.

- Moderate-sized diverticula (1 to 4 cm) are treated with a myotomy plus either **inversion of diverticulum or suspension of diverticulum** cephalad along prevertebral fascia.
- Diverticula 3 cm–5 cm in size in an elderly/frail patient, may be treated with **rigid endoscopic stapling diverticulotomy.**
  - The stapling device is introduced trans-orally with one jaw in the diverticulum and one in the esophageal lumen. When fired, this allows the diverticular lumen to have confluence with the esophageal lumen.
  - Since the diverticulum must be large enough to accommodate one limb of the stapling device, this technique is not used on smaller outpouchings.
  - Flexible endoscopy can also accomplish such a diverticulotomy, but is less well studied.
- A diverticulum >5 cm can be treated with **myotomy plus diverticulectomy**, which breaches and closes the esophageal mucosa by transecting at the neck of the diverticulum.
- For epiphrenic pulsion diverticula:
  - Treatment includes a left thoracic approach for **diverticulectomy** with a 56F bougie in place to prevent luminal narrowing. The muscle is closed at the site of the diverticulectomy, and a **contralateral myotomy** is performed to address the underlying distal hypertonicity. One may buttress the repair with a tongue of gastric fundus.
- For traction diverticula, the treatment is aimed at the underlying fibrotic process. The diverticula can be resected as well, but no myotomy is needed.

HIATAL HERNIA

- **PATHOPHYSIOLOGY**
  - These occur when the stomach or other abdominal organs herniate through the esophageal hiatus of the diaphragm into the mediastinum.
  - This is due to stretch of the crural fibers resulting in dilation of the hiatus and laxity and redundancy in the phrenoesophageal ligament.
- **DIAGNOSIS**
  - Patients may be asymptomatic or have symptoms of GERD that are already being treated with empiric PPI and well controlled. In these patients, the diagnosis of a hiatus hernia may occur incidentally on imaging.
  - If a patient has GERD refractory to PPI, or they have additional warning signs, then further workup should follow to exclude esophageal motility disorders, esophagitis, ulcers, stricture, dysplasia, neoplasia, or a hiatus hernia.
    - **Warning signs** include dysphagia, regurgitation of ingested liquid or particulate matter, chest pain, aspiration, early satiety, weight loss, or anemia secondary to mucosal ulceration within the herniated stomach.
    - Patients may have **obstructive symptoms of severe pain and nausea, but *without* the ability to vomit,** if the herniated stomach has undergone organoaxial (along long axis) or mesenteroaxial (along short axis) volvulus.
  - **UGI contrast study** is the main diagnostic modality. The patient ingests barium liquid or a barium-laden soft food bolus in Trendelenburg or prone position. UGI contrast can outline gastric herniation into the mediastinum.
  - **EGD** may demonstrate the end of rugal folds (GEJ) as >2 cm cephalad of the diaphragmatic pinch (hiatus). This would be diagnostic of a hiatal hernia.
    - Given that many hiatal hernias are sliding, they may not be present above the diaphragm pinch at the moment of an EGD. However a **Hill grade** can be determined via a retroflexed view of the GEJ. This helps determine the laxity in the crural area in the absence of an obvious hiatus hernia.
    - EGD can visualize strictures, or esophageal or gastric ulcerations. It must also exclude metaplasia, dysplasia, or neoplasia.
  - **24 hours pH monitoring** quantifies the severity of reflux with a DeMeester score.
  - **Esophageal manometry** evaluates peristalsis and LES tone.
  - **Chest x-ray (CXR)** may show the gastric cardia, fundus, or other hollow viscous or solid organ silhouettes in the mediastinum, posterior to the heart. However, since many hiatal hernias are sliding, they may not be present or visible at the time of a CXR.
- **TREATMENT AND OUTCOMES**
  - Management is guided by the presence of symptoms and the type of hiatus hernia. *See Figure 7.8a-d.*
  - **Type 1 sliding hiatal hernias** are when >2 cm of rugae reside above the diaphragmatic pinch (hiatus) on EGD. Often only the cardia of the stomach slides into and out of the mediastinum via the esophageal hiatus.
    - It is associated with GERD and may contain **Cameron's ulcers** within the herniated portion of the stomach. These occur from friction between opposing gastric mucosal surfaces within the mediastinum.
    - If asymptomatic, type 1 hernias do not require treatment.
    - If GERD is the only symptom, but controlled on a PPI, then no further therapy is needed.
    - For refractory GERD or any warning signs, a **laparoscopic fundoplication and cruroplasty** with or without biologic mesh in horseshoe fashion is the preferred treatment.
    - Patients may have an associated reflux related **stricture or Schatzki ring** at the squamocolumnar junction.

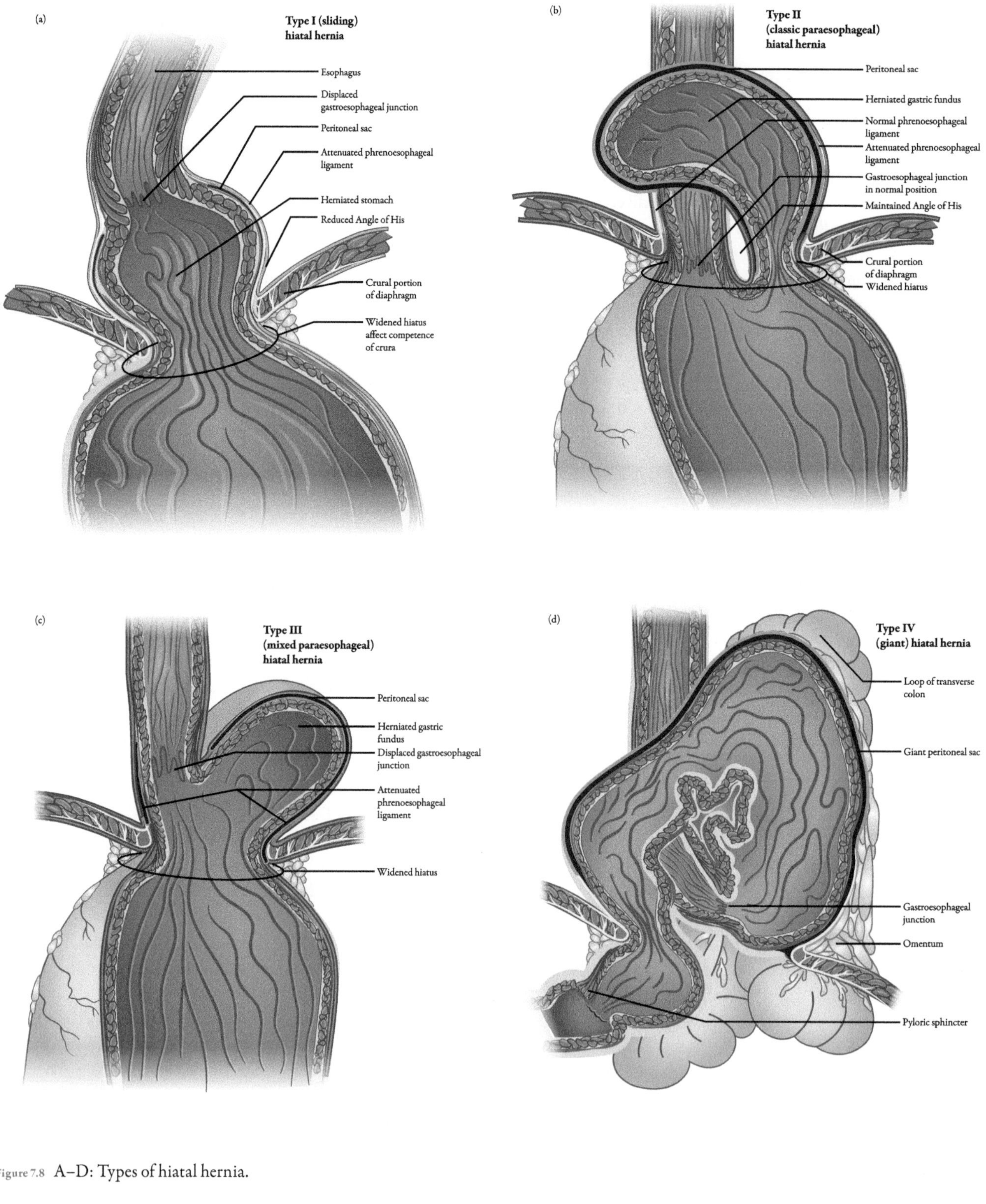

Figure 7.8 A–D: Types of hiatal hernia.

Either can be **dilated** and managed with PPI therapy, provided biopsies are negative for dysplasia or neoplasia.

- **Type 2 hiatal hernias** occur when a portion of the gastric fundus herniates through the hiatus and resides adjacent to the esophagus within the mediastinum. The GEJ is unchanged in position within the abdominal cavity.
    - Chest pain or dysphagia can occur from fundal incarceration and compression of the neighboring

esophagus. Patients are less likely to have reflux symptoms alone.
  - Treat most with **laparoscopic fundoplication and cruroplasty** with or without biologic mesh in a horseshoe fashion, after dissection of the phrenoesophageal ligament from the crura and mediastinum and reduction of herniated contents.
  - May **observe** small, asymptomatic paraesophageal hernias if the patient is elderly or comorbidly ill and not believed to be a good candidate for surgery.
  - Symptoms abate after surgery and rarely recur, even when there is radiographic evidence for hernia recurrence.
- **Type 3 hiatal hernias** are a combination of types 1 and 2. The GEJ and combined cardia and fundus reside in the mediastinum. Surgical treatment is the same as type 2. Patients may have lost their angle of His and have a very short intra-abdominal esophagus length after reduction, necessitating a Collis gastroplasty.
- **Type 4 hiatal hernias** are when the entire stomach plus the colon or spleen reside in the mediastinum. Surgical treatment is the same as types 2 and 3.
- **Organoaxial or mesenteroaxial volvulus of the stomach** can complicate types 3 and 4 hernias, effectively producing a gastric or esophagogastric obstruction with severe pain and the inability to vomit. This needs emergent laparoscopic or open surgical reduction and cruroplasty, with or without a wrap, depending on the health of the gastric body and fundus. Infarcted portions of herniated stomach or other visceral organs may need to be resected. Hemodynamically compromised patients may simply undergo reduction and gastrostomy to pexy the stomach within the abdominal cavity, provided the gastric wall has not necrosed. Formal cruroplasty and fundoplication can be performed at a later date.

CAUSTIC INJURY

- **PATHOPHYSIOLOGY**
  - Alkali ingestions cause **liquefactive necrosis** and deeper burns. They can be neutralized with vinegar, lemon, and orange.
  - Acid ingestions induce **coagulative eschar** at the esophageal mucosa and are neutralized with milk, egg, and antacids. At times, acids may pass through the esophagus and incur gastric injury.
- **DIAGNOSIS**
  - Symptoms include chest pain and odynophagia after either an intentional or accidental ingestion. Patients who have perforated may be systemically ill.
  - Perform a **CT chest and abdomen** to look for signs of perforation, and also assess extent and depth of injury.
  - **Endoscopy** should be completed within 48 hours to assess the depth of injury.
    - The striated and smooth muscle interface is the area of poorest clearance of a luminal irritant, and hence tends to be the site of greatest mucosal injury. Do not try and scope past the most cephalad area of injury. Injuries are classified by grade.
      - Grade 1: hyperemia only.
      - Grade 2: hemorrhage, erosion, and superficial ulcerations that may be circumferential.
      - Grade 3: **necrosis, charring, deep ulcerations**, and a narrowed lumen.
        - An ensuing **fibrotic scar** may shorten the esophagus and result in GERD or a stricture that narrows the lumen.
        - Strictures may become symptomatic as early as three weeks post injury.
      - Grade 4: **perforation.** This may be visualized at presentation, or may occur as late as three weeks after the ingestion.
- **TREATMENT**
  - Patients should be managed conservatively as nil per os **(NPO) and intravenous fluids.**
  - Grade 3 or 4 injuries should receive **antibiotics.**
  - A **caustic perforation with systemic illness** is best managed with **esophagectomy, proximal diversion, and J tube placement** rather than primary repair, since the general health of the esophageal wall in this setting is poor. Future conduits include right colon, left colon, tubularized gastric greater curvature, or the entire stomach.
  - **Strictures** should be managed with **endoscopic dilation** antegrade, or retrograde via a G tube. Resection is reserved for cases refractory to dilation.
  - Self-expanding plastic (SEPS) or self-expanding metal stents (SEMS) may be used in strictures or perforation with minimal mediastinal or pleural contamination. Stents should be 3 cm–4 cm longer than the stricture. It is ideal if the stent does not have to cross the CP or GEJ.
  - Button batteries produce alkaline injury and must be removed immediately by endoscopy, even if they have passed through the esophagus and already made it into the stomach.
  - Patients with high-grade injuries have a risk of future malignancy and may need surveillance endoscopies and biopsies.
  - Some patients never regain meaningful esophageal function and remain morbidly ill with a strictured and dysfunctional esophagus. They require an esophagectomy and autologous conduit.

OBSTRUCTION

- **PATHOPHYSIOLOGY**
  - Complete obstruction is often from a **food bolus.** However, three-fourths of patients also have underlying baseline esophageal pathology including stricture, esophagitis, malignancy, or motility disorder.
  - Less often, the patient has ingested a non-digestible object that has caused the esophageal obstruction. The most common example of this would be a **pediatric coin ingestion.**
  - Occasionally, complete obstruction is solely due to a **stricture or malignancy** (see Carcinoma section below).
- **DIAGNOSIS**
  - Patients will be continuously **spitting up saliva** and any liquids they attempt to take. They will not have the ability to wretch or vomit. They generally appear well, unless a secondary perforation has occurred.
  - Diagnosis can be based on these clinical findings and an **EGD.** UGI contrast can be utilized to evaluate whether the obstruction is complete or partial.
  - If the object is not a food bolus, and is believed to be radiopaque, then a **CXR** can visualize the position of the object and level of obstruction.
- **TREATMENT AND OUTCOME**
  - **Glucagon** may be tried for a food bolus, as it can relax the LES and allow the bolus to pass.
    - If the food bolus or obstructing object passes into the stomach, patients can be followed up.
    - If the object is radio-opaque and has made its way into the stomach, serial x-rays at follow-up can ensure complete passage over the ensuing several weeks.
  - 15% of patients will need urgent **flexible or rigid endoscopic retrieval** at the time of presentation, because:
    - The object **does not pass into the stomach** with glucagon.
    - The object has passed into the stomach, but it is **sharp, large, or angulated** in such a way that successful passage through the remainder of the gastrointestinal tract is suspect and possibly dangerous.
    - The object is a **button battery.** Whether esophageal or gastric in position, these need immediate removal with endoscopic evaluation of any potential mucosal injury from its alkaline discharge.
    - 95% of retrieval attempts are successful, and the complication rate is <5%.
    - Rigid esophagoscopy may be needed for objects lodged proximally in the esophagus.
  - **Three-fourths of patients with a food bolus obstruction have underlying esophageal pathology** such as a stricture, esophagitis, malignancy, or a motility disorder that requires further workup and treatment after the obstruction is resolved.

PERFORATION

- **PATHOPHYSIOLOGY**
  - **Two-thirds are iatrogenic in the cervical esophagus** at the cricopharyngeus muscle after an EGD.
  - Perforations can also occur during thyroidectomy or anterior approach to cervical spine surgery.
  - **Boerhaave syndrome** describes a transverse laceration of the left lateral esophageal wall at 3 cm–5 cm above the GEJ secondary to forceful vomiting.
  - A **Mallory Weiss** tear is 0.5 cm–2.5 cm, longitudinally oriented, and on the right side of the distal esophagus extending down the lesser curvature of the stomach. Results in bleeding that usually spontaneously resolves. It is not a true full thickness perforation.
- **DIAGNOSIS**
  - Generally patients have pleuritic pain and odynophagia. Symptoms may be preceded by vomiting if the perforation was not iatrogenic.
  - The degree of systemic illness will parallel the volume of mediastinal or pleural mesothelial contact with extruded esophageal contents like saliva, acidic gastric secretions, or recently ingested food.
    - Heavy mediastinal contamination or free flow of esophageal contents into the pleural space will generally result in systemic illness and abnormal vital signs.
  - Patients may have Hamman's sign on auscultation, or subcutaneous air on neck palpation.
  - A **CXR** can show pneumomediastinum with or without a pleural effusion. The effusion would be on the left side if the distal esophagus is perforated. Otherwise it would likely be on the right side.
  - **Gastrografin UGI contrast study** in the right lateral decubitus position will demonstrate a perforation, and whether it is contained in the mediastinum or freely flowing into the pleural space. If negative, the study may be repeated with **dilute barium** for enhanced sensitivity.
  - **CT chest and abdomen** can show extraluminal air cluttered around the area of esophageal perforation as well as a fluid collection, either contained to the mediastinum or collecting in the right or left pleural space.
    - If oral contrast was used, this can be seen extruding into the mediastinum or pleural space in some instances.
- **TREATMENT AND OUTCOME**
  - 1. **Conservative management includes NPO, spit, antibiotics, and PPI.** This strategy is reserved

for patients without signs of systemic illness who have a contained perforation confined to the mediastinum.

- In such contained circumstances, the extraluminal contents will drain back into the lumen of the esophagus.
- Repeat contrast study can be done at 72 hours to confirm the perforation remains contained, after which one may start a liquid diet.

- 2. If <24 **hours** since perforation and the patient has systemic illness, pleural spillage, and no mediastinal containment, one may perform **open or video assisted thoracoscopic surgery (VATS) debridement and drainage of mediastinum and pleura with primary repair of the esophagus in *two layers*, and leave a drain.**
  - A longitudinal myotomy can be performed to see the full extent of mucosal injury.
  - May use **strap muscles** as a buttress of repair if perforation is in cervical esophagus.
  - May use **intercostal muscle** or pericardium as a patch to buttress the repair if perforation is in the thoracic esophagus. *See Figure 7.9.*
  - If the perforation is in the setting of achalasia previously requiring dilation, then perform a Heller myotomy contralateral to the repair site.
  - **Endoscopic esophageal stent with VATS debridement and drainage** is an option in select cases where the perforation does not cross cricopharyngeus (CP) or GEJ. However, the mediastinal or pleural contamination would still need to be washed out if the patient is systemically ill.
- 3. **If >24 hours** since the event, the pleural space is contaminated, and the patient has signs of systemic illness, then **do not attempt primary repair of the esophagus.**
  - In the cervical esophagus, just leave **drains**.
  - In the thoracic esophagus, debride the pleural and mediastinal spaces via thoracotomy or thoracoscopy, and do one of the following:

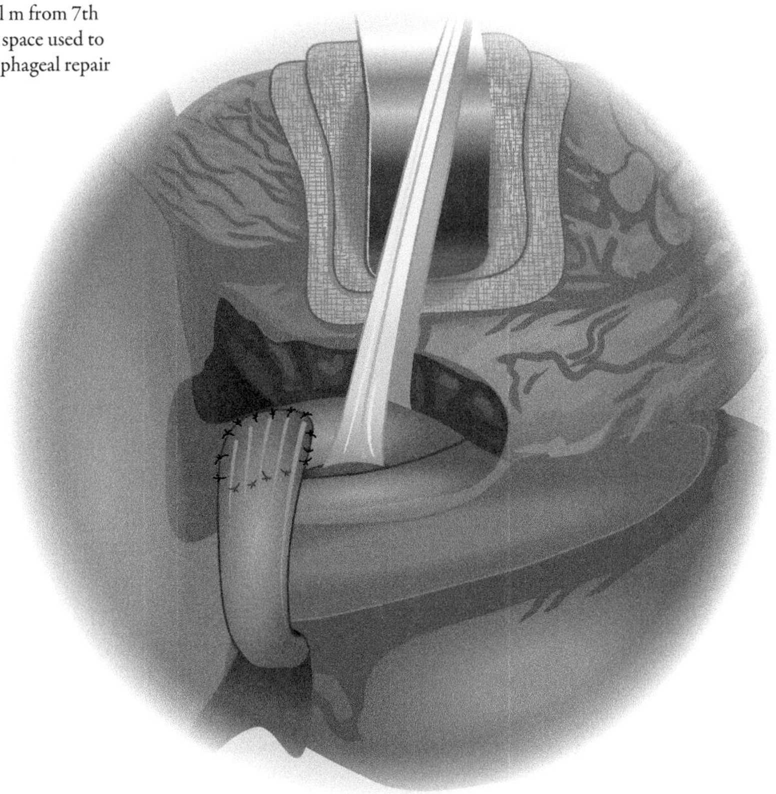

Figure 7.9 Esophageal perforation repair.

- a) **Place an endoscopic self-expanding stent** for the perforation if the neighboring esophagus is otherwise healthy and the tear is not crossing the CP or GEJ.
  - Stents may also be used for leakage or perforation at a prior anastamotic site, but should not sit solely within the body of a gastric conduit since the larger diameter of the conduit may not house the stent appropriately.
- b) An alternative option is to **staple across the esophagus proximal to the injury and perform a proximal diverting cervical loop or end esophagostomy** if the esophagus is not healthy enough to accommodate a stent and the neighboring contamination is heavy.
  - May need to resect the diseased portion of thoracic esophagus or conduit if it is necrotic. Leave pleural drains.
- Place a **gastrostomy or jejunostomy tube.** The latter is preferred if you believe the stomach will need to function as a future conduit.

- 4. If a patient has minimal pleural contamination and minimal systemic signs of illness, then an **endoscopic self-expanding stent and pleural drains may be placed without a thoracotomy or thoracoscopic debridement or washout.**
  - Again, for a stent to work, the tear should not cross the CP or GEJ, and shouldn't be within the body of a gastric conduit since the larger diameter of such a conduit may not house a stent appropriately. Stents may reside across a leaking anastomosis, however.
- 5. Perforation or **leakage at the site of a previous esophagogastric anastomosis can be managed with a stent** if the conduit is healthy or resection and diversion if the gastric conduit is necrotic.
- 6. **Benign or malignant perforations can be managed with covered self-expanding plastic (SEPS) or metal stents (SEMS)** that have exposed struts at margins to prevent migration. Mediastinal or pleural contamination may still need operative debridement/drainage, or simply pleural drains.
  - Stents are best tolerated if they are not near the CP or crossing the GEJ, since there is an increased risk of migration. However, the latter is not an absolute contraindication, as a stent may remain the best temporizing option in the patient with a malignant etiology to their perforation.
  - SEPS have increased radial force, no exposed struts, and hence an increased risk of migration.
- **Esophagectomy** may eventually be needed if there is significant intrinsic disease from caustic injury, achalasia with megaesophagus, or carcinoma. However, the above measures should be used first in the acute period.
  - A J tube should be placed so the stomach is preserved as a future conduit.
  - Other conduit options include colon or jejunum, which may be kept on its mesentery for distal esophageal replacement or used as a free flap anastomosed to the superior thyroid arteries for cervical esophageal replacement.
- After any perforation, **fistulization** with the tracheal lumen should be suspected in the chronic setting if a patient consistently coughs after swallowing.
- **Mallory Weiss tears** can be endoscopically banded with or without adjunctive epinephrine injection. If bleeding is refractory to endoscopic attempts, then a left gastric artery embolization or a gastrotomy with oversewing of the vessel may be necessary.

BENIGN NEOPLASIA

- **Leiomyoma** is the most common benign tumor of the esophagus. It originates from the muscularis propria or muscularis mucosa of the distal esophagus and is smooth and well encapsulated.
  - Diagnosis is by **esophagram** showing a smooth, well-encapsulated mass that may be mobile. There is no need for a biopsy.
  - Treatment is with **enucleation** if symptomatic or large.
  - If asymptomatic and small, it may be **observed** provided the imaging characteristics are diagnostic.
- **Esophageal polyps** are the second most common benign esophageal tumors.
  - They normally reside in the cervical esophagus and present with dysphagia or hematemesis.
  - Treatment is with endoscopic or surgical resection.
- **Esophageal cysts** may be congenital or acquired from a secretory gland obstruction. They should be resected.

CARCINOMA

- **PATHOPHYSIOLOGY**
  - Risk factors include smoking, alcohol, nitrosamines, caustic injury, achalasia, and long-standing acid exposure.
  - **Squamous cell carcinoma (SCC)** is the most common type of esophageal cancer worldwide**.** It is more common in Asians and Africans.
    - SCC generally resides in the **mid to upper esophagus** and is drained by thoracic and cervical nodes.
    - The **lung** is the most common metastatic site.
  - **Adenocarcinoma** is the second most common esophageal cancer worldwide, but outnumbers SCC 15:1 in the United States. It has an increasing prevalence related to **obesity and GERD.** *See Figure* 7.10.
    - Adenocarcinoma develops in the **lower esophagus**, and its lymphatic drainage heads inferiorly into the celiac nodes.
    - The **liver** is the most common metastatic site.

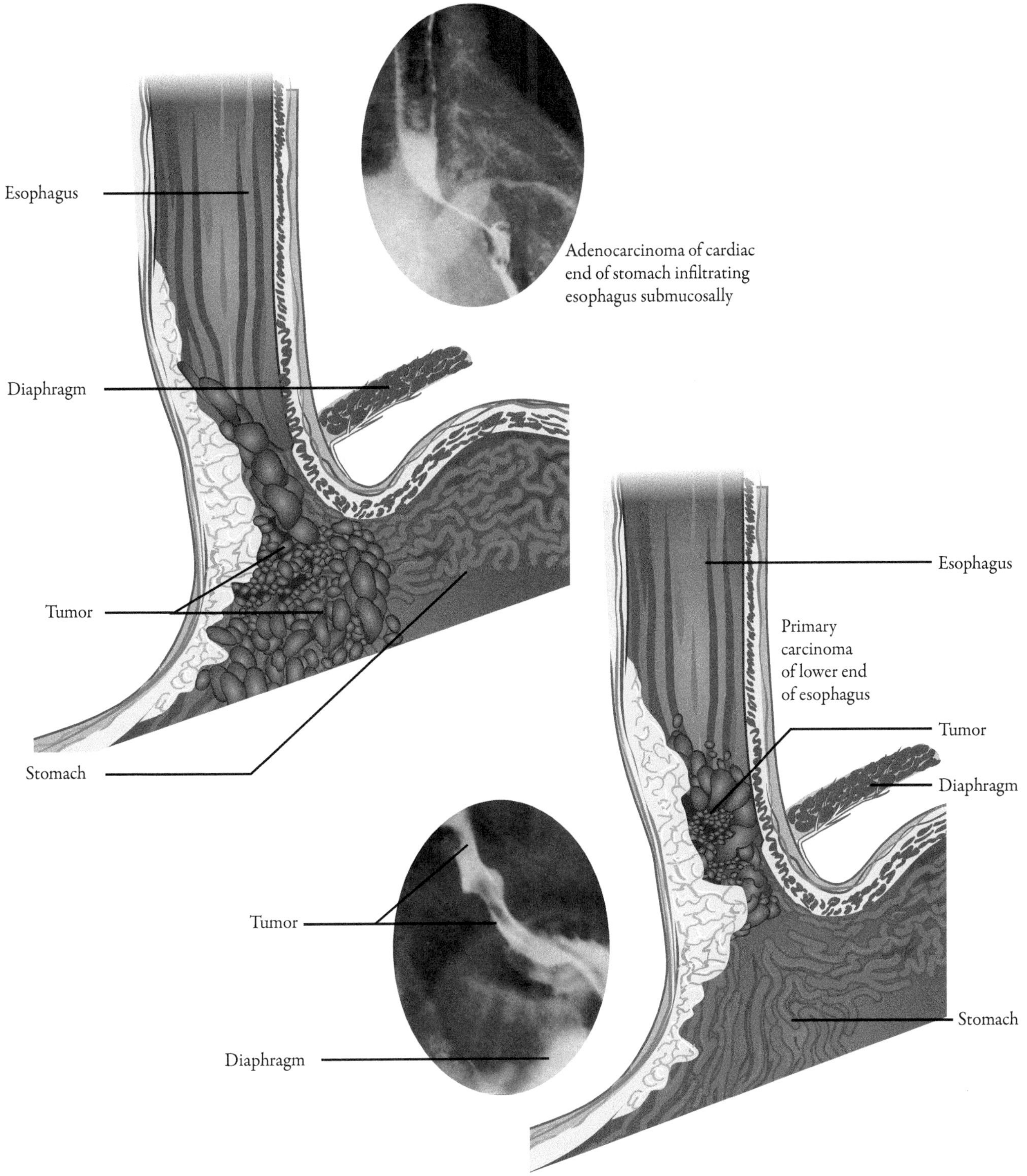

Figure 7.10 Distal esophageal adenocarcinoma.

- **DIAGNOSIS**
  - Symptoms include weight loss and dysphagia to solids that progresses to liquids.
  - Hoarseness can develop from vagal recurrent nerve invasion or Horner's syndrome from cervical sympathetic chain invasion. Both imply unresectability.
  - **UGI contrast study** can show a filling defect corresponding to the tumor's encroachment into the lumen. However, further characterization as benign or malignant is limited.
  - **EGD and biopsy** will determine if an identified lesion has dysplasia, carcinoma in situ, or invasive disease.

- **EUS can determine T stage and local paraesophageal or celiac node status.** This can be accompanied by a transesophageal fine needle aspiration (FNA) of suspicious nodes.
- **CT chest and abdomen** evaluates for nodes and metastases in the setting of a T1b or greater lesion. Positron emission tomography (PET) may be added to delineate the significance of borderline lesions.
- **Bronchoscopy** can be used to exclude unresectable tracheobronchial invasion in proximal esophageal tumors. This can be accompanied by a transbronchial FNA.
- **Laparoscopy** is a staging adjunct for tumors around the GEJ that may have liver lesions or local nodes outside the radius of EUS detection.
- N1 status is 1–2 paraesophageal nodes, and N2 status is 3–6 paraesophageal, cervical, or celiac nodes.

- TREATMENT
  - SCC
    - **Chemoradiation** is the primary treatment if the tumor is within 5 cm of the upper esophageal sphincter or larynx. 25% have a complete response and therefore may not need surgery. In these cases, the chemoradiation is definitive therapy.
    - Surgery is reserved for those proximal tumors that fail chemoradiation or have subsequent mechanical complications.
      - It also can be employed after chemoradiation for distal tumors, as in the case of adenocarcinoma (see below).
  - Adenocarcinoma
    - In the absence of metastases, the sequence of interventions depends on T stage and local extent of disease. *See Figure* 7.11.
    - **HGD or carcinoma in situ** are both contained in the mucosal layer and do not extend into the lamina propria. So, either can be managed with the following:
      - **Radiofrequency ablation (RFA)** to ablate the superficial 0.5 mm of mucosa or
      - **Endoscopic mucosal resection (EMR)** utilizing a submucosal saline lift injection, snare polypectomy, and/or a suction cup. EMR is preferred if the lesion is nodular and <2 cm.
    - **T1a lesions** extend through mucosa and lamina propria and into, but not through, the muscularis mucosa. This can be treated by **EMR if <2 cm, provided the EUS and CT are not suspicious for lymph node involvement or metastases.** Lymph node involvement is unlikely, since the lamina propria has minimal lymphatic network.
    - **T1b lesions** either extend through muscularis mucosa and into the submucosa, or appear to respect the muscularis mucosa but have evidence for lymphovascular invasion on biopsy.
      - These have been treated by **endoscopic submucosal dissection (ESD)** if the tumor is limited to the superficial submucosa (SM1).
      - There is a 20%–40% risk of lymphatic invasion with deeper submucosal involvement (SM2 and SM3) due to the extensive submucosal lymphatic network.

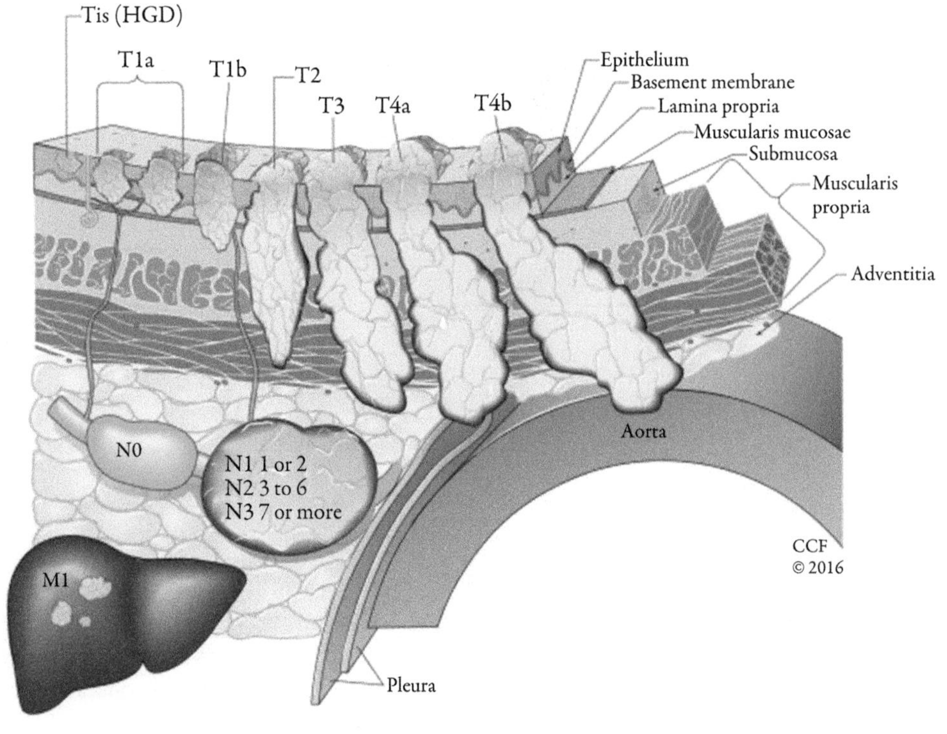

Figure 7.11 T stages of esophageal adenocarcinoma.

So esophagectomy may be a better option for these lesions.
- **Vagal sparing esophagectomy** is an option for T1a or T1b lesions or high-grade dysplasia with high risk features including multifocality, ulcerations, or extensive nodularity.

- **T2 lesions** extend through the submucosa into, but not through, the muscularis propria. These should receive **neoadjuvant chemoradiation** followed by resection 5–7 weeks later.
- **T3 lesions** extend beyond the muscularis propria into the sub-serosa. These should receive **neoadjuvant chemoradiation** followed by resection 5–7 weeks later.
- **T4a lesions** encroach into resectable adjacent structures like the pleura or diaphragm. These should receive **neoadjuvant chemoradiation** followed by resection 5–7 weeks later.
- Any T stage tumor with **FNA positive paraesophageal nodes** should receive **neoadjuvant chemoradiation** and then esophagectomy 5–7 weeks later. This will decrease local relapse rates.
- Neoadjuvant chemoradiation most often includes **5FU, cisplatin, and 35–50 Gy radiation.**
  - One may perform PET imaging early within the neoadjuvant chemoradiation regimen, or after the regimen is complete, to determine response and the foreseen benefit of completing therapy versus an early resection.
    - Progression of disease at such interval evaluations could imply that surgery should be abandoned altogether.
  - Chylothorax, and possibly thrombosis, rates may be higher postoperatively in patients who received neoadjuvant chemoradiation therapy.

- **T4b lesions invade into unresectable neighboring structures** like nerve, tracheobronchial tree, aorta, or vertebral bodies. The presence of a **malignant effusion** is also a criterion to be categorized as T4b.
  - These lesions are treated with **definitive chemoradiation.** Radiation may be external beam or brachytherapy.
- **M1 status** is ascribed to any patient with metastases. These are treated with **chemotherapy** alone. Isolated metastatic lesion resection in esophageal cancer is not an option.
- Radiation can be used to palliate bleeding or obstruction near the GEJ if the patient is not an operative candidate (T4b or M1).
- **Malignant stricture, obstruction, perforation, or fistulization** can be managed with a covered SEMS that has exposed struts at margins to prevent migration.
  - SEPS have increased radial force, no exposed struts, and hence may migrate.

- Stents are best tolerated if they are not close to the cricopharyngeus (CP) and do not cross the GEJ. The latter has an increased risk of migration but is not an absolute contraindication. Often a stent is still the best temporizing measure for obstructing or perforated malignancy despite having to cross the GEJ.
- Once the stricture or perforation is excluded by the stent, the patient may initiate chemoradiation treatment and thereafter be considered for resection provided they do not have a T4b lesion or metastases.
- **Open Esophagectomy**
  - **5 cm–10 cm longitudinal margins** are needed due to the tendency for carcinoma to extend along submucosal lymphatics. **1 cm radial margins** are considered R0.
  - **Transhiatal esophagectomy** via upper abdominal and cervical incisions. This is a total esophagectomy and cervical esophagogastric anastomosis after the stomach is pulled through the mediastinum.
    - The gastric conduit is perfused by the right gastroepiploic artery, since the left gastroepiploic, left gastric, and often the right gastric artery are all ligated.
    - There is a limited intrathoracic lymphadenectomy up to the inferior pulmonary veins.
    - This approach may be better for patients with baseline pulmonary comorbidities.
  - **Ivor Lewis esophagectomy** is performed via an upper abdominal and right thoracic incision. It is a subtotal esophagectomy with an intrathoracic anastomosis of the gastric conduit to the proximal or mid-portion of the remaining esophagus. The thoracic lymph node harvest may be more complete than a transhiatal approach.
    - Leak rates may be lower than those with a cervical anastomosis. However if it should occur, a leak will result in mediastinitis and greater morbidity.
  - **McKeown esophagectomy** is performed via an upper abdominal, right thoracic, and cervical incision. This combines the transhiatal and Ivor Lewis approaches and is a total esophagectomy with cervical esophagogastric anastamosis and thoracic lymph node harvest.
  - A pyloromyotomy may be performed but is not mandatory, as it does not seem to alter the rate of postoperative gastric outlet obstruction (GOO).
    - If it should occur, **post-op GOO is treated with balloon dilation.**
  - Colonic interposition can be utilized instead of a gastric conduit.
  - A **jejunostomy tube** may be placed 40 cm distal to ligament of Treitz (LT), with tacking stitches to the anterior abdominal wall and a percutaneous Seldinger technique into the jejunal lumen. May place an "anti-torsion" stitch distally along jejunum affixing this area

to the anterior abdominal wall to prevent rotation. Alternatively, a Witzel tunnel technique can be employed.

- **Minimally invasive esophagectomy (MIE)** is associated with decreased length of stay (LOS) and pulmonary complications. The steps of the procedure include
  - Open the gastrohepatic ligament. Open the gastrocolic ligament to free the greater curvature, taking care to **preserve the right gastroepiploic artery** as the primary blood supply left intact for perfusion of the gastric conduit.
  - Visualize the left crus of the diaphragm, free the retroesophageal space and the posterior portion of stomach, and take the left gastric artery and vein.
  - Then staple across cardia or body of the stomach at incisura after taking the right gastric artery. Proceed cephalad with a second staple fire at a right angle to the first fire. This should be aimed cephalad and parallel to the greater curvature. This tubularizes the greater curvature of the stomach in preparation as a conduit.
    - One may keep the superiormost part of the greater curvature in continuity with the esophageal specimen to prevent rotation of the conduit as it is pulled through the thorax.
  - Gastric conduit length can be increased by transection of the right gastric artery and a Kocher maneuver if necessary.
  - Right thoracic dissection begins inferiorly with incision of the inferior pulmonic ligament, taking all nodes, and using heavy cream via the J tube to accentuate the thoracic duct.
    - Mediastinal dissection proceeds cephalad close to the esophagus as the mediastinal reflection of the pleura is opened. The azygous vein is ligated. The esophagus is freed from the hiatus inferiorly, heading superiorly on its anterior surface to free it from the membranous posterior walls of the tracheobronchial bifurcation.
    - Subcarinal nodes are harvested.
    - The specimen is transected at least **5 cm** cephalad of the tumor based on EGD.
    - A functional end–end **stapled intrathoracic esophagogastric anastomosis** is accomplished with an EEA circular stapler. Alternatively, the anastomosis may be **thoracoscopically hand sewn.**
    - **Omentum** from the greater curvature can be used to buttress the anastomosis to decrease the leak rate, and the conduit can be tacked to the diaphragm.
  - If it is chosen to resect the entire esophagus, one may pull the gastric conduit up into the neck for a cervical hand sewn or stapled esophagogastric anastomosis.
  - A jejunostomy tube can be placed via a percutaneous Seldinger technique with laparoscopic assistance.
- **POSTOPERATIVE COURSE AND OUTCOMES**
  - **Adjuvant chemotherapy** may be considered in cases without an R0 resection or cases that had positive lymph nodes in either the preoperative workup or in the operative specimen.
  - **Postoperative chylothorax** persistently >1000 ml/day via pleural drain output despite a trial of medium chain triglyceride (MCT) diet, NPO, and TPN can be managed with right thoracic approach **duct ligation or lymphatic embolization.**
  - A **postoperative uncontained anastomotic leak** is preferably managed with an esophageal stent and washout. Other options include direct repair with buttressing, resection and revision of the anastomosis, or resection of the anastomotic segment and diversion of the proximal esophagus if the conduit is unhealthy or contamination is excessive.
    - Covered SEMS have exposed struts at margins to prevent migration.
    - SEPS have increased radial force, no exposed struts, and hence may migrate.
    - Stents are best tolerated if they are not close to the CP and do not cross the GEJ. The latter has an increased risk of migration, but is not an absolute contraindication as a stent may still be the best temporizing option in the postoperative state.
  - Chronic **anastomotic strictures** can be dilated or stented before having to resort to resection and re-anastomosis
  - **Postoperative recurrence** should be treated with **chemotherapy** alone. There is no role for resection of a metastatic recurrence in esophageal cancer.

8.

STOMACH

ANATOMY AND PHYSIOLOGY

- The stomach is lined by columnar mucosa that invaginate into the wall forming gastric pits. Deep to the mucosa lies lamina propria, muscularis mucosa, submucosa, and three muscular layers (oblique, circular, and longitudinal). There is an outer serosal covering. *See Figure 8.1.*
- Gastric perfusion is via the left and right gastric and gastroepiploic arteries and veins.
- Perigastric lymphatics drain to celiac nodes.
- **Cardia** is the superiormost portion of the stomach in continuity with the esophagus. It contains mucous secreting glands.
- **Fundus and body** of the stomach are responsible for **receptive relaxation** under vagal stimulation.
  - **Parietal cells** situated here produce intrinsic factor (IF) and hydrochloric acid (HCl) under the influence of vagal acetylcholine (ACh), histamine, and gastrin. Gastric acid production has three phases:
  - **Cephalic phase** of acid production is initiated by the thought, sight, smell, or chewing of food. It is mediated by descending vagal nerve output and is responsible for 30% of HCl.
  - **Gastric phase** of acid production starts once food enters the stomach, is mediated by the hormone gastrin, and is responsible for 60% of total HCl.
  - **Intestinal phase** of acid production starts when chyme enters the small bowel.
  - Acid production is inhibited by proton pump inhibitors (PPIs) and histamine receptor blockers (H2B). PPIs have greater efficacy against upper gastrointestinal bleeding (UGIB), but higher C. difficile rates. Both agents increase nosocomial PNA.
  - **Chief cells** in the fundus and body release **pepsinogen** into the gastric lumen for protein digestion and **leptin** into the circulation to generate satiety.
  - **Endocrine cells** release **ghrelin** into circulation. This is responsible for appetite stimulation.
- The **antrum** is responsible for **churning of contents and motility.**
  - Churning is vagally mediated during the **migrating motility complex (MMC) phase 2**. It consists of high amplitude, non-propulsive contractions with a closed pylorus for mixing food within the stomach
  - Peristalsis is mediated by motilin during **MMC phase 3** and results in propulsive contractions with an open pylorus, pushing contents forward and out of stomach.
    - Gastric peristalsis is atop, and at the same frequency of, the basal electrical rhythm of pacemaker Cajal cells.
    - 50% of gastric contents should empty by 2 hours, with nearly complete emptying by 4 hours.
  - **D cells** in the antrum release somatostatin into circulation, which inhibits gastrin, secretin, motilin, insulin, and glucagon.
  - **G cells** release gastrin into circulation. They are stimulated by amino acids, ACh, calcium, alcohol intake, luminal distension, and increased luminal pH. The result is **increased parietal hydrochloric acid (HCl) and intrinsic factor** (IF) production as well as chief **cell pepsinogen release.**
  - Mucous and bicarbonate producing cells are also present in the antrum. *See Figure 8.2.*
- In the **duodenum**, Brunner's glands produce **protective mucus**, and endocrine cells release **secretin (from S cells) and cholecystokinin (CCK, from I cells)** into the circulation to inhibit gastric acid production.

ESOPHAGOGASTRODUODENOSCOPY (EGD)

- This is the main diagnostic modality for the stomach and is used for **diagnosis, treatment, and surveillance of gastritis, varices, or ulcers.**
- EGD may also be performed **intraoperatively** to confirm anastamotic integrity with a leak test, mucosal integrity after a submucosal tumor resection, luminal patency after any intervention, or as a guide for margins during laparoscopic or open resection.
- For **epigastric abdominal pain, early satiety, vomiting, or nausea**, an EGD can reveal acute or chronic gastritis, ulceration, stricture, obstruction, or malignancy (See the appropriate sections below.) Biopsies should be taken from the pathologic site, as

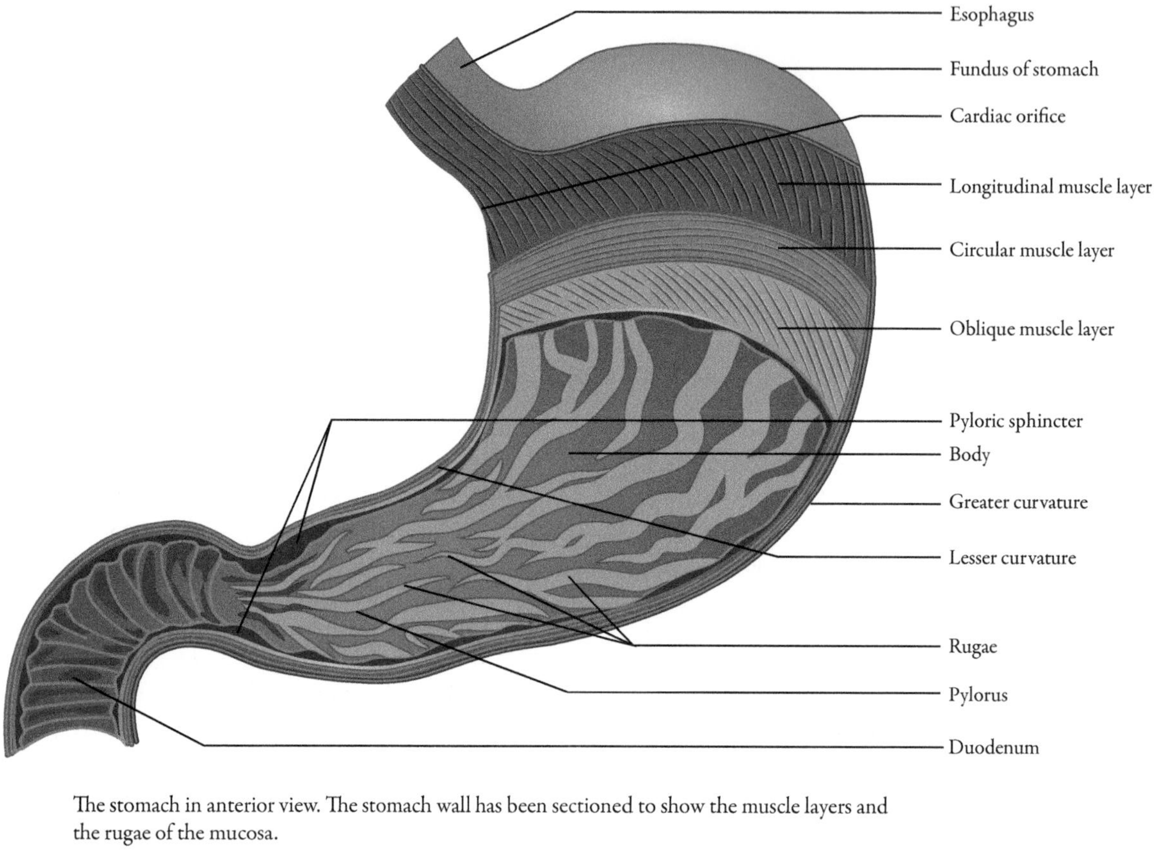

Figure 8.1 Anatomy of stomach.

well as the antrum, incisura, and body for H. pylori testing.

- For **UGIB**, the EGD is performed with the right side down to prevent blood pooling in the fundus and to see a Dieulafoy lesion (an exposed submucosal vessel).
  - Rotate tip posterior and to the right to traverse the C loop of duodenum if needed.
  - One must treat **active bleeding**, since such lesions are the most likely to re-bleed. A **visible non-bleeding vessel** and **adherent clot** are also likely to re-bleed. Any **active ooze** from the mucosal surface needs treatment as well.
  - **Hemostatic measures** include submucosal epinephrine, banding varices, clips, balloon tamponade, or coaptive pressure and heat. Argon plasma coagulation treats superficial bleeding.
  - Perform an EGD **q6 weeks** thereafter until any treated esophageal, gastric, duodenal ulcers, or varices have healed.
- EGD may show esophageal pathology as the source for pain or dysphagia (See Chapter 7, Esophagus, for complete details).
  - **Barrett's metaplasia** of the esophagus requires surveillance every two years for any metaplastic segment <3 cm, or yearly for a metaplastic segment >3 cm. **Biopsies** are performed every 1–2 cm along the segment.
  - **Cold forceps or snare** is adequate for most lesions. The snare may be heated.
  - May resect a submucosal lesion endoscopically provided it elevates with saline injection. The lift indicates the lesion is not invading the muscularis propria.
  - Surveillance frequency is increased to q6 months for any **low-grade dysplasia (LGD)** within the Barrett's metaplastic segment seen on biopsy. Persistence of LGD despite PPI may warrant anti-reflux surgery.
- **High-grade dysplasia or carcinoma in situ** on biopsy needs ablation or resection.
  - **Ablation** of pathology is usually accomplished with radiofrequency (RFA). Other options include Nd:YAG laser, alcohol, photodynamic therapy, or argon plasma coagulation.
- Focal carcinoma in situ or T1a or T1b malignant lesions can undergo **endoscopic mucosal resection (EMR) or endoscopic submucosal dissection (ESD)** provided one is confident in the T stage by endoscopic ultrasound (EUS), and there is no evidence for perigastric nodes or distant spread. *See Figure 8.3.*
  - Resection may occasionally be assisted by laparoscopic ports placed trans-abdominally and trans-gastrically. After resection, the ports can be pulled back into the peritoneal cavity and gastrotomy sites closed.

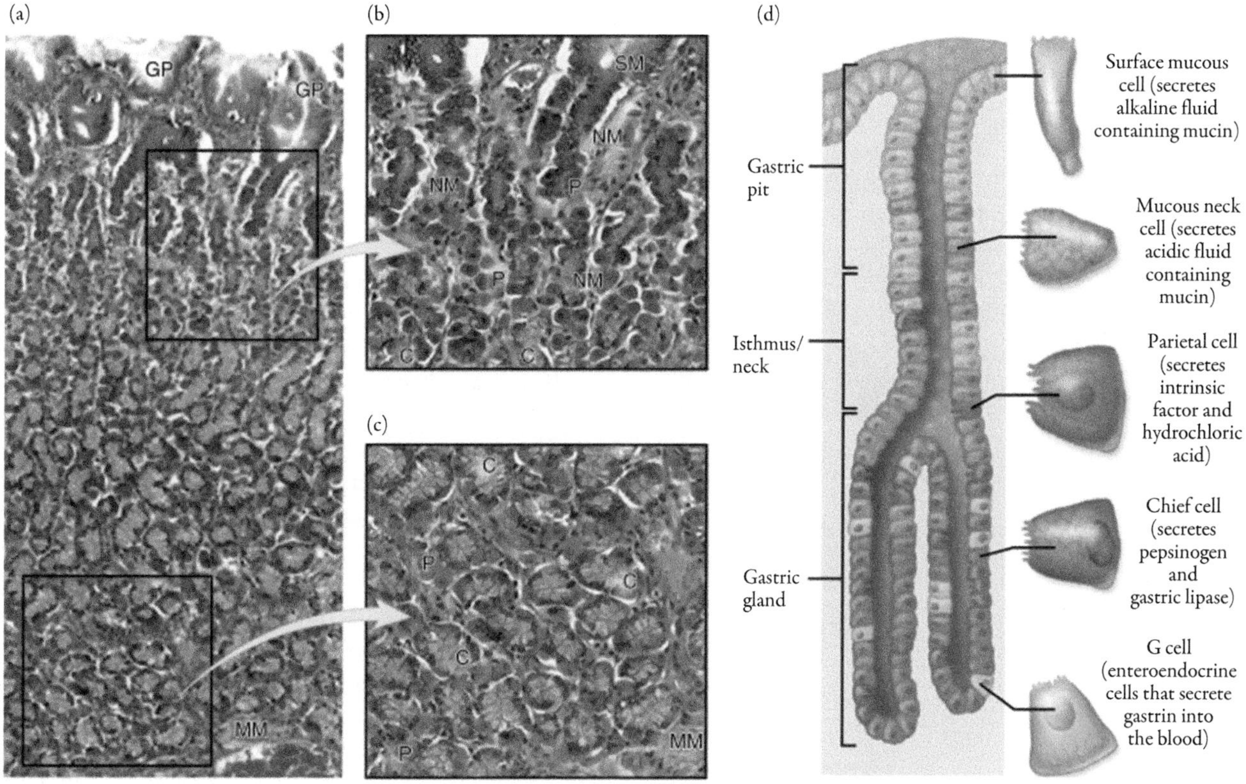

Figure 8.2 Resident cells of the gastric mucosa.

- Rigid EGD is used for removal of a proximal esophageal foreign body (FB) and flexible EGD for removal of a distal esophageal FB or gastric foreign body.
- **Esophageal stricture dilation** is accomplished with successively larger bougies until resistance is felt or blood is noted coating the device upon withdrawal. Generally need >40F to treat dysphagia. Balloon dilation under fluoroscopy is equivalent to bougie dilation. There is an approximately 3% perforation rate in the setting of achalasia.
- **Self-expanding metal esophageal stents (SEMS)** may be covered and have flanges on the ends for anchoring. These can be used for malignancy, obstruction, or perforation. They should not be placed near the cricopharyngeus muscle.

GASTROSTOMY TUBE

- This is indicated in a patient with **intact enteric function who cannot nourish themselves orally** and is perceived will remain so for weeks to months. Their underlying pathology is often a neurologic, pharyngeal, or esophageal pathologic process. Temporary feeding can be initiated with a nasoenteric tube, however if the patient is deemed to need long-term feeding, then a gastrostomy or jejunostomy tube should be considered.
- Preoperative antibiotics should be given to prevent wound site infection.
- The endoscopic **pull technique** is the most commonly employed. It may introduce oral flora into the surgical site.
  - The procedure is referred to as a percutaneous endoscopic gastrostomy (PEG) and starts with endoscopic visualization of the anterior gastric wall and transillumination across this, with the light noted externally on the dermal abdominal wall surface.
  - Discrete external finger pressure indentation of the abdominal wall should be noted endoscopically as well.
  - Then, percutaneous access of the gastric lumen is obtained and a wire advanced.
  - Any bubbles aspirated percutaneously before visible entry into the gastric lumen on endoscopy indicates an intra-colonic position of the needle. The needle should then be withdrawn, repositioned, and gastric access re-attempted.
  - Once adequate percutaneous gastric lumen entry is obtained and the wire advanced, it is snared via the endoscope and pulled retrograde out the esophagus and mouth.
  - The gastrostomy tube is attached to the wire at the head of the bed and this is pulled antegrade down the

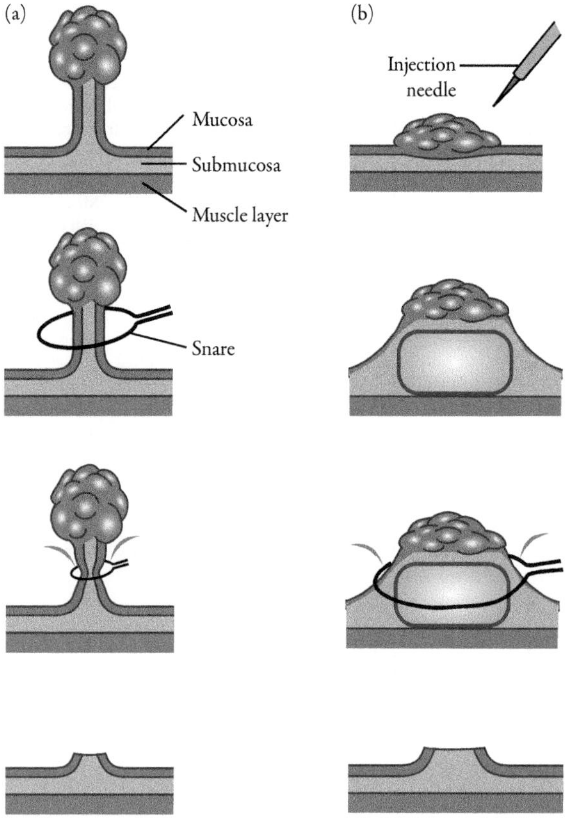

Figure 8.3 Endoscopic mucosal resection by snare polypectomy and submucosal saline lift.

esophagus and out through the anterior abdominal wall puncture site.

- The phlange should sit nicely apposed to the gastric mucosa on endoscopy. The external bumper should be secured at the abdominal wall dermal surface, and the centimeter marking noted. *See Figure 8.4.*
- The bumper should not be excessively tight against the dermis, or else the inner phlange will slowly begin to erode through the gastric mucosa under tension.

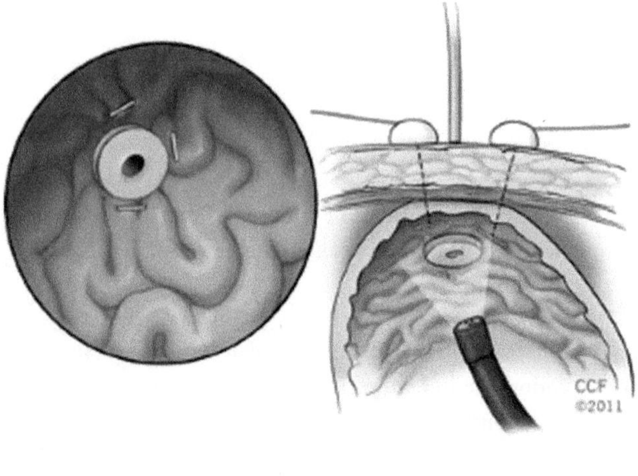

Figure 8.4 Percutaneous endoscopic gastrostomy (PEG) tube.

- An alternative technique is the endoscopic or image guided **push technique.**
  - Both place the gastrostomy tube from the percutaneous access point over a sheath and into the gastric lumen, thereby avoiding oral transit of the tube.
  - T fasteners may be used under endoscopic or radiologic guidance to secure the gastric wall to parietal peritoneum.
- Percutaneous gastrostomy by any technique is **contraindicated in massive ascites**, since ascites will prevent apposition of the gastric wall to the peritoneal lining, and hence tract epithelialization around the tube may not occur.
- The main complications of percutaneous gastrostomy tubes include **leakage** of gastric contents or **hemorrhage.** Either of these can occur atop the dermis, within the soft tissue of the abdominal wall along the tube tract, or intraperitoneally. Hemorrhage may also occur into the gastric lumen.
  - Any **peri-tube leakage** overtop the dermis can be managed by loosening the external bumper to allow the wound to contract.
  - **Dislodgement** of the tube in the early postoperative period may result in gastric leakage into the peritoneal cavity and require operative washout and replacement of the tube, since there had not been time for tract epithelialization.
    - At >14 days, an epithelial lined tract is mature, and a dislodged tube can be replaced at bedside.
  - A **gastrocolic fistulae** from trans-colonic placement of a percutaneous gastrostomy can be managed with removal of the tube, repair of the colon wall, and replacement of the gastrostomy in an open or laparoscopic fashion.
    - Alternatively, the patient can be managed conservatively, allowing time for a tract to mature, provided that:
      - The patient is stable and not having significant diarrhea from tube feed passage into the colonic lumen.
      - There are no signs of a colonic obstruction from the anterior and posterior transverse colon walls being skewered and sandwiched between the stomach and parietal peritoneal lining.
    - If not recognized initially, upon replacement of G tube weeks later after the tract has matured, patients may have the tip of the gastrostomy tube seated within the transverse colon, and resultant intracolonic delivery of feeds causing intolerable diarrhea. Fecal halitosis via the gastrocolic fistula can also be observed at that time.
  - **Buried bumper syndrome** occurs when the outer bumper is affixed too tightly to the abdominal wall, and the inner phalange consequently induces pressure ischemia and bores its way through the gastric mucosa and into its wall.

    - This is a chronic complication and results in trouble with administering tube feeds, often with concomitant abdominal pain.
    - Treatment includes loosening the bumper, provided it is still visible on endoscopy and the stomach wall has not necrosed during the preceding period. Otherwise if the bumper is completely outside of the gastric lumen, the tube should be removed and replaced endoscopically.
  - **Gastric outlet obstruction** can occur from the inner phalange or balloon, obstructing the pyloric lumen distally.
  - Gastric **mucosal ulceration** can result from irritation by the tip of the gastrostomy tube at a contralateral or neighboring site.
- **Laparoscopic gastrostomy** tube placement has similar complications as percutaneous endoscopic placement. The gastric wall at the tube entry site may be enclosed with a purse-string suture and affixed to the anterior peritoneal wall to prevent intraperitoneal leakage should the tube be dislodged in the early postoperative period.
- A **Stamm open gastrostomy** is through a midline incision with placement of gastric tube, purse-string of gastric wall at the tube's entry site, and affixing the gastric wall to the anterior peritoneal wall as protection against intraperitoneal leakage.

GASTROPARESIS

- **PATHOPHYSIOLOGY**
  - Gastroparesis describes impaired gastric motility and prolonged emptying. It is more common in **females** than males.
  - The most common etiology is **idiopathic**, in half of cases. The second most common cause is **diabetes** in one-quarter of cases. **Medication side effects** are the third reason. Rarely, it is due to an **iatrogenic vagal injury**.
- **DIAGNOSIS**
  - Symptoms include bloating, dyspepsia, early satiety, nausea, or delayed vomiting of undigested food.
  - The first step is to **exclude a mechanical obstruction** with an EGD or contrast study. These modalities normally will only show a dilated gastric lumen in the case of gastroparesis.
  - Then perform a nuclear medicine study by having the patient ingest a **radiolabeled solid meal** (e.g., albumin egg yolk) and if >50% is retained at 2 hours or >10% is retained at 4 hours, then gastroparesis is present.
  - If gastroparesis is due to an iatrogenic vagal nerve injury, there is increased pyloric, fundal, and body tone in addition to delayed solid emptying.
    - In these cases, liquid emptying may actually be accelerated due to loss of vagally mediated migrating motility complex (MMC) phase 2 antral churning.
    - On a contrast study, the stomach may not be as dilated due to the loss of vagally mediated relaxation capabilities.
    - Also, a blunted increase in pancreatic polypeptide (PP) may be seen after a sham feed in the setting of vagal nerve injury.
- **TREATMENT**
  - Dietary modification should include **small, frequent, liquid meals that are low fat and low fiber.**
  - Glucose control for diabetics and thyroid supplementation for hypothyroidism are both essential.
  - One can add **metoclopramide**, or low-dose erythromycin can improve gastric motility.
  - Patients refractory to medical management may need a **gastric stimulator.**
  - A venting G tube and feeding J tube can be placed, provided there is no associated SB dysmotility.
  - Pyloroplasty or subtotal or total gastrectomy are last line surgical options.

GASTRIC ULCERS

- **PATHOPHYSIOLOGY**
  - In an ulceration, the mucosal layer is eroded down through the lamina propria and into the muscularis mucosa or deeper into the submucosa and muscularis propria.
  - The main risk factor is **H. pylori, which is present in 70% of ulcers.**
  - Other risks include male sex, tobacco, alcohol, nonsteroidal anti-inflammatories (NSAIDs), uremia, physiologic stress (e.g., Cushing's or Curling's ulcers), glucocorticoids, or immune suppression.
  - There are five types of gastric ulcers designated by location. *See Figure 8.5.*
    - Type 1 ulcers are located inferiorly along the lesser curve (near the incisura). They result from a weakening of the gastric mucosal barrier.
    - Type 2 ulcers include one lesser curve ulcer and one duodenal ulcer. These result from high levels of acid secretion.
    - Type 3 ulcers are pre-pyloric and also from high acid secretion.
    - Type 4 ulcers reside superiorly along the lesser curvature and are related to a weakened mucosal barrier.
    - Type 5 ulcers are NSAID-induced and diffuse.
  - Any ulcer may harbor intestinal metaplasia, dysplasia, or malignant neoplasia. Giant ulcers >3 cm and ulcers on the greater curvature have a higher risk of containing malignancy on biopsy.

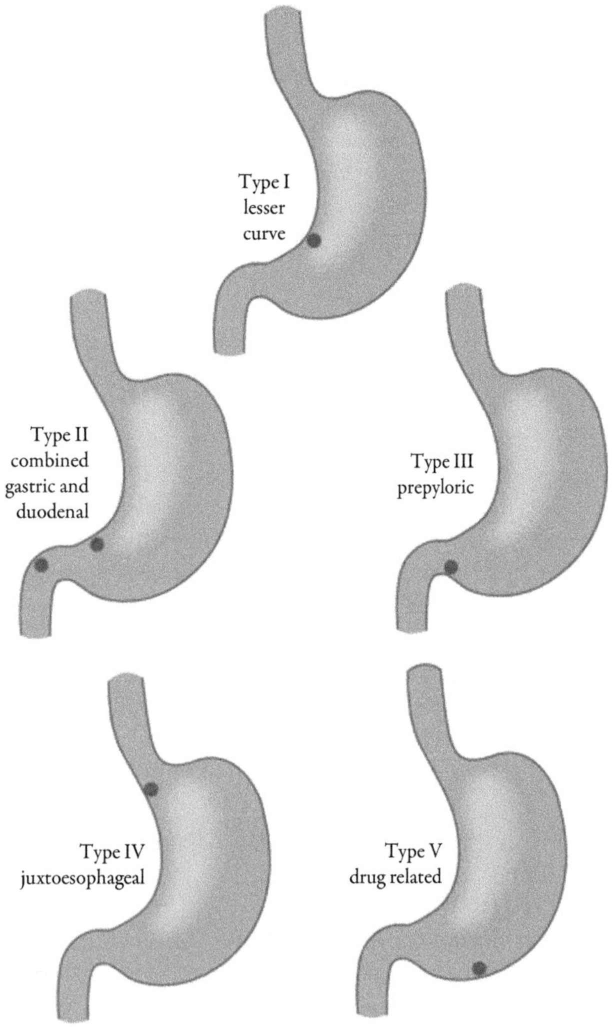

Figure 8.5 Types of gastric ulcers and their locations. Types 2 and 3 are related to acid hypersecretion and would benefit from a vagotomy procedure in addition to resection, if refractory to medical management.

- **DIAGNOSIS**
  - Symptoms include epigastric, post-prandial pain radiating to the back, and possibly guaiac positive stools.
  - If the ulcer has extended deep to the muscularis propria and through serosa of the stomach wall, then the patient can have more severe or widespread pain with peritoneal irritation due to leakage of gastric contents. Such leakage could be contained by neighboring omentum, or it could be uncontained resulting in peritonitis.
- **TREATMENT**
  - 1. **Proton pump inhibitor (PPI)**: Most patients can be **empirically started** on a PPI for a clinical suspicion of PUD and the absence of any signs of perforation. This treatment can be continued for weeks, provided the patient continues to improve. Other agents used include sucralfate and misoprostol.
  - 2. **EGD**: For patients **not responding to empiric PPI**, or for those with more severe symptomatology like **upper gastrointestinal bleeding (UGIB)**, the diagnosis can be formally made by endoscopy.
    - **The ulcer should be biopsied to exclude malignancy, and random biopsies of the antrum, incisura, and body should be taken** to test for H. pylori colonization. This is done with either tissue culture or the rapid urease (CLO) test of each specimen. Biopsies can have a reduced sensitivity during an UGIB.
    - If positive, H. Pylori should be treated with **triple therapy**: omeprazole, clarithromycin, and either amoxicillin or metronidazole for 14 days. Some recommend quadruple therapy with bismuth and triple antibiotics.
    - After completion of a two week course, H. Pylori eradication is tested by a urease breath test or stool antigen.
    - **Bleeding ulcers** are generally diagnosed and managed by EGD as well.
      - The Blatchford score includes pulse, blood pressure (BP), hemoglobin (HGB), urea, age, and other comorbidities. It can be used to stratify the need for urgent EGD in UGIB.
      - **Hemoclips and epinephrine injection** are the main modalities for hemorrhage control. Diffuse bleeding or type 5 ulcerations that cannot be managed medically may need endoscopic topical control or angiographic embolization. Occasionally a total gastrectomy is required.
      - **Active pulsatile bleeding, a visible non-bleeding vessel, and a clot adherent to a vessel are in descending order of re-bleeding risk after endoscopic treatment.**
      - Diffuse oozing and a clean non-bleeding ulcer base are at lower risks of re-bleeding.
      - A **Dieulafoy lesion** is an exposed, superficial submucosal arteriole on the lesser curvature that bleeds easily in the setting of overlying mucosal erosion by gastric acid.
  - 3. **Surgical treatment** is reserved for patients refractory to medical management.
    - Patients receiving >3 months of PPI therapy without endoscopic improvement may be classified as **"intractable" ulcerations** and qualify for consideration of surgical therapy.
    - Other indications for surgery include the following:
      - Ulceration complicated by **perforation.**
      - Ulceration complicated by **acute bleeding refractory to multiple attempts at endoscopic or angiographic control.**
      - Ulcerative **gastric outlet obstruction (GOO)** that is not improved by a trial of nasogastric tube (NGT) decompression or balloon dilation.
      - Presence of **malignancy** on a biopsy.

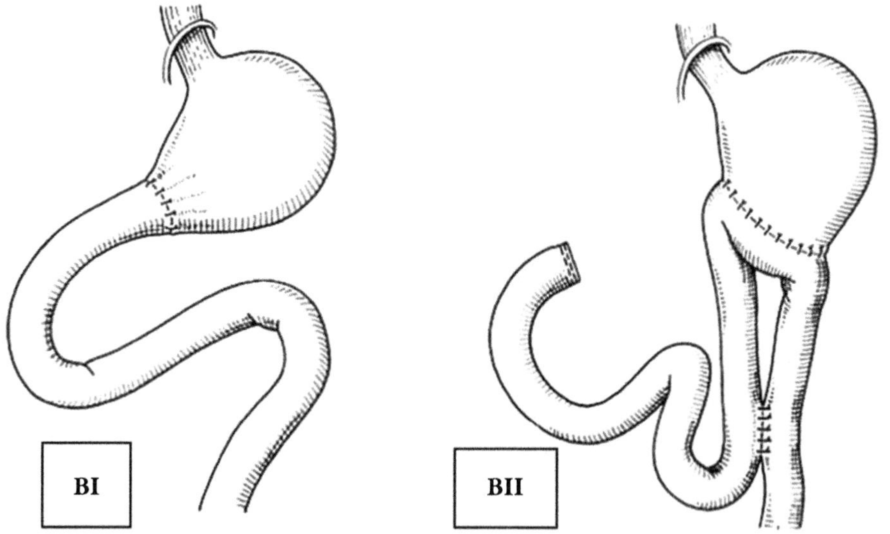

Figure 8.6 Billroth I and II reconstructions. Billroth I is preferred, as it does not leave an end duodenal stump. However, the first part of the duodenum must be healthy, and the remaining gastric pouch must reach to this point. Otherwise a Billroth II is used.

- In all instances, one must resect the ulcer and send for frozen section.
- a) **Resection of the ulcer bed and primary closure of gastrotomy** is the simplest approach and used in the unstable, acutely perforated patient.
- b) **Resection and reconstruction** is the most aggressive option for the stable patient. It has the lowest recurrence rate, but highest morbidity. Resection options and addition of a vagotomy are dependent on ulcer type. See *Figure 8.6.*
  - **Type 1 ulcer resection includes antrectomy and Billroth I (BI) reconstruction.** A vagotomy is not mandatory since the ulcer is not related to high acid secretion.
  - **Type 2 ulcer resection includes antrectomy, vagotomy, and reconstruction with Billroth II (B II) and 20 cm afferent limb, or Roux-en-Y (RxY) and a 40 cm Rx limb.**
    - B II is used for larger gastric remnants since it has a higher risk of bile reflux gastritis and esophagitis if the gastric remnant is small.
    - RxY is used when the post-resection gastric remnant is small, since there is no gastrin-producing antrum and very few acid-producing parietal cells. Both of which protect against peri-anastamotic jejunal ulcers in the Roux limb.
    - No B I reconstruction can be done in type 2 ulcers due to the presence of the concomitant duodenal ulcer.
  - **Type 3 ulcer resection includes an antrectomy, vagotomy, and B I, B II, or RxY reconstruction.**
  - **Type 4 ulcers require an extended resection along the lesser curvature due to their proximal location.** Subtotal gastrectomy, Pauchet or Cesendes, which has a resection margin closer to GEJ, are various options.
    - Reconstruction should proceed by a B II or RxY format. The latter can be used if the gastric remnant is very small. No B I reconstruction can be done, since the gastric remnant in extended resections often does not reach the duodenum.
    - Vagotomy is not needed since the ulcer is not acid related.
  - **Type 5, diffuse ulcerations refractory to medical management should be treated with a gastrotomy with oversewing or wedge resection of culprit bleeding sites, leaving the non-bleeding ulcers alone.**
    - These cases may require total gastrectomy with esophagojejunostomy if refractory to endoscopy, angiography, or prior conservative resections.
  - **Gastric outlet obstruction (GOO):** In the obstructed patient refractory to balloon dilation, do either of the following:
    - Perform vagotomy and antrectomy including the ulcer bed, with a B2 reconstruction; or
    - Perform vagotomy and bypass with gastrojejunostomy if there are concerns that a B II would result in poor duodenal stump healing.
- c) **Vagotomy** should be considered for types 2, 3, and duodenal ulcers that were **unresponsive to medical management provided the patient is stable intraoperatively**. See *Figure 8.7.*
  - Loss of vagal tone afterward leads to decreased parietal HCl production and a reflexive increase in gastrin release with G cell hyperplasia.
  - Loss of receptive vagal relaxation of fundus, body, and pyloric regions results in slower transit time for solids and quicker transit times for liquids.

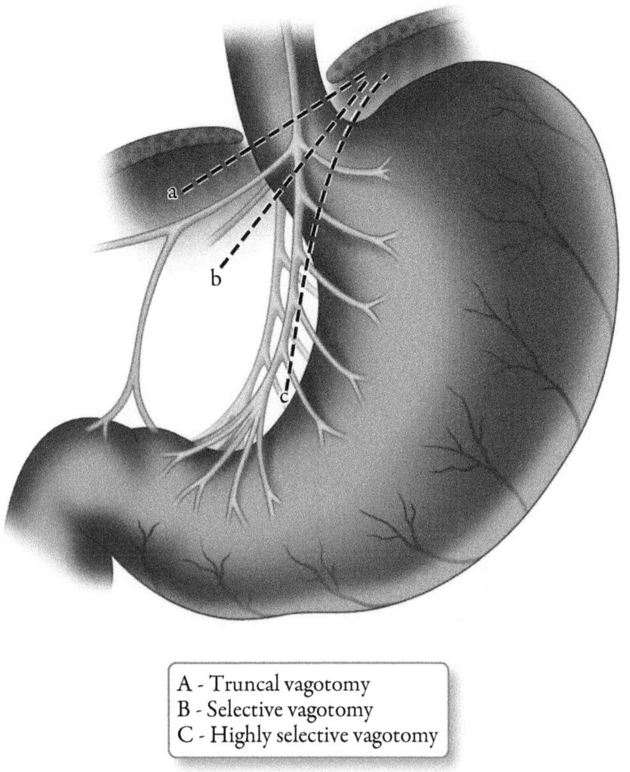

Figure 8.7 Types of vagotomy: truncal, selective sparing the celiac and hepatic branches, and highly selective sparing the crow's feet as well.

- **Truncal vagotomy** describes transection of the nerve about 4 cm proximal to GEJ, cephalad of the biliary neural branch from the anterior trunk, and the pancreatic/celiac neural branch from the posterior trunk.
  - This results in decreased biliary and pancreatic excretion with increased frequency of small bowel migrating motility complexes (MMC). The latter produces post-vagotomy diarrhea as the most common side effect.
  - Increased pyloric tone makes for increased solid transit time. So an accompanying **pyloroplasty or antrectomy with B I, B II, or RxY reconstruction** should be completed as described elsewhere.
- **Selective vagotomy** describes transection of the nerve just inferior to the takeoff of biliary and pancreatic/celiac neural branches. This spares the patient from alterations in biliary, pancreatic, and small bowel MMC frequency.
  - However, the patient will still need a drainage procedure like **pyloroplasty or antrectomy with reconstruction to counteract** the increased pyloric tone.
  - The **criminal nerve of Grassi** may bifurcate cephalad of the pancreatic/celiac neural branch from the posterior vagal trunk, and so must be securely transected, otherwise fundal parietal cell acid production will not decrease.
- **Highly selective "proximal" vagotomy (HSV)** describes transection of the anterior and posterior branches of Latarjet nerves along the lesser curvature of the stomach, while sparing the crow's feet at a point about 6 cm from the pylorus. HSV has the lowest morbidity but the highest ulcer recurrence rate.
  - **No drainage procedure is needed,** since pyloric relaxation is preserved and solid emptying time is unchanged. This can be used **when not resecting the ulcer bed,** as in the case of duodenal ulcers.
  - It may be used after a Graham patch of a duodenal ulcer perforation in a stable patient, since antrectomy and Billroth I is not ideal with a diseased D1 segment, and Billroth II or RxY would leave the Graham patch at the site of an end-duodenal stump.
  - **A variation of HSV** includes an anterior seromyotomy with posterior proximal vagal trunk transection. This may be accomplished laparoscopically.
- d) **Pyloroplasty** may be performed in a Heineke-Mikulicz (HM), Finney, or Jaboulay fashion. See *Figure 8.8*.
  - This allows for pyloric relaxation in the setting of truncal or selective vagotomies. It is not necessary in HSV.
  - May use interrupted gambee sutures for mucosal inversion.
  - In the setting of a medically refractory posteriorly penetrating duodenal ulcer (DU) that is persistently bleeding despite multiple endoscopic attempts, a surgical exploration and duodenotomy will allow one to oversew the gastroduodenal artery branches. The **horizontal duodenotomy can be extended proximally** to the patients' left across the pylorus and reconstruction can follow in an HM or Finney fashion. This accomplishes a pyloroplasty with minimal extra operative time. A truncal vagotomy can then be performed provided the patient remains stable.

ACUTE GASTRITIS

- **PATHOPHYSIOLOGY**
  - This is broad inflammation of the gastric mucosa. The inflammation can extend through mucosa into lamina propria or muscularis mucosa, however the overlying mucosal cells remain intact and without any areas of erosion or ulceration.
  - It is due to H. pylori, NSAIDs, alcohol, smoking, cocaine, bile reflux, or systemic physiologic stress in the critically ill patient.

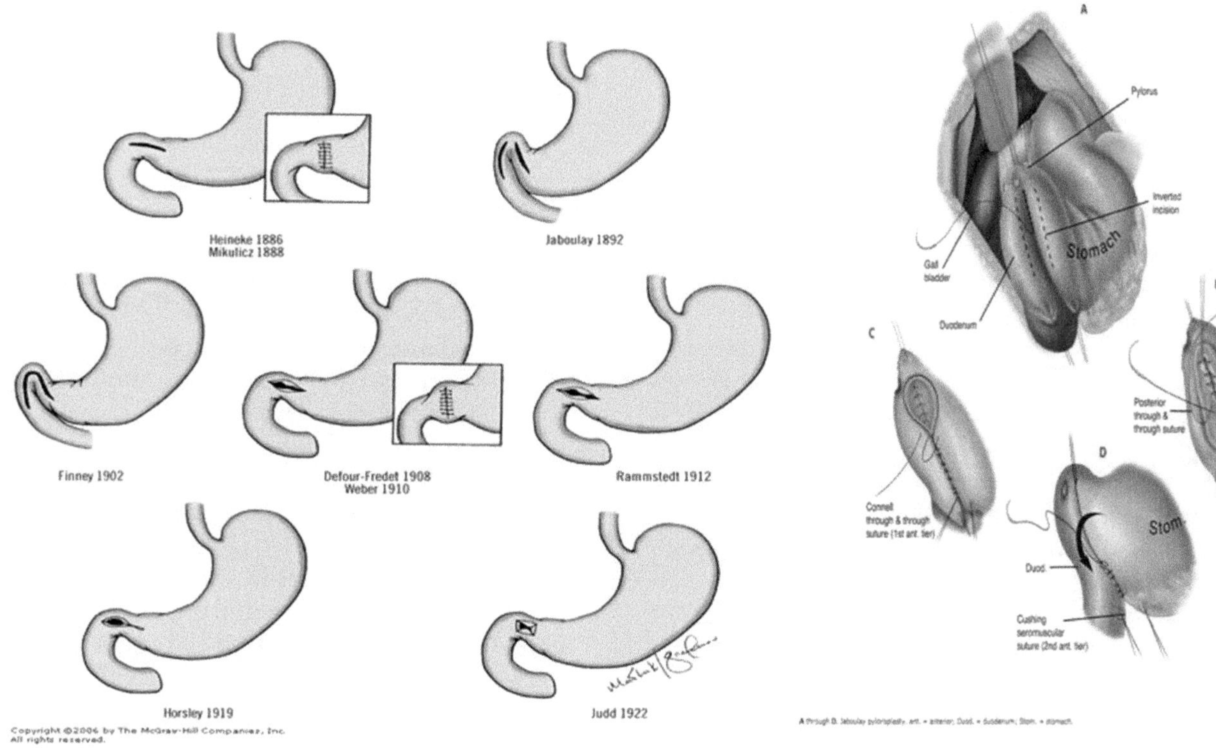

Figure 8.8 Types of pyloroplasty.

- **H. pylori-induced acute gastritis** can be diffuse or localized to the antrum or body.
- **Stress gastritis and stress ulceration** risk factors include **mechanical ventilation** for more than 48 hours, **coagulopathy**, burns, brain injury, spinal cord injury, and **lack of an enteral diet**. It occurs 3–10 days after the inciting event and may begin with localized fundal lesions. Stress gastritis can be associated with H. pylori colonization at baseline.

- **DIAGNOSIS**
  - based on symptoms of epigastric pain, nausea, and anorexia with or without vomiting that prompts an EGD. Some patients will have an UGIB. Critically ill patients may not demonstrate symptoms other than those relatable to an UGIB and anemia.
- **TREATMENT**
  - triple or quadruple therapy for H Pylori, PPI, sucralfate, stopping nsaids, and initiation of an enteral diet if the patient is not already on one. EGD with cautery should be THE initial approach for bleeding complications.

CHRONIC GASTRITIS

- This is chronic broad, superficial mucosal inflammation that can again extend into the lamina propria or muscularis mucosa. However, there should be no erosion, ulceration, or other break in the mucosal cell layer.
- **Fundal, type A** is associated with pernicious anemia or autoimmune disease. Treatment is aimed at the underlying immune activity.
- **Antral, type B** is associated with H. pylori. Treatment is with triple or quadruple therapy, including a PPI.
- Chronic gastric mucosal inflammation in either case can eventually lead to **atrophy and intestinal metaplasia**, both of which are risk factors for gastric adenocarcinoma (see below).
- In the setting of atrophic gastric mucosa, low parietal cell HCl production results in higher gastrin levels and resultant stimulation of enterochromaffin-like (ECL) cells. This chronic stimulation can occasionally produce **carcinoid features of the ECL cells**. Low parietal cell production of intrinsic factor leads to **B_{12} malabsorption and macrocytic anemia.**

GASTRIC ADENOCARCINOMA

- **PATHOPHYSIOLOGY**
  - The primary risk factor is **intestinal metaplasia associated with H. pylori colonization**. This intestinal type adenocarcinoma is the most common

worldwide and is prevalent in elderly patients from East Asia. It is usually located in the antrum and associated with areas of ulceration.
- The **diffuse type of adenocarcinoma** is the most common in the United States. It occurs in young patients and is often poorly differentiated with signet ring cells. It spreads laterally in the submucosal layer. Early cases may be missed on endoscopy. When present, an associated mass may be polypoid and fungating or intramural and ulcerating.
  - Some patients have an autosomal dominant hereditary syndrome producing this cancer. Genetic testing for **CDH1 cell adhesion molecule** should be considered if multiple young family members have diffuse type adenocarcinoma. In this setting, prophylactic total gastrectomy may be needed before the age of 40.
- Other risks include atrophic gastritis with or without pernicious anemia, adenomatous polyps, tobacco, pickled, salted, or smoked food, nitrosamines, and type A blood.

- **DIAGNOSIS**
  - Symptoms include unrelenting epigastric pain and weight loss. Occasionally patients will have an upper gastrointestinal bleed (UGIB).
  - Diagnosis is by **EGD with biopsy.** The cancer may present as a mass lesion or within an ulcer bed.
    - **An EUS** can provide an accurate T stage. This can be accompanied by a transgastric fine needle aspiration (FNA) of any suspicious perigastric lymph nodes seen.
    - **CT abdomen and pelvis** should be obtained for staging.
    - **A CT chest should be added if near the GEJ**, since tumors within 5 cm of it are considered similar to esophageal carcinomas and staged accordingly.
    - **PET-CT** is optional to evaluate for distant METS and node status.
    - **Diagnostic laparoscopy** in addition to standard imaging is a useful adjunct for T3 or greater lesions. This can determine if nodes outside of the perigastric area are involved, or if the patient has small liver metastases or other peritoneal implants not visible on standard imaging. Malignant cells in the peritoneal wash do not necessarily preclude resection, but are formally classified as M1.
- **TREATMENT**
  - In the absence of metastases, the surgical intervention depends on T stage and local extent of disease.
    - Biopsies showing **T-is (in situ) or T1a lesions** may be treated with **endoscopic mucosal resection (EMR) or endoscopic submucosal dissection (ESD)**. However, the specimen should have negative margins and no high-grade histologic features or lymphovascular invasion.
    - **T1b lesions** extend deep to the muscularis mucosa into the submucosa where lymphatic channels reside, so they have 20% chance of positive nodes. If there are no suspicious nodes and the staging workup is otherwise negative, **ESD** can be a consideration. Again, the specimen should have negative margins and no high-grade histologic features or lymphovascular invasion.
  - Any T1b with aggressive features on biopsy or any higher T stage should be treated with a laparoscopic or open subtotal or total gastrectomy, D1 lymphadenectomy, and reconstruction.
    - **T2 lesions** extend into the muscularis propria with a 50% chance of positive nodes.
    - **T3 and T4 lesions** have a 25% chance of peritoneal spread so one may consider **neoadjuvant chemoradiotherapy** after the staging workup is completed.
    - N1 status is 1–2 positive nodes and N2 status is 3–6 positive nodes. Neoadjuvant therapy can be considered in these cases.
- **Subtotal gastrectomy**, ideally with 6 cm margins when feasible, is best for an antral tumor. Reconstruction is with Billroth II (B II) if the gastric remnant is large, or Roux-en-Y (RxY) gastrojejunostomy if the remnant is small.
- **Total gastrectomy** and RxY esophagojejunostomy is the treatment of choice for a proximal gastric or esophagogastric tumor or any extensive diffuse type adenocarcinoma.
- An **antecolic or retrocolic B II or RxY gastrojejunostomy or esophagojejunostomy** can be constructed in an anti- or iso-peristaltic arrangement. See *Figure 8.9*.
- A **jejunostomy tube** may be placed distally as access for enteral feeding should an anastamotic complication occur.
- A **complete omentectomy** is performed with a plane created by cephalad retraction on omentum and inferior retraction on the transverse colon to enter the lesser sac.
- **D1 lymphadenectomy** includes perigastric nodes along greater and lesser curvatures and requires a **minimum of 16 nodes** for staging.
- **D2 lymphadenectomy** includes left gastric, celiac, superior mesenteric artery (SMA), and splenic artery nodes with or without a distal pancreatectomy or splenectomy. This extent of a lymphadenectomy is performed most often.
- **D3 lymphadenectomy** includes para-aortic nodes and is not mandatory.
- **Adjuvant chemotherapy should be considered for positive nodes or aggressive features.** Regimens include a combination of topoisomerase inhibitors (epirubicin, etoposide), cyclophasphamide, 5-FU, capacetabine, or oxaliplatin. Trastuzamab is added if the tumor is HER2 positive.
- Adjuvant radiation may also be considered for positive margins not manageable with re-excision.

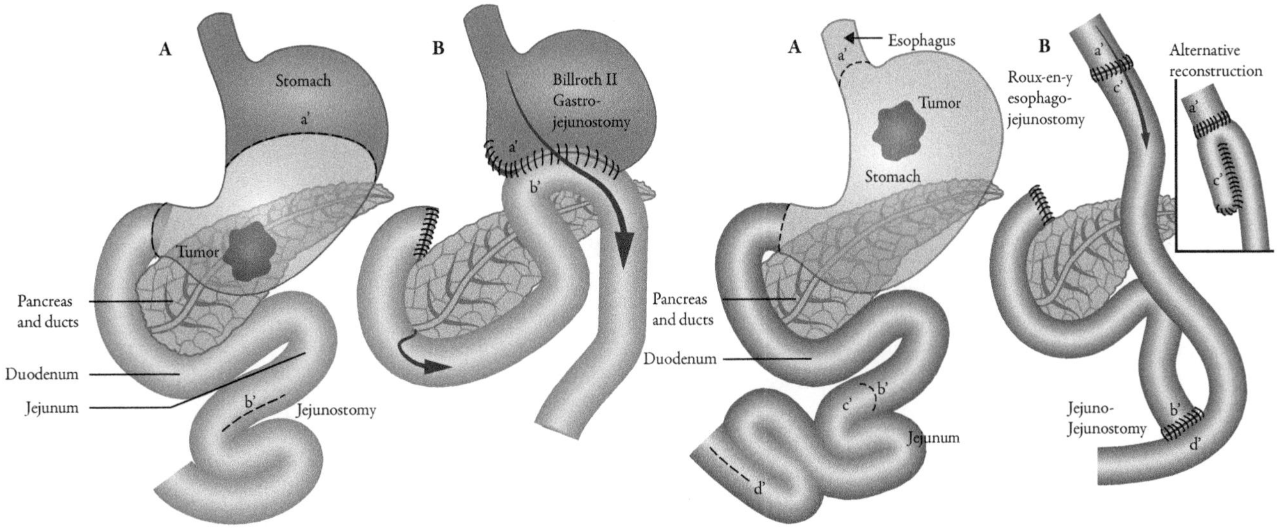

Figure 8.9 Resection and reconstruction based on gastric carcinoma location. Billroth II and Roux-en-Y esophagojejunostomy.

- **OUTCOMES AND PALLIATION**
  - The esophagojejunal anastomotic **leak rate is 1%–5%.**
  - **5-year overall survival** is in the range of **20%**. Half of patients have a recurrence after surgery.
  - Palliation of advanced tumors or recurrences can include:
    - An endoscopic **stent** for a proximal obstructing gastric adenocarcinoma.
    - Endoscopic stent or **gastrojejunostomy** for a distal gastric obstructing lesion.
    - **Radiation** to palliate bleeding.
    - Partial or total gastrectomy **with or without** isolated metastasectomy may occasionally be employed in fit patients as an adjunct to other palliative measures. There is no need for an omentectomy or an extended lymphadenectomy in palliative surgery.

POST-GASTRECTOMY COMPLICATIONS

- **Marginal ulcers** reside on the jejunal side of a gastrojejunostomy. They are more common **after an RxY** than B2 reconstruction. The main causes include ischemia, aspirin, NSAIDs, tobacco, or persistence of G cells within an antrum and parietal cells in the gastric remnant, resulting in high HCl production.
- A **gastrogastric fistula** between the gastric pouch and remnant may develop >1 year post-op of an RxY gastric bypass. This may impair weight loss, since the gastric remnant can now serve as a reservoir for ingested liquids or food, preventing pouch distension and the associated sense of satiety.
- **Early dumping syndrome** occurs in 5%–10% of patients. Symptoms include nausea, bloating, abdominal pain, lightheadedness, palpitations, and diaphoresis. They occur **within minutes** of ingestion and are from a carbohydrate load rapidly entering the small bowel. **Late dumping syndrome** is from insulin release in response to an enteric carbohydrate load rapidly entering the proximal small bowel. This results in symptoms similar to early dumping, but occurring **hours after** ingestion.
  - Neither type of dumping syndrome necessarily has diarrhea.
  - Treatment for both types is medical with **small, high protein, high fat, high fiber, low carbohydrate solid meals taken slowly followed by lying down for 30 minutes. Liquids may be taken thereafter.**
  - **Octreotide** can be initiated if symptoms are refractory to dietary modification.
  - Surgical options to create a jejunal pouch reservoir or use a reversed jejunal segment are rarely needed.
- **Post-vagotomy diarrhea** occurs after a **truncal vagotomy** disrupts the posterior vagal trunk neural branches to celiac ganglia that are responsible for regulating small bowel motility. The resulting sustained migrating motility complexes result in rapid transit of bile acids into the colon. This produces an osmotic diarrhea at 30–60 minutes after a meal. There is no actual malabsorption, just rapid transit. Treatment is with **small, high protein, fat, and fiber meals as well as cholestyramine.** Octreotide does not seem to work. Surgical treatment with a reversal of a jejunal segment, to slow the migrating motility complexes, is rarely used.
- **Alkaline reflux gastritis** produces symptoms of postprandial epigastric pain, nausea, and vomiting. It is more common after **B2** than B1 reconstruction. Diagnosis is by **EGD showing gastritis and a HIDA scan** showing radiotracer entering the gastric remnant. Treatment includes cholestyramine, PPI, or metoclopramide. However, this often requires definitive

management by conversion to an RxY reconstruction with a biliopancreatic limb **jejunojejunostomy anastomosed 60 cm** distal to the gastrojejunostomy. This leaves a 60 cm Roux limb, up which bile is unlikely to reflux and make its way to the gastric mucosa.

- **Gastric atony** results in symptoms of nausea, vomiting, and early satiety. It is diagnosed by a gastric emptying study. Treatment is with a high carbohydrate, low fat, low protein diet and small, liquid meals. Metoclopramide can be added. Unlike standard gastroparesis, gastric stimulators will not work if the antrum was removed. Rarely a pyloroplasty or a gastrectomy with RxY reconstruction is necessary to remove the dysmotile body of the stomach.
- **Dilated gastric pouch with or without a gastrogastric fistula** to the remnant may impair weight loss. The same can happen with a gastrojejunal fistula. Fistulae generally result from a staple line or anastomotic breach in integrity at some point in the patient's postoperative course.
- **Roux stasis syndrome** occurs in the setting of a long Roux limb within which there is retrograde peristalsis emanating from a jejunal pacemaker downstream. Treatment is with pro-motility agents, or rarely Rx limb shortening or gastrectomy.
- **Blind loop syndrome** has symptoms of malabsorption and steatorrhea due to bacterial overgrowth in a small bowel diverticula, a bypassed segment of small bowel, or a poorly mobile afferent limb of a B2 or biliopancreatic (BP) limb of an RxY.
  - Bacterial activity on enteric contents results in **deconjugation of bile acids (producing deoxycholic and lithocholic bile acids) and destruction of pancreatic enzymes**, both of which result in steatorrhea.
  - Also concomitant **impairment of intrinsic factor-B_{12} complex** results in a macrocytic anemia.
  - Treatment is with an **EGD and culture of the afferent B2 or BP limb**, then **antibiotics** and metoclopramide for motility. A medium chain triglyceride diet can help with fat absorption. Reconstruction of a shorter (40 cm) afferent limb is occasionally required.
  - If the syndrome is emanating from a diverticulum, then a clinical diagnosis and a d-xylose test can be enough to justify initiation of antibiotics. The diverticulum may eventually need to be resected.
- **Afferent loop syndrome** produces symptoms of abdominal pain, nonbilious vomiting, and malabsorption due to obstruction of the afferent limb in B2 or BP limb in RxY. Increased duodenal stump back pressure can cause backwash pancreatitis or even stump blowout.
  - Diagnosis is by **CT scan.**
  - Treatment is with an **EGD and balloon dilation if stenotic. If completely obstructed, then surgical resection and reconstruction of a shorter (40 cm) afferent limb is indicated.**
  - For a duodenal stump blowout, treatment includes external drainage and retrograde transanastomotic intraluminal duodenal drain.
  - On occasion, pain may resolve spontaneously with the subsequent appearance of bile in vomitus.
- **Efferent loop obstruction** has symptoms of abdominal pain and
  - **Nonbilious** vomiting if RxY with an efferent limb obstruction proximal to the jejunojejunostomy; or
  - **Bilious** vomiting if it is a B2 efferent limb obstruction or RxY common channel obstruction distal to the jejunojejunostomy.
    - Diagnosis is by **CT scan.**
    - Treatment is with **EGD and dilation of a chronic stenosis.** If the efferent limb is acutely and completely obstructed, then **surgical treatment** is indicated since the efferent limb may be internally herniated posterior to an anastomosis. This needs reduction and possibly resection and reconstruction depending on the health of this limb. If it is discovered the efferent syndrome is simply adhesive in origin at the time of operation, it can be managed with lysis of adhesions similar to other SBOs.

BARIATRIC SURGERY

- Two-thirds of the American population has a BMI of >25 and is considered overweight.
- Obese refers to a BMI >30, severe obesity is a BMI >35, and morbid obesity a BMI >40.
  - **Visceral (android) obesity** is deep to the fascia and associated with greater vascular and metabolic sequelae.
  - There is enhanced production of adipokines (TNF-a, IL-6, leptin, adiponectin, resistin) with paracrine and endocrine effects that manifest as chronic low-grade inflammation and insulin resistance.
  - Hepatocytes store free fatty acids as triglycerides, after they are released from insulin-resistant peripheral adipose tissue. This is believed to be protective for the body, but results in **hepatic steatosis and eventually nonalcoholic steatohepatitis (NASH).**
  - The **metabolic syndrome** is defined by having 3 of the 5 following criteria: central obesity, hypertriglyceridemia, low HDL, hypertension, and hyperglycemia. This syndrome correlates with a x2 increase in CV disease and x5 increase in type 2 DM within 5–10 years.
- **Indications for bariatric surgery** are severe obesity (BMI >35) with multiple comorbidities or morbid obesity (BMI >40) without comorbidities. The patient must have tried medical management and diet for 6

months, not partake in alcohol or other drugs, and be psychologically stable.

- **Laparoscopic vertical sleeve gastrectomy** is the most common procedure. It consists of a stapled resection from a point 6 cm proximal to the pylorus, across from incisura, cephalad toward the angle of His, with a 40F bougie in place along the lesser curvature. *See Figure 8.10.*
  - This removes epsilon ghrelin-producing fundal cells that mediate hunger, growth hormone production, and insulin release.
  - By stapling across the sling fibers of the cardia, this procedure may worsen gastroesophageal reflux (GER).
  - **Early leakage** of gastric contents can occur from the distal staple line, and **late leaks** generally come from the proximal staple line. The latter are due to a narrowing and stricturing at the distal staple line near the incisura, which increases resistance to outflow of gastric contents. This can cause the proximal staple line to leak under higher pressure.
  - Leaks are treated with either an **endoscopic stent** if there is minimal peritoneal contamination suspected, or **laparoscopic washout and repair** (and possible later revision to RxY) if significant peritoneal contamination is suspected based on systemic signs of illness.
  - Strictures can be treated with multiple **balloon dilation** attempts and conversion to RxY if those fail to relieve the stenosis.
- **Roux-en-Y gastric bypass (RxY GB)** is the second most commonly employed procedure. It is both restrictive and malabsorptive and results in the greatest amount of weight loss. See *Figure 8.10.*
  - Start by opening the hepatogastric ligament and area around the angle of His. A stapled gastric transection proximally should leave a 20 ml pouch. The jejunum is then transected at 40 cm beyond the ligament of Treitz and an **antecolic/antegastric** gastrojejunostomy (GJ) is performed using the distal segment of jejunum. A side-to-side jejunojejunostomy (JJ) is constructed **75–150 cm distal to the jejunal transection** point. The greater the distance of this anastomosis, the longer the Rx limb length and greater the weight loss. There is an increased risk of Rx limb syndrome at longer lengths.
  - Complications of RxY GB include:
    - **Marginal jejunal ulcers** from retained parietal cells in the gastric pouch producing HCl that injures the neighboring peri-anastamotic jejunal mucosa. 90% of these respond to PPI.
    - **Dilation of the gastric pouch** or development of **gastrogastric fistula** between the pouch and remnant stomach. Both effectively increase the reservoir capacity and hence diminish feelings of satiety and can result in return of any lost weight.
    - **Anastomotic leak** is due to ischemia or tension at GJ or JJ. In some instances, the GJ could potentially be endoscopically stented if there is minimal peritoneal contamination in a stable patient. However, if the

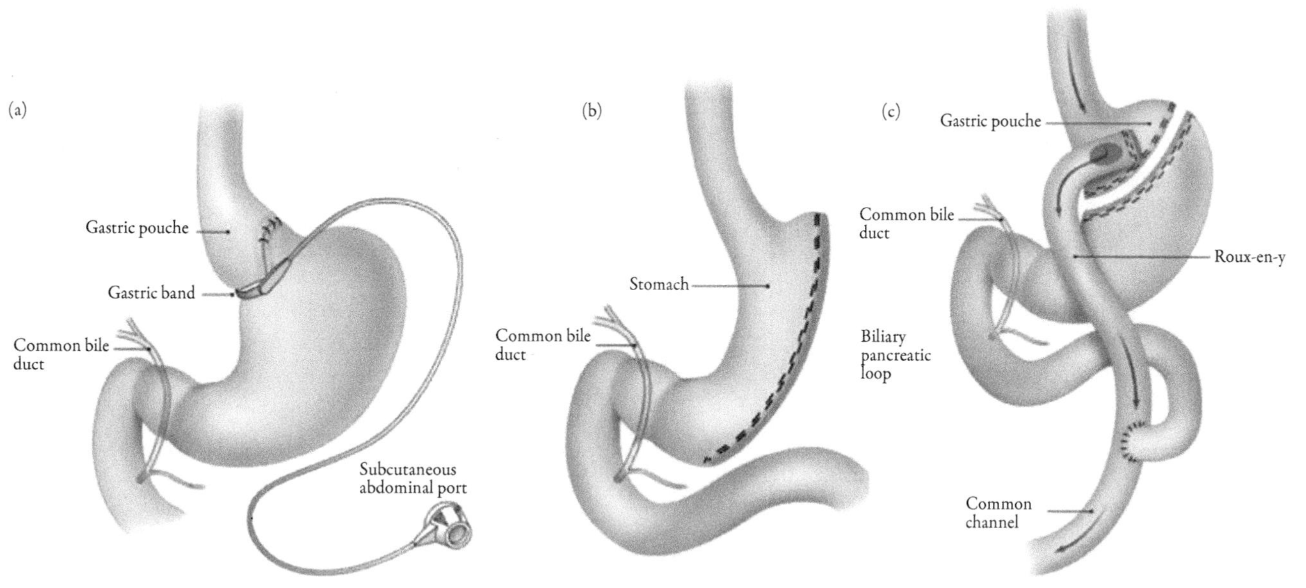

Figure 8.10 The two most common weight loss procedures. Roux-en-Y gastric bypass (C), which is restrictive and malabsorptive. Gastric sleeve resection (B), which is restrictive. (A) is the laparoscopic adjustable gastric band. Note the appropriate oblique position of the band, and the fundal suture to the cardia for security. The subcutaneous port can be filled incrementally. This is a restrictive procedure and is no longer used very often.

patient is systemically ill, then it should be presumed there is more significant contamination, and laparoscopic washout and repair should be considered. Leakage at the JJ may need a laparoscopic approach, as it is less likely to be endoscopically accessible.

- **Anastamotic stricture** at GJ is treated by endoscopic balloon dilation.
- **Y (biliopancreatic) limb obstruction or stenosis** with dilation of the gastric remnant may need surgical revision, as it may not be endoscopically accessible for dilation or decompression. An endoscopist may attempt access across the pouch and into the remnant stomach, creating a gastro-gastric connection in certain instances.
- **Roux (Rx) limb obstruction** may need to be ballooned or surgically revised.
- Small bowel obstruction can occur from **internal herniation** through Peterson's space (between the mesocolon and antecolic jejunal limb) or the mesojejunal space near the jejunojejunostomy. It also can occur through the mesocolic window if a retrocolic GJ was constructed. Internal herniation is more common than adhesive SBO after RYGB because almost all cases are laparoscopic, and therefore have few adhesions and a resultant easy mobility of bowel.
- In the open format of RYGB, **adhesive SBO** is more common. Adhesions also limit the mobility of the SB, so internal herniation is relatively less common than the laparoscopic format.
- **Iron deficiency, B_{12} deficiency, and calcium malabsorption** occur because ingested contents bypass stomach acid, intrinsic factor, and the duodenum respectively.
- Gallstones can develop during rapid weight loss. One should consider a prophylactic cholecystectomy at the time of surgery if there was any history of symptomatic cholelithiasis.

- **Laparoscopic gastric band** is much less common than a sleeve gastrectomy or RxY GB, and has largely fallen out of use. It can however be used for select patients with a BMI 30–35. See *Figure 8.10.*
  - The pars flaccida technique describes placement of the proximal gastric band around the stomach, but also including part of the gastrohepatic ligament. The fundus inferior to the band can be sutured to the cardia superior to the band in order to prevent slippage.
  - The first fill of the port should occur at **3–6 weeks post-op** with multiple incremental fills of 0.5–1 ml over the ensuing 2 years in order to lose >2 lbs/week and feel a restriction to intake of excess solid foods.
  - Position and tightness of the band are assessed with fluoroscopy. The band should appear **vertical on radiograph** with only a small cardia pouch of stomach cephalad of the band.
  - Symptoms of **obstruction or gastroesophageal reflux (GER)** could indicate that the band has been overfilled and should be treated by taking fluid out of port.
  - If the **band has slipped caudal**, patients can have an obstruction at the gastric pouch. Rather than appearing vertical, the band will appear horizontal on a radiograph. This requires immediate desufflation of the port.
    - If the patient is systemically ill or there is any concern of a compromised gastric wall, consider laparoscopy for removal of the band altogether and assessment of the pouch for ischemia, perforation, or erosion.
  - The band may chronically **erode into stomach** lumen, after which the subcutaneous port site may appear inflamed. A UGI study would show contrast lateral to the band margins. This is treated with removal of the band and repair of the stomach.
  - If a band needs removal, one may consider conversion to another weight loss surgery.
- **Duodenal switch-biliopancreatic diversion (DS-BPD)** is the least employed weight loss procedure. It is occasionally used for the super obese.
  - The first stage is a gastric sleeve using a 40F–48F bougie.
  - After a year of weight loss from the gastric sleeve alone, the patient can undergo the second stage. The jejunum is transected 250 cm proximal to the ileocecal (IC) valve, and the proximal segment anastomosed to a point on the ileum 100 cm proximal to IC valve. The distal transected jejunal end is brought up as a 150 cm Rx limb to the proximal end of a transected proximal duodenum. This leaves the native duodenum and biliary/pancreatic tracts as part of a Y limb draining into the ileum 100 cm proximal to the IC valve.
  - DS-BPD results in the greatest amount of weight loss, but also the greatest magnitude of metabolic derangements.
- **Postoperative Considerations**
  - All patients should be maintained on a **high protein, pureed diet for one month** after surgery. Solid food can be gradually reintroduced thereafter.
  - MVI, calcium, iron, and B_{12} supplementation is needed. Fat soluble vitamin supplementation can be considered in DS-BPD patients.
  - Ursodiol may help prevent gallstone formation.
  - **Low molecular weight heparin should be given for 10–30 days postoperatively**. A preoperative dose is also indicated.
  - **Mortality is from pulmonary emboli or anastomotic leakage**. Both are about 1%.
  - Most procedures produce a **50%–60% loss of excessive weight at 2 years** post-op. RxY GB is slightly higher at 60%–70%.
  - Pseudo-tumor resolves in nearly 100%, sleep apnea in approximately 85%, and nonalcoholic fatty liver disease (NAFLD) is reduced in 75%–85%.

- DM has a 50%–70% complete remission rate, compared to approximately 15% in medical management.
- Hyperlipidemia and hypertension improve in 70%.
- GERD, venous stasis, arthralgias, depression, polycystic ovarian syndrome, infertility, menstrual irregularities, and stress incontinence also improve.
- Cardiovascular events decrease by one-third, and cardiovascular mortality decreases by half.

OTHER SURGICAL PATHOLOGY

- **Benign lesions** that reside anteriorly or posteriorly in the subserosal position can undergo stapled wedge resection. If a lesion is submucosal and posterior, it can be approached by laparoscopic ports placed transgastrically with endoluminal resection after tumescence, and then endoscopic extraction.
- **Polyps** are generally benign; however, if they are >1 cm they should be removed.
  - **Fundic polyps** are small and sessile. They are the most common polyp. Most often they are secondary to PPI use, high gastrin, or familial adenomatous polyposis (FAP). Fundic polyps have no malignancy risk.
  - Heterotopic polyps may contain pancreatic tissue and are benign.
  - Hamartomatous polyps contain disorganized native tissue and are benign.
  - **Hyperplastic polyps** have a 2% malignancy risk. They are located in the antrum and are associated with H. pylori.
  - **Adenomatous polyps** have a 10%–20% malignancy risk if >2 cm or villous rather than tubular architecture. They may be associated with FAP.
- **Gastrointestinal stromal tumor (GIST)** originates from pacemaker Cajal cells residing high on the greater curvature. They are encapsulated and smooth, so often do not grow into adjacent organs. Gastric is the most common location and has the best prognosis. Small bowel is the second most common locale. Most tumors are asymptomatic, but some may erode the overlying mucosa and bleed or, less commonly, obstruct.
  - **Diagnosis** is by **EUS FNA or biopsy** showing spindle cells.
    - **The tumor is considered malignant if it is >5 cm or has >5 mitoses/50 HPF.** 75% have a **KIT mutation**, and some types of KIT mutations confer a better response to medical therapy, generally in gastric origin GIST. 10% have a PDGFRa mutation. This is also indolent and gastric, but imatinib-resistant. KIT exon 9 mutations require higher doses of imatinib.
    - **CT with or without PET** is done to exclude peritoneal or hepatic metastases. GISTs generally do not spread to lymph nodes.
  - **Treatment** is with **wedge resection and 1 cm margins**, with bougie in place if needed. Rupture or spillage can result in peritoneal recurrence. There is no need for an R0 resection if the tumor is low grade.
    - **May observe lesions that are <1 cm in size and without any concerning features on biopsy.**
    - **Imatinib** is a tyrosine kinase inhibitor that can be added if the GIST is believed to be malignant and **CD117, cKIT positive.** It may be given as neoadjuvant, adjuvant, or both. Also it may be used to palliate nonresectable lesions or after an R1 resection. Treatment resistance can be managed with increasing the dose or switching to sunitinib.
    - Resection is an option for metastases or recurrences previously controlled on imatinib.
- **Neuroendocrine tumor (carcinoid)**: Gastric carcinoids come in three types.
  - **Types 1 and 2** are from increased gastrin in the setting of MEN, gastrinoma, or a baseline low parietal HCl pdxn, for example in atrophic gastritis. Elevated gastrin stimulates enterochromaffin-like (ECL) cells to make more histamine and can result in an atypical carcinoid syndrome. Lesions may be multifocal. **Treatment is with endoscopic polypectomy, wedge resection, or formal gastric resection including the antrum to remove the source of hypergastrinemia.**
  - **Type 3 gastric carcinoids** are aggressive, ulcerated, grow independent of gastrin, and also may make histamine. Treatment can include wedge resection or subtotal or total gastrectomy. A D1 or D2 lymphadenectomy can be included. Supplemental octreotide can be added.
- **Leiomyoma** is a benign tumor of smooth muscle origin. It may be enucleated if <2 cm and undergo wedge resection if >2cm.
- **Gastric diverticula** are acquired, pulsion, pseudo-diverticula generally on the posterior cardia or fundus. Symptomatic diverticula should be resected.

OTHER NONSURGICAL PATHOLOGY

- **Mucosa-associated lymphoid tissue (MALT) lymphoma** is related to H. Pylori colonization. Treatment for low-grade lesions is triple therapy against H. Pylori. Most regress within a year. Radiation therapy can be added if refractory to H. Pylori treatment. Chemotherapy is reserved for high-grade lesions, deeper lesions by EUS, or presence of perigastric nodes.
- **Gastric lymphoma** is the second most common primary gastric malignancy after adenocarcinoma. Generally it is non-Hodgkin's diffuse large B cell. Diagnosis is by EGD and biopsy. Treatment is with chemotherapy and radiation. Surgery is reserved for an obstruction, perforation, or intractable bleeding.
- **Menetrier** is characterized by giant rugal folds in the fundus and body of a middle-aged male. These folds

represent **hyperplasia of mucosal cells**, and their increased secretion can result in a **protein losing state**. There is an associated diminution of parietal cells with consequent low HCl production. Treatment is with triple therapy against H. Pylori, high protein diet, anticholinergics, and rarely gastric resection if refractory.

- **Watermelon stomach** describes dilated, ectatic antral mucosal vessels. Generally this is in an elderly female. Treatment is with topical laser therapy, hormonal therapy, or antrectomy if refractory.
- **Trichobezoars** are conglomerates of ingested hair in a psychiatric patient. These generally need a **gastrotomy for removal.**
- **Phytobezoars** are conglomerates of fiber, generally in a diabetic with gastroparesis. They may respond to **enzymatic degradation** with Coca Cola or cellulase, followed by **EGD fragmentation and removal.**

9.

SMALL BOWEL

ANATOMY AND PHYSIOLOGY

- **Extracoelomic herniation** of the bowel occurs at the 5th week of gestational age. Then at 10 weeks of gestational age, there is a 270 degrees **counterclockwise rotation** upon retraction of the bowel back into the coelomic cavity. **Organogenesis** is completed by the 12th week. Intestinal ischemia prior to the completion of organogenesis results in atresia.
- The entire small bowel (SB) is 4–6 meters in length in an adult.
- The **basal electrical rhythm** is set by the interstitial Cajal cells and forms the basis for post-prandial **peristaltic contractions** as well as the **migrating motility complex (MMC)** during the fasting state. The phases are below:
  - 1 = bowel is at rest.
  - 2 = acceleration and gallbladder contraction that is vagally mediated.
  - 3 = peristalsis that is motilin mediated.
  - 4 = deceleration.
- SB, specifically jejunum, is the major site of absorption for PO intake (about 2 L/day) and any intrinsically secreted fluid, including the following:
  - saliva up to 1.5 L/day, gastric secretions accounting for 2.5 L/day, pancreatic production of 1.5 L/day, biliary output of 0.5 L/day, and small bowel secretions totaling 1 L/day.
- The lining of the SB is arranged in **villi**. Each is composed of a fingerlike projection of **epithelium and underlying lamina propria**, within which lies a central lacteal. Multipotent stem cells differentiate and multiply at the depth of each villus, and migrate to the tip over 2–5 days. Epithelial cell types include:
  - **Absorptive epithelial** cells across which active or passive transport of nutrients may occur transcellular. Passive transport can sometimes be paracellular.
  - **Goblet** cells secreting mucin.
  - **Paneth** cells secreting proteins with local paracrine activity for mucosal defense.
  - **Enteroendocrine** cells secreting proteins with systemic endocrine activity. These include:
    - Enterochromaffin cells (EC) produce serotonin.
    - Enterochromaffin-like (ECL) cells make histamine.
    - S cells release secretin, which increases pancreatic ductal HCO_3 release and decreases gastrin.
    - I cells produce CCK, which increases gallbladder contraction, relaxes the sphincter of Oddi, and increases pancreatic acinar enzyme release.
    - G cells make gastrin, which increases parietal cell HCL production and ECL cell histamine release.
    - D cells are responsible for somatostatin.
    - Other endocrine compounds include motilin, which stimulates phase 3 of the MMC, and bombesin, which increases pancreatic and gastric secretions.
  - **Brunner's gland** cells have alkaline secretions to neutralize acidic stomach contents entering the SB.
  - **M cells** are present in the epithelium for antigen processing and presentation. Peyer patches are conglomerates of lymphocytes in the underlying lamina propria of the ileum. 70% of all immune cells are in the enteric system.
- **Muscularis mucosa** allows for villi motility and contains interstitial Cajal cells.
- **Submucosa** surrounds muscularis mucosa. Inflammatory processes or injury deep to this level can produce strictures.
- **Muscularis propria** surrounds submucosa and consists of circular and longitudinal layers.
- **Serosa and mesothelium** form the outer covering.
- **Duodenum**
  - The duodenum is perfused by anterior and posterior branches of superior and inferior pancreaticoduodenal arteries.
  - It is the major site of **iron and calcium absorption**.
  - Duodenal mucosal cells and Brunner's glands help manage acidic gastric contents exiting the pylorus.
  - Lymphocytes reside in the lamina propria underlying the mucosa.
  - **Celiac sprue** is secondary to a lymphocytic response to gluten antigens in the duodenum. It can result in upper gastrointestinal (UGI) symptoms and weight loss. Over the long term, this can lead to the development of a T-cell lymphoma.
- **Jejunum**
  - The superior mesenteric artery perfuses the jejunum by branching into long vasa recta.

- Valvulae conniventes, also known as plica circularis, are the luminal mucosal folds.
- The jejunum is the **major site of absorption of sodium chloride (NaCl) and most all nutrients.**
  - Carbohydrates normally compose 45% of the diet.
    - **Salivary and pancreatic amylase** produce oligo- and disaccharides including maltose (glucose-glucose), sucrose (glucose-fructose), and lactose (glucose-galactose).
    - Brush border enzymes **maltase, sucrase, lactase, and limit dextrinase** produce the monosaccharides glucose, galactose, and fructose.
      - Glucose and galactose move into the epithelial cells from the bowel lumen by **secondary active transport** coupled to the movement of a sodium (Na) molecule. This is possible because a Na gradient is created by a basolateral sodium-potassium (Na/K) pump.
      - Fructose is transported by **facilitated diffusion.**
      - Fructooligosaccharides that are undigested move along into the colon as short chain fatty acids to fuel colonocytes.
  - Protein comprises 15% of diet.
    - **Gastric pepsin** produces polypeptides that are converted to oligopeptides by **endopeptidases** like trypsin. Trypsin is activated by enterokinase in the bowel lumen.
    - **Brush border enzymes** produce oligo, tri, and dipeptides all of which can be absorbed into the lumen of the epithelial cell.
  - Lipids comprise 40% of the diet.
    - **Gastric and pancreatic lipase** produce three lengths of fatty acids
      - **Short chain fatty acids** result from metabolization of fructooligosaccharides and continue into the colon to fuel colonocytes.
      - **Medium chain fatty acids** are absorbed directly into portal venous tributaries.
      - **Long chain fatty acids** are micellized and vesicularized via bile salts with transport into the epithelial cells. Intracellular construction of chylomicrons are then transferred into lacteals draining to the cisterna chyli and eventually into the brachiocephalic vein. *See Figure 9.1.*
      - Gastric acid hypersecretion, pancreatic exocrine insufficiency, luminal bacterial overgrowth destroying pancreatic enzymes, and terminal ileal mucosal insufficiency or resection all interfere with fat absorption and produce clinical or sublinical steatorrhea. Treatment is with proton pump inhibitor (PPI), restriction of fat intake, pancreatic enzyme supplements, antibiotics for bacterial overgrowth, and diphenoxylate-atropine.
- **Ileum**
  - It is perfused by the superior mesenteric artery branching into short vasa recta.
  - The ileum is the major site of absorption of **non-conjugated bile acids** (passively transported at any point along the ileum) and **conjugated bile acids** (actively transported at the terminal ileum) as well as **folate and B_{12}-IF complex.**
    - B_{12} is ingested and bound to salivary R protein, then hydrolyzed by pancreatic enzymes in the proximal small bowel and bound to intrinsic factor (IF), which was produced by gastric parietal cells.
    - Normal ileal output is 500–800 ml/day entering the colon or exiting out a stoma.

DUODENAL ULCERS (DU)

- **PATHOPHYSIOLOGY**
  - These are erosions that extend beyond the lamina propria into the muscular mucosa. *See* Figure 9.2.
  - They present with epigastric or right upper quadrant (RUQ) pain that abates with eating but returns a few hours later, sometimes in the middle of the night. This is because in the non-prandial state, gastric acid heading into the duodenum is unbuffered by ingested food or liquid. So the low pH irritates an area of duodenal mucosal weakness.
  - Ulcers are often in the anterior wall of duodenal segment 1 (D1). Posterior ulcers may penetrate into the gastroduodenal artery (GDA) leading to an upper gastrointestinal bleed.
  - 90% of ulcers are associated with ***H. pylori* in the gastric antrum or duodenal mucosa. There may also be gastric metaplasia at the duodenal site.** *H. Pylori* decreases D cell somatostatin production, which in turn increases gastrin.
- **DIAGNOSIS**
  - **EGD with biopsy of the gastric antrum if a DU is seen. Culture** of this specimen should show *H. pylori* or a **CLO test** can be done by adding urea to the sample and noting any ammonia production and increased pH. This would indicate the presence of urease producing *H. Pylori.*
  - EGD can also help exclude a duodenal malignancy or inflammatory disorder as the cause to the UGI symptoms.
- **TREATMENT**
  - **Empiric initiation of a PPI and avoidance of tobacco and NSAIDs.** If *H. pylori* testing is positive, add antibiotic therapy to PPI.
    - Eradication can be confirmed with **stool antigen or urea breath test,** where ingested labeled urea is converted to labeled CO_2 if *H. pylori* is still present. This is absorbed into the portal circulation

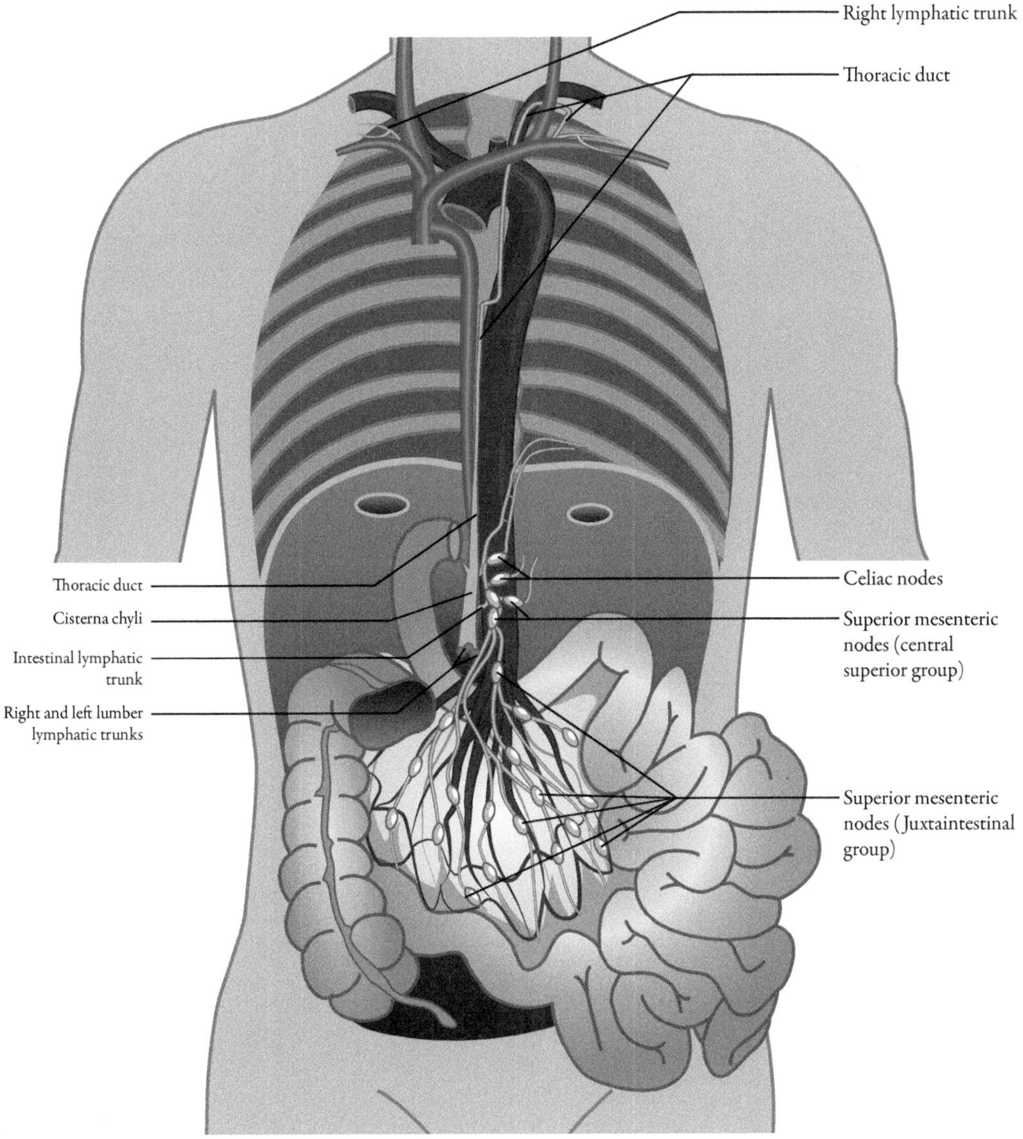

Figure 9.1 Lymphatic drainage of the small bowel.

and exhaled from the pulmonary circuit. Serology remains positive, so it is not as useful to confirm eradication.

- **Surgical treatment** is reserved for obstruction, perforation, or an intractable ulcer refractory to medical management. If a patient is NSAID-dependent or cannot be compliant with *H. pylori* treatment, then they can also be considered for surgery.
  - **Anteriorly perforated ulcers** in D1 should undergo a **Graham patch** with or without the use of a jejunal loop as a serosal patch. *See Figure 9.3.*
    - Another option if the perforated ulcer bed is large is to bring a Roux limb up for a duodenojejunal anastamosis and then, if stable, a highly selective vagotomy (HSV) since nonselective vagotomy would require a pyloroplasty, and this is not ideal when adjacent to an ulcer bed. Also antrectomy with reconstruction is not suitable in the acute setting.
- **Posteriorly perforated bleeding DUs** refractory to multiple attempts at **endoscopic control** with epinephrine, cautery, or hemoclips can undergo **angioembolization of the GDA.** If refractory to this or if the patient is unstable from hemorrhage, one can perform a **duodenotomy with oversewing of the GDA and transverse pancreatic artery branch.** *See Figure 9.4.*

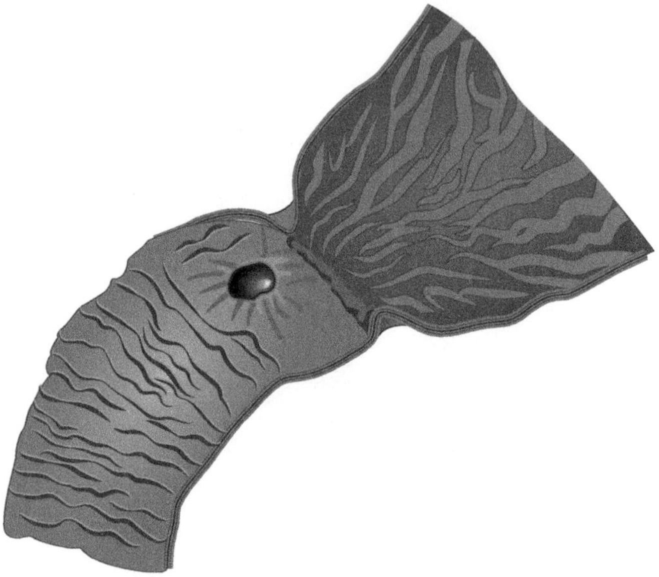

Figure 9.2 Duodenal ulcer.

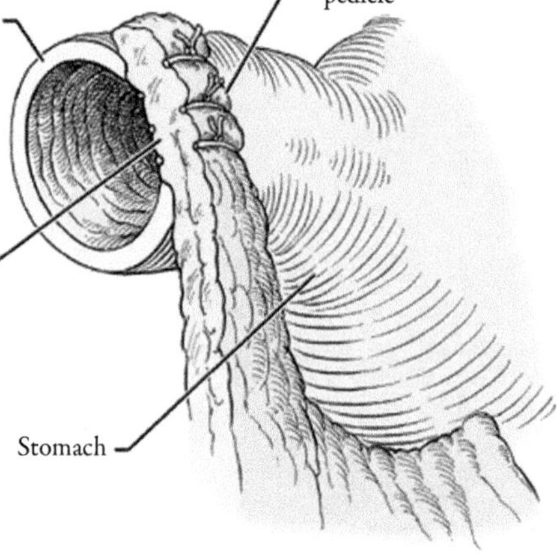

Figure 9.3 Anterior perforation repair.

  - If the patient remains stable intraoperatively, this can be followed by extension of the duodenotomy proximally, across the pylorus, and closure in a transverse Heineke-Mikulicz (HM) or Finney fashion. This functions as a pyloroplasty, and the procedure can be completed with a truncal vagotomy.
- **Malignant or intractable ulcers** failing >3 months of medical therapy should be resected, along with antrectomy, vagotomy, and Billroth-II (B2) or Roux-en-Y reconstruction. See Chapter 8, Stomach, for details on these reconstructions.
- **Obstructing ulcer beds** in which malignancy has been excluded by biopsy can be managed with nasogastric tube (NGT) decompression in the acute setting, but may require **serial balloon dilations** in the chronic setting. *See Figure 9.5.*

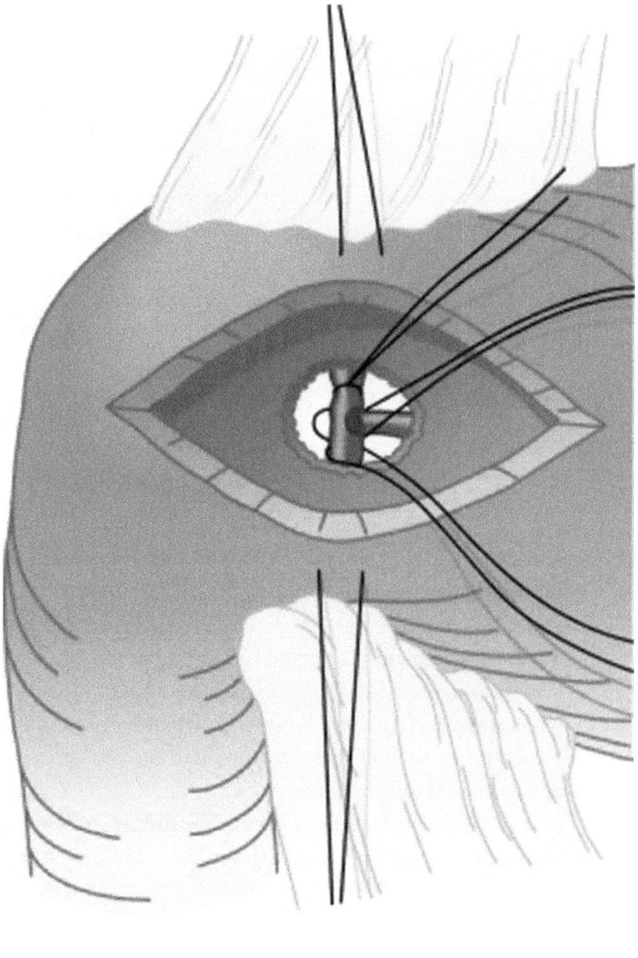

Figure 9.4 Posterior duodenal ulcer repair.

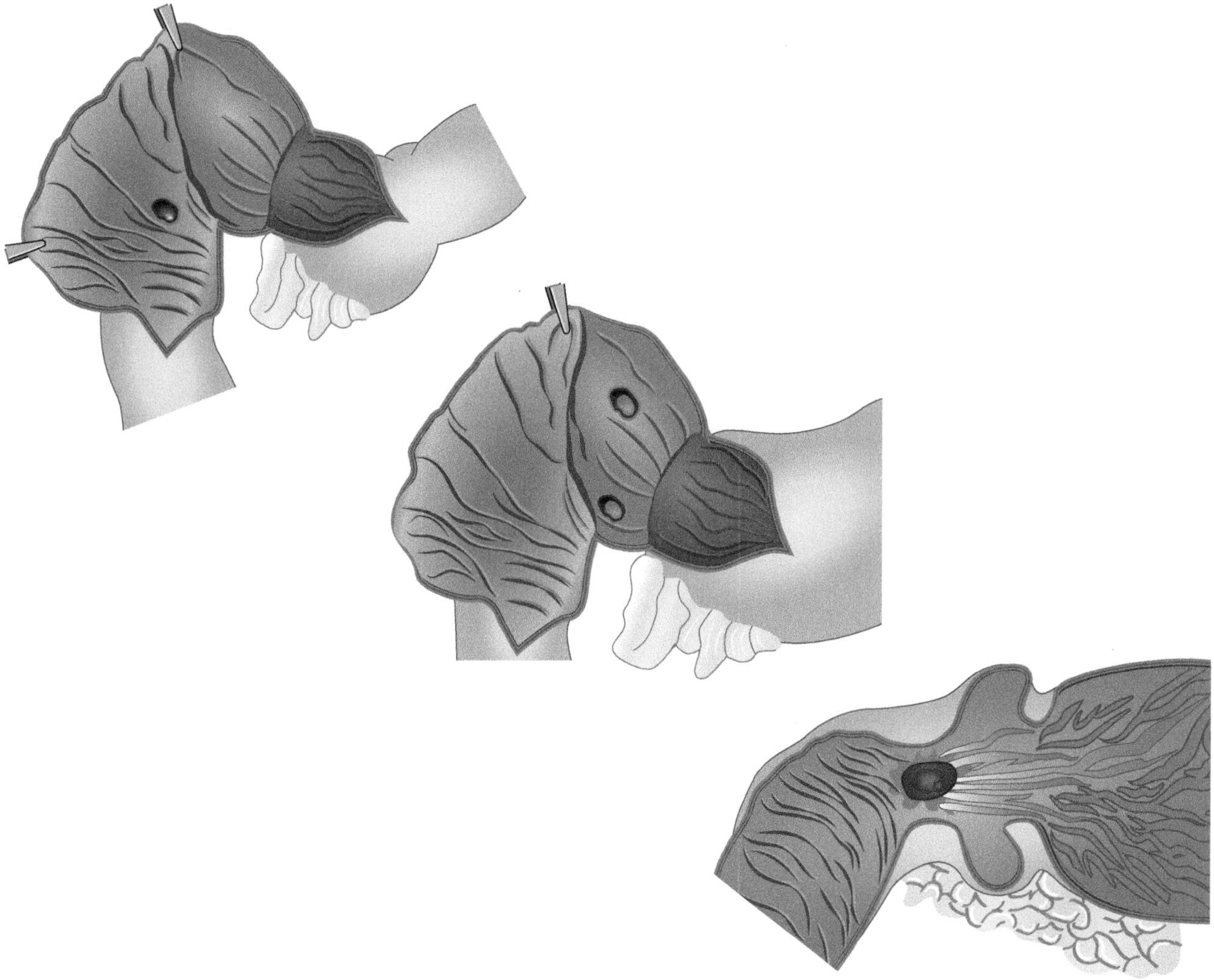

Figure 9.5 Duodenal ulcer complications.

    - Ultimately, **vagotomy and resection of the ulcer bed** along with antrectomy plus B2 reconstruction may be required if serial dilations are inadequate.
    - An HSV and bypass with gastrojejunostomy, avoiding ulcer or antrum resection, is an alternative option if it is believed that the duodenal bed would be a difficult to close after resection.
  - **HSV** without directly addressing an ulcer has a 10%–15% ulcer recurrence rate.
- **Vagotomy and pyloroplasty** has a 5%–10% recurrence.
- **Vagotomy with antrectomy** has a 1%–2% recurrence but the highest mortality of the three.

CROHN'S DISEASE

- **PATHOPHYSIOLOGY**
  - Crohn's is mostly a disease of 15- to 35-year-old smokers in the higher socioeconomic class. Some patients have a first-degree relative with the disease.
  - It is characterized by full thickness, patchy inflammation of the **ileum (40% of presentations), colon (20%–35%), and anus (5%–10%).** The process usually **spares the rectum.** *See Figure 9.6.*
  - Any other segment from the mouth, where aphthous ulcers are the most common, to the anus, where large skin tags are the most common finding, can be involved.
  - The chronic inflammation characteristic of Crohn's disease increases SB and colon adenocarcinoma risks.
  - **Steatorrhea** is secondary to ileal malabsorption of bile salts or their destruction by pancreatic enzymes in a blind loop. This results in **ADEK vitamin deficiency and calcium oxalate ureteral stones.**
  - **Decreased B12** is from malabsorption at the ileum or destruction of the IF-B12 complex by blind loop bacterial overgrowth in an entero-enteral fistula. Folate deficiency can also occur.

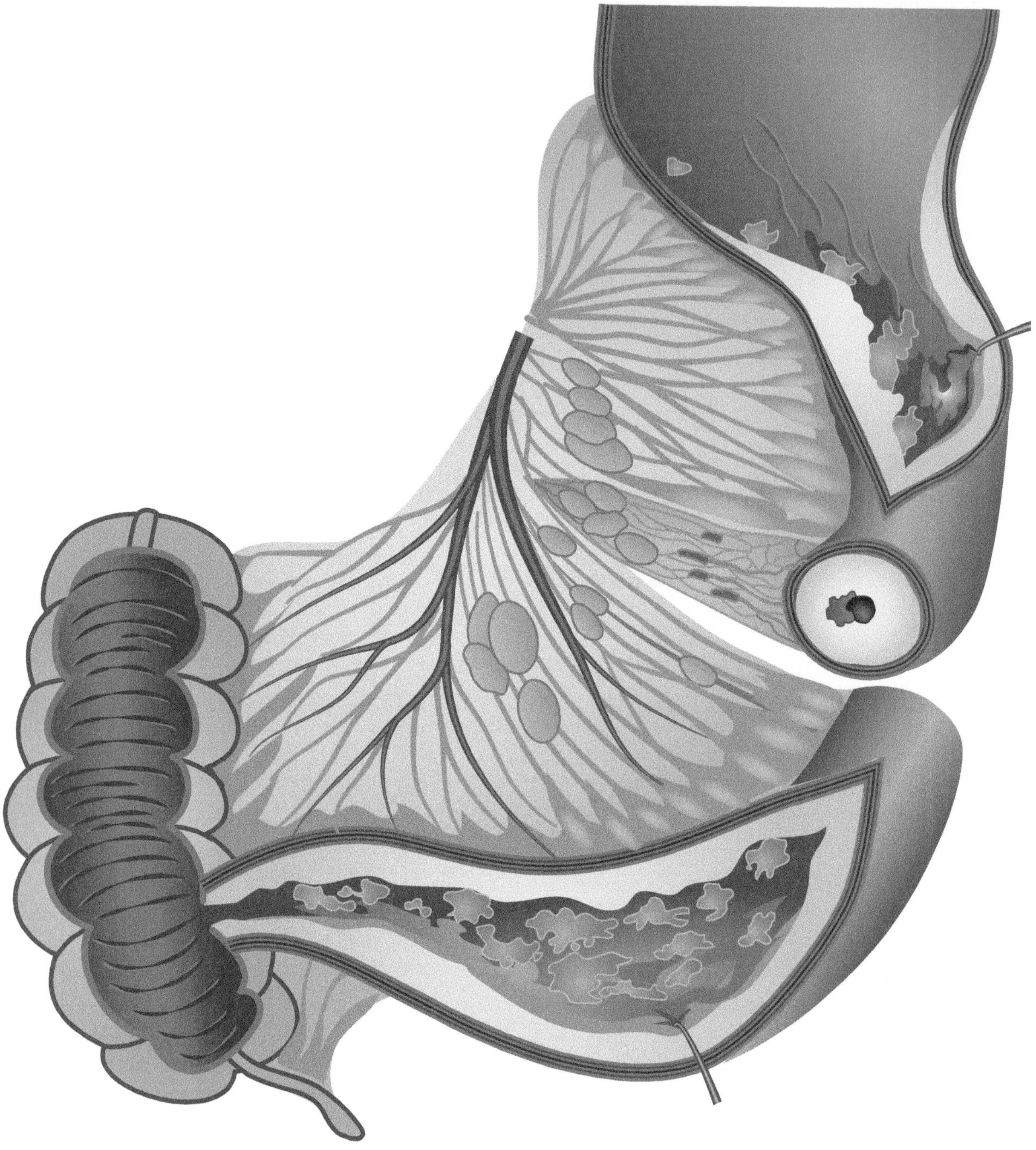

Figure 9.6 Crohn's ileitis.

- **Gallstones** result from lack of bile salt reabsorption in the terminal ileum, and hence an imbalance between cholesterol and bile salts within the gallbladder.
- **DIAGNOSIS**
  - Symptoms include chronic, recurrent abdominal pain, diarrhea, and weight loss. **Extraintestinal manifestations** include arthritis, pyoderma gangrenosum, erythema nodosum, primary sclerosing cholangitis, or ocular disease such as episcleritis or uveitis.
  - Diagnosis is by **colonoscopy with ileal intubation** showing skip lesions, cobblestoning, pseudopolyps, deep ulcerations, and strictures. Biopsies reveal granulomas.
- **Characteristic imaging findings** include strictures, deep ulcers, fistulae, and creeping mesenteric fat. These may be seen on CT abdomen and pelvis or MRI of the abdomen and pelvis. Either may be accompanied by enterography.
  - Small bowel follow-through (SBFT) with or without enteroclysis is another modality to view the mucosal side of bowel.

- **TREATMENT**
  - Medical management is the initial approach.
    - Topical anti-inflammatories include **sulfasalazine** for mild disease or **mesalamine** if colonic involvement.
    - **Systemic glucocorticoids** are for an exacerbation or extraintestinal manifestations. **Budesonide** is a steroid limited to luminal activity.
    - TNF-alpha inhibitors include **Infliximab IV q6 weeks or adalimumab subcutaneous q2 weeks.** They are for moderate or severe disease with or without fistulae.
    - **6 Mercaptopurine (6MP), Azathioprine, and Methotrexate** can help reduce glucocorticoid dependence over the long term.
    - **TPN** can assist in closure of proximal small bowel fistulae in the setting of malnutrition.
    - Image-guided drainage is preferred for perienteric or pericolonic abscesses. Abscesses that are very small may be treated with antibiotics without drainage.
  - **Surgery** is required at some point in 90% of patients.
    - Indications include obstruction, fibrotic strictures, non-resolving fistulae, blind loop syndrome, perianal abscesses or fistulae needing setons or an advancement flap (see Chapter 7, Anorectum).
      - Glucocorticoid dosage should be decreased or stopped prior to surgery.
      - There is no need to hold TNF-alpha inhibitors.
      - At laparotomy, **resect only grossly involved small bowel or large bowel**. There is no need for clear microscopic margins. If the terminal ileum (TI) is resected and only a short segment of normal TI exists before the IC valve, it is best to take that segment and the cecum in the specimen.
      - One must **resect the diseased source of a fistula**. The innocent neighboring SB, colon, or bladder into which the fistula traversed may simply be oversewn with omentum placed between it and the site of resection.
      - A Foley balloon can be passed antegrade and retrograde to determine the extent of luminal strictures proximally and distally to the site of any enterotomy.
      - Take care with thickened mesentery so as to not create a lymphatic leak. After mobilization laparoscopically, it may be best to transect the mesentery extracorporeally via the specimen extraction site.
      - If colonic inflammation is noted and diffuse, a total proctocolectomy with end ileostomy is preferred because anal disease may complicate any attempt at ileoanal anastomosis.
      - One may remove a normal appendix provided the cecum is not inflamed. This helps prevent future diagnostic uncertainty.
- **Strictureplasty** *See Figure 9.7.*
  - This is a better option than resection for preserving bowel length in the setting of multiple strictures, a known diagnosis of Crohn's enteritis, and prior operative resections and biopsies negative for malignancy.
  - The most common complication of a strictureplasty is bleeding.
  - The bowel may undergo inflammatory regression at the strictureplasty site.
  - It should not be used **in the colon, at sites of prior anastomoses, chronic radiation enteritis, or suspected malignancy.**
  - **Heineke-Mikulicz format for strictures <7–10 cm.**
  - **Finney or Jaboulay strictureplasties if <15–20 cm.**
  - If the diseased area is longer than 20 cm, one may divide the SB and mesentery at the midpoint of this long stricture, align the strictured halves to overlap side to side with normal SB proximally and distally,

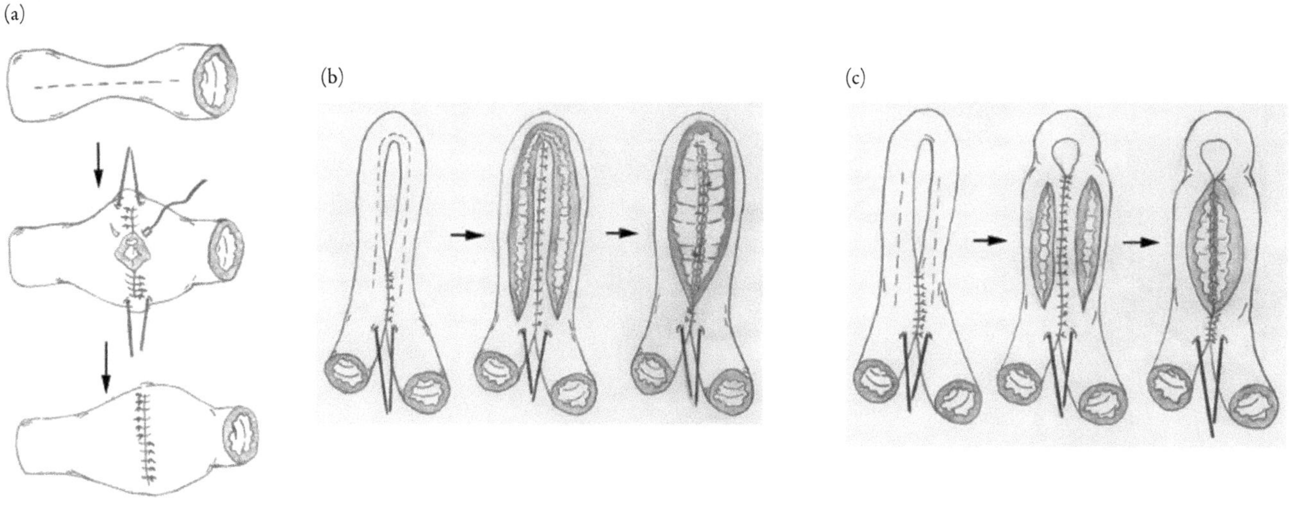

Figure 9.7 a–c: Crohn's strictureplasty.

complete longitudinal stricturoplasties of each half, and anastomose the diseased segments to the paired segment of normal proximal or distal SB in a side-to-side fashion.

MALIGNANT SB NEOPLASMS

- There are a variety of etiologies, and the presentation can vary from an incidentally diagnosed mass to obstruction, bleeding, perforation, or fistulization. *See Figure 9.8.*
- **Metastases** to the SB are from melanoma, lung, breast, or renal primary. If obstructing, these may be resected or palliatively bypassed.
- **NEUROENDOCRINE TUMORS**
  - are the most common primary SB malignancy.
- **PATHOPHYSIOLOGY**
  - They arise from enterochromaffin cells (carcinoid) that make serotonin.
  - **The terminal ileum** is more often involved than the jejunum. Some neuroendocrine tumors are located in the appendix or rectum. .

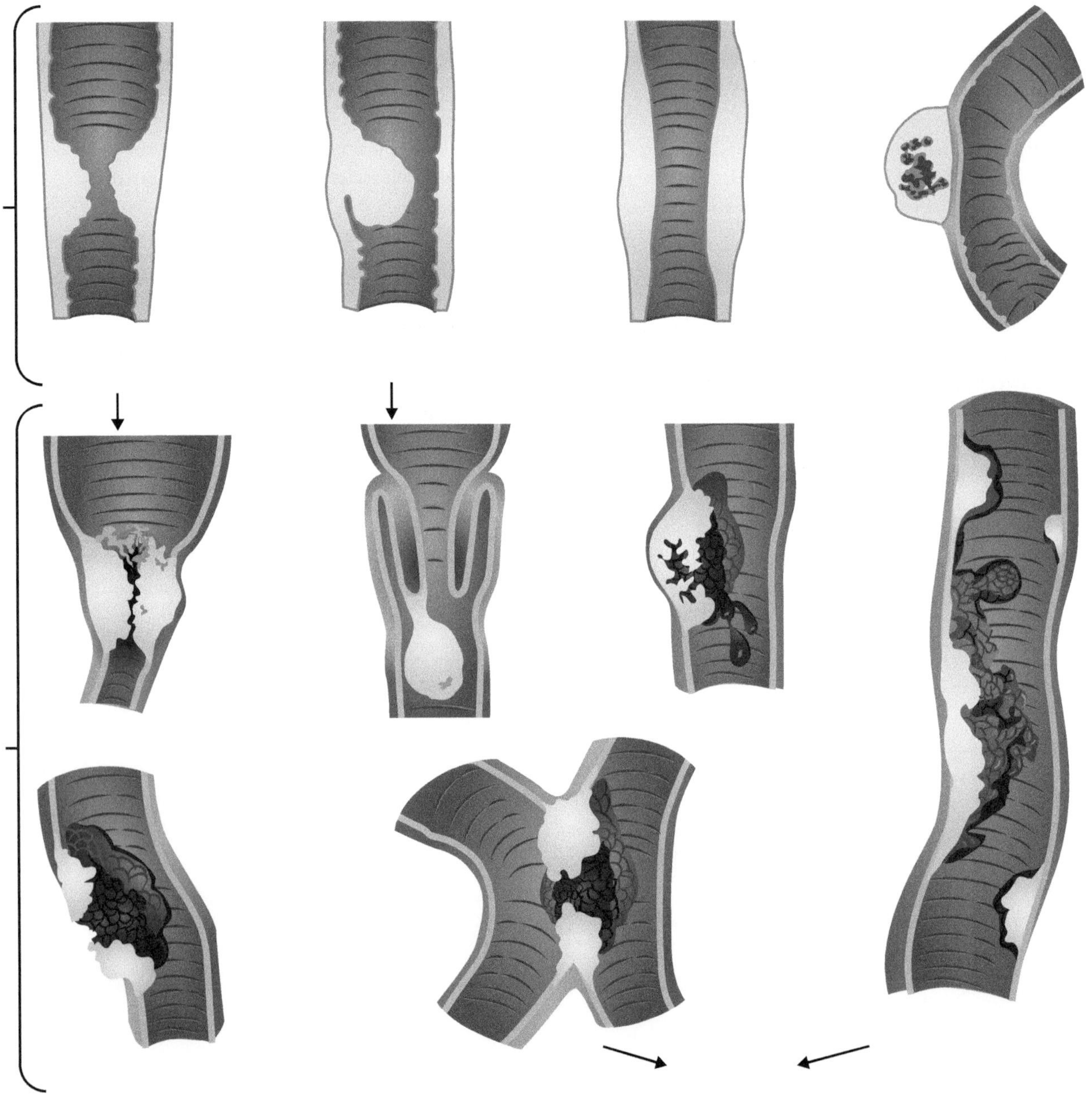

Figure 9.8 Morphologic variations of malignant bowel tumors.

- Tumors can result in mesenteric fibrosis causing a **Small Bowel Obstruction (SBO) or gastrointestinal bleeding (GIB)**. The latter is more common if rectal origin.
- About **25% of cases have multifocal involvement, and 50% have metastases.**
- The risk of lymphatic spread is related to the size of the primary tumor.
- Well-differentiated tumors are low or intermediate grade and have <10 mitoses/HPF and a Ki-67 <20%. Poorly differentiated tumors exceed these limits and are considered high grade.
- **Carcinoid syndrome** occurs if the tumor makes products that bypass the portal circulation; for example in bronchial carcinoid, liver metastases, bone metastases, or a retroperitoneal tumor.
  - **Serotonin causes diarrhea, kallikrein causes flushing, and bradykinin causes bronchiolar spasm.**
  - Products also injure the tricuspid valve causing **tricuspid regurgitation.**
  - Large-scale shifting of tryptophan to serotonin production can result in a relative deficiency of the tryptophan metabolite, niacin, and hence symptoms and signs of pellagra.

- DIAGNOSIS
  - Diagnosis is by **CT scan**, octreotide scan, positron emission tomography (PET) imaging, and chromogranin A levels. The latter have the highest sensitivity.
  - For carcinoid syndrome, the serotonin breakdown product **5-HIAA** can be measured.
- TREATMENT
  - **Endoscopic resection** is adequate if the tumor is duodenal and <1 cm.
  - **Wide local excision (WLE)** is best if the tumor is between 1 and 2 cm.
  - **Segmental resection and lymphadenectomy** should be performed if the tumor is >2 cm, since there is a higher risk for lymphatic spread. Any tumor with mesenteric fibrosis would also need segmental resection including the mesentery and its lymph nodes.
  - If on the tip of appendix and <2 cm, **appendectomy** is appropriate. Otherwise, a neuroendocrine appendices tumor not meeting these two requirements will need a **right colectomy.**
  - **Octreotide** is an adjunct for patients without complete resection.
  - Standard, effective chemotherapy protocols do not yet exist, so **aggressive surgical debulking** may need to be undertaken. Most applicable to liver metastases.
    - Hepatic artery chemoembolization, cryoablation, or radiofrequency ablation are other options for liver involvement.
    - A cholecystectomy is done prophylactically if practitioners are planning to use octreotide postoperatively.
    - A liver transplant would be definitive management.
  - For **carcinoid syndrome**, one should pursue resection of the culprit lesion(s) and adjunctive **octreotide and everolimus.**
  - Postoperative surveillance includes q3- to 6-month follow-up **imaging and chromogranin A levels.**
- **Other neuroendocrine tumors** are from the proximal bowel and more often metabolically active.
  - Duodenal **gastrinomas** (from G cells) are associated with MEN1 and present with peptic ulcer disease (PUD) and diarrhea. See pancreas Chapter 14, Pancreas, for details.
  - **Somatostatinomas** (from D cells) are associated with neurofibromatosis and present with glucose intolerance, diarrhea, and cholelithiasis. See Chapter 14, Pancreas, for details.
  - **Gastric neuroendocrine tumors.** Types 1 and 2 arise from histamine-producing enterochromaffin-like (ECL) cells under the stimulation of increased gastrin. Type 3 is gastrin independent and more aggressive, with ulceration and metastases. See the Chapter 8, Stomach, for details.

- ADENOCARCINOMA
  - PATHOPHYSIOLOGY
    - Most tumors reside in the **duodenum.**
    - Crohn's patients may develop adenocarcinoma of the ileum.
    - Some cases are associated with a familial syndrome like Familial Adenomatous Polyposis (FAP), Peutz-Jehger syndrome (PJS), HNPCC, or neurofibromatosis.
  - DIAGNOSIS
    - Many cases are asymptomatic, and so present late or are incidentally found on imaging.
    - **CTAP or CT enterography** can diagnose the primary site and mesenteric or nodal involvement. CT enteroclysis may also be performed.
    - CT chest for metastases.
    - A formal diagnosis can be made by biopsy via **EGD** if duodenal, double balloon or push **enteroscopy** if jejunal, or **colonoscopy with ileal intubation** if in TI.
    - **Capsule endoscopy** is useful if the tumor is located in the mid-distal jejunum or ileum, and can also detect synchronous lesions. Capsule studies are not used if there is a stricture or risk of obstruction.
    - Upper GI contrast series with SBFT or enteroclysis can also detect luminal irregularities.
  - TREATMENT
    - **Resection and lymphadenectomy with a minimum 8–10 nodes** is the standard approach. If there are 1–3 nodes positive, the patient is N1 status.
    - A **pancreaticoduodenectomy** may be necessary if the tumor is in the **2nd portion of the duodenum.**

- There are no standard adjuvant chemotherapy protocols.
- A **palliative stent** is used for an unresectable obstructing duodenal tumor in a patient with <2 months lifespan.
- **Palliative Gastrojejunostomy** is used for an unresectable obstructing duodenal tumor in a patient with >2 months projected lifespan, or if the obstruction is first discovered intraoperatively.

- **LYMPHOMAS**
  - **PATHOPHYSIOLOGY**
    - Most commonly B-cell in origin and located in the ileum where there is a high density of lymphocytes in Peyers patches. Celiac sprue patients can develop duodenal T-cell lymphoma. The majority of patients do not have B symptoms of fever, night sweats, and weight loss.
  - **DIAGNOSIS**
    - **Diagnosis is by CT** imaging demonstrating nodal burden with or without splenomegaly.
  - **TREATMENT**
    - **chemotherapy with or without a bone marrow transplantation.** Segmental resection and mesenteric lymphadenectomy for a minimum of 12 nodes is utilized only in those minority of patients who suffer from obstruction or bleeding.
- **GASTROINTESTINAL STROMAL TUMOR (GIST)**
  - **PATHOPHYSIOLOGY**
    - GISTs arise from pacemaker Cajal cells at various depths within the SB wall. Tumors are encapsulated and smooth and do not grow into adjacent organs.
    - **Gastric** is the most common location and has the best prognosis, with **small bowel** being second.
    - GISTs generally do not spread to lymph nodes.
    - **DIAGNOSIS**
      - Tumors are usually asymptomatic, but can erode the overlying mucosa and bleed or obstruct.
      - **EUS-FNA or biopsy** is the main modality for diagnosis. It can show spindle cells.
      - Lesions are considered malignant if the tumor is **>5 cm or has >5 mitoses/50 HPF** on pathology.
      - 75% of lesions have a KIT mutation, and some types of KIT mutations confer a better response to medical therapy (generally in gastric origin GISTs).
      - 10% of lesions have a PDGFRa mutation which is also indolent and gastric, but imatinib resistant.
      - **CT imaging** is needed if the tumor is beyond the reach of endoscopy. CT-PET can be used to exclude peritoneal or hepatic metastases.
    - **TREATMENT**
      - Segmental resection with any negative margin is adequate.
      - No lymph node resection is needed.
      - Rupture or spillage of the primary lesion can result in peritoneal recurrence.
      - Imatinib is an adjunct tyrosine kinase inhibitor employed if the GIST is believed to be malignant and CD117/cKit positive.
        - It may be given as neoadjuvant, adjuvant, or both.
        - TK inhibitors may be used to palliate non-resectable lesions.
        - Treatment resistance can be managed with increasing the dose or switching to sunitinib.
        - Resection of metastases or recurrences previously controlled on imatinib is allowable.
- **LEIOMYOSARCOMAS**
  - These are malignant tumors of small bowel wall smooth muscle and occur in the jejunum and ileum. They appear like leiomyomas (see below) except histologically have >5 mitosis/HPF and atypia or dysplasia on the specimen. Check for cKit to exclude GIST. Treatment is with **segmental resection without lymphadenectomy**.

BENIGN SB NEOPLASMS

- Benign lesions more often present with **bleeding** rather than pain.
- **POLYPS**
  - **PATHOPHYSIOLOGY.**
    - Most are **adenomatous.** Villous architecture portends an increased cancer risk compared to tubular architecture.
    - Some may be hyperplastic or hemartomas with mucuslaris mucosa in the polyp.
  - **DIAGNOSIS**
    - Suggested by a "soap bubble" or "paintbrush sign" on a UGI study.
    - Formal confirmation would require an endoscopic resection by **EGD** if duodenal, **double balloon or push enteroscopy** if jejunal, or **colonoscopy** with ileal intubation if in the TI.
    - Capsule endoscopy can image polyps elsewhere, but will not provide a tissue diagnosis.
  - **TREATMENT**
    - **Endoscopic resection** with or without laparoscopic assistance is sufficient if the polyp is <1 cm and there is no high-grade dysplasia or neoplasia in the specimen.
    - **Surgical resection** may be needed if the polyp is >2 cm.
  - Polyps may be associated with an underlying syndrome:
    - 1. Familial Adenomatous Polyposis (FAP)
      - Duodenal polyps progressing to adenocarcinoma are the #2 cause of death. **EGD screening should be performed q2 years starting at age 25–30.**
      - **NSAIDs** can decrease polyp growth.

- Pancreaticoduodenectomy is indicated if a polyp is at the ampulla or if there are multiple, high-grade/ dysplastic, villous duodenal polyps (Spigelman 4 classification or above). Q6-month EGDs should be done postop.
- 2. **Peutz-Jeager syndrome (PJS)** is characterized by multiple SB and LB hamartomas, with smooth muscle in the submucosa of polyp.
  - Patients have a family history of similar polyps and mucosal pigmentation that fades by their 4th decade of life.
  - **EGD and colonoscopy screening should begin at 10–20 years of age.** Patients are at increased risk for breast, uterus, cervix, and testicular cancer.
  - Treatment is by **removing all polyps >1.5 cm and biopsy others. Formal resection** is reserved for any **adenomatous change** found on biopsy or for any obstruction, or bleeding.
- **HEMANGIOMAS** come from submucosal vessels and **lipomas** move with the SB wall during peristalsis on fluoroscopy, since they are noninvasive. Both can be left alone unless symptomatic.
- **LEIOMYOMAS** are well encapsulated and should be resected to exclude leiomyosarcoma or GIST.

ENTEROCUTANEOUS FISTULA

- **PATHOPHYSIOLOGY**
  - This is an epithelial lined tract from the small bowel lumen to the dermal surface.
  - Enteroatmospheric fistulae connect an enteric lumen to a granulating wound bed without mature, intact dermis.
- **DIAGNOSIS**
  - Most fistulae are postoperative in origin and generally diagnosed after an **abdominal CT** shows an intraperitoneal or subcutaneous postoperative abscess. Once the abscess is drained, enteric contents follow shortly thereafter and persistently drain from the site.
  - Low-output fistulae are <200 ml/day, moderate 200–500 ml/day, and high-output >500 ml/day.
  - Output tends to persist in the setting of **FRIENDS** (foreign body, radiation treatment, inflammatory bowel disease, epithelialization, neoplasm, distal obstruction, or sepsis).
- **TREATMENT AND OUTCOME**
  - Surgical or radiologic **drainage of any infection**.
  - Surgical **removal of any foreign body**.
  - **Skin care** with a stoma protection device or sump suction device to keep the edges clean.
  - Electrolyte replacement.
  - **PPI, loperamide, codeine, psyllium. and octreotide.**
  - **Regular diet can be continued if output is <200 ml/day.**
  - **NPO and distal enteric feeding should be instituted if >500 ml/day output** during attempts at proximal enteric feeding. If distal enteric feeding cannot be done, then **TPN** should be utilized.
  - One-third of fistulae will close by 18 weeks. Colonic fistulae may close simply with a low-residue diet.
  - **Small bowel resection and primary anastamosis** is reserved for failure of conservative care at 6–18 weeks.
  - Mortality is secondary to complications from sepsis and malnutrition.
  - Enteroatmospheric fistulae may be managed with negative pressure wound therapy and split thickness skin grafts, in addition to standard conservative measures, until an enterocutaneous fistula results.

SHORT BOWEL SYNDROME (SBS)

- **PATHOPHYSIOLOGY**
  - SBS generally occurs after a single operative resection of a long segment of small bowel resulting in <180–200 cm remaining length. However, the diagnosis is not dependent on the exact length of remaining bowel.
  - Remaining SB adapts over a 2-year period with bowel lumen widening, villi lengthening, and an increased number of enterocytes.
  - Severe adaptive dilation can independently result in malabsorption or bacterial overgrowth.
- **DIAGNOSIS**
  - **Made by malabsorptive signs** including diarrhea, steatorrhea that is Sudan red stain positive, and nutritional deficiencies of B12/ADEK without another cause, and in the appropriate setting of a known limited length of small bowel.
- **TREATMENT**
  - Institute a **high carbohydrate, high protein, fat-restricted diet.** Medium chain triglycerides are used for direct portal venous absorption, since bile acid–mediated ileal absorption of long chain triglycerides may not be possible. Oxalate is also restricted from the diet.
  - **PPI** helps treat elevated G cell gastrin secretion and subsequent increased parietal cell HCl production.
  - **Cholestyramine** binds unabsorbed bile salts.
  - **Octreotide** to decrease excretions and SB volume.
  - **Fiber, codeine, loperamide, and diphenoxylate-atropine** are useful for diarrhea.
  - **TPN** may be needed if enteral feeding cannot meet nutritional needs.
    - Patients can survive with only short-term TPN at >60 cm adult length provided the ileocecal (IC) valve is competent. They may need >75 cm of length if IC valve

is not competent. At >120 cm adult or >60 cm in a child, they are unlikely to need any supplemental TPN at any time.

- Surgical therapy includes strictureplasty, tapering of dilated segments (if patient has adequate remnant length), interposition of reversed segments, or **serial transverse enteroplasties.** *See Figure 9.9.*
  - The Bianchi procedure is a longitudinal mesenteric and enteric transection creating two parallel segments of narrower SB that can then be anastomosed in a serial fashion.
- If the patient has an ileostomy, reestablishment of continuity would provide more potential absorptive area (via the colon), but also would expose the patient to colonic absorption of unbound oxalate and the perianal complications of frequent diarrhea.
- **SB transplant** should be considered for patients with complications from TPN over the long term. If a patient has irreversible TPN-related cholestatic liver disease as well, then both **SB and liver transplant** can be done.

SMALL BOWEL OBSTRUCTION (SBO)

- **PATHOPHYSIOLOGY**
- **Extrinsic** causes to SBO are most common and include **adhesions** (from open pelvic or colonic surgery more so than laparoscopy), **hernia,** or neighboring **colonic pathology.** *See Figure 9.10.*
- **Intramural** causes include Crohn's disease, neoplasm, or intramural hematoma.
- **Intraluminal** causes include intussusception, foreign body or bezoar, and gallstone ileus.
  - Gallstone ileus is a fistula between a chronically inflamed gallbladder fundus and the duodenum, sometimes due to the weight of a non-mobile eroding gallstone. The stone traverses the fistula into the duodenum and ends up obstructing the small bowel at the ileocecal valve. Air from the duodenum can ascend into the gallbladder and biliary tree. Patients have an increased risk for ascending cholangitis.
- Ischemia can ensue if intraluminal and intramural pressure exceeds capillary pressure in the setting of a complete SBO, closed loop with or without volvulus, or a patient with baseline mesenteric arterial insufficiency.
  - Ischemia starts at the inner mucosa and progresses through the thickness of the bowel wall, eventually involving the serosa and producing peritoneal irritation by exam at that time.
- **DIAGNOSIS**
  - Pain may be constant or colicky and associated with peristaltic waves against the resistance of the obstruction.
  - Eventually. accompanying nausea and vomiting ensue as the foregut and stomach distend. Vomiting is a more prominent component of proximal SBO. and dissension is more prominent in distal SBO.

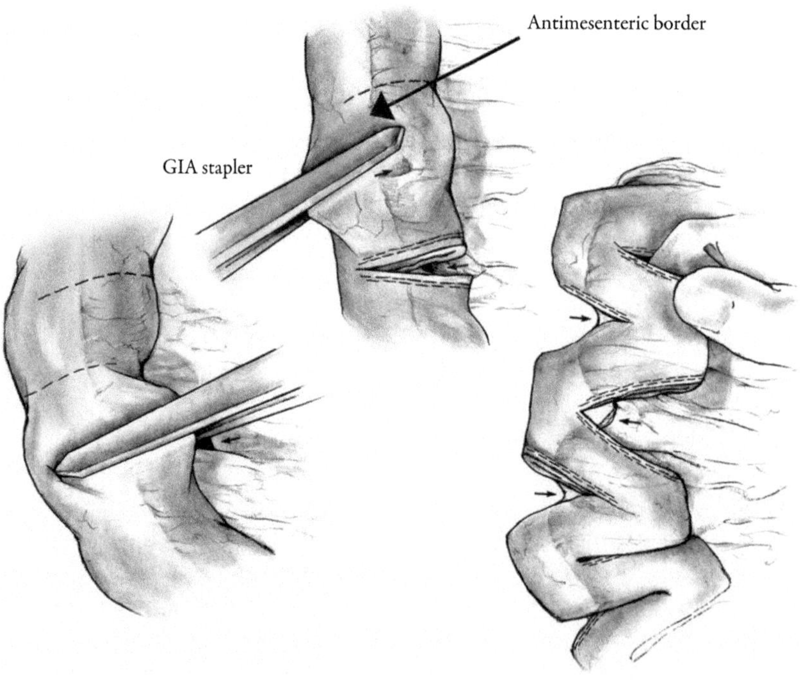

Figure 9.9 Serial transverse enteroplasty.

- Complete SBO implies the patient is obstipated and cannot pass intraluminal gas.
- Partial SBO refers to the continued passage of flatus or stool at > 6 hrs after onset of pain.
- **DIAGNOSIS**
  - is by **CTAP** with or without oral contrast.
  - Consider a **gastrografin small bowel follow-through** if there is no improvement as an inpatient with conservative management, NPO, NGT decompression, and IV fluids. The contrast can exclude a complete SBO by demonstrating entrance into the cecum. It can also hasten the resolution of a partial, adhesive SBO. If at **8 hours** post -ngestion there is no contrast in cecum on serial radiographs, then likely the patient will need surgery.
  - Contrast studies can include enteroclysis with 250 ml barium and 1–2 L water plus methylcellulose through a nasojejunal tube. This enhances SB mucosal detail in the setting of an intramural or intraluminal etiology to the obstruction.
- **TREATMENT**
  - For an adhesive SBO in a stable patient without peritonitis, the mainstay of management is **npo, fluid and electrolyte replacement, and NGT decompression** for up to 72 hours. This may cure 40-80% of such cases.
  - NGT decompression may be prolonged for 2-4 weeks if the scenario is an early postoperative SBO (within 30 days of a procedure) or an acute duodenal hematoma. In the latter case you operate only for those patients progressing to a chronic, fibrotic state or for those who have another indication to be in the OR in the acute setting.
  - **Exploratory laparotomy** is reserved for **suspicion of high intraluminal pressure causing ischemic mural sequelae**, closed loop obstruction, a non-reducible hernia, suspicion for **malignancy, a non-progressing and endoscopically non-retrievable foreign body, or failure of conservative management** in presumed adhesive disease.
  - Exploratory **laparoscopy** is an option if a single adhesive band is suspected and there is minimal proximal dilation. There needs to be enough peritoneal cavity working space to insufflate and visualize.
  - Gallstone ileus is treated by a **laparotomy, enterotomy and gallstone removal** if the patient is systemically ill in the setting of SBO. Rarely would one pursue cholecystectomy and closure of the duodenal fistula even in a stable patient.
  - If malignant and not surgically correctable, the patient may benefit from a **decompressive gastrostomy.**

ILEUS

- Gastrointestinal inactivity on postoperative days 1-4 is actually due to gastric and colonic dysmotility since they take longer to reanimate than the small bowel.
- If **beyond 4 days** of GI inactivity, then the patient may have a true adynamic ileus of the entire tract due to sepsis, hypokalemia, peritonitis, trauma, or medications (ex, anticholinergics, calcium channel blockers, or narcotics)
  - Methods to prevent an ileus include
    - **Epidural amide anesthetics for pain control**
    - **Non-steroidal anti-inflammatory drugs (nsaids)**
    - **Decreasing the use of opioids**
    - The **peripheal mu receptor blocker alvimopan** can also be used to prevent an ileus. However this does carry a risk of myocardial ischemia.
  - Chronic idiopathic intestinal pseudo-obstruction is a familial enteric and genitourinary myopathy, neuropathy, or a combination of both that presents insidiously in adolescence with waxing and waning symptoms of adynamic ileus and eventually results in progressive dilation of the esophagus, duodenum, or colon.
  - Chronic SB dysmotility is managed with decompressive gastrostomy, small bowel resection of the dysmotile segment, or small bowel transplant

OTHER BENIGN PATHOLOGY

- **FALSE DIVERTICULA** are also known as pulsion divertula.
- **PATHOPHYSIOLOGY**
  - Diverticula are protrusions of the mucosa either into the periampullary area in the duodenum or into the mesentery if located in the jejunum or ileum. *See Figure 9.11.*
  - Some large diverticula may be complicated by **blind loop bacterial overgrowth** within them. The bacteria destroy pancreatic enzymes and impair the mechanism of B_{12} absorption.
- **DIAGNOSIS**
  - Many diverticula are asymptomatic.
  - If symptomatic, the presentation can include pain from inflammation, upper or lower gastrointestinal bleeding, or biliary or pancreatic duct obstruction.
  - The diagnosis is made by **CT imaging.** Occasionally a diverticulum may be encountered on endoscopy.
  - Blind loop bacterial overgrowth can produce abdominal pain and malabsorption diarrhea. It is diagnosed

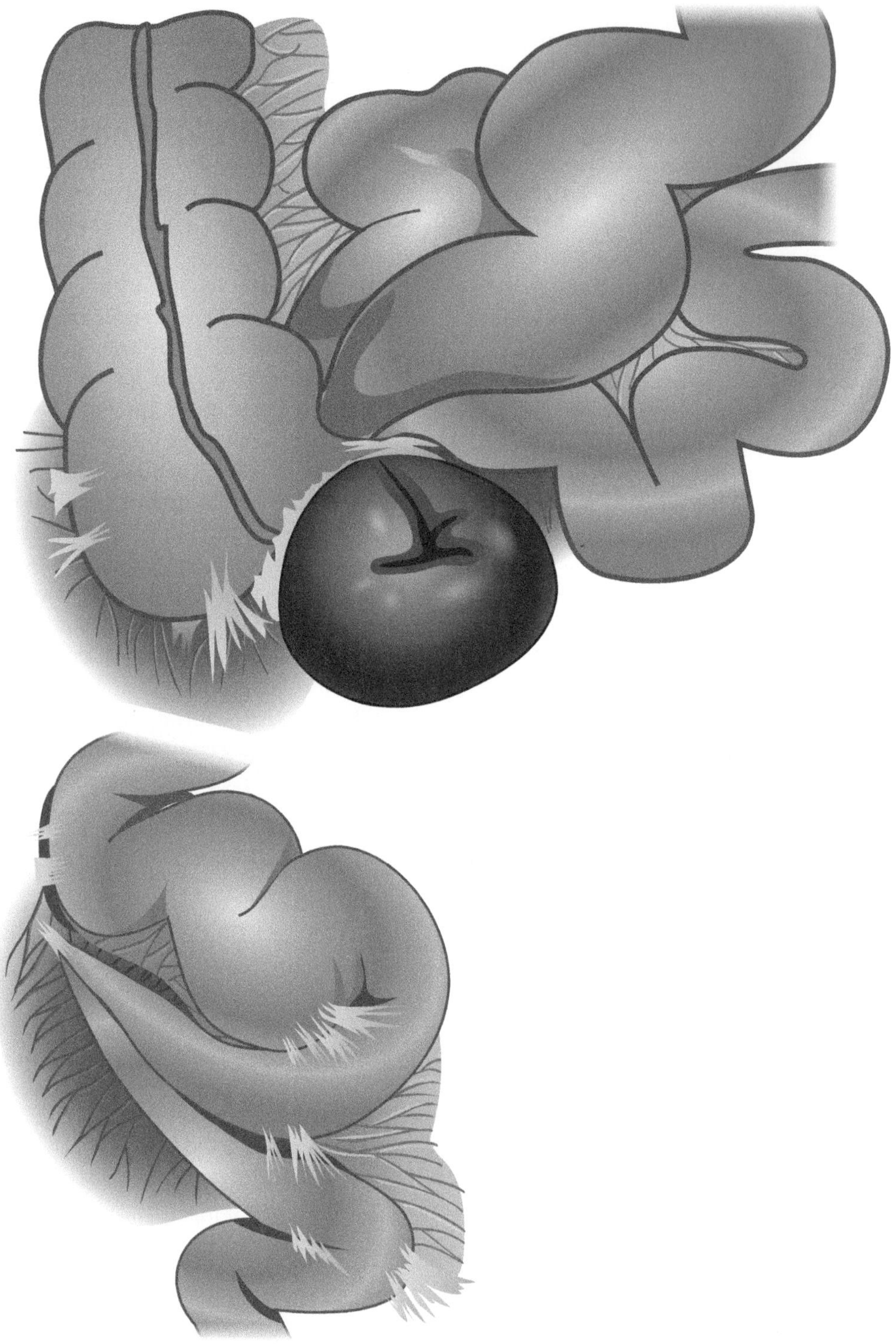

Figure 9.10 Small bowel obstruction.

by a radiolabeled D-xylose test producing exhaled radiolabeled CO_2. A Schilling test without radiolabeled B_{12} in the urine would also suggest this diagnosis.

- **TREATMENT**
  - observation if asymptomatic.
  - For an obstructing duodenal diverticulum, **ERCP with biliary or pancreatic stents** may be necessary
    - In the event that endoscopic therapy is not possible or fails, surgical treatment includes **segmental resection** of the duodenum provided the anatomy permits.

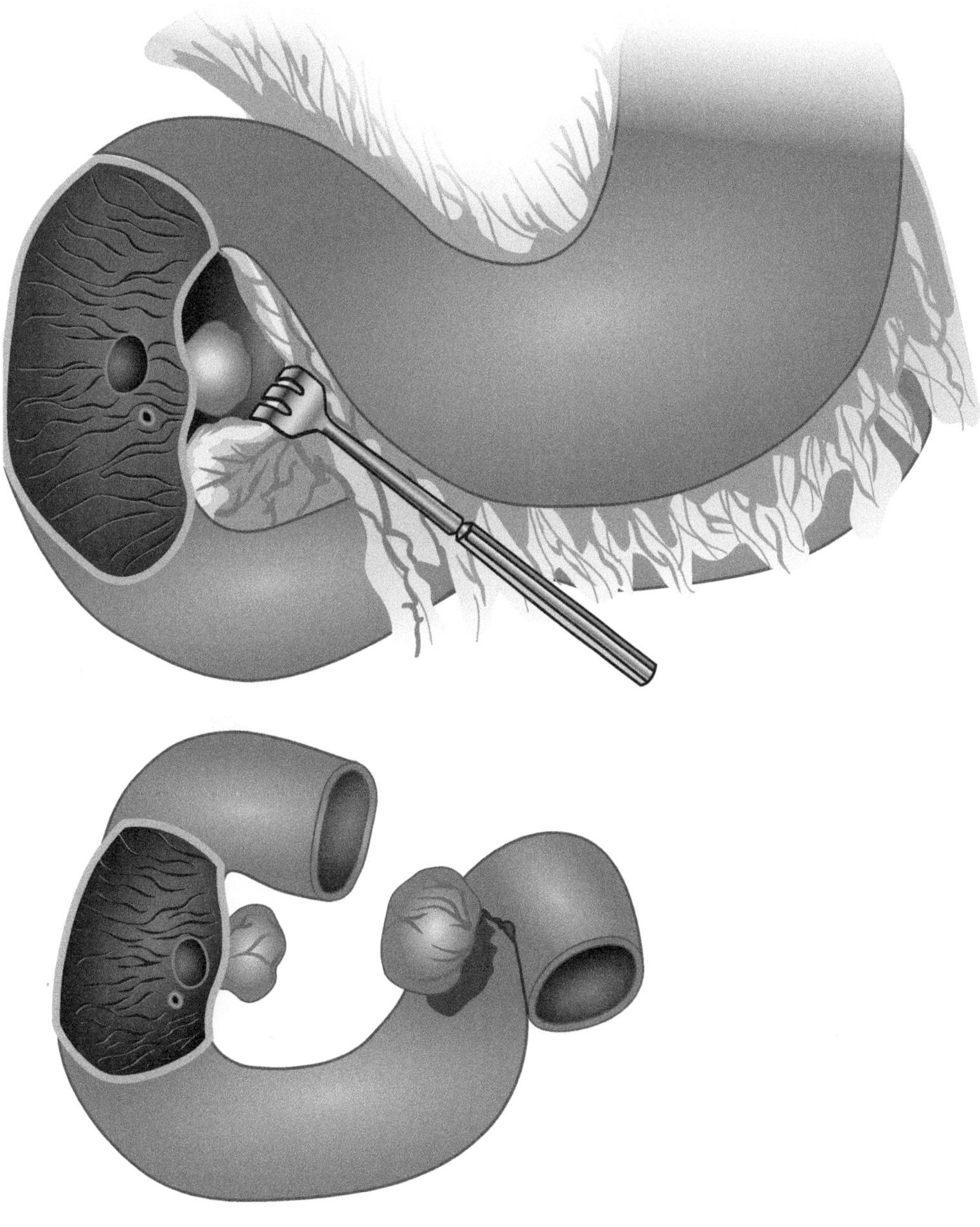

Figure 9.11 Duodenal diverticula.

- If obstructing and periampullary, consider **hepatico or choledochojejunostomy** accompanied by an ERCP with pancreatic duct stent.
- Treatment of symptomatic jejunoileal mesenteric pulsion diverticula is initially **antibiotics and nonoperative management for diverticulitis.** This can be followed by **segmental resection** when the inflammation has subsided.
- Any bleeding diverticulum should be resected.
- Bacterial overgrowth in blind loop syndrome is treated with **oral antibiotics and a medium chain triglyceride diet**, which avoids long chain malabsorption that would otherwise occur secondary to bacterial destruction of pancreatic enzymes.

- **TRUE DIVERTICULA** of the small bowel refers to a Meckel's diverticulum. *See Figure 9.12.*
- **PATHOPHYSIOLOGY**
  - This is a remnant of the omphalomesenteric duct and resides within 2 feet of the ileocecal valve. It contains two types of ectopic mucosa, either pancreatic or gastric.
  - It is symptomatic in 2% of the population, and presents within 2 years of life. The most common **pediatric presentation is bleeding from ulceration** on the mesenteric mucosal side of the ileum due to ectopic gastric mucosa within the diverticulum.
  - Ectopic pancreatic tissue can result in **diverticulitis.**
  - Meckel's diverticulum can present as an **SBO in an adult** secondary to the following:

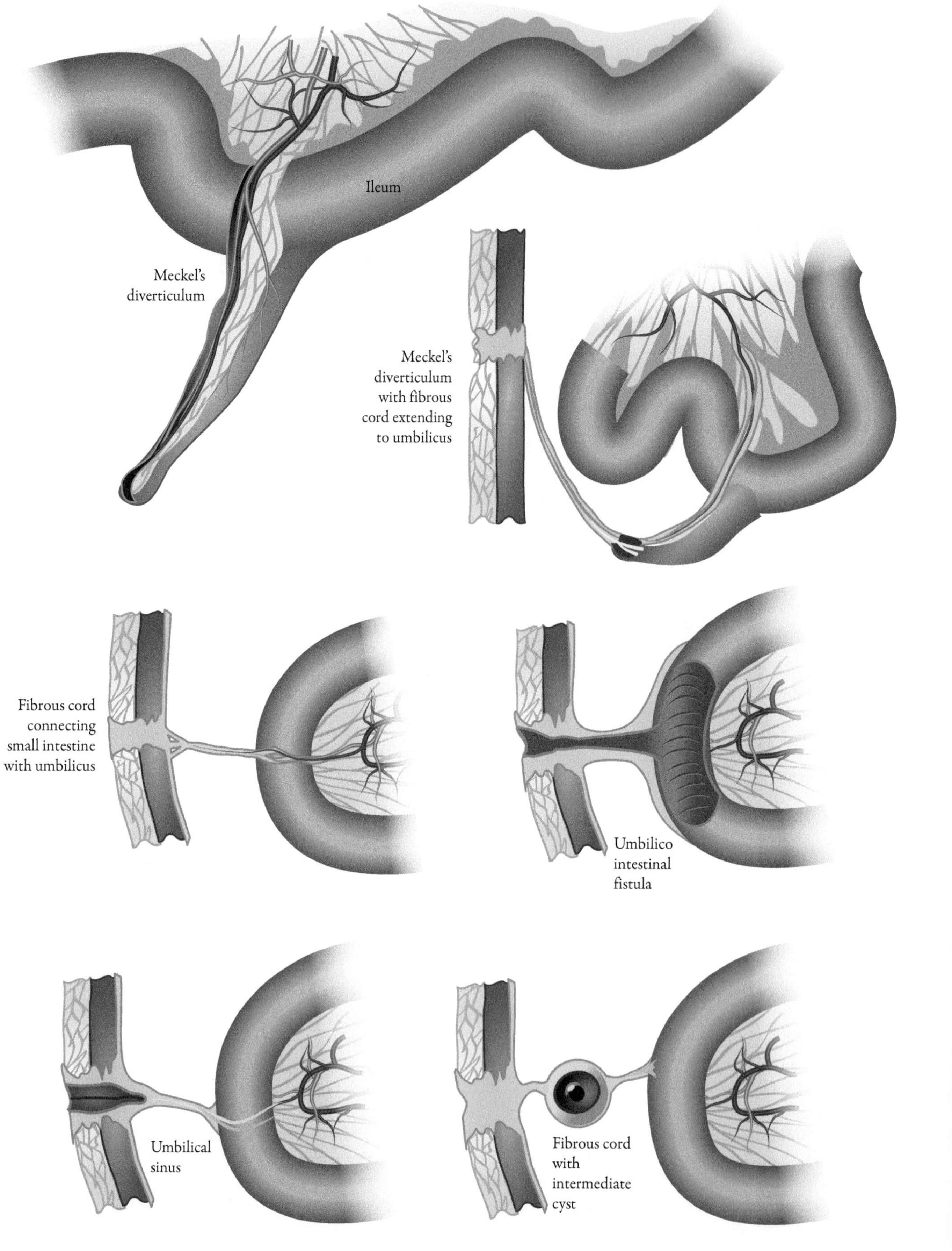

Figure 9.12 Anatomic variations of Meckel's diverticulum.

- A fibrous band remnant from the diverticulum or anti-mesenteric wall of the ileum to the abdominal wall.
    - A *meso*diverticular fibrous band; or
    - Intussusception with the diverticulum as the lead point.
  - **DIAGNOSIS**
    - Obtain a **technetium 99m-pertechnaetate nuclear medicine scan** if it is a pediatric patient presenting with a gastrointestinal bleed (GIB). This enhances the ectopic gastric mucosa within the diverticulum at the same time as the gastric mucosa in the stomach.
    - If presenting with diverticulitis or obstruction, diagnosis will likely be made via **CTAP** or at the time of surgery.
  - **TREATMENT**
    - Diverticulectomy is adequate if the base is <2 cm.
    - Segmental resection of the ileum housing the diverticulum is a better option for broad based diverticulitis, perforation, ulceration, or bleeding from the mesenteric mucosal side of the adjacent ileum.
    - One may also perform an appendectomy at the time of surgery
    - If found incidentally, consider a diverticulectomy if the patient is a child, has had prior symptoms that could have been related to the diverticulum, has a mesodiverticular band, or has thick gastric mucosa palpable within the outpouching. Otherwise simply **observe.**
- **SB GASTROINTESTINAL (GI) HEMORRHAGE**
  - **PATHOPHYSIOLOGY**
    - From an **arteriovenous malformation** (75% of cases) or **neoplasm** (10%). The most common cause in children is ectopic gastric mucosa in a Meckel's diverticulum causing peridiverticular mucosal ulceration and bleeding
  - **DIAGNOSIS**
    - **esophagogastroduodenoscopy** identifies the source if duodenal or a **push enteroscopy** if jejunal. **Capsule endoscopy, tagged RBC scan,** or **laparoscopic assisted enteroscopy are other options to evaluate the entire SB.**
  - **TREATMENT**
    - laparoscopic or open SB **resection.**
- **CHYLOUS ASCITES**
  - **PATHOPHYSIOLOGY**
    - Secondary to malignancy, cirrhosis, or trauma or surgery involving the draining mesenteric lymphatics, retroperitoneal lymphatics or cisterna chyli.
  - **DIAGNOSIS**
    - Fluid sample showing triglyceride >110 and WBC >500 with a predominance of lymphocytes.
  - **TREATMENT**
    - 1. A **high protein low fat diet with medium chain triglycerides** for direct portal vein absorption, bypassing the lacteal system. If a peritoneal drain is in place and <500 ml/day, then one may continue this diet.
    - 2. **NPO with TPN** supplementation is indicated if the drain has **>500 ml/day** of chylous output.
    - 3. Lymphoscintigraphy to localize the source followed by **lymphatic embolization or surgical ligation of the culprit lymphatics** should be pursued if the drain has >**1000 ml/day** chylous output despite the above measures. If a focal area of obstructed and congested SB lacteals proves to be the source, then **SB resection** may be curative.
- **INTUSSUSCEPTION** in adults is generally ileocolic or cecocolic. Often there is an underlying malignant lead point in the intussusceptum of adults, so **resection** of the involved area is necessary. If found incidentally on CT scan and without signs of obstruction, one may **observe** and perform serial exams.
- **RADIATION ENTERITIS**:
  - **Acute mucositis** and sloughing at > 4500 cGy. O2 radicals are produced and disrupt chromosomes during metaphase as well as inducing inflammatory cytokine production. Patients have acute symptoms of malabsorption and pain usually **within weeks.** Treatment is supportive care. The appearance of acute mucositis does not necessarily portend the development of chronic injury.
  - **Chronic endarteritis** is injury to microvasculature that leads to ischemia and a secondarily indurated, fibrotic, ulcerated, and strictured submucosa. It generally presents within two years of exposure but can occur anywhere **from 6 months to 30 years after exposure.** Injury is compounded by preexisting microvascular disease, vascular insufficiency upstream, or chemotherapy. Treatment is first to address any diarrhea from malabsorption or bacterial overgrowth. Resection or bypass of the involved segments may be needed.
  - Prevention of radiation enteritis is by focused application of the radiation beam. A neoadjuvant, rather than adjuvant, manner of treatment may also be beneficial.
- **SMA SYNDROME** is due to obstruction of the third portion of the duodenum by the overlying SMA. Compression is secondary to dissipation of the retroperitoneal fat pad in the setting of weight loss, allowing the SMA to angle posteriorly. Classically it is seen in a female after scoliosis surgery. Obstruction is relieved by the prone or right lateral decubitus position. Diagnosis is by **CT angiogram** with or without a contrast bowel study. Treatment is by providing **enteral nutrition or performing duodenojejunostomy bypass** around the obstructing overlying SMA.
- **PNEUMATOSIS INTESTINALIS** refers to air in the bowel wall. Its significance requires clinical correlation.
  - First one must **rule out ischemic bowel.** Generally this is done with a CT angiogram to look for an embolic or thrombotic occlusion of the SMA. If the

CT angiogram (CTA) is not diagnostic of an arterial occlusion, then an endoscopy or proctosigmoidoscopy can be a useful adjunct to visualize mucosal ischemia.

- **Laparoscopic or open exploration** is the best strategy for systemically ill patients or those with peritonitis or a LGIB presumed to be due to sloughing, infarcted mucosa.
- **Non-ischemic pneumatosis** may be due to a mucosal breach in a setting of immunosuppression, a distal obstruction with increased intraluminal pressures allowing air to dissect into the bowel wall, chronic obstructive pulmonary disease (COPD), or gas-forming bacterial growth within the wall. Treatment in these settings is determined by the systemic state of the patient and their abdominal exam. If both are benign and non-ischemic pneumatosis is likely based on the clinical picture, then **nonoperative management** is an option.
- **Primary pneumatosis intestinalis cystoides** is a benign condition in which air filled intestinal wall pockets can rupture into the peritoneal cavity. However, there is no significant bacterial contamination since the **mucosa is not breached. In this particular condition, surgery is not needed.**

- ACUTE MESENTERIC ISCHEMIA
  - PATHOPHYSIOLOGY
    - 1. Thrombosis atop a ruptured mesenteric vascular plaque. This is the **most common** cause of mesenteric ischemia, and the second highest mortality.*See Figure 9.13.*
    - 2. Embolus to the superior mesenteric artery.
    - 3. Ischemia related to overall low flow in mesenteric vasculature due to low cardiac output (nonocclusive ischemia). This has the **highest mortality** of all types.
    - 4. Mesenteric venous thrombosis producing SB mucosal congestion and secondary inflow ischemia. *See Figure 9.13.*
  - **DIAGNOSIS**
    - Presents with upper gastrointestinal (UGI) symptoms of pain and nausea or vomiting.
    - Diarrhea or hematochezia are less common.
    - The abdominal exam may be soft and nontender early on, when the ischemia is limited to the mucosal lining. The patient will still have severe pain from mucosal oxygen deprivation, however.
    - Eventually the ischemia extends full thickness to the serosal covering. At that time, palpation of the abdominal wall will bring the parietal peritoneum up against ischemic serosa and result in involuntary guarding.
    - The diagnosis is confirmed by a CTA showing the following:
      - 1. **Tapering of superior mesenteric artery** within 1-2cm of its origin if thrombosis atop plaque.
      - 2. **Abrupt meniscus cutoff of SMA** several centimeters from origin if embolic.
      - 3. Patent SMA and IMA if nonocclusive ischemia; or
      - 4. Thrombosis somewhere along portal venous system.
    - For patients who are unstable or who have an equivocal CTA, a **diagnostic laparoscopy** or exploratory laparotomy may be required for diagnosis and therapy.
  - **TREATMENT**
    - Start a heparin drip for any suspicion of thrombosis or embolus, provided there are no contraindications.
    - Embolus is treated by laparotomy and exposure of the SMA.
      - The SB is retracted medially and the transverse colon cephalad.
      - The ligament of Treitz is transected and the 4th portion of duodenum retracted to the patient's right side.
      - Then the retroperitoneal reflection at the root of the mesentery is opened and a transverse SMA arteriotomy proximal to the takeoff of the middle colic artery is performed.
      - Distal and proximal Fogarty balloon embolectomies can retrieve clot.
      - Finally, the transverse arteriotomy is closed primarily.
    - **Thrombosis atop a plaque** is treated by laparotomy, SMA exposure as above, a longitudinal arteriotomy, and **thrombectomy.**
      - A retrograde wire and stent of the proximal SMA may be performed.
      - The longitudinal arteriotomy can be closed by patch angioplasty.
      - Some patients may need **autogenous vein bypass** from the supraceliac aorta, distal aorta, or right common iliac artery (CIA) to the proximal SMA, just inferior to the pancreas.
      - A few patients will need an additional bypass to the celiac artery, common hepatic artery, or proximal splenic artery.
      - Pure endovascular approaches for embolus or thrombosis in situ are not regularly used if there is need for assessment of bowel viability.
    - **Nonocclusve ischemia** is treated by supportive measures, fluid resuscitation, optimization of cardiac output, antibiotics, or less commonly SMA catheterization, and papaverine infusion (30–60 mg/hour for 24 hours).
      - A laparotomy may be needed to assess bowel viability, but is not routinely done.
      - **Ischemic colitis** is one type of nonocclusive ischemia. It manifests at Griffiths point, Sudeks point, and the distal ileocolic watershed areas. 80% of these cases are successfully managed medically provided the ischemia is limited to superficial mucosa. See colon chapter.

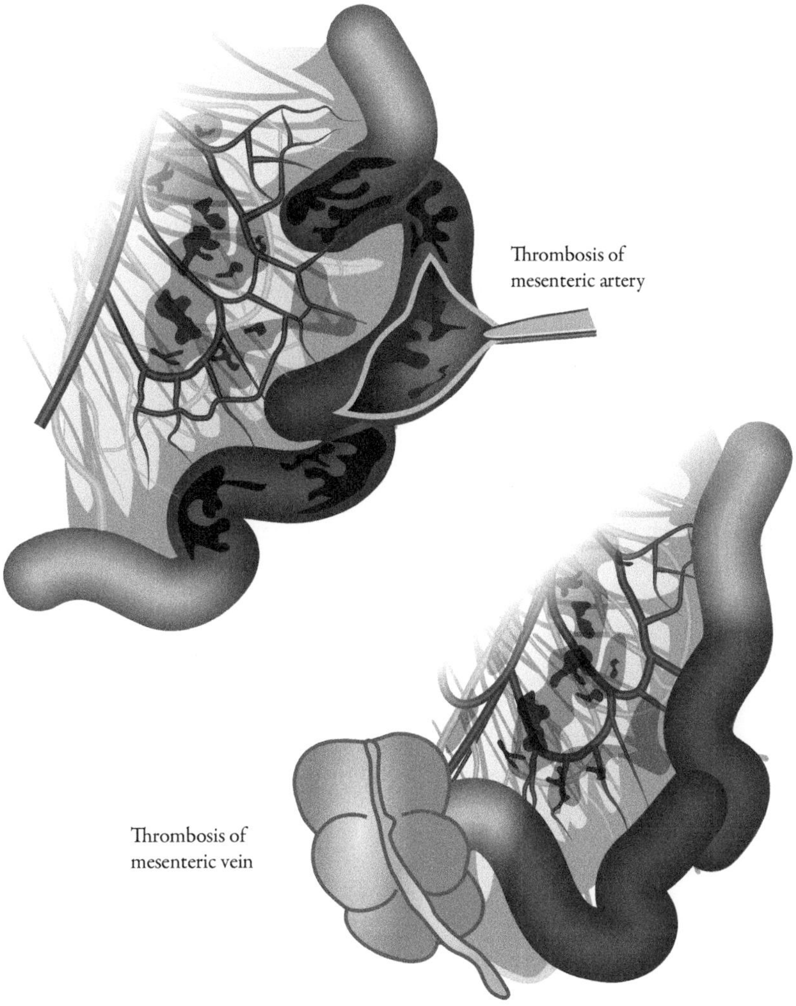

Figure 9.13 Acute mesenteric ischemia from arterial or venous thrombosis.

- **Mesenteric vein thrombosis** is treated with a heparin drip.
  - Intra-arterial thrombolytics via the SMA or splenic artery have been used if ischemia is refractory to anticoagulation.
  - There is no role for venous thrombectomy.
  - Patients may need a laparotomy and bowel resection if the clinical picture is consistent with full thickness necrosis of the bowel wall. But this is not common. Strong consideration should be given to a mandatory second look laparotomy in all operative cases.
- **CHRONIC MESENTERIC ISCHEMIA**
  - **PATHOPHYSIOLOGY**
    - Longstanding, progressive celiac and SMA artery atherosclerotic stenosis is the hallmark of this condition.
    - The metabolic demands of food intake, luminal metabolization, and peristalsis overwhelm oxygen supply in the setting of impaired perfusion. This results in mucosal ischemia.
    - **Median arcuate ligament syndrome** is a less common cause of chronic ischemia due to an extrinsic impingement of the celiac artery and impaired foregut perfusion.
  - **DIAGNOSIS**
    - Patients with chronic ischemia have food fear and post-prandial pain with weight loss. Their abdominal exam is usually benign, however.
    - Suspicious cases who have concomitant renovascular disease can receive a **duplex US of mesenteric vasculature.** This can also be used as **surveillance** in known cases. Concerning features include:
      - Peak systolic velocity (PSV) >275 cm/s or end diastolic velocity (EDV) >45 cm/s
      - PSV:aorta ratio ratio >3 is indicative of >70% diameter reduction.
      - PSV <40 is indicative of critical stenosis.
      - A test meal of 8 oz. protein drink can be given and then a repeat US evaluation of the SMA to note any gradient change. Gradient changes in the celiac or IMA indicate this vessel is providing critical collateral blood flow.

- **Diagnostic angiography or CTA** should follow for patients with a positive ultrasound. Alternatively, if there is a high clinical suspicion for the diagnosis and the patient does not have renovascular disease, CTA can be the initial modality.
- **TREATMENT**
  - An endovascular **balloon expanding stent** can be placed across a 1–2 cm calcified, ostial or proximal stenotic lesion.
  - A **self-expanding stent** may be better for lesions distal along the artery. Stents have a 30% failure rate at 1 year.
  - The definitive treatment is **revascularization** of the SMA with or without revascularization of the celiac or IMA as well
    - **An antegrade PTFE** bypass goes from the supraceliac aorta (transcural exposure) to the end of the celiac artery beyond the diseased segment. A second graft can traverse from the supraceliac position in a retro-pancreatic course to the side of the SMA.
    - **A retrograde PTFE** bypass goes from the side of the infrarenal aorta or right iliac artery to the side of the SMA or side of the common hepatic artery.
    - An aortic endarterectomy extending 2 cm into the ostia of the celiac, SMA, and renal arteries requires medial visceral rotation and is less commonly done.
    - Median arcuate ligament syndrome is treated with transection of the ligament and arterial reconstruction if needed.
- **Postoperative duplex surveillance** of an SMA stent or bypass graft is done every 6 months to detect myointimal hyperplasia. This can be followed by **angiography** if criteria indicate a >70% stenosis. **Endovascular intervention** is done to maintain assisted patency in these cases, even if the patient is asymptomatic.

10.

COLON

ANATOMY AND PHYSIOLOGY

- **COLONIC ARCHITECTURE**
  - The peritoneum extends over the colonic surface inferiorly and covers the upper rectum and anterior surface of mid-rectum. The entire rectum is 5–15 cm long.
  - The plica semilunaris are haustra formed by the stripes of incomplete longitudinal muscles known as tenia. The tenia merge at the rectosigmoid junction.
  - The **colonic mucosal lining** secretes potassium (K) and absorbs sodium (Na) and 1–5 liters of water per day. Most of this absorption occurs in the cecum and right colon.
    - Crypts of Lieberkühn are epithelial invaginations at the base of villi that contain mucous-secreting goblet cells.
    - The mucosa metabolizes bilirubin, bile acids, estrogen, cholesterol, and proteins.
    - Carbohydrate oligosaccharides are fermented by bacteria to short chain fatty acids, and these are used as fuel for the colonic mucosa.
    - Epithelium produces vitamin K, H, and CH4.
    - At baseline the colon is colonized with billions of bacteria.
- **COLONIC PERFUSION AND DRAINAGE.** *See Figure 10.1.*
  - Ascending and two-thirds of the transverse colon are perfused by the **superior mesenteric artery (SMA)** and drained by the **superior mesenteric vein (SMV)**.
  - The distal transverse and descending colon, sigmoid, and upper and mid-rectum are perfused by the **inferior mesenteric artery (IMA)** tributaries. These include the left colic, sigmoid arteries, and superior rectal artery. These areas are drained by the **inferior mesenteric vein (IMV).**
    - The **marginal artery** and **arc of Riolan** provide collaterals between the SMA and IMA.
    - Griffiths' point is a watershed area near the splenic flexure that lies between the SMA and IMA vascular territories.
  - Inferior rectum and anal canal are perfused by the **internal iliac artery tributaries (middle rectal)**, and anal margin and verge are perfused by the **pudendal artery tributaries (inferior rectal)**.
    - Sudek's point is a watershed area that lies between superior rectal and middle rectal vascular territories.
  - **Lymphatic drainage** parallels arterial perfusion. **Epicolic** nodes are immediately adjacent to the colon wall; **pericolic** nodes are in the mesentery near arcades; **intermediate** nodes are near named arterial branches; **main** nodes are near takeoff of SMA or IMA. Inguinal nodes drain the anal margin.
  - **Superior rectal veins** drain the upper and mid-rectum into the IMV and then splenic vein and portal system. **Middle rectal veins** drain inferior rectum and anal canal (internal hemorrhoidal plexus), and **inferior rectal veins** drain the anal verge (external hemorrhoidal plexus). Both lead into the internal iliac venous system and vena cava.
- **NEURAL INNERVATION**
  - Sympathetic lumbar and sacral plexi lie posterior to the sigmoid and rectum. Parasympathetic nerves come from sacral foramina (nervi erigentes).
  - Meissner's (submucosal) and Auerbach (intramuscular) myenteric plexi form the enteric nervous system.
  - Pudendal nerves to the external sphincter comprise the somatic nervous system.
- **Denonvilliers fascia** separates the anterior rectum from the bladder and vagina.
- **Waldeyer's fascia** separates the retrorectal space superiorly from the supralevator space inferiorly on the posterior side of the rectum. **It** fuses with the fascia propria of the rectal wall at the anorectal junction. The retrorectal space may be the site of pediatric malignancies including teratoma, chordoma, enteric cysts, dermoid cyst, myelomeningocele, meningocele, or other neural or osseous tumors.

DIAGNOSTICS

- **Anoscopy** visualizes up to 8 cm from the anal verge, **proctoscopy** up to 25 cm, and **flexible sigmoidoscopy** up to 60 cm from the anal verge.
- **Endoscopic rectal ultrasound (US)** is used for determining the T stage and presence or absence of mesorectal nodes in a low rectal tumor.

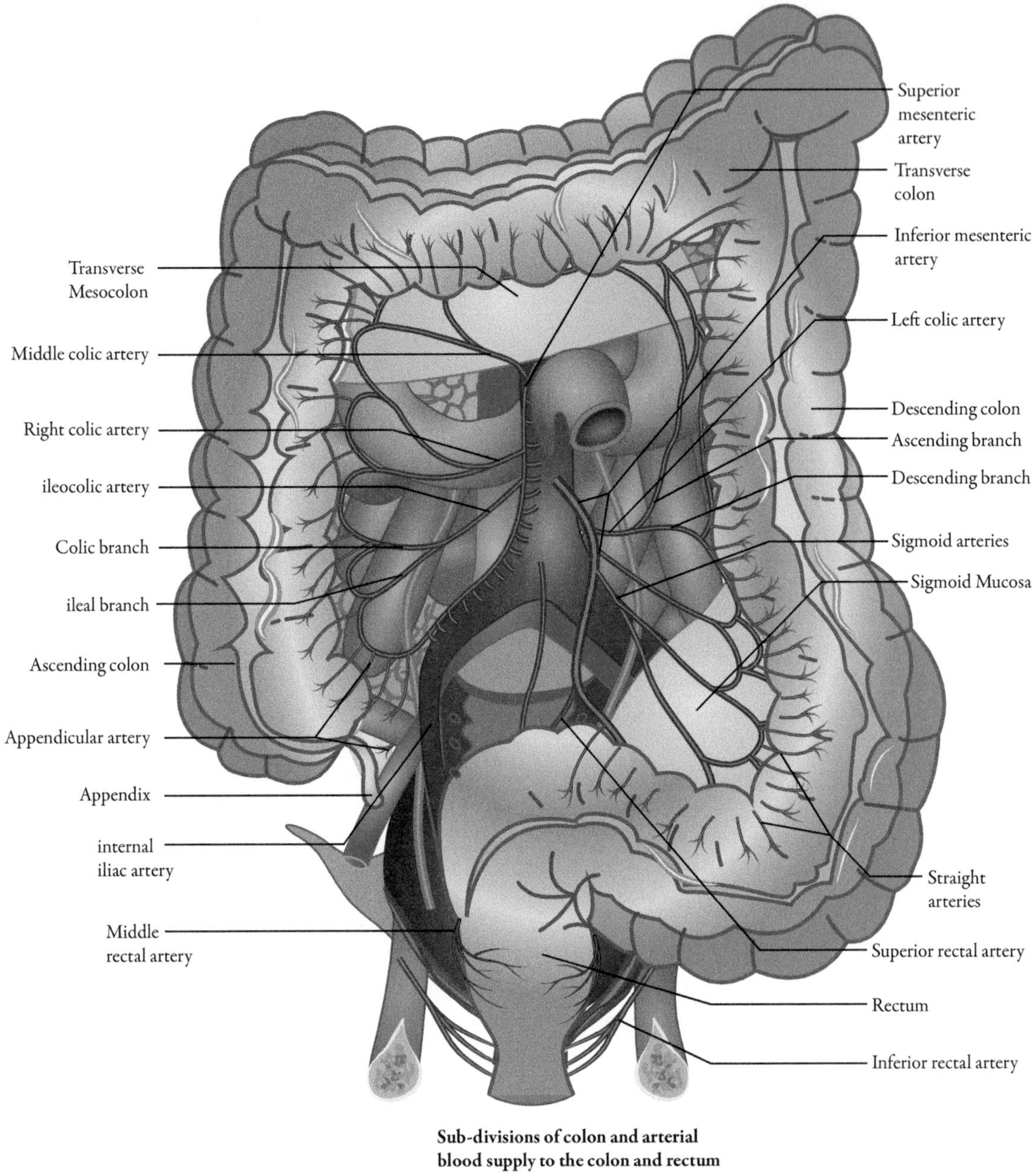

Figure 10.1 Colon arterial supply.

- **Manometry** measures resting and squeeze pressures at the internal and external anal sphincters, plus it can evaluate adequacy of the inhibitory sampling reflex.
- **Defecography** involves contrast enhancement of the rectal lumen to diagnose rectal prolapse, intussusception, cystocele, enterocele, or a rectocele into the vaginal vault.
- **Pudendal nerve latency** checks for a pudendal neuropathy as a cause to sphincter dysfunction and fecal incontinence.
- **Sitzmark test** is for colonic inertia or pelvic floor dysfunction. Markers are ingested and serial x-rays performed to see where along the colon, rectum, or anus the delay in transit is occurring.
- **Fecal occult blood test (FOBT)** detects human heme. If partially digested, it is upper gastrointestinal (UGI) in origin. If undigested, it is lower gastrointestinal (LGI) in origin. Nonhuman heme can be found in certain food products. All three cases would produce a positive FOBT.
- **Fecal immunohistochemical testing (FIT)** tests 3 stool samples at home. It detects only undigested human heme, which is specific for a lower gastrointestinal bleed (LGIB). Some tests add **multi-target DNA testing**

of the sample. This increases sensitivity but decreases specificity.

- **Colonoscopy** screening starts at age **45** for average-risk patients.
  - For those with a family history of colon cancer, start screening either:
    - 1. At 40 years of age if relatives with colon cancer were diagnosed after the age of 50;
    - 2. Ten years prior to the member of the family who was youngest when diagnosed with colon cancer. Use this method if it allows for earlier screening than 40 years of age.
    - In familial cases, screening should be repeated every 5 years. See below for screening of familial genetic syndromes.
  - Other screening options for those patients considered average risk include yearly FIT or a mixture of **FIT/FOBT and flexible sigmoidoscopy. CT colonography** is approaching the sensitivity of colonoscopy for lesions >10 mm, and can be done q5 yrs.
  - An adequate colonoscopy program needs the following:
    - A high completion rate.
    - A withdrawal time of >6 minutes.
    - Post-procedure pain in <10%.
    - Bleeding <1%.
    - Perforation <0.1%.
    - Cecal intubation >90%–95% of the time; and
    - An adenoma detection rate of 15%–30%.
  - Hold warfarin 5 days before a colonoscopy with the potential for needing biopsies. Hold any direct oral anticoagulants (DOACs) for 48 hours beforehand.
  - Hold dual antiplatelet therapy (DAPT) for 5–7 days prior to the colonoscopy provided that it has been >6–12 months since a drug eluting stent was placed or >1 month since a bare metal stent was placed.
  - Aspirin as a sole agent may be continued throughout the colonoscopy.
  - **Mechanical prep** with polyethylene glycol (PEG) requires a larger volume intake but does not alter electrolytes. Magnesium-sulfate or sodium-phosphate are smaller volume alternatives, but they do alter electrolytes.
    - Dosing may be split into half on the night before and half on the morning of the procedure. This regimen possibly improves visualization.
    - Generally these osmotic agents are combined with **oral, luminally active antibiotics** given the day prior to the procedure.
  - Flexible sigmoidoscopy can be performed after only **enemas** as a prep the morning of the procedure.
  - Moderate sedation for colonoscopy is associated with a higher morbidity and mortality than general anesthesia. Small caliber scopes need less sedation.
  - Navigate the splenic flexure of the colon by having the patient lie on their back or on the right lateral decubitus position.
  - Polypectomy can be performed via a **cold or hot forceps biopsy or a snare** if >4 mm size.
    - Polypectomy has a 0.1% perforation rate overall.
    - **Endoscopic mucosal resection (EMR)** is for superficial lesions >2 cm extending only into the lamina propria. It is performed with submucosal injection to lift the lesion prior to snaring, and has a 1% perforation rate.
    - **Endoscopic submucosal dissection (ESD)** is for lesions extending deeper than the lamina propria and into the muscular mucosa or submucosa. It is performed by a submucosal injection and dissection of the lesion. It has a 4% perforation rate.
    - If the lesion does not raise, then it may penetrate the muscularis propria layer and hence not be amenable to ESD. These cases will need a segmental surgical resection.
    - If a polyp cannot be suctioned into the trap, one may suction-withdraw the scope and specimen together all the way out of the anus. Alternatives include straining the stool post-procedure or using a retrieval net.
    - **Bleeding** is the most common complication of a polypectomy and occurs in 1%–2% of cases. If absolutely necessary, biopsies can still be taken in a coagulopathic patient provided the INR is <2.5.
    - **Post-polypectomy syndrome** is characterized by fever and pericolonic inflammation, but no free air on imaging. It is secondary to electrical current or heat injuring the muscularis or serosa. It occurs in 0.3% of polypectomies. Management is conservative with NPO and IV antibiotics.
    - Tattoo suspicious lesions that are not entirely removed by biopsy, at 4 circumferential points distal to the lesion.
  - The most common non-biopsy perforation site is at the rectosigmoid junction.
  - **6-month** follow-up colonoscopy is needed if there was **neoplasia** or **piecemeal** removal of an adenomatous villous polyp.
  - **3-year** follow-up colonoscopy for any **dysplasia** on the biopsy specimen, removal of a single **villous** adenoma, or a **single large tubular adenoma or multiple small tubular adenomas.** This interval is adequate provided all specimens were obtained in single attempts.
  - **5-year** screening colonoscopy for patients with a strong family history of colon cancer, provided pathologic findings do not necessitate it be done sooner.
  - **10-year** follow-up colonoscopy if a **single small tubular adenoma** or any number of **hyperplastic** polyps were removed.
  - Postoperative colonoscopy should be performed at 3 months for patients who received emergency cancer surgery, or at 1 year if they received elective cancer surgery.

POLYPS

- **Hyperplastic polyps** have epithelial cell division at the crypt only, and no malignancy risk. Most are sessile, <10 mm, and in the rectosigmoid area. Follow-up colonoscopy should be done in 10 years.
- **Hamartomatous polyps** have a disorganized collection of normal tissue elements (aka "juvenile polyps") and may be part of a syndrome like familial juvenile polyposis, Peutz-Jeghers, or Cowden. All syndromes have an increased colon cancer risk.
- **Adenomatous polyps** have epithelial division at their surface in addition to the crypt. Tubular architecture confers a 5%–10% malignancy rate, tubulovillous 20%, and villous up to 40% chance of malignancy.
  - Serrated versions of sessile adenomas can be found in the right colon, and they have an increased rate of malignant transformation due to hypermethylation.
  - **Repeat colonoscopy should be done in 6 months** if a large benign sessile polyp had to be removed **piecemeal**, or any **neoplasia** is noted on a polypoid specimen and further resection is not being performed.
  - **Repeat colonoscopy should be done in 3 years** if a single large benign tubular adenoma or multiple smaller tubular adenomas are removed. Any **villous** adenoma or any **dysplasia** would also warrant a colonoscopy in 3 years. In all cases, the specimens must have been removed in single attempts and not piecemeal.
  - **Repeat colonoscopy can be done in 10 years if a small benign tubular adenoma <1 cm** is removed in single attempt, or small distal hyperplastic polyps are the only findings. An entirely negative colonoscopy also warrants a repeat in 10 years.
- For polyps with **carcinoma in situ** (superficial to muscularis mucosa), or invasion deep to muscularis mucosa **T1**, a **polypectomy alone could be sufficient treatment** provided:
  - 1. There is **>2 mm margin.**
  - 2. Neoplastic cells are **well differentiated.**
  - 3. Invasion of muscular mucosa is at **Haggitt level 1-3** along the head, neck. or stalk of a pedunculated polyp. *See Figure 10.2.*
  - 4. Invasion is to a depth of <1 mm or defined as **SM1-2,** which refers to the upper or middle third of submucosal involvement.
  - 5. There is no lymphatic or vascular invasion.
  - For in situ disease, since there is neoplasia, a repeat colonoscopy should be performed in **3 to 6 months.**
  - If the depth of submucosal invasion is **SM3** or occurring at **Haggitt level** 4 (invasion at the base of a pedunculated polyp or in any sessile polyp), then the patient has a much higher node-positive risk at 10%–25%. In these instances, one should **consider a segmental resection and lymphadenectomy.**
- If a **low rectal sessile adenoma** is encountered, it may be **removed transanally** provided the lesion is <3 cm size, <30% of the canal circumference, and <8 cm from anal verge.
  - This may be adequate for in situ disease or invasion into the submucosa, provided the margins are >2 mm, cells are well differentiated, and no lymphatic or vascular invasion exists.

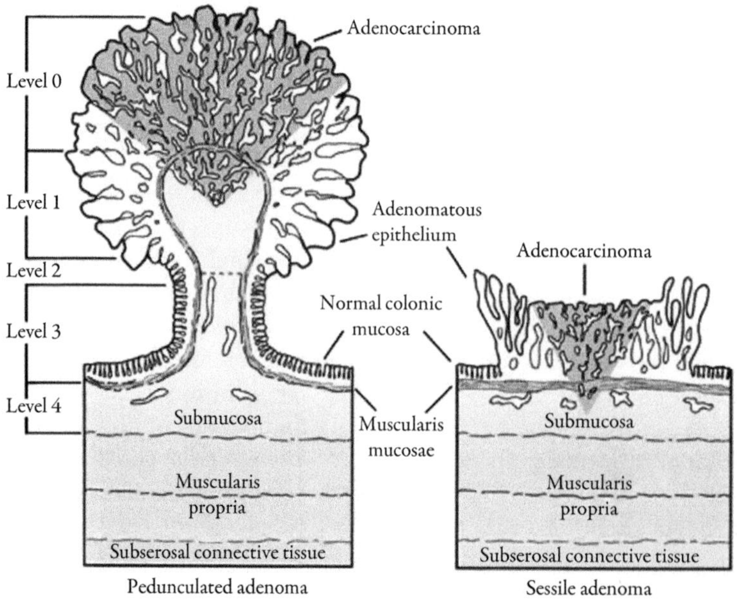

Figure 10.2 Colon polyps—Haggitt level.

- Otherwise the patient may need neoadjuvant therapy, low anterior resection (LAR), or abdominoperineal resection (APR) depending on the lesion size and proximity to sphincter mechanism (see below).

COLON ADENOCARCINOMA

- **PATHOPHYSIOLOGY**
  - This is the 2nd leading cause of cancer-related death in the United States.
  - 20% of cases are hereditary.
  - Risk factors include dietary fat, alcohol, tobacco, and *Clostridium septicum* infection.
  - It is due to unregulated epithelial division in a polyp that progresses to neoplasia and basement membrane invasion. Most cases are in the distal left colon or the cecum. *See Figure 10.3.*
  - Underlying abnormalities to the DNA regulatory process includes **APC (tumor suppressor gene), KRAS (G protein downstream of EGFR), DCC, and p53.**
  - Mucoepidermoid features, signet ring cells, and DNA aneuploidy are all associated with aggression and a worse prognosis.
  - Colon cancer classically spreads to the liver via portal veins, and rectal cancer spreads to the lung and bones via the middle rectal veins, draining to the inferior vena cava and Batson's plexus respectively.
- **DIAGNOSIS**
  - Malignancy presents with altered bowel habits, bleeding, or pain. Alternatively, it can be discovered on a screening colonoscopy.
  - **Colonoscopy and biopsy** will confirm the diagnosis. Patients may have a single tumor or multiple synchronous lesions.
  - **CT abdomen and pelvis (CTAP)** can demonstrate the local extent of disease.
  - For low rectal cancers, a **pelvic magnetic resonance image (MRI)** is better for overall local extent of disease, T stage, and nodal status.
  - An **endorectal ultrasound (US)** can distinguish T1 from T2 lesions. A T3 primary extends through the muscular propria but not outside the serosa. This qualifies as stage 2. Endorectal US can also visualize perirectal lymph node involvement and facilitate biopsy.
  - N1 status is 1–3 nodes, and this qualifies as stage 3.
  - **CT chest** can evaluate for metastases.
  - **Positron emission tomography (PET)-CT** is employed for indeterminate lesions in possible metastatic disease.
    - FDG is taken up by metabolically active cells, but not metabolized.
    - Comparison of standardized uptake values (SUV) before and after neoadjuvant or adjuvant chemotherapy or chemoradiation can be used to prognosticate, but cannot be used to definitively diagnose complete pathologic response.
  - Liver metastases can be diagnosed by an **intraoperative ultrasound,** which is the most sensitive. Preoperative diagnosis is by abdominal MRI, abdominal CT with or without PET, or transabdominal US. 20% of patients will have metastases at the time of presentation.
- **TREATMENT**
  - **Neoadjuvant chemoradiotherapy** is for rectal cancer that is clinically **stage 2 (T3 primary) or stage 3 (1–3 lymph nodes).** It may also be used for **T2 rectal tumors with aggressive features** or to downstage a **T2 lesion near the sphincter muscles** in a patient wishing to preserve continence. Neoadjuvant chemoradiotherapy decreases local recurrence and increases overall survival, with decreased rates of anastamotic stricture and chronic enteritis.
    - The regimen includes **45–50 Gy and a radiosensitizing dose of 5-FU or capecitabine** concomitantly for 6 weeks. Surgery is conducted 3–4 weeks afterward.
    - At that time, one may consider a low anterior resection (LAR) instead of an APR if there was a positive response and at least 1 cm distal margin obtainable below any residual tumor or scar, and above the levator muscles.
    - If the patient had stage 4 disease initially, then surgery need not always be performed depending on the distribution of metastases and peritoneal involvement.
  - **Neoadjuvant chemotherapy** is less often used for colon cancer, but can be considered for stage 4 patients perceived to be resectable and certain tumors with high-risk features like poor differentiation, lymphovascular invasion, perineural invasion, a T4 primary, or microsatellite instability (MSI) negative. Regimens are generally 5-FU based and can include, leucovorin, and oxaliplatin (FOLFOX) or irinotecan (FOLFIRI).
- **Surgical Options**
  - A **mechanical bowel** prep precedes surgery. Either use 4 L or 2 L PEG or magnesium citrate, as an osmotic agent. The latter can result in electrolyte changes. The osmotic agent is combined with **oral topical antibiotics** (neomycin, erythromycin, metronidazole). Finally, pre-op systemic IV antibiotics are given within one hour of the incision.
  - The primary tumor should be resected. This can be done en bloc with other organs if necessary.
    - **5 cm proximal and distal margins are ideal.** The distal margin may be as little as 1–2 cm for low rectal tumors where sphincter preservation is desired after neoadjuvant therapy.
    - **1 mm radial margin** is adequate.

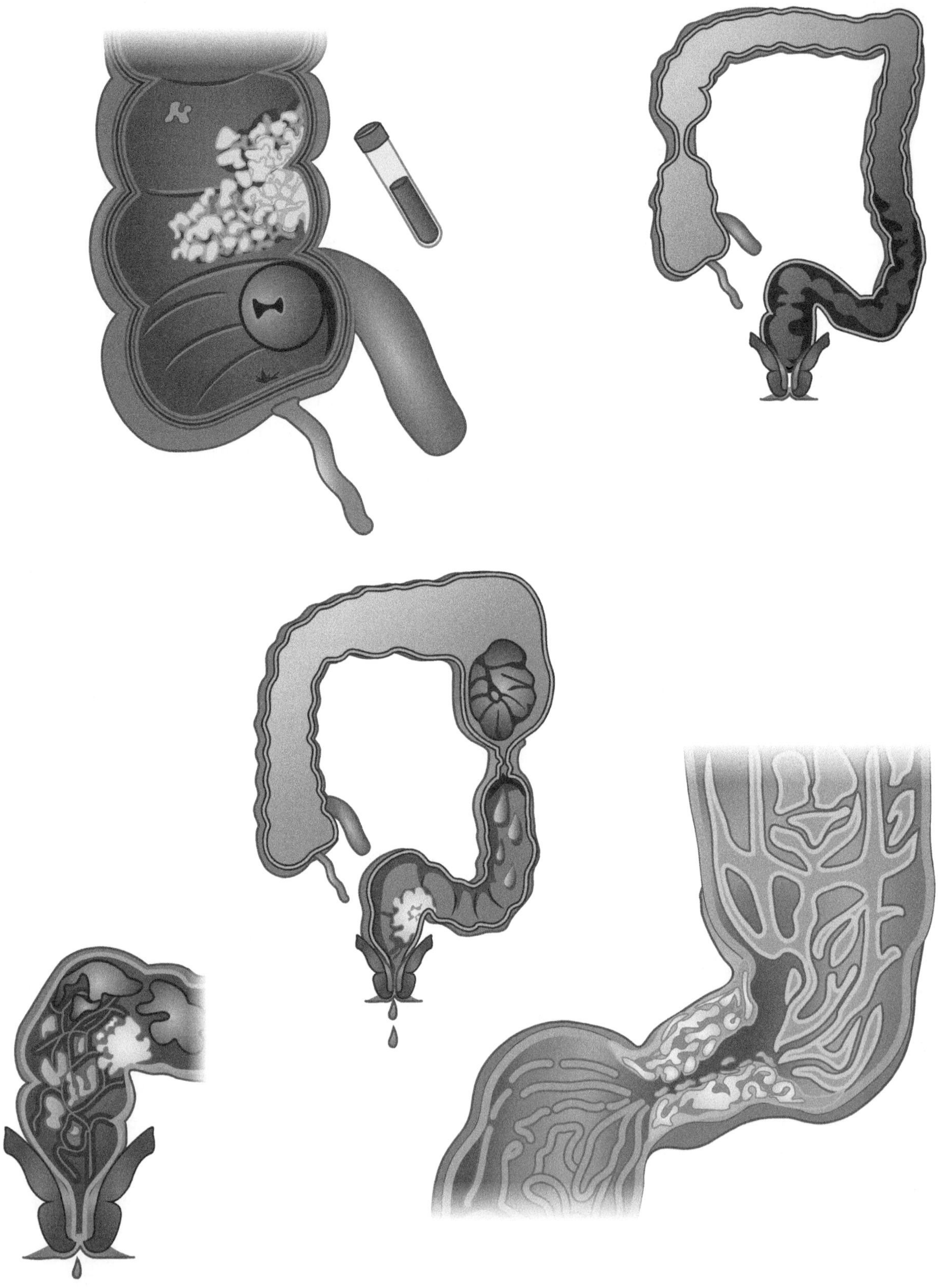

Figure 10.3 Presentations of colon adenocarcinoma.

- A minimum of **12 lymph nodes** should be retrieved.
- **Cytoreductive surgery,** generally after systemic chemotherapy, can be used for T4 tumors with peritoneal carcinomatosis and <5 liver mets (see below).
- The nodal status in the pathologic specimen is the most important prognostic factor.
- There are a number of segmental colon resections listed below. *See Figure 10.4.*
- **Right colectomy** involves taking the ileocolic artery and right branch of the middle colic artery, and removing the ileocecal junction, cecum, ascending colon, and proximal transverse colon.
  - Laparoscopic approach is with 10 mm infraumbilical, LLQ, LUQ, and RLQ ports. The ileocolic vessels are tented up to open the mesentery and begin a medial to lateral dissection.
  - An **extended right colectomy** takes both R and L branches of middle colic, and hence more of the transverse colon is removed.
- **Transverse colectomy** takes middle colic artery at its origin,
- **Left hemicolectomy** involves taking the left colic artery, but spares the sigmoid and superior rectal arteries coming from the IMA. This approach is used for a splenic flexure or proximal descending colon tumor. It allows for a side-to-side transverse to sigmoid colon anastomosis.
- **Sigmoid colectomy** takes the sigmoid artery and superior rectal artery, but spares the left colic artery.
- A formal **left colectomy** is a combination of a left hemicolectomy and sigmoid colectomy, and essentially takes the IMA at its origin.
  - A laparoscopic approach is with a 10 mm port infraumbilically, and 12 mm RLQ, RUQ, and LLQ ports. A medial to lateral dissection is begun in a plane directed under the proximal IMV. Splenocolic, renocolic, and gastrocolic ligaments are transected for a full mobilization of the splenic flexure of the colon.
- **High anterior rectal resection** takes upper rectum that lies above the peritoneal reflection, as well as its sacral attachments
- **Low anterior rectal resection** is a total mesorectal excision (TME) for tumors residing within 10 cm of the anal verge. *See Figure 10.5.* The rectum is excised below the peritoneal reflection down to Denonvilliers and Waldeyer's fascia, to the sacral fascia at the level of the coccyx.
  - A laparoscopic approach uses 10 mm infraumbilical and 12 mm RLQ, RUQ, and LLQ ports. The rectum is retracted cephalad and anterior while the mesorectum is bluntly dissected off the posterior surface. Dissection is extended through Waldeyer's fascia posteriorly, and up to Denonvilliers fascia anteriorly.

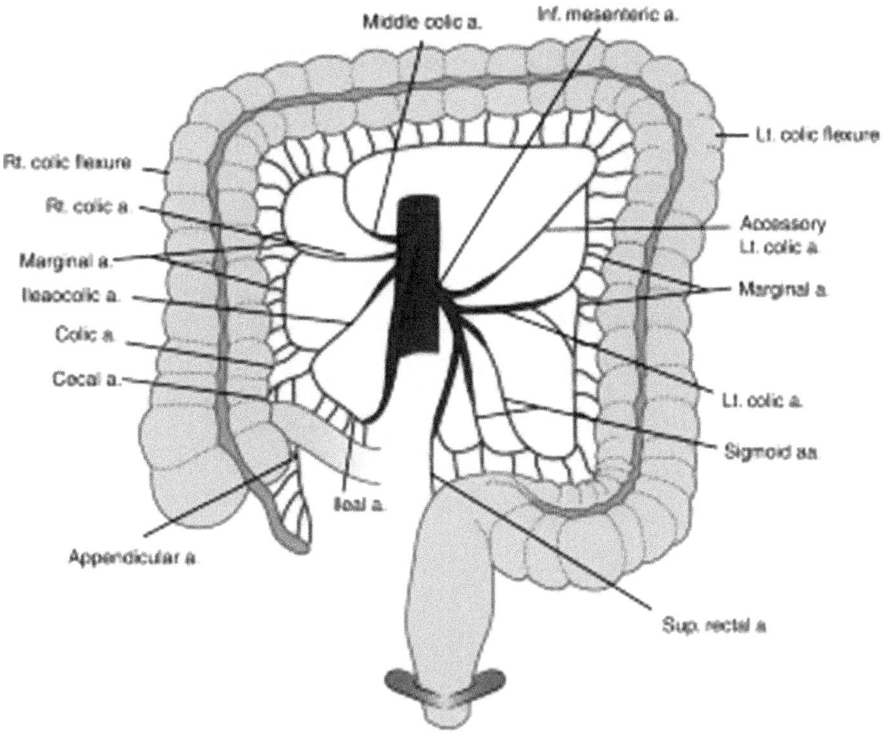

Figure 10.4 Arterial territories guiding segmental colon resection.

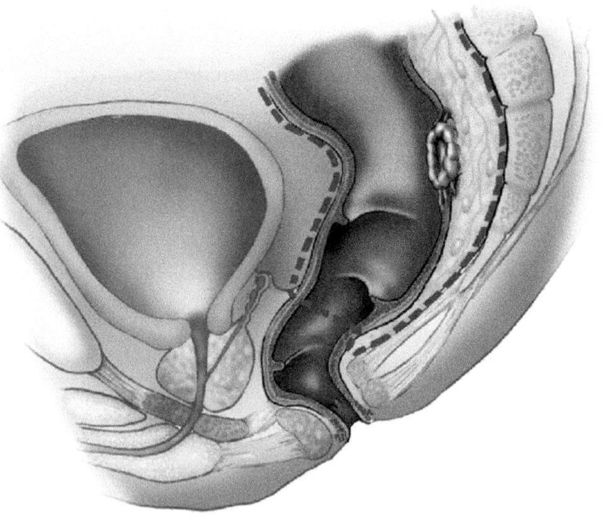

Figure 10.5 Total mesorectal excision.

- A size 28 anvil is placed in the proximal descending colon or sigmoid when extracorporeal for an EEA coloanal anastomoses. One may include a proximal diverting loop ileostomy constructed at the RLQ port site to protect a low coloanal anastomosis.
- An **extended low anterior resection** is a TME below the peritoneal reflection and below Denonvilliers fascia anteriorly.
- An **abdominoperineal resection (APR)** is a TME extending below the anal canal and taking an ellipse of perianal skin as well. Posterior dissection is carried down to the coccyx and the anococcygeal ligament divided. Lateral dissection is carried through the ischiorectal fossa, and the levator muscles are transected laterally. The anterior dissection is done last. The dissection is completed from a perineal approach, and primary closure of the perineum is performed. A permanent end colostomy or ileostomy is constructed. The extralevator version of APR takes a wide berth around the lateral margins of levator muscles to include them with the specimen and creates a larger perineal wound that requires tissue flap coverage from the rectus abdomens. Gracilis flaps can be used to cover smaller defects.
- Anastomotic options include:
  - 1. **Colorectal or coloanal,** either with or without a colonic J pouch.
  - 2. **Ileorectal** after a total colectomy.
  - 3. **Ileoanal pouch** after a total proctocolectomy.
    - An endoanal stapled anastomosis is commonly done.
    - Alternatively, a hand sewn anastomosis can be employed after a transanal mucosectomy of the transition zone above the dentate line.
    - Either technique can be used. provided the patient has normal preoperative anal function and no plans for postoperative radiation to this area.
    - Neoadjuvant chemoradiation does not preclude this anastomosis.
    - The patient should expect 8–10 bowel movements per day and 50% nocturnal incontinence.
    - Complications include pouchitis, pouch-fistula, and anal stricture.
    - A permanent end ileostomy is done if they do not meet the above criteria, or the patient does not request continence.
  - **Risk factors for an anastomotic leak** include patients in ASA class 4, a narrow male pelvis, tumor size >5 cm and located <7cm from the anal verge, increased OR time, increased blood loss, and if an excessive number of staples had to be fired.
    - In such cases the anastomosis may be protected with a proximal diverting ileostomy.
    - The ileostomy can be taken down in 3 months, or after adjuvant chemotherapy. **Digital exam and enema** at that time must show that the distal anastomosis is patent prior to ileostomy reversal.
  - Surveillance postoperative proctoscopy should be done to exclude inflammation, dysplasia, or carcinoma of any retained rectal mucosa
- **Transanal excision (TAE)** is an option for a low rectal sessile adenoma, Tis (in situ), or T1 that is <3 cm size, <30% circumference of the rectal wall, and located <5 cm–8 cm from the anal verge. 1 cm gross margins are required. This may be adequate for in situ disease or invasion into the submucosa, provided the histologic margins are >2 mm, cell types are well differentiated, and no lymphatic or vascular invasion is seen.
  - **Transanal endoscopic microsurgery (TEM)** utilizes instruments through an endoscope port for more proximal, cephalad rectal Tis or T1 lesions
  - **Transanal minimally invasive surgery (TAMIS)** uses a single site laparoscopic (SILS) adaptor placed within the anal canal and laparoscopic instruments for resection. This is another option for a more proximal, cephalad rectal Tis or T1 lesion.

- **METASTASECTOMY**
  - About 20% of patients will have metastases at presentation. Both liver and lung mets can be resected before or after chemotherapy, so long as they are isolated and other prognostic factors are positive, including:
    - Original disease was N0.
    - <5 mets are detected.
    - Largest metastasis is <5 cm.
    - CEA level of <200.
    - If a metastatic recurrence is noted, this can be treated if >12 months of a disease-free interval had elapsed after the initial primary treatment, and before the metastasis is detected.

- Recurrence rates after metastasectomy are >50%. Survival is about 50% at 5 years in patients qualifying for, and undergoing, a metastasectomy. This is compared to a <10% survival at 5 years if no therapy is conducted for metastases.
- **POSTOPERATIVELY**
  - **Adjuvant chemotherapy** for 4-6 months may be given for **stage 3 or above colon cancer,** or select stage 2 cases that have high risk features such as poor differentiation, lymphovascular invasion, perineural invasion, a T4 primary, <12 nodes recovered, or are MSI negative. Regimens are generally 5-FU based, and can include leucovorin, and oxaliplatin (FOLFOX) or irinotecan (FOLFIRI).
  - Adjuvant chemotherapy is also indicated for stage 3 or above rectal cancer and select stage 2 cases with aggressive features.
  - **Bevacizumab** can be used for stage 4 disease, and can be associated with delayed wound healing and colonic perforation in the setting of a stent.
  - If the tumor is KRAS wild type, then the **EGFR inhibitor cetuximab** may be used for stage 4 disease.
  - **Serial CEA levels** should be drawn q3–6 months to assess response to treatment, and **surveillance colonoscopy** should be done within 1 year to look for recurrence or a second primary. Then colonoscopies can be spaced out to 3–5 years, provided they remain negative.
    - For rectal cancer most recurrences are within 2 years and at the anastomotic line. Otherwise, if beyond 2 years, they may be seen in the bone or lung.
    - For colon cancer most recurrences are in the liver, whether within or beyond 2 years.
  - **CT chest and abdomen pelvis** can be done every 6–12 months. PET-CT also can be used to detect recurrences.
  - Isolated recurrences at the anastamotic line of low rectal cancer, with or without extension into perirectal pelvic tissue, are treated with an APR provided there are no other metastases. Neoadjuvant chemoradiation, if not used initially, can improve resectability.
- Peritoneal metastases are graded by the Peritoneal Surface Disease Severity Score (PSDSS), which includes symptoms and (a) histopathology of the primary neoplasm, (b) histopathology of its peritoneal implants, and (c) the extent and volume of peritoneal dissemination.
  - **Peritoneal colorectal cancer implants found at laparoscopy** should be biopsied and the procedure aborted unless obstruction, perforation, or bleeding mandates resection.
    - The patient should then receive systemic chemotherapy before consideration of future surgery.
    - Occasionally, in patients discovered to be PSDSS 1 at the time of laparoscopy and without distant metastases, cytoreductive surgery (CRS) can be done at the time of the index operation. But this is uncommon.
  - **Peritoneal colorectal cancer implants found preoperatively** are managed by:
    - **Systemic chemotherapy alone** if the patient also has distant metastases (other than liver or lung oligometastases). In these two cases, the patient may still be a surgical candidate.
    - **Systemic chemotherapy followed by cytoreductive surgery and hyperthermic intraperitoneal chemotherapy (HIPEC)** is the ideal regimen for PSDSS 2–3 *without* distant metastases.
    - CRS and HIPEC without neoadjuvant chemotherapy can be considered for PSDSS 1 and no metastases.
    - CRS and HIPEC have improved survival compared with either alone.
- **5 years survival for colorectal cance**r is estimated at 70%–95% if T1/2, 55%–65% if T3/4, 40%–60% if N1/2, and 0%–16% if M1, unless the metastases are resectable liver or lung mets (see above). In that case, the 5-year survival can be up to 50%.

FAMILIAL GENETIC COLON ADENOCARCINOMA SYNDROMES

- **FAMILIAL ADENOMATOUS POLYPOSIS (FAP)**
  - **PATHOPHYSIOLOGY**
    - This syndrome results from an autosomal dominant mutation in the **APC tumor suppressor gene.** 20% of mutations are spontaneous.
  - **DIAGNOSIS**
    - It can present with **hematochezia in a child or young adult**, prompting a sigmoidoscopy or colonoscopy demonstrating adenomas. Otherwise it is diagnosed by **yearly flexible screening sigmoidoscopies starting at 10–12 years of age** in a patient with a family history of FAP or positive APC mutation. Any adenomas detected in this situation would be diagnostic. At times, the diagnosis is made later in life when a screening colonoscopy shows >100 adenomas in an adult.
    - **Duodenal polyps** can be detected on **esophagogastroduodenoscopy (EGD) every 2 years starting at age 25–30.** These can progress to adenocarcinoma and are the second most common cause of death for FAP patients.
    - **Desmoid tumors** are located in the abdominal wall, retroperitoneum, or mesentery, and are the third most common cause of death in FAP patients.
    - Osteomas, retinal pigment epithelial hypertrophy, and epidermoid cysts are other extraintestinal manifestations
    - **Variants**
      - The **MYH variant** and **attenuated FAP** present with **10-100 polyps in the right colon.** Malignancy

can develop by the age of 55 without extraintestinal manifestations. This variant requires screening with a **formal colonoscopy** yearly starting at age 20.

- **Gardner's Syndrome** is APC-associated colon adenocarcinoma, sebaceous cysts, osteomas, and desmoid tumors.
- **Turcot Syndrome** is APC-associated colon adenocarcinoma and cerebellar medulloblastomas or another brain tumor.

- **TREATMENT.**
- Any adenomatous polyps detected on colonoscopy should be treated with a colectomy because of the high risk of developing adenocarcinoma at that site or elsewhere. Surveillance anoproctoscopy will still be needed postoperatively in most cases. Resection options include the following:
  - 1. **Total colectomy and ileorectal anastomosis (IRA)** if there is minimal rectal polyposis. Lifetime endoscopic surveillance of the rectal mucosa is needed q6–12 months.
  - 2. **Total proctocolectomy with stapled ileal pouch to anal anastamosis (IPAA)** if there is rectal polyposis, but no anal transition zone polyps are noted.
  - 3. **Total proctocolectomy with mucosectomy of the transition zone and hand sewn ileal pouch to anal anastomosis** is the preferred treatment if there is polyposis at the anal canal or transition zone.
  - 4. **Total colectomy, abdominoperineal resection, and a permanent end ileostomy** is the best option if there is malignant anal pathology or baseline sphincter dysfunction.
- If no polyps are detected on screening endoscopies, the patient may still need a prophylactic total colectomy once reaching age 20, as malignancy is very likely to occur before age 40.
- A **pancreaticoduodenectomy** may be needed if endoscopy shows multiple high-grade, dysplastic, villous duodenal polyps (Spigelman classification 4) or if a polyp is at the ampulla. EGD every 6 months should be done postoperatively. One may use nonsteroidal anti-inflammatories (**NSAIDs**) to decrease polyp growth whether or not surgery is indicated.
- **Desmoid tumors** may be treated with **NSAIDs or tamoxifen** if asymptomatic and mesenteric or retroperitoneal in location. One may **resect** the tumor if it is symptomatic, in the abdominal wall or retroperitoneum, or along the periphery of the mesentery and not near its root.

- **FAMILIAL JUVENILE POLYPOSIS**
  - **PATHOPHYSIOLOGY**
  - This is an autosomal dominant syndrome producing multiple **colonic hamartomas.**
  - **DIAGNOSIS**
  - It presents as **hematochezia in a child** or is discovered on a **screening colonoscopy starting at 10–15 years of age** for patients in whom the syndrome is known to exist. Biopsy or resect any polyps encountered.
    - **EGD screening should start at 25 years of age** to address the gastric and duodenal malignancy risk.
  - **TREATMENT**
  - Biopsied polyps that are **hamartomas require no further therapy.**
  - Any **adenomatous polyp** discovered requires a **total colectomy and IRA or IPAA** because of the high risk of adenocarcinoma developing at that site or elsewhere. Partial colectomies are generally not performed.

PEUTZ-JEGHERS SYNDROME

- **Patients have** multiple **hamartomas** that contain smooth muscle in the submucosa of the polyp. These are spread throughout the small and large bowel.
- They may have a family history and mucosal pigmentation that fades by 4th decade of life. They should receive **EGD and colonoscopy screening at 10–20 years of age.** There is also an increased risk for breast, uterus, cervix, and testicular cancer. Biopsy all polyps and resect those that are adenomas, bleeding, or obstructing.
- **HEREDITARY NON-POLYPOSIS COLON CANCER (HNPCC)**
  - **PATHOPHYSIOLOGY**
    - Secondary to mutations in the following genes responsible for **DNA mismatch repair protein** production: MLH, MSH2, MSH6, PMS2, EPCAM.
    - If they are mutated, the result can be **microsatellite instability (MSI)** and alterations in nucleotide sequences in the colonic epithelial genome. Single or multiple colorectal adenocarcinomas can ensue.
    - All new colorectal tumors are **evaluated for MSI in the DNA of the tumor specimen.** They are considered **high MSI if >2** different DNA sequences are affected.
    - This is followed by **immunohistochemistry (IHC)** to identify the true presence or absence of one of the above listed mismatch repair proteins.
  - **DIAGNOSIS**
    - This is often made at the time of evaluation of a tumor specimen and determining high MSI and the absence of the associated mismatch repair protein in serum.
      - **Amsterdam criteria** may suggest HNPCC if three first-degree relatives over two generations were diagnosed with colorectal carcinoma. At least one of them should have been diagnosed at age <50.
    - In patients with known HNPCC, perform **colonoscopy screening yearly starting at 25 years of age** or 10 years prior to the youngest relative's diagnosis of adenocarcinoma. Patients may have multiple primaries and metachronous lesions.

- **Transvaginal ultrasound (TVUS) and endometrial biopsy should start at 25–35 years of age** if Lynch 2, given the risk of gynecologic malignancy.
- Patients also have an increased risk of pancreatic, biliary, and small bowel malignancy, so they should receive **screening EGD at age 50.**
- **TREATMENT**
- **total colectomy and IRA** plus endoscopic surveillance of rectal mucosa thereafter for any non-rectal adenocarcinoma identified. Partial colectomies are generally not performed. A concomitant proctectomy may be needed for any rectal dysplasia or neoplasia.
- Total abdominal hysterectomy and bilateral salpingo-oophorectomy (TAH BSO) may be needed for a gynecologic malignancy. It can also be considered prophylactically.

- **SERRATED POLYPOSIS SYNDROME**
  - **This** results from **hypermethylation** and generally presents as a **right-sided sessile serrated polyp** which can be easily missed on colonoscopy. Diagnosis is by noting multiple such polyps proximal to the sigmoid colon with or without a family history. Surveillance colonoscopy starts at 40 years old, or 10 years prior to the youngest age at which a relative was diagnosed, whichever is earlier. Surveillance should be repeated at a minimum of every 5 years.

NEUROENDOCRINE TUMORS

- The **rectum** is the most common hindgut site of neuroendocrine tumors. The small bowel and appendix are more common overall. Diagnosis is made by colonoscopic biopsy.
- Treatment is with **local excision** if <2 cm, or **segmental resection (LAR or APR) with lymphadenectomy** if >2 cm or there is invasion of muscularis propria (T2 lesion) on biopsy.
- Octreotide and tumor debulking can relieve symptoms even if not entirely resectable. There are no standard chemotherapy protocols.

DIVERTICULITIS

- **PATHOPHYSIOLOGY**
  - Colonic diverticula result from high intraluminal pressures causing pulsion of the mucosal layer in between taenia musculature and neighboring vasa rectae. In that sense, they do not contain all layers of the intestinal wall, and so are technically pseudo-diverticula.
  - If the outpouching becomes occluded or perforated, it can result in infection and inflammation termed diverticulitis. Perforations may be microscopic and contained or may be larger or uncontained.
- **DIAGNOSIS**
  - Patients most commonly present with localized abdominal pain and tenderness, often in the LLQ over the sigmoid colon where diverticula are most prevalent. They can have a history of chronic constipation or may have had prior colonoscopies or imaging showing colonic diverticula. They may or may not have signs of systemic inflammation.
  - For stable patients without major comorbidities who have a history of diverticulitis and now present similarly with a short period of focal peritoneal irritation and no systemic illness, no further imaging is necessary. The practitioner may proceed to treatment (below).
  - All other patients, including those with multiple comorbidities, immune suppression, or a longer course of symptoms, should receive an abdominal pelvic CT scan (APCT) to confirm the diagnosis and define any Hinchey grade complication should it exist.
  - Patients with uncontained perforation and intraperitoneal spillage can have diffuse abdominal pain, tenderness, and guarding with shallow palpation or even percussion. This would be indicative of fecal irritation of the parietal peritoneum. They may have systemic signs of illness. Patients in this category may proceed directly to treatment (below).
  - **CTAP is used primarily to diagnose any Hinchey grade complication.** *See Figure 10.6.*
    - Hinchey **1a** is pericolic phlegmon or an abscess <4 cm.
    - Hinchey **1b** is a pericolic abscess >4 cm.
    - Hinchey **2** is a pelvic or retroperitoneal abscess.
    - HInchey **3** is purulent peritonitis and often will only be diagnosed at laparotomy.
    - Hinchey **4** is feculent peritonitis, also diagnosed at laparotomy.
  - **Colovesicular fistula** present with complaints of pneumaturia or fecaluria and are secondary to perforated sigmoid colon diverticula breaching the bladder wall over a prolonged course of time. They are diagnosed by CTAP with or without cystogram or rectal contrast.
- **TREATMENT**
  - Mild cases with focal tenderness and no systemic signs can be treated with oral **antibiotics and bowel rest as an outpatient.**
    - **Colonoscopy should be done 6–8 weeks after** an acute episode has subsided in order to exclude malignancy. There is an increased risk of finding malignancy if the original episode had a Hinchey grade complication. **This should be done before any elective surgery.**
    - Future **elective surgery should be considered on a case by case basis** for two or more episodes of uncomplicated diverticulitis, especially if the episodes are lifestyle limiting, or the patient has diabetes or an immune compromised state. In these instances,

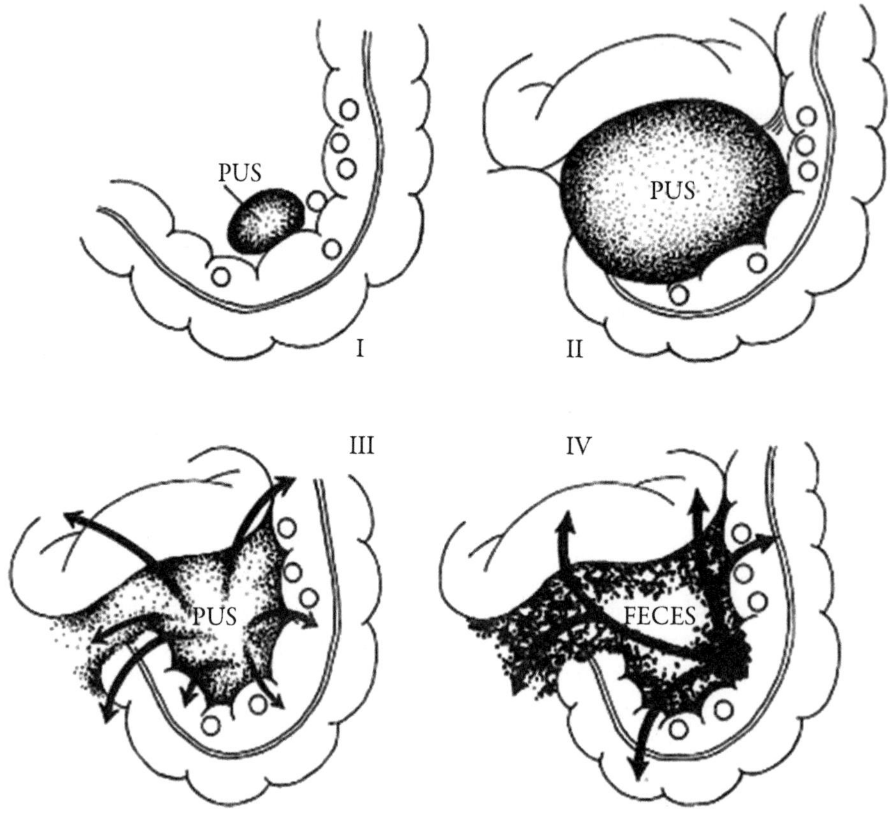

Figure 10.6 Diverticulitis—complications by Hinchey classification.

resect the involved colon down to the rectum at the level where the tenia splay, and then re-anastomose the colon. A proximal protective diverting loop ileostomy may be utilized, but will require a second surgery for reversal.

- **Hinchey 1a** cases can be treated with **inpatient IV antibiotics** converted to PO before discharge. The patient may undergo colonoscopy and elective surgery as an outpatient.
- **Hinchey 1b or 2** should receive **radiologic guided drainage** in addition to IV antibiotics. The patient may undergo colonoscopy and elective surgery as an outpatient.
- **Hinchey 3 or 4** generally require **emergent laparotomy**.
  - Resection of the involved colon down to the rectum, washout, and anastomosis (with or without a proximal diverting loop ileostomy) is acceptable for the clinically stable patient who does not have long-standing peritoneal contamination.
  - Resection of the involved colon down to the rectum, with an end stoma and Hartmann's pouch, or resection and leaving the patient in discontinuity with an open abdomen, are more radical options that may be required based on the degree of contamination and clinical status of the patient respectively. For those left in discontinuity, a planned second-look laparotomy will be needed to determine if further resection is needed or if the patient can be converted to an end stoma or anastomosis.
  - Select Hinchey 3 cases may undergo only a **laparoscopic washout without resection.** This can shorten the length of stay (LOS), but is associated with an increased need for subsequent radiologic drainage.
  - Drains are left in place until pain, fever, and leukocytosis improve, and daily output is under 30 ml.
- **Colovesicular fistulae** are treated by elective resection of the involved colon down to the rectum with re-anastomosis. The bladder is managed simply by closure of the cystotomy.
- Regardless of type of resection, patients will still need a future colonoscopy of their remaining colon for malignancy surveillance as an outpatient.

COLONIC BLEEDING

- **PATHOPHYSIOLOGY**
  - Most patients present with **hematochezia** if the source is the left colon, anorectal area, or there is rapid transit form a more proximal source. Otherwise, patients will present with **melena.** An increased BUN and bilirubin would be indicative of mucosal

reabsorption of blood products in luminal bleeding that originated from the upper gastrointestinal tract (UGIB), not a colonic source
  - **Diverticula** cause arterial bleeding from vasa recta disrupted by pulsion of the inner colonic wall through them. Other causes include **colitis and neoplasia**, and are less common than diverticulosis. **Angiodysplasia** is generally venous and of right colon origin and improves with valve replacement if the patient has aortic stenosis.
- **DIAGNOSIS**
- Rule out an UGIB with **nasogastric drainage (NGT)**, followed by:
  - 1. **Proctoscopy or colonoscopy** after a bowel prep. Preparation may be truncated to 4–6 hours if the patient is continually bleeding and practitioners do not want to wait a full 24 hours for a prep. A complete colonoscopy should ideally intubate the terminal ileum as a comprehensive evaluation of a lower gastrointestinal bleed (LGIB). Control of any hemorrhage encountered can be accomplished with **hemoclips, epinephrine injection, cautery, or argon plasma coagulation (APC).**
  - 2. **Angiography** can detect an UGIB or LGIB occurring at a rate of >0.5ml/minute and allows for **angioembolization** of bleeding sites if they are arterial in origin. It can be particularly useful in angiodysplasia of the right colon. Generally, angiography is only done if colonoscopy is not diagnostic or therapeutic.
  - 3. **CT angiography (CTA)** can detect an UGIB or LGIB occurring at a rate of >0.2 ml/minute. This may be done before or after a colonoscopy.
  - 4. A **nuclear medicine tagged RBC scan** can give a general sense of locale of bleeding in a stable patient who is slowly oozing at >0.1 ml/minute.
- **TREATMENT**
  - For **unstable patients** or those who cannot undergo endoscopy or imaging, **intraoperative colonoscopy with or without rectal irrigation and an enteroscopy** may need to be performed urgently and without a bowel preparation. At that point, if a bleeding site cannot be identified or controlled via endoscopic measures and the patient remains unstable, **emergent colectomy** (right, left, or total) may be performed, provided rectal irrigation does not suggest an anorectal source.
  - For stable diverticular bleeding, 75% stop on their own and do *not* require future surgery. Many cases will recur within a year. Colonoscopic control with **hemoclips, epinephrine injection, cautery, or APC** is usually sufficient for those cases that are not self-limited. Angioembolization is occasionally required, and surgical resection is a last resort for failure of the above measures, or if there is recurrence of bleeding within a week.
  - Colitis and neoplasia can be temporized colonoscopically, but ultimately require medical or surgical treatment for the underlying pathology.
  - Angiodysplasia can be treated with colonoscopic hemoclips or APC. Angioembolization is a second-line option. Resection is reserved for refractory cases.

APPENDIX

- **PATHOPHYSIOLOGY**
  - The appendix is perfused by the appendicular artery from the ileocolic artery. It contains its own mesentery, the mesoappendix. The appendix secretes IgA and is possibly protective against ulcerative colitis (UC) in patients <20 years of age.
  - Inflammation occurs when it is obstructed at its base from an appendicolith, fecalith, lymphadenopathy, neoplasm, or a parasite.
    - This stretches the appendiceal lumen and activates visceral afferent fibers which produce poorly localized, periumbilical pain indicative of midgut origin pathology.
    - Once the inflammation proceeds transmural to the appendiceal serosa, this inflamed serosa abuts the parietal peritoneum and activates somatic afferents in the local area. At this time, the patient's pain "migrates" to the right lower quadrant (RLQ).
  - Perforation typically occurs at the midpoint of the antimesenteric border due to high intraluminal pressure and wall tension at this site. *See Figure 10.7.*
- **DIAGNOSIS**
  - **Pain** is the first symptom.
    - It begins periumbilical during the visceral distension phase, and migrates to the RLQ once the serosa is inflamed and abutting the somatically innervated parietal peritoneum.
  - **Nausea and anorexia generally follow the onset of pain.** Patients may have vomiting, but only occasionally have diarrhea.
  - Obturator sign is pain on internal rotation of the right hip and can be seen in a pelvic orientation of the appendix.
  - Psoas sign is pain on extension of the hip and is associated with a retrocecal appendix lying atop the psoas muscle.
  - Rovsing's sign is pain in the RLQ when one pushes on the left lower quadrant (LLQ). This is due to a shift of enteric contents over to the patient's right side and secondarily displacing an inflamed appendiceal serosa against the parietal peritoneum on the right side.
  - The **Alvarado score** is based on clinical parameters and can help quantify a patients' likelihood of appendicitis:

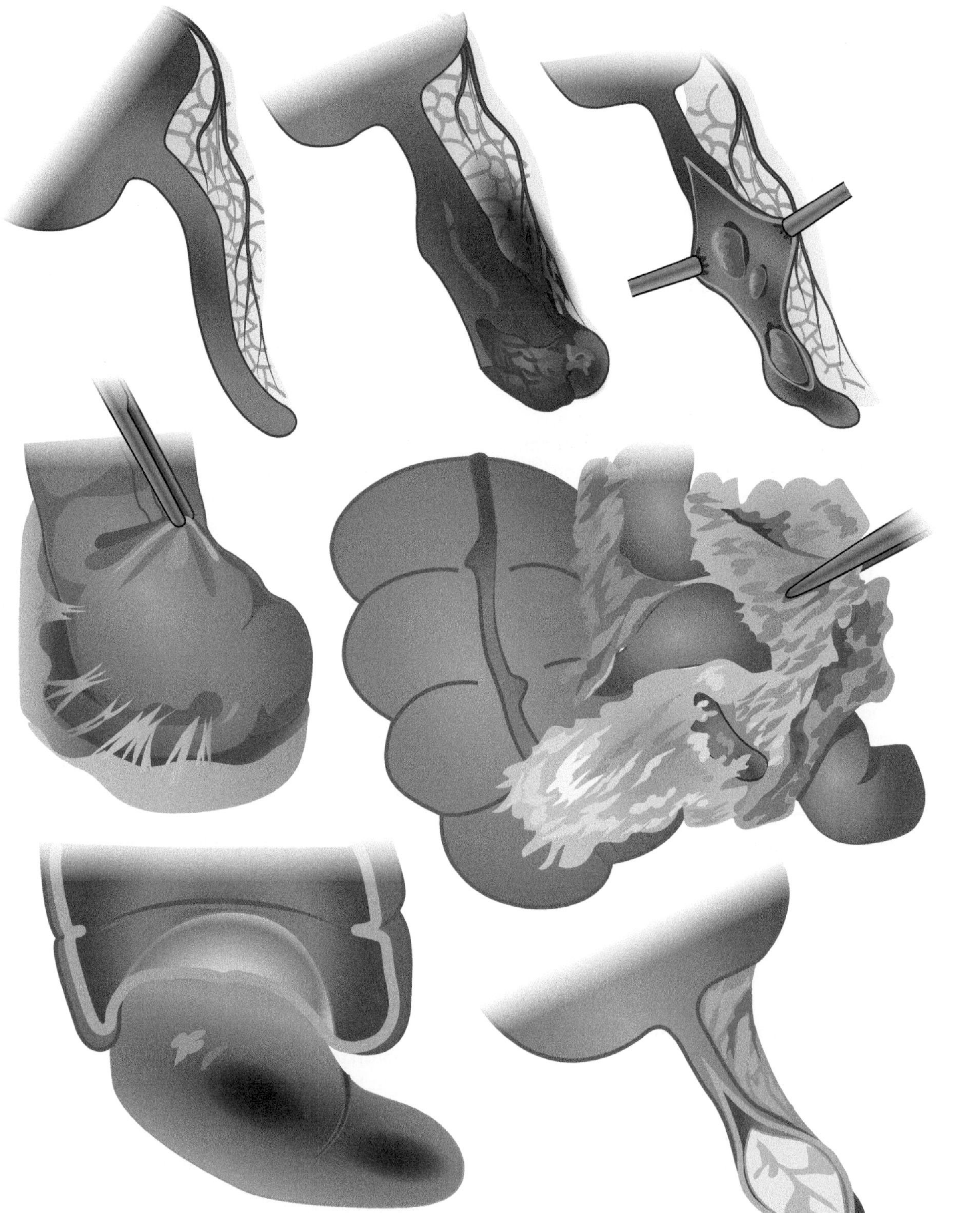

Figure 10.7 Appendicitis, perforation, and neoplasm.

- <3: likely no appendicitis These patients do not need imaging and can be discharged from the emergency department for a repeat abdominal exam in the near future.
- 4–6: indeterminate. These patients should have imaging performed.
- >7: very likely appendicitis. These patients can proceed to the operating room without imaging. If imaging was

obtained and the results equivocal or the appendix not visualized, these patients can still proceed to diagnostic laparoscopy and appendectomy.

- Initial imaging is with **abdominopelvic CT scan (APCT)** in the majority of adults.
  - **Ultrasound (US)** is the initial study for pediatric or pregnant patients. This is followed by **MRI** if the diagnosis remains unclear in these patient populations.
- **TREATMENT**
  - **Non-operative management** is acceptable if the patient is >5 years old, not systemically ill, and the appendix is perforated without an appendicolith. Intravenous (IV) antibioticsare given for phlegmon, and a drain should be placed for any sizable abscess. **Pediatric patients may receive an interval appendectomy** in 6–8 weeks, and adult patients generally do not. All **adult patients should receive an outpatient colonoscopy**, since they may have an appendiceal malignancy or an incidental colon carcinoma elsewhere. If an interval appendectomy is planned for an adult, the colonoscopy should be completed first.
    - Some cases of simple, non-perforated appendicitis can also be managed non-operatively with antibiotics alone. These cases have a slightly longer LOS, the same pain and complication profile as operative patients, but lower costs. However, there is a 25% recurrence needing appendectomy at 1 year and up to 40% recurrence at 5 years.
  - **Laparoscopic appendectomy** is the standard of care for operative management.
    - It has a decreased wound infection rate compared to open appendectomy, but may be associated with an increased surgical site abscess rate if the appendicitis was complicated or perforated.
    - There are a number of variations, but ports are generally placed suprapubically, in the LLQ, and periumbilically. The appendix and its mesentery are liberated from the base of the cecum, leaving a short appendiceal stump. One must ensure that the terminal ileum is not inadvertently injured or narrowed during the appendiceal resection.
    - A postoperative **cecal fistula** will close on its own. A stump >5 mm can lead to **stumpitis** within the ensuing decade.
    - During **pregnancy** the best time to operate is the 2nd trimester. There is a 5% fetal loss from appendicitis, but **25% fetal loss if the appendix is ruptured.** The 1st trimester has a higher fetal loss rate, and the 3rd trimester has a higher preterm labor rate. Avoid the suprapubic port, and use bilateral lower quadrant (BLQ) working ports if 2nd trimester. Place both working ports on the right side if the patient is in the 3rd trimester, since the uterus will not allow for ports on the left side.
- **NEOPLASM.** *See Figure 10.7.*
  - **Neuroendocrine tumor**s at the tip that are <2 cm can receive a simple appendectomy. All other neuroendocrine tumors (including ones with goblet features or adenocarcinoid features) need a right colectomy and lymphadenectomy.
  - **Lymphoma** isolated to the tip can receive an appendectomy, otherwise a right colectomy can be done if disease burden is truly isolated to the colon. Most patients, however, have multiple sites, so chemotherapy is the mainstay of treatment for lymphoma.
  - All **adenocarcinomas, high-grade appendiceal mucinous neoplasms,** or any low-grade lesion involving the base of the appendix need a right colectomy. **Low-grade appendiceal mucinous neoplasms (cysts, mucoceles, cystadenomas) confined to the tip** are managed with an appendectomy alone, provided the margins are negative and no invasive tumor is seen on pathology. Any assumed benign neoplasm resected by appendectomy that turns out to have carcinoma anywhere in the specimen needs a right colectomy.
  - **Peritoneal implants** are categorized by PSDSS, which encompasses the histopathology of primary tumor, histopathology of implants, and extent and volume of dissemination by CT or MRI. There is often associated mucinous ascites and a high recurrence rate after treatment.
    - **Low-grade mucinous adenocarcinoma implants** may undergo cytoreduction and hyperthermic intraperitoneal chemotherapy (HIPEC).
    - **High-grade mucinous adenocarcinoma implants** need neoadjuvant systemic chemotherapy first, and cytoreduction with HIPEC if a good response to chemotherapy is noted. These cases have a high recurrence and a high mortality.

ULCERATIVE COLITIS (UC)

- **PATHOPHYSIOLOGY**
  - Risk factors include a low fiber, high preservative, high sugar diet. Cigarettes may be protective.
  - **Mucosal and submucosal inflammation** starts at the rectum and progresses proximally, sparing the anus. Inflammation is generally confluent. and there are no skip lesions. *See Figure 10.8.*) Many patients are p-ANCA positive.
  - Characteristic pathologic findings include crypt abscesses, pseudopolyps, backwash ileitis, and eventual loss of haustra with a narrow caliber, shortened, noncompliant colon wall.
- **DIAGNOSIS**
  - Symptoms include bloody diarrhea, abdominal pain, fever, and weight loss of subacute or chronic duration. There is a bimodal age distribution, with presentations either in the early 20s or in the 50s.

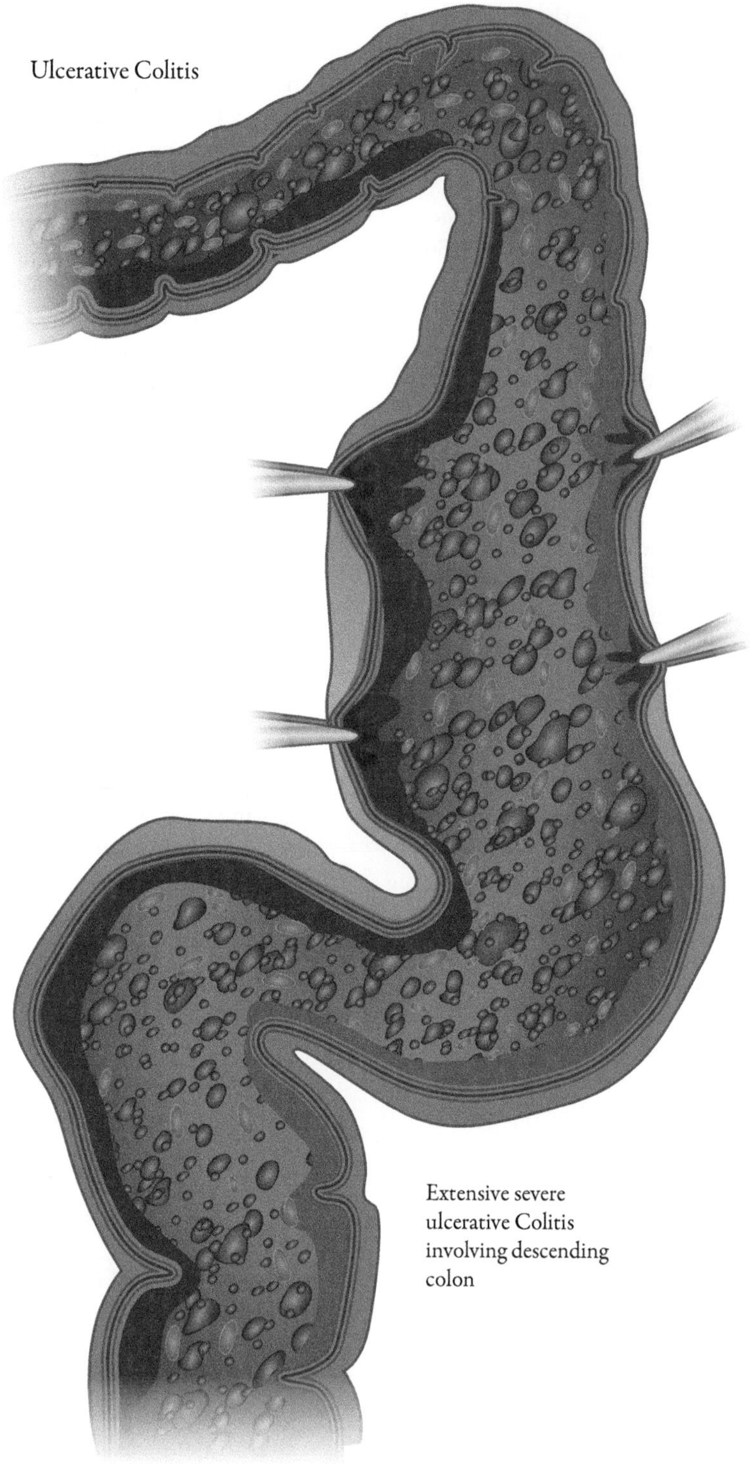

Figure 10.8 Ulcerative colitis.

- **Colonoscopy** will show confluent inflammation beginning at the rectum and advancing proximally.
- **Extraintestinal manifestations** include ocular inflammation (episcleritis or uveitis), inflammatory arthritis, primary sclerosing cholangitis (PSC), ankylosing spondylitis, and pyoderma gangrenosum. A new diagnosis of PSC without a history of UC requires colonoscopy and multiple biopsies to evaluate for UC and dysplasia, regardless of symptoms

- The risk of cancer is 10% at 20 years and 50% or more after 30 years. So **surveillance colonoscopies** should be performed every 1-3 years beginning within 8 years of the IBD colitis diagnosis. Four quadrant random biopsies are done every 10 cm at 32 sites or more. Any suspicious-appearing areas should also be biopsied. A polyp without neighboring active inflammation can be managed as per polyp biopsy characteristics in a non-IBD patient. See the diagnostics section above.

- **TREATMENT**
  - **Medical treatment** is with a combination of 5-ASA, sulfasalazine, mesalamine suppositories, luminally active glucocorticoids (budesonide), systemic glucocorticoids (hydrocortisone), TNF alpha inhibitors, and 6MP or azathioprine. The latter two are for sparing the patient from high glucocorticoid dosages.
  - **Indications for surgery include:**
    - 1. Refractory disease with continued pain or bleeding not responding to a medical regimen, NGT, bowel rest, and antibiotics.
    - 2. Refractory extracolonic manifestations of erythema nodosum, arthritis, or ocular disease.
    - 3. *Any* dysplasia, malignancy, or stricture on colonoscopic biopsy.
    - 4. Toxic colitis characterized by signs of systemic inflammation and guarding on exam that is not responding to 48 hours of medical management.
    - 5. Toxic megacolon (defined as toxic colitis with colonic diameters >8 cm).
    - 6. Hemorrhagic shock, colonic obstruction, or perforation (which is most commonly in the transverse colon).
    - 7. Growth restriction in children.
    - **Total proctocolectomy with ileal pouch anal anastomosis (IPAA).**
      - A stapled anastomosis is common. But if low rectal dysplasia was noted, a hand sewn anastomosis after perianal mucosectomy of the canal may be a better option. However, this can impair the anal sensing reflexes, accentuating postoperative incontinence, urgency, and frequency.
      - **Pouch complications** include pelvic sepsis, anastamotic leak, bleeding, small bowel obstruction (SBO), impotence or infertility, fistulae to the vagina or elsewhere, strictures, or pouchitis.
        - **Pouchitis** is diagnosed by endoscopy and biopsy to exclude Crohn's disease of the pouch. Treatment then follows with PO ciprofloxacin and metronidazole. Daily suppressive antibiotics may be employed in patients with multiple bouts. Probiotics and salicylate or glucocorticoid enemas are additional options. Severe cases may require diversion, exclusion, or resection for dysplasia, carcinoma, or refractory inflammation. Strictures can be treated with dilation.
      - **End ileostomy** rather than an anastomosis can be considered if the remainder of the terminal ileum cannot reach the anal canal despite mobilizing the SMA, transecting the ileocolic artery or SMV, and scoring the mesentery to release more length. Ileostomy can also be utilized if the patient has baseline anal incontinence or a high suspicion for Crohn's colitis.
    - **Total colectomy with ileostomy and Hartmann's pouch**, and an intra-rectal drain or mucous fistula construction from the distal end, can be employed if the patient is acutely ill and not able to tolerate an anastomosis. In the future an ileorectal anastomosis (IRA) or ileal pouch anal anastomosis (IPAA) with or without a protective diverting ileostomy can be performed
    - Total colectomy with ileorectal anastomosis (IRA) and rectal mucosal surveillance afterward can be used for the rare UC patient with minimal rectal inflammation, or those that are chronically ill and a poor operative candidate for the proctectomy component.
    - Perineal intersphincteric proctectomy leaves the external sphincter to buttress the perineal closure. It can be used as a component of total proctocolectomy with permanent end ileostomy. Complications include perineal wound healing issues and impotence.
    - **Postoperative surveillance proctoscopies** are necessary to exclude inflammation, dysplasia, or carcinoma of any retained rectal mucosa.
    - Inflammatory arthritis and ocular inflammation (episcleritis or uveitis) both improve after colectomy. Pyoderma gangrenosum may improve. Primary sclerosing cholangitis (PSC) and ankylosing spondylitis do not improve after colectomy.

CROHN'S COLITIS

- **PATHOPHYSIOLOGY**
  - Subtypes include nonstricturing-nonpenetrating, penetrating, or stricturing, a small percentage of which may have occult adenocarcinoma. Patients may be p-ANCA negative and have ASCA antibodies.
- **DIAGNOSIS**
  - Colonoscopy with ileal intubation is the main method of diagnosis. Inflammation can be in discrete locales (skip lesions) or be confluent and hence difficult to distinguish from UC. The latter case is an "indeterminant colitis."
  - Biopsy will show **granulomatous inflammation** and possibly extension through the deeper layers of the colon wall. The ileum may be involved as well. These findings help distinguish Crohn's disease from UC.
  - **Colonoscopic screening for dysplasia** is similar to UC.

- **TREATMENT**
  - Medical management is similar to UC above.
  - **Surgical indications** are similar to UC above.
    - Unlike UC, a **segmental colectomy** is an option in Crohn's. Margins are determined by reaching a grossly normal appearing colon serosa. Fistulization is more common with Crohn's colitis than UC. In this case the culprit segment of colon should be resected, and the bystander (recipient) small or large bowel, vagina, or bladder may be repaired primarily.
    - **Total proctocolectomy with end ileostomy** is the traditional, conservative option, since Crohn's disease often affects the anus. The proctectomy may be intersphincteric to save the external sphincter for better perineal wound healing. Total colectomy with IRA can be employed if the anorectum is spared of Crohn's disease.
    - Due to the transmural penetrating inflammation and resulting fibrosis, preoperative ureteral stent placement may assist in identifying any intraoperative ureteral injuries incurred.

OTHER TYPES OF COLITIS

- 1. **ISCHEMIC COLITIS.**
- **PATHOPHYSIOLOGY**
  - Most often it is due to a low flow state with watershed ischemia to specific sites. *See Figure 10.9.*
  - Griffiths point is at the splenic flexure **between the SMA and IMA vascular territories.**
  - Sudek's point is at the upper rectum **between the superior rectal and middle rectal vascular territories.**
  - There is also a right colon watershed area along the **distal ileocolic artery.** If this site is involved, patients often have concomitant mesenteric ischemia to the distal small bowel. Pain is more prominent than bloody diarrhea in this vascular distribution.
- **DIAGNOSIS**
  - Patients present with abdominal pain followed by bloody diarrhea
  - Diagnosis is by **colonoscopy** showing patches of erythema, ulcerations, or gray discoloration associated with varying levels of ischemia. There is a sharp transition between ischemic and non-ischemic segments.
  - **CTA** could show segmental colonic wall thickening, however its sensitivity for the diagnosis is generally lower than colonoscopy. Most cases are not associated with a discrete arterial defect. In a minority of circumstances, ileocolic artery cutoff may be visualized in patients with right-sided ischemic colitis and accompanying mesenteric ischemia.
- **TREATMENT**
  - **Bowel rest, IV fluid, and IV antibiotics** is successful in 80% of cases, since episodes are self-limited.
  - **Surgery** is needed only in a minority of patients suffering from perforation with pneumoperitoneum, toxic colitis with or without megacolon, or necrosis seen on colonoscopy or as pneumatosis intestinalis or portal venous gas on imaging.
  - Chronic ischemia without acute necrosis can result in strictures that may require dilation or resection.
- 2. **NEUTROPENIC ENTEROCOLITIS**
  - Can be seen in patients with an absolute neutrophil count <1000. Inflammation involves the cecum and ascending colon as well as the distal ileum. Diagnosis is by **CT abdomen pelvis** showing thickening in these areas, with or without pneumatosis intestinalis. Colonoscopy is an alternative diagnostic modality. Treatment is with **NPO, IVF, and IV antibiotics.** Conservative management is indicated even if pneumatosis intestinalis is present on imaging. Colony stimulating factor can be used. Surgery is reserved for patients with perforation.
- 3. **DIVERSION COLITIS**
  - It is from distal colonic mucosal deprivation of short chain fatty acids (SCFAs), including acetate, butyrate, and propionate. This can be secondary to decreased dietary intake or a proximal diverting colostomy.
- 4.***CLOSTRIDIUM DIFFICILE* COLITIS PATHOPHYSIOLOGY**
  - It is due to a change in colonic microbial flora allowing for a proliferation of the organism. Host immunosuppression from HIV or inflammatory bowel disease (IBD) therapy also plays a role.
- **DIAGNOSIS**
  - Symptoms include non-bloody diarrhea within 3 weeks after antibiotics, hospitalization, or endoscopic intervention. *C. difficile* may occasionally cause an enteritis or pouchitis in a patient who has had a total colectomy.
  - Confirmation **toxin assay** will detect active disease. The organism produces toxins A and B. **PCR** has a higher sensitivity than toxin assay, but detects prior disease, current disease, and simple colonization without toxin production.
  - **Sigmoidoscopy** can be utilized if the diagnosis is in question. It can show pseudomembranes as well as neutrophils in the mucosa and lamina propria on biopsy.
- **TREATMENT**
  - **First step is to stop antibiotics** and consider starting probiotics.
    - **PO vancomycin 125 mg qid** is first-line for the first episode and first recurrence. It also may be reused for a second recurrence or if the patient has underlying IBD.
    - **IV Flagyl or PR vancomycin** can be added for proctosigmoid disease or concomitant ileus.
    - **PO fidaxomicin 200 bid or IV tigecycline** are options if the above modalities are insufficient.
    - **Fecal microbial transplant can be via NGT or colonoscope,** and is for patients with >3 recurrent

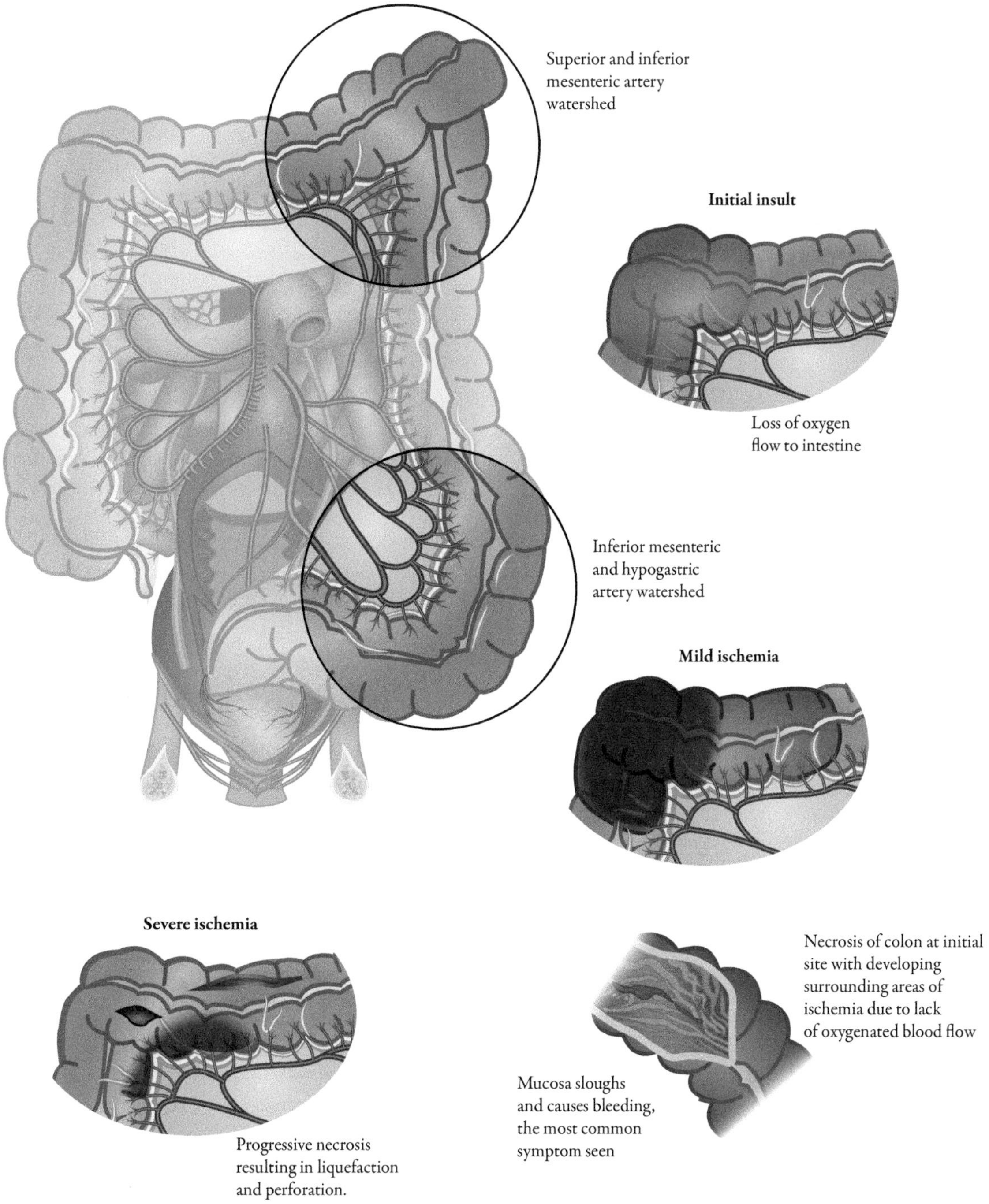

Figure 10.9 Ischemic colitis.

episodes, or no response to 1 week of the above treatment.

- Up to 50% of episodes recur, and some patients have a chronic carrier state, while others are colonized.
- Surgical therapy with **colectomy, end ileostomy, and Hartmann's pouch** is reserved for patients who progress to toxic colitis that is vasopressor-dependent or associated with organ failure. A less commonly employed alternative to colon resection involves constructing a loop ileostomy with antegrade PEG followed by vancomycin colonic irrigation for 10 days.

- 5. **AMOEBIC COLITIS**
  - Results from *Entamoeba histolytica* infecting travelers and alcoholics.
  - Diagnosis is by **serology antibody testing** and colonoscopy showing **ulcerations**. A **stool sample** may show trophozoites.

- Treatment is with **Flagyl** 500 tid for 10 days for acute dysentery. Add diiodohydroxyquin 650 tid for 20 days if chronic. Some patients may have an amoebic liver abscess without stool trophozoites.
- **6. CYTOMEGALOVIRUS (CMV) COLITIS**
  - Generally affects HIV positive patients. They present with abdominal pain more so than diarrhea. Penetrating ulcerations can result in perforation of the colon. Treatment is with acyclovir or ganciclovir, and possibly surgical resection if perforated or refractory to medical therapy.
- 7. **ACTINOMYCES**
  - Anaerobic Gram positive rods (GPR) in the cecum that form abscesses and sometimes fistulae. Yellow sulfur granules are seen on pathology. Treatment is with **abscess drainage and penicillin.**
- 8. **OTHER INFECTIOUS COLITIDES**
  - *Salmonella, Shigella, Campylobacter, Yersinia,* and *Giardia* can be treated with PO or IV antimicrobials.
- 9. **TOXIC COLITIS**
  - This is a general term applied to systemic derangements resulting from colonic inflammation. The etiology could be any of the above, but is most commonly secondary to IBD or *C. difficile.* Patients have transmural inflammation impairing muscularis propria function and resulting in dilation and bacterial translocation. If colonic diameter is >8 cm, the term **toxic megacolon** is applied.
  - The etiology of colitis should be diagnosed by clinical presentation and imaging. The patient should also have evidence of systemic illness and possibly end organ dysfunction to qualify as "toxic". Although colonoscopy is also diagnostic, it carries a higher risk of perforation.
  - Treatment starts with bowel rest, volume and electrolyte support, **IV glucocorticoids for IBD, and broad antibiotics, unless it is believed that the colitis is due to *C. difficile*.**
  - **Total colectomy, end ileostomy, and Hartmann's pouch** is reserved for those patients not responding to 48 hours of medical therapy or with progressive sepsis and increasing colonic diameters. Additional management of the distal rectal stump can include placement of an intra-rectal drain, construction of a mucous fistula within the midline wound, or burial of the stump in the subcutaneous tissue below the dermal incision.

OTHER PATHOLOGY

- **COLONIC OBSTRUCTION**

- **PATHOPHYSIOLOGY**
  - Most commonly, this is secondary to colon adenocarcinoma, volvulus, or fibrotic complications of diverticulitis. *See Figure 10.10a, b, c.*
  - **Sigmoid volvulus** occurs in debilitated, neurologically limited psychiatric patients with a high fiber diet.
  - **Cecal volvulus** occurs in 20–30-year-old patients with a congenitally poor fixation of the cecum. 20%–30% of these cases just have a cecal bascule.
  - Other causes include IBD, extrinsic compression, fecal impaction, foreign body, or intraluminal stricture from chronic inflammation, ischemia, or malignancy.
  - All complete colonic obstructions have the potential for closed loop physiology if the ileocecal valve remains competent.
  - The cecum is the most common site for perforation, usually when its diameter reaches 12 cm or greater.
- **DIAGNOSIS**
  - AXR shows a dilated colon with air fluid levels within it. A transition point to decompressed, distal colon or rectum may be identifiable. Sigmoid volvulus appears as a bent inner tube with the concave side facing the LLQ. Cecal volvulus appears as a coffee bean sign with the concave side facing the RLQ.
  - A CT abdomen pelvis can be obtained for confirmation of the diagnosis and its etiology. Any pneumatosis intestinalis or air in the portal venous system indicates ischemic colon wall possibly accompanied by bacterial infection and translocation.
  - A contrast enema can delineate the site of a distal obstruction, but may not necessarily provide an etiology. A bird's beak tapering of the distal colon or rectum is an exception, as it can be diagnostic of volvulus.
  - Colonoscopy or sigmoidoscopy can also confirm the obstruction when difficulties are encountered trying to pass the scope proximally.
- **TREATMENT**
  - **Diverting loop colostomy** is used for a patient who is acutely ill with an obstruction low in the rectum, where a stent cannot be placed (see below),
  - **Segmental colectomy** and a Hartmann's pouch, or a segmental colectomy with primary anastomosis with or without a proximal diverting loop ileostomy, are both options for the stable patient.
  - **Self-expanding metal stents** can be placed across left colonic, non-rectal obstructions from a unifocal, short stricture that is <10 cm or an intraluminal tumor without a component of extrinsic compression, provided a wire can be navigated proximal to the lesion.
    - Their benefits include:
      - 1. Palliation.
      - 2. To facilitate colonoscopy or completion of neoadjuvant therapy for cancer.
      - 3. To facilitate resection, by relieving an obstruction and allowing for a bowel prep. If the patient improves in their physiologic condition with this, then a single staged resection and anastomosis can be accomplished and a stoma avoided.

(a)

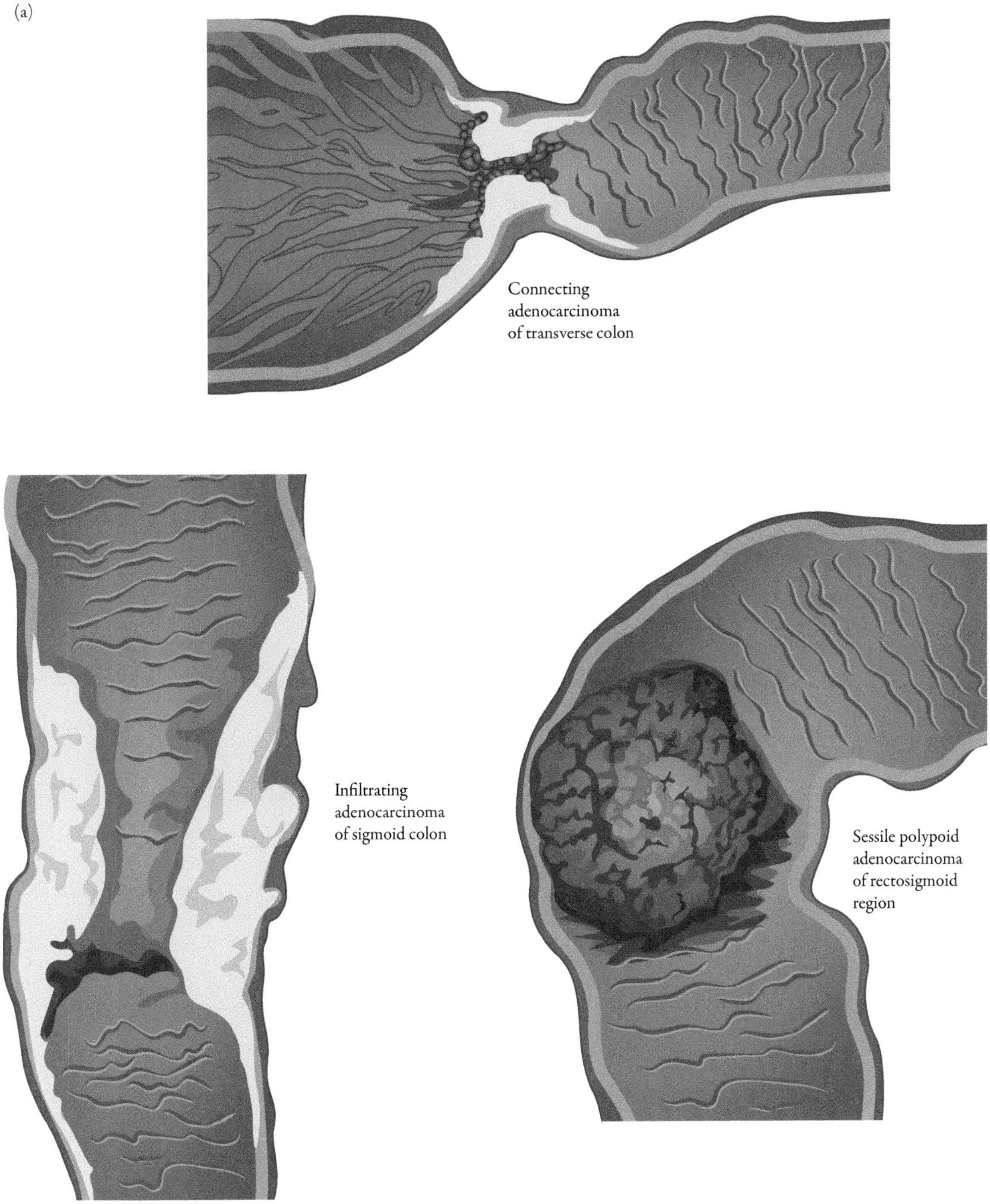

Figure 10.10 (a) Colon cancer obstruction.

(b)

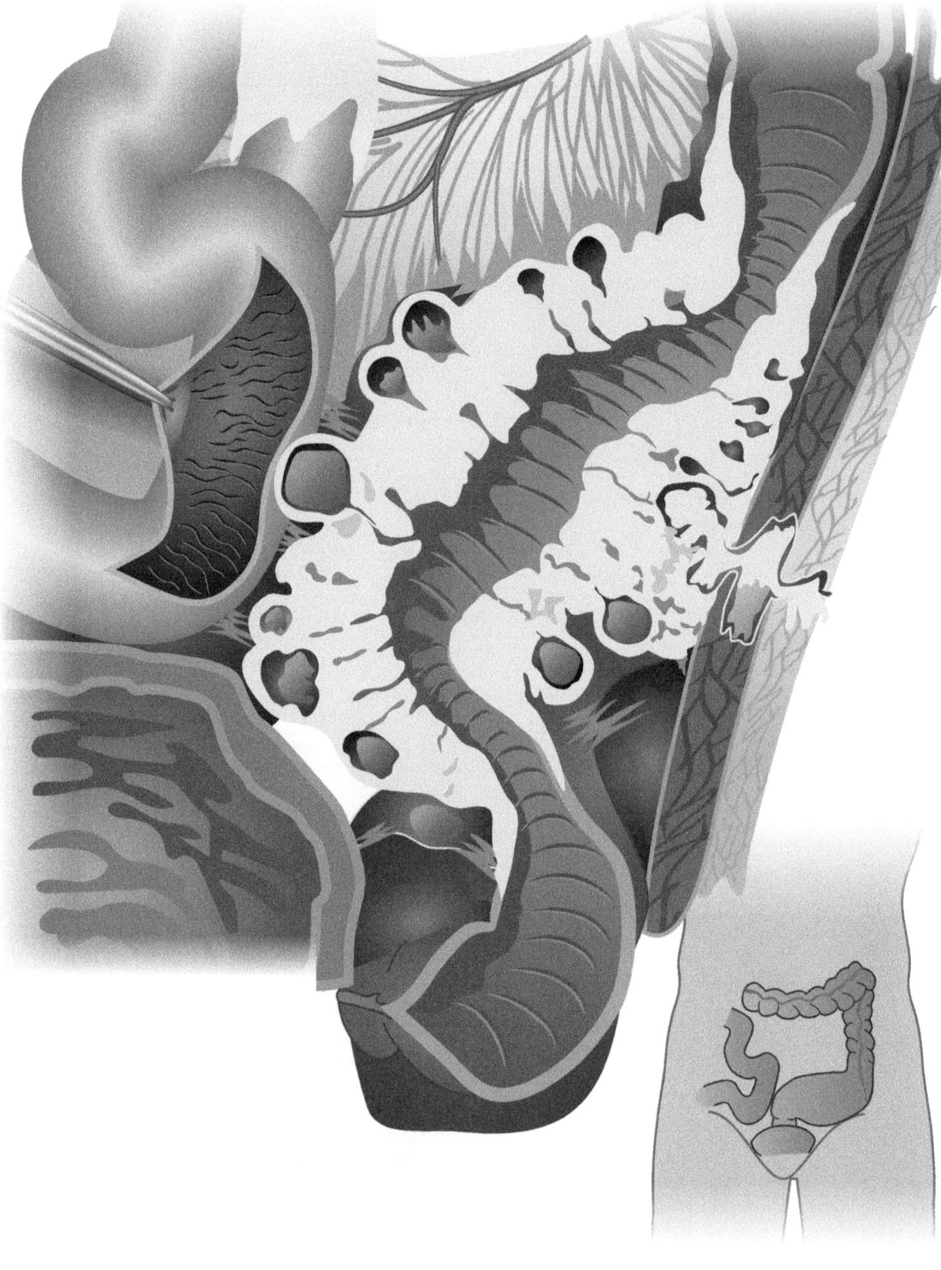

Figure 10.10 (b) Colon diverticular obstruction and fistulisation.

- Preparation is with **enemas** for a near-complete obstruction, or **PEG** by mouth for a partial obstruction. Use the scope and guidewire to pass proximal to the lesion. Usually, no dilation of the lesion is attempted because of the perforation risk. Self-expanding metal stents have decreased migration but increased tumor ingrowth. Covered stents have increased migration rates but decreased tumor ingrowth.
    - Stents are **not ideal for obstructions low in the rectum**, since they will extend to the sensate infra-dentate anal margin.
    - In the setting of cancer, a history of bevacizumab use or radiation to the area can complicate stent placement or result in perforation.
- Treatment of sigmoid volvulus is by decompression with a **rigid or flexible colonoscope and leaving a**

(c)

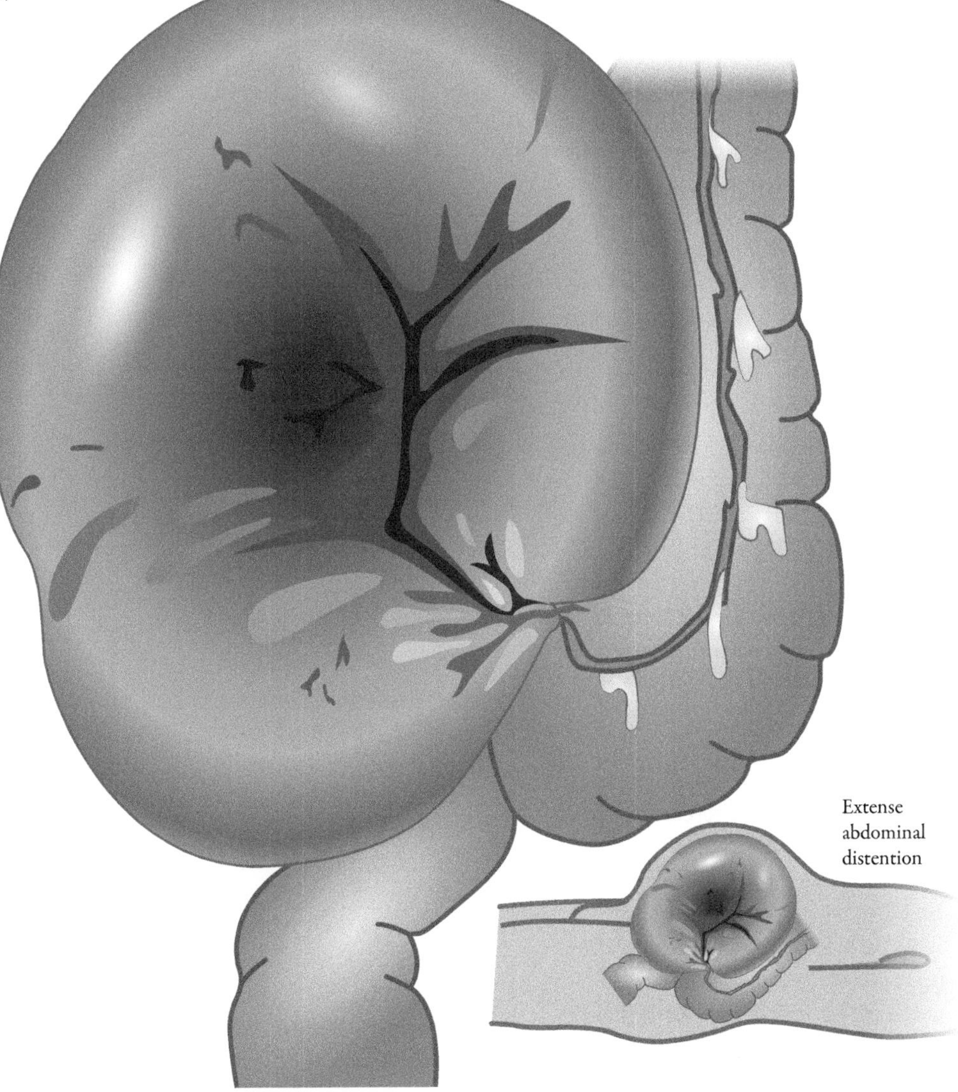

Figure 10.10 (c) Sigmoid volvulus.

colonoscopic tube in place beyond the torsed segment. This prevents repeat torsion in the short term and should be followed by **sigmoid colectomy and re-anastomosis** as an inpatient. If gangrene is suspected, then just perform colectomy as initial therapy and forego endoscopy.

- Treatment of cecal volvulus is with **ileocecectomy with or without right colectomy**, and establishment of continuity with an ileocolic anastomosis. This is preferred over cecopexy or a cecostomy tube. The latter is reserved for frail patients who cannot tolerate surgery.

- **OGILVIE'S COLONIC PSEUDO-OBSTRUCTION**
  - **PATHOPHYSIOLOGY**
    - Medications that slow colonic motility are one of the main culprits. Other than this, any general physiologic insult or retroperitoneal process can cause pseudo-obstruction.
    - Patients are often elderly, bedridden, on opioids and post-op, septic, or recovering from a recent trauma.
  - **DIAGNOSIS**
    - AXR shows air throughout the colon and rectum. Distension may roughly parallel the SMA vascular distribution, with the ascending and transverse colon dilated and the descending and sigmoid decompressed.
    - Confirmation is with CT, colonoscopy, or contrast enema to **definitively exclude a colonic obstruction.**
  - **TREATMENT**
    - Start with NGT decompression, correction of electrolytes, and cessation of causative medications.
    - **Neostigmine** 2 mg can be given if there is no improvement after 24–48 hours of the above conservative measures, or if the cecum is approaching 12 cm diameter. Obstruction must have been definitively

excluded before using neostigmine. The dose may be repeated in 24 hours.
    - **Colonoscopic decompression** involves suctioning out air, advancing the scope to the hepatic flexure (unless ischemia is encountered prior to this), and leaving a decompressive colonic tube in this position. The tube is placed to intermittent suction for up to 72 hours.
    - Surgical colostomy tube or resection is a last resort for those cases suspected to have progressed to a necrotic wall or perforation. Percutaneous cecostomy has occasionally been employed as an alternative decompression strategy in a frail patient with recurrent episodes.
- **STERCORAL ULCER**
    - It is secondary to pressure necrosis of the colonic wall from hard stool. It generally occurs at the rectosigmoid junction in an elderly patient on NSAIDs. The ulcer may perforate and result in fecal peritonitis.
- **MELANOSIS COLI**
    - A term given to pigment-harboring macrophages that reside in the lamina propria. This is simply a clinical finding without known sequelae. It is secondary to cathartics and laxatives, and requires no treatment other than stopping the offending agent.
- **RARE TUMORS**
    - Lipomas may undergo colonoscopic resection or enucleation if the patient is symptomatic.
    - Leiomyomas or sarcomas can be resected without the need for a lymphadenectomy.
    - Large B cell lymphoma may be resected if it is large, isolated, and is impending obstruction. Otherwise, the majority of cases are treated with chemotherapy.

COLOSTOMY AND ILEOSTOMY COMPLICATIONS

- **Local Wound Care**
    - A properly fitted appliance around the stoma orifice prevents content leakage.
    - Silver nitrate can be applied to granulation tissue.

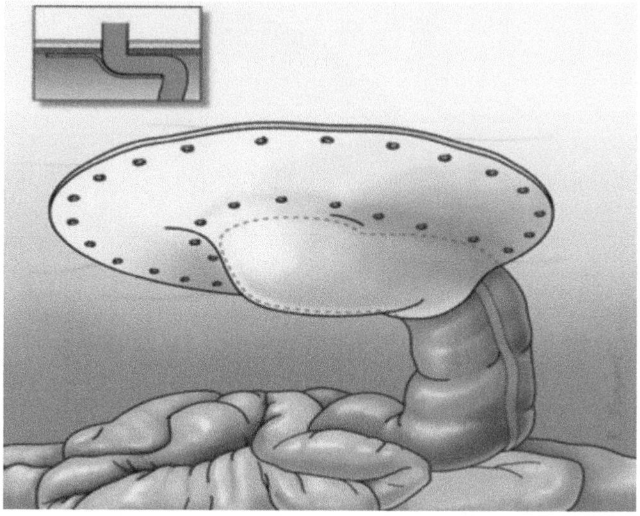

Figure 10.11 Laparoscopic parastomal hernia repair with mesh in Sugarbaker fashion.

    - Zinc oxide can be placed over areas of skin irritation from enteric contents.
- **Parastomal Hernia**
    - Asymptomatic patients can be treated with a support belt and weight loss.
    - Symptomatic parastomal hernias need repair with mesh in a sugar-baker fashion. *See Figure 10.11.* This is more common with colostomies than ileostomies. One may consider prophylactic mesh at the time of stoma construction to prevent this.
- **Stomal prolapse** should be treated with revision of the stoma and resection of redundant bowel. This can all be done without entering the peritoneal cavity.
- **Pyoderma gangrenosum** of the peristomal area in an IBD patient can managed with infliximab or glucocorticoids.
- **Stomal mucosal necrosis** can be evaluated using a clear test tube within the lumen. If the depth of necrosis remains superficial to the level of fascia, the stoma may be observed and mucosa allowed to slough. If the necrosis extends deep to the fascial level along the mucosa, then the stoma will need to be revised.
- Most temporary stomas can be reversed after 12 weeks.

11.

ANORECTAL

ANATOMY AND PHYSIOLOGY

- The **anal canal is from the anorectal junction, marked by the puborectalis muscle, down to the dentate line.** It is lined by **columnar epithelium** and is 5 cm in length. It is perfused by the middle rectal arteries and veins. The columns of Morgagni house the intersphincteric glands in the canal. The transition zone to the anal margin crosses the dentate line. *See Figure 11.1.*
- The **anal margin is from the dentate line to the anal verge.** It is lined by non-keratinized **squamous epithelium** and is perfused by the pudendal arteries and veins from the inferior rectal vessels.
- The **anal verge** is the transition to **keratinized** squamous epithelium of the perianal skin. This area is also perfused by the pudendal vessels.
- The **internal sphincter** is an extension of the muscularis propria of the rectum. It has a baseline pressure of 40–80 mmhgn and is under autonomic control. It normally relaxes when the rectum superiorly is accommodating stool, liquid, or gas. This allows contents to descend and be sensed by the anal canal, a process called the sampling reflex. Loss of this reflexive relaxation of the internal sphincter can be seen in Hirschsprungs disease.
- The **external sphincter** is an inferior extension of the puborectalis muscle and levator complex. It is innervated by the pudendal nerve. During the sampling reflex when rectal contents descend, the external sphincter contracts and increases pressure by 40–80 mmhg in order to prevent incontinence.
- The crypts of lieberkuhn line the rectum and contain mucous secreting goblet cells.

DIAGNOSTICS

- **Anoscopy and proctoscopy** can define mucosal lesions, fistulae, fissures, inflammation, rectal bleeding sources, and tumor.
  - **Biopsy** is performed for any suspicious anal lesion, to exclude HSV (#1 cause of anal ulcers), CMV, B cell lymphoma, Kaposi's sarcoma, melanoma (pigmented or nonpigmented), low-grade or high-grade squamous dysplasia, Squamous cell carcinoma in situ ("Bowens disease"), Squamous cell carcinoma (SCC), or adenocarcinoma.
  - Biopsy can also help diagnose Hirschsprung disease in the setting of an inadequate inhibitory reflex.
- **Endoscopic anorectal ultrasound** is used to check sphincter muscle integrity for incontinent patients and also delineate T1a, b, or T2 status and perirectal nodes in low rectal tumors.
- **Manometry** can determine the internal sphincter resting pressure (normally 40–80 mmHg) and external sphincter squeeze pressure (normally is an increase of 40–80 mmHg).
- ENG **pudendal nerve latency** can check for a neuropathic component to symptoms of incontinence.

CONSTIPATION

- **PATHOPHYSIOLOGY.**
  - Stasis of stool in the colon and rectum results in continued water absorption and desiccation of contents, hardening of stool, and potential irritation and mechanical trauma to the mucosa.
- **DIAGNOSIS.**
  - By definition, patients should have 3 or less defecations per week, and at least one-fourth of defecations have straining, hard stools, incomplete evacuation, a sensation of obstruction, or require manual assistance.
  - Next step is to **rule out rectal prolapse on exam, and rule out obstruction** with colonoscopy, flexible sigmoidoscopy, proctoscopy, or a contrast enema.
- **TREATMENT AND FURTHER DIAGNOSTICS**
  - First trial **fiber bid** and add an **osmotic agent** like polyethylene glycol (PEG), magnesium citrate, lactulose, or glycerin. Then try adding **stimulants** like bisacodyl or senna, as well as **stool softening** agents like docusate.
  - If there is no improvement after 6 months of this therapy, proceed to testing below.
    - First check for **pelvic floor dysfunction (PFD)** with:
      - **1. Manometry** to test the inhibitory reflex. The internal sphincter should relax with

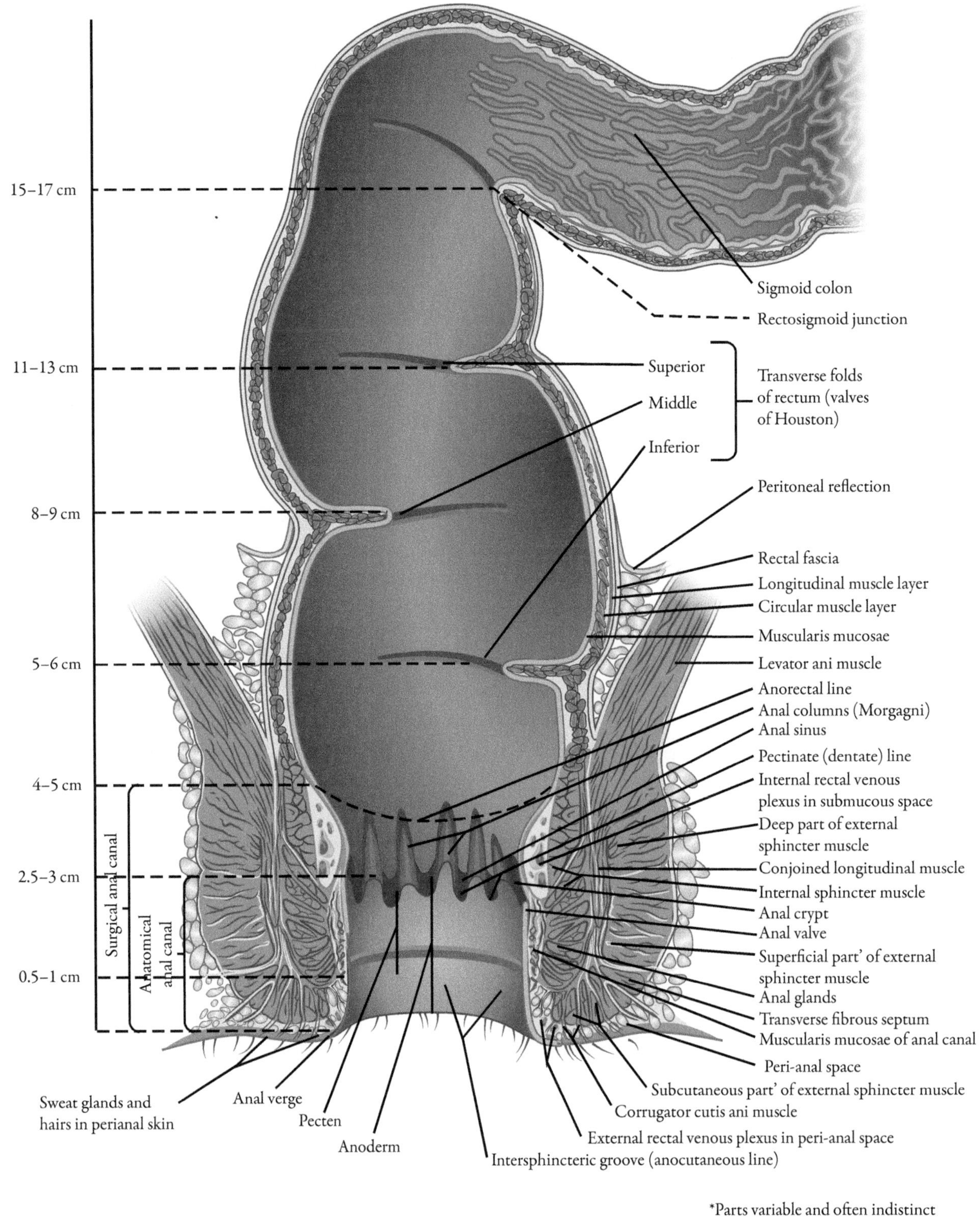

Figure 11.1 Anal canal anatomy.

intra-rectal balloon insufflation, unless the patient has Hirschsprung disease, Chagas, or an inflammatory sphincter disorder.

- **Hirschsprung disease** is usually congenital and diagnosed as a neonate; however, if a patient has only a **short segment of aganglionosis** in the rectosigmoid, they may remain constipated and undiagnosed until adulthood. The acquired version is most often associated with ***trypanosoma cruzi*** infection (Chagas disease). Diagnosis first requires noting an **impaired anorectal inhibitory reflex,** and then a **biopsy should show aganglionosis.** Treatment for select cases with a short segment in an adult patient is with **anorectal myotomy.** If that fails, or the aganglionic segment is long, the patient may need a Duhamel, Swenson, or Suave procedure (see Chapter 25, Pediatrics).
- **2. Balloon Expulsion Test**
- **3. Cinedefecography** using rectal barium with or without contrast into the vagina or bladder as well. **MRI defecography** provides superior imaging.
- Initial treatment for paradoxical puborectalis contraction or a descending perineum is with **biofeedback.**
- Treatment for obstructed defecation with an associated rectocele, sigmoidocele, or enterocele involves **plication of the anterior rectal muscles and rectovaginal septum** via a transanal, transvaginal, or transperineal incision. A levatorplasty can be added.
- Treatment for obstructed defecation with intussusception or prolapse of the rectum is with **resection and rectopexy with or without a ventral biologic mesh.**

- If PFD is not confirmed on the above studies, then check for **slow transit constipation (STC)** with a **Sitzmark study:**
  - 24 radio-opaque markers are ingested, and retention is noted by imaging at 5 days. Normal transit through the colon should be 30–40 hours.
  - If the markers are scattered throughout the colon, then the patient has generalized inertia.
  - If markers are clustered at the distal sigmoid or proximal rectum, then the patient has pelvic floor dysfunction.
  - Treatment for confirmed STC is a **total colectomy and ileorectal anastomosis.** This is complicated by postoperative **diarrhea** more commonly than recurrent constipation. A subtotal colectomy with ileosigmoid or cecosigmoid anastomoses are alternatives. Stoma construction is a last resort for either STC or PFD.
  - Some patients have a mixed STC and PFD disorder.

PROCTITIS

- **PATHOPHYSIOLOGY**
  - Rectal mucosal inflammation from a variety of etiologies. **Infectious** proctitis is usually gonorrhea, or less commonly chlamydia. **Mechanical and chemical** causes are externally introduced. An **inflammatory** etiology is usually ulcerative proctocolitis, or occasionally Crohn's.
  - **Radiation**-induced proctitis can occur after therapy for pelvic, prostatic, or rectal tumors.
    - Diarrhea and bleeding are secondary to **acute mucosal sloughing** within weeks.
    - **Chronic endarteritis and ischemic injury** occur months to years afterward. Fibrosis and poor rectal wall compliance in this situation can lead to urgency and incontinence.
- **DIAGNOSIS**
  - Proctoscopy will demonstrate lesions that may appear ulcerated, but should not be biopsied.
- **TREATMENT**
  - Start with sucralfate enemas, argon plasma coagulation, and formalin.
  - Surgery is less commonly employed. Options include a diverting colostomy, proctectomy with coloanal anastomosis, or APR in the most severe, refractory cases.
  - Radiation-induced rectal fistulae that do not close with conservative management can be resected and re-anastomosed with interposition of a healthy tissue graft in between. Upstream fecal or urinary diversion may be needed.

RECTAL PROLAPSE

- **PATHOPHYSIOLOGY**
  - Predisposing factors include a deep pouch of Douglas, multiple previous pregnancies, lack of rectal fixation via lateral tissue, a redundant sigmoid colon, laxity of pelvic floor muscles and anal sphincter, pudendal neuropathy, straining during defecation, and chronic diarrhea.
  - Rectal invagination starts 6–7 cm from the anal verge.
    - Radial folds indicate **mucosal prolapse** of the hemorrhoidal plexus alone.
    - Circular folds indicate **full thickness prolapse.** *See Figure 11.2.*
- **DIAGNOSIS**
  - Perform a **physical exam on commode** while the patient is straining to mimic a bowel movement. **Manometry** can show a weakened sphincter.

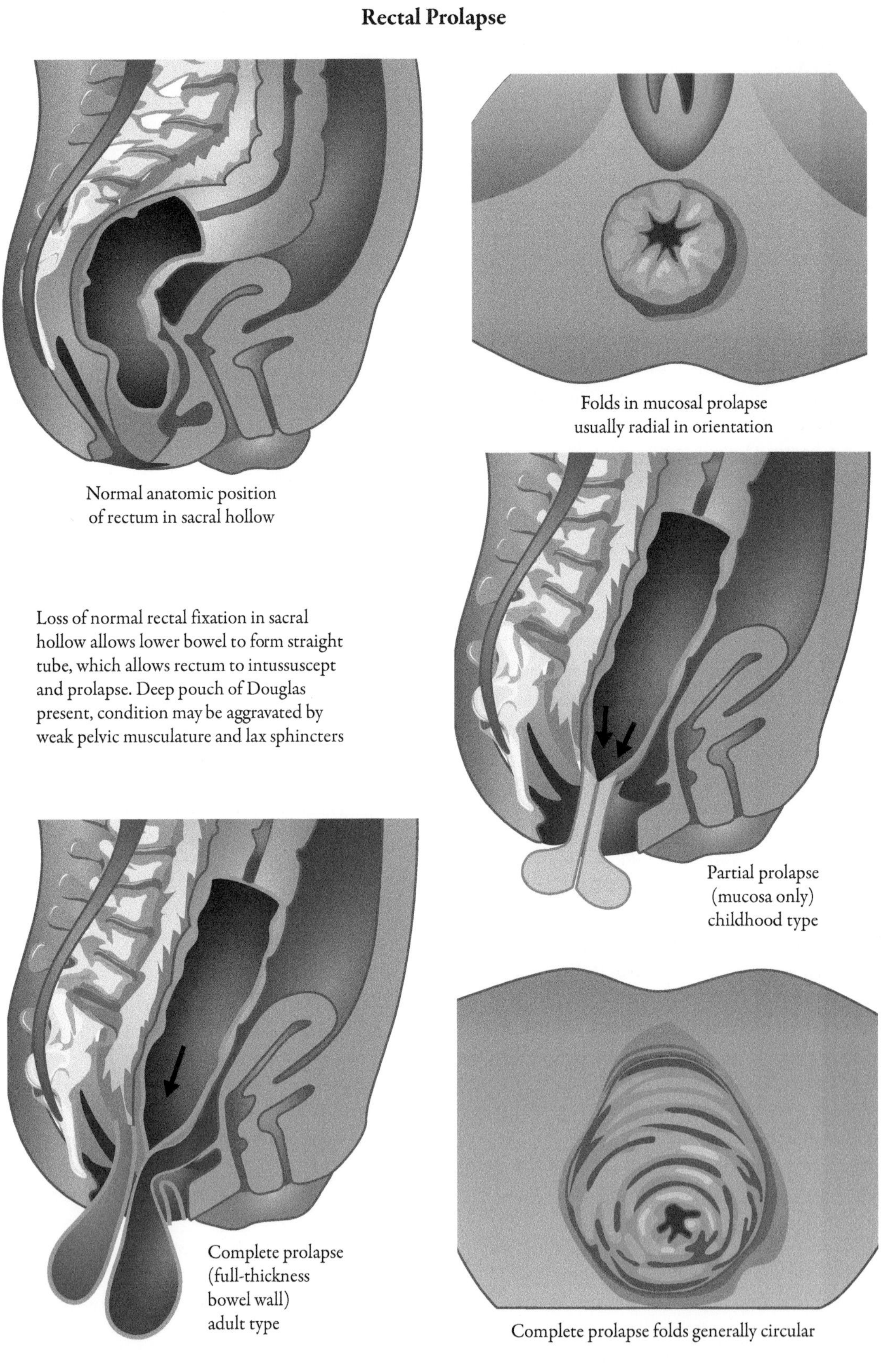

Figure 11.2 Rectal prolapse—full thickness and mucosal.

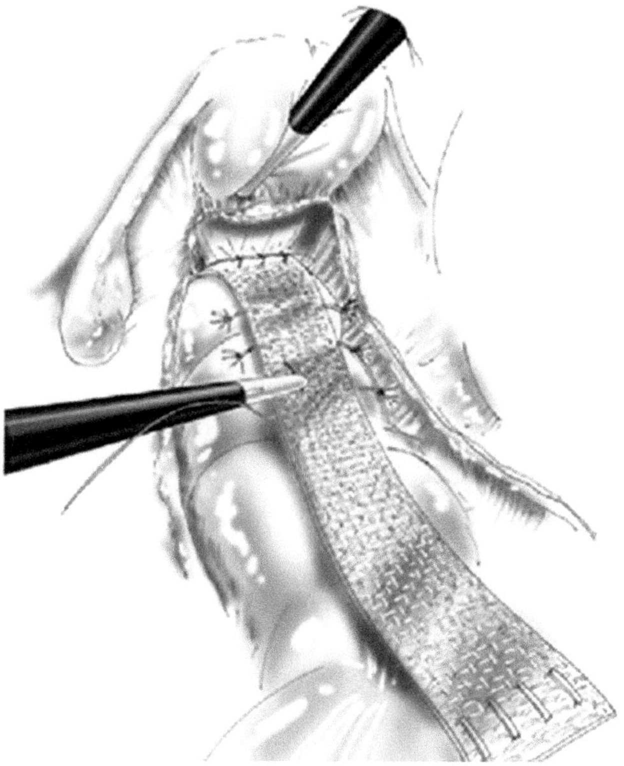

Figure 11.3 Rectal prolapse repair—ventral rectopexy.

- Contrast defecography or MRI **defecography** can diagnose an internal rectal intussusception that fails to prolapse below the anal verge ("occult prolapse").
- Patients must be evaluated for an associated cystocele, enterocele, or uterine prolapse.

- **TREATMENT**
  - Employ manual reduction in an acute presentation. Then a **high-fiber diet** followed by either transabdominal or transanal surgery.
  - Abdominal approaches have **higher complication rates but lower recurrence rates**, and should be used for patients who are good surgical candidates.
    - **1. Ventral rectopexy to the sacral promontory with or without a biologic mesh.** This requires minimal posterior rectal dissection, since there is no resection. May be complicated by future constipation. *See Figure 11.3.*
    - **2. Low anterior resection or sigmoidectomy with rectopexy with or without biologic mesh.** This is a better option if they have a history of constipation. It can be complicated by mesh infection and erosion. Transecting the lateral rectal stalks decreases recurrence rates but worsens constipation.
  - Transanal approaches have higher recurrence rates, but lower complication rates. They are reserved for the frail, elderly, and poor surgical candidates.
    - 1. **Altemeier procedure** is a full thickness resection of the redundant rectal length, with or without a levatorplasty. A stapled transanal full thickness resection is an alternative. This procedure is useful in an acutely strangulated, non-reducible, necrotic rectal prolapse.
    - 2. **Delorme procedure** is a mucosal sleeve resection extending from just proximal to the dentate line up to the most cephalad extent of redundant mucosa. The two mucosal edges are plicated to include bites of the underlying circularis muscle, which is "accordionized."
  - Recurrent rectal prolapse can be treated by:
    - 1. A second Altemeier procedure if their first procedure was perineal; or
    - 2. Rectopexy alone if there was a perineal or abdominal approach resection at the time of their first procedure.
    - 3. Rectal resection and repeat rectopexy if there was a transabdominal rectopexy alone at the time of their first procedure.

SOLITARY RECTAL ULCER

- **PATHOPHYSIOLOGY.**
  - These ulcers are found **10–12 cm from the anal verge** on the anterior rectal wall. They are associated with chronic constipation and straining.
  - The ulceration is secondary to friction from **internal intussusception of the rectum** in the setting of paradoxical puborectalis muscle contraction during defecation.
- **DIAGNOSIS.**
  - Patients present with blood and mucous output prompting colonoscopy at which time the ulceration is visualized. **Biopsy** is needed to exclude malignancy. The ulcer specimen usually has thickened muscularis mucosa and fibrosis.
  - **Defecography or endoanal US** can diagnose the internal intussusception.
  - **EMG** may suggest puborectalis hyperactivity.
  - **ENG** evaluates for pudendal nerve latency.
- **TREATMENT**
  - Start with **fiber, laxatives, enema, and biofeedback** for puborectalis hyperactivity.
  - Since there is an underlying internal intussusception ("occult prolapse"), an **abdominal resection of the redundant colorectum and ventral mesh rectopexy or a perineal resection and anastomosis** are options for cases that are refractory to conservative management.
  - There is no need to actually resect the segment of rectum containing the ulcer, provided the biopsy was negative for malignancy.

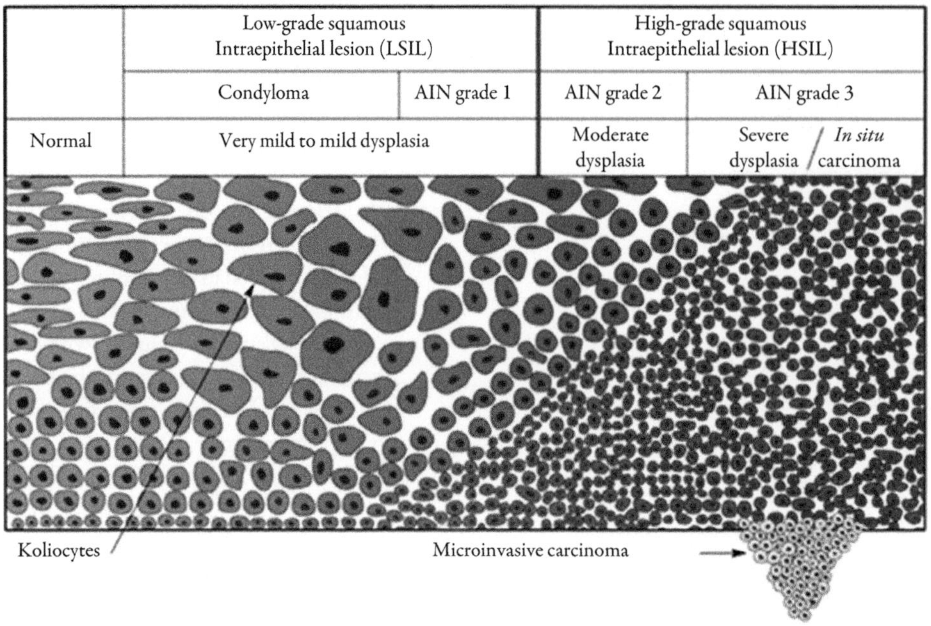

Figure 11.4 Spectrum of anal intraepithelial neoplasia.

ANAL NEOPLASIA

- **CONDYLOMA ACCUMINATA (ANAL WARTS)**
  - **PATHOPHYSIOLOGY**
    - Lesions are secondary to **HPV types 6 or 11**, and considered pre-malignant.
  - **DIAGNOSIS**
    - The majority of patients with HPV are asymptomatic, yet can transmit the virus.
    - If condylomas develop, it is usually years after the transmission event. Patients will complain of a painless growth around their anus.
    - Always check for anal canal, pharyngeal, laryngeal, vulvar, or cervical HPV, dysplasia, or neoplasia. A biopsy may be necessary to exclude dysplasia or neoplasia.
  - **TREATMENT**
    - Topical agents for small lesions.
      - **Imiquimod** induces cytokine production. It is not used in the anal canal.
      - **Podophylin and 5-FU** are anti-mitotic.
      - **Interferon alpha**, **chemical cautery** with trichloroacetic acid, **cryotherapy** with liquid nitrogen, **argon plasma coagulation**, or **fulguration** with fine tip cautery are other topical options.
    - **Tangential or local excision** is for larger, palpable lesions. This can also help exclude dysplasia or neoplasia.
    - **Recurrence rates are as high as 20%–50%**, especially if the patient is immune compromised or has HIV. Recurrences can occur at perianal or non-anal sites.
    - **Prevention** is with the **HPV vaccine** for females aged 11–26 years old and sexually active males.
    - **Buschke-Lowenstein lesion** (giant condylomata accuminata, GCA) is a pathologic intermediate between accuminata and squamous cell carcinoma (SCC). Treatment is with wide local excision (**WLE**) **with 1 cm margins.** Nigro protocol may be needed if the lesion contains malignancy.
- **ANAL INTRAEPITHELIAL NEOPLASIA (AIN)**
  - **PATHOPHYSIOLOGY**
    - These are precancerous lesions where the basement membrane remains intact.
    - Risk factors include HPV, HIV, homosexuality, immune suppression, radiation, and smoking. Lesions may be flat or ulcerated; the latter implies an increased risk of eventually finding an invasive component on biopsy.
    - **Low-grade squamous lesion** (AIN 1) has dysplastic cells limited to the lower third of the epithelium.
    - **High-grade squamous lesion** (AIN 2 or 3) has dysplasia extending into the more superficial epithelium. There is an increased risk of progression to SCC, especially in those who are immune compromised. *See Figure 11.4.*
  - **DIAGNOSIS**
    - A **pap smear** will reveal dysplastic cells. This prompts **high-resolution anoscopy (HRA) with acetic acid, or Lugol's staining to direct biopsies.**
    - In AIN, biopsies will demonstrate low- or high-grade dysplasia, but no actual malignant transformation or breach of the basement membrane.
  - **TREATMENT**
    - 1. Following with **serial examinations** and HRA and acetic acid. This option is chosen for small lesions

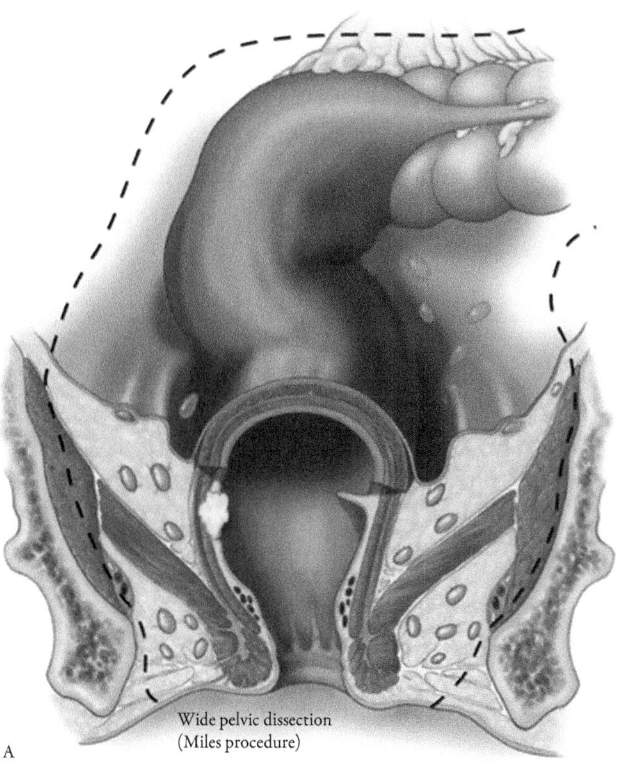

Figure 11.5 Abdominoperineal resection and boundaries of total mesorectal excision.

or patients with diffuse disease who are immune compromised, implying a high risk of recurrence should one attempt any treatment.
- 2. **Cautery ablation or topica**l therapy with imiquimod or 5-FU.
- 3. **Excision.**

- **SQUAMOUS CELL CARCINOMA IN SITU (SCC IN SITU)**
  - This is AIN 3 with cells that have progressed in appearance to malignant, but are confined by an intact basement membrane. Lesions are erythematous, scaling, crusted, burning, or pruritic. Treatment includes local excision. Some practitioners may complete four quadrant mapping biopsies at the dentate line, anal verge, and perianal skin to determine the extent of disease before proceeding to resection.

ANAL CANAL MALIGNANCIES

- These reside above the dentate line but below the rectal transition zone. Presenting symptoms include pruritus, bleeding, or a palpable mass. A lesion may also be found incidentally on sigmoidoscopy or colonoscopy.
- **ADENOCARCINOMA**
  - **PATHOPHYSIOLOGY**
    - This remains the most common anal canal cancer and has a similar etiology as rectal adenocarcinoma.
  - **DIAGNOSIS**
    - Start with a physical exam and biopsy. A **complete colonoscopy** may identify other synchronous lesions or assist in biopsy of a lesion extending into the rectum.
    - Workup may include a pelvic MRI, endoanal US, or CT abdomen pelvis or chest to define the local extent of disease and any metastases.
  - **TREATMENT**
    - **Transanal WLE** with 2–3 mm margins is adequate for low lying T1 lesions that are <3 cm in size, occupy <1/3rd the circumference of the anus, and have no high-grade features or lymphovascular invasion on biopsy.
    - **Neoadjuvant chemoradiotherapy** is used for T3 lesions or N1 status, then low anterior resection (LAR) or WLE should follow if anatomically feasible.
    - **Abdominoperineal resection (APR)** is needed if a lesion does not meet sphincter preserving resection criteria after neoadjuvant therapy. *See Figure 11.5.*
    - Adjuvant chemotherapy use is based on the same criteria as rectal adenocarcinoma.
- **SQUAMOUS CELL CARCINOMA**
  - **PATHOPHYSIOLOGY**
    - Subtypes include epidermoid, mucoepidermoid, cloacogenic, and basaloid.
  - **DIAGNOSIS**
    - Imaging workup are similar to adenocarcinoma, including a complete colonoscopy, pelvic MRI, endoanal US, or CT abdomen pelvis. This helps define the local extent of disease and metastases.
    - T2 lesions are 2 cm–5 cm in size. N1 status is ascribed to positive perirectal nodes, and N2 status given to positive inguinal or internal iliac nodes. Any concerning nodes on exam or imaging should undergo FNA.
  - **TREATMENT**
    - The **Nigro protocol** is first-line therapy. It includes 5-FU, mitomycin, and 30 Gy radiation. It may induce a complete pathologic response. Best results are obtained if the primary is <5 cm in size.
    - Neoplasia may linger from 8 weeks to 6 months post-treatment; however, this does not necessarily represent treatment failure.
    - Follow-up digital rectal exam (DRE), anoscopy, and clinical inguinal node surveillance should occur at regular intervals. About 20% of patients have a recurrence at the primary site.
    - Additional **radiation can be directed at nodal beds** outside of the initial field, or metachronous lesions at these sites. Inguinal lymphadenectomy can be considered for recurrent or nonresponsive

inguinal nodes after chemoradiation. A suprainguinal incision is used for the deep external iliac nodes, and an infrainguinal incision is used for superficial inguinal nodes.
  - **APR** is indicated if the above measures fail to control the primary site, or if SCC recurs near the primary site.
  - Metastases are treated with chemotherapy, including cisplatin and 5-FU.
- **MELANOMA**
  - Melanoma in the anal canal may not be pigmented. This is the 3rd most common site of occurrence, after skin and ocular melanoma. One-third of cases have already spread to mesenteric nodes or distant sites, including the liver and lungs, at the time of the initial presentation. Therefore, treatment is with **WLE** rather than APR, since neither would be curative. Adjuvant therapy may follow (see melanoma section in Chapter 16, Skin and Soft Tissue).

ANAL MARGIN MALIGNANCIES

- These lie distal to the dentate line, behave like other skin cancers, are treated in a similar manner to skin cancers, and have a better prognosis than canal lesions.
  - **SQUAMOUS CELL CARCINOMA**
    - This is the most common. Treatment is with **WLE and 5 mm–10 mm margins with inguinal lymphadenectomy for clinically positive nodes.** Neoadjuvant chemotherapy can be used to downstage lesions in proximity to the anal sphincter. Follow-up resection can then possibly spare continence.
  - **BASAL CELL CARCINOMA**
    - These are treated with WLE and 3 mm margins
  - **ADENOCARCINOMA in situ ("Paget's disease")**
    - Can exist either as a solitary lesion from apocrine glands or as a marker for another enteric visceral malignancy. It may presents as perianal pruritis. Workup should focus on finding a synchronous malignancy with **EGD, colonoscopy, and a genitourinary evaluation.** Treatment is with **Mohs surgery or mapping biopsies at the perineum** to determine the extent of disease, followed by **WLE with 1 cm margins.** Adjuvant radiation and chemotherapy may be required.

HEMORRHOIDS

- **PATHOPHYSIOLOGY**
  - These are dilated venous complexes that are a normal part of the submucosa of the anal canal. They contribute to the resting tone and the sensation of rectal contents. Columns reside at 3, 7, and 11 o'clock locations in the lithotomy position.
- **DIAGNOSIS**
  - Diagnosis is by physical exam and anoscopy.
  - **External** hemorrhoids are distal to the dentate line and covered by sensate squamous epithelium.
  - **Internal** hemorrhoids are proximal to the dentate line, insensate, and covered by columnar epithelium. There are 4 grades. *See Figure 11.6.*
    - **Grade 1** hemorrhoids prolapse below the dentate line, and **grade 2** prolapse below the anal verge, but reduce spontaneously.
    - **Grade 3** hemorrhoids prolapse below the verge and require manual reduction, and **grade 4** are incarcerated below the verge.
- **TREATMENT**
  - First-line therapy for enlarged hemorrhoidal columns includes **fiber, stool softeners, and Sitz baths.**
  - For external hemorrhoids that are acutely thrombosed, an **elliptical incision** at the bedside under local anesthesia is indicated. This is best done at <4 days of symptoms.
    - When a large external hemorrhoid is associated with an internal hemorrhoid proximally, the entire column may need to be excised in the operating room.
  - Grade 1, 2, and 3 internal hemorrhoids are treated by **banding or infrared coagulation.** Bands can be placed at bedside in an office visit and should encompass the mucosa associated with the internal hemorrhoid, well above the dentate line.
    - Excessive pain implies the band is close to the sensate dentate line
    - Banding may be complicated by **bleeding at 7–10 days post-op due to a necrotic slough.** This is treated with ice packs, Foley balloon, or suture ligation.
  - Grade 4 internal hemorrhoids, hemorrhoids refractory to banding, or those associated with a large external component will require **excisional hemorrhoidectomy.** After excision, the mucosal bed may be closed or left open. *See Figure 11.7.*
    - Resection may be complicated by urinary retention secondary to perineal muscle spasm, thrombophlebitis, and pelvic sepsis. The latter requires an exam under anesthesia and debridement of any necrotic mucosa.
    - Bleeding in the early post-op period should be re-explored, as it is not likely due to slough.
  - A **stapled hemorrhoidopexy** removes a circumference of mucosa proximal to a 360-degree grade 2–4 internal hemorrhoid plexus using a circular stapling device. This devascularizes the submucosal venous network and lifts everything cephalad. There is less pain, since the staple line is far from the dentate line, but there is a higher recurrence rate.

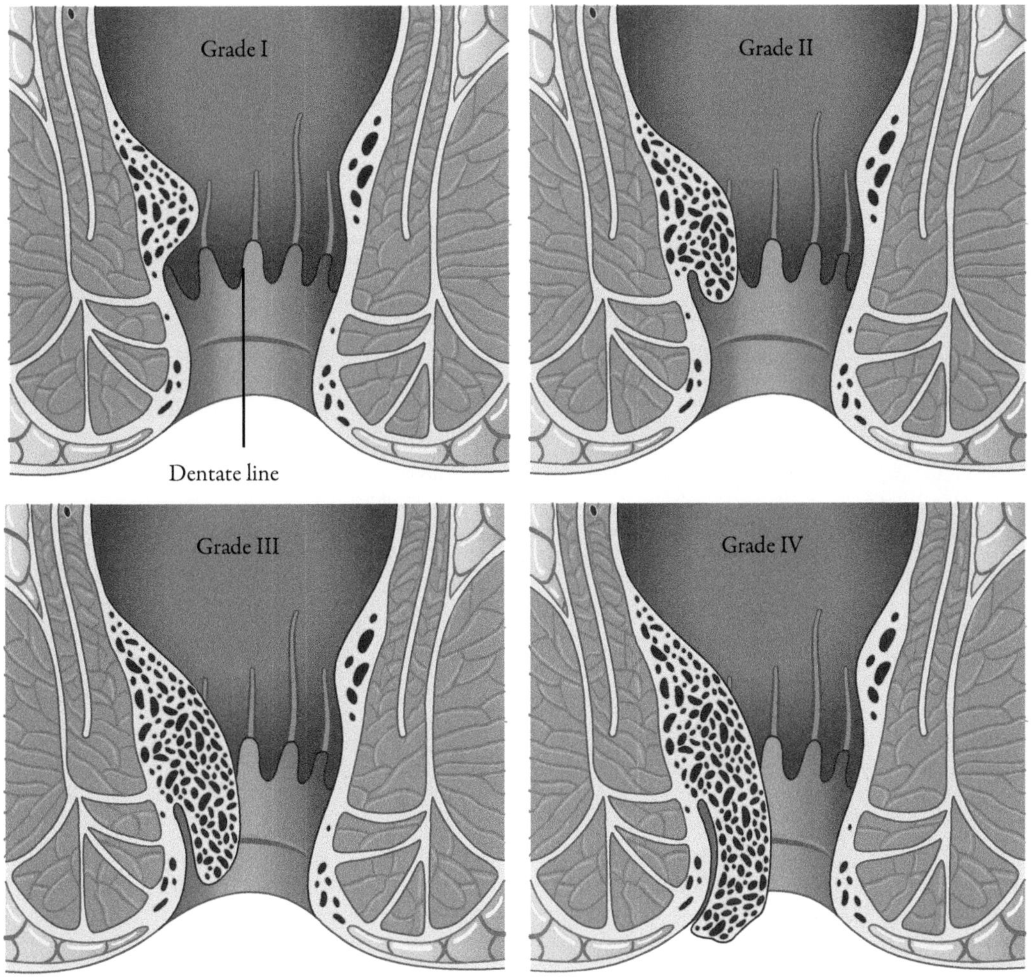

Figure 11.6 Internal haemorrhoids—Grade 1-4.

ANORECTAL ABSCESS AND FISTULAE

- **PATHOPHYSIOLOGY**
  - Abscesses arise from anal glands in the intersphincteric space when there is obstruction of their natural drainage pathways between mucosal columns into the crypts of Morgagni.
  - Pockets of purulence may extend to the dermal surface in a serpiginous manner between, through, or around sphincter muscles. If such a connection persists to be lined by epithelium, it is termed a perianal fistula.
- **DIAGNOSIS**
  - Patients present with perianal pain, induration, and erythema with or without signs of systemic inflammation.
  - If there is a subanodermal component, fluctuance may be palpable, and one may notice purulent discharge. If the patient has no dermal findings but can tolerate a DRE, fluctuance may be palpable in the anorectum.
  - CT of the pelvis with IV contrast is used to delineate confines of the abscess if it is believed to be complex and extending cephalad of any palpable evaluation at the anorectum. *See Figure 11.8.*
  - **Perirectal fistulae** are diagnosed by **probing,** endoanal ultrasound, pelvic MRI, or, less commonly, fistulography. *See Figure 11. 9.* Goodsall's rule predicts the direction a fistula would take from its external opening toward the anal canal. If the external opening is >3 cm from the anal verge, then the fistula tract takes a circuitous route to the posterior midline, regardless of the exact position of the external opening.
- **TREATMENT**
  - Abscesses
    - **Transrectal drainage** is sufficient for submucosal and intersphincteric abscesses bulging into the anorectal canal.
    - Perianal subanodermal, intersphincteric, and ischiorectal abscesses can all be **drained transdermally** via incision over the involved area, but as close as possible to the anal margin.

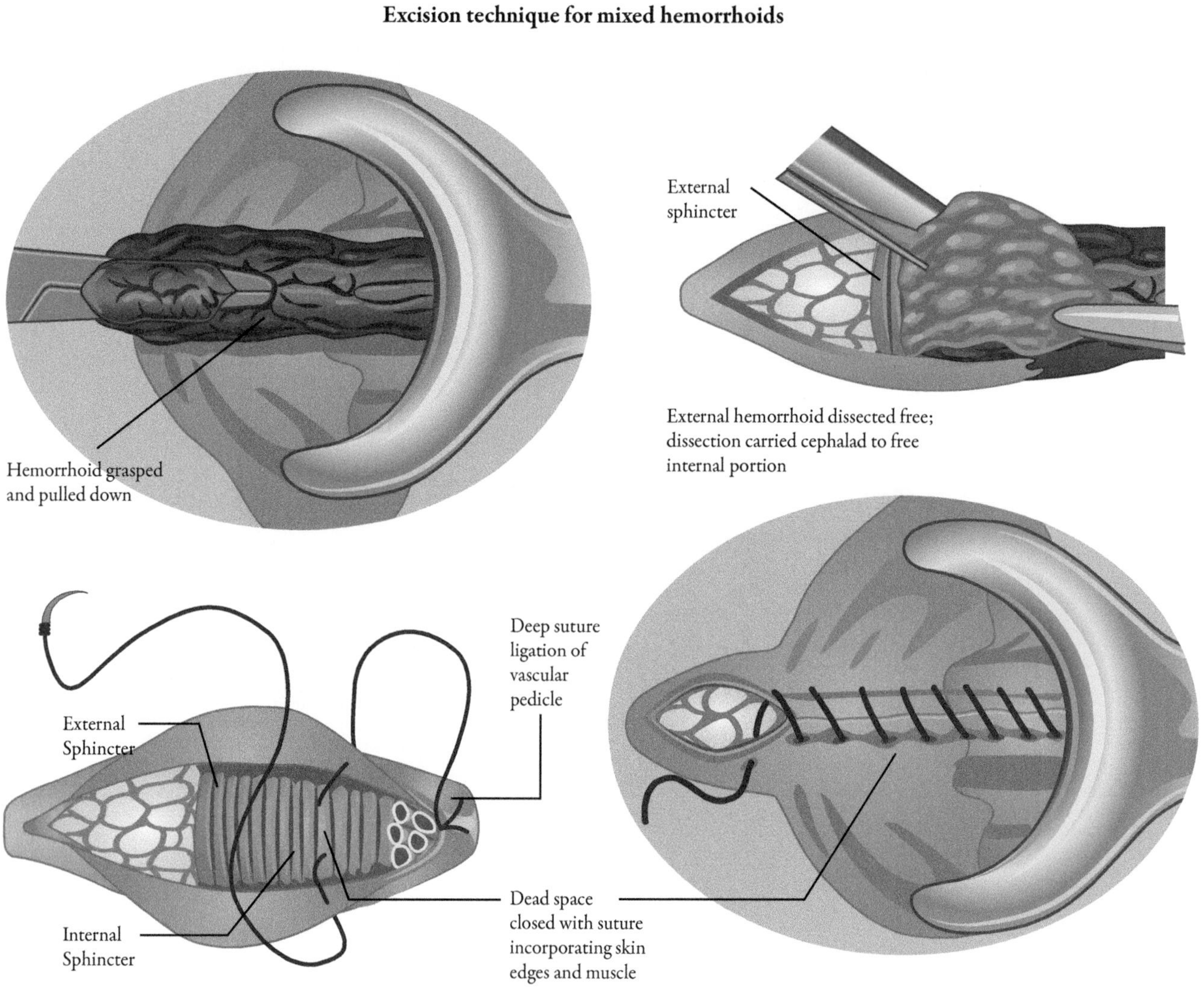

Figure 11.7 Excisional hemorrhoidectomy.

- **Supralevator abscesses** can be drained transrectally if intersphincteric in origin, transdermally if ischiorectal in origin, or transabdominally if peritoneal in origin.
- **Horseshoe abscesses** are either intersphincteric or ischiorectal and wrap around the posterior margin of the rectum. One can drain the postanal space in the midline, incising low through the sphincter mechanism and dividing the anococcygeal ligament transdermally in the posterior midline. Separate counter-incisions in each ischiorectal fossa may also be required.
- Antibiotics should be added for patients who are immune compromised or have a surrounding cellulitis. Practitioners may start with antibiotics alone if no fluid collection is present in a neutropenic patient. Empiric incision can be done thereafter if the patient is not improving.
- Fistulae
  - **Fistulotomy or fistulectomy** with marsupialization is the best option if the fistula course is subanodermal, intersphincteric, or transphincteric across the lower one-third of the external sphincter, and the patient has no other risks for incontinence.
  - If a fistula is low transphincteric, but the patient has a risk of not retaining post-op continence should a fistulotomy be done, then a **dermal advancement flap** would be a better option.
  - If the course of a fistula is transphincteric through the upper two-thirds of the external sphincter, or if it is suprasphincteric or extrasphincteric, then perform **curettage and Seton placement** with a braided thread, vessel loop, or, less commonly, a cutting suture.
    - A **rectal mucosal advancement flap** can be attempted 6–10 weeks after Seton placement, should the tract have not closed by this time. The flap should have a broad base of mucosa and internal sphincter cephalad of the internal opening of the

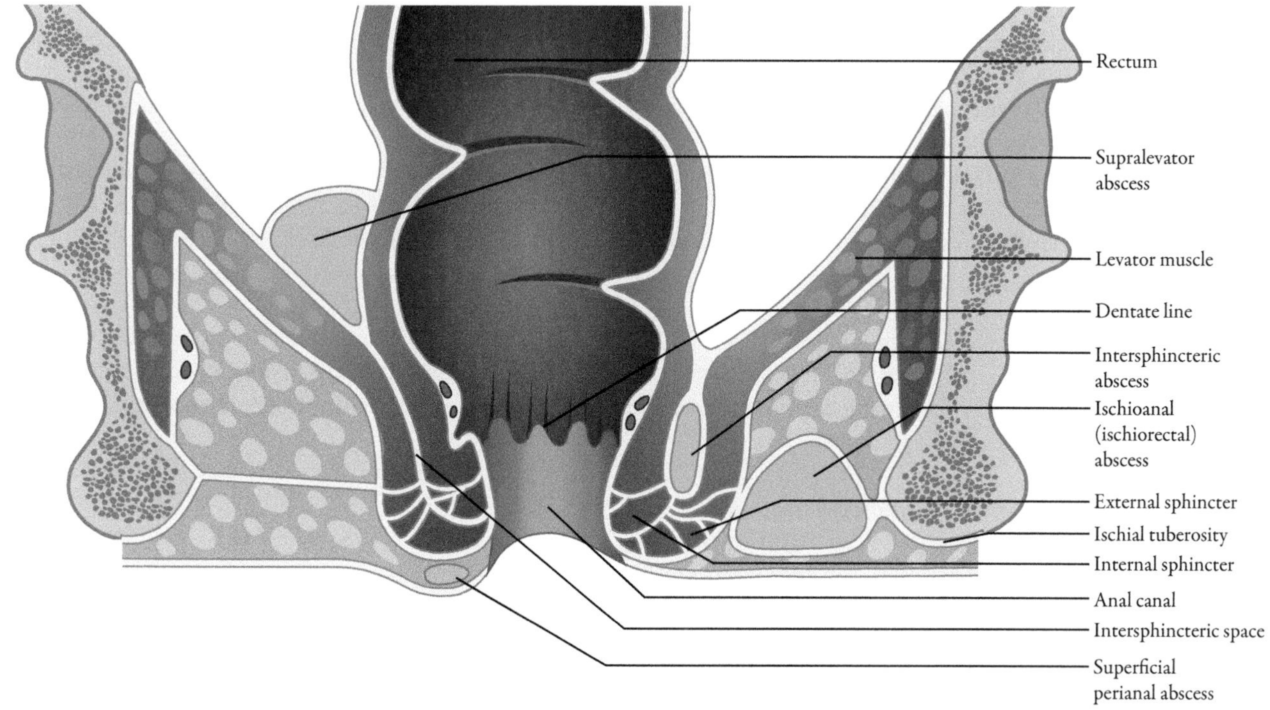

Figure 11.8 Types of anorectal abscesses.

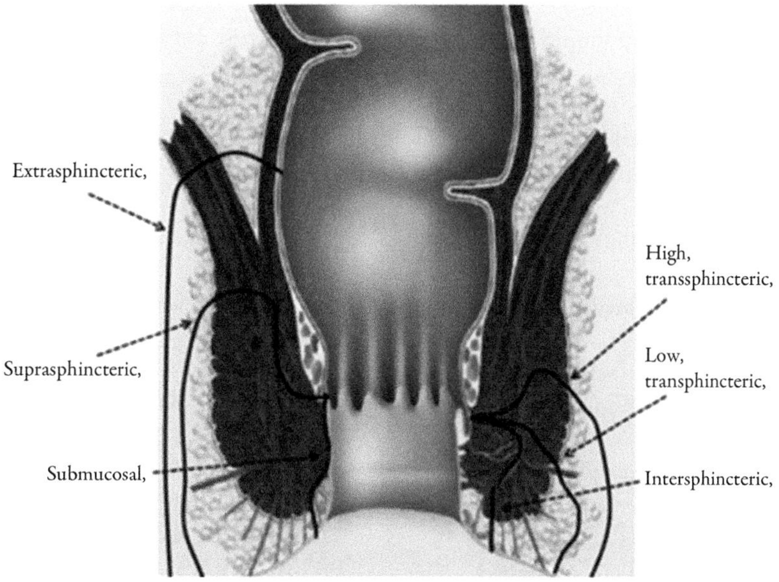

Figure 11.9 Types of anal fistulae.

fistula. Widening the dermal or vaginal opening assists with drainage.

- **Ligation of intersphincteric tract (LIFT)** is an alternative for a high transphincteric fistula. Other options for upper fistulae include inserting a fibrin plug or filling the tract with a fibrin sealant.

RECTOVAGINAL FISTULA (RVF)

- **PATHOPHYSIOLOGY**
  - RVF is a pathologic communication between the anorectal canal and the vaginal vault. It most commonly results from obstetric trauma. Other

etiologies are Crohn's disease, infection, neoplasm, or non-obstetric iatrogenic injury.

- **DIAGNOSIS**
  - This is confirmed by physical exam demonstrating rectally administered **methylene blue** contaminating a vaginal tampon. **Hydrogen peroxide** can also be used in a similar test.
  - **Proctoscopy** can evaluate the mucosa for a neoplastic, inflammatory, infectious, or radiation etiology after a methylene blue tampon test confirms the presence of an RVF.
- **TREATMENT**
  - **Start with conservative management for the first 2–3 months.** This can be extended as long as 6 months if the fistula is in the setting of prior radiation therapy. Occasionally a Seton or plug may be trialed during this time.
  - If the rectovaginal communication is persistent, and located low in the vaginal canal and near the level of sphincter mechanism on the anal side, then a **rectal mucosal advancement flap** can be attempted. This should include a broad -ased flap of mucosa and internal sphincter muscle cephalad of the anal fistula opening that is mobilized to cover the defect. The lateral edges of the internal sphincter and external sphincter should be apposed so as to provide an overlapping sphincteroplasty. Leave the vaginal opening to drain.
  - A **transvaginal advancement flap** should be considered if the patient has anorectal stenosis or fibrosis, Crohn's disease refractory to medical therapy, or if a previous rectal flap has failed.
  - One may add a formal anterior sphincteroplasty to either type of flap if needed.
  - A **perineal approach** can be attempted for failure of the above.
    - 1. An **episioproctotomy with layered closure** involves transperineal division of all tissue between the rectum and vagina, and repair of each layer separately. An interposition gracilis flap, bulbocavernosus flap (Martius flap), or mesh can be added.
    - 2. A **transperineal ligation of intersphincteric fistula (LIFT) procedure with or without anterior sphincteroplasty** is another transperineal option. This can also include an interposition gracilis flap, bulbocavernosus flap, or mesh.
    - Consider a **proximal diverting stoma** in both cases of flap interposition.
  - If a low opening is persistent despite the above interventions, or if the opening is high in the vaginal canal, constituting a complex RVF, then perform:
    - 3. A **transabdominal primary repair of both rectal and vaginal openings** with omental interposition; or
    - 4. **Transabdominal resection and anastomosis of rectum** and omental interposition between the anastamotic line and the vaginal opening.
      - Consider a proximal diverting stoma in either case.
  - If the RVF is **radiation induced, wait 6 months** before an attempt at any perineal or abdominal repair with interposition.
  - If secondary to **Crohn's disease,** assure **medical therapy** has obtained quiescence of the disease process, and then consider a **vaginal flap** instead of a rectal flap. Proctectomy with an end stoma is a last resort.
  - If secondary to **squamous carcinoma,** start with the **Nigro chemoradiation protocol**, and perform **APR or pelvic exenteration** if the patient is a candidate afterward.

CROHN'S OF THE ANUS

- Skin tags at the anal verge and sentinel piles near the dentate line are the most common manifestations of Crohn's disease of the anus. Do not excise either of these.
- Another presentation is with an off-midline fissure. These should be managed non-operatively.
- Any abscesses that develop still need to be drained.
- Fistulae secondary to Crohn's disease can be treated with a Seton, antibiotics, and infliximab.

ANORECTAL FIBROTIC STRICTURE

- **PATHOPHYSIOLOGY**
  - Strictures are secondary to trauma, malignancy, inflammatory disorders, or prior surgery with loss of anoderm (excisional hemorrhoidectomy).
- **DIAGNOSIS**
  - Symptoms include pain and resistance with defecation. DRE may be limited by a narrow aperture. Exam under anesthesia (EUA) relaxes the sphincter and opens the aperture if the stenosis was primarily from muscular hypertrophy. However, it will not facilitate DRE in true fibrotic stricture.
- **TREATMENT**
  - Includes a trial of fiber, increased water intake, stool softeners, and **Hegar dilation. Transanal strictureplasty** can be done for anastomotic strictures. A rectal mucosal **advancement flap** is used for a proximal stricture. A dermal (V-Y, Y-V, house, or S) advancement flap is used for a distal stricture. Radiation, DM, Crohn's, immune suppression, and smoking are associated with poor outcomes despite intervention.

ANAL INCONTINENCE

- **PATHOPHYSIOLOGY**
  - Etiology includes laxity of the levator muscles and sphincter complex. This is more common in obese patients and multiparous females. Abnormalities to neural input to the sphincter muscles or obstetric trauma with discontinuity to the anterior component of the sphincter complex are other causes.
- **DIAGNOSIS**
  - This first requires **exclusion of rectal prolapse** or a canal mass by Valsalva physical exam and defecography. An **endoanal US** can then evaluate if the sphincter mechanism is intact.
- **TREATMENT**
  - If the sphincter muscles are intact, then treatment includes **fiber** for stool bulk plus **constipating agents** like codeine, loperamide, or diphenoxylate atropine. **Biofeedback** is also useful for patients with an intact sphincter and no prolapse. A **sacral nerve stimulator** with a tined lead at the S3 nerve root is reserved for patients refractory to the above who have passed a percutaneous trial of this therapy.
    - Other less commonly used options include gel injections into the submucosa above the dentate line or implantation of an artificial sphincter.
  - If the sphincter mechanism is discontinuous secondary to obstetrics, surgery, or trauma, then an anterior **overlapping sphincteroplasty** with or without a posterior intersphincteric levatorplasty can be done 3–6 months post-injury.

ANAL FISSURE

- **PATHOPHYSIOLOGY**
  - Fissures are a posterior midline split in the anoderm due to relative ischemia in this area. 20% of fissures are anterior. Women are more commonly affected than men.
- **DIAGNOSIS**
  - Symptoms include pain and bleeding with defecation. A posterior midline breach in the dermis may be visible on inspection. Digital rectal examination can be limited by pain.
    - When present for >6 weeks, fissures can be associated with hypertrophic papillae at the dentate line, and a skin tag or sentinel pile at the anal verge.
- **TREATMENT**
  - Start with **Sitz baths, fiber, and stool softeners for the first 6 weeks.**
  - **Topical** lidocaine, diltiazem, or nitroglycerin can be added to relax the sphincter muscles and ease the tension on the fissure with passage of stool.
  - If the above conservative measures fail, then a **lateral subcutaneous internal sphincterotomy** is the next option. This addresses any hypertonicity to the sphincter that is contributing to stool retention, stool desiccation, and subsequent exacerbation of the fissure.
    - An open sphincterotomy incises through the anoderm along with internal sphincter.
    - A closed sphincterotomy leaves the anoderm and mucosa intact while incising through the internal sphincter up to the cephalad point of the fissure. *See Figure 11.10.*
    - There is a 5%–15% fecal incontinence rate postoperatively.
  - A **chemical sphincterotomy** is when botulinum toxin is injected at the 10 o'clock and 2 o'clock intersphincteric positions anteriorly. It is used if surgical sphincterotomy poses a high risk for incontinence, or if the patient may have Crohn's disease. Relaxation of the sphincter muscles can last up to 3 months, during which time the fissure may heal.
  - If refractory to the above measures, one must **rule out Crohn's disease and malignancy**. Thereafter, consideration can be given to either:
    - **Repeat lateral internal sphincterotomy** on the opposite side, if the sphincter tone is still high after the initial sphincterotomy.
    - **Fissurectomy with dermal advancement flap** if the sphincter was hypotonic after the initial sphincterotomy.

OTHER PATHOLOGY

- **PRURITIS ANI**
  - Perianal pruritus with or without visible lesions. Causes include the following:
    - **Idiopathic** from pruritogenic foods or poor anal hygiene leading to fecal anodermal irritation.
    - **Secondary** from infection (including skin bacteria, sexually transmitted disease, fungal organisms, or pinworm), contact or allergic dermatitis, neoplasm (including squamous carcinoma or adenocarcinoma in situ), any anorectal anatomic or inflammatory pathology resulting in fecal irritation of the anoderm, or lichen sclerosis which classically presents in a perimenopausal woman with vulvovaginal symptoms as well.
  - Treatment is with **hygiene**, avoidance of irritants, application of topical glucocorticoids for up to 4 weeks, and treatment of any underlying contributing disorder.
- **HIDRADENITIS SUPPURATIVA (HS)**
  - **PATHOPHYSIOLOGY**
    - HS is secondary to obstruction of follicles in the perianal skin causing pilosebaceous unit rupture and adjacent inflammation. It does not involve the

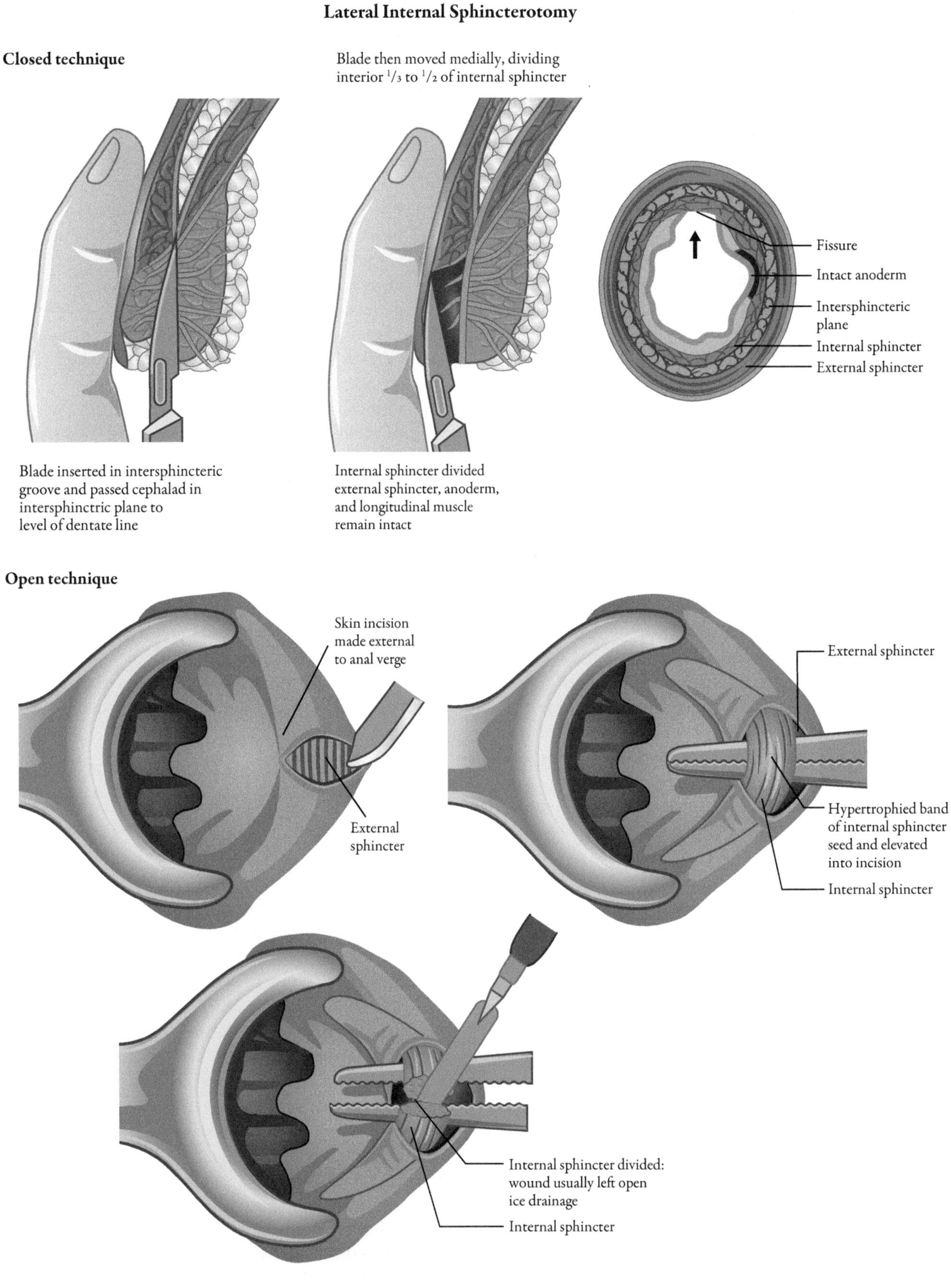

Figure 11.10 Lateral internal sphincterectomy--open and closed techniques.

anal margin or canal, since pilosebaceous units are not present here. Risk factors include smoking, alcohol, and infection with staphylococcus or streptococcus.

- **DIAGNOSIS**
  - Diagnosis is by physical exam. Staging is based on the extent of abscess and sinus formation:
    - Hurley 1: singular abscesses that are not connected.
    - Hurley 2: multiple separate abscesses with sinus tracts.
    - Hurley 3: confluence of multiple abscess cavities and their sinus tracts.
- **TREATMENT**
  - Start with antibiotics, antiandrogens, zinc, biologic modifiers (adalimumab), and incision and drainage of any acutely infected abscess. Advanced cases need **WLE** with healing by secondary intention or coverage with a skin graft or flap.
- **PILONIDAL CYST**
  - This is cyst formation in the midline, or just off midline, of the superior gluteal cleft from an ingrown hair. There may be a deeper sinus connection to sacrococcygeal structures
  - Treatment for **mild cases** includes shaving or laser hair removal, curettage of small sinus tracts, or excision of small pits and closure in the midline or off midline. If secondarily infected, incision and drainage via an off-midline incision should be performed.
  - For **extensive disease** there are two options:
    - 1. Excise and leave the wound open, or marsupialize for secondary intent healing. This has the lowest infection rate and hence lowest recurrence rate.
    - 2. Excise and construct an off-midline flap with tension-free closure which flattens the natal cleft (cleft procedure, rhomboid flap).

12.

LIVER

ANATOMY AND PHYSIOLOGY

- The inferior border of the **falciform ligament** is called the **round ligament,** and contains the obliterated umbilical vein. As a fetus, the umbilical vein coursed through this area and met with the left portal vein. Posteriorly, the vascular channel continued as the ductus venosus, draining into the left hepatic vein. As an adult, the residua of the ductus venous is known as the **ligamentum venosum.**
- **Hepatic Arteries (HA)**
  - A replaced right hepatic artery (RHA) may come from the superior mesenteric artery and run posteriorly within the portal tract. If this artery exists in addition to the normal right and left hepatic arteries, it is known as an accessory right hepatic artery.
  - A replaced left hepatic artery (LHA) may come from the left gastric artery in 20% of patients.
  - The middle hepatic artery branches off the LHA.
- **Bile Ducts.** Aberrant anterior or posterior divisions of the right hepatic bile duct may arise inferiorly along the left hepatic duct. Alkaline phosphatase comes from the biliary canalicular membrane and has a half-life of 7 days.
- **Portal vein (PV) branches run horizontally.** The left courses extrahepatic and feeds left lobe segments 2, 3, and 4. It does not perfuse the caudate. The right runs intrahepatic and feeds right lobe segments 5, 6, 7, and 8.
- **Hepatic vein (HV) branches run vertically.** The middle HV drains 4b and 5, and merges with the left HV, which drains 2, 3, and 4a. This merger occurs very close to their confluence with the inferior vena cava (IVC). *See Figure 12.1.* Accessory veins, draining the medial right lobe, and inferior phrenic veins may be present and drain directly into IVC.

- Cantlie's line runs from the gallbladder fossa to the IVC and marks the right and left lobe borders.
  - **Left Lobe Segments**
    - Caudate 1 has its own arterial, portal and venous branches.
    - Superolateral 2 and inferolateral 3: Both 2 and 3 lie lateral to the left hepatic vein. Segment 2 is superior to the left portal vein, and segment 3 sits inferior to it.
  - **Right Lobe Segments**
    - Medial segment contains 4a superiorly and 4b inferiorly. They lie between the middle and left HV.
    - Inferior anteromedial segment 5 sits below the right portal vein, and the superoanteromedial segment 8 resides above the right portal vein. Both lie between the right and middle HV.
    - Inferior posterolateral segment 6 is below the right portal vein and superior posterolateral segment 7 sits above it. Both lie lateral to the right HV.
- **Hepatocytes**
  - Drug and compound metabolism is in 2 phases. Phase 1 is **cytochrome p450 oxidation, reduction, and hydrolysis.** Phase 2 includes **conjugation and subgroup addition** to make hydrophilic compounds for excretion.
  - Nutrient uptake occurs across the sinusoidal membranes. **Kupffer cells** line the sinusoidal membrane. **Stellate cells** reside in the space of Disse, between sinusoid and the hepatocyte membrane. They mediate liver response to injury and fibrosis. *See Figure 12.2.*
  - Hepatocytes store glycogen and release glucose for systemic needs. They also store fat-soluble vitamins and water-soluble B_{12}. They synthesize urea and all coagulation factors except VIII and VWF, which are in the endothelium. Hepatocytes can use ketones for energy.

IMAGING

- **Ultrasound (US)** may be contrast enhanced or accompanied by elastography and is the primary modality for screening for liver pathology. **Intraoperative US** has the highest sensitivity for hepatic lesions <5 mm, even if such lesions reside deep within the parenchyma.
- **Abdominal CT** with IV contrast is done in three phases when evaluating the liver. -0\Hepatic artery (HA)

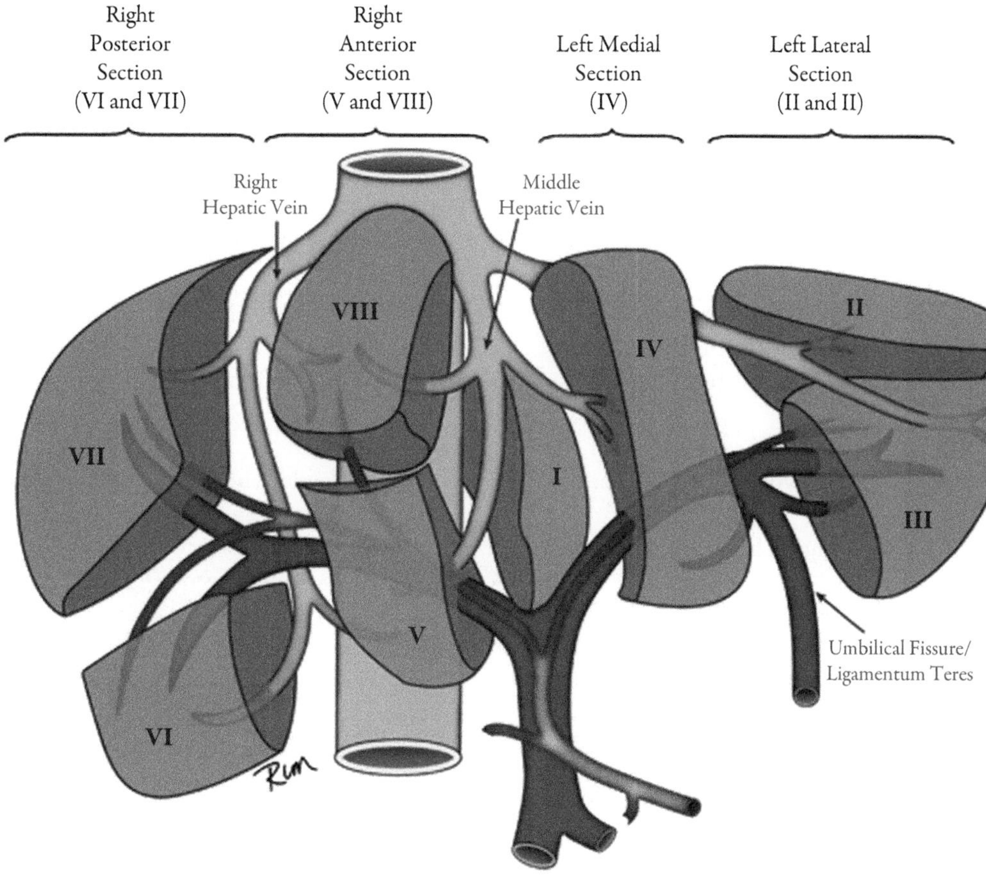

Figure 12.1 General hepatic anatomy including bile ducts, portal vein, and hepatic artery branches.

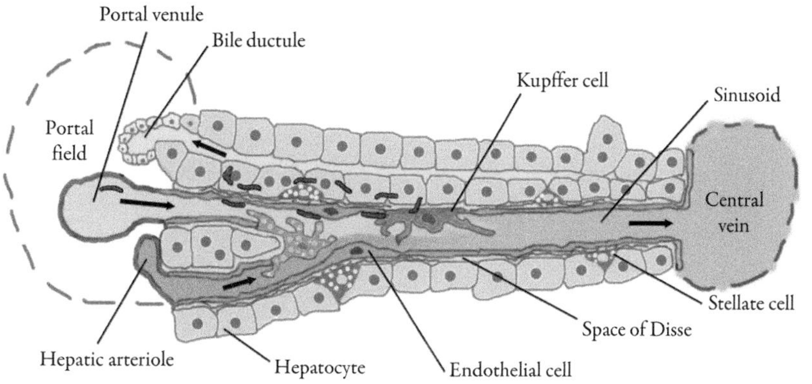

Figure 12.2 Hepatocyte anatomy showing cellular architecture and relation to venous sinusoids and biliary canaliculi.

phase is 20–30 seconds after injection. This enhances hepatocellular carcinoma (HCC), adenoma, and the periphery of a hemangioma. Portal vein (PV) phase is 60–70 seconds after injection. This washes out HCC, enhances metastases, and also enhances the central component of a hemangioma.

- **Magnetic resonance imaging (MRI)** includes T1 and T2 weighted images. It may be accompanied by gadolinium, which enhances Kupffer cells. **Magnetic resonance cholangiopancreatography (MRCP)** is a version of T2 imaging that defines the biliary tree. No contrast is used. It may be accompanied by elastography. **Gadoxetate (Eovist)** is excreted into, and enhances, biliary ducts at 20-minute delayed imaging. It also accentuates normal liver to differentiate it from colorectal metastases and to confirm the diagnosis of FNH.

- **Technetium 99m colloid scan** (aka, liver scan) accumulates in Kupffer cells in the sinusoidal lining.

HEPATITIS

- **DIAGNOSIS**
  - Patients generally present with right-sided abdominal pain, tenderness over the liver margin, and accompanying nausea or vomiting. Jaundice or icterus may be noticeable. Transaminases are elevated, but liver synthetic function (INR, bilirubin, albumin) should be preserved.
- **HEPATITIS B** is a DNA virus and responsible for the majority of infectious hepatitis worldwide. Most cases resolve on their own.
  - The presence of surface antigen (HBsAg) indicates active or chronic infection.
  - The antibody to core antigen (Anti-HBc) is elevated first in a patient's immune response, and indicates initiation of recovery from acute Hepatitis B infection.
  - The antibody to surface antigen (Anti-HBs) is detectable after recovery from an acute Hepatitis B infection or after vaccination.
  - **Treatment** of select patients is with pegylated interferon (PEG-IFN) and nucleoside analogs (tenofovir, entecavir). Immune globulin (HBIG) and vaccination are given for workplace exposures in employees who have inadequate titers or an insufficient vaccination history.
- **HEPATITIS C** is an RNA virus and is more aggressive than Hepatitis B. 70% of individuals end up with a chronic infection, some of which progress to cirrhosis or hepatocellular carcinoma (HCC).
  - Polymerase chain reaction (PCR) is positive within days of infection. Antibodies can be detected 2–6 months afterward.
  - For those with chronic infection, practitioners may screen for HCC with serial US.
  - **Treatment**
    - Historically PEG-IFN, ribavirin, and protease inhibitors were used. Now the combination drug Ledipasvir-Sofosbuvir (Harvoni) has had greater success.
    - In the setting of a workplace exposure, there is no post-exposure prophylaxis. Instead, the exposed worker should be monitored with PCR and antibody testing. If they turn positive and symptoms develop and fail to resolve, then treatment may be initiated.
- Fulminant hepatic failure may result from either **hepatitis D,** which can coinfect with B, or **hepatitis E,** which afflicts women in the third trimester.
- **Acetaminophen** toxicity is the most common cause of drug-induced hepatitis. It can be treated with N-acetylcysteine. If fulminant failure ensues, the patient will need a liver transplant.
- **Acute Fatty Liver of Pregnancy (AFLP)** is a deficiency of B-oxidation of fatty acids resulting in failure of hepatic synthetic function. Patients will have an increased conjugated bilirubin and decreased glucose. Treatment is delivery of the fetus.

LIVER FAILURE COMPLICATIONS

- **ACUTE FULMINANT HEPATIC FAILURE (FHF)**
  - The rapid loss of liver synthetic function in less than 8 weeks.
  - **DIAGNOSIS**
    - Patients develop **encephalopathy, coagulopathy, and jaundice.** They do not necessarily generate portal hypertension, ascites, or dermal vascular prominence. Encephalopathy will generally not respond to standard therapy.
    - Causes include acetaminophen overdose, idiosyncratic drug reaction, Hepatitis B, D, E, or an autoimmune attack on hepatocytes.
  - **TREATMENT**
    - Treatment requires emergent liver transplantation unless herniation and irreversible neuronal damage has already set in. N-acetylcysteine (NAC) 140–150 mg/kg bolus followed by 50–70 mg/kg q4 for 12–17 doses can be used in the setting of acetaminophen toxicity.
- **SUBFULMINANT HEPATIC FAILURE**
  - Loss of synthetic function over 8–26 weeks.
- **CHRONIC CIRRHOTIC HEPATIC FAILURE**
  - Associated with portal hypertension, ascites, episodes of encephalopathy, and dermal vascular prominence (varicosities or telangiectasias). Etiologies include nonalcoholic fatty liver disease (NAFLD) progressing to nonalcoholic steatohepatitis (NASH) and then to cirrhosis. Hepatitis C, alcohol, and autoimmune diseases are other causes
- **HEPATIC ENCEPHALOPATHY**
  - Due to a buildup of ammonia compounds and false neurotransmitters from enteric bacterial production, enteric intake, or skeletal muscle breakdown. This can be exacerbated by volume contraction alkalosis, hypokalemia, benzodiazepines, or other sedatives that increase GABA transmission.
  - **DIAGNOSIS**
    - Patients may present with confusion, asterixis, or a decrease in level of consciousness in the setting of acute or chronic liver disease. Ammonia levels may be elevated.
  - **TREATMENT**
    - **Embolization** of any shunts.
    - **Lactulose** 30 ml PO every few hours, or a 300 ml retention enema, to acidify colon contents and trap

NH4+ promoting its catharsis in the form of 2–3 soft bowel movements daily.

- **Rifaximin** 550 mg PO BID can be used in combination with lactulose to decrease enteric flora ammonia production.
- Polyethylene glycol (PEG) 4 liters PO helps with catharsis.
- Adjunct therapy includes limiting protein intake and using branched chain amino acids, which are metabolized by muscle instead of hepatocytes.

- **ASCITES OR HEPATIC HYDROTHORAX**
  - Transudative sequelae of pre-, intra-, or post-sinusoidal obstruction.
  - **DIAGNOSIS**
    - Patients will have abdominal distension, shifting dullness to percussion, and a fluid wave. They should lack focal abdominal pain or guarding.
  - **TREATMENT**
    - **sodium restriction** (1.5–2 g/day, versus the normal intake of 4–6 g/day) concurrently with a combination **diuresis** of furosemide (40–160 mg/day) and spironolactone (100–400 mg/day) to counteract the hyperaldosteronism common to liver failure.
    - **Free water restriction** is reserved for patients that are hyponatremic.
    - A **paracentesis** is the most rapid way to remove large volumes of ascites and diagnose peritonitis. If truly a large-volume paracentesis, it can be accompanied by albumin infusions of 1gm per 100 ml of ascites removed.
    - A **transjugular intrahepatic portosystemic shunt (TIPS)** can be placed if a patient is needing a large-volume paracentesis more often than every 2–4 weeks. (see details of the procedure in the varices section below). *See* Figure *12.3*. It may take 2–4 weeks before ascites resolves after shunt placement. Patients should receive spontaneous bacterial peritonitis (SBP) prophylaxis.
    - **Surgical shunts** are less commonly used for ascites. These include H-type mesocaval shunts or a splenorenal shunt.
    - Refractory ascites is defined as either:
      - Resistant to a maximal dosage of diuretics and sodium restriction; or
      - Intractable in patients who cannot tolerate the side effects of diuretics including azotemia, hyponatremia, and encephalopathy.
    - Repair any umbilical hernia after medical optimization and minimization of ascites. This should be done before tension from the protruding hernia causes a dermal breach that drains externally and increases the infectious risk. The latter is a surgical urgency and should be primarily repaired expeditiously.
- **SPONTANEOUS BACTERIAL PERITONITIS (SBP)**
  - This is generally monomicrobial from *E. coli* or *Klebsiella*. Staphylococcus and Streptococci are more common in nephrotic children.
    - Risk factors include a history of SBP, low protein ascites (<1 mg/dl), or an upper gastrointestinal bleed (UGIB) secondary to varices.
  - **DIAGNOSIS**
    - Patients will have abdominal pain, distension, and tenderness with or without a fever. A paracentesis demonstrating **>250 polymorphonuclear leukocytes (PMNs)** is diagnostic. Cultures should eventually grow a single organism.

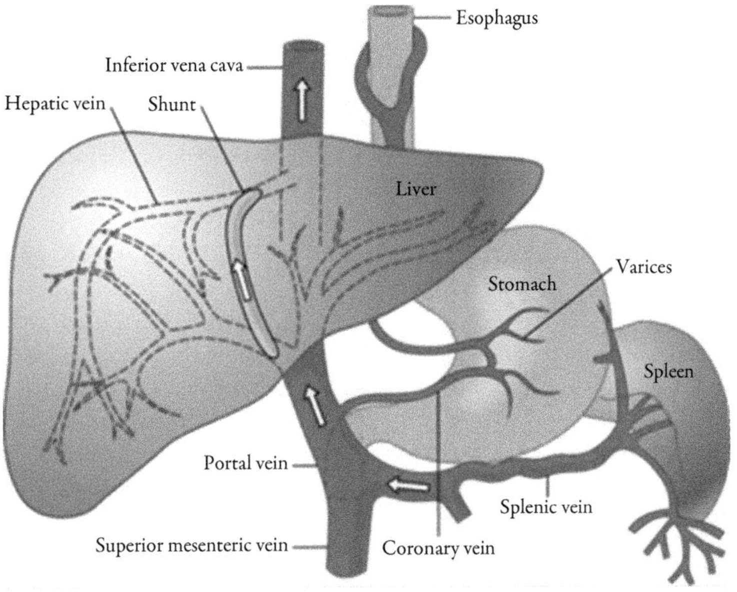

Figure 12.3 Transjugular intrahepatic portosystemic shunt (TIPS) bypassing liver parenchyma.

- If the patient has guarding or a paracentesis, demonstrates enteric material, polymicrobial growth, any anaerobe, or an increased ascitic protein, high LDH, or low glucose, one should consider **secondary bacterial peritonitis.**
- **TREATMENT**
  - 3rd-generation cephalosporin with or without albumin infusion.
  - **Prophylaxis** is with a quinolone antibiotics for those with a low ascitic albumin, or ceftriaxone for patients with active variceal bleeding.
  - Any patient not responding to therapy should have further testing for secondary peritonitis and possible enteric inflammation or perforation that has inoculated the patient's baseline ascites.
  - If a **peritoneal dialysis catheter** is in place, then **>100 PMNs** is positive for SBP. Staph epidermidis is the most common organism in the setting of a foreign body. Treatment is with **intraperitoneal antibiotics,** rather than systemic. Removal and replacement of the catheter is only needed if there is fungal growth, multiple episodes of bacterial peritonitis, or a soft tissue infection along the tract that will not clear. Repeated infection is a reason to consider conversion to hemodialysis.

- **PORTAL HYPERTENSION**
  - Pressure within this venous system exceeding the normal range of 3–7 mmHg. Patients can present with dermal vascular prominence, esophageal varices, prominent upper rectal veins, ascites, or splenomegaly.
    - Portal pressure is measured by a transjugular catheter transducing free hepatic vein pressure, and then balloon occlusion of the hepatic vein to transduce pressure in continuity with the portal vein inferiorly, devoid of the inferior vena cava (IVC) back pressure superiorly. This value is termed the **wedged hepatic vein pressure** and equals the portal venous pressure.
    - The difference between the balloon and non-occluded pressures is the hepatic vein pressure gradient, which equals the gradient between portal veins and IVC.
  - A hypertensive portal system normally decompresses into the systemic venous system via the following:
    - 1. Coronary veins and right gastric vein into the azygous.
    - 2. Left gastroepiploic vein or short gastric veins into azygous; and
    - 3. Recanalization of the umbilical vein.
  - Categories of portal hypertension are as follows.
  - **Presinusoidal**
    - **Portal vein thrombosis** presents with a dull ache and low volume ascites. It is due to an underlying hypercoagulable condition or myeloproliferative state, stasis from cirrhosis, or a prior neonatal umbilical vein catheter. Treatment is with anticoagulation with or without TIPS, or a superior mesenteric vein (SMV) to left portal vein (LPV) shunt.
    - **Schistosomiasis** is another cause of presinusoidal hypertension via fibrosis of intrahepatic portal tributaries.
  - **Intrasinusoidal** portal hypertension is most commonly from cirrhosis.
  - **Post-sinusoidal** portal hypertension is from right heart failure or Budd Chiari syndrome (BCS).
    - BCS is hepatic venous thrombosis in two of the three veins, usually due to an underlying myeloproliferative disorder (polycythemia, essential thrombocytosis, paroxysmal nocturnal hemoglobinemia) or hypercoagulable state (factor V Leiden or antiphospholipid syndrome).
    - Patients may present with an enlarged and tender liver with or without ascites. They usually have preserved hepatocyte synthetic function. **Doppler US** demonstrates hepatic vein thrombus. CT with triple phase contrast or MRI are other options. Caudate hypertrophy is common, since venous flow heads into this lobe to drain directly into the IVC. Regenerative nodules may be seen if chronic. A transjugular hepatic venogram would reveal a spider-like appearance of multiple collaterals with no discrete hepatic vein.
    - Treatment is with **anticoagulation** initially. Local thrombolytics can be considered in the acute setting.
      - If the vein obstruction persists, and the patient is not in acute fulminant liver failure and has a MELD <15, the next step would be **TIPS** or DIPS (a transjugular, transcaudate, direct shunt between the IVC and main portal vein). If this fails, a **side–side mesocaval shunt** can be constructed.
      - If the patient has fulminant liver failure from BCS or has underlying cirrhosis with MELD >15, then **transplantation** is the best option.
- **ESOPHAGEAL VARICES**
  - Seen in 50% of cirrhotics. One-third of these patients will have an UGIB. *See Figure 12.4.*
  - **DIAGNOSIS**
    - Made either at endoscopy for UGIB, or via screening endoscopy done for chronic liver disease or other indications.
  - **TREATMENT**
    - Endoscopic **suction cup banding** is primary or secondary prevention. Sclerotherapy is used if bleeding is refractory to banding. **Octreotide** drip should be started in conjunction with endoscopy for active bleeding.
    - **SBP prophylaxis antibiotics** are indicated if the patient also has ascites.
    - A **Sengstaken-Blakemore tube or Minnesota tube** (which has an extra proximal port to siphon oral secretions) are used for intraluminal compression in the emergent setting. Inflate the gastric port and place it on traction. The distal port can siphon any residual bleeding. The esophageal port can be inflated

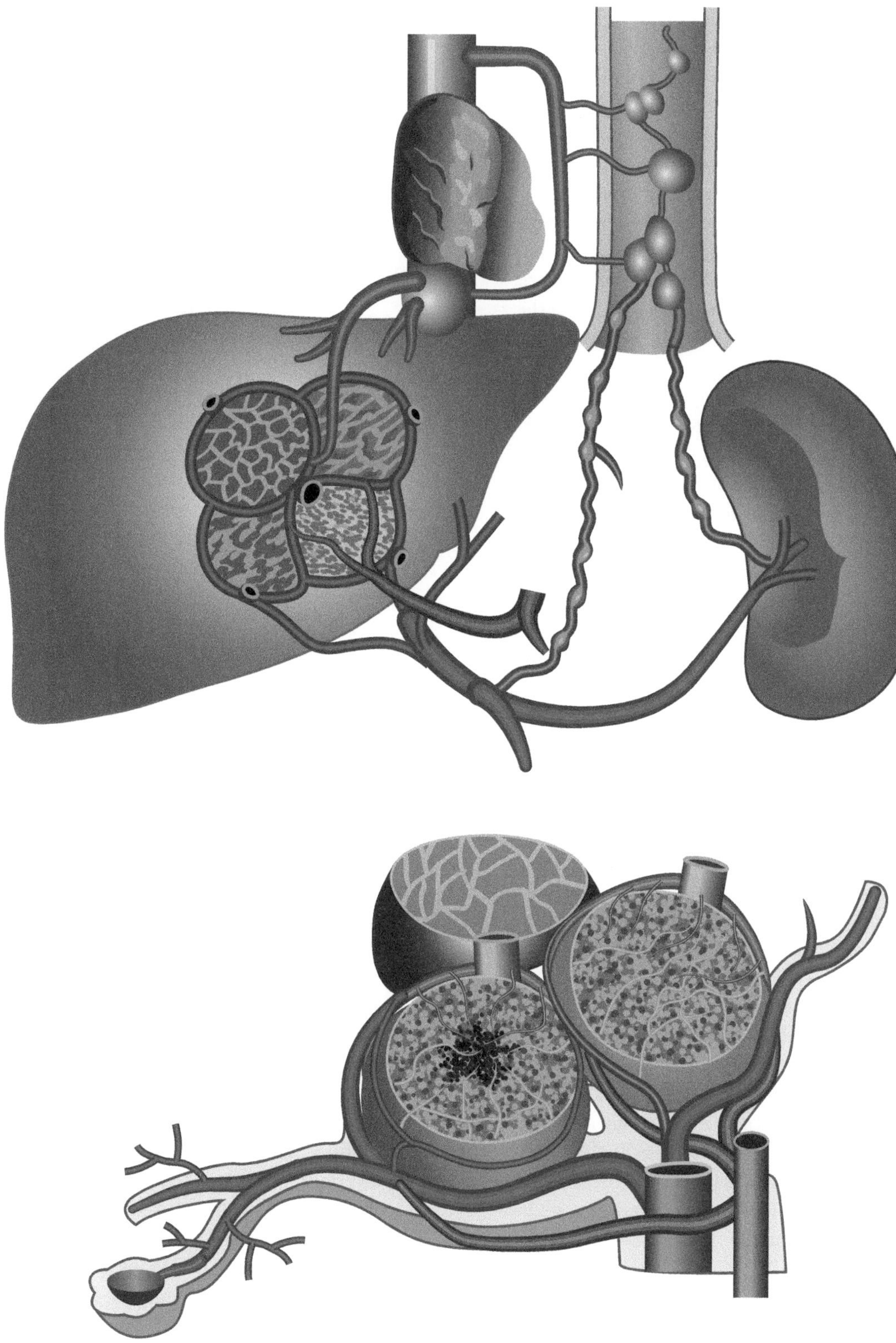

Figure 12.4 Cirrhotic liver and esophageal varices.

if needed. The tube may be left in position for up to 36 hours.

- A **self-expanding covered metal stent** can also provide luminal tamponade. It may be left in place for up to 2 weeks.
- **Propranolol** is given for secondary prevention and surveillance; esophagogastroduodenoscopy (EGD) should be performed every 6 months.
- **Transjugular intrahepatic portosystemic shunt (TIPS)** is a nonselective shunt for refractory variceal

bleeding or ascites. It also helps reduce bleeding in portal hypertensive gastropathy. *See* Figure 12.3.

  - It may be performed transjugular with portography via a wire crossing the liver parenchyma. Alternative routes include image-guided direct portal venous system access via the umbilical vein, or retrograde via the femoral to renal vein, then traversing a splenorenal collateral. Such routes may be needed if a transjugular, retrograde, transhepatic route is not possible. Sinusoids within fibrotic hepatic parenchyma should be balloon dilated until the pressure gradient drops to <12. Then a stent is placed.
  - The procedure can be complicated by thrombosis or intimal hyperplasia with stenosis within 1 year. The reported 50% patency at 1 year may be improved with covered PTFE stent grafts.
  - TIPS is **contraindicated** in right heart failure or, relatively, if the patient is a Childs C cirrhotic. In the latter case, shunting of portal venous blood away from hepatocytes may worsen liver failure and hepatic encephalopathy.
- **Balloon retrograde transvenous obliteration (BRTO)** is accomplished by advancing a catheter from the femoral to renal vein and then retrograde through a gastrorenal or splenorenal collateral. The culprit varix can then be sclerosed. *See Figure 12.5.*
- A **mesocaval interposition nonselective shunt (H-graft)** is occasionally needed for emergent control of bleeding in the instance of failed medical management and TIPS. It is a graft anastomosed from the side of the SMV to the side of the IVC. This approach leaves the portal vein (PV) unaltered for potential future transplantation.
- A **distal, selective, splenorenal shunt** is the transected end of the splenic vein (SV) anastomosed to the side of the left renal vein. All renal vein tributaries are ligated *near* their confluence with the renal vein, and all SV tributaries are ligated *far from* their takeoff. This helps encourage the maximal amount of portal blood to flow through the shunt into the systemic circuit.

- **GASTRIC VARICES**
  - Can exist and cause an UGIB independent of the presence of esophageal varices. Treatment is with **cyanoacrylate injection,** not banding.
  - If non-bleeding and discovered incidentally after pancreatitis complicated by splenic vein thrombosis, one may observe the varices without anticoagulation for the splenic vein. Once bleeding ensues, splenectomy is indicated.
- **HEPATORENAL SYNDROME**
  - Characterized by splanchnic vasodilation and renal vasoconstriction producing a prerenal picture. Type 1 is rapid onset and high mortality. Type 2 is slow onset. Treatment is with **midodrine or norepinephrine, octreotide, and albumin.** Renal function will only improve after hepatic transplantation.
- **HEPATOPULMONARY SYNDROME**
  - Involves pulmonary vasodilation and pulmonary arterial to pulmonary venous shunt development leading to hypoxemia. This improves after transplantation.

LIVER CYSTS

- Cysts are lined by epithelium and can be found incidentally on imaging done for other purposes, or they may produce abdominal pain and nausea. Intrahepatic biliary compression is a less common way to present.
- **SIMPLE SEROUS CYSTS**
  - Originate from biliary epithelium and have lost communication with the biliary tree.
- **DIAGNOSIS**
  - They are generally asymptomatic and visualized on CT imaging. Large cysts may cause atrophy of adjacent hepatic parenchyma, biliary obstruction, or portal vein obstruction and portal hypertension.
- **TREATMENT**
  - Reserved for symptomatic cases or those in which imaging suggests the cyst could be malignant. Start with **aspiration** to confirm clinical improvement with desufflation of the cyst cavity. If improved, one may proceed to **laparoscopic fenestration** by removing the anterior cyst wall and sending it to pathology to rule out cystadenoma or cystadenocarcinoma. Then fulgurate the remaining cyst lining and place an omental flap into the cyst bed. An intraoperative cholangiogram (IOC) can be performed to exclude biliary communication.
  - A **cystojejunostomy** can be done if there is biliary communication and pathology is consistent with a benign simple serous cyst. Alternative therapy includes US-guided intracystic injection of contrast to exclude biliary communication, and then instillation of alcohol or minocycline to **ablate or sclerose** the epithelial lining. Recurrence is higher than fenestration.
- **POLYCYSTIC LIVER DISEASE**
  - A condition marked by multiple simple, serous cysts that are asymptomatic and can be **observed.** Most often there are associated polycystic kidneys. Open or laparoscopic fenestration may be done for a dominant, symptomatic cyst. Transcatheter hepatic arterial embolization, aspiration and sclerosis, and medical therapy with octreotide are all less effective alternatives.
- **CYSTADENOMAS**
  - Have a collagenous pseudocapsule housing ovarian mesenchymal stroma. The cyst lining is cuboidal

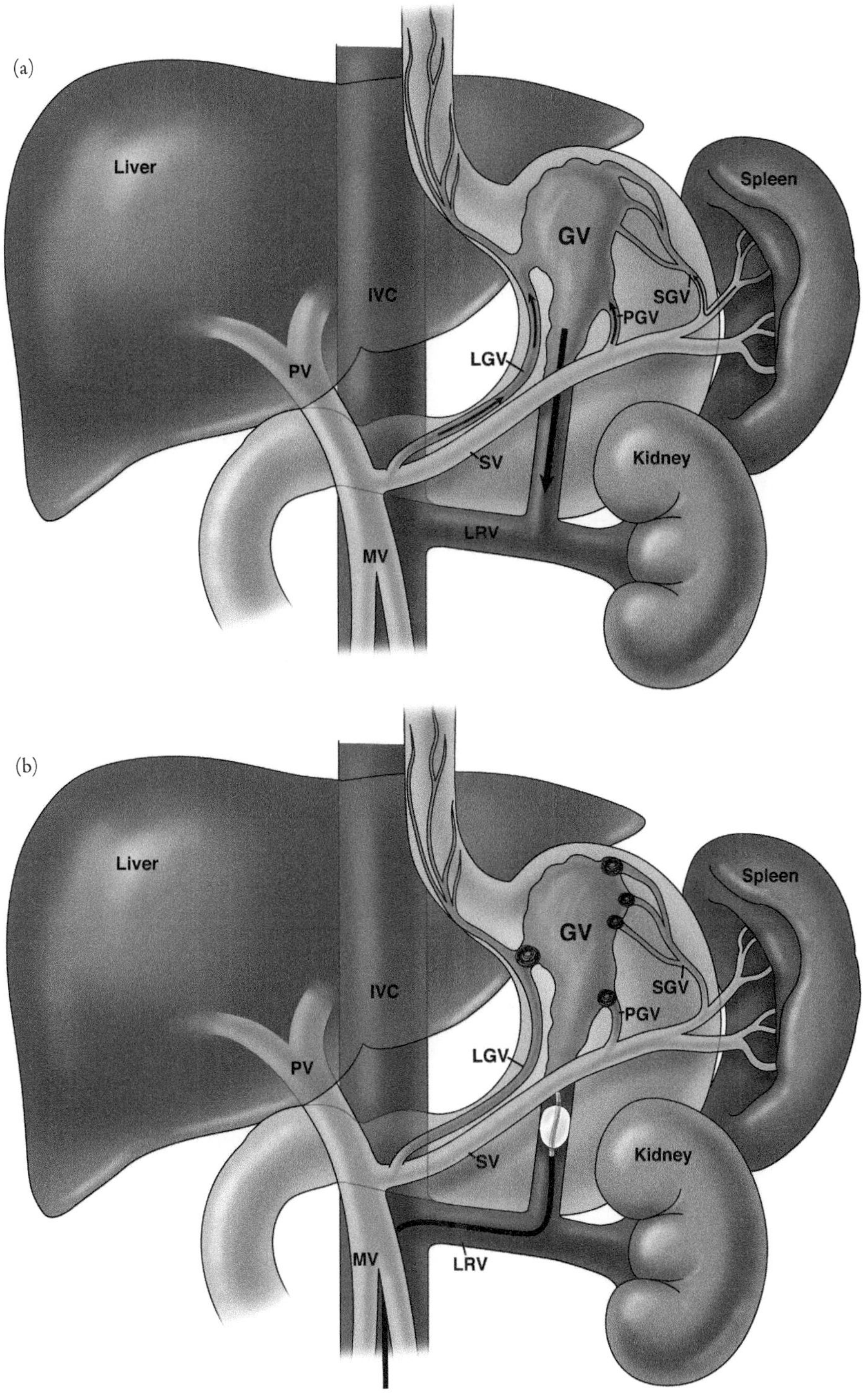

Figure 12.5 Balloon retrograde transvenous obliteration procedure (BRTO).

or columnar, and contents are serous or mucinous (mucinous cystic neoplasm). An MRCP, MRI with gadoxetate, or IOC can exclude biliary communication, and an intraductal papillary mucinous neoplasm of biliary origin (IPMN-B). Treatment is **resection or enucleation.**

- **CYSTADENOCARCINOMAS**
  - Harbor increased septations, wall nodularity, and intracystic hemorrhage on imaging when compared to cystadenomas. Intracystic CEA and CA 19-9 levels are similar to those in a cystadenoma. They may have serous or mucinous contents (mucinous cystic

neoplasm with malignant transformation). Exclude an intraductal papillary mucinous carcinoma of biliary origin (IPMN-B with malignant transformation) with MRCP, MRI with gadoxetate, or IOC. Treatment is with a **partial hepatectomy or wide local excision (WLE).**

- **INTRADUCTAL PAPILLARY MUCINOUS NEOPLASMS OF BILIARY ORIGIN (IPMN-B)**
- Have a communication with the biliary tree on ERCP, IOC, MRCP, or MRI with gadoxetate. There is no ovarian stroma. They are the hepatic counterpart to pancreatic IPMN, may cause biliary obstruction or cholangitis, and have a high malignant potential. Treatment is with **partial hepatectomy or WLE.**
- **TRAUMATIC HEPATIC PSEUDOCYSTS**
  - Also known as hematoceles, lack an epithelial lining and are not true cysts.

FOCAL LIVER INFECTIONS

- Patients have abdominal pain and nausea, sometimes accompanied by fever.
- **AMOEBIC ABSCESSES**
  - fecal–oral transmission of *Entamoeba histolytica.* They classically occur in the right lobe of a male alcoholic. Hepatocytes in the center of the lesion undergo liquefactive necrosis. Protozoa exist on the periphery of the lesion,
  - **DIAGNOSIS**
    - ***Entamoeba* serology** and an abdominal **CT** showing a fluid collection without an enhancing rim. A **culture from the peripheral rim** will show protozoa, but is not mandatory.
  - **TREATMENT**
    - **Flagyl 750 mg PO TID.** Paromomycin, diiodohydroxyquin, or diloxanide furoate will treat remaining enteric luminal trophozoites to prevent a relapse. Aspiration is undertaken only if the collection is refractory to the above measures. or imaging suggests possible impending rupture.
- **ECHINOCOCCAL CYSTS**
  - Result from human exposure to sheep or dogs, and have a right lobe predominance. There are two main types.
  - In the **cystic version,** a pericyst adventitia of reactive host cells lines the periphery. Within that lies the calcified ectocyst wall. An inner endocyst wall is the lining from which protoscolices bud, and intact daughter cysts detach.
    - **DIAGNOSIS**
      - The cyst may be asymptomatic or compress adjacent hepatocytes causing atrophy. Other structures that can be compressed include the biliary tree, portal vein, or hepatic veins. Confirmation is with **serology and US** imaging. There are four types.
      - **Type 1:** anechoic unilocular cyst.
      - **Type 2:** multilocular material within cyst.
      - **Type 3:** detached endocysts within a liquid cavity, with or without solid components.
      - **Type 4:** inactive, calcified wall with solid interior. There is generally no active infection.
      - An ERCP can exclude biliary communication in patients with jaundice, biliary insufficiency, or cholangitis.
    - **TREATMENT**
      - Start **albendazole with or without praziquantel.** This is sufficient for small, <5 cm type 1 cysts.
      - **Percutaneous transhepatic aspiration and alcohol or hypertonic saline injection with re-aspiration (PAIR)** can be added for small type 1 cysts that have no biliary or pulmonary communication on imaging.
      - **Complete pericystectomy or conservative endocystectomy** with an omental pedicle placed inside the residual cavity are both options for type 2 and 3 cysts. Generally the cyst is aspirated and injected or sclerosed intraoperatively prior to resection.
      - **Liver resection** is reserved for cystic disease nestled deeply within hepatic parenchyma.
      - Laparoscopic resection can be performed in accessible liver segments.
      - Open resection is preferred if the cyst is large, multiloculated, complex, or has biliary communication or compressive effects.
      - Post-procedural albendazole should be continued in all cases.
  - In the **alveolar version** the lesion is invasive and has a non-cystic growth pattern, so diagnosis requires **histopathology and PCR of a biopsy specimen.** Treatment is with laparoscopic or open liver resection.
  - **Postoperative imaging** includes serial US for cystic disease or serial CT or MRI for alveolar disease.
- **SCHISTOSOMIASIS**
  - Starts with a rash (swimmers itch), progresses to fever and systemic symptoms (Katayama fever), and then the organism nestles into either the bladder or colon. Petechia, ulceration, and hemorrhage of the mucosal lining can be seen. Eventually, organisms gain access to the hepatic parenchyma and incite a periportal presinusoidal fibrotic response producing portal hypertension. Treatment is with **praziquantel.** Secondary varices are managed in the standard fashion.
- **PYOGENIC ABSCESS**
  - Results from bacteria of biliary, diverticular, or appendiceal origin. In the latter two cases, organisms have migrated within the portal venous system and

established infection within the liver parenchyma. Most commonly they are GNR and afflict the right lobe.
    - Some abscesses are associated with hepatic transplant recipient arterial insufficiency causing hepatic necrosis or ischemic cholangiopathy and biliary stricture. The sequelae of this pathology is abscess formation.
  - **DIAGNOSIS**
    - CT imaging in a symptomatic, sometimes septic, patient.
  - **TREATMENT**
    - **CT guided aspiration and systemic antibiotics** for 4–6 weeks. Aspiration will provide the causative organisms and can help tailor therapy. If fungal, treatment is with antifungals in addition to drainage. In all instances, surgical debridement is the last resort.
- **ASCARIS**
  - Generally inhabits the enteric lumen and occasionally the biliary tree. Filling defects can be seen on ERCP. Conglomerates of the organism can serve as a nidus for intrahepatic biliary gallstones. Treatment is with anti-helminth medication and biliary decompression.

BENIGN LIVER MASS

- **HEMANGIOMA**
  - The most common benign mass. It has a female predominance and is composed of a conglomerate of vasculature with an endothelial lining surrounded by a thin connective tissue pseudocapsule. *See Figure 12.6.*
- **DIAGNOSIS.**
  - Most are asymptomatic. Three-phase liver **CT or MRI** will show asymmetric, **slow nodular peripheral (hepatic artery) heading to central (portal vein) enhancement**.
- **TREATMENT**
  - **Observation** if asymptomatic, regardless of size, since most remain stable. Otherwise, symptomatic patients can receive **enucleation** with or without preoperative embolization. The latter is utilized if the hemangioma has ruptured or there is a consumptive coagulopathy. Radiofrequency, microwave, or alcohol **ablation** is second-line therapy. Embolization and corticosteroids, with or without chemotherapeutics, are the primary treatment modalities in pediatric patients. Follow-up MRI should be done at 3–6 months.
- **FOCAL NODULAR HYPERPLASIA**
  - The second most common benign mass and also has a female predominance. It is due to an embryonic blood flow disturbance leading to a central vascular scar with surrounding **hyperplasia of normal hepatic structures**.
  - **DIAGNOSIS**
    - The majority of lesions are **asymptomatic** and incidentally found on the **periphery** of the liver by three phase liver CT or MRI showing a **hypervascular, enhancing central stellate scar.** A **positive Tc-99m colloid scan** can be confirmatory, since Kupffer cells are present. The lesion will also enhance on MRI with Eovist, since it has normal hepatobiliary components.
  - **TREATMENT**
    - **Observation** alone. If there is diagnostic uncertainty, or the differential includes fibrolamellar carcinoma, which classically has a non-enhancing central scar on imaging, then resection may be needed for a definitive diagnosis.
- **HEPATIC ADENOMA**
  - An overgrowth of sheets of hepatocytes fed by larger peripheral arteries, often in the right lobe of the liver. It has a female predominance and is related to oral contraceptive (OCP) or anabolic steroid use. **Hemorrhagic potential** is the main concern when the lesions are >5 cm. Malignant potential is the main concern otherwise.
  - **DIAGNOSIS**
    - 80% of lesions are asymptomatic. Otherwise, patients can have abdominal pain. Three-phase liver **CT** scan shows a heterogeneous mass and **rapid arterial phase contrast uptake with delayed washout.** Lesions will also be hypervascular on MRI and negative on Eovist. A **Tc-99m colloid scan** will be negative, since there are no Kupffer cells.
  - **TREATMENT**
    - Generally **observation and stopping OCPs** if the lesion is <4 cm and asymptomatic. In this situation, regression can be noted. **Excision with a narrow margin** is the best option for lesions that are symptomatic or >4 cm, lesions that are beta catenin positive, or any lesion occurring in a male patient. Excision can also rule out malignancy.
      - **Ablation** has been used for multiple adenomas.
      - **Angioembolization** should be first line for rupture and intraperitoneal hemorrhage in a patient who demonstrates stability after resuscitation. This should be followed by resection during the same hospitalization.
- **BILE DUCT HAMARTOMAS**
  - Seen on the liver surface at laparoscopy. No intervention is needed.

MALIGNANT LIVER MASS

- **METASTASES**
  - These are more common than primary liver tumors.
  - **Colorectal** metastases are the most common malignancy found within the liver in the United States. **Other metastases** include neuroendocrine, ocular melanoma, renal, breast, and sarcoma.

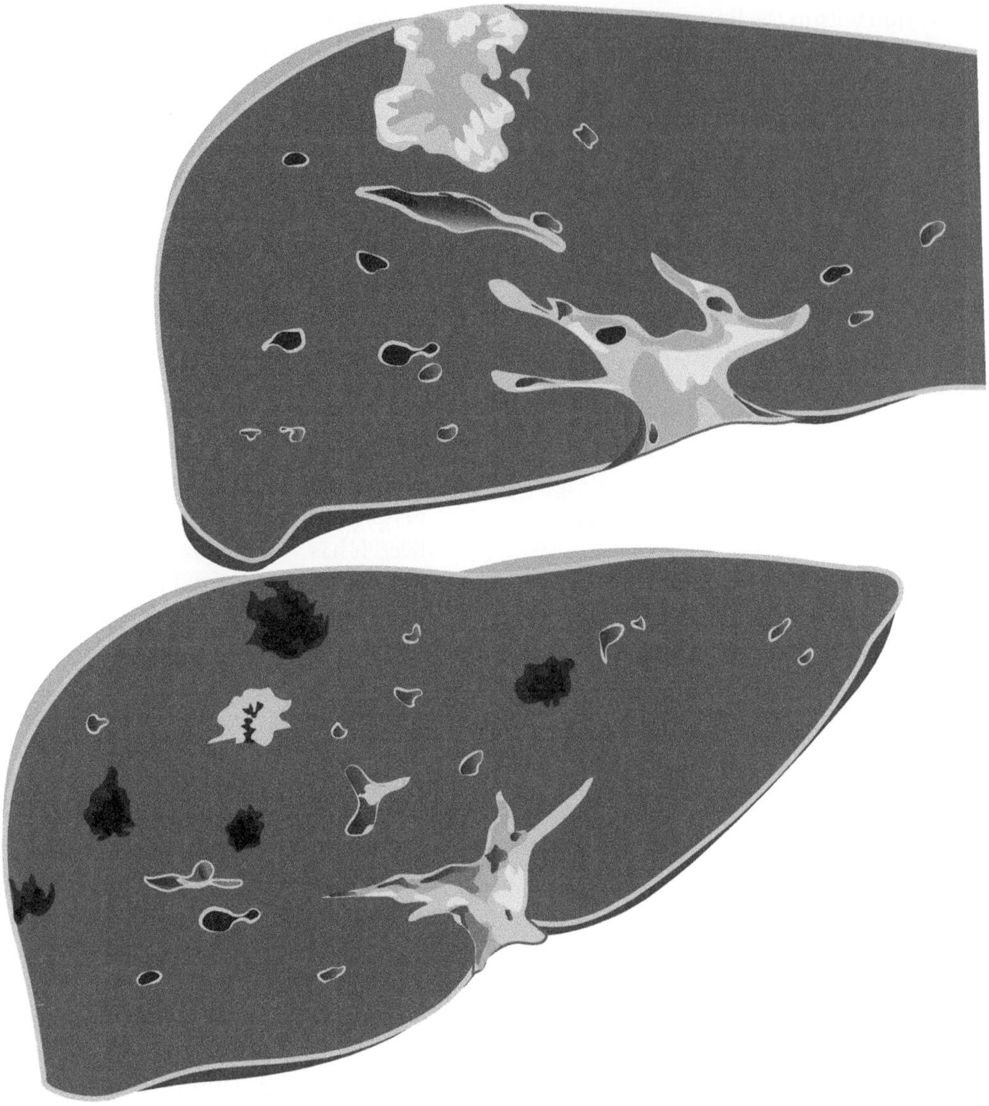

Figure 12.6 Hepatic adenoma (above) and hemangiomas (below).

- **DIAGNOSIS**
  - They can be discovered on staging workup for known primary malignancy, or occasionally a heavy metastatic burden to the liver can present with right upper quadrant abdominal pain.
    - **Triple phase CT** can provide a definitive diagnosis. Metastases do not enhance on the hepatic artery phase, so they appear **hypodense**. No biopsy is needed if lesions are categorized as LI-RADS 4 or 5, and the history and overall clinical picture provide a likely origin for the metastases.
    - **Intraoperative US** has the highest sensitivity for hepatic lesions <5 mm, even if such lesions reside deep within the parenchyma.
- **TREATMENT**
  - Options include chemotherapy, resection of the primary and metastases, or ablative techniques. The order of any particular treatment algorithm will vary based on factors below.
  - Resection of the primary tumor can be followed 8 weeks later by wedge resection of the metastatic liver tumor. This can be done before or after chemotherapy.
    - One may **resect a liver lesion first** if this appears it may border unresectability in the near future as it grows and encroaches on vascular structures. Laparoscopy is acceptable for all hepatic segments other than 7 or 8. Minimally invasive technique decreases blood loss, length of stay, and cirrhotic decompensation. They also allow for easier future liver transplantation. The primary can then be resected at a future date.
    - **Resecting synchronous hepatic lesions at the time of a colectomy** is an option for patients <70 years old who have a low operative risk and a small

metastatic burden. Both the primary and hepatic metastases must be resectable to R0 status in order to proceed. If one or the other is not resectable, or if there are extensive extrahepatic metastases, the patient should receive chemotherapy alone.
    - **Metachronous lesions** have the best prognosis if an R0 metastasectomy can be accomplished, the largest metastasis is <5 cm, there are 3 or fewer metastases, CEA level is <200, or if the initial diagnosis was N0.
    - Repeat resections are acceptable if a single metastasis is diagnosed after >12 months of a disease-free interval.
    - Neuroendocrine hepatic metastases may be debulked aggressively, since there are no good chemotherapeutic alternatives.
  - For **chemotherapy** one must keep in mind the following:
    - 1. Some agents are hepatotoxic, and if given preoperatively will necessitate a larger hepatic remnant (30%) be present after any planned resection.
    - 2. Some treatment regimens will make the metastasis difficult to localize during future resection.
  - **Radiofrequency (RFA) or microwave ablation** are alternatives for small lesions that are not resectable. Either may be used as sole therapy or in conjunction with chemotherapy, or in conjunction with resection of a dominant lesion elsewhere.
    - Microwave ablation has less char and less heat sink effect, so it can perform better than RFA for lesions near blood vessels.
    - Intraoperative approaches have lower recurrence rates than percutaneous approaches due to more precise application of the ablator, and the opportunity to dissect lesions away from heat sink structures or resect the neighboring gallbladder.
    - Neither should be used if the patient has untreatable extrahepatic disease.
    - **Follow-up imaging should be done in 1–3 months** and regularly thereafter.
    - CT PET is useful to monitor response to therapy after 3 months, when inflammation has subsided.
  - **Transarterial chemoembolization (TACE),** either conventional or DEB-TACE, or **transarterial radioembolization (TARE)** are third-line options for multiple lesions clustered together in a hepatic arterial neighborhood.
  - **Poor prognostic findings** include metastases diagnosed within 12 months of treatment of the primary, >3 metastases, >5 cm size single metastasis, or CEA level >200.
- **HEPATOCELLULAR CARCINOMA**
  - The most common primary liver malignancy in the United States, and the most common solid tumor worldwide. *See Figure 12.7.*
- **PATHOPHYSIOLOGY**
  - Carcinoma is secondary to Hepatitis B, C, EtOH, hepatic adenoma, glucocorticoids, hemochromatosis, or cirrhosis of any cause, including primary biliary cirrhosis. Malignant cells spread along portal venous tributaries. A fibrolamellar variant is seen in young patients without cirrhosis or Hepatitis B. It appears as a non-enhancing central scar on imaging, and increased serum neurotensin levels.
- **DIAGNOSIS**
  - Screening in high-risk patients is done with **q6 months US, possibly alternating with MRI**. AFP levels are not needed.
  - Confirmation is by **MRI with gadolinium or gadoxetate (Eovist) or a triple phase liver CT.** CT imaging will show hypervascular mass(es) on the hepatic artery phase with washout on the portal venous phase. T stages are as follows:
    - **T1:** single tumor <2 cm or any single larger tumor **without vascular invasion.**
    - **T2:** >2 cm and radiographic **vascular invasion** (other than the main PV or HV) or **multiple tumors each <5 cm** diameter.
    - **T3:** multiple tumors with one of them >5 cm diameter.
    - **T4:** any single or multiple tumors with **main portal or hepatic vein involvement** or growing out of the confines of the liver to visceral peritoneum or adjacent organs.
- **TREATMENT**
  - 1. **Resection with 2 cm margins** is the preferred treatment.
    - The patient should be non-cirrhotic, or at most Child's A with a MELD <9, and be free of portal hypertension.
    - The tumor can be of any size, but should be solitary and have **no macrovascular invasion on imaging.** Only a minority of cases will be at a resectable stage.
    - There should be an **adequate predicted post-resection liver remnant.** A 20% remnant is fine if the patient has normal liver function. A 30% remnant may be required for patients who have received, or will receive, hepatotoxic chemotherapeutics. Up to a 40% remnant may be needed if the patient is Child's class A.
    - If needed for maximization of a future liver remnant, one may use ipsilateral transcutaneous **portal vein embolization (PVE)** to induce apoptosis within the area to be resected. A compensatory hypertrophy of other segments then follows. **Repeat volumetric imaging analysis (3d CT or MRI)** is done to determine the size of the new predicted remnant. Surgical resection can then be planned for 4–6 weeks later. Avoid PVE if the patient is Childs B or C.

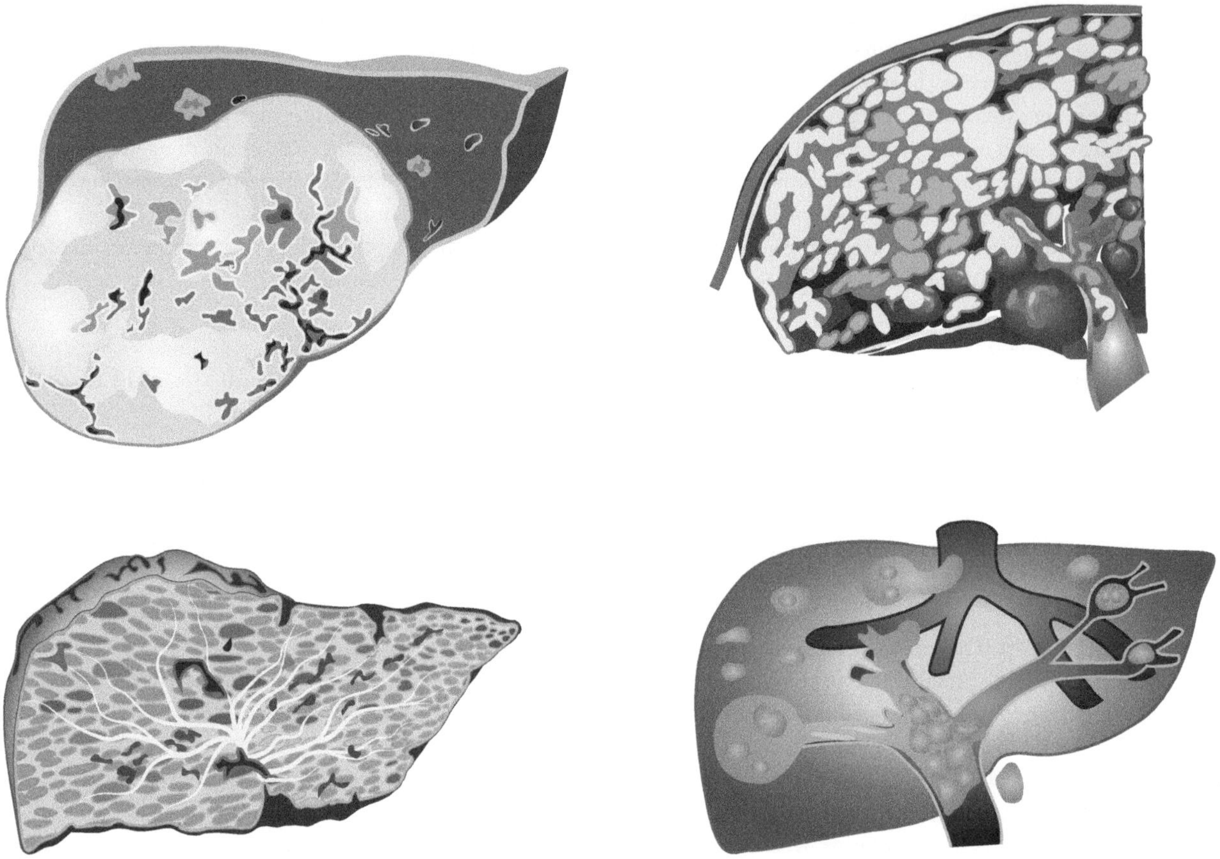

Figure 12.7 Hepatocellular carcinoma.

- **Intrahepatic recurrences** are common, and as high as 30% in 5 years.
- 5 years survival is 30% after resection.
- 2. **Liver transplantation** is the second-line treatment for patients who do not qualify for resection based on **Childs class B or C** or an **inadequate predicted liver remnant.**
  - The tumor should have **no macrovascular invasion** and must meet criteria listed below.
    - **Milan criteria:** single tumor <5 cm diameter, or three or fewer tumors each <3 cm diameter with total diameters of all tumors <9cm
    - **UCSF criteria:** single tumor <6.5 cm, or multiple tumors the largest of which is <4.5 cm, with total diameters of all tumors <8 cm.
    - May use **TACE and ablation to meet either of the above criteria.**
- **3. Radiofrequency or microwave ablation** is the third-line treatment option if the patient is not a candidate for either of the above based on tumor burden, small predicted liver remnant, or underlying cirrhosis.
  - Tumors should be <3 cm–5 cm and not near a vascular structure.
  - A percutaneous or intraoperative needle is placed into the tumor and coagulative necrosis induced at up to 100°C.
  - RFA may be equivalent to surgical resection for tumors that are <2 cm.
  - It can be used to improve candidacy for either resection or transplantation.
  - The intraoperative approach has a lower recurrence rate than the percutaneous approach, due to a more precise application of the ablator and the opportunity to dissect lesions away from heat sink structures or to resect the neighboring gallbladder.
- 4. Bland transarterial embolization (**TAE**), transarterial chemoembolization (**TACE-** conventional chemotherapeutics emulsified in oil with embolic material infused afterward), drug eluting beads with TACE (**DEB-TACE**), or **transarterial radioembolization (TARE** with radioactive microspheres) are options to treat multiple tumor sites in Childs A or B patients that are not candidates for resection.
  - TACE can be combined with RFA in order to improve candidacy for transplantation.

- Embolic methods are contraindicated if there is biliary obstruction, because a superimposed ischemic cholangiopathy ensues.
- PV thrombosis and advanced liver disease with elevated total bilirubin (TB) are relative contraindications to transarterial therapy.
- **Sorafenib** can be used adjuvantly, or as sole therapy if a patient is not a candidate for any of the above surgical or adjunctive options.

- **INTRAHEPATIC CHOLANGIO-CARCINOMA (CCA)**
  - The second most common primary liver malignancy.
- **DIAGNOSIS**
  - Patient presentation can include signs of insidious biliary obstruction and right upper quadrant (RUQ) abdominal pain.
  - **Triple phase liver CT** will demonstrate a **low attenuation lesion** with biliary obstruction and **liver capsule retraction.** This is diagnostic and distinguishes it from HCC, adenoma, hemangioma, or FNH. If no resection is planned or imaging is equivocal, biopsy can be performed and will show adenocarcinoma.
    - T stages are:
      - **T1:** solitary lesion without vascular invasion.
      - **T2:** multiple tumors or a solitary with vascular invasion.
      - **T3:** lesion extends to visceral peritoneum.
      - **T4:** lesion has direct invasion into local extrahepatic structures.
  - **Diagnostic laparoscopy** is done to confirm resectability. Any nodes outside of the porta hepatis contraindicates resection.
- **TREATMENT**
  - Resection to negative margins and portal hilar lymphadenectomy. Transplant is not an option for intrahepatic CCA.
    - Preoperative PV embolization can help maximize the liver remnant.
    - Adjuvant chemotherapy and radiotherapy may be added for an R1 resection or N1 status.
    - Palliative options include chemotherapy, RFA, TAE, TACE, or TARE.
    - **Perihilar cholangiocarcinoma** may be resected along with a portal lymphadenectomy. These patients can alternatively be considered for neoadjuvant chemotherapy and transplantation, especially if they have underlying primary sclerosing cholangitis (PSC).
- **Sarcoma and angiosarcoma** are less common than either HCC or CCA. Risk factors include exposure to polyvinylchloride, thorotrast, or arsenic. Treatment is palliative.

HEPATECTOMY PRINCIPLES

- **PREOPERATIVE ASSESSMENT**
  - The **model for end stage liver disease score (MELD)** includes INR, creatinine, TB, and sodium (Na). It is used to prognosticate after TIPS or any elective surgery, as well as prioritize patients with chronic disease on the transplant list.
    - Scores range from 6–40. In emergency surgery, every point equates to an increase in mortality of 14%.
    - Scores >40 are associated with a 70% mortality at 3 months after elective surgery. Scores <10 have a 2% mortality at 3 months after elective surgery.
    - Scores >15 mean the mortality from liver disease is greater than that of a transplant, so the patient can be considered for transplantation.
  - **Child-Pugh score** includes albumin, ascites, encephalopathy, international normalized ratio (INR), and TB. It predicts mortality after surgery. Scores <7 are class A (10% mortality), 7–9 are class B (30% mortality), and >9 are class C (80% mortality).
  - **Volumetric analysis (3d CT or MRI)** is used to predict the liver remnant after a planned resection.
    - A minimum of 20% functional liver remnant (FLR) is all that is needed in most *non*-cirrhotic patients undergoing hepatectomy.
    - A 30% FLR is desired if chemotherapeutic hepatic injury has occurred or is anticipated.
    - A 40% FLR is required if a Child's A cirrhotic is planning to undergo resection.
  - **Functional testing with indocyanine green (ICG)** injection depends on intact portal venous flow to deliver the tracer to hepatocytes. If this inflow is intact, then the amount of remaining compound in the serum can be used to estimate hepatocyte function and thereby determine the minimum FLR needed.
    - <10% of ICG left in serum at 15 minutes is normal and implies one can resect 3 liver segments and the patient will still be left with an adequate FLR.
    - 10%–20% of ICG remaining in the serum at 15 minutes means one should resect only 2 segments.
    - 20–30% ICG remaining suggests that the patient will tolerate only a single segment resection.
- **INTRAOPERATIVE CONSIDERATIONS**
  - Maintaining a **low central venous pressure** reduces hepatic vein bleeding. Trendelenburg position reduces the chance of a right atrial air embolus.
  - Pneumoperitoneum during laparoscopy will increase SVR, decrease portal flow and renal perfusion, and decrease venous return. This can lead to an increased rate of venous thrombosis.

- A right subcostal incision or thoracoabdominal incision can be utilized if segments VI, VII, or VIII need access.
- One can mobilize the falciform ligament over the liver dome, right up to the IVC.
- **Mobilizing** the right liver includes transecting the triangular, coronary, and retrocaval ligaments to expose the IVC and right hepatic vein (HV) confluence. Then the right HV can be encircled with umbilical tape. Short hepatic veins draining directly into the inferior cava can be ligated.
- During the dissection, one may employ portal triad occlusion of 15–20-minute episodes in order to control hemorrhage. Five minutes of reperfusion rest should be instituted in between portal occlusive periods.
- Hepatotomies near the gallbladder (GB) fossa and anterior portion of the caudate allows one to control the takeoff of right portal structures with umbilical tape.

- **Nonanatomic wedge resection** is an option for lesions in segments 2–6. Margins should be 1–2 cm.
- Anatomic resections respect the liver segments. *See Figure 12.8.*
- A **left lobectomy** involves ligation of the left hepatic artery and portal vein posterior to it. The next step is intrahepatic control of the left hepatic vein with or without the middle hepatic vein(s). The left hepatic ducts are transected intrahepatically.
  - A **left lateral segmentectomy** can be done without taking the middle hepatic artery or vein.
  - In an **extended left hepatectomy taking segments V and VIII,** the right anterior HA, PV, and HV tributaries should be ligated, with preservation of their posterior counterparts.
- A **right hepatectomy** starts with extrahepatic ligation of the right HA and PV posterior to it, using traction on the falciform and round ligaments as well as the cystic duct stump after cholecystectomy.
  - The right hepatic duct is transected intrahepatically. Branches of the middle hepatic vein to the patient's right should be clipped or stapled. The right HV can be ligated intra- or extrahepatically.
  - In an **extended right hepatectomy taking segment IVa,** the middle HA and HV should be ligated.
- Once the vasculature is controlled, **parenchymal transection** can be done with finger fracture, hydrotherapy, or ultrasonic devices.

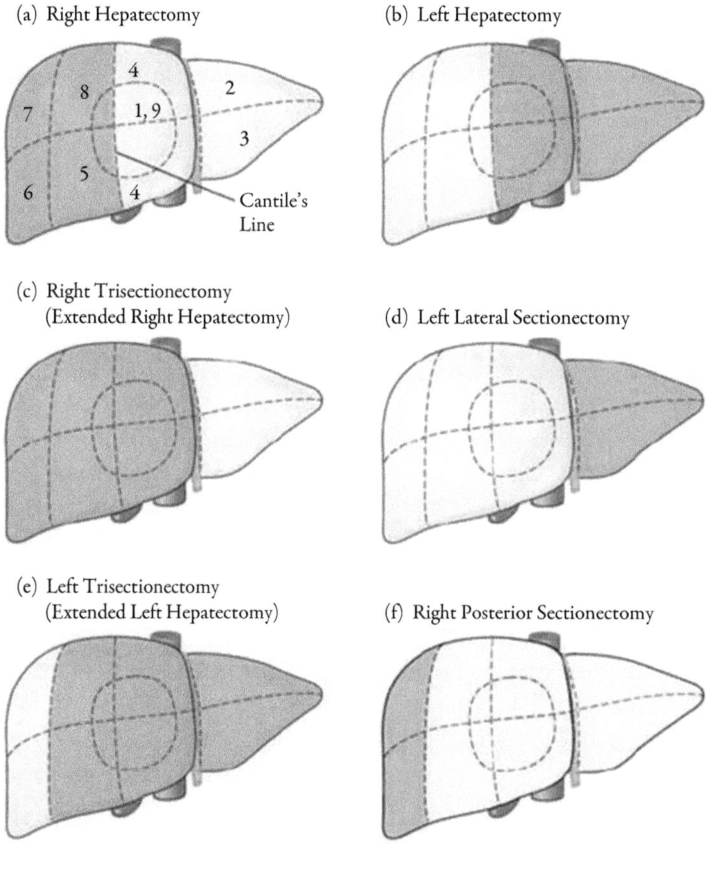

Figure 12.8 Hepatectomy options.

- Vascular structures encountered that are <7 mm can be sealed thermally or ultrasonically. Otherwise, clips or a stapler can be used.
- The transection can be done purely laparoscopically with or without hand port assistance from midline, or may be accomplished in a hybrid manner where the liver is mobilized laparoscopically and then transected in an open fashion.
- Laparoscopic resection is reasonable in segments 2–6 unless the mass is large or near major vascular structures. Segments 7 and 8 over the dome of the liver are better approached as an open resection.

LIVER TRANSPLANTATION PRINCIPLES

- **PREOPERATIVE CONSIDERATIONS**
  - Indications include:
    - 1. **Fulminant hepatic failure (FHF).**
    - **2. Cirrhosis** with MELD >15 or hepatopulmonary or hepatorenal syndrome.
    - 3. **Hepatocellular carcinoma** in a patient who does not meet criteria for resection or has Childs B or C status cirrhosis, but meets either Milan or UCSF criteria.
      - **Milan criteria:** single tumor <5 cm diameter, or 3 or fewer tumors each <3cm diameter with total diameters of all tumors <9 cm.
      - **UCSF criteria:** single tumor <6.5 cm, or multiple tumors the largest of which is <4.5 cm, with total diameters of all tumors <8 cm.
    - 3. **Hilar cholangiocarcinoma** that is not resectable, especially if the patient has concomitant PSC.
    - **4. In conjunction with a small bowel transplant** conducted for complications of TPN during the management of short bowel syndrome.
  - **Contraindications** include donor sepsis, donor malignancy, donor CJD, recipient active EtOH use, intracranial hemorrhage, acute ulcerative colitis, or severe pulmonary hypertension with elevated IVC pressures refractory to prostaglandins.
  - Other considerations that are not necessarily contraindications are described below.
    - **ABO compatibility is not mandatory.**
    - **Recipient portal vein thrombosis** is not a contraindication to transplantation.
    - If the donor is **Hepatitis B** positive, the recipient can be treated with HBIG and lamivudine to prevent reinfection. Reinfection rates can be reduced to <20%.
    - If the donor is **Hepatitis C** positive, the recipient can be treated with ribavirin, IFN, or ledipasvir-sofosbuvir (Harvoni).
    - If the donor is **HIV** positive, the organ is restricted to HIV positive recipients.
    - Living donors may be used. A right lobe or two donor left lobes are needed for a recipient adult. A single left lobe is sufficient for a recipient child.
- **INTRAOPERATIVE CONSIDERATIONS**
  - **Recipient hepatectomy** involves the following:
    - 1. Transection and ligation of the common hepatic artery as it splits into the GDA and proper hepatic artery. This widens the lumen available for anastomosis.
    - 2. Transection of common hepatic duct (CHD) near the hilum so there is plenty of recipient common bile duct (CBD) remaining for an anastomosis; and
    - 3. Transection of the PV only after the liver has been completely mobilized off the retrohepatic IVC.
  - The donor organ can endure up to **24 hours cold ischemic time** after procurement.
  - For implantation, the **piggyback technique** involves a side–side anastomosis of donor and recipient inferior cava. This avoids venovenous bypass.
  - The **PV** is anastomosed end to end, ensuring there is no kink from excessive length.
  - The **hepatic artery donor carrel patch** includes the donor celiac trunk and part of the neighboring aorta or takeoff of the splenic artery. It is anastomosed to the recipient common hepatic artery (CHA).
    - In cases where the recipient CHA is stenosed, procurement may need to include the donor iliac artery fashioned as an interposition graft from the donor Carrell patch to the recipient supraceliac or infraceliac aorta.
  - The donor **CBD is anastomosed end to end or side to side** with the recipient CHD. *See Figure 12.9.*
  - A **Roux-en-Y hepaticojejunostomy** can be used if the recipient has a diseased CBD or the recipient is a pediatric patient with a small CBD.
- **POSTOPERATIVE CONSIDERATIONS**
  - AST can rise up to 2500, peaking on postoperative day 2. Cholestasis can last up to 12 days. Both of these findings are normal.
  - **Primary nonfunction** of the hepatic graft occurs in 3%–4% of recipients. It presents within 2–4 days post-op with decreased bile production, elevated prothrombin time (PT), and ensuing Signs of fulminant hepatic failure. It is more common if the warm ischemic time was >25 minutes or if the donor liver has macrosteatosis. Treatment requires **re-transplantation.**
  - **Hepatic artery complications** are diagnosed by Doppler ultrasound.
    - **Thrombosis** occurs in 1%–5% at the anastomotic site, and can result in graft loss requiring re-transplantation.
      - Intraoperative US to assess arterial flow <200 ml/minite at the time of construction of the anastomosis can help predict future thrombosis. Low flow can also predict primary nonfunction of the graft in the absence of arterial thrombosis.

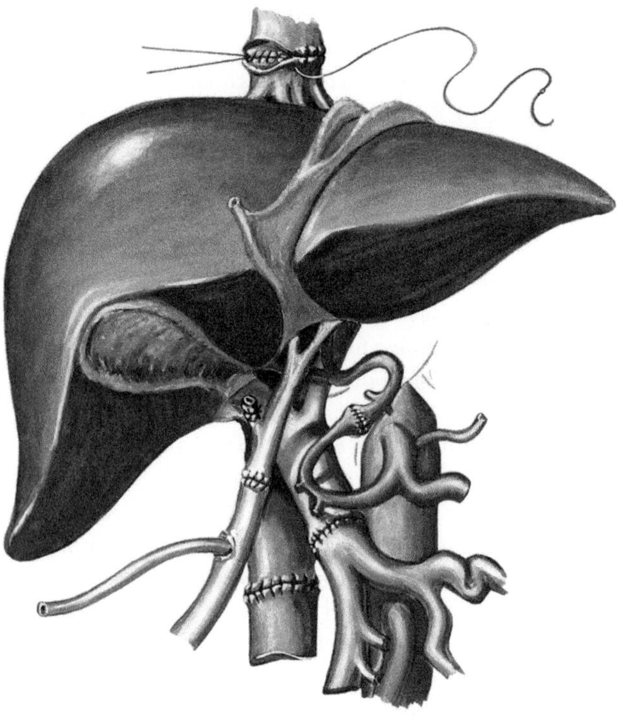

Figure 12.9 Hepatic transplantation.

  - **Stenosis** usually presents within 3 months of transplantation, and can be managed by angioplasty with or without a stent. Hepatic artery insufficiency from either stenosis or late-onset thrombosis results in biliary compromise, strictures, and intrahepatic abscesses of biliary origin.
- **Portal vein complications** are diagnosed by Doppler ultrasound.
  - **Thrombosis** may be due to kinking of the anastamosis, recipient portal vein disease, or hypercoagulability. If it occurs early post-op, it can result in graft loss. Treatment is with **thrombectomy, anastamotic revision, or re-transplantation**.
    - PV thrombosis can be predicted by an intraoperative US showing flow <1 L/min.
  - **Stenosis** occurring early can be revised surgically, and those occurring late can be managed with **angioplasty** via transjugular or percutaneous transhepatic routes.
  - TIPS may be used for chronic portal vein thrombosis or stenosis, or delayed graft function. It has higher infectious complications than the pre-transplant patient.
- **Biliary complications** can be diagnosed by RUQ ultrasound, cholangiopancreatography (ERCP or MRCP), or a nuclear medicine study.
  - A **bile leak** is the most common and is managed with an external drain, ERCP, and stent.
  - **Biliary strictures** can be **stented** if at the anastamotic site. If not at the anastamotic site, they are often due to arterial insufficiency in the setting of vasopressor use, chronic hepatic artery thrombosis or stenosis, prolonged cold ischemic time, or a prolonged warm ischemic time in donation after cardiac death (DCD).
    - In these cases, the stricture can be **stented, resected if extrahepatic, or undergo re-transplantation if intrahepatic.**
- **Cholangitis** can present with fever, abdominal pain, and signs of biliary obstruction. It characteristically has neutrophil infiltrates around biliary tracts. Treatment is with broad antibiotics and ERCP to relieve any obstruction.
- **IVC stenosis or thrombosis** is rare and treated with thrombectomy, thrombolysis, stent, or anastamotic revision.
- **Acute rejection** is T cell mediated with a **lymphocytic infiltrate** around bile ducts and an **endotheliitis**. It presents within 2 months with fever, jaundice, decreased bile output, increased TB and PT.
- **Chronic rejection demonstrates portal fibrosis with loss of bile ducts** and a rise in alkaline phosphatase. Both cellular and humoral processes mediate these findings.
- **CMV infection** occurs in the setting of CMV positive organs received by negative recipients. Patients present with malaise, leukopenia, low platelets, and a transaminitis. Treatment is with antivirals. The incidence is decreasing due to sensitive screening tests.
- **"Small for size syndrome"** results in cholestasis and ascites in a recipient who has received a graft that is an insufficient volume of hepatocytes relative to their body size.

13.

BILIARY TRACT

ANATOMY AND PHYSIOLOGY

- **BILE COMPOSITION**
  - 1. **<u>80%–85% bile salts</u>**
    - **Primary bile salts include cholic and chenodeoxycholic acid.** They are formed from cholesterol via 7 alpha-hydroxylase. This enzyme has decreased activity in the fasting state.
    - Primary bile salts are **conjugated** with glycine and taurine for water solubility, released into the biliary ducts, and then into the duodenum.
    - Bile salts serve to emulsify ingested fat and form micelles and vesicles to help with long chain fatty acid absorption across the jejunum mucosa. Remaining bile salts are actively reabsorbed in the terminal ileum.
    - Bacterial deconjugation produces **secondary deoxycholic and lithocholic acids** that may be passively reabsorbed along the remainder of the small bowel or colon.
    - About 5% of bile salts are not reabsorbed, and pass into stool for excretion.
    - **Cholestyramine** binds bile salts in the GI lumen and prevents reabsorption. This results in increased production of bile salts from cholesterol by hepatocytes, and hence decreases serum cholesterol.
    - A buildup of bile salts in the serum can result in acute kidney injury (AKI) and pruritus.
  - 2. **10%–15% phospholipids and lecithin** which help solubilize cholesterol and prevent cholesterol stone formation.
  - 3. **5% cholesterol,** which is formed from HMG-CoA reductase. This enzyme has increased activity in obesity and is inhibited by statins and niacin.
  - 4. Bilirubin and other proteins make up the remainder of bile.
    - **Unconjugated bilirubin** load comes from the breakdown of biliverdin and hemoglobin.
    - This is **conjugated to glucuronic acid** by hepatocytes and transferred across canaliculi into the bile ducts.
    - Then the bilirubin is either **deconjugated** by bacteria in the common bile duct (CBD) or intestinal lumen, or **converted to urobilinogen.**
    - Urobilinogen is either reabsorbed across the intestinal lining into the bloodstream and excreted in urine, or it is converted to **stercobilinogen** to remain in stool for excretion.
    - **Hyperbilirubinemia** can result from biliary obstruction (conjugated hyperbilirubinemia) or impaired hepatocyte conjugation (unconjugated hyperbilirubinemia). It decreases by 50% in 1–2 days and normalizes by one week after resolution of the obstruction.
    - Delta bilirubin is an exception to this, as it builds up in prolonged jaundice and remains bound to albumin. It has a half-life of 18 days, so it takes much longer to clear.
    - Hyperbilirubinemia at levels >40 is not from obstruction alone. There must be a component of hemolysis, hepatocyte dysfunction, bile duct to hepatic vein fistula, or concomitant AKI or chronic kidney disease.
- **Gallstones** are present in 10%–15% of the population. **80% are asymptomatic.** 20% develop biliary colic. About 2%–3% per year develop symptoms, and 1%–4% develop an indication for an intervention.
  - **Cholesterol stones** are the most common in the United States. They result from stasis within the gallbladder (GB), increased water reabsorption, calcium nidus, decreased lecithin, or increased cholesterol with decreased conversion to bile acids. The latter is from increased HMG-CoA reductase activity with decreased 7 alpha-hydroxylase activity. Most cholesterol stones are radiolucent on imaging.
  - **Black calcium bilirubinate stones** are the most common worldwide. They result from an increased **unconjugated bilirubin load** secondary to hemolysis, impaired hepatocyte conjugation, total parenteral nutrition (TPN), or increased calcium. Treatment must be aimed at the underlying cause for elevated unconjugated bilirubin.
  - **Brown calcium bilirubinate stones** are from bacterial **deconjugation of previously conjugated bilirubin** residing within the common hepatic duct (CHD) or common bile duct (CBD). They are often associated with **distal common duct pathology** promoting stasis and infection, which leads to deconjugation. Brown stones are a **compilation of calcium bilirubinate and bacterial debris.** They are considered primary CBD stones because they do not originate in the GB, so cholecystectomy is not curative. Treatment is aimed at the underlying biliary pathology and includes

antibiotics and endoscopic sphincterotomy or surgical CBD drainage.

- **BILE EXCRETION**
  - 500–1000 ml of bile are produced per day. The final concentration is determined by gallbladder reabsorption of NaCl and water, as well as the secretion of mucous and hydrogen ions by the GB. Such secretion helps solubilize calcium.
  - **Postprandial GB contraction and emptying** starts within 40 minutes of a meal, and is influenced by cholecystokinin (CCK) and secretin. 80% of the GB should be emptied at 2 hours after a meal.
    - **Non-prandial GB contraction and emptying** occurs during phase 2 of the migrating motility complex in the fasting state.
    - Both are under left (anterior) vagal coordination as well as T7–10 sympathetic input via the celiac and splanchnic ganglia.
  - A **HIDA scan** involves an intravenous radiotracer injection that transitions from hepatocytes into the biliary system, and then duodenum and gallbladder by 60 minutes.
    - If the GB does not fill in **4 hours**, then the patient has acute cholecystitis.
    - If it fills at **1–4 hours**, then the patient may have chronic cholecystitis.
    - 80% of gallbladder contents should empty within 2 hours after a meal or CCK administration. **Biliary dyskinesia** is suggested by a GB that fills normally, but **empties <35%–40% at 20 minutes**. Patients who suffer from true dyskinesia should have 30-minute or longer episodes of unrelieved postprandial pain.
- **Common hepatic duct (CHD)** is 1 cm–4 cm in length and 4 mm in diameter. It is perfused by branches from the right hepatic artery (RHA). The cystic artery is an RHA branch.
  - The anterior right hepatic ducts drain segments 5 and 8. The posterior right hepatic ducts drain segments 6 and 7. Either may insert aberrantly low on the left hepatic duct.
- The **cystic duct** contains Heister valves and normally drains into the CHD. However, it may on occasion drain into on an aberrant right posterior hepatic duct.
- **Gallbladder (GB)** components include the following. See *Figure 13.1.*
  - Mucous secreting glands in the infundibulum.
  - Elastic fibers within the body for receptive capabilities.
  - Muscular elements in the fundus for contraction and emptying.

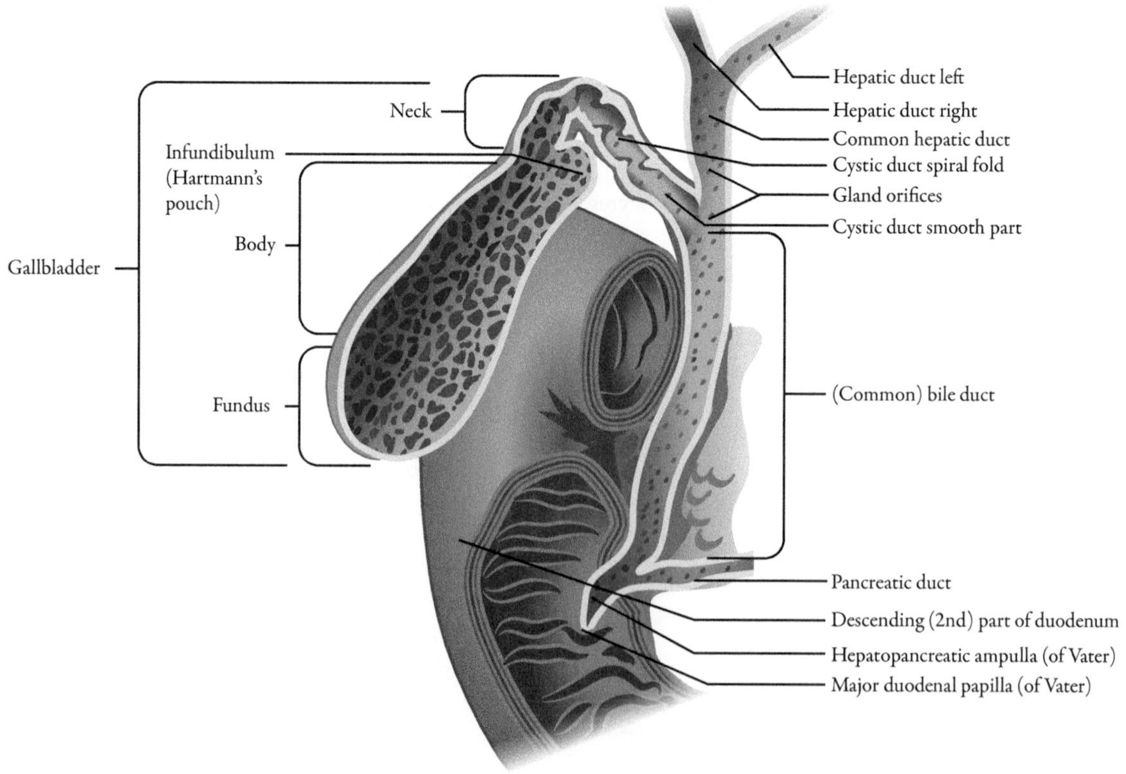

Figure 13.1 Anatomy of the gallbladder and biliary tree.

- Columnar epithelial mucosa with Rokitansky-Aschoff sinuses. These are invaginations from intraluminal pressure, and may progress to adenomyomatosis.
- There is no submucosal layer.
- Ducts of Luschka drain bile from the hepatic parenchyma directly into the gallbladder.
- Cystic veins drain from the gallbladder directly into the right portal vein.

- The **common bile duct (CBD)** is 7 cm–10 cm in length and 5 mm–10 mm in diameter, and is lined by mucous secreting glands. It has a longitudinal blood supply the sits at the 3 o'clock and 9 o'clock positions. Inferiorly, the blood supply comes from branches of the gastroduodenal artery.
  - The **common channel of the CBD and pancreatic duct** (PD) is 1 cm–2 cm in length, and their merger occurs outside the duodenal wall in the majority of patients, ultimately draining through a single ampulla of Vater.
    - Occasionally, the two may merge within the duodenal wall or drain via separate openings into the duodenal lumen.
  - The **sphincter of Oddi** is 4–6 mm in length and has a baseline pressure that is 13 mmHg above that of the duodenal lumen. Its tone is controlled by Cajal cells, and it modulates the patency of the ampulla of Vater.
    - Morphine causes contraction, and glucagon causes relaxation of the sphincter.
  - **Lymphatics** reside to the patient's right of the CBD, within the hepatoduodenal ligament.

SYMPTOMATIC CHOLELITHIASIS AND CHOLECYSTITIS

- **SYMPTOMATIC CHOLELITHIASIS**
  - transient GB obstruction due to stones that impact in the neck as the GB attempts to contract and empty. Patients have acute onset of pain, which resolves when the stones drop back into the GB lumen. Episodes are generally short enough that acute inflammation does not set in. These episodes are an indication for an elective laparoscopic cholecystectomy. Symptomatic cholelithiasis also encompasses stones transiently or permanently obstructing the common bile duct (see below).
- **ACUTE CHOLECYSTITIS**
  - Gallbladder inflammation secondary to pressure overload from non-resolving obstruction at its neck or the cystic duct. This is most commonly due to a cholesterol or pigmented gallstone. See *Figure 13.2.*
    - Obstruction may be acalculous and secondary to stasis, sludge, or edema at the gallbladder neck. Inciting factors for acalculous cholecystitis include burns, trauma, major surgery, or prolonged use of TPN.
    - Whether calculous or not, less than 30% of cases are actually infected. Organisms include E. Coli, Klebsiella, and Enterobacter.
    - Up to 10% may have GB wall necrosis due to pressure overload, infection, and poor arterial inflow.
    - Patients may have increased systemic alkaline phosphatase levels during cholecystitis. If the CBD or CHD are not obstructed, the serum alkaline phosphatase is generally elevated via cystic vein drainage or via inflamed ducts of Luschka.
- **CHRONIC CHOLCYSTITIS**
  - multiple long episodes of GB neck or cystic duct obstruction over time that resulted in recurrent inflammation and eventual scar formation.
- **DIAGNOSIS**
  - The classic patient is female, forty, fertile, and obese with cholesterol stones. The most common symptom is episodic or constant postprandial RUQ pain that worsens with inspiration.
  - **Ultrasound (US)** is the initial modality to confirm the diagnosis. Findings include a gallbladder wall thickness >4 mm, pericholecystic fluid, sonographic Murphy's sign, or simply cholelithiasis in the correct clinical presentation. A CBD diameter >8 mm on US should suggest concomitant choledocholithiasis.
  - **CTAP** is a second-line imaging option. Signs of cholecystitis, with or without visible stones, can be sufficient.
  - A **HIDA** scan can be used in cases that remain equivocal despite US and CT.
    - If the GB and duodenum fill at <60 min, then there is no neck obstruction or CBD obstruction.
    - A GB that does not fill by 4 hours in the correct clinical picture is acute cholecystitis.
    - A GB that slowly fills over 1–4 hours can be chronic cholecystitis.
    - HIDA with CCK injection after the GB is full can be used for the diagnosis of biliary dyskinesia. This is suggested by a GB ejection fraction <35%–40% at 20 minutes post injection of CCK. Biliary dyskinesia can be the cause of colic, even in the absence of stones.
- **TREATMENT**
  - 1. **Laparoscopic cholecystectomy** can be elective for symptomatic cholelithiasis, or urgent for acute or chronic cholecystitis.
    - In the case of acute inflammation, the procedure should ideally be completed within 72 hours of presentation. This was historically associated with a lower open conversion and complication rate. However, laparoscopic cholecystectomy may still be attempted beyond 72 hours.
    - Dissection should stay anterior to Rouviere's sulcus, which runs to the right of the hepatic hilum, anterior to the caudate lobe, and contains the right portal pedicle.

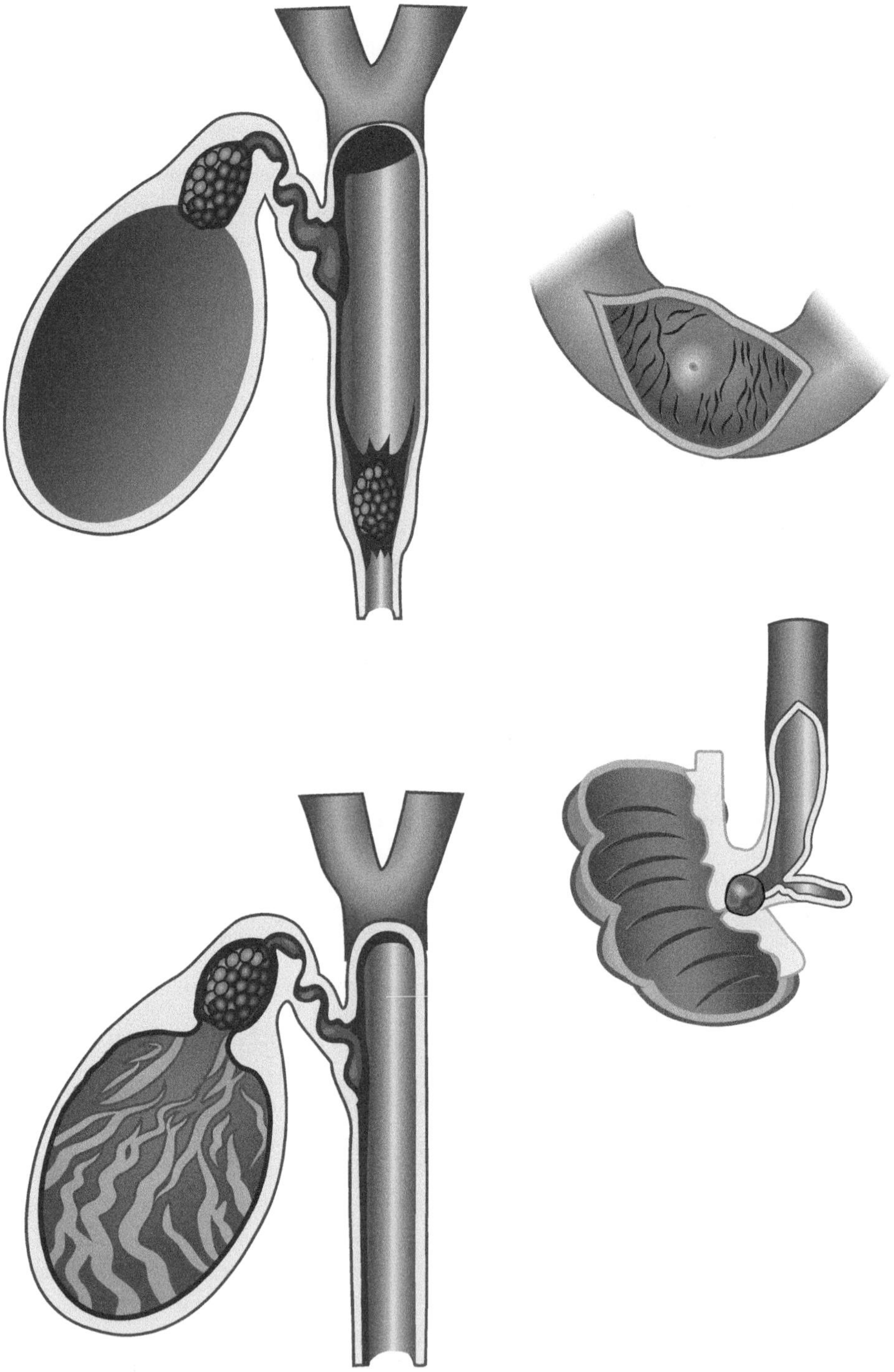

Figure 13.2 Complications of gallstones.

- The **critical view of safety** includes dissecting the GB off the cystic plate, visualizing a single cystic duct and single cystic artery entering the GB, and noting liver parenchyma posteriorly.
- An **intraoperative cholangiogram (IOC)** can be performed via the cystic duct or gallbladder infundibulum in order to clarify biliary anatomy if the critical view of safety cannot be achieved. It is also useful if choledocholithiasis was suspected preoperatively. See *Figure 13.3*.
- One should see the distal CBD draining into the duodenum caudally, plus both the right and left hepatic ducts and their respective subdivisions cephalad.
- If the cephalad hepatic ducts are not seen, the possibility of an inadvertently clamped or injured

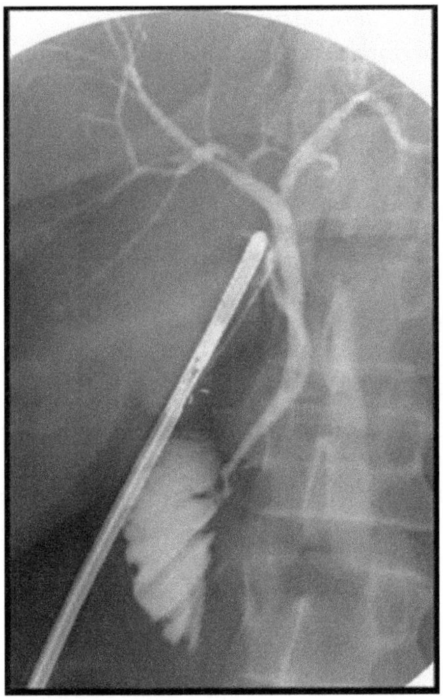

Figure 13.3 Intraoperative cholangiogram showing normal cephalad hepatic ducts and distal duodenal filling.

CHD should be entertained. One may repeat the IOC after giving morphine to constrict the sphincter of Oddi and encourage cephalad flow of contrast. Visualized hepatic ducts can then ensure there has been no injury.
  - If the duodenum does not fill, one may try glucagon to relax the sphincter of Oddi. Visualization of duodenal filling then excludes CBD obstruction, ligation, or transection.
- Once the cystic duct and artery are clipped and transected, the specimen can be removed in an endoscopic bag, and one should irrigate all bile spillage and remove all free-floating stones to prevent abscess formation.
- **Laparoscopic transcystic CBD exploration** is an option for an IOC suggesting obstruction and a CBD that is <8 mm. See *Figure 13.4*.
  - A **saline flush** via the cystic duct, plus **glucagon 1–2 mg,** can help clear stones that are <3 mm in size
  - A wire can be passed all the way through the ampulla into the duodenum for instrument and sheath exchange, and **balloon dilation** of the cystic duct, CBD, and ampulla.

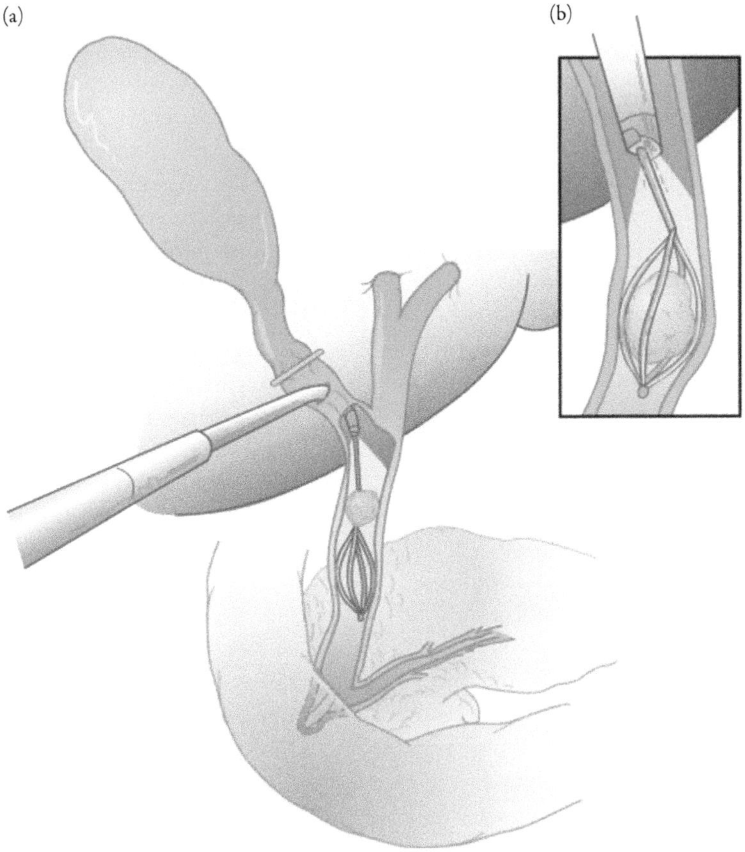

Figure 13.4 Laparoscopic trans-cystic common bile duct exploration.

- A **choledochoscope** can be passed over the wire, then **basket retrieval**, or laser or shock wave **lithotripsy** can be performed.
- A completion cholangiogram should note that the CBD is patent and there are no cephalad CHD stones.
- Then the cystic duct should be clipped or ligated.

- **Laparoscopic transcholedochal CBD exploration** is an option for an IOC suggesting obstruction and a very dilated CBD >8 mm diameter, a visualized CBD stone >10 mm, or any stones seen cephalad of the cystic duct residing in the CHD.
  - Compared to cystic duct access, direct CBD access allows retrieval of larger stones and cephalad stones, resulting in an increased clearance rate.
  - A **1–2 cm longitudinal choledochotomy** is made with direct insertion of a choledochoscope and basket retrieval of stones.
  - The choledochotomy is closed with running absorbable monofilament, and a **14F T-tube** can be placed if desired.
  - A completion cholangiogram can be obtained via the cystic duct to ensure the closure is secure and there are no more stones.
  - If a T-tube was used, a post-op cholangiogram via the T-tube can be performed 3–4 weeks later, and the tube gradually backed out.
- **Open CBD exploration** is an option if the above interventions fail to relieve a common duct obstruction. A Kocher maneuver allows for palpation of the distal CBD.
  - A 1.5 cm longitudinal choledochotomy is made, and stones are manually milked back. Randall stone forceps, flexible spoons, duct irrigation, choledochoscopy, and lithotripsy are used to assist with extraction.
  - Rarely, a duodenotomy and transduodenal sphincterotomy at the 11 o'clock position can help remove a large impacted stone. The CBD wall can then be sewn to the duodenal mucosa to maintain patency.
  - A diverting choledochoduodenostomy or choledochojejunostomy can bypass a large distal obstructing stone that cannot be removed.
- A **subtotal laparoscopic cholecystectomy** can be performed if there is dense scar and inability to successfully isolate the cystic duct and artery or clear the GB off the cystic plate.
  - A portion of the GB is left adhered to the liver, the exposed mucosa is cauterized, the cystic duct aperture is ligated, and a drain is left in place. In the future, only 5% need a completion cholecystectomy.
- Conversion to an **open cholecystectomy** is needed in <10% of acute cholecystitis cases, and <5% of elective cases for symptomatic cholelithiasis.

- 2. **Endoscopic retrograde cholangiopancreatography (ERCP)** is the primary method of removing common bile duct stones after US or MRCP suggests their presence. ERCP is acceptable to perform in the 3rd trimester of pregnancy.
  - It should be completed prior to cholecystectomy in patients who are **jaundiced (TB>4), have cholangitis, harbor a large CBD stone seen on imaging, or have non-resolving gallstone pancreatitis.** A laparoscopic cholecystectomy can then be completed during the same hospital admission.
  - If CBD obstruction is suspected but the above criteria are not met, the obstruction may be transient in nature and associated with **microlithiasis or sludge** only. In these cases, it may **resolve on its own** and the patient can move on to laparoscopic cholecystectomy with or without an intraoperative cholangiogram, without incurring the risks of an ERCP (see above).
  - **Magnetic resonance cholangiopancreatography (MRCP)** should be performed when CBD obstruction is suspected clinically and the above criteria are not met, yet the etiology of the obstruction is unclear. In such cases, MRCP can provide a definitive diagnosis of stricture or malignancy, as opposed to stones. It also avoids the complication profile of an ERCP when no immediate intervention is needed.
  - **ERCP** is performed with the patient in a prone position using a side viewing scope. In this orientation, the pancreatic duct (PD) sits at 1 to 3 o'clock within the ampulla of Vater, and the common bile duct (CBD) is at 11 o'clock. See Figure 13.5.
  - Maneuvers to clear the CBD of debris include sphincterotomy, balloon sphincteroplasty, shock wave or laser lithotripsy, cholangioscopy, and balloon or basket stone retrieval.
  - A **plastic stent** can be placed and may remain in the transpapillary position for 3–6 months if needed. A transpapillary nasobiliary tube can also be left to function as an external biliary drain.
  - **Uncovered metal stents** are used for longer term patency in the setting of strictures or malignancy.
  - A stent without sphincterotomy can be performed if the patient has an elevated INR.
  - Fine needle aspiration (**FNA**) used for suspicious masses can be diagnostic for malignancy up to 30% of the time.
  - **Brushings** are diagnostic up to 60% of the time.
  - **Post-procedure pancreatitis** occurs in 3%–5%. Generally these patients are young and had a normal pre-procedure bilirubin. Prophylactic pancreatic duct stents or rectal indomethacin have been used to try and prevent pancreatitis.

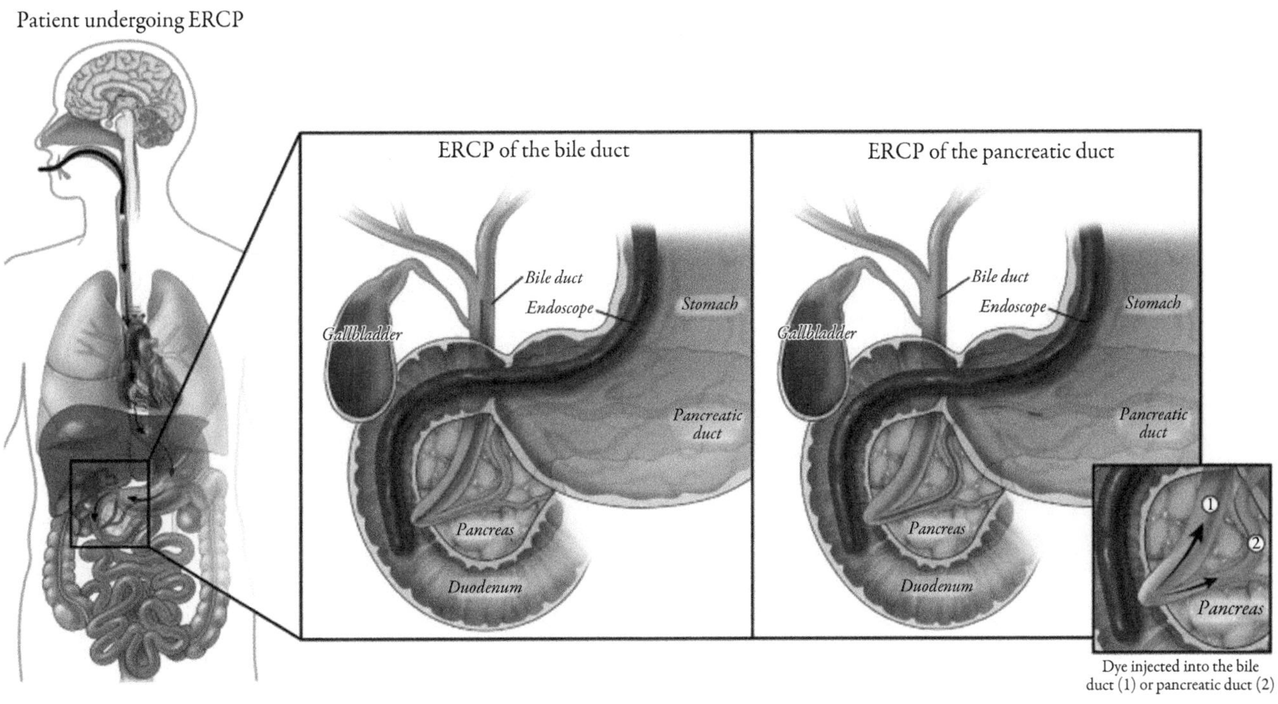

Figure 13.5 Endoscopic retrograde cholangiography with sphincterotomy for stone retrieval.

- A less likely complication is **perforation** of the duodenum, ampulla, or CBD.
- If after an EGD a patient has pain and retroperitoneal air of unclear origin on CT scan, this is likely an occult perforation from the EGD and can be managed conservatively, provided the patient is clinically stable.
- **Hemobilia** is from iatrogenic injury to the biliary mucosa.

- 3. **Antibiotics alone for 7–14 days** is an option for the treatment of cholecystitis in a high surgical risk patient. If not improving in 72 hours, consider cholecystostomy tube or a cholecystectomy. An ERCP can be added if the patient is suspected of having concomitant cholangitis.
- 4. A **transhepatic cholecystostomy tube** is used for those who are unable to tolerate laparoscopic surgery, have multiple comorbidities, or have failed a trial of antibiotics. This may be definitive treatment in some cases, since recurrent cholecystitis occurs in <30%.
  - Consider GB necrosis if the patient has no clinical improvement after an appropriately positioned cholecystostomy tube draining the lumen of the GB.
  - A **cholangiogram should be done via the cholecystostomy tube at 2–4 weeks.** One may clamp the tube if the cystic duct is patent on cholangiogram. Shortly thereafter, the tube may be removed. Then one can make a case-by-case decision as to whether or not to perform a future cholecystectomy.
    - If the cystic duct is still obstructed at the time of cholangiogram, continue leaving the tube open to drain. If the patient is in better clinical condition at this juncture, one should consider performing a cholecystectomy.
    - For a persistent cystic duct obstruction in a patient who still remains a poor surgical candidate, an endoscopic transcystic duct stent can be placed. This extends from the GB lumen through a sphincterotomy at the ampulla, and into the duodenum. This internalizes GB drainage and allows one to remove the cholecystectomy tube.

- **COMPLICATED CHOLECYSTIC SCENARIOS**
  - A **hydropic gallbladder** can occur in elderly diabetic men from cystic duct obstruction and increased mucous secretion by the GB infundibular epithelium. The GB may be very large, tense, and difficult to handle intraoperatively, requiring aspiration of contents before dissection can be undertaken.
  - **Emphysematous cholecystitis** is due to **C. perfringens,** usually in an elderly male diabetic. Gas will be noted in the wall of the GB, so the artifact may result in a negative US. CT will depict the gas, however. Treatment is with emergent laparoscopic or open cholecystectomy.
  - **Mirizzi syndrome** is caused by a stone chronically lodged in the neck of the GB producing extrinsic compression of the CHD, or a CHD stricture due to neighboring, chronic cystic duct or GB inflammation. See *Figure 13.6.* A laparoscopic cholecystectomy should be attempted; however, subtotal resection, open conversion, or a Roux-en-Y hepaticojejunostomy may be needed.

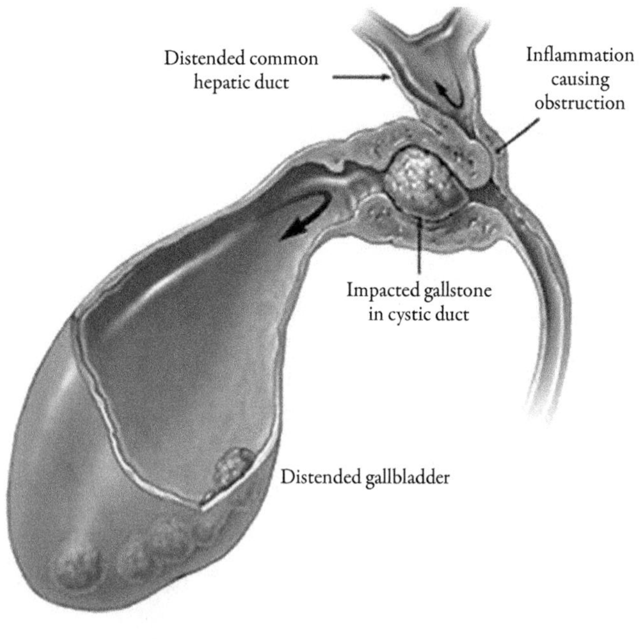

Figure 13.6 Mirizzi syndrome.

- **Gallstone ileus** is due to erosion of a large stone through the GB fundus into the duodenum. The stone then progresses through the SB lumen and ultimately obstructs at the ileocecal valve. See *Figure 13.7*. Patients may have secondary pneumobilia as air rises from the duodenal lumen into the biliary tract. The stone could alternatively erode into the jejunum, hepatic flexure of the colon, or the stomach, causing gastric outlet obstruction. Treatment is by **laparotomy and enterotomy** to remove the stone, and run the bowel to exclude other stones. Most patients are left with a persistent cholecystoenteric fistula and do not have issues with this. Cholecystectomy and takedown of the cholecystoenteric fistula to prevent cholecystitis, cholangitis, or recurrent ileocecal obstruction from a large stone is not routinely done at the index operation.
- **Cholecystitis in pregnancy.** 1st and 2nd trimester patients can receive laparoscopic cholecystectomy, as it reduces the risk of fetal death and improves quality of life. 3rd trimester patients should be treated with antibiotics with or without a cholecystostomy tube. Laparoscopic or open cholecystectomy is reserved for failure of conservative management.
- Indications for **prophylactic cholecystectomy** in asymptomatic patients include the following:
  - 1. **Polyps** >1 cm, polyps of any size with concomitant gallstones, or polyps of any size with primary sclerosis cholangitis.
  - 2. Patients who have had extensive small bowel resection and **short bowel syndrome.** The subsequent prolonged TPN use and impairment to bile salt reabsorption predispose to future gallstone formation.
  - 3. **Carcinoid syndrome** expected to require octreotide, since this will cause biliary stasis and gallstone formation.
  - 4. **Focal mucosal calcifications** that are suspicious for carcinoma. "Porcelain" diffuse intramural calcifications are no longer an indication for prophylactic cholecystectomy.
  - 5. **Patients who have a Roux-en-Y reconstruction (RxY) after gastrectomy or gastric bypass.** With this anatomy, future access of the CBD via ERCP would be difficult should any choledocholithiasis ensue.

OTHER BENIGN GB PATHOLOGY

- **Adenomyomatosis** is a benign, fundal thickened nodule of mucosa and invaginated muscle derived from Rokitansky-Aschoff sinuses. Treatment is with cholecystectomy if symptomatic or there is diagnostic uncertainty.
- **Adenomatous hyperplasia** is a benign overgrowth of the mucosal lining.
- **Cholesterolosis** are cholesterol deposits within mucosal macrophages, producing a strawberry gallbladder appearance.
- **Polyps** are most commonly composed of cholesterol. See *Figure 13.8*. Other polyps may be inflammatory or adenomatous in morphology. A single polyp is more concerning for cancer than multiple small polyps. A CT scan can exclude invasive malignancy beyond the confines of the gallbladder in an older patient with a sessile adenoma. Treatment is with **cholecystectomy** if the patient is >50 years of age or symptomatic, if the polyp is >1 cm, sessile, or associated with stones or primary sclerosing cholangitis (PSC). No biopsy is needed. If there is no indication for a cholecystectomy, then serial US can be done q6–12 months.

GALLBLADDER ADENOCARCINOMA

- **PATHOPHYSIOLOGY**
  - GB carcinoma has a female predominance and is more common than cholangiocarcinoma.
  - Neoplasia develops secondary to chronic irritation of the GB mucosa from:
    - 1. A large stone; or
    - 2. A long common channel between the CBD and pancreatic duct allowing pancreatic refluxate to ascend to the GB.
  - T-1a lesions are limited to the superficial lamina propria.
  - Since there is no muscularis mucosa or submucosa, deeper lesions immediately encroach upon the muscularis propria and the deeper subserosal layer.

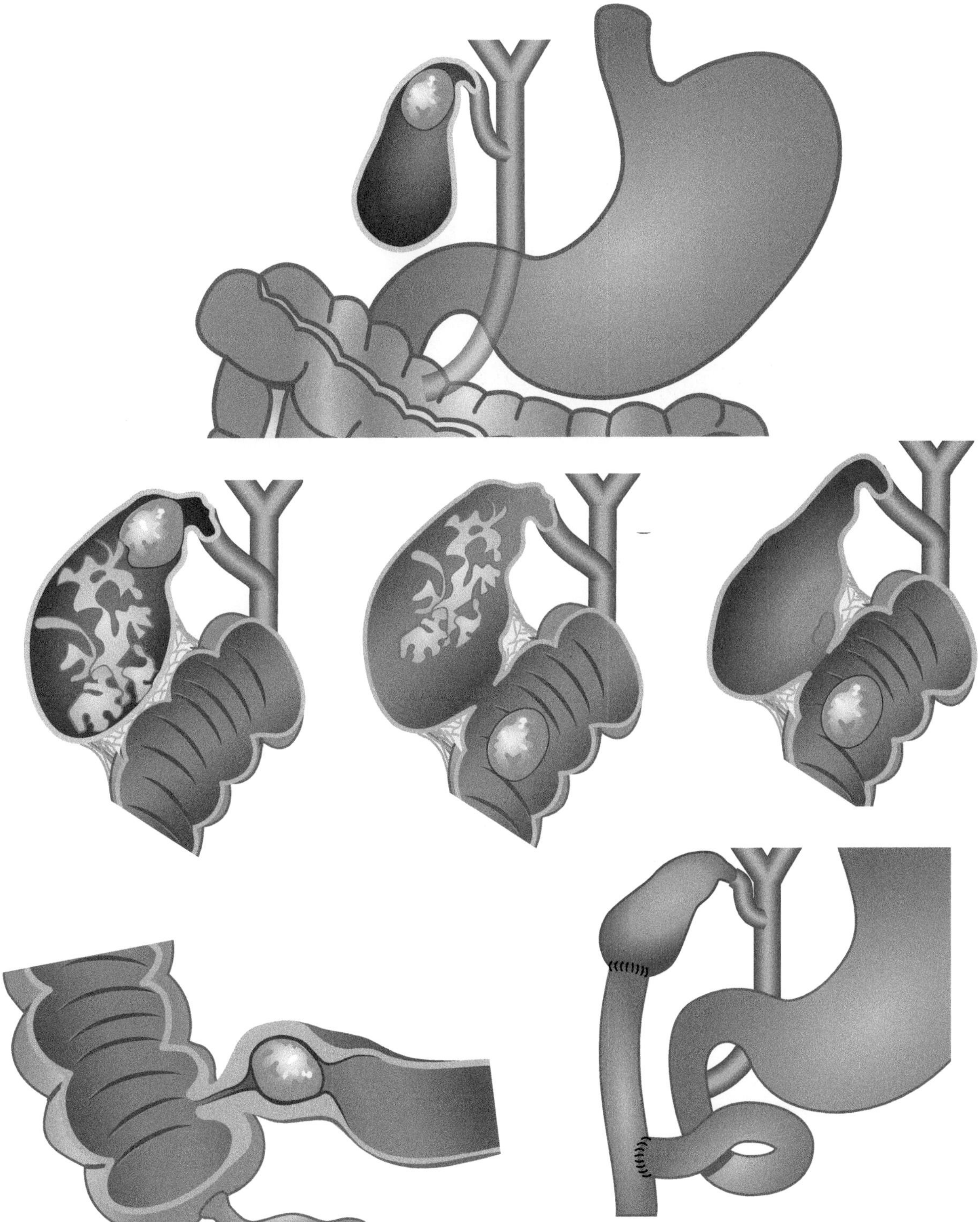

Figure 13.7 Etiology of gallstone ileus as a cholecystoduodenal fistula.

- Advanced disease includes the following:
  - Stage 3a: local invasion of adjacent structures.
  - Stage 3b: lymph nodes positive.
  - Stage 4a: vascular invasion by the primary.
  - Stage 4b: distant metastases with or without vascular invasion by the primary.
- **DIAGNOSIS**
  - Most tumors are **discovered incidentally on pathology of a cholecystectomy (CCY) specimen.** About 1% of all CCY specimens have carcinoma.
  - On the other hand, tumors may be larger and seen **incidentally on imaging** performed for

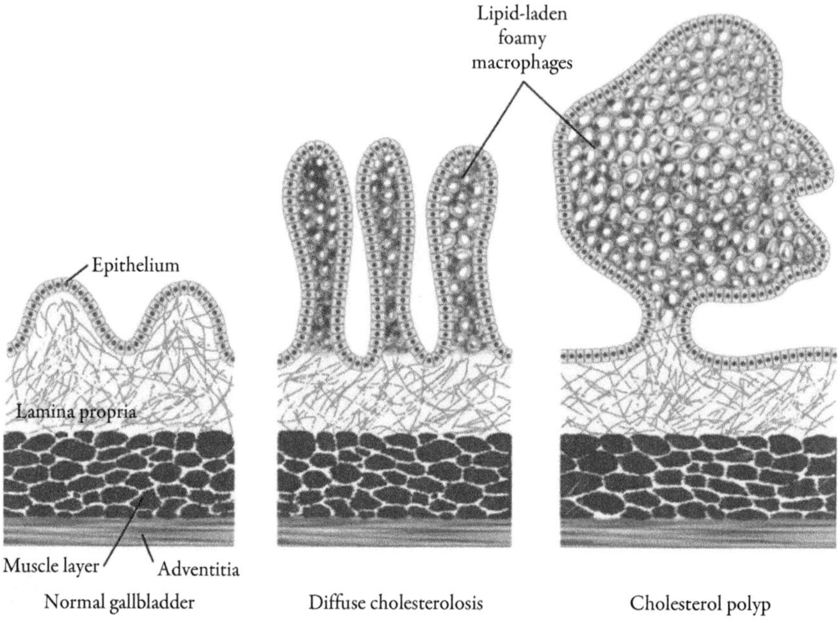

Figure 13.8 Genesis of a cholesterol polyp.

another reason. 60% of these are located in the fundus.

- Some patients present with **RUQ pain, with or without jaundice from CHD obstruction. Subsequent imaging then suggests the diagnosis of GB malignancy.**
- **US** is the initial imaging modality, followed by **MRCP, or CT** to delineate the extent of disease.
- **CT chest, abdomen, and pelvis is done for staging**, and PET imaging can be added for any suspicious nodes or lesions seen.

- **TREATMENT**
  - **T1a** lesions are newly **discovered at pathologic examination** of a CCY specimen and are treated completely by **CCY alone.** No further imaging or therapy is needed.
  - **T1b–T2** lesions noted on a CCY specimen need a re-operation for cystic, portal, and hepatic artery **lymphadenectomy, and re-resection of the cystic duct stump margin.**
    - Extrahepatic duct resection with a **RxY HJ** would need to be added if the the cystic duct margin is positive.
    - A non-anatomic wedge or a complete **resection of IVb and V** may also be necessary. Liver resection is important because the normal laparoscopic cholecystectomy dissection plane is subserosal, and hence leaves GB serosa on the undersurface of the liver. This will result in recurrence and positive nodes if the original lesion was T2.
  - **T3–4** lesions need the above and also possibly an extended **right hepatectomy (segments IVa, IVb, V–VIII)** to achieve an R0 resection.
  - Port recurrence can occur at extraction or non-extraction port sites in up to 10% of cases. However, resection of port sites at the time of the initial operation does not seem to help prevent recurrence and is not recommended.
  - **Adjuvant chemotherapy** can be added for a primary that is T1b or greater, N1 status, or an R1 resection. Tumors are generally chemoresistant, though.
    - 5-year survival is almost 100% for T1a lesions, otherwise there are stepwise decrements at higher T stages and node positivity.

BENIGN COMMON BILE DUCT PATHOLOGY

- **CHOLEDOCHOLITHIASIS**
  - A stone in the common bile or common hepatic ducts. See *Figure 13.2*. This can be the primary component of the clinical presentation, or can be seen secondarily in 5% of patients undergoing a cholecystectomy.
    - If a patient has had their GB removed within the past 2 years and they present with choledocholithiasis, the stone is considered a **secondary stone (cholesterol or black)** that was originally generated in the GB lumen.
    - If it has been > 2years since their CCY, then the stone is a **primary brown stone** generated within the CBD itself, and associated with downstream obstruction, stricture, or CBD diverticula.
  - **DIAGNOSIS**
    - **US** showing CBD or CHD dilation in the appropriate clinical setting. Occasionally, but not always, the stone itself may be seen on US.

- **MRCP.** This is a T2 image focused on intra- and extrahepatic bile tracts and can identify the cause of CBD or CHD dilation. Stones of varying sizes, a stricture, or malignancy can all be directly visualized by MRCP.

- **TREATMENT**
  - **Endoscopic retrograde cholangiopancreatography (ERCP)** is the primary method of removing CBD stones after US or MRCP suggests their presence. ERCP is acceptable to perform in the 3rd trimester of pregnancy.
    - It should be performed prior to cholecystectomy in patients who are **jaundiced (TB>4), have cholangitis, harbor a large CBD stone seen on imaging, or have non-resolving gallstone pancreatitis.** A laparoscopic cholecystectomy can then be completed during the same hospital admission.
    - If CBD obstruction is suspected but the above criteria are not met, the obstruction may be transient in nature and associated with **microlithiasis or sludge** only. In these cases, it may **resolve on its own** and the patient can move on to laparoscopic cholecystectomy with or without an intraoperative cholangiogram, without incurring the risks of an ERCP.
    - **MRCP** should be performed when CBD obstruction is suspected clinically and the above criteria are not met, yet the etiology of the obstruction is unclear. MRCP can provide a definitive diagnosis of stricture or malignancy, as opposed to stones. It also avoids the complication profile of an ERCP when no immediate intervention is needed.
    - **ERCP** is performed with the patient in a prone position using a side viewing scope. In this orientation, the pancreatic duct (PD) sits at 1 to 3 o'clock within the ampulla of Vader, and the CBD is at 11 o'clock.
    - Maneuvers to clear the CBD of debris include sphincterotomy, balloon sphincteroplasty, shock wave or laser lithotripsy, cholangioscopy, and balloon or basket stone retrieval.
    - A **plastic stent** can be placed and may remain in the transpapillary position for 3–6 months if needed. A transpapillary nasobiliary tube can also be left to function as an external biliary drain.
    - **Uncovered metal stents** are used for longer term patency in the setting of strictures or malignancy.
    - A stent without sphincterotomy can be performed if the patient has an elevated INR.
    - Fine needle aspiration (**FNA**) used for suspicious masses can be diagnostic for malignancy up to 30% of the time.
    - **Brushings** are diagnostic up to 60% of the time.
    - **Post-procedure pancreatitis** occurs in 3%–5%. Generally these patients are young and had a normal pre-procedure bilirubin. Prophylactic pancreatic duct stents or rectal indomethacin have been used to try and prevent pancreatitis.
    - A less likely complication is **perforation** of the duodenum, ampulla, or CBD. If after an EGD a patient has pain and retroperitoneal air of unclear origin on CT scan, this is likely on occult perforation from the EGD and can be managed conservatively provided the patient is clinically stable.
    - **Hemobilia** is from iatrogenic injury to the biliary mucosa.
  - **Laparoscopic or open transcystic or transcholedocho CBD and CHD explorations,** transduodenal sphincterotomy, or choledochoduodenal or choledochojejunal bypass can be used if ERCP is not feasible (see cholecystitis section above).
  - **Percutaneous, transhepatic biliary drains** may be needed if ERCP is not sufficient to recanalize (see cholangitis section below).

- **CHOLANGITIS**
  - This is a bacterial infection of the CBD or CHD after inoculation retrograde via the ampulla in the setting of choledocholithiasis, biliary duct injury, biliary stricture, primary sclerosing cholangitis (PSC), biliary diverticulum, choledochal cyst, malignancy, or an obstructed or kinked stent.
    - E. Coli and Klebsiella are the most common organisms.
    - Pneumobilia can be secondary to cholangitis with gas-forming organisms, or may be the benign sequelae of prior ERCP or gallstone ileus.
    - Once choledochal pressure exceeds 20 cm H20, there is reflux into the venous system and potential bacteremia.
  - **DIAGNOSIS**
    - Patients present with fever, RUQ pain, and jaundice. Mental status changes and shock indicate an advanced infection.
    - **US** is the initial imaging study and can be diagnostic if it shows CBD dilation in the correct clinical picture. Often the etiology of the obstruction will not be directly visible on US.
    - **MRCP** may be used to diagnose the etiology of CBD obstruction, if it is not evident by US and the clinical picture alone.
  - **TREATMENT**
    - Start with **broad antibiotics.** In a majority of patients, the obstruction is due to sludge or microlithiasis that will pass on its own.
    - **ERCP** is needed in **15%** of patients to clear a CBD obstruction that doesn't resolve spontaneously. An accompanying cholangioscopy, sphincterotomy, or stent may be needed.
    - **Percutaneous transhepatic cholangiography (PTC) and percutaneous biliary drain (PBD)** are useful for patients with a cephalad biliary obstruction or CBD transection that ERCP cannot traverse. The PBD can be

replaced with an internal-external drain if the lesion can be crossed at a later date.

- **Surgical CBD and CHD exploration,** transduodenal sphincterotomy, or choledochoduodenal or choledochojejunal bypass can be used if ERCP and PBD do not adequately address underlying distal biliary pathology.
- **Recurrent cholangiohepatitis** is due to brown stones in the setting of bacterial or parasitic infection. Patients may have liver abscesses as well. Treatment is with antimicrobials, ERCP and PBD, then a RxY hepaticojejunostomy.

- **BILE DUCT INJURIES**
  - These most commonly occur from direct injury or traction during a cholecystectomy.
  - **DIAGNOSIS**
    - Symptoms include postoperative nausea, vomiting, and jaundice.
    - **US** will show either a **biloma, dilated intra- or extrahepatic ducts, or both.**
      - **An US or CT guided percutaneous drain** can be placed into the biloma for drainage. A subsequent **bilogram or sinogram** can be performed by injection of contrast through the drain and into the fluid collection to define the original ductal injury.
      - **An US or CT guided percutaneous transhepatic biliary drain** can decompress the cephalad dilated ducts. Contrast may be injected to define the injury.
  - **TREATMENT**
    - **Intraoperatively** discovered **minor duct injuries can be ligated,** and sidewall injuries of the common duct that are **<1 cm can be repaired over a T-tube** placed nearby or within the injury.
    - If **discovered within 72 hours post-op,** perform image guided biloma drainage and bilogram. If a sidewall, cystic duct, or minor duct injury is noted, then **ERCP and stent** can be definitive treatment.
    - Any more significant injury discovered intraoperatively or within 72 hours post-op will need a **RxY HJ or CDJ with or without a silastic stent** placed across the anastomosis and exiting transhepatic and transcutaneously.
    - If **discovered late,** drain the biloma, drain the cephalad dilated biliary tree with PBD, define the injury with contrast injection, and **delay surgical repair for 6–12 weeks** in case there is any concomitant vascular injury. This allows the true healthy margin of biliary ends and liver parenchyma to be declared.
    - Specific management by injury type A–E is as follows. See *Figure 13.9.*
      - **A: cystic duct or duct of Luschka leak.** Treatment is with ligation if intraoperatively discovered, or biloma drainage, ERCP, sphincterotomy, and covered stent if discovered postoperatively.
      - **B: ligated aberrant R hepatic duct.** No treatment unless symptomatic, in which case the patient needs a right posterior hepatic sectionectomy.
      - **C: transected aberrant right hepatic duct.** No treatment unless symptomatic and patient needs biloma drainage with or without a posterior sectionectomy.
      - **D: <50% sidewall injury of CHD or CBD.** Treatment if discovered intra-op is repair and T-tube placement at the injury site or nearby. If discovered postoperatively, then ERCP, sphincterotomy, and stent across site. Thermal injuries may need repair with T-tube placed at a separate site, or a RxY HJ or CDJ.
      - **E: transection of main ductal system.** Needs RxY HJ or CDJ if discovered within 72 hours postoperatively. Delay repair 6–12 weeks if discovered >72 hours postoperatively.
        - **E1**: injury >2 cm inferior to duct bifurcation.
        - **E2:** <2 cm from duct bifurcation.

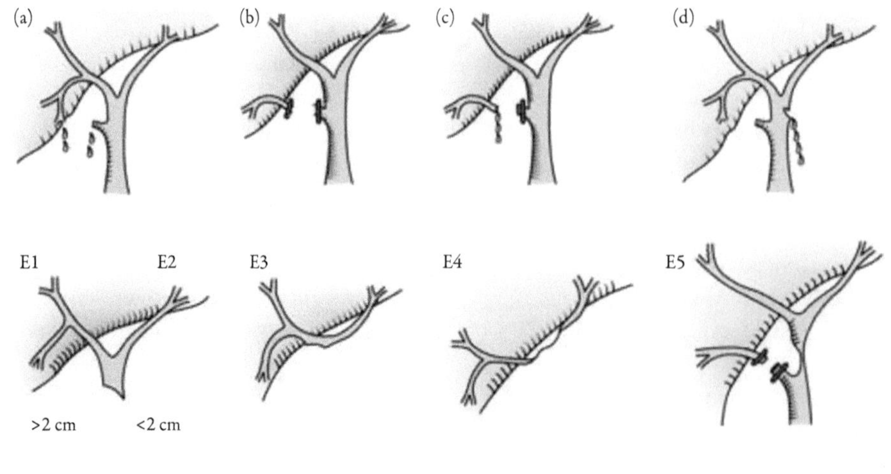

Figure 13.9 Types of biliary tract injuries and strictures.

- **E3**: injury at bifurcation, but R&L ducts still communicate. Future RxY HJ can be anastomosed to this common site.
- **E4**: stricture of both R&L ducts preventing their merger. RxY HJ needs separate anastomoses and an intrahepatic dissection to access duct divisions.
- **E5**: aberrant anterior or posterior right hepatic duct injury plus a CHD stricture.

•BILIARY STRICTURES

- Most commonly secondary to **prior surgery** resulting in traction-related arterial insufficiency and ductal ischemia, or direct injury to the bile duct.
- Other causes include pancreatitis (often a distal stricture in this case), prior biliary–enteric or biliary–biliary anastomosis, erosion from biliary stones, cholangitis (pyogenic or primary sclerosing), or malignancy.
- Depending on the etiology, strictures can be intrahepatic or extrahepatic and classified like biliary injuries (see specific injuries, "E" above).
- In the chronic setting, patients often have **positive bile bacterial cultures.**
- **DIAGNOSIS**
  - Patients present with biliary stasis or obstruction in the setting of low suspicion for a gallstone etiology. This prompts imaging by **MRCP**, or performance of an **ERCP with brushings or biopsy** to exclude malignancy.
- **TREATMENT**
  - a) **Resection and RxY HJ or CDJ** with or without a temporary transhepatic silastic stent. Choledochoduodenostomy (CDD) or a pancreaticoduodenectomy are options if a distal stricture exists with concerns for malignancy or duodenal obstruction. This is most often encountered in the setting of chronic pancreatitis.
  - b) Palliation can be with **ERCP, balloon or bougie dilation, and a plastic or self-expanding metallic stent** if the lesion can be crossed endoscopically. One may also obtain brushings, biopsy, and cholangioscopy to evaluate for malignancy at the stricture site.
    - Endoscopic treatment works better for anastomotic, rather than non-anastomotic, strictures.
    - If the ampulla is not accessible, practitioners may use EUS to directly cannulate the distal CBD transduodenally, or cannulate dilated intrahepatic biliary ducts in the left lobe of the liver via a transgastric-transhepatic route, to place an antegrade stent.
    - Alternatively, the endoscopist may perform a rendezvous procedure with a percutaneous transhepatic wire passed through the ampulla and recovered endoscopically within the duodenal lumen. This enables the endoscopist to use traction to enter the ampulla.
  - c) If the stricture cannot be crossed with ERCP, the next option for palliation is with a **percutaneous biliary drain (PBD) placed via a transhepatic route into the biliary tree cephalad of the stricture.**
    - This approach is **contraindicated if the patient is coagulopathic or has significant ascites.** In the latter case, peri-drain ascitic leakage and biliary liver leakage will occur, as the hepatic parenchyma cannot remain apposed to the parietal peritoneum due to the interposed ascites.
    - The PBD may be **converted to an internal-external transhepatic drain** if the lesion can eventually be crossed. The external component can later be capped, allowing for internal drainage and reestablishment of biliary–enteric circulation so the patient may utilize their bile salts for lipid absorption.
    - Via the percutaneous transhepatic route of the drain, a **permanent** (uncovered metallic), **long-term** (covered metallic), **or short-term** (plastic) **biliary stent** can be placed.
      - If a plastic stent is used, then at 3–12 months it can be removed and the drain pulled back cephalad of the prior stricture site.
      - A **contrast study** can be performed via the drain to show patency through the previously stented stricture site, and **manometry** can demonstrate cephalad pressure <20 cm H20. These two findings in a patient who is asymptomatic for 1–2 weeks without the stent portend that the drain can be removed, and likely the strictured area will remain patent at one year.
  - d) Consider **liver transplantation** for non-treatable intrahepatic strictures or primary sclerosing cholangitis.
- **HEMOBILIA**
  - This presents as an upper gastrointestinal bleed (UGIB) and is due to a direct communication between a hepatic vein, portal vein, or hepatic artery and an intrahepatic bile duct.
  - Hepatic arterial connections with the biliary tree are usually in association with a **pseudo-aneurysm** that developed post PBD, ERCP, liver biopsy, or trauma.
  - **DIAGNOSIS**
    - **Endoscopy** will show blood emanating from the ampulla, or **CTA** showing an intrahepatic arterial pseudo-aneurysm in the setting of an UGIB.
    - Biliary tracts may be obstructed by coagulated blood, leading to an increased TB and alkaline phosphatase.
  - **Treatment** is by **angioembolization** of any hepatic artery source.
    - If secondary to a PBD, then **upsizing** the drain may tamponade any portal or hepatic vein source.
- **CHOLEDOCHAL CYST**
  - This is due to an abnormally **long common channel between the distal pancreatic and common bile ducts.** Generally there is >1 cm

length of extraduodenal merger of the two ducts. The abnormality is present since infancy, but only diagnosed in young adulthood. **Women** are more often affected than men. Stasis, chronic inflammation, and epithelial dysplasia can eventually result in cholangiocarcinoma.

- **DIAGNOSIS**
  - **Patients have** symptoms of biliary insufficiency and alkaline phosphatase elevation, prompting an **MRCP.** There are multiple types.
- **TREATMENT**
- Depends on the type of choledochal cyst, as follows. *See Figure 13.10.*
  - **Type 1: extrahepatic focal or fusiform dilation primarily of the CBD**
    - This carries a higher malignancy risk within the dilated segment.
    - Treatment is with **resection** of the cystic area and a **Roux-en-Y hepaticojejunostomy** (RxY HJ).
  - **Type 2:** diverticulum off the CBD sidewall. Treatment is with **resection and CBD sidewall closure.**
  - **Type 3:** choledochocele within the duodenal wall. Treatment is with **sphincterotomy, marsupialization, or choledochoduodenostomy.**
  - **Type 4a** is type 1 dilation plus intrahepatic cysts. Treatment is with resection, including the intrahepatic component, and RxY HJ.
  - **Type 4b** has multiple extrahepatic cystic dilations. This has a high malignancy risk and is treated with resection and RxY HJ.
  - **Type 5** are intrahepatic choledochal cysts only. There is no extrahepatic component. This can present with pain secondary to Glisson's capsule stretch. Treatment is with **hepatic resection if focal, or transplantation** if bi-lobar and complicated by cholangitis, portal fibrosis, or cholangiocarcinoma.
- **PRIMARY SCLEROSING CHOLANGITIS (PSC)**
  - An autoimmune-mediated inflammation resulting in **heterogeneous fibrosis of both intrahepatic and extrahepatic biliary ducts.** See *Figure 13.11.* **Men** are more often affected than women.
- **DIAGNOSIS**
- Patients present with signs of biliary obstruction and increased **alkaline phosphatase** preceding total bilirubin (TB) elevation. Some patients will have concomitant IBD or pouchitis, colon cancer, GB polyps, GB malignancy, Reidel fibrosing thyroiditis, or retroperitoneal fibrosis.
- Confirmation is with **MRCP** followed by **ERCP** with brushings, biopsy, and cholangioscopy. Serum **autoimmune profiling** can be done thereafter. A **liver biopsy** may be necessary if only the smaller intrahepatic biliary ducts are involved.
- Concomitant **cholangiocarcinoma can be seen in the strictured areas** in 15% of cases. An elevated CA 19-9 and ERCP brushings and biopsy can confirm malignancy.
- PSC can ultimately result in **cirrhosis** secondary to widespread biliary occlusion, stagnation, and fibrosis.

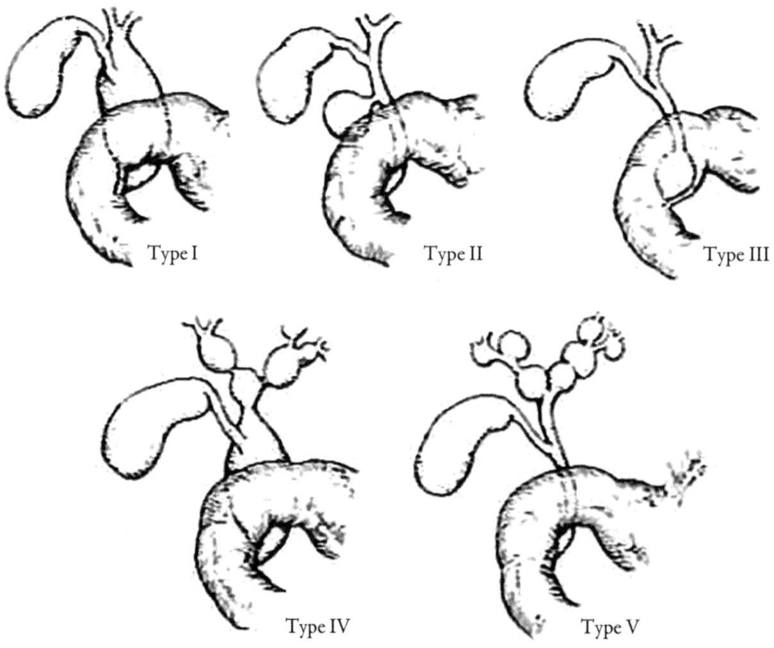

Figure 13.10 Types of choledochal cysts.

- **"Secondary" sclerosing cholangitis** is due to stones, trauma, recurrent infection, hepatic artery insufficiency, or surgery, and is not related to an underlying autoimmune process.
- **TREATMENT**
  - A dominant **extrahepatic stricture can be resected, and RxY right and left HJ** can be performed over transhepatic silastic stents.
  - Palliative options are similar to management of strictures in general, and include **ERCP, PBD, internal-external drains, and stents (see above)**.
  - **Liver transplant** is an option for concomitant cholangiocarcinoma or cirrhosis. In this case the entirety of the recipient native CBD should be resected, and a donor choledocho to recipient RxY construct of jejunum should be performed. One-fourth of transplant recipients have recurrent PSC in the graft. Recipients must be screened aggressively for colon cancer, since they are immune suppressed.
- **PRIMARY BILIARY CIRRHOSIS (PBC)**
  - This is intrahepatic biliary stenoses in a female patient.
  - **DIAGNOSIS**
    - PBC presents with indolent obstructive symptoms, pruritus, and xanthomas. MRCP and anti-mitochondrial antibodies will confirm the diagnosis.
    - Patients have an increased risk for cholangiocarcinoma.
  - **TREATMENT**
    - **ursodeoxycholic acid** and medical management of symptoms. Liver transplant is reserved for severe, refractory cases.

CHOLANGIOCARCINOMA (CCA)

- **INTRAHEPATIC CCA**
  - Accounts for 5%–10% of cases. Details are included in Chapter 12, Liver.
    - **DIAGNOSIS**
      - **Triple-phase liver CT** can help distinguish CCA from hepatocellular carcinoma (HCC), hepatic adenoma, hemangioma, or focal nodular hyperplasia (FNH). Imaging shows a **low attenuation lesion with biliary obstruction and liver capsule retraction.** This can be diagnostic and, if surgically resectable, no biopsy is needed.
        - T1 lesions are solitary with no vascular invasion.
        - T2 lesions are either solitary with vascular invasion, or multiple tumors without invasion.
        - T3 lesions extend to the visceral peritoneum.
        - T4 lesions have local invasion of periductal tissue.
      - **Diagnostic laparoscopy** can determine resectability. Any nodes outside of the porta hepatis contraindicates resection.
      - If no resection is planned, or imaging is equivocal and the diagnosis is uncertain, then a **biopsy** is indicated. This will show adenocarcinoma.
- **TREATMENT**
  - **Resection to negative margins and portal hilar lymphadenectomy.**
  - One may use preoperative portal vein embolization to maximize the functional liver remnant (FLR).
  - Chemotherapy and radiotherapy may be added for an R1 resection or any positive lymph nodes.
  - Adjunctive treatment and palliative options include ablation, TAE, TACE, or TARE.
  - Transplant is generally not an option for intrahepatic cholangiocarcinoma.
- **EXTRAHEPATIC CCA**
  - Accounts for 90% of cases.
  - Most cases are sporadic. A minority will be from chronic infection (recurrent cholangitis or Clostridium sinensis) or chronic inflammation associated with ulcerative colitis, primary sclerosing cholangitis, or choledochal cysts.
  - Neoplastic cells spread longitudinally along biliary tracts via lymphatics.
  - **Two-thirds of extrahepatic tumors are perihilar (Klatskin tumors)**. See *Figure 13.12.)*
    - **Type 1** tumor resides at CHD just inferior to bifurcation.
    - **Type 2** tumor resides at the bifurcation.
    - **Type 3a** tumors extend into the right hepatic duct.
    - **Type 3b** tumors extend into the left hepatic duct.
    - **Type 4** tumors involve both ducts.
  - **One-third of extrahepatic tumors are in the middle or lower common bile duct.**
  - **DIAGNOSIS**
    - Patients present with painless jaundice, pruritus, weight loss, and increased TB and alkaline phosphatase. Most cases have an increased CA 19-9.
    - **MRCP** can suggest the diagnosis by showing signs of biliary obstruction secondary to a stricture or mass. Sometimes the mass is very small, so a transition point between distended and collapsed biliary segments is all that is seen.
      - **T1** lesions are confined to the duct mucosa.
      - **T2** lesions extend beyond the wall of the duct.
      - **T3** lesions involve liver parenchyma, GB, pancreas, or unilateral HA or PV.
      - **T4** lesions invade the main PV, proper HA, or a hollow viscous.
    - **ERCP** with brushings, biopsy, and cholangioscopy may be needed for confirmation if a mass cannot be appreciated on MRCP.
  - **TREATMENT**
    - Perihilar Klatskin tumors are treated with **extrahepatic biliary resection and hepatic lobe resection** (if needed) to obtain a negative cephalad

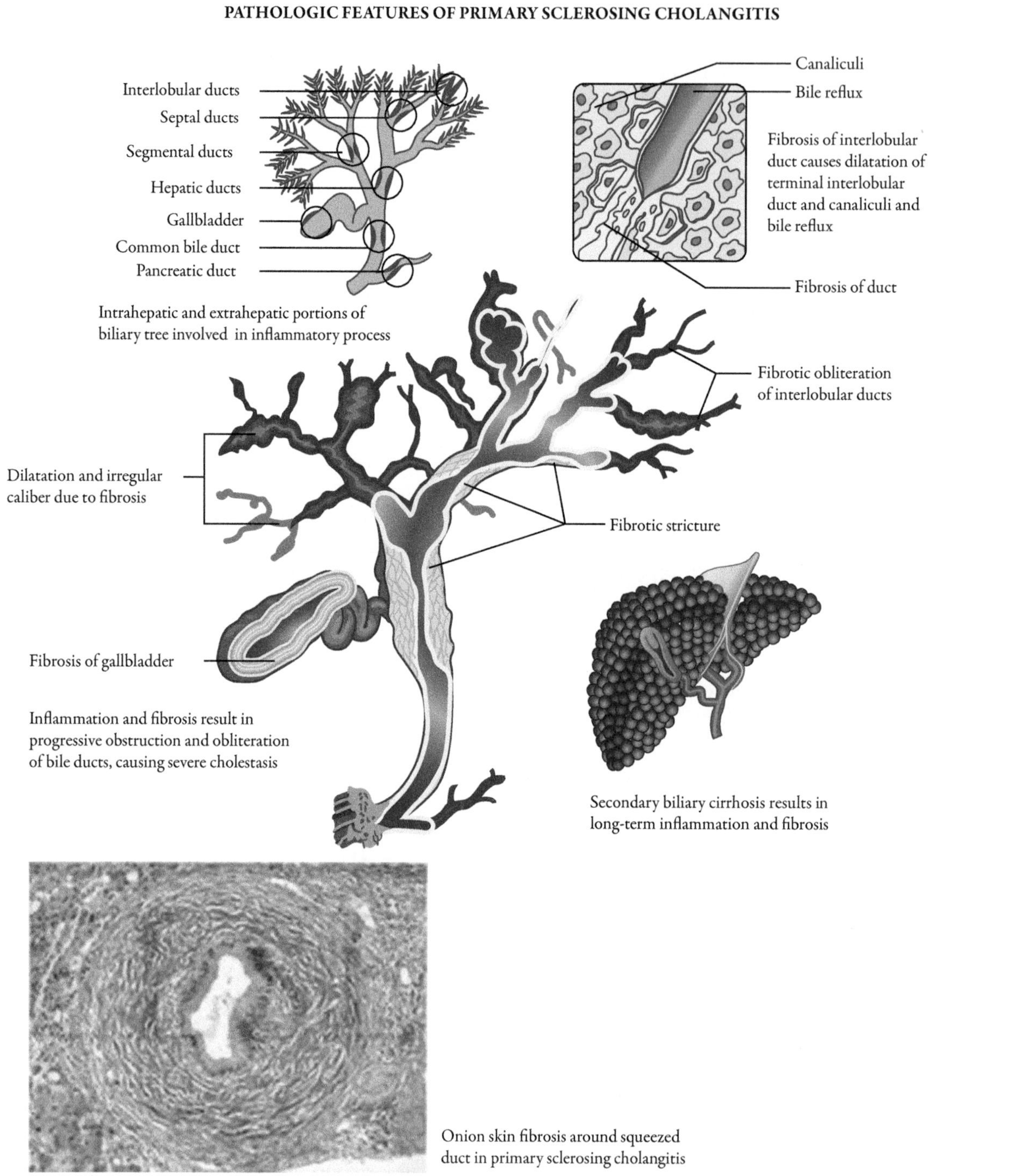

Figure 13.11 Sclerosing cholangitis.

duct margin on one side. If the caudal CBD margin was positive, one may extend the resection inferiorly as far as a pancreaticoduodenectomy. A **portal lymphadenectomy** should be done in all cases. Reconstruction is with a **RxY HJ**.

- **Preoperative ERCP or PBD** can decompress the biliary system and get TB <7. This is especially important for obstructing perihilar tumors that may need a hepatic resection or chemotherapy.
    - Neoadjuvant or adjuvant **chemotherapy** is used for any positive surgical margin or any positive node. Tumors are characteristically chemoresistant, though.
    - For nonresectable extrahepatic Klatskin CCA, **transplant after neoadjuvant therapy** is the best option for treatment, provided there is no nodal involvement or metastases. This is especially important in patients who have underlying PSC.
- Middle and lower CCA are treated with neoadjuvant therapy and **resection with RxY HJ or**

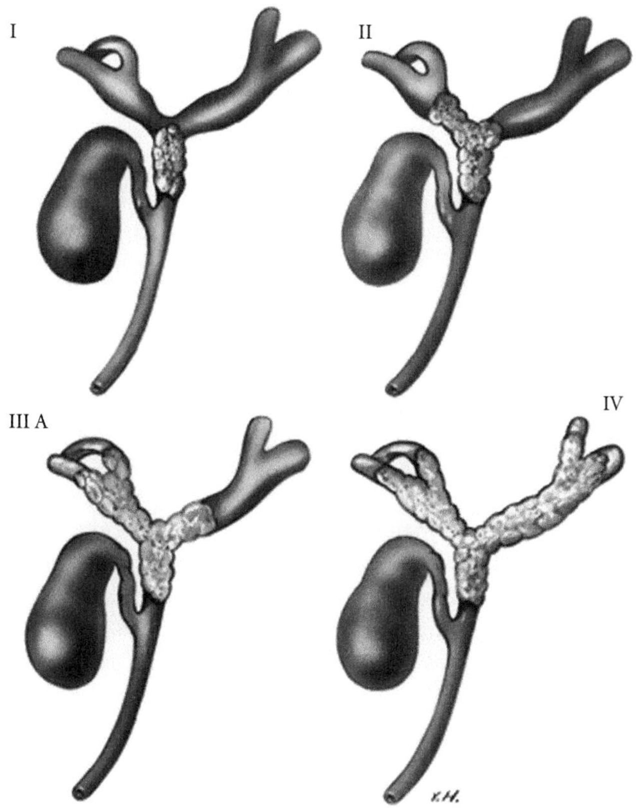

Figure 13.12 Klatskin type of extra-hepatic cholangiocarcinoma.

pancreaticoduodenectomy, respectively. A **portal lymphadenectomy** should be included in all cases.

- For those patients receiving a pancreaticoduodenectomy, use of an omental flap, octreotide, or a Braun enteroenterostomy can decrease the incidence of a pancreaticojejunal anastomotic leak.
- There is no need for preoperative biliary decompression in middle or lower CCA, since a hepatic resection is not needed.

- **Palliative options** include the following:
  - **Segment III biliary–enteric bypass.**
  - **Percutaneous biliary drain (PBD), internal-external drainage, or endobiliary** uncovered or covered metal **stents.** These can remain patent for 6–12 months before needing recanalization or replacement if the patient happens to live beyond this span of time. Bile or ascites leakage around a PBD can be managed by upsizing the drain for tamponade, or by placement of a pericatheter dermal suture.
  - **ERCP and a metal stent** is an option to prevent occlusion, and even deliver interstitial brachytherapy in middle or lower CCA.
  - **Photodynamic therapy.**
  - **Systemic chemotherapy** includes 5-FU, mitomycin, or doxorubicin for nonresectable lesions.

14.

PANCREAS

ANATOMY AND PHYSIOLOGY

- The organ develops from a ventral and dorsal bud.
  - The **ventral bud** contains the **Wirsung duct**, which drains the head and uncinate process
    - 70% of ducts merge with the common bile duct (CBD) outside of the duodenal wall and converge into a 1–2 cm common channel. 20% of Wirsung ducts merge with the CBD within the duodenal wall, and 10% open at a separate duodenal site from the CBD.
    - An **annular pancreas** results from failure of the dorsal bud to rotate clockwise. This leads to a second segment duodenal (D2) obstruction from the encircling pancreas. It is more common in Down syndrome patients. Treatment is with duodenoduodenostomy (DD) or duodenojejunostomy (DJ) to bypass the encirclement.
  - The **dorsal bud**: contains **Santorini duct,** which can drain through a separate opening within the duodenum.
    - **Pancreas divisum** results from failed fusion of Wirsung and Santorini ducts. The latter drains the majority of the organ during fetal life.
      - The blind tail of Wirsung should normally meet with the side of Santorini, and this combination should form the main pancreatic duct for adult life. The residual Santorini duct that drained into the minor papilla either regresses or remains as an accessory duct. *See Figure 14.1.*
      - If this does not happen, then Wirsung remains blind-ended, and Santorini remains independently draining the dorsal majority of the organ into the minor papilla sitting 2–5 cm proximal to the major papilla.
      - Most patients are **asymptomatic**, but some have pancreatitis from an insufficient Santorini. Individuals may have an associated genetic, enzymatic, or cellular predisposition to pancreatitis. Confirmation is by **MRCP, pancreatic protocol CT (PPCT) or EUS.**
      - Treatment involves **minor papilla sphincterotomy** using the pull technique over a wire, or needle-knife technique over a 3–5 Fr pancreatic duct stent that will self-extrude. Preventative measures for post-ERCP pancreatitis include a PD stent and rectal indomethacin.
      - Puestow longitudinal pancreatojejunostomy is an option for fibrosis and diffuse ductal dilation.
      - Pancreaticoduodenectomy is used if the fibrosis and calcification are limited to the head and the distal CBD, and ampulla is involved and obstructed.
      - A combination of a limited pancreatic head resection and longitudinal pancreatojejunostomy (LR-LPJ) is useful if the head is involved but the CBD remains uninvolved.
- A **heterotopic pancreas** in the duodenum is seen as a submucosal nodule with central umbilication on endoscopy.
- Arterial perfusion first goes to endocrine islet cells, and then portal channels take blood to ductal and acinar exocrine cells.
  - The inferior head and uncinate process receive perfusion from **pancreaticoduodenal vessels.**
  - Superior head/neck/body/tail are perfused by the **inferior pancreatic artery,** which runs parallel and inferior to the splenic artery. This is augmented by the **dorsal, great, and caudal pancreatic arteries,** which are tributaries off the splenic artery that connect it to the inferior pancreatic artery. *See Figure 14.2.*
- The superior mesenteric vein (SMV) resides below the pancreatic neck. The middle colic vein, right gastroepiploic vein, and inferior pancreaticoduodenal veins drain into it.
- **Endocrine Function**
  - **Alpha cells release glucagon**, which decreases intestinal motility and relaxes smooth muscle. It decreases gastric acid, pancreatic secretions, and biliary secretions. Glucagon increases glycogenolysis, gluconeogenesis, proteolysis, lipolysis/ketogenesis. Glucagon is also a positive inotrope and chronotrope.
  - **Beta cells release insulin.** They are stimulated by oral and intravenous glucose, vagal nerve activity, cholecystokinin (CCK), and B agonists.
  - **Delta cells release somatostatin** which decreases exocrine and endocrine activity. Delta cells are stimulated by fat intake and inhibited by vagal activity.
  - **Epsilon cells release ghrelin** to stimulate appetite. These cells are also present in the gastric fundus.

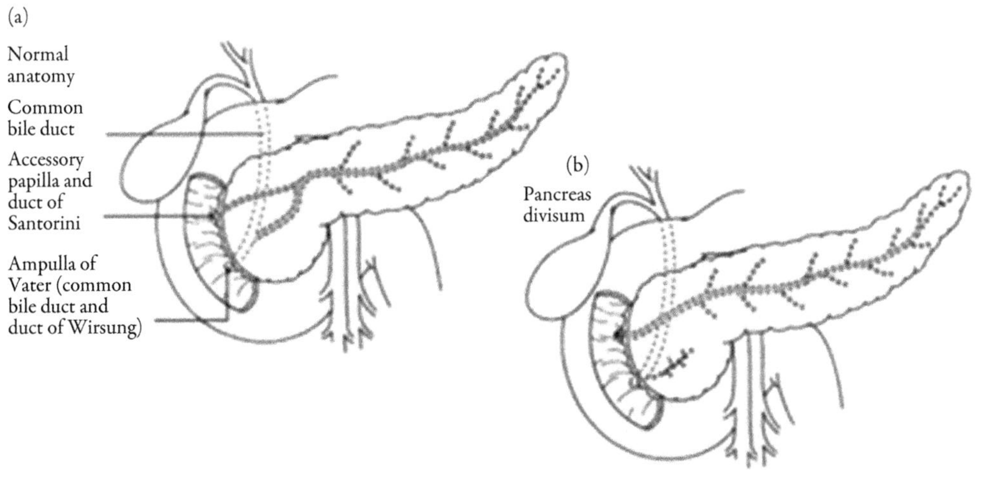

Figure 14.1 Normal pancreatic anatomy and pancreas divisum.

- **F cells release pancreatic polypeptide** to decrease pancreatic and biliary exocrine secretions and increase hepatic insulin sensitivity. This **decreases circulating serum glucose.** F cells reside mainly in the pancreatic head. Therefore, patients may have impaired glucose tolerance or brittle type3 DM after a pancreatic head resection.
- **Exocrine Function**
  - **Acinar cells release amylase and lipase** in their active forms into the pancreatic duct. **Trypsinogen** is also transferred into the duct, but is activated to trypsin by enterokinases in the duodenal lumen. **Chymotrypsinogen** is activated to chymotrypsin by trypsinogen. **Exopeptidases** are released in response to vagal activity and CCK.
  - **Ductal cells release bicarbonate** into the lumen in response to vagal activity and secretin. The latter normally produces increased flow rates, with increased HCO3 and less chloride in the ductal secretions.

ACUTE PANCREATITIS

- **PATHOPHYSIOLOGY**
  - Acute inflammation is characterized by autodigestion of pancreatic tissue and interstitial edema. This is secondary to zymogens fusing with lysozymes and activating within pancreatic cells and parenchyma. *See Figure 14.3.*
  - Inflammatory mediators spill into lymphatics and go systemic, bypassing the liver. This can lead to multi-organ dysfunction and adult respiratory distress syndrome (ARDS).
  - The **etiology** is most commonly **gallstones (70%) or alcohol (10%–20%).**
  - Alcohol increases proteinaceous secretions, duct cell permeability, and damage to acinar cells. Simultaneously, acinar perfusion is decreased and the sphincter of Oddi contracted, resulting in ductal stasis.
  - Other triggers include trauma, hyperlipidemia, hypercalcemia, viral infections, medications (diuretics, glucocorticoids), hypoperfusion, scorpion venom (which increases pancreatic secretions), pancreas divisum, or post-ERCP pancreatitis.
  - **Hereditary pancreatitis** has increased intra-pancreatic trypsin activity. It presents in childhood. Calcifications develop from recurrent, chronic inflammation. Malignancy can occur in up to 40% by the age of 50. This is most commonly associated with **PRSS1 and Spink1 mutations** inherited in an autosomal dominant fashion.
  - **Familial pancreatitis** refers to families that carry a high incidence of pancreatitis unrelated to obstructive or environmental causes, cystic fibrosis, or diagnosed hereditary pancreatitis. Again, there is conversion of intra-pancreatic trypsinogen to trypsin. Patients are predisposed to alcohol-related pancreatitis.
  - **Tropical pancreatitis** also has intra-pancreatic trypsin activity. It is associated with cassava root ingestion and can result in both chronic exocrine and endocrine dysfunction.
  - **Cystic fibrosis** is a receptor defect in ductal and acinar cells that produces thickened secretions which occlude the pancreatic duct. Baseline pancreatic insufficiency is common.
- **DIAGNOSIS**
  - Clinical presentation includes **epigastric and periumbilical pain** along with a **serum lipase that is greater than three times the upper limit of normal.**
  - **Right upper quadrant ultrasound** (**RUQ US**) helps diagnose a gallstone etiology.

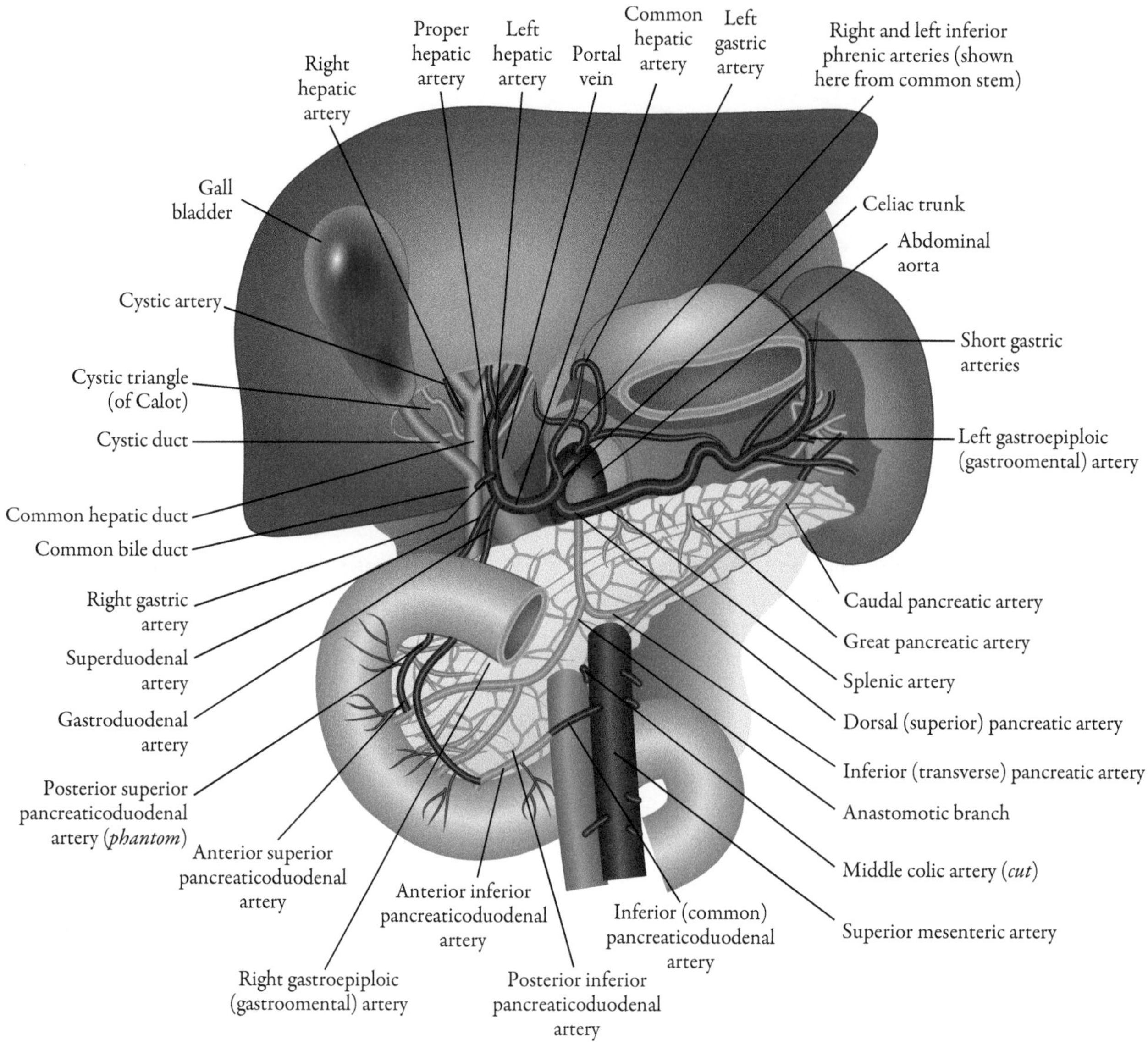

Figure 14.2 Arterial perfusion of the pancreas.

- **CT with IV contrast at 72–96 hours** can determine the severity of structural injury. It allows one to distinguish between simple interstitial pancreatitis, an **acute peripancreatic fluid collection (PPFC)**, or an acute **necrotic liquid or semisolid collection.**
  - A PPFC that persists >4 weeks and is encapsulated by a rind is known as a **pseudocyst.** It has homogenous liquid components but no true cellular lining. Pseudocysts occur in <15% of patients, only half of whom are symptomatic.
  - **Necrotic areas** do not uptake IV contrast and occur in **15%** of patients, particularly the obese. Such collections can have gas within them if secondarily infected. Otherwise, secondary infection is diagnosed by the clinical picture and a CT guided aspirate.
  - Necrotic liquid or semisolid collections that are >4 weeks in age are known as **walled off necrosis** and have a thick rind with either liquid or semisolid heterogeneous contents.
- The **state of the pancreatic duct** seems to correlate with the presence and size of fluid collections, and possibly even necrosis.
  - The grade of ductal injury can be defined by **MRCP, ERCP, EUS, or injection of contras**t into a percutaneously placed drain. Any of these methods can be enhanced with secretin stimulation.
  - **Type 1:** no injury.
  - **Type 2:** stricture that may need a PD stent.
  - **Type 3:** disruption that may need surgical intervention.

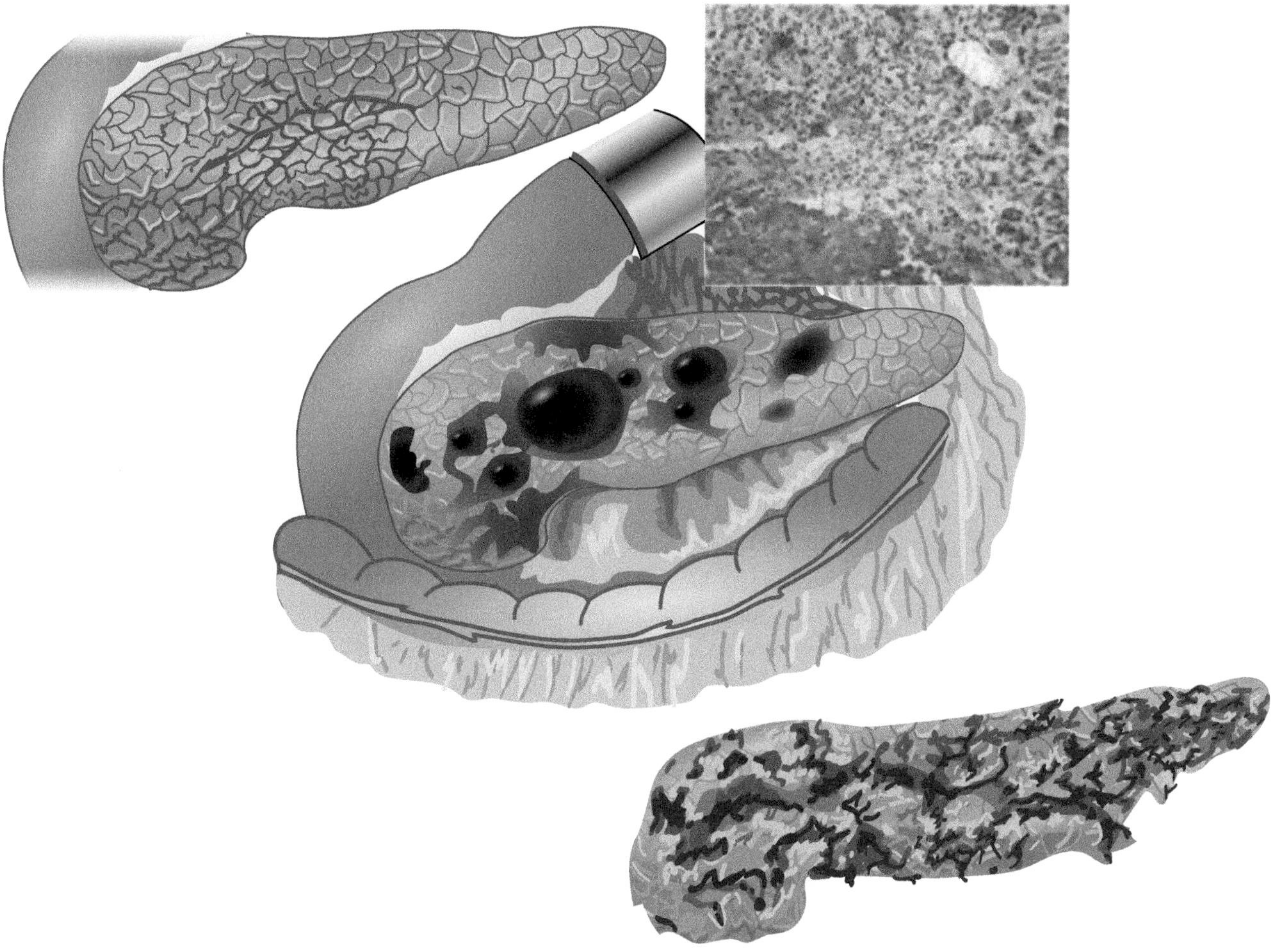

Figure 14.3 Acute pancreatitis.

- In addition to imaging, the degree of parenchymal damage, and the state of the pancreatic duct, clinical severity can be graded by organ dysfunction or a point system.
  - **Moderate** pancreatitis is defined as transient organ dysfunction with or without a local complication on imaging.
  - **Severe** pancreatitis is defined as organ dysfunction that persists beyond 48 hours with or without a local or systemic complications.
  - **BISAP** system includes one point for each of the following: **B**un >25, **I**mpaired mentation**, S**irs criteria, **A**ge >60, and **P**leural effusion. A score of 3 or more correlates with a mortality between 5%–15%.
  - **Ranson's Criteria** include:
    - GLC >200, AST >250, LDH >350, Age >55, WBC >16k at presentation, and
    - Calcium <8, hematocrit dropoff >10, PaO2 <60, Base deficit >4, Bun rise >5, Sequestration of fluid >6L at 48 hours.
    - The total points at 48 hours are used to determine mortality. A score of 3–5 has a 10%–15% mortality. A score of 6 or greater has a 40% mortality.
- **TREATMENT**
  - 1. Therapy starts with intravenous **volume resuscitation** plus NPO at admission. **Oral or gastric feeding can be started within 24 hours for mild disease.** Post-pyloric feeding can be used for those intolerant of gastric feeds.
    - Broad IV antibiotics are only used if there is concern for infected necrosis based on clinical signs, CT showing a necrotic collection containing gas, or a diagnostic CT guided aspirate Gram stain and culture. Antibiotics are **not** used in a prophylactic fashion.
  - 2. **ERCP** with sphincterotomy and stone extraction is needed if pancreatitis is believed to be secondary to an immobile stone at the ampulla of Vater, and the patient demonstrates **jaundice (TB >4), a dilated CBD, a large stone lodged in the CBD on imaging, a clinical picture consistent with**

cholangitis, or a lack of improvement with medical management.

- If the patient meets none of the above criteria, likely the pancreatic inflammation is due to sludge or microlithiasis in the common channel that will resolve on its own. Serial abdominal exams and bilirubin draws may demonstrate improvement. At that time, the patient may proceed directly to a cholecystectomy (see below), having avoided any of the morbidity of an ERCP.
- **Sphincterotomy** alone (without subsequent cholecystectomy) results in a 5% recurrence rate at 1 year. This strategy is employed for patients who are not candidates for early cholecystectomy due to severe pancreatitis, PPFC, or necrotic collections. Delayed cholecystectomy with or without drainage of any accompanying mature fluid collections can be accomplished in 6 weeks.

- 3. **Laparoscopic cholecystectomy** and intraoperative cholangiogram (IOC) with or without a transcystic or transcholedochal CBD exploration is performed at the same admission to prevent recurrence in mild pancreatitis.
  - If this is not done at the same hospital admission, the recurrence risk is up to 30% within 6 weeks.
  - One may omit a CBD exploration and simply perform a postoperative ERCP and sphincterotomy if a persistent obstruction is suggested by the IOC.
  - Cholecystectomy may be **delayed 6 weeks** in patients with a severe episode of gallstone pancreatitis complicated by PPFC, or sterile liquid or semisolid necrosis. These patients usually have already had an ERCP with sphincterotomy at the time of index episode to decrease their recurrence risk. When eventually performed, the cholecystectomy may be combined with any laparoscopic pancreatic drainage or debridement needed at that time (see below).
  - Pregnant patients with pancreatitis in the 1st and 2nd trimester should receive a laparoscopic cholecystectomy with or without an ERCP. Lower insufflation pressures are used in pregnancy.
    - If in the **3rd trimester,** patients receive an **ERCP followed by a postpartum laparoscopic cholecystectomy.**
- 4. **Sterile, asymptomatic** PPFC, or a sterile liquid or semisolid necrotic collection should be **left alone**.
- **5. PPFC** that is **symptomatic or infected** and nonresponsive to antibiotics should be **drained.** The drain may be upsized sequentially, and multiple drains may be placed. This process is known as "step-up" drainage.
  - Drainage can be done **percutaneously, either:**
    - **a) directly into the collection; or**
    - **b) percutaneously via a transgastric route** through the skin, anterior and posterior stomach walls, and into the collection. In this case the anterior wall gastrotomy site can be converted into a G or GJ tube weeks later, when the posterior gastric wall in communication with the collection has been encapsulated and drainage is effectively internalized.
- 6. A **liquid acute necrotic collection that is infected** and nonresponsive to antibiotics should be **drained percutaneously** with a step-up approach. An endoscopic transgastric or transduodenal approach with a stent is an option later in the chronic phase when the collection has been encapsulated.
- 7. A **semisolid acute necrotic collection that is infected** and nonresponsive to antibiotics needs **debridement.**
  - **Transgastric endoscopic access**, stent, and serial necrosectomy can be done with or without a post-procedure nasogastric or nasojejunal drain into the cavity. This approach is used for collections adjacent to the stomach wall, and later in their clinical course, such that a capsule has formed.
  - **Video assisted retroperitoneal debridement (VARD)** can be performed unilaterally or bilaterally after percutaneous retroperitoneal drains are placed and a wire, sheath, port. and zero-degree camera are used to endoscopically debride the area. This approach is used for collections in the appropriate, lateral retroperitoneal location.
    - As a last resort, one can attempt a **laparoscopic or open necrosectomy** and serial takebacks with or without continuous drain irrigation in between operative visits. This is only utilized if all of the above measures fail or are not anatomically feasible.
  - Whenever possible, intervention on all infected liquid or semisolid acute necrotic collections should be **delayed 4–6 weeks for adequate encapsulation** so that internal drainage can be accomplished (see below).

- **Late Complications**
  - 1. **Pancreatic pseudocysts develop from PPFC in <15% of patients,** only half of whom are symptomatic.
    - May **observe for up to 3 months if asymptomatic.** Up to 70% resolve if <5cm. Patients may need TPN or post-pyloric feeding during this time.
    - **Drainage** is employed for any **symptomatic** collection, or any collection that **persists >3 months** and is still >5 cm. An underlying type 2 or 3 pancreatic duct (PD) injury is likely in these cases. Options include:
      - Internal drainage via **endoscopic cystgastrostomy.** *See Figure 14.4.*

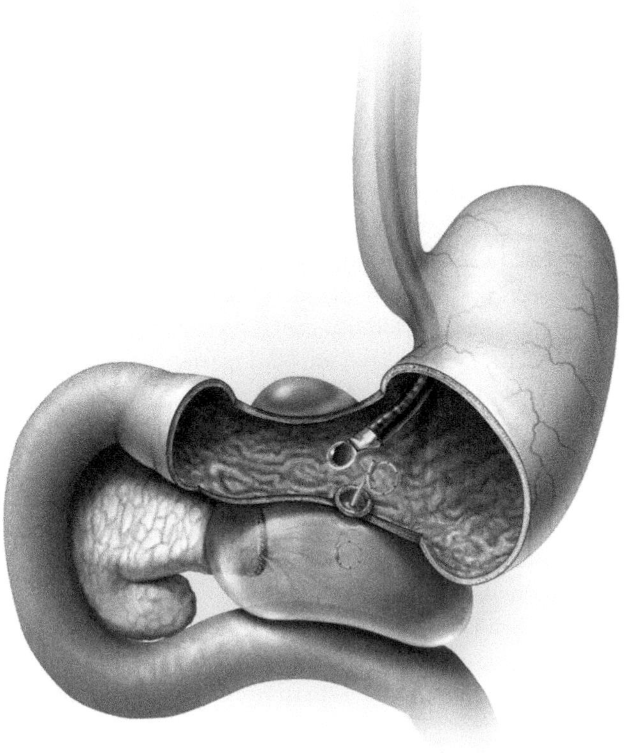

Figure 14.4 Endoscopic drainage of pancreatic fluid collection.

- **Endoscopic transpapillary PD stent**, which may extend through the PD breach and into the cyst.
- **Laparoscopic** drainage via anterior and posterior gastrotomy into collection or via RxY format into jejunum if the collection is not adjacent to the stomach.
- **Open** cyst-gastrostomy or enteric drainage in RxY format.
- Surgical drainage options are favored if the patient needs a cholecystectomy as well.
- May resect a pseudocyst if it is refractory to the above, or there is concern for a mucinous neoplasm. The latter may be suspected in the setting of a growing cyst or one incidentally noted without a definitive history of pancreatitis.
- **Roux-en-Y (RxY) pancreatojejunostomy** is the optimal management for the distal segment in the setting of a type 3 pancreatic ductal disruption at the watershed area of the pancreatic neck (between the vascular territories of the pancreaticoduodenals and splenic tributaries). Other options to manage this distal segment include endoscopic transgastric drainage or a distal pancreatectomy.

- 2. **Walled off pancreatic liquid or semisolid necrosis** may be **debrided and drained in an endoscopic transgastric, laparoscopic, or open** manner. *See Figure 14.5.*
  - Surgical approaches may be transgastric into the collection or an RxY enteric format.
  - In the chronic setting, the thick, mature capsule allows for anastomotic drainage and not just debridement.
  - Surgical drainage is favored if they need a cholecystectomy as well.

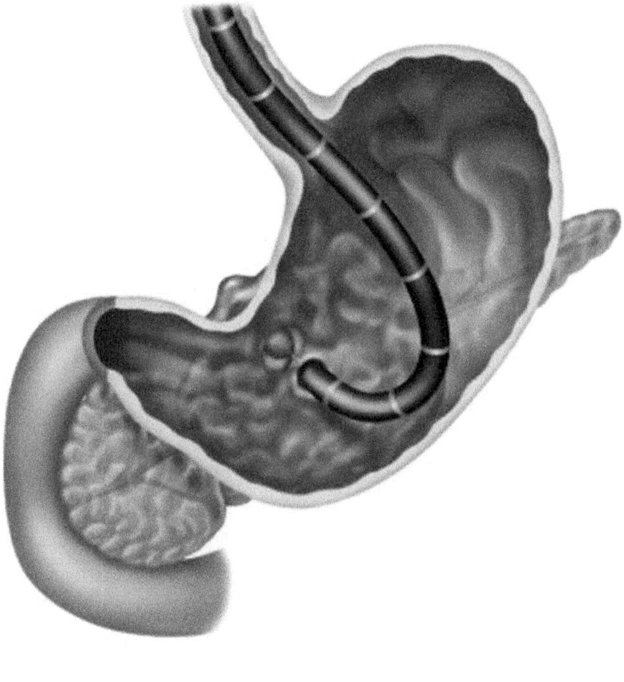

Figure 14.5 Endoscopic drainage of peri-pancreatic infection.

- 3. **Pancreatic ascites** is from a ruptured pseudocyst, and a **pleural effusion** can accumulate from the retroperitoneum through pleuroperitoneal foramina.
  - The fluid protein is >25 g/L, fluid amylase is also high, and serum amylase and lipase may be elevated from reabsorption of these enzymes across the mesothelium and into the bloodstream.
  - Treatment is with **paracentesis or thoracentesis** to appose the parietal and visceral linings and encourage scar formation.
    - **NPO, TPN, octreotide** are adjuncts.
    - **ERCP sphincterotomy and PD sten**t may help decrease resistance and thereby encourage pancreatic fluid to flow into the duodenum rather than into the peritoneal or pleural spaces.
    - **Distal pancreatectomy or distal segment RxY pancreatojejunostomy** are other options to manage the distal segment of an underlying type 3 PD injury.
- 4. **Pancreatico-cutaneous fistula** management includes the same principles of NPO, TPN,

octreotide, ERCP sphincterotomy, and PD stent. Most fistulae will close if there is <200 ml/day output.

- Measures to prevent fistulae after distal pancreatectomy include suture ligation of the PD, in addition to the staple line, and placement of a PD stent.

CHRONIC PANCREATITIS

- **PATHOPHYSIOLOGY**
  - Chronic inflammation is characterized by a monocytic infiltrate, hypertrophy of neural structures, and stellate cells transforming to myofibroblasts.
  - Acinar and ductal cells are lost, resulting in exocrine dysfunction. *See Figure 14.6.*
  - Islet cells are preserved, or even hypertrophy initially in chronic pancreatitis.
  - **Alcohol** use is the most common etiology.
    - Otherwise, chronic pancreatitis can be idiopathic, hereditary, familial, or tropical.
    - Impaired drainage can produce chronic pancreatic inflammation. Causes include pancreas divisum, an obstructing tumor, or a pancreatic duct stricture after trauma, iatrogenic injury, or an episode of acute, severe pancreatitis.
    - Hyperparathyroidism, hyperlipidemia, and toxins are other potential etiologies.
    - Autoimmune pancreatitis produces an inflammatory head mass causing biliary obstruction. It is associated with serum IgG4 elevation. Treatment is with glucocorticoids (see benign pancreatic mass below).
- **DIAGNOSIS**
  - **Pain** is the main symptom. It may be postprandial if chronic inflammation is related to ductal insufficiency. Pain may be continuous if the origin of chronic inflammation is primarily cellular or enzymatic, and not secondary to inadequate ductal drainage. Anorexia, weight loss, and malabsorption are other presenting features.
  - **Exocrine dysfunction** and steatorrhea may be **subclinical** early on. So generally, an empiric trial of non-enteric coated pancrelipase is started early if the diagnosis is suspected (see below). Once exocrine function drops to less than 10% of normal, clinical signs of malabsorption may be evident.
    - Exocrine dysfunction can be further evaluated by **EUS and ERCP** showing a beaded appearance and chain of lakes to the pancreatic ductal system.
    - An **endoscopic pancreatic function test** uses a tube or endoscope with a gastric port for decompression, and a duodenal port to measure HCO3, enzymes, and volume produced by the pancreas after secretin stimulation.
      - Low volume in response to secretin stimulation represents duct obstruction or malignancy.
      - Low bicarbonate but normal volume is seen in chronic pancreatitis.
      - Low enzymes but normal volume can indicate malnutrition.
      - If volume, bicarbonate, and enzyme levels are all low, then the patient has end-stage pancreatitis.
    - Fecal chymotrypsinogen and elastase levels, the PO bentiromide test, Sudan stain of stool, CCK stimulation test measuring luminal pancreatic output, and low ADEK vitamin levels are other means of assessing pancreatic exocrine function.
  - **Endocrine dysfunction** occurs later in the disease process, compared with exocrine dysfunction.
    - It is evaluated by low **pancreatic polypeptide (PP) levels** after a sham feed. Other etiologies to low PP levels include a prior vagal nerve injury or vagal neuropathy secondary to DM.
    - When islet cells are eventually lost, **brittle type 3 DM** results from sharp glucose drops (secondary to no glucagon) and sharp rises (secondary to no insulin or PP). Ketoacidosis, however, does not develop often.
  - **Adenocarcinoma** is 30 times more likely in chronic pancreatitis. It can be seen in up to 10% of patients, so screening with EUS and CA19-9 should be offered.
- **TREATMENT**
  - Initial measures include **pain control**, a high protein, high carbohydrate, **low fat diet, octreotide, PPI, and non-enteric coated pancrelipase.** The latter should start early for subclinical steatorrhea, and increased from 40,000 to 90,000 U as needed for weight gain and normalization of bowel movements.
  - Surgery is reserved for **intractable pain, distal biliary stricture, or inability to exclude malignancy**.
    - **Puestow** longitudinal pancreatojejunostomy is performed if the duct is diffusely dilated >6 mm and there is no pancreatic head involvement. This will drain any associated pseudocysts as well.
    - **Beger resection** is a duodenal and CBD preserving pancreatic head resection with pancreatic neck transection, and:
      - 1. A side–side jejunal anastamosis of the back wall of the pancreatic head in RxY format, plus
      - 2. An end–side jejunal anastamosis of the distal pancreas.
    - **Frey resection (LR-LPJ)** is a limited pancreatic head resection (core-out) with a lateral jejunal anastomosis extending down the length of the pancreatic duct. *See Figure 14.7.*

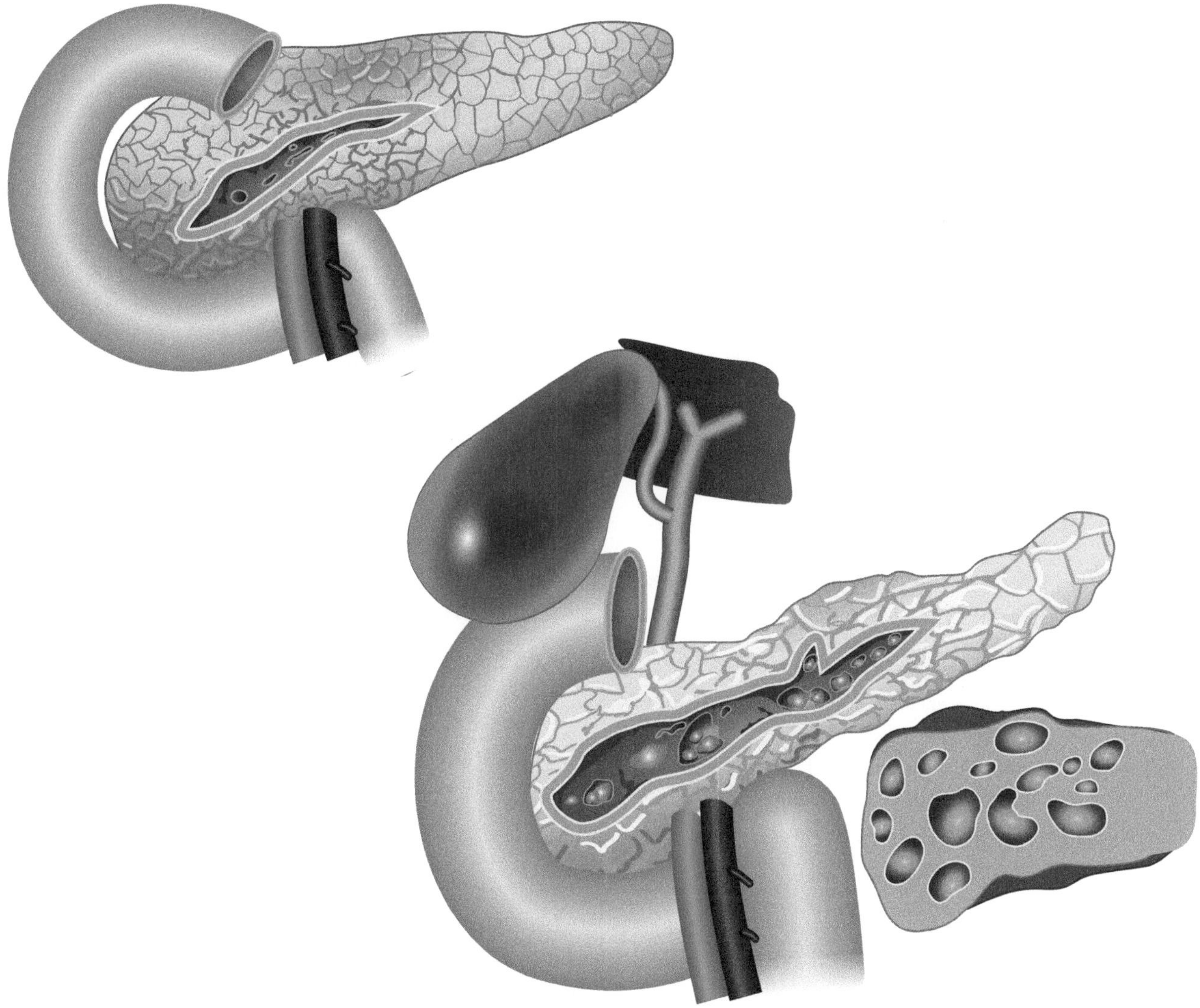

Figure 14.6 Chronic pancreatitis.

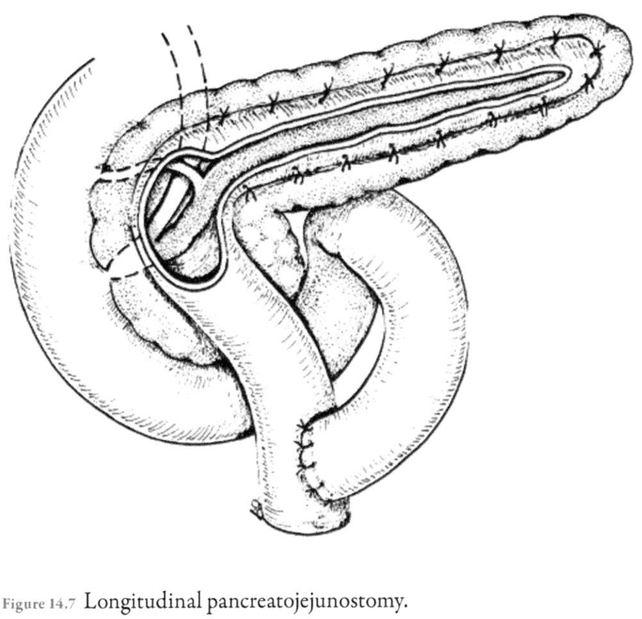

Figure 14.7 Longitudinal pancreatojejunostomy.

- A **pancreaticoduodenectomy** is used for a calcified head with CBD involvement, obstruction, stricture, or concern for malignancy.
- **Distal pancreatectomy** is an option for inflammation and is limited to this area. *See Figure 14.8.*
- **Total pancreatectomy** includes removal of D2–D4 in addition to the entire pancreas with or without the spleen. If the spleen is preserved, the splenic artery (SA) and splenic vein (SV) are ligated at the hilum and the organ resides on short gastric vessels. Enteric reconstruction is with a DD, DJ, or B2 GJ. Biliary reconstruction is via choledochojejunostomy.
    - Digestion of pancreatic tissue and **transportal islet cell *auto*-transplantation** can be performed at the time of laparotomy or the next day, in order to try and reintroduce pancreatic endocrine function.

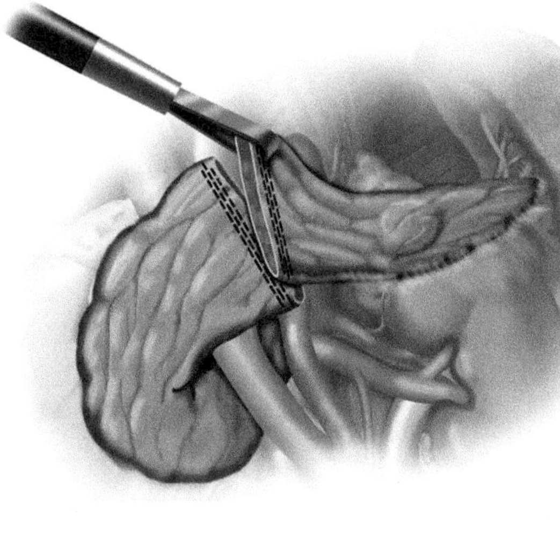

Figure 14.8 Distal pancreatectomy.

- The infusion should be stopped if portal pressures exceed 25. Post-op tight glucose control on an insulin drip is instituted for 48 hours.
- It can take months for islets to neovascularize and engraft in the liver, and result in normoglycemia.
- Clinical endpoints include reduction in chronic pain, improvement in quality of life, and insulin independence or at least improvement in glucose control.

- **Adjuvant Procedures and Palliation**
  - Occasionally, the cause of repeated inflammation is an isolated stricture. If so, this can be treated by **ERCP sphincterotomy, stone extraction, and PD stent** placement. The stent can be replaced every 3 months to prevent restenosis.
  - A palliative bilateral thoracic splanchnicectomy, or endoscopic **celiac block or ganglionectomy** are options for pain control. *See Figure 14.9.* Each lasts about 6 months.
  - Any associated splenic artery (SA) pseudo-aneurysm should be ligated intraoperatively.
  - A splenectomy should be done for SV thrombosis complicated by bleeding gastric varices. Otherwise, SV thromboses without bleeding gastric varices can be left alone without anticoagulation.

PANCREATIC MALIGNANT MASS

1. DUCTAL ADENOCARCINOMA

- **PATHOPHYSIOLOGY**
  - **Tumors arise from the main PD epithelium. 70% are in the head of the pancreas.** 50% invade vascular structures, have nodes at the celiac or SMA origins, peritoneal implants, or distant metastases at the time of diagnosis. *See Figure 14.10.*

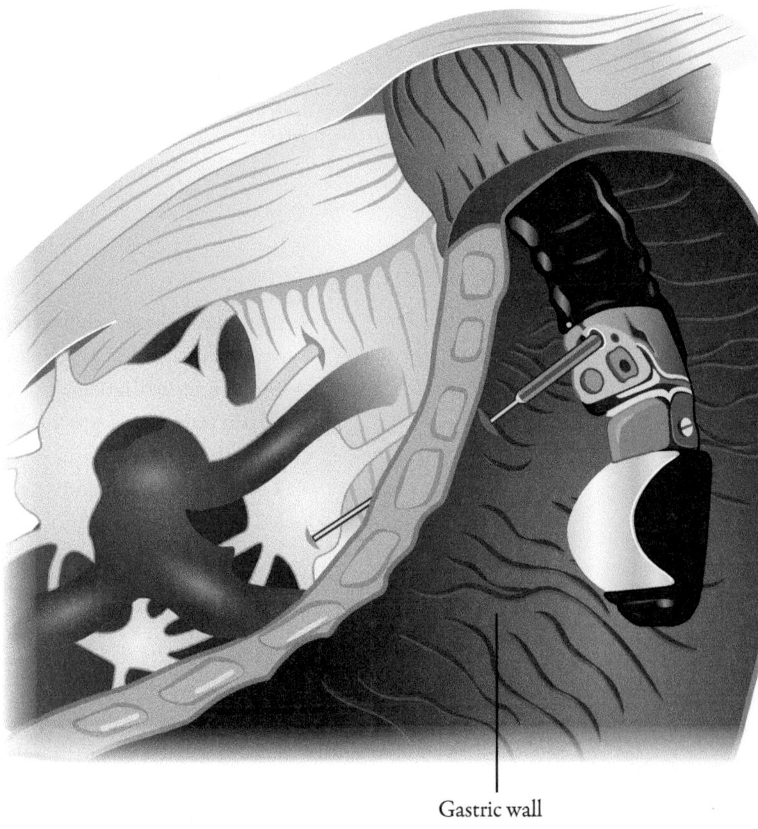

Figure 14.9 Endoscopic blockade of the celiac ganglia.

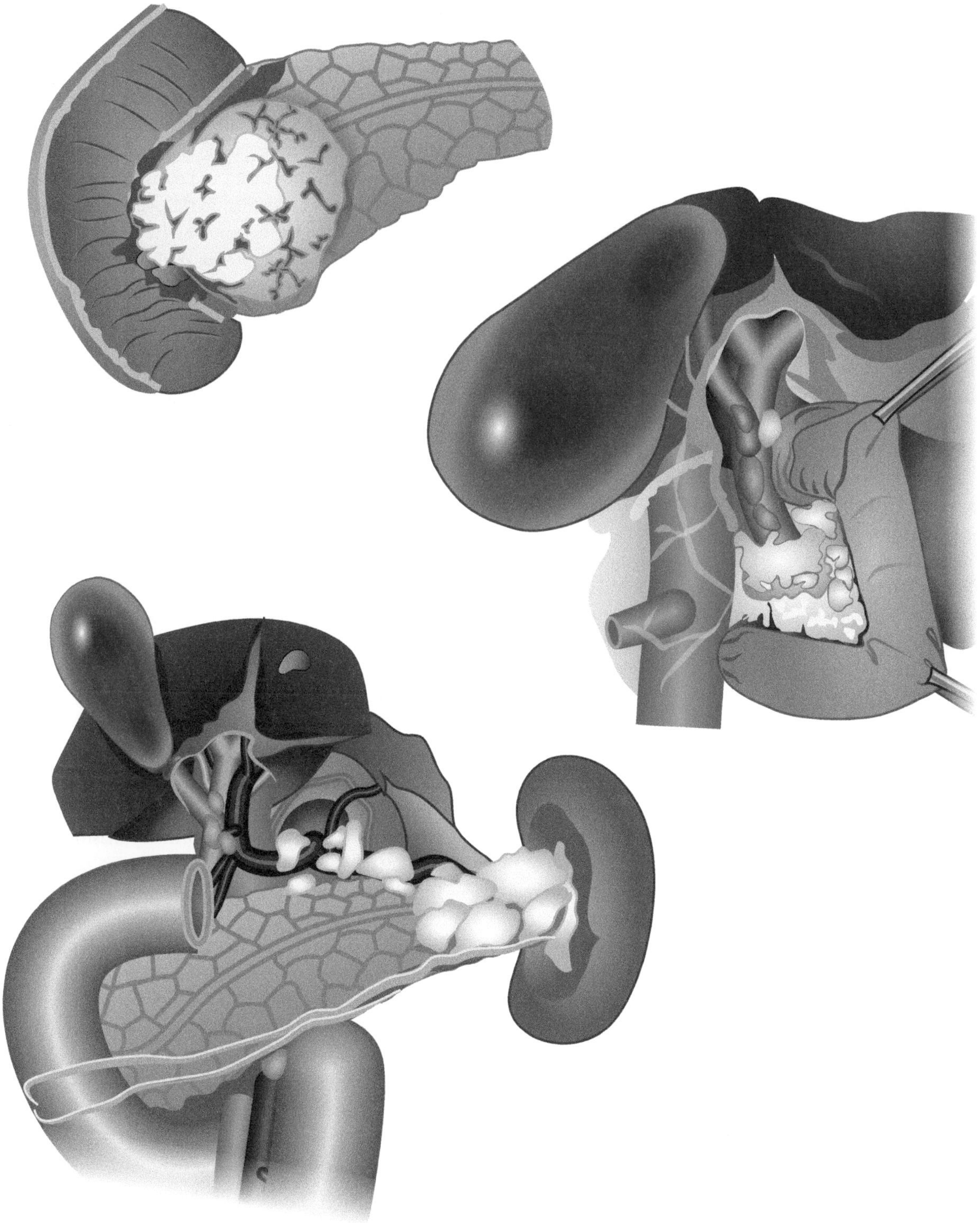

Figure 14.10 Locales of pancreatic adenocarcinoma.

- Risk factors include tobacco, obesity, a high fat diet, chronic pancreatitis, DM, familial predisposition, and possibly a history of cholecystectomy.
- **Men** are more commonly affected than women.
- Genetic associations include mutations in **KRAS** or the absence of p16 or DPC4 tumor suppressor gene. Also associated with Her2/neu, FAP, HNPCC.

- **DIAGNOSIS**
  - Symptoms include **painless jaundice** from CBD obstruction, or episodes of **pancreatitis** from PD obstruction. Weight loss and cachexia are from high TNF levels.
  - **CT scan pancreatic protocol** or a magnetic resonance cholangiopancreatography (MRCP) will confirm the diagnosis.
    - The T stage of the lesion is defined as **T1** <2 cm mass, **T2** >2 cm but confined to the pancreas, **T3** extending outside of the pancreas into adjacent nonvascular structures, **T4** extending into the PV, SMA, or peritoneum.
  - There is no need for a biopsy unless the diagnosis is in doubt in spite of imaging, or if the mass is borderline resectable or unresectable (see below). In these cases, **EUS-FNA** pathology may be needed to guide medical care.
- **TREATMENT**
  - Centered around pancreaticoduodenectomy with or without neoadjuvant therapy.
  - **Neoadjuvant therapy** is utilized for borderline resectable masses.
    - a) Tumor **abutment** of the SMA around <180 degrees of the artery circumference.
    - b) Tumor **encasement** of the **common hepatic artery or a venous structure around >180 degrees of its circumference.**
    - **c)** A tumor causing **venous indentation** or contour irregularity.
    - The regimen is 5-FU based for 2 months, then restaging by imaging. Thereafter the patient may receive chemoradiotherapy over 6 weeks and a final restaging.
    - It is likely that there will be no downstaging of the primary tumor, but occult micrometastatic sites are being sterilized. These can be followed with CA 19-9 levels.
    - Surgery can be performed **4 weeks after completion of neoadjuvant course.**
    - This sequence may reduce postoperative pancreatic anastomotic leaks and fistula.
    - Tumors that encase **>180 degrees of the SMA** are considered **locally advanced and unresectable.** They may be treated with chemoradiation alone.
  - There is no need to relieve a distal biliary obstruction with an endoscopic stent pre-op, as this may increase the wound infection rate.
  - **Pancreaticoduodenectomy Principles**
  - The procedure is most commonly performed open, but laparoscopic and robotic approaches are also used.
  - Mobilize the proximal transverse colon, Kocherize the duodenum, dissect the plane between the pancreas and SMV, control the right gastric artery and GDA, remove the gallbladder, and isolate and transect the CBD. Transect distal or proximal to the pylorus. Transect the jejunum distal to the ligament of Treitz.
  - Transect the pancreas anterior to the course of SMV and PV, and shift the entire specimen to the patient's right. Complete the dissection of the uncinate process and any remaining vascular connections from the pancreatic head to the SMV and SMA.
  - Ensure there is no replaced right hepatic artery (RHA) in the hepatoduodenal ligament, and test clamp the GDA while feeling for a proper hepatic artery (PHA) pulse to confirm that the liver is not dependent on retrograde flow from the pancreaticoduodenal arteries in the setting of celiac trunk stenosis. In that case, the patient may need a bypass graft to the PHA or celiac artery, or lysis of the median arcuate ligament.
  - If a portomesenteric venous resection is necessary, primary transverse venorrhaphy and end-to-end anastomosis can be done for defects <3 cm in length. Larger defects may require an autologous greater saphenous vein (GSV), left renal vein, or left internal jugular vein interposition graft.
  - A limited resection and reconstruction of a short segment of the common hepatic artery (CHA) along with the GDA can be done by primary anastomosis or GSV interposition graft.
  - The **enteric reconstruction** involves taking the distal free jejunal end to the pancreas, common bile duct, and either the proximal duodenum or distal stomach (depending on if a pylorus-preserving resection was done). *See Figure 14.11.*
  - A pancreatojejunostomy (PJ) is first completed as a two layered, end to side duct to mucosa anastomosis. The other option is intussuscepting the jejunal end over the transected distal pancreases.
  - An end to side hepaticojejunostomy (HJ) is performed along the jejunum distal to the PJ.
  - The transected end of the duodenum or stomach is then anastomosed to the side of the jejunum at a point distal to the HJ in a Billroth II (B2) fashion.
  - About 25% of resected patients are alive at 5 years post-op.
  - **Complications**
    - **Delayed gastric emptying** is more likely if a pylorus-preserving approach was taken. Treatment is with metoclopramide after first excluding a leak or fluid collection with imaging.
    - **Marginal ulceration** of the jejunal limb is treated with a PPI.

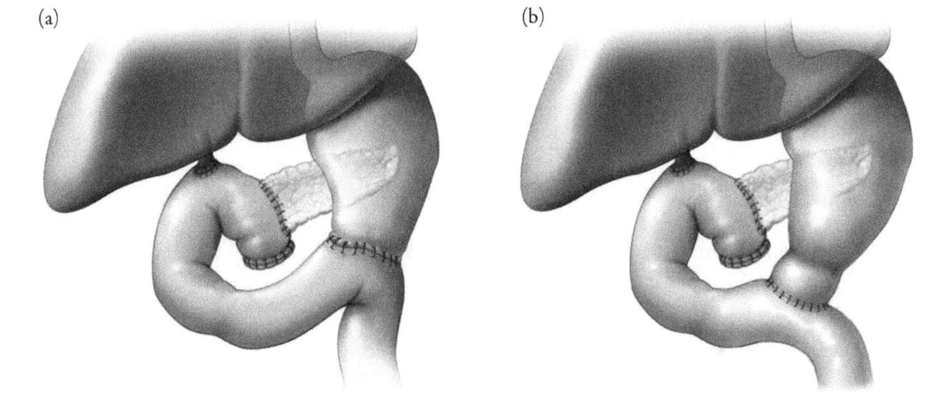

Figure 14.11 Pancreaticoduodenectomy reconstruction without and with pylorus preservation.

- **Intraperitoneal hemorrhage** from the GDA stump is treated with angioembolization. This can be prevented by wrapping the GDA stump in the falciform ligament at the time of resection.
- **Pancreatic anastomotic leak** or fistula is diagnosed by a POD 3 drain amylase level that is greater than three times the serum level. It occurs in 10%–25% of cases and is more common if the PD diameter was <3 mm.
  - **Type A** leak is low output (<200 ml) and clinically inconsequential.
  - **Type B and C** leaks need percutaneous drainage and possibly distal enteral feeding or TPN. Generally >500 ml/day output necessitates NPO and TPN.
  - Prophylactic measures to prevent or decrease a leak include octreotide and intraoperative consideration to performing the gastrojejunostomy (GJ) or duodenojejunostomy (DJ) further distally, so contents do not reflux back to a fragile PJ.
- **Adjuvant chemotherapy** is given in all cases. The regimen is 5-FU and gemcitabine based, with or without radiation or NanoKnife.
- **Palliation** for unresectable lesions may include any of the above medical therapies, as well as:
  - A covered metal **duodenal stent** for gastric outlet obstruction if the patient's lifespan is perceived to be <2 months.
  - **A B2 isoperistaltic antecolic or retrocolic GJ** can be done if preoperative imaging suggests duodenal encroachment, and the patient's life expectancy is perceived to be >2 months. If palliation is decided upon intraoperatively at the time of a diagnostic laparoscopy, a GJ is a better option than a stent. Also, it can be used for patients in whom the duodenal stent fails.
  - Biliary obstruction can be palliated with an endoscopic self-expanding **metal stent (covered or uncovered) or a percutaneous biliary drain (PBD)**.
  - A **Roux-en-Y biliary bypass (RxY CDJ or HJ)** can be employed if a stent is insufficient.
    - First transect the jejunum 30 cm–50 cm beyond the ligament of Treitz and then perform a jejunojejunostomy 75 cm distal to that. Then, complete the procedure with a CDJ or HJ to the staple line of leading edge of distal jejunum.
    - If the patient already had a GJ, then the jejunum can be transected distal to GJ, a JJ performed at that site, and the distal stapled end of jejunum can be taken to the CBD or CHD for a CDJ or HJ.
    - A cholecystojejunostomy is an option for those patients with an inaccessible common duct.
  - **Celiac plexus block or ablation** can be performed endoscopically via the stomach, laparoscopic, or open. The latter two are used if this palliative technique is decided upon at the time of a diagnostic laparoscopy or attempted resection.
  - A palliative splenectomy can be done for SV thrombosis causing bleeding gastric varices, hypersplenism, sequestrative thrombocytopenia, and intolerance of chemotherapy.

2. **ACINAR CARCINOMA** is rare. It is derived from acinar cells, not ductal cells, and produces lipase. **Lipase levels** may be followed as a tumor marker. It is associated with paraneoplastic arthralgias, subcutaneous fat necrosis, erythema nodosum, and eosinophilia. Treatment is with resection, and it has a better prognosis than ductal carcinoma.
3. **NONFUNCTIONAL NEUROENDOCRINE TUMORS** are generally **malignant,** have an indolent course, and represent **one-third** of the endocrine neoplasms occurring in the pancreas.
   - **DIAGNOSIS**
     - Image with a **pancreas protocol CT**, MRI, or an octreotide scan. **Half of patients have metastases** at the time of diagnosis. CgA or PP levels are also helpful.

- **TREATMENT**
  - **Enucleation** is adequate if the lesion is small and low grade.
  - Otherwise, **WLE with 2 cm margins is recommended** since the size of the lesion correlates with its biologic aggressiveness. May add a cholecystectomy if using postoperative octreotide.
  - Metastases should be debulked with surgery, ablation, or TARE.
  - Follow CgA levels postoperatively.
  - Palliative or postoperative **octreotide** can be used for symptom control.
  - Chemotherapy regimens are of limited utility.
  - >20 mitosis/HPF correlates with approximately 3 years OS.

4. **GLUCAGONOMAS**
   - Generally large and located in the **tail** of the pancreas. They result in **diabetes,** stomatitis, glossitis, diarrhea, **thrombosis of deep veins**, and a fascial lower abdomen, and perineal dermatitis with central healing (aka **necrolytic migratory erythema**).
   - **CT or MRI** imaging and a fasting **glucagon level >500 will make the diagnosis.** Sometimes a biopsy of presenting skin lesions may be useful.
   - Treatment is with a **distal pancreatectomy, debulking of metastases, octreotide**, and IFN alpha. Zinc, amino acids, and fatty acids can assist with resolution of rash.
5. **VIPOMAS**
   - Large pancreatic **tail** tumors that produce **achlorhydria, watery diarrhea, and secondary hypokalemia.**
   - **CT or MRI** imaging and **VIP levels >200 will make the diagnosis.** Other causes of diarrhea should be excluded. Metastases are common at time of diagnosis.
   - Treatment is with a **distal pancreatectomy, debulking metastases, and octreotide.**
6. **SOMATOSTATINOMAS**
   - Large pancreatic **head** tumors that produce **diabetes, gallstones, steatorrhea**, and hypochlorhydria.
   - **CT or MRI** imaging and fasting **somatostatin levels >10 will make the diagnosis.** Metastases are common at time of diagnosis.
   - Treatment is with **pancreaticoduodenectomy and cholecystectomy plus debulking metastases.**
7. **PERIAMPULLARY AND AMPULLARY TUMORS**
   - Four cellular etiologies. Pancreatic adenocarcinoma is #1, cholangiocarcinoma #2, a true ampullary epithelial tumor #3, and duodenal adenocarcinoma #4. The latter have better prognoses. All have early presentations with obstructive jaundice.
   - Occasionally, benign adenomas or polyps are obstructing the ampulla (see Chapter 9, Small Bowel). These can be managed with endoscopic or transduodenal WLE with 2–3 mm margins or a pancreaticoduodenectomy if there are multiple dysplastic polyps and a history of familial adenomatous polyposis (FAP).
8. **PRIMARY PANCREATIC NON- HODGKIN'S LYMPHOMA**
   - A bulky mass with lymphadenopathy. Diagnosis is by FNA and treatment is with chemotherapy (CHOP).
9. **METASTASES TO THE PANCREAS**
   - Include renal cell carcinoma, melanoma, or breast cancer. They are hypervascular on imaging and may be resected if isolated.

PANCREATIC BENIGN MASS

1. **SEROUS CYSTADENOMAS**
   - These have a **female** predominance and a low malignancy risk at <1%.
   - Cysts are associated with APLD, APKD, and VHL. If biopsied, pathology normally shows **benign PAS positive cuboid epithelium.**
   - **DIAGNOSIS**
     - **APCT** showing a central stellate scar or a honeycomb, cluster of grapes appearance. EUS guided **aspiration demonstrates a low cyst CEA and amylase** because there is no communication with the pancreatic duct.
   - **TREATMENT**
     - **Observation.** Resection is reserved for the rare instances where lesions are enlarging >0.5–2.0 cm per year on serial imaging, are compressive, or have a clinical or radiographic concern for malignancy prompting an FNA whose results are not consistent with a benign serous cystadenoma.
2. **MUCINOUS CYSTADENOMAS**
   - Also known as mucinous cystic neoplasms and are located in the **body or tail** of the pancreas in a middle-aged female.
   - **DIAGNOSIS**
     - **APCT** and **EUS** guided needle **aspiration showing a high cyst CEA** but a **low cyst amylase and lipase,** since lesions are not in communication with the pancreatic duct. Cyst CEA levels >200 correlate with mucin, but not necessarily malignancy.
   - **TREATMENT**
     - **Resection** for all cases because 15%–30% of lesions may be a mucin producing adenocarcinoma.
     - If carcinoma is confirmed after resection, negative margins are mandatory, and patients may need adjuvant therapy in a fashion similar to a ductal adenocarcinoma.
     - Otherwise, pathology normally shows benign columnar, mucin producing epithelium with **ER/PR positive ovarian stroma and peripheral eggshell calcifications.** Clear margins in this setting generally mean no recurrence.

3. INTRADUCTAL PAPILLARY MUCINOUS NEOPLASMS

- Found in elderly patients. Lesions may be located in the pancreatic head and in communication with the **main duct** or, more commonly, a **branch duct.** The neoplasm spreads lengthwise along the pancreatic ductal system with skip lesions, multifocality, and a high recurrence rate. Some cases can have malignant transformation over the course of 10–20 years.
- **DIAGNOSIS**
  - **APCT** imaging with or without **EUS** guided aspiration showing **elevated cyst CEA and amylase.** The latter is secondary to ductal communication.
  - Other diagnostic modalities include ERCP, intraductal ultrasound, and pancreatoscopy revealing a patulous ampulla with mucin extruding from it.
- **TREATMENT**
  - **Resection** for all main duct origin neoplasms.
  - Resection is also used for branch duct origin neoplasm that are >3 cm size or associated with >10 mm dilation of the main pancreatic duct (PD), or lesions with enhancing mural nodules on CT.
  - If a branch duct origin lesion is associated with a 5–10 mm main PD or non-enhancing mural nodules on APCT, then an EUS and aspiration can be performed. Resection can then be based on any worrisome pathologic features on the EUS.
  - **Observation and serial imaging** is best for branch duct origin neoplasms in the absence of any of the above features. The development of worrisome radiographic features or rapid growth on future imaging may warrant resection at that time.
  - **Non-anatomic enucleation** can be employed for low-grade and small lesions millimeters away from the main duct.
  - **Pancreaticoduodenectomy** is used if the lesion is located in the head.
  - Perform a **distal pancreatectomy** for tail lesions, or **total pancreatectomy** if multifocal.
  - Pathology will show mucin producing columnar ductal cells that are one of four different subtypes. The gastric subtype is most common and least likely to recur.
  - Resect to **negative margins with respect to high-grade dysplasia or invasive disease.** R1 resection is adequate if only low-grade or moderate dysplasia are noted.
  - **Adjuvant therapy** should be given if there is a carcinoma in the specimen with extensive local disease or nodal involvement.
  - Surveillance options include **MRI or EUS q3–6 months** if the pathology showed borderline features or if the patient has a family history of pancreatic cancer.
  - MRI can be done q1–3 years if the pathologic specimen was entirely benign.

4. SOLID PSEUDO-PAPILLARY TUMORS

- Present in **young females** with abdominal discomfort from a large mass in the **body or tail** of pancreas. Solid vascular papillary stalks may have liquefied central necrotic degeneration and appear "cystic" on imaging. FNA shows epithelial cells. NSE and beta-catenin levels may be positive. Most are benign, but all should be **resected** due to a minority having malignant potential.

5. LYMPHOEPITHELIAL CYSTS

- Present in **males** as a large, usually asymptomatic cyst containing keratin and cholesterol with stratified squamous epithelium and lymphoid stroma. Contents appear bright on T1 MRI imaging. There may be cyst fluid CEA or CA 19-9 level elevation. If cytology is definitely benign and the patient asymptomatic, then one may **observe** and not resect the lesion.

- **6. AUTOIMMUNE PANCREATITIS**
  - A conglomerate of lymphocytes and eosinophils in an inflammatory mass at the pancreatic **head** causing **biliary obstruction**. Diagnosis is by **MRCP.** The majority have elevated serum **IgG4** levels. Some patients have autoimmune disease elsewhere. Treatment is with **corticosteroids** and repeat imaging at 2–6 weeks while following CA 19-9 levels.
- **7. INSULINOMAS**
  - **Evenly distributed** throughout the pancreas. Most are small, benign, and may be multifocal and difficult to localize.
  - **DIAGNOSIS**
    - First demonstrate **Whipple's triad**: fasting hypoglycemia with symptoms relieved by glucose administration.
    - This is followed by measurement of **insulin:glucose ratios.** Ratios exceeding 0.4 imply that insulin is being made without negative feedback from low glucose levels.
    - Additional testing should reveal positive **c-peptide and proinsulin levels** and urine that is negative for **sulfonylurea metabolites.** These tests confirm that the cause of the hyperinsulinemia is endogenous.
    - Once the above tests determine the presence of an insulinoma, imaging of the pancreas should follow with APCT **(pancreas protocol), MRI, EUS,** or a radiolabeled octreotide scan. The latter has a lower sensitivity for insulinoma than for other neuroendocrine tumors.
    - Angiography with calcium stimulation and venous sampling for insulin levels is less commonly used. It can help localize the veins that are draining the insulinoma, and hence focus the resection.
    - Since they are small and possibly multifocal, **intraoperative pancreas US** is useful for detecting insulinomas not seen by the above methods.

- **TREATMENT**
  - **Enucleation** for lesions <2 cm and at least 2–3 mm away from main PD by intraoperative US.
  - **WLE** is best for lesions >2cm, since size correlates with biologic aggressiveness.
  - **Diazoxide** inhibits insulin release to help palliate symptoms for non-resectable lesions.
  - 5-FU, streptozocin, glucocorticoids, and octreotide can also be used for metastatic disease or symptom relief.
  - For the rare patient with a malignant insulinoma, the most common site for metastases is the liver. One may **debulk** liver mets via resection, ablation, or hepatic artery embolization.
- **8. GASTRINOMA**
  - A tumor of G cells producing excess gastrin, resulting in prominent gastric folds, increased histamine, and increased parietal cell HCl production. The sequelae include peptic ulcer disease (PUD) and diarrhea.
  - 25% of gastrinomas are associated with **MEN1.** Gastrinoma is the most common functional neuroendocrine tumor in this syndrome. Half of patients have multiple gastrinomas, and half of tumors are malignant.
  - **Passaro's triangle** is the area between the cystic duct–CBD junction, the pancreatic neck, and the third portion of the duodenum. Gastrinomas are most commonly found here. *See Figure 14.12.*
  - **DIAGNOSIS**
    - **Serum gastrin levels should be >200.** A level >1000 increases specificity.
    - The differential diagnosis of elevated gastrin includes baseline PPI use, pernicious anemia, CKD, and a retained or excluded antrum after gastric surgery.

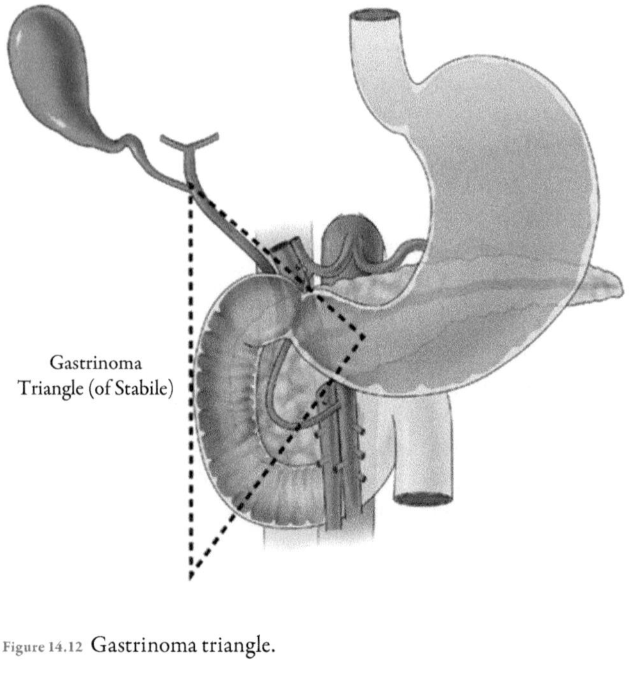

Figure 14.12 Gastrinoma triangle.

    - A confirmatory **secretin stimulation test** will result in an increase in gastrin by 200, measured at various intervals after an intravenous bolus of secretin. This is specific for gastrinoma.
      - Secretin may be selectively injected into celiac artery tributaries while checking hepatic vein gastrin levels to aid in localization.
      - A calcium stimulation test can be used if secretin is negative.
    - A radiolabeled **octreotide scan** is used for localization.
    - **EUS or intraoperative US** can be employed to find pancreatic tumors not seen on octreotide scan.
    - EGD with transillumination and **duodenotomy** are last resorts if all of the above are non-localizing.
  - **TREATMENT**
    - PPI (60–120 mg/day) and **medical management for MEN1 or multiple tumors provided they are all <2 cm.**
    - **Enucleation** is an option if there is a single tumor <2 cm and it is a few mm from the pancreatic duct.
    - **WLE** for any tumor >2cm, since size generally correlates with biologic aggressiveness.
    - May **debulk** symptomatic liver mets with resection, ablation, or hepatic artery embolization if PPI therapy is not adequate.

LAPAROSCOPIC DISTAL PANCREATECTOMY

- This can be employed for benign or malignant lesions of the distal pancreas. *See* Figure 14.8.
- Endoscopic ultrasound (EUS) should be done first and the lesion tattooed.
- Mobilize the splenic flexure of the colon, open the gastrocolic ligament, dissect the inferior and superior pancreatic margins, and ligate the short gastrics.
- Then staple ligate the SA followed by a second firing across the pancreatic parenchyma and SV.
- Complete the procedure with a splenectomy, leaving the organ hanging on the phrenosplenic ligament until the end. This helps with traction and assists with the dissection. Remove the specimen as pancreas and spleen en bloc, morcellizing the spleen manually.
- If pathology is definitively benign preoperatively, then splenic preservation is possible about 70% of the time. In these cases, the pancreatic parenchyma is transected with a stapler, but the SA is preserved and the SV is dissected free of the pancreatic specimen.

PANCREAS TRANSPLANTATION

- Indications include **brittle diabetes mellitus** with hypoglycemic episodes, **nephropathy**, neuropathy, and retinopathy.

- Transplantation reverses all of the above but does not reverse macrovascular disease.
- 80% of recipients receive a **simultaneous pancreas and kidney (SPK) transplant.**
- 15% receive a pancreas after kidney transplant and 8% receive a solo pancreatic transplant.
- The donor organ consists of the **C loop of the duodenum and the entirety of the pancreas.** The gastroduodenal artery (GDA) is ligated.
  - The donor external and internal iliac artery are fashioned into a **Y graft** anastomosed to the proximal SMA and proximal splenic artery of the donor pancreas and duodenal C loop complex.
  - The donor common iliac artery containing the new **Y graft complex is then anastomosed to the recipient common iliac artery (CIA) or external iliac artery (EIA).**
  - **Donor portal vein is then anastomosed to the IVC (if transplanting to the right iliac fossa) or the left external iliac vein (EIV)** for outflow. *See Figure 14.13.*
  - Either are more commonly done than a donor portal to recipient portal or SMV implantation.
  - One may use an extension donor iliac vein graft anastomosed to the proximal donor portal vein in order to gain enough venous length to place the organ onto the recipient's left EIV. Take care to avoid donor splenic vein thrombosis. Consider perioperative anticoagulation.
- **Duodenal C loop is anastomosed to the side of recipient jejunum** at a point 30–50 cm beyond the ligament of Treitz
- An alternative would be to anastomose the duodenum to the bladder. This has increased leak rates, but less deleterious effects on graft loss if it does leak. Urine amylase can be used as a marker for rejection. Otherwise rejection is normally monitored by following the function of the renal graft. Oral enzymes would have to be taken by the patient to supplant exocrine pancreatic function. This technique is less commonly used.
- For a **living donor,** the organ is procured as a distal pancreatectomy with the donor splenic artery and vein anastomosed to the recipient EIA and EIV. The pancreatic duct (PD) is anastomosed directly to the bladder, since there is no enteric portion of the graft in this case.
- Complications include **donor portal or splenic venous thrombosis,** peripancreatic abscess, hemorrhage in 5%–10%, pancreatic fistulae, peripancreatic fluid collection (PPFC), or an arteriovenous fistula. Ultrasound or CT scan can be diagnostic, and graft failure may require explantation.
- **Islet cell *allo*-transplantation** may be performed after a kidney transplant or in isolation**.**
  - The recipient selection and donor procurement are similar to whole organ transplantation.
  - One needs to harvest 10,000 cells per kg of the recipient. This may require up to four donors.

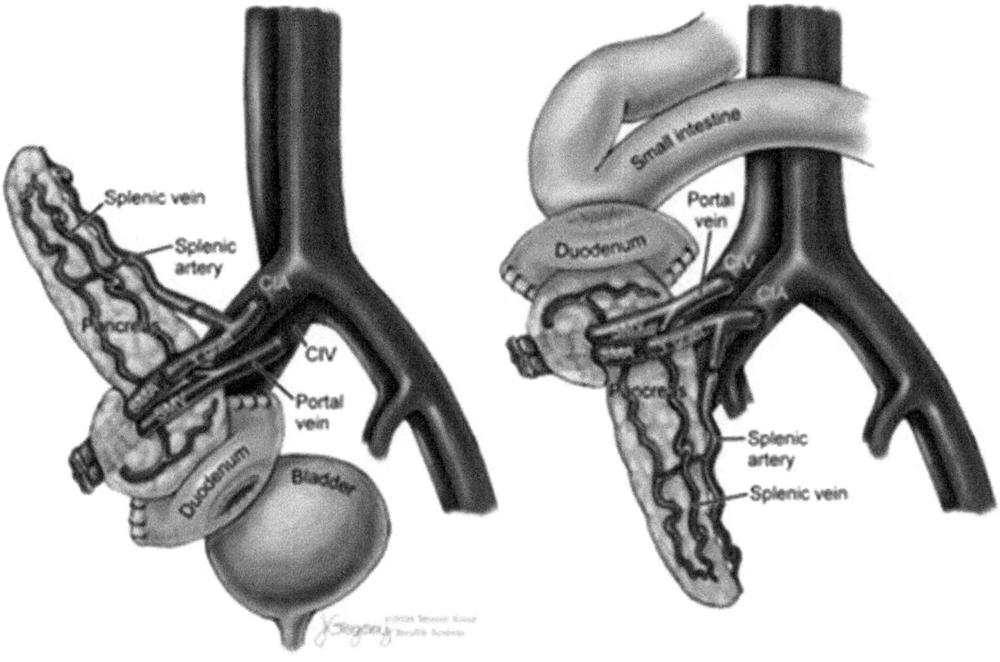

Figure 14.13 Pancreatic transplantation into the right iliac fossa.

    - The cells are injected into the portal venous system with the hope that they engraft in liver parenchyma and begin producing insulin.
    - Induction is with antithymocyte globulin (ATG), alemtuzumab, basiliximab.
    - Tacrolimus is used for maintenance despite its potential islet cell toxicity.
    - >50% of recipients achieve insulin independence at 5 years.
    - Valganciclovir is for CMV-positive organ or cells given to a CMV-negative recipient, or for any CMV-positive recipient.
    - Trimethoprim-sulfamethoxazole is for pneumocystis prophylaxis postoperatively.
- **Total pancreatectomy with transportal islet cell *auto*-transplantation** is for patients with chronic pancreatitis causing pain and endocrine insufficiency.
    - The procedure entails removal of the patient's D2–D4, pancreas, and possibly the spleen.
    - If the spleen is preserved, the SA and SV are ligated at the hilum, and the organ resides on the short gastrics.
    - Enteric reconstruction is with a DD, DJ, or B2 GJ. Choledochojejunostomy establishes biliary reconstruction.
    - The harvested pancreatic tissue is then digested and undergoes transportal islet cell *auto*-transplantation either at the same OR visit or the next day. The infusion should be stopped if portal pressures are elevated. Post-op tight glucose control requires an insulin GTT for 48 hours.
    - It can take months for islets to neovascularize and engraft and result in normoglycemia.
    - Endpoints are reduction of chronic pain, improvement in quality of life, and insulin independence or at least improvement in glucose control.

15.

SPLEEN

ANATOMY AND PHYSIOLOGY

- Hematopoiesis occurs in the fetal spleen from 2 weeks to 5 months of gestational age.
- Splenic artery and vein reside in the splenorenal ligament. 70% of patients have distributive arterial anatomy, and 30% are magistral. *See* Figure 15.1.
- Short gastric vessels and the left gastroepiploic vessel reside in the gastrosplenic ligament.
- **White pulp** makes up 15% of the spleen volume and consists of arterioles surrounded by **periarteriolar lymphatic sheaths (PALS)**, within which are T and B cell follicles where macrophages present antigens, and lymphocytes produce IgM. *See Figure 15.2.* White pulp is the site of phagocytosis of poorly pre-opsonized bacteria or debris. **Tuftsin** assists with opsonization. It is also the site of alternate complement pathway activation by **properdin.**
- The **marginal zone** lies between the white and red pulp,
- **Red pulp** makes up 85% of the spleen volume and consists of non-endothelialized cords of fibroblasts, reticular fibers, and macrophages leading into endothelialized sinuses. Actin-myosin complexes control entry of red blood cells (RBCs) into the sinuses. Within a sinus, both **pitting** (removal of RBC membrane abnormalities) and **culling** (removal of non-deformable RBCs in their entirety) take place.
- The spleen normally holds one-third of circulating platelets at any one point in time. **Indium labeled autologous platelets** injected into the patient's bloodstream can quantify the contribution of sequestration to thrombocytopenia prior to considering splenectomy.
- **Hypersplenism** is defined as decreased circulating RBC, leukocyte, or platelet counts with a normal bone marrow response. The abnormalities should correct after splenectomy. Not all hypersplenism patients have splenomegaly.
- **Spontaneous rupture** of the spleen can occur due to mononucleosis, malaria, sepsis, sarcoidosis, leukemia, or a myeloproliferative disorder.
- A **wandering spleen** resides on a pedicle, is mobile, and can torse. It can be treated by splenopexy provided the spleen hasn't already infarcted from torsion. In the latter case, a laparoscopic splenectomy is needed.
- A **technetium labeled colloid or RBC injection** can diagnose a **splenosis, an ectopic spleen, or an accessory spleen** (which usually resides near the splenic hilum). Increased uptake may be noted due to splenomegaly in the setting of cirrhosis or non-alcoholic steatohepatitis (NASH).

SPLENIC MASS

- A **hemangioma** is the most common benign mass. Lymphangioma and hamartoma are less common.
- **Lymphoma** involving the spleen is the most common malignant mass.
- **Angiosarcoma** is the most common primary malignancy.
- **Cysts.** The majority are **asymptomatic, posttraumatic or post-infectious pseudocysts** with no true cellular lining.

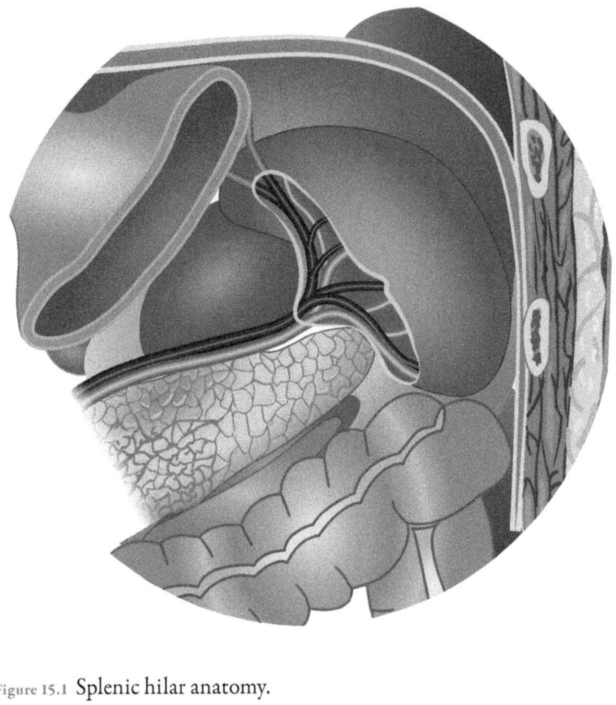

Figure 15.1 Splenic hilar anatomy.

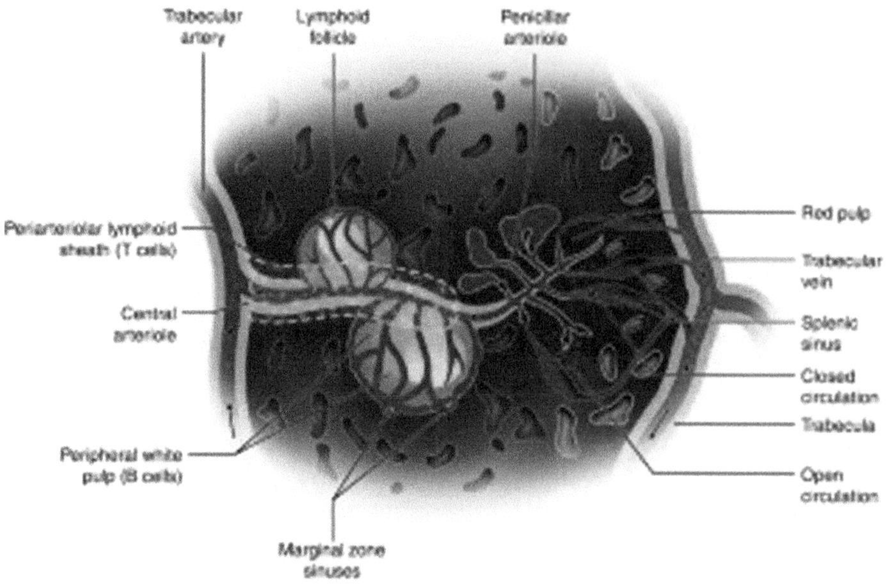

Figure 15.2 Splenic parenchymal architecture.

    - Cysts that are >5 cm are more likely to have symptoms.
    - True cysts make up only 10% of all cysts and have a benign squamous cellular lining. Some may have CA 19-9 or CEA elevations.
    - One may **aspirate** to diagnose a true cyst, evaluate for malignancy (rare), and to see if the patient's symptoms improve with desufflation of the cavity. If truly symptomatic based on the aspiration test, definitive treatment is with **drainage, laparoscopic fenestration, or splenectomy**.
    - An **echinococcal cyst** is treated with splenectomy or unroofing and marsupialization of the cavity. Albendazole or praziquantel should accompany any intervention.
- An **abscess** may be posttraumatic, hematogenous, or from a contiguous organ. Treatment is with **antibiotics and a percutaneous drain** if the abscess is unilocular. **Splenectomy** is reserved for a multilocular abscess or a refractory unilocular abscess.
- Peliosis describes multiple cyst-like cavities within the splenic parenchyma which have the potential for rupture.
- "Idiopathic" splenomegaly may need a splenectomy for histopathologic diagnosis and to exclude lymphoma.

PLATELET DISORDERS

- **IDIOPATHIC THROMBOCYTOPENIC PURPURA (ITP)**
- **PATHOPHYSIOLOGY**
    - This is the most common acquired reason for non-traumatic splenectomy.
    - Splenic lymphoid production of **IgG directed against platelet receptors gp2b3a or gp1b** coats their cell surface, which then results in phagocytosis within the spleen.
    - A peripheral smear with megakaryocytes represents an inadequate bone marrow response.
- **DIAGNOSIS**
    - **Based on exclusion of other causes of thrombocytopenia**. Serum testing for IgG cannot rule out the diagnosis.
- **TREATMENT**
    - **Observation** is appropriate in an adult with platelet counts above 30,000 and children with platelet counts above 20,000. ITP is more often self-limited in children.
    - **Glucocorticoids** are indicated for platelet counts <20,000–30,000ooo
    - **IVIG** will provide short term platelet increases needed for active hemorrhage or procedural risks. This effect can last up to 6 weeks.
    - **Platelet transfusions** are only used in the setting of active bleeding or if the count is <10,000.
    - **Eltrombopag** is a non-peptide thrombopoietin receptor agonist for refractory cases or for relapse after the above measures.
    - **Splenectomy** is reserved for failure of medical management at 6–8 weeks, with platelet counts continuing to drift below 10,000. Platelet transfusions may be given intraoperatively after splenic artery ligation.
- **THROMBOTIC THROMBOCYTOPENIC PURPURA (TTP)**
    - This is due to an abnormal activity of von Willebrand factor that causes platelet clumping, obstruction of

microcirculation, shearing of RBCs coursing through the platelet and fibrin aggregates, and end-organ ischemia and dysfunction. This is a medically treated disease with **plasma exchange and glucocorticoids.** Splenectomy is only rarely used in refractory cases.

HEMOLYTIC DISORDERS

- **WARM AUTOIMMUNE HEMOLYSIS**
  - **From IgG directed against RBC membranes.** If refractory to glucocorticoids or other immune suppression, the patient may need a splenectomy.
  - Splenectomy is **not used for cold, IgM mediated hemolysis**, as RBC destruction in these cases is intravascular or intrahepatic, and not mediated by splenic macrophages.
- **HEREDITARY SPHEROCYTOSIS**
  - The **most common congenital reason** to require a non-traumatic splenectomy.
  - A spectrin deficit to the RBC wall results in a high mean corpuscular hemoglobin concentration (MCHC), high red cell distribution width (RDW) and low mean corpuscular volume (MCV). Due to their abnormal shape, RBCs are sequestered and destroyed in the spleen.
  - **DIAGNOSIS**
  - Confirmation is by peripheral smear and the appropriate RBC indices (above).
  - **TREATMENT**
  - **splenectomy and cholecystectomy,** given the potential to develop future pigment stones. Partial splenectomies have been used successfully in children.
- **OTHER ERYTHROCYTE DISORDERS**
  - **Hereditary elliptocytosis** has oval shaped RBCs, some of which can slip through splenic cords and sinuses, so there is less hemolysis than in spherocytosis.
  - **Pyruvate kinase deficiency** is the most common *non-membrane* congenital reason for non-traumatic splenectomy. Impaired glucose metabolism shortens RBC lifespan and results in their destruction within the spleen.
  - **B-thalassemia** is a hereditary autosomal recessive disorder. The minor version is heterozygous and asymptomatic. The major version is homozygous with symptoms starting at 2–6 months of age when the body is attempting to switch to adult hemoglobin (A2B2 chains). Signs include anemia, growth limitations, head enlargement, and leg ulcers. Microcytic, hypochromic, target cells can be seen on a peripheral smear. Treatment is with blood transfusions and iron chelation. Splenectomy is for patients with symptomatic splenomegaly and persistently high transfusion requirements.
  - **Sickle Cell Disease (HbS)** is an amino acid substitution resulting in an abnormal conformation to the hemoglobin molecule in the deoxygenated state. RBC deformation and clumping occurs when oxygen levels are low on the venular side of a capillary. This results in venular obstruction and organ ischemia. Most patients have autoinfarcted their spleen by childhood. **Treatment** can start with **hydroxyurea** to increase HbF production. **Exchange transfusions** are needed for severe crises, including the following:
    - CVA or pulmonary arterial vaso-occlusion.
    - Other end organ vaso-occlusion or splenic sequestration.
    - Aplastic crisis; or
    - Dactylitis and digital inflammation when switching from HbF to the less oxygen-avid HbS as an infant.
  - Patients rarely require **splenectomy** for sequestration or splenic auto-infarction with abscesses in early childhood.
- **G6PD deficiency** results in decreased NADPH and reduced glutathione. This imbalance causes oxidized hemoglobin to build up within RBCs as Heinz bodies. Splenectomy is rarely ever needed.
- **Felty's syndrome** describes patients with rheumatoid arthritis and antibodies against white blood cells that result in coating and leukopenia from splenic sequestration and phagocytosis. Patients generally have leg ulcers. Splenectomy is reserved for those with refractory neutropenia and infections despite medical therapy.

LYMPHOID DISORDERS

- **Non-Hodgkin's lymphoma.** A systemic malignancy of B cell origin and the most common metastatic malignancy in the spleen. Splenectomy is needed if patient has refractory cytopenia, symptoms of splenomegaly, inadequate tissue obtainable elsewhere for a diagnosis, or not responding to treatment as expected.
- **Hodgkin's lymphoma.** Stage 1 is lymphadenopathy in one area or contiguous sites. Stage 2 is lymphadenopathy in noncontiguous areas on the same side of the diaphragm. Stage 3 is disease on both sides of the diaphragm. Stage 4 liver, lung, or bone involvement. Splenectomy was previously part of staging, but now it is less commonly used.
- **Lymphocytic and lymphoblastic leukemias** can also cause splenomegaly. But splenectomy is not indicated.

MYELOID AND OTHER DISORDERS

- Splenectomy is primarily reserved for hypersplenism with cytopenias or splenomegaly symptoms.

- **Polycythemia Vera** is an increase in all cell lines. Treatment is with aspirin, phlebotomy, and chemotherapy.
- **Essential thrombocytosis** presents with thrombosis or hemorrhage. Patients may have vasomotor symptoms.
- **Chronic myelogenous leukemia** is treated with imatinib.
- **Acute myelogenous leukemia** may be secondary to blastic transformation from CML.
- **Agnogenic myeloid metaplasia with myelofibrosis** has a peripheral smear with teardrop RBCs. Patients have extramedullary hematopoiesis with splenomegaly.
- Sarcoidosis, amyloidosis, and Gaucher storage disease rarely need splenectomy.

SPLENIC TRAUMA

- Injuries are graded by the circumference involved, depth of laceration, or size of intraparenchymal hematoma (IPH). *See* Figure 15.3.
- **Grade 1 injury** is a <10% circumference subcapsular hematoma or a <1 cm laceration.
- **Grade 2**: 10%–50% circumference subcapsular hematoma or 1–3 cm parenchymal laceration or <5 cm IPH.
- **Grade 3**: subcapsular hematoma >50% circumference or laceration >3 cm or IPH >5 cm. Any extravasation or pseudo-aneurysm is also considered grade 3. This may be managed with **selective** angioembolization if the patient is hemodynamically stable.
- **Grade 4:** >25% of the spleen is devascularized. This can be managed with observation or **nonselective** angioembolization if the patient is hemodynamically stable and a nonoperative course is chosen.
- **Grade 5**: a shattered spleen with hilar devascularization. This can be managed with observation or **nonselective** angioembolization if the patient is hemodynamically stable and a nonoperative course is chosen.
- Atraumatic ruptures can occur in the setting of hematologic malignancies or splenic infection.
- A **nonoperative trial** is acceptable regardless of the grade of injury, provided that hemodynamics are stable after the initial IVF or pRBC load, and there are no other major injuries. 10% of low-grade injuries and 75% of high-grade injuries fail a nonoperative trial.
- **Splenectomy** is reserved for the following:
  - Hemodynamic instability despite the initial resuscitation.
  - Continued product transfusion requirement during a nonoperative trial with or without selective or nonselective angioembolization; or
  - A high-grade injury with other confounding injuries needing laparotomy.
  - The splenocolic, gastrocolic, gastrosplenic, and phrenocolic ligaments are taken down. The splenorenal ligament contains the splenic artery and vein and can be taken en masse, or after isolation of each vessel.
  - **Immunizations should be given prior to discharge.** The highest risk for post-splenectomy sepsis would be within 2 years post-op in a child whose hematologic disorder prompted a splenectomy (see below).
  - **Delayed hemorrhage** can occur up to 6 weeks post-injury. After 6 weeks the parenchyma has healed.

SPLENORRHAPHY AND PARTIAL SPLENECTOMY

- **Splenorrhaphy** can be attempted if one-third of the organ can be saved.
  - Topical cautery, argon coagulation, microfibrillar collagen, gelatin sponge (gelfoam), methylcellulose (Surgicel) are topical maneuvers for splenic salvage.
  - Horizontal mattress suture plication can be performed with or without Dacron or organic pledgets, or a tongue of omentum included.
  - A mesh wrap of the entire spleen may be held in place by a purse-string suture. *See Figure 15.4.*
- **Partial splenectomy** is an option for deeper lacerations provided one-third of the organ can be preserved. Ligation of inferior or superior pole arterial or venous tributaries is followed by staple or cautery excision of devitalized splenic component, and then oversewing margin with pledgets or an omental inclusion.
  - Partial splenectomy may result in persistent hemolysis and cholelithiasis if used to manage extravascular hemolysis. Regrowth of the remainder of the spleen may occur within one year, but rarely is completion splenectomy necessary.

SPLENECTOMY PRINCIPLES

- Generally try to delay a non-emergent splenectomy until after the age of 5 whenever possible. This will decrease the risk of overwhelming post-splenectomy sepsis (OPSI).
- One may perform **pre-op gelfoam or coil splenic artery embolization.**
- **Laparoscopic** splenectomy is an option if dimensions are <20–25cm. *See Figure 15.5.* One may use a hand assist as well.

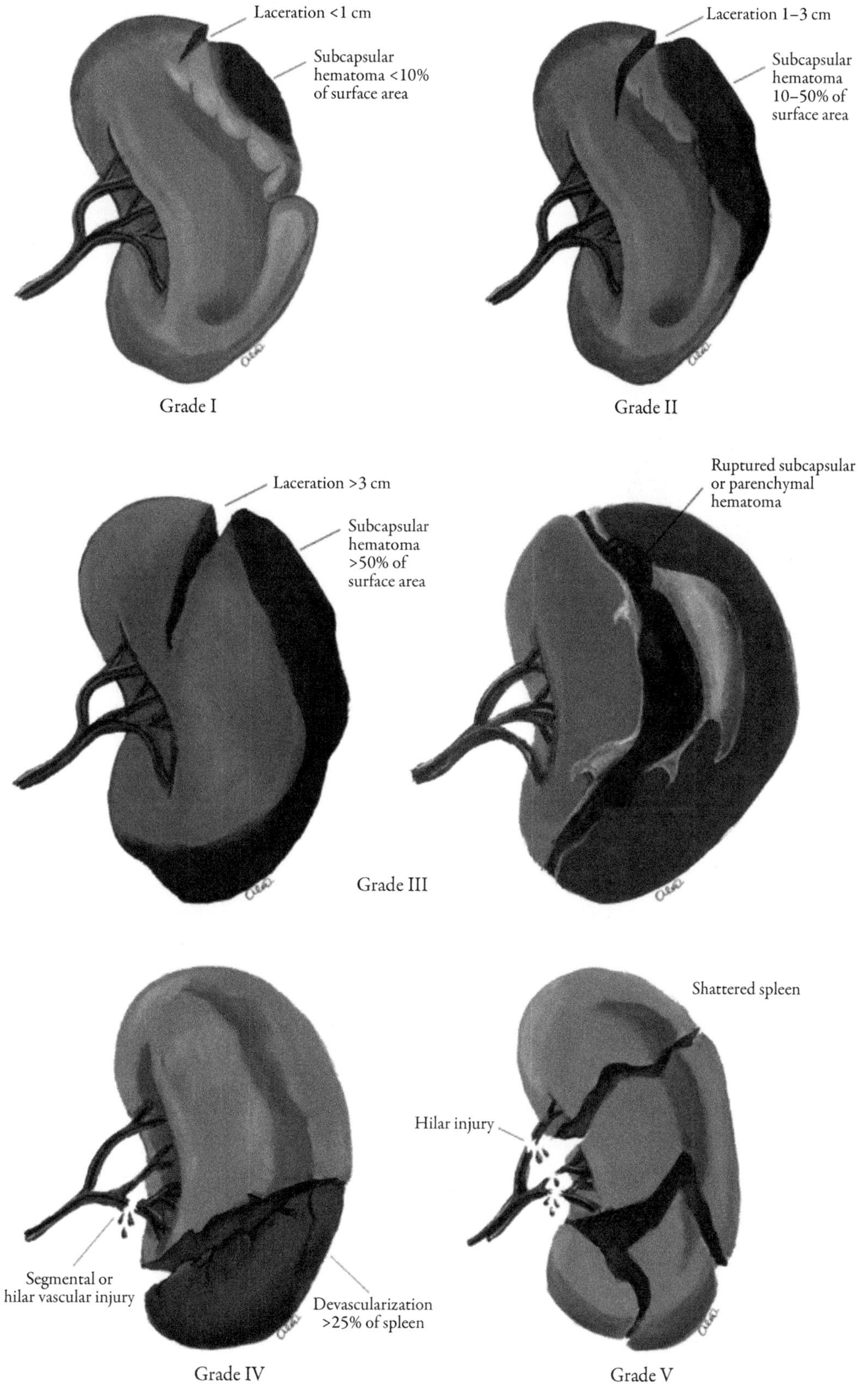

Figure 15.3 Grades of splenic injury.

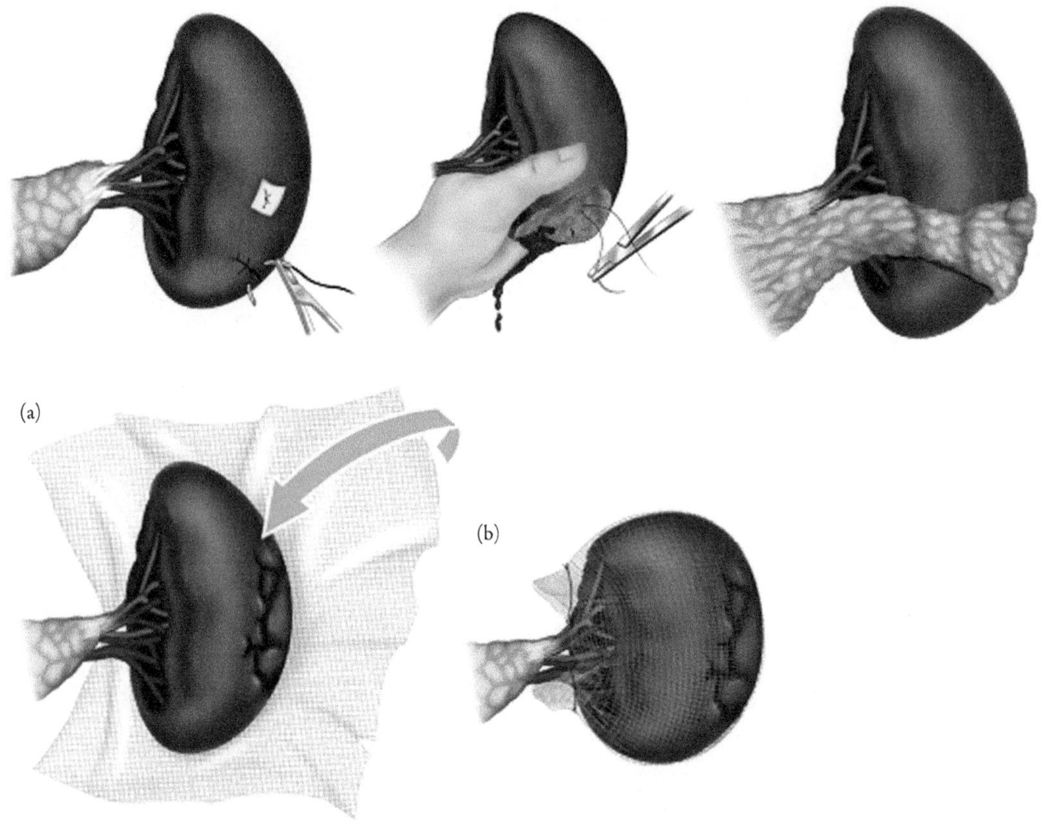

Figure 15.4 Options for splenorrhaphy.

- Take the splenocolic ligament and the lateral portion of the gastrocolic ligament. Move cephalad along the gastrosplenic ligament, taking the short gastrics.
- Leave the phrenosplenic and phrenocolic ligaments to maintain cephalad retraction of the spleen while the patient resides in the right lateral decubitus position.
- Then take the splenorenal ligament with splenic artery first and vein second.
- Morcellize spleen in a bag to prevent spillage and future splenosis.
- May extract spleen without morcellization via a mini midline laparotomy or Pfannensteil incisions.
- **Smear changes** after splenectomy include Howell Jolly bodies (nuclear fragments), Heinz bodies (oxidized Hgb deposits), Pappenheimer bodies (iron deposits), target cells, and acanthocytes.
- Transient thrombocytosis will peak in weeks.
- Transient leukocytosis can persist for months, with a lymphocytic and monocytic differential.
- Consider that an accessory spleen may be present if RBC smear changes are not visible post-op. The diagnosis can be pursued with a nuclear colloid or tagged RBC scan.
- Complications after splenectomy include the following:
- Postoperative **portal vein, splenic vein, or renal vein thrombosis**.
- **Deep venous thrombosis (DVT)** can be due to post-splenectomy thrombocytosis or the patient's underlying hematologic disorder. So consider prophylactic enoxaparin up to 4 weeks post-op.
- 2% of patients have a **pancreatic tail injury**. Drain any peripancreatic collection and consider endoscopic retrograde cholangiopancreatography (ERCP) and pancreatic duct stent if there is persistent leakage.

POST-SPLENECTOMY INFECTION (OPSI)

- There is a 1%–2% risk of OPSI. It occurs in the absence of IgM against *S. pneumo*, *H. flu*, and *N. meningitides* and has a 40%–70% mortality.
- It most often presents in patients **<5 years old** who have had a splenectomy for hematologic malignancy or myelodysplasia within the past 2 years
- **Vaccination** should be done 2 weeks prior to an elective splenectomy, or at the time of discharge after an emergency splenectomy.

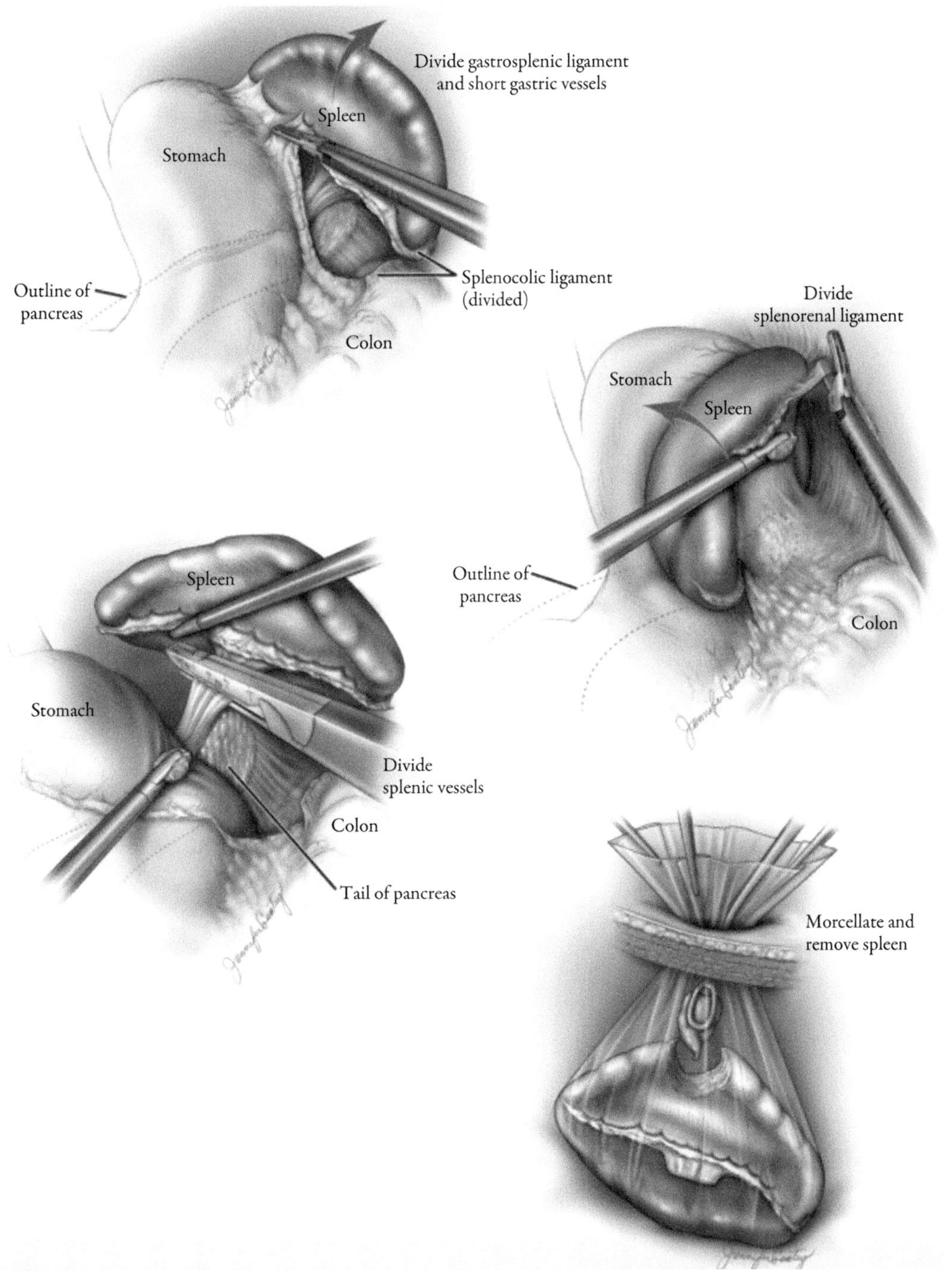

Figure 15.5 Laparoscopic splenectomy.

- SP13-valent and 23-valent vaccine and the *Neisseria meningitides* vaccine are each given for two doses, and this is repeated every 5 years.
- *H. influenza* vaccine is generally given just once.
- There is a possible benefit to prophylactic postoperative antibiotics for 2 years or longer.

SOFT TISSUE AND ENDOCRINE

16.

SKIN AND SOFT TISSUE

ANATOMY AND PHYSIOLOGY

- **Epidermis** consists of mostly **squamous keratinocytes.**
  - **Melanocytes** reside in the basal layer and transfer melanin to neighboring keratinocytes as an umbrella cap protecting their nuclei from ultraviolet light.
  - **Immune cells** are interspersed in the epidermal layer and include lymphocytes and Langerhans antigen presenting cells.
  - **Mechanoreceptors (Meissner's corpuscles)** are in the lower epidermis and superficial dermis.
  - **Eccrine and apocrine sweat glands** are innervated by the sympathetic system. They are present in the palms, soles, axillae, anogenital region, and nipple
  - Hair follicles are nestled within a **pilosebaceous unit** occupying invaginations of the epidermis that dip into the dermal layer.
  - Type 4 collagen is at the basal lamina of the epidermal to dermal junction.
  - Type 7 collagen anchors the epidermis to dermis.
- **Dermis** sits below the epidermis and is composed of **papillae** interdigitating with the epidermis and containing loose structural proteins and a superficial vascular network plus a deeper **reticular** dermis with thick horizontal proteins and its own vascular network. *See Figure 16.1.*
  - **Type 1 collagen** is the most predominant type in the dermis. A branching pattern to the collagenous network endows elasticity and memory to regain shape. Type 3 collagen is in vasculature.
  - **Pacinian** corpuscles detect pressure, **Ruffini** endings detect warmth, and **Krause** bulbs detect cold.
  - The dermis is more resistant to pressure necrosis than the underlying muscle. In muscle, ischemia develops if pressure exceeds 30 mmHg. Body weight produces a pressure at the ischial spine of >300 and at the sacrum of >150 mmHg.
- The **subcutaneous tissue** runs below the dermis and is of varying thickness depending on body locale and patient habitus. Deep to the subcutaneous tissue is **fascia and muscle.**
- **Radiation** damages basal epidermal cells, hair follicles, and melanocytes. Inflammation sets in over weeks including capillary thrombosis. Re-epithelialization begins at 2 weeks. This is followed by chronic fibrosis and possible permanent pigmentation.
- **UVA light** is low energy but **penetrates deeper**.
- **UVB light** is responsible for sun damage and produces **superficial erythema.**

INFECTION AND INFLAMMATION

- **DERMAL ULCERS**
- Due to immobility with pressure ischemia at bony prominences. *See Figure 16.2.*
  - **Grade 1** is erythema.
  - **Grade 2** has a dermal injury, but no exposed subcutaneous (SQ) fat, Treatment is with a moist dressing and Silvadene.
  - **Grade 3** is a dermal breach that exposes SQ. Treatment is with sharp debridement and negative pressure wound therapy (NPWT) if the wound is clean and not ischemic.
  - **Grade 4** is down to fascia, muscle, bone, or joint. Treatment is with debridement of the wound and any bony prominence, followed by soft tissue local flap coverage.
- **HIDRADENITIS SUPPURATIVA**
  - Secondary to obstruction of hair follicles in the axilla or inguinal regions. Obstruction causes pilosebaceous unit rupture and adjacent inflammation.
    - Risk factors include smoking, alcohol, and infection with *staphylococci* or *streptococci.*
    - **Hurley stage 1**: singular abscesses without sinus tracts or connections.
    - **Hurley 2**: multiple separate abscesses with sinus tracts.
    - **Hurley 3**: confluence of multiple abscess cavities and their sinus tracts.
  - Treatment is with **antibiotics, antiandrogens, zinc, and biologic modifiers (adalimumab)**.
  - Incision and drainage should be done if there is an acute infection.
  - Advanced cases need **wide local excision (WLE)** with healing by secondary intention or coverage with a skin graft or flap.

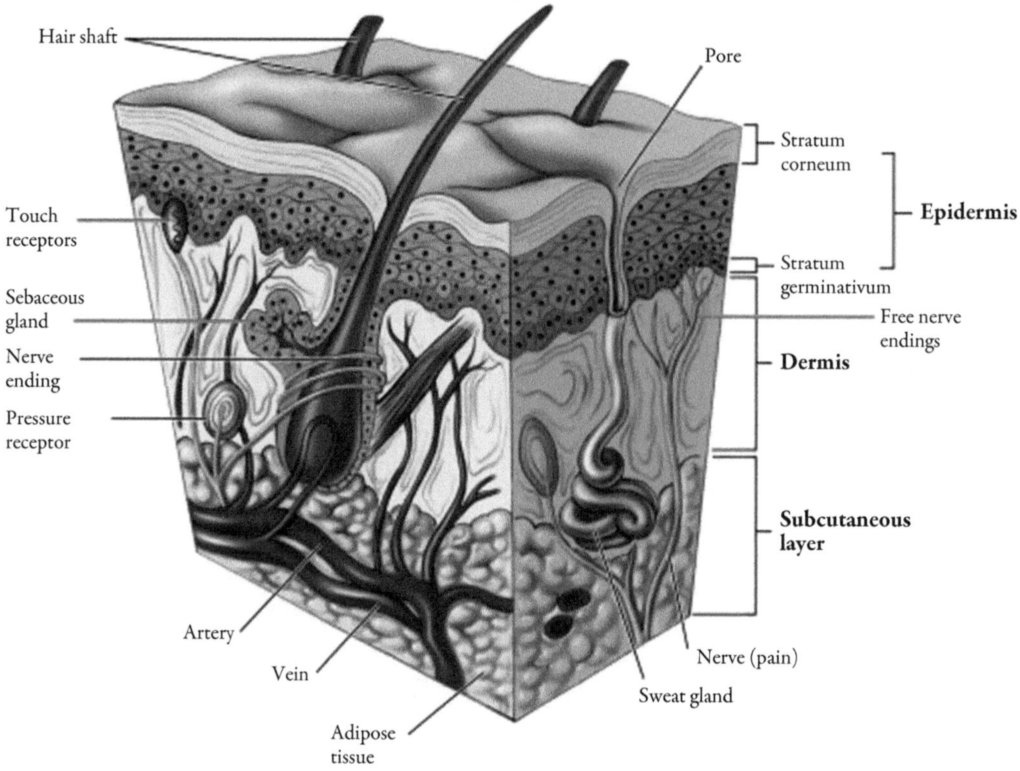

Figure 16.1 Dermal architecture.

- **PILONIDAL CYST**
  - This is cyst formation in the midline, or just off-midline, of the superior gluteal cleft from an ingrown hair. There may be a deeper sinus connection to sacrococcygeal structures.
  - Treatment for **mild cases** includes shaving or laser hair removal, curettage of small sinus tracts, or excision of small pits and closure in the midline or off-midline. If secondarily infected, incision and drainage via an off-midline incision should be performed.
  - For **extensive disease** there are two options:
    - 1. Excise and leave the wound open or marsupialize for secondary intent healing. This has the lowest infection rate and hence lowest recurrence rate.
    - 2. Excise and construct an off-midline flap with tension-free closure, which flattens the natal cleft (cleft procedure, rhomboid flap).
- **NON-NECROTIZING INFECTIONS**
  - **Impetigo** is a non-tender bacterial dermatitis treated topically.
  - **Erysipelas** is an infection of the deeper dermis and superficial SQ layer.
  - **Cellulitis** is an infection of the deeper SQ. All are generally streptococcal or staphylococcal and treated with antibiotics alone. Less common organisms include Eikenella bacteria from human bites and Pasteurella from dog and cat bites.
  - **Lymphangitis** is due to a *Streptococcus pyogenes* inoculation after minor trauma, and presents as red streaks running along a limb. Treatment is with antibiotics alone.
  - **Folliculitis** is inflammation at the base of a hair follicle. It can progress to a localized intradermal abscess centered around the follicle (**furuncle**) or a complex, loculated connection of intradermal abscess cavities (**carbuncle**). Furuncles and carbuncles need incision and drainage.
- **NECROTIZING SOFT TISSUE INFECTIONS**
  - **Type 1** is polymicrobial, including *bacteroides*. It presents on the trunk or limb of an immune compromised patient, diabetic, or intravenous drug user (IVDU).
  - **Type 2** is monomicrobial. Organisms include *streptococci*, *vibrio* from seafood or saltwater, *aeromonas* from freshwater, or *clostridia* from environmental inoculation.
    - **Clostridial myonecrosis** is also known as gas gangrene. It is secondary to muscular inoculation from trauma, surgery, IVDU, or translocation across the colonic mucosa.
      - Organisms include *C. perfringes* from the environment, *C. sordelli* from gynecologic procedures and black tar heroin, *C. novyi* in IVDU patients, or *C. septicum* translocated across the colon.
      - **Alpha and theta exotoxins** disrupt soft tissue and muscle cell membranes, reduce myocardial

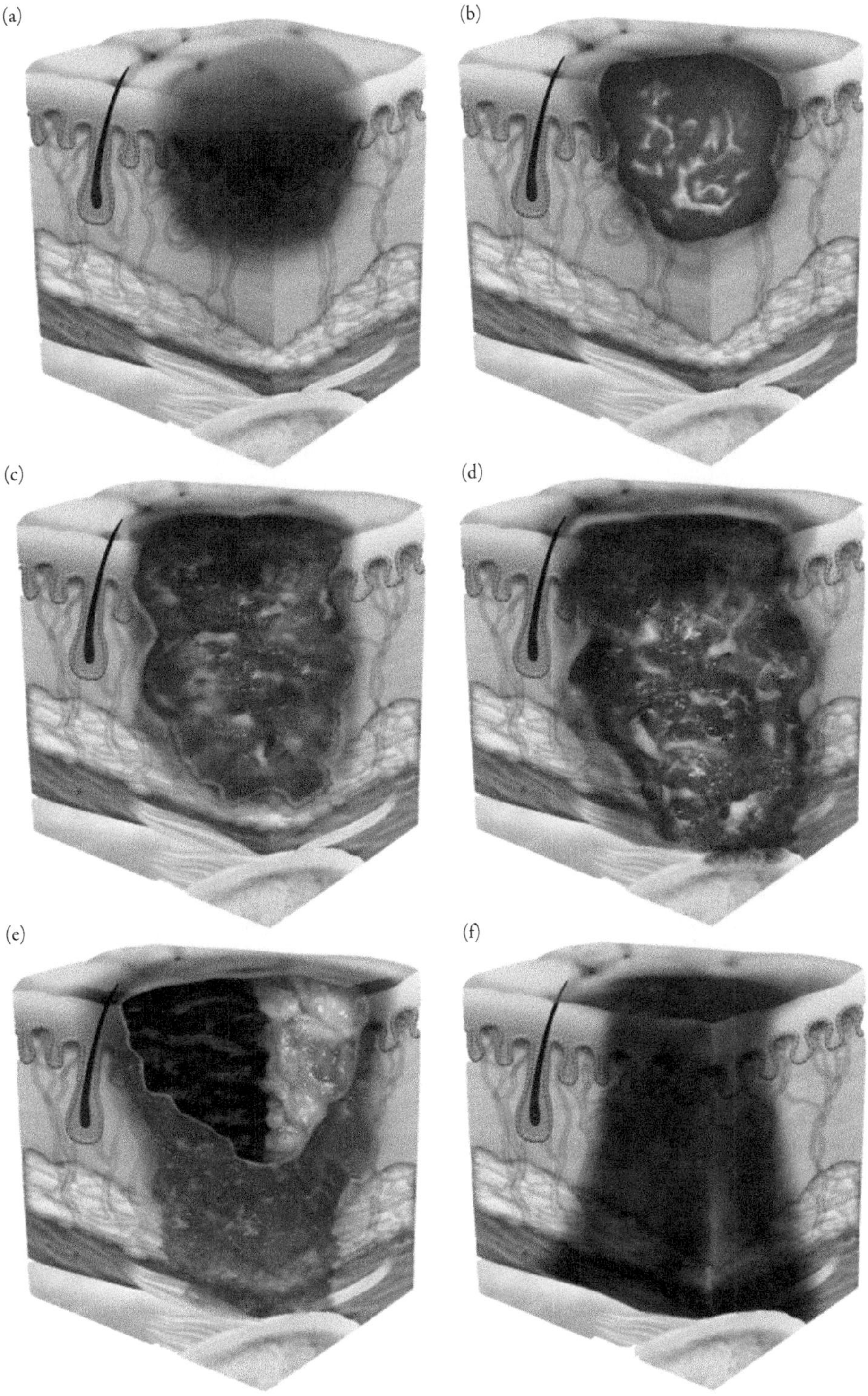

Figure 16.2 Degrees of dermal pressure necrosis a) grade 1 b) grade 2 c) grade 3 d) grade 4 e & f) grade 4 with necrotic infections.

contractility, and decrease systemic vascular resistance (SVR). The result is distributive shock with impaired cardiac compensation.

- **DIAGNOSIS**
  - Based on the physical exam. Patients have localized pain and a systemic response that usually starts with tachycardia. Since the infection is often deep to the dermis, skin changes lag behind. Early on, there will be just erythema-mimicking simple cellulitis,
  - The topology of tenderness usually extends beyond the visible skin changes.

- In time there can be a bluish hue secondary to necrotic disruption of subcutaneous vasculature and percolation of bloody exudate to the skin surface.
- Hemorrhagic bullae are a late finding, and form because a dome of superficial keratin remains intact, encasing necrotic sanguineous components from below.
- Any weakness of a muscle compartment suggests myonecrosis, not just fasciitis. Fever, elevated WBC, and depressed sodium are often, but not always, present.

- **TREATMENT**
  - Start with crystalloid resuscitation and **broad antibiotics** that can be narrowed if a monomicrobial organism is ultimately identified. Penicillins may be adequate for streptococcal or clostridial infections. Clindamycin is added to inhibit exotoxin production.
  - **Debridement should occur within 24 hours of presentation.** Repeat debridements may be needed. Mortality is up to 25%. Hyperbaric oxygen (HBO), may be used as an adjunct in a patient who has stabilized after debridement.
- **DISORDERS OF SCAR FORMATION**
  - Dermal scar tissue is made up of proteoglycans, hyaluronic acid, and type 1 collagen.
    - **Hypertrophic scars** are characterized by an excess of type 1 collagen within the confines of the original scar. It is due to wound infection, burns, or impaired healing, but not genetics. Treatment is with **lidocaine and steroid injection** early on, silicone sheeting, pressure garments, or re-excision.
    - **Keloid scars** are characterized by excess type 1 collagen that extends beyond the confines of the original scar. There is a **genetic component to keloids**. Treatment is with lidocaine and steroid injections early on, intra-lesional chemotherapeutics, silicone sheeting, pressure garments, radiation, or excision and Z-plasty closure.

WOUND COVERAGE

- Skin grafts are the most common means of coverage for wounds that cannot be primarily closed. They may be placed directly over muscle, granulation tissue, peritendineum, or periosteum.
  - After placement of a graft, **imbibition** of nutrients starts within 48 hours. **Inosculation** and capillary alignment occur at 2–4 days. Arterial and venous **revascularization** begins at 5–7d. Sensation develops at 10 weeks to 2 years post grafting.
  - The main cause of graft failure is a **sub-graft fluid collection** or hematoma that disrupts imbibition, inosculation, and revascularization.
  - **Split thickness skin grafts (STSG)** have decreased primary graft contraction at 10%, compared to full thickness skin grafts (FTSG), which have 40% primary contraction. STSG have increased recipient site contraction compared to FTSG. Thinner grafts are perfused and vascularize more easily than thick ones.
  - **FTSG** are taken from retroauricular and preauricular spaces, supraclavicular sites, inguinal fossa, or the forehead. Subdermal fat is removed.
- Local tissue rearrangement **random flaps** are based on intradermal and subdermal vascular networks for perfusion. It may be a VY advancement, rotation, transposition around a pivot point (rhomboid flap or Z-plasty), or interpolation.
  - Recipient site **tissue expanders** may be used for 6–8 weeks for dermal stretch and skin hyperplasia in order to accommodate a future random flap construction and coverage.
- **Axial flaps** are based on a named vessel. They may be **pedicled,** remaining attached to the donor site, with or without skeletonization. The most common example is a contralateral superior epigastric artery–based transverse rectus abdominis muscle (TRAM) flap.
  - Axial flaps may also be constructed **free** with a transected parent vessel, with or without the associated muscle. One may leave some muscle at the donor site (muscle sparing free flap) or leave all muscle at the donor site (perforator free flap).
    - The donor site may be preconditioned by ligating other vessels such that the donor parent vessel angiosome is enlarged and more robust prior to harvest of the flap.
    - The donor site artery must be sewn to a vessel at the recipient site with or without an interposition free vein graft.
    - Vasospasm is mitigated with a limited adventitiectomy of the donor parent vessels, avoidance of caffeine and smoking, and the addition of lidocaine or papaverine topically for vasodilation.
  - Venular injury will lead to edema and venous occlusion. This results in secondary impaired inflow and graft loss. Leeches or topical heparin can prevent venular congestion.
  - Keep the flap warm and the intravascular space replete with colloid or blood infusions. Serial Doppler evaluations can ensure arterial inflow is adequate.

AESTHETICS

- **Augmentation mammoplasty** with silicone implants has a decreased incidence of capsule contracture or rupture. A submuscular position of implants prevents rippling.

- **Reduction mammoplasty** can be performed for nipple ptosis or associated muscular discomfort, and can be complicated by nipple island hypoesthesia and necrosis.
- **Blepharoplasty** involves removal of eyelid adipose tissue.
- **Facelift** tightens the superficial muscular aponeurosis, most often complicated by hematoma.
- Suction lipectomy, autologous fat grafting, brachioplasty, transverse abdominoplasty, or total lower body lift are all methods of body contouring.

MELANOMA

- **PATHOPHYSIOLOGY**
  - Melanocytes are neural crest in origin and reside in the basal layer of the epidermis. They produce melanin from tyrosine. Melanin is shared with neighboring keratinocytes.
  - Melanoma occurs most commonly on the back in men and legs in women. 10% of melanoma patients get a second primary.
  - Risk factors include dysplastic, atypical, or large congenital nevi and xeroderma pigmentosum. Familial atypical multiple mole syndrome has the highest risk.
  - Nevi may be limited to the epidermis (junctional nevi), dermis, or they may have both epidermal and dermal components (compound nevi).
- **DIAGNOSIS**
  - Patients complain of a new, enlarging, painless pigmented lesion.
  - Clinical features include **A**symmetry, irregular **B**orders, **C**olor change within the lesion, and increasing **D**iameter or **E**volution over time.
    - Suspicious lesions should undergo **excisional biopsy of the entire lesion if <2 cm in a non-cosmetic area.** This may be accomplished with an automated punch biopsy for very small lesions. Otherwise, a formal excisional biopsy whose orientation would facilitate future WLE (if needed) should be done. All biopsies should extend down through the reticular and deep dermis into the SQ or fascia.
      - For lesions larger than 2 cm or those in cosmetically sensitive areas, a full thickness **punch biopsy** can be performed at the **thickest area of darkest pigment, ideally containing part of a margin with normal skin.** Multiple such biopsies can be taken.
      - Samples should be evaluated for thickness, mitoses, ulceration, lymphovascular invasion, and margins. BRAF testing should be performed.
  - Check for tumor along the primary site's lymphatic drainage route, known as **disease in-transit.**
  - Check for nodal involvement by exam and **ultrasound (US). Fine needle aspiration (FNA)** can confirm if suspicious nodes actually harbor melanoma. These would be clinically positive nodes.
  - **In situ disease (thin lentigo)** is limited to the superficial papillary dermis.
  - **Lentigo maligna** have radial more so than nodular growth. It makes up 10% of all melanomas and is common in the elderly.
  - **Superficial spreading** melanoma is the most common overall.
  - **Acral lentiginous** melanoma presents on the palms and soles of African Americans. It is generally diagnosed in a delayed fashion, and therefore has the worst overall prognosis.
  - **Nodular** melanoma is the most aggressive and is staged by depth. *See Figure 16.3.*
    - **Stage 1** is < 2 mm without ulceration or <1 mm with ulceration.
    - **Stage 2** is >2 mm with or without ulceration or 3–4 mm depth without ulceration
    - **Stage 3** is clinical or sentinel lymph node positive. N1 (1 node) and N2 (2–3 positive nodes). N status is the best predictor of survival. If the patient has multiple nodes or bulky lymphadenopathy, then consider staging imaging or a PET scan to evaluate for distant metastases.
    - **Stage 4** patients have metastases. Lung metastases are via the lymphatics, and SB metastases are hematogenous. Ocular melanoma can metastasize to the liver.
- **TREATMENT**
  - In situ melanoma (thin lentigo) can be excised with **0.5 cm** margins.
  - < 1 mm depth melanoma is excised with **1 cm margins**. If the lesion is **>0.8 mm or has ulceration, a mitotic rate >1, or aggressive features**, then lymph nodes may be involved. So a sentinel lymph node biopsy (SLNB) should be considered, provided there are no clinically evident nodes. See below.
  - 1–2 mm depth lesions are excised with **1–2 cm margins.** 10%–20% of tumors at this depth will have positive lymph nodes, so SLNB should be done unless the patient has clinically positive nodes.
  - >2 mm depth lesions are excised with **2 cm margins.** 30%–40% of these tumors will have positive lymph nodes, so SLNB should be done unless the patient has clinically positive nodes.
  - The depth of excision must extend **through the superficial fascia of the SQ.** One may take the deeper muscular fascia in the excision if there is an associated desmoplastic reaction.
    - On the sole of the foot, excise to plantar fascia and cover with a split thickness skin graft (STSG), or a pedicled

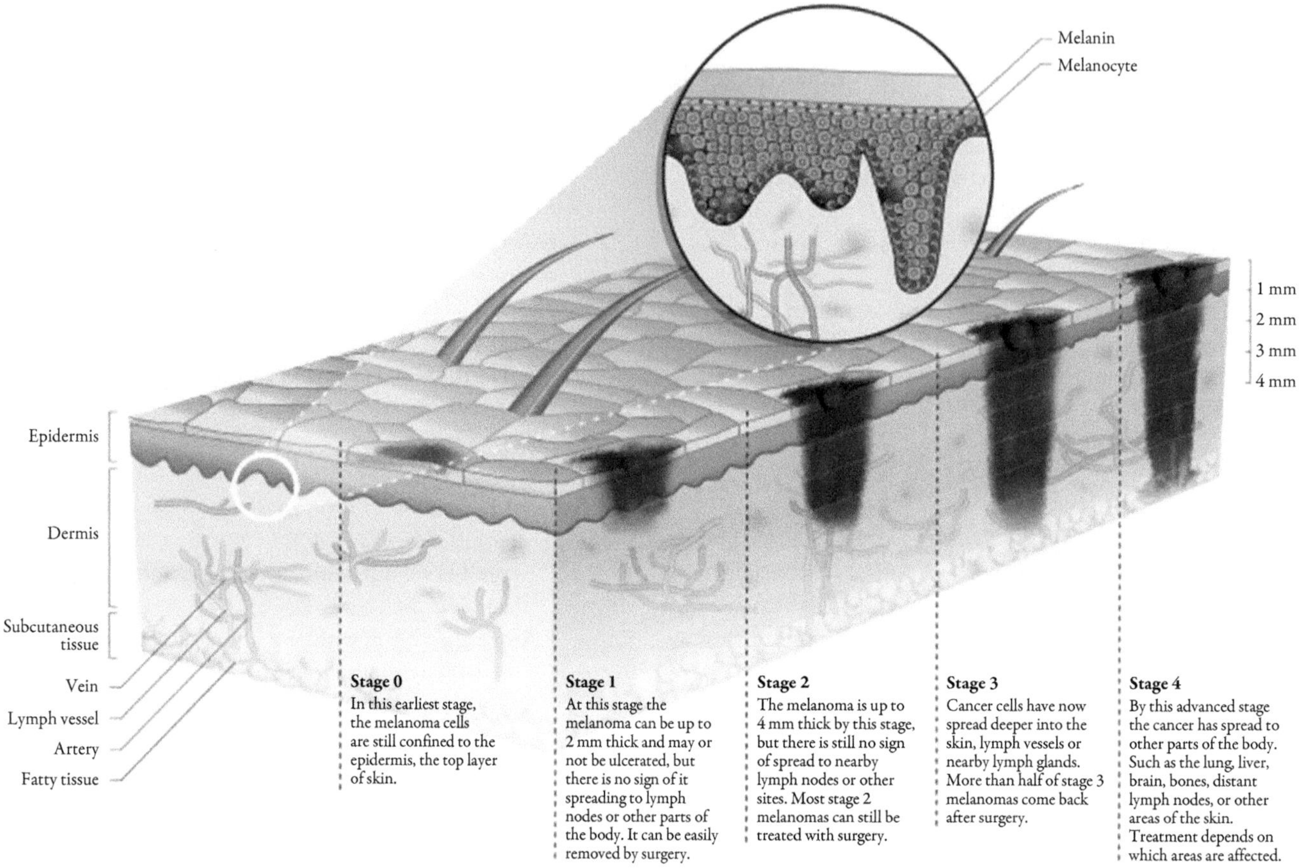

Figure 16.3 Nodular melanoma depth.

flap or free flap if the excision is over a weight-bearing bony prominence.
- If the melanoma is subungual, treatment can be either wide local excision (WLE) or amputation just proximal to the interphalangeal (IP) joint.
- In advanced cases, the resection site may benefit from adjuvant radiation.

- One may close the resection site primarily in a longitudinal or oblique orientation if on a limb.
  - Orientation of closure should be along skin lines if the resection was on the trunk, head, or neck.
  - A random flap, pedicled flap, free flap, or graft can be used if primary closure is not possible. The source of coverage should not come from an area lying within the lymphatic transit pathway of the primary lesion.
- **Sentinel lymph node biopsy (SLNB)** involves injection of a radiotracer and blue dye into the peritumoral dermal area in a patient that is clinically node-negative, but meeting any of the above mentioned criteria. In certain anatomic locales, the radiotracer can determine which nodal basin the tumor initially drains into.
  - Take the sentinel node demonstrating the highest radioactivity as well as all nodes >10% of that radioactivity. If any are positive for melanoma, the patient needs a completion lymphadenectomy of that entire basin. See below.
  - **Complications** of SLNB include seroma, hematoma, lymphedema, and wound infection. All at rates of **<5%.**
  - Lymphazurin blue anaphylaxis risk is <1%.
- **Completion lymphadenectomy** should be performed for:
  - **1. A heavy positive nodal burden or aggressive features on SLNB without distant metastases.** For limited nodal involvement, observation and serial US without completion lymphadenectomy is an option.
  - **2. Any clinically positive node by US guided FNA, provided there are no distant metastases.** If the primary is not identified by physical exam, a PET-CT may be needed to identify it.
  - **3. Any clinically positive node by US guided FNA is the first manifestation of recurrent disease.** The exception would be if the patient has multiple metastases identified during the workup of the recurrence. In this case, the benefit of a completion lymphadenectomy may not be realized.
    - **Cervical** completion LN dissection for an anterior face or neck primary melanoma. A superficial

parotidectomy is included for facial melanoma, even if periparotid nodes are negative. One may consider additional neck radiation if nodes were clinically positive. There is a 20%–30% recurrence rate.
    - **Axillary** completion LN dissection for an arm melanoma includes zones 1–3.
    - **Inguinal** completion LN dissection via an infrainguinal oblique or vertical incision is employed for a thigh melanoma meeting depth criteria.
        - The area of dissection is bordered by the pubic tubercle medially, anterior superior iliac spine (ASIS) laterally, apex of femoral triangle inferiorly, and 5 cm above the inguinal ligament superiorly. The dissection may be performed open or endoscopically.
        - Sartorius muscle coverage can be considered for a large residual defect.
        - There is a **20% lymphedema risk,** so consider postoperative stockings.
    - **Iliac** completion LN dissection via a suprainguinal oblique or vertical incision is performed if there are positive iliac nodes on imaging or >4 nodes positive on a preceding inguinal dissection. One may transect the inguinal ligament for exposure, and take all nodes cephalad along the external iliac vessels up to the common iliac bifurcation. Any defects can be closed with a Sartorius muscle flap or mesh.
    - **Popliteal** LN dissection is for a lower leg melanoma with SLNB localizing to this area or clinically suspicious nodes here. If positive, one may advance cephalad to a superficial inguinal lymph node dissection.
- One may **resect single, dual, or isolated metastases** to distant soft tissue, distant nodes, lung, small bowel, or liver. Also, it can be beneficial to resect isolated metastatic recurrences after a long disease-free interval.
- **Adjuvant Therapy**
    - **Interferon alpha (IFN-a)** was indicated for bulky primary disease (stage 2b, 2c) or node positive (stage 3). It is less commonly utilized now.
    - **Ipilimumab** is an anti-CTLA antibody that prevents down-regulation of T cell function. It can be used for positive lymph nodes, multiple metastases, or an unresectable primary. **Pembrolizumab** is an anti-PD-1 preventing programmed cell death of T cells, and is another option for advanced disease. Neither require a specific melanoma cell mutation to be effective, since they both work by augmenting T cell function.
    - **BRAF inhibitors** can be used if the melanoma itself is wild type positive.
    - **Isolated limb perfusion** (open surgical vessel access with an oxygenator and bypass circuit) or **isolated limb infusion** (percutaneous access and a hypoxic limb environment) are both considerations for treating locally advanced limb disease, before resorting to an amputation.
    - **Radiation therapy** can be added to any lymph node basin demonstrating advanced disease (multiple, large nodes with extracapsular spread of tumor) in order to decrease local recurrence.

BASAL CELL CARCINOMA (BCC)

- **PATHOPHYSIOLOGY**
    - BCC is the **most common malignancy in the United States.** It is four times more common than squamous cell carcinoma (SCC). Tumors arise from the basal epithelial layer.
        - Neoplasia is secondary to **UVB exposure.**
        - Basal cell nevus syndrome can lead to multiple BCCs.
    - Over 80% are located on the **head and neck,** including the eyelid and upper lip. Some patients have palmar and plantar pits.
    - **Nodular** lesions are the most common and have a pearly appearance, rolled borders, peripheral palisading of nuclei, and a stromal reaction. Superficial and morpheaform lesions are less common.
- **DIAGNOSIS**
    - **Incisional punch biopsy or excisional biopsy** (for smaller lesions that are not in cosmetically sensitive areas).
- **TREATMENT**
    - **Mohs micrographic surgery with negative margins or wide local excision with 3 mm–5 mm margins.** This involves serial resections with increasing margins, and real time evaluation of clearance of that margin. This technique is more time consuming but produces equal results to WLE with better cosmesis. *See Figure 16.4.*
    - If Mohs surgery is not being done, one may perform frozen section of the margins or leave the wound uncovered or temporarily grafted until final pathology confirms the margins as negative. At that time, the wound may be formally closed or covered.
    - **High-risk features** include location on the central face, lesions >6 mm diameter, a desmoplastic reaction around the tumor, indistinct borders, a morpheaform subtype, and underlying immunosuppression. These cases may need up to 6 mm margins.
    - **Regional lymphadenectomy** should be done for clinically FNA positive nodes.
    - Radiation may be added for close margins or aggressive tumors with neurovascular or lymphovascular invasion. Other adjuncts include imiquimod, 5-FU, and photodynamic therapy.
    - **Radiation and chemotherapy** can be employed as primary therapy for non-operable disease and metastases.

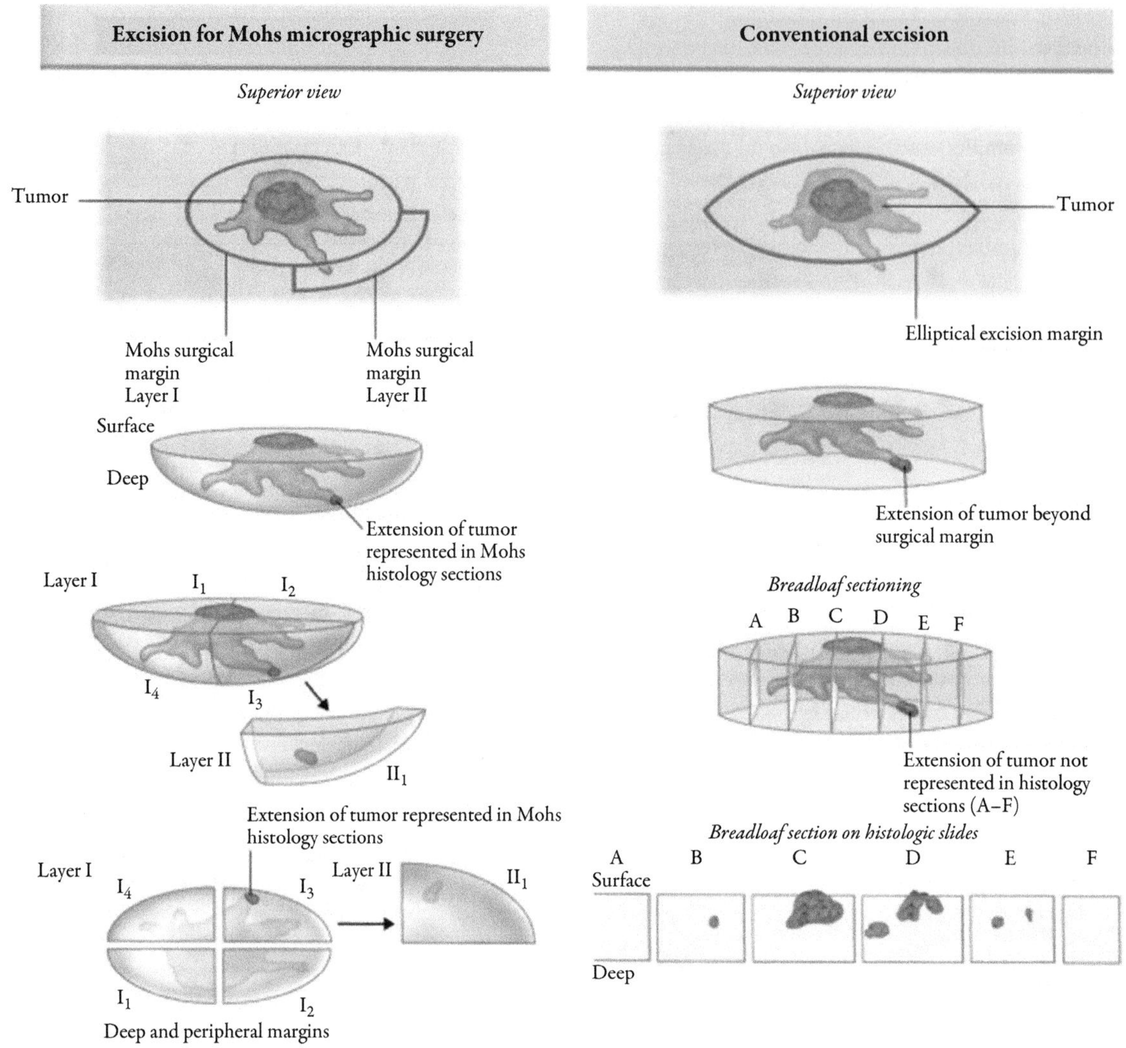

Figure 16.4 Mohs micrographic surgery versus conventional excision.

SQUAMOUS CELL NEOPLASIA

- **PATHOPHYSIOLOGY**
  - **Actinic keratoses (AKs)** are precursor lesions with nuclear pleomorphism. There is a 6%–15% risk per year of developing SCC. Transformation to SCC can also occur years later during immunosuppression. Precursor AK lesions may be treated with **cautery, topical therapy,** or excision if there is concern for concomitant SCC.
  - **Bowen's disease (BD)** is in situ squamous carcinoma lacking basement membrane invasion. It may be associated with human papilloma virus (HPV). 10% of BD become invasive SCC. It appears as a red-brown scaly plaque on physical examination. **Treatment** is with **topical imiquimod, 5-FU, or excision with 5 mm margins.** The latter may be preferred in the perianal location. Surveillance exams and biopsies should be performed regularly. High resolution anoscopy is used as an adjunct for surveillance in the perianal area.
  - **Squamous cell carcinoma (SCC)** risk factors include AK, BD, HPV types 16 and 18, prior radiation, UV light exposure, xeroderma pigmentosum, epidermal atrophy, arsenics, hydrocarbons, chlorophenols, or a chronic wound or ulcer. SCC within such a chronic wound is known as a Marjolin ulcer.
    - HPV types 1, 2, and 4 are associated with cutaneous condyloma and 6 and 11 with mucous membrane condyloma. None of these types cause SCC. Such benign HPV disease can be treated topically or with cryotherapy or fulguration.
- **DIAGNOSIS**
  - It appears as a flesh colored hyperkeratotic papule or nodule with surrounding induration, erythema, and possibly ulceration or satellitosis. Most presentations are on the face, forearms, and hands.

- Lesions on the lips, genitalia, legs, perianal area, or within a chronic scar can behave more aggressively.
- The keratoacanthoma variant of SCC grows rapidly and should be biopsied to distinguish it from a true benign keratoacanthoma crater filled with keratin.

- TREATMENT
  - **Resection with 5 mm–10 mm margins from the edge of erythema.**
  - Use the higher end margins if the patient has had prior radiation, recurrent disease, indistinct borders, immunosuppression, a high-risk location of SCC, or perineural involvement noted on biopsy.
  - Frozen section of margins can be done to confirm they are negative. Alternatively, the wound may be left uncovered or temporarily grafted until final pathology confirms negative margins, at which time the wound can have formal closure or coverage.
    - **Mohs surgery** can be considered for high-grade aggressive lesions, perineural involvement, large lesions, indistinct margins, or if the lesion is in a cosmetic area.
  - Regional **completion lymphadenectomy** should be done for clinically FNA positive nodes. Certain aggressive cases that are clinically node-negative may benefit from sentinel lymph node biopsy (SLNB) on a case by case basis, to determine if a completion lymphadenectomy is needed.
  - **Radiation and chemotherapy** are used for non-operable disease, high-grade lesions, perineural, neurovascular, or lymphovascular invasion, metastases, or instances where adequate margins are not achievable. LN basin radiation may be applied on a similar basis.

SARCOMA

- PATHOPHYSIOLOGY
  - Sarcomas are considered **idiopathic or secondary to prior radiation** to the breast, ovary, cervix, testes, or lymphoma. Post-radiation sarcomas are considered more aggressive and can have a p53 mutation. Other causes include herbicides or chronic lymphedema. Thorotrast, polyvinylchloride, or arsenic can give rise to hepatic angiosarcoma.
  - Syndromes associated with sarcomas include the following:
    - **Neurofibromatosis**: CNS tumors, pheochromocytomas, peripheral nerve sheath neurofibromas ranging to malignant peripheral nerve sheath tumors.
    - **Li-Fraumeni syndrome**: childhood rhabdomyosarcoma in addition to brain tumors, breast cancer, sarcoma, osteosarcoma, nephroblastoma, neuroblastoma, adrenal tumors, and leukemia.
    - **Tuberous sclerosis**: angiomyolipomas, cardiac myxomas.
    - **Gardner's syndrome:** Familial adenomatous polyposis plus desmoids and osteomas.
    - Hereditary retinoblastoma and cardiac rhabdomyomas can also be familial.
  - The most common type of sarcoma is an **unclassified pleomorphic sarcoma**, also known as "malignant fibrous histiocytoma." Spindle cells will be seen on histology.
  - 50% of sarcomas occur on the **extremities.**
  - 15% occur in the **retroperitoneum** as liposarcomas or leiomyosarcomas. *See Figure 16.5.*
    - **Lipomatous tumors** are the most common sarcoma in the retroperitoneum, and the second most common sarcoma overall. They may range from atypical lipomatous tumors to poorly differentiated liposarcomas.
    - **Leiomyosarcoma** is the second most common retroperitoneal sarcoma. They may also be found in the uterus or arising from vasculature.
  - **Hematogenous spread** is the predominant route for metastases. Lymph node involvement is uncommon except for Clear cell, Alveolar, Rhabdomyosarcoma, Embryonal, and Synovial subtypes.

- DIAGNOSIS
  - Extremity sarcomas present as rapidly growing, painless, masses. Often times, patients attribute the mass to an incidental traumatic episode. Retroperitoneal sarcomas can have vague abdominal or flank pain. This may radiate in a lumbar or sciatic distribution, depending on what nerves lie adjacent. Quite commonly, retroperitoneal sarcomas are asymptomatic and incidentally discovered on abdominal imaging.
  - **Magnetic resonance imaging (MRI)** is the first step for extremity sarcomas, and helps define the extent of the tumor beyond physical palpation.
  - **Excisional biopsy** is reasonable for tumors that are <3 cm in size, superficial, and mobile.
  - **Core needle biopsy (CNBx)** is more commonly used, since most lesions do not meet criteria for excisional biopsy. Core samples can be followed by a **formal incisional biopsy via a longitudinal incision** if the tissue sample was inadequate.
  - **CT abdomen-pelvis (CTAP)** is a good first step for retroperitoneal tumors. It is also used to evaluate metastases in lower extremity sarcomas, like myxoid liposarcoma in particular. This is followed by a **core needle biopsy (CNBx).**
  - **CT chest** should be done to evaluate for hematogenous lung metastases
  - Fine needle aspiration (FNA) can be used to confirm the identity of metastases or evaluate any suspicious

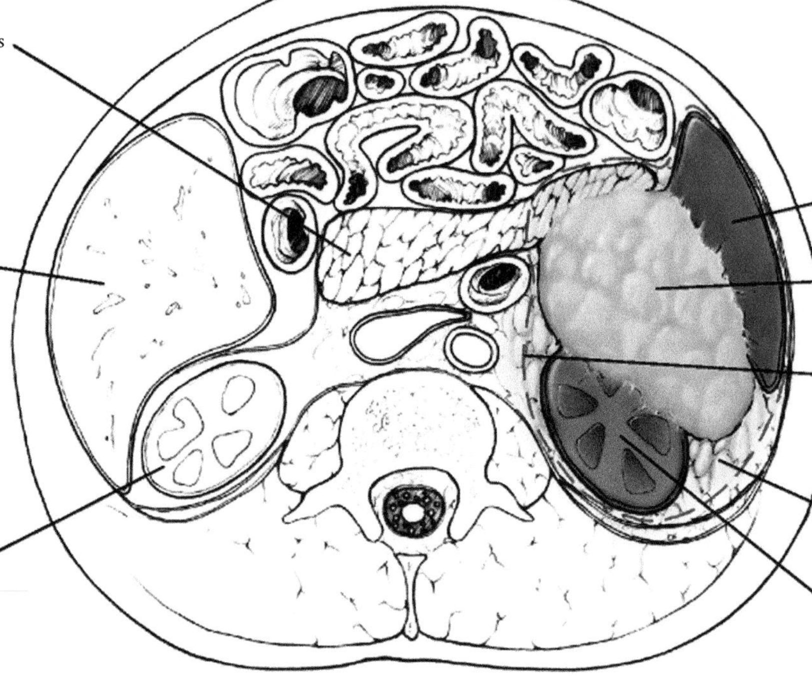

Figure 16.5 Retroperitoneal sarcoma.

lymph nodes on physical exam or imaging. However, lymphatic spread is rare.

- **Histologic grade** is more important than size.
  - G1 is well differentiated, <50% necrosis, <10 mitoses per HPF, and G4 is poorly differentiated, >50% necrosis and >19 mitoses per HPF.
  - Increased Ki67, cadherin, catenin, or CD100 are all poor prognostic indicators.
- **T1** lesions are <5 cm. T1a is superficial to fascia, and T1b is deep to fascia. **T2** lesions are >5 cm.
- N1 is any positive node and considered stage 3.
- M1 is any metastasis and considered stage 4.

- **TREATMENT**
  - Wide local excision (**WLE) with 1–3 cm margins** and at least one uninvolved fascial plane. The tumor should be resected **en bloc** with arteries, veins, nerves, tendons or perineurium, periosteum, or adventitia. Bone may be prophylactically plated.
  - If on an extremity, all attempts should be made to try and spare the limb.
  - If in the retroperitoneum, clear the entire compartment of fat, since frozen section intraoperative evaluation of adipose margins is generally not accurate.
  - **Narrow margin resection** is sufficient for low-grade lesions like atypical lipomatous tumors.
  - **Palliative debulking** can be sufficient if the tumor is low grade, and curative resection is not possible.
  - Lymphadenectomy of the entire basin would only be required if FNA confirmed a clinically positive node.
  - **Neoadjuvant external beam radiation 50 Gy** over 4–6 weeks can be considered for large retroperitoneal tumors or extremity tumors with borderline resectability for limb salvage, especially if abutting critical nerves or arteries. The upper extremity tolerates radiation better than the lower extremity, with decreased fractures, fibrosis, contractures, and lymphedema. Surgery is performed 4–8 weeks after completion of radiation. Neoadjuvant chemotherapy is less often utilized.
  - **Postoperative adjuvant external beam radiation 60–70 Gy** over 4–6 weeks should be considered for high-grade histology, close or positive margins (R1), or dedifferentiated areas within an atypical retroperitoneal lipomatous tumor. Other options include the following:
    - Intensity modulated radiation therapy (IMRT), which provides greater precision.
    - Brachytherapy, which can be given for one week starting after postoperative day 5, via catheters left in place at the time of surgery.
  - **Postoperative doxorubicin-based adjuvant chemotherapy** should be considered if the tumor was >10 cm, high grade, or if the sarcoma developed within a previously irradiated field. Such tumors are often aggressive and multifocal.

- Hyperthermic isolated limb infusion or perfusion with melphalan has been used as local adjuvant treatment to help spare a limb from amputation. Tcm99 radiolabeled albumin is given to confirm there has been no systemic leak of the infusion.
- **Follow-up imaging** after surgery can be performed every 3 months for the first two years and every 6–12 months thereafter.
  - **Recurrences** are common and often asymptomatic, with rates of 10%–30% on the extremity and up to 50% in the retroperitoneum.
  - One may resect local or distant, isolated recurrences if there is a long disease-free interval, <4 lung mets, or an isolated liver metastasis.
  - Other options for treatment of metastases include doxorubicin, ablation, chemoembolization, or palliative radiation at >60 Gy.

OTHER SARCOMA TYPES

- **DESMOID**
  - Benign but locally invasive. Some tumors have a **beta catenin** gene defect.
  - **DIAGNOSIS**
  - Patients present with a slow-growing mass on their abdominal wall, often at an old scar site secondary to prior trauma or surgery. Sometimes the mass is diagnosed during pregnancy. Abdominal or back pain with gastrointestinal symptoms can be the presenting symptoms if the tumor is arising in the mesentery or retroperitoneum.
    - Confirmation is by **MRI** showing infiltration of neighboring structures and a **core needle biopsy (CNBx)** showing fibroblasts and collagen.
  - **TREATMENT**
    - **Excision** if symptomatic and located in the abdominal wall or peripheral mesentery. Microscopically positive resection margins (R1) are adequate.
    - **Pre-op or post-op radiation 50–56 Gy** can occasionally be added for local control.
    - If not resectable, treat with **sulindac, tamoxifen,** imatinib, or doxorubicin-based chemotherapy.
- **DERMATOFIBROMA**
  - 1–2 cm benign nodule of collagen and fibroblasts.
- **DERMATOFIBROSARCOMA PROTUBERANS (DFSP)**
  - A red, brown, or purple plaque on the trunk or proximal extremity. It has slow growth and fingerlike projections. Diagnosis is by **CNBx.** Treatment is with **Mohs micrographic surgery or WLE with 3–4 cm margins** to accommodate for the heterogeneous projections. May delay closure of the wound bed until final pathology confirms adequate margins. Then proceed with staged wound closure, graft, or flap coverage. Adjuvant radiation is an option.
- **MYXOFIBROSARCOMA**
  - Presents in the extremity of an elderly patient. It infiltrates in a similar fashion to DFSP. Treatment is with WLE with 2–4 cm margins. May re-resect and add radiation for a recurrence.
- **BREAST ANGIOSARCOMA**
  - Presents as an erythematous plaque or patch.
    - **Primary** tumors arise de novo in the breast and are treated with mastectomy.
    - **Secondary** tumors occur 5–10 years after radiation or from long-standing lymphedema. These are treated with WLE including the dermis. There is no need for lymph node dissection. One may use neoadjuvant chemoradiation.
- **KAPOSI'S SARCOMA**
  - Endothelial spindle cell in origin and associated with human herpes virus 8 (HHV 8). Most patients are HIV positive. Lesions may present on the skin or the oral or pharyngeal mucosa. Treatment is focused on management of the underlying HIV. Resection is reserved for cases complicated by bleeding.
- **CHILDHOOD RHABDOMYOSARCOMA**
  - Presents in the head, neck, genitourinary area, extremity, or trunk. The latter has the worst prognosis. Embryonal subtype is most common. Alveolar subtype has a poor prognosis. Unlike most sarcomas, pediatric rhabdomyosarcoma spreads to lymph nodes.
- **GASTROINTESTINAL SARCOMAS**
  - These are either stromal tumors (GIST) or, less commonly, a leiomyosarcoma. Treatment is with WLE without an LN resection.
- **UTERINE LEIOMYOSARCOMA OR STROMAL HIGH-GRADE SARCOMAS**
  - Both treated with total abdominal hysterectomy and bilateral salpingo-oophorectomy (TAH-BSO).
- **OSTEOSARCOMA**
  - Comes from metaphyseal stromal cells around the knee or in the thoracic cage. Metastases are to the lung. Treatment may include neoadjuvant chemotherapy followed by resection. Isolated pulmonary metastases can also be resected if cure is deemed a possibility.

OTHER SOFT TISSUE TUMORS

- **MERKEL CELL CARCINOMA**
  - An aggressive neuroendocrine tumor arising from mechanical touch receptors, called Meissner's corpuscles, of an immune compromised individual. Neuron-specific Enolase (NSE), cytokeratin, and neurofilament protein may be present.

- **DIAGNOSIS**
  - It presents as red or purple papules, nodules, or plaques. Diagnosis is confirmed by **CNBx,** a **CT** to evaluate for metastases, and an **octreotide scan**. US guided FNA can determine if the patient has clinically positive nodes.
- **TREATMENT**
  - **WLE with 1–5 cm margins, SLNB (if no clinical positive nodes), and completion lymphadenectomy** for any positive SLNB or FNA positive nodes. **Postoperative radiation** can be given to the resection site and the LN basin for positive nodes.
- **GLOMUS TUMOR**
  - A benign growth of vessels and nerves at the tip of a digit. Treatment is tumor excision.
- **ADENOCARCINOMA IN SITU OF APROCRINE GLANDS (EXTRAMAMMARY PAGET'S)**
  - This can occur in the perianal margin or, rarely, elsewhere on the skin. Check for an underlying gastrointestinal or uro-gynecologic malignancy with endoscopy and imaging.
- **VASCULAR TUMORS**
  - **Hemangiomas, angiosarcomas, and hemangioendotheliomas** may regress by the age of 5 to 7 years. Treatment is with prednisone, interferon alpha, and beta-blockade.
  - **Vascular Malformations**
    - Capillary malformations like Port-wine stain are treated with pulsed dye laser.
    - Lymphangiomas or hygromas are treated with sclerotherapy and serial excisions.
    - Arteriovenous malformations are treated with embolization and resection.
    - Venous malformations should be excised.
- **NEURAL TUMORS**
  - **Nerve sheath origin** tumors may be spinal, intradural, or dumbell morphology. Most are benign. Neurilemmomas are also known as schwannomas. Neurofibromas may be associated with underlying neurofibromatosis. Some of these are malignant neurofibrosarcomas. **Malignant peripheral nerve sheath tumors (MPNST)** are resected with a 4 cm margin
  - **Neuronal origin** tumors arise from ganglion cells of the sympathetic chain and may be benign ganglioneuromas, ganglioneuroblastomas, or a malignant neuroblastoma presenting in infancy.
  - **Neural crest cell origin** tumors are from tissue adjacent to the sympathetic chain and include malignant paraganglioneuromas, pheochromocytoma from the adrenal medulla, or chemodectomas. Diagnosis is by CT or MIBG scan. Treatment is resection.
- **CYSTS**
  - **Epidermal inclusion cysts** have a granular layer of mature epidermal cells lining a cavity filled with keratin. The cyst resides within the dermis on the scalp, face, neck, or trunk.
  - **Trichilemmal cysts** are located on the scalp and have an epidermal lining without the granular layer.
  - **Ganglion cysts** are on the dorsal wrist, over tendons, and filled with joint fluid or collagen. Half resolve in 2 years.
  - **Dermoid cysts** contain tissue originating from multiple cell lines. Most are benign, with a minority having malignant potential. Treatment is with excision.

17.

HERNIA

ABDOMINAL WALL ANATOMY

- The **external oblique muscle** ends laterally, and its medial extension is an aponeurosis up until the recuts sheath. Contraction results in contralateral torso rotation.
- The **internal oblique muscle** has a longer aponeurosis superiorly and a shorter fascial extension inferiorly before meeting the rectus sheath. Contraction results in ipsilateral torso rotation.
- **Transversus abdominus muscle** extends medial to the linea semilunaris up to the posterior rectus sheath in the upper abdomen. As it progresses down the mid- and lower abdomen, the muscular component is shorter and the fascial extension longer. The fascia fuses with the internal oblique aponeurosis and forms the posterior rectus sheath. Inferior to the **arcuate line,** this fascial layer resides entirely anterior to the rectus muscle.
- The **ilioinguinal nerve** runs posterior to the external oblique, and on top of spermatic cord, to the mons and the inner thigh. The **iliohypogastric nerve** runs overtop the internal oblique, cephalad and lateral to the conjoined tendon.
- **T7-L1 neural and vascular bundles** run posterior to the internal oblique and superficial to the transversus abdominis. They penetrate, through the posterior sheath if above the arcuate line, into the rectus muscles medially.
- The genital branch of the **genitofemoral nerve (GFN)** heads to the scrotum or labia, and the femoral branch of the GFN goes to the fossa ovalis.
- The **deep inferior epigastric artery** runs in the retrorectus position as the main blood supply medially. It collateralizes with lateral vascular tributaries as they penetrate into the rectus muscle.
- The **transversals fascia** runs posterior to the transverses abdominis laterally, the posterior sheath superiorly, and posterior to the rectus muscle itself inferiorly below the arcuate line.
  - The space of Retzius is between the transversalis and bladder. The space of Bogros is lateral to the space of Retzius
- The **parietal peritoneal membrane** is posterior to the transversals fascia. It has fenestrated lymphatics and can transport solutes and ions between the peritoneal cavity and the vascular space.

INGUINAL HERNIA

- **PATHOPHYSIOLOGY**
  - The iliopubic tract runs from the anterior superior iliac spine (ASIS) to the pubic tubercle.
  - The myopectineal orifice encompasses direct, indirect, and femoral hernia spaces. *See Figure 17.1.* Femoral hernias are inferior to the inguinal ligament and mentioned separately below.
  - **Indirect hernias** are the most common and occur within a patent processus vaginalis that runs lateral to the inferior epigastric vessels.
  - **Direct hernias** occur medial to the inferior epigastric vessels through the triangle marked by the medial border of the rectus muscle and the inguinal ligament. They have higher recurrence rates than indirect hernias.
  - **Sliding hernias** are when any of the above contain the ovary, fallopian tubes, cecum, sigmoid colon, or bladder confluent with part of the sac wall.
  - Hernia risk factors include **obesity, smoking, chronic obstructive pulmonary disease (COPD),** prostatic hypertrophy (BPH), ascites, pregnancy, constipation, and connective tissue disorders. Some patients are predisposed to hernias based on their family history or high levels of type 3 collagen in soft tissue (rather than type 1).
  - **Littre hernia** have a Meckel's diverticulum in the sac.
  - **Amyand hernia** has the appendix in the sac.
- **DIAGNOSIS**
  - Patients may have a symptomatic or asymptomatic bulge around the area of the inguinal ligament. Clinical exam is usually sufficient for diagnosis, although distinction between an indirect and direct hernia can be difficult. Contents may descend into the scrotal sac in men with indirect hernias. A bulge below the inguinal ligament is a femoral hernia.
  - Imaging with ultrasound, CT, or MRI may be needed in equivocal cases.

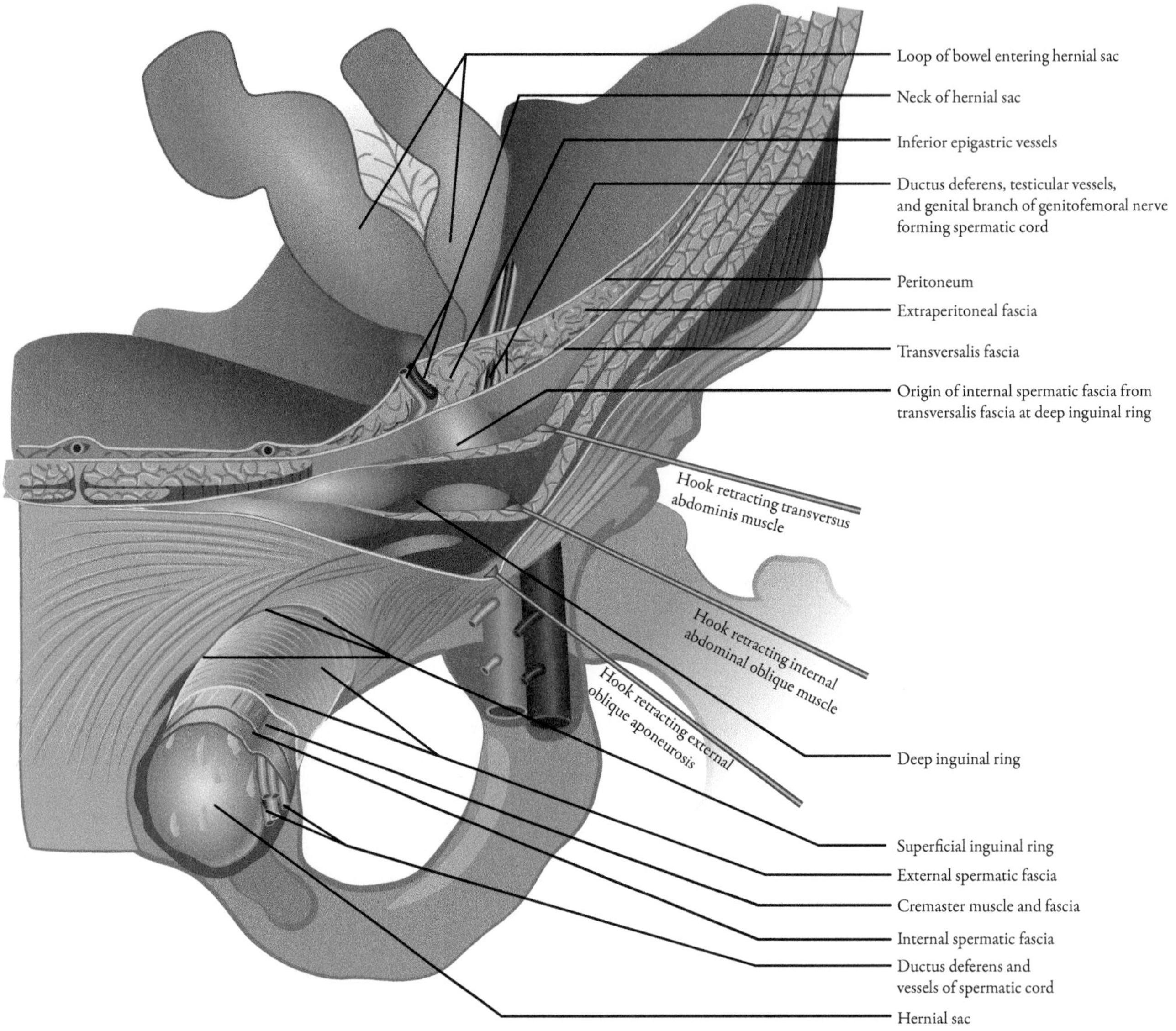

Figure 17.1 Hernia wall: Indirect inguinal hernia and abdominal wall layers.

- **TREATMENT**
  - Manual reduction if needed, followed by surgical repair. This can be done on an elective, outpatient basis in most instances.
  - Emergent repair is reserved for patients with acute incarcerated bowel or other visceral organs that cannot be reduced manually at the bedside after appropriate sedation and analgesia. If there is any concern for ischemia or strangulation, the surgery should be performed in an open format.
  - A male patient who is minimally symptomatic from in inguinal hernia may decline surgery and choose **observation**, but should be aware that symptoms will progress and ultimately require surgery over the ensuing years.
  - Pediatric inguinal hernias are treated by **high ligation of the sac** only. There is no tissue repair directed at the inguinal ring or the direct space.
  - A primary **tissue repair** is for emergencies with strangulated, compromised bowel precluding mesh placement.
    - **Bassini repair** involves opening the transversalis and then sewing the conjoined tendon and superior transversalis to the inguinal ligament. The recurrence rate is **10%–15%**.
    - **Shouldice repair** is done in multiple overlapping layers of the superior transversalis and conjoined tendon with the inferior transversalis and the external oblique. This has a reported recurrence rate of **1%–2%** in an experts' hands.

- **McVay repair** brings the transversalis and conjoined tendon to Cooper's ligament medially to cover the femoral space. Lateral to the femoral vessels, this transitions to a Bassini repair and the conjoined tendon is sewn to the inguinal ligament. A **relaxing incision** in the anterior rectus sheath may be needed for approximation. This is the only open repair (tissue or prosthetic) that routinely covers the femoral space.
- **Marcy repair** is a sutured closure of the internal ring.

- **Lichtenstein repair** is the most common way to address an indirect or direct hernia in the absence of compromised bowel. It does not cover the femoral space. *See Figure 17.2.*
  - Functionally, it is an inlay synthetic mesh placed over the direct space and sewn to the conjoined tendon superiorly and inguinal ligament inferiorly. The tails of the mesh wrap around the internal ring to help prevent a future indirect hernia. Ideally, the mesh should overlap medial to the pubic tubercle and cephalad of the conjoined tendon, and should not be under any tension.
  - A plug may be inserted into the internal ring, or a posterior component to the mesh can be tucked into the internal ring or posterior to the conjoined tendon or transversalis in an underlay fashion.
  - Light or heavyweight synthetic non-absorbable mesh seem to be equivalent, with about a **1% recurrence rate**. A biologic mesh is not used for a Lichtenstein repair, since the inlay position will be associated with a recurrence.
  - Mesh may increase postoperative chronic pain (see below).
- **Laparoscopic repair** is an option for bilateral inguinal hernias or a recurrent unilateral inguinal hernia after a prior open mesh repair. Laparoscopy allows visualization and coverage of the entire myopectineal orifice, including the femoral space, and allows one to address a contralateral occult hernia.
  - However, if there are strangulated contents, a large hernia, prior lower abdominal surgery, or the patient is high-risk for general anesthesia, an open repair should be done.
  - **Total extraperitoneal repair (TEP)** is the preferred laparoscopic method. One port is at the umbilicus, a second port 1–2 cm above the symphysis, and a third port sits halfway between the two. The preperitoneal space is entered with the assistance of a balloon dissector and cleared to make space for a synthetic non-absorbable mesh to cover the entire myopectineal orifice, including the direct, indirect, and femoral spaces.
  - **Transabdominal preperitoneal repair (TAPP)** is an option in place of TEP if the patient has had prior preperitoneal or prostate surgery, or a large sac needing reduction.
    - Ports are placed to enter the peritoneal cavity, and the parietal peritoneal lining is incised and dissected off the transversals to create a preperitoneal space.
    - Synthetic non-absorbable mesh should cover the entire myopectineal orifice, including the direct, indirect, and femoral spaces.

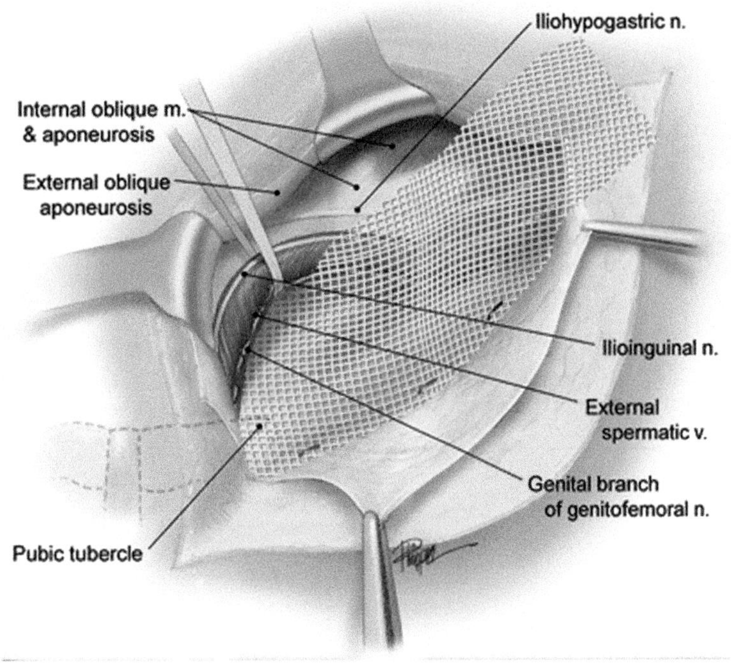

Figure 17.2 Lichtenstein open inguinal hernia repair.

- Absorbable or permanent **tacks** can be placed over the portion of Cooper's ligament medial to the femoral space and over the anterior abdominal wall medial to the ASIS, so the mesh overlaps past midline, down to inguinal or Cooper's, and lateral all the way to the ASIS.
  - Tacks are not placed laterally along Cooper's, in the abdominal wall lateral to the ASIS, or inferior to the inguinal ligament in the triangle of doom or triangle of pain. *See Figure 17.3.*
- There is no advantage to heavyweight mesh.
- Pain, length of stay, and return to normal activities are all improved with a laparoscopic approach. **Recurrence rate of <5% is similar to open repair.**
- Patients may return to labor at 2 weeks post-op.

- **Complications**
  - **Urinary retention** postoperatively is secondary to anesthetic effects.
  - **Seroma** formation can occur within the potential space anterior to mesh. No intervention is needed.
  - The **wound infection** rate is 1%. This may be at a pre-fascial level, or deeper at the surgical site. **Mesh infection** presents with pain and drainage from the wound. Imaging will show a fluid collection posterior to the mesh, which should be presumed to represent infection and possible occult bowel injury. Treatment includes a **trial of antibiotics before mesh explantation**. For infection of prior open mesh repairs, one may explant and place a new mesh in the preperitoneal space laparoscopically.
  - **Chronic postoperative pain** (aka inguinodynia) occurs in up to 30% of patients and lasts for greater than 3 months. Severe pain is present in 4%.
    - The first step is to **exclude recurrence** of the hernia or a missed femoral hernia with imaging.
    - Vague, **dull, non-radiating**, positional discomfort may be due to the mesh itself. Start with a medication regimen and physical therapy for scar release. If the pain remains refractory, consider explantation.
    - **Shooting or burning** pain is neuropathic, and may be from intraoperative nerve injury, entrapment, impingement, postoperative perineural inflammation, or ingrowth of the mesh into a nerve. Diagnosis is by an anesthetic infiltration trial noting improvement in pain. Treatment is with NSAIDs, nerve modulating medications, anesthetic blocks, ablation, or steroid injections. A triple neurectomy, including the iliohypogastric, ilioinguinal, and genital branch of the

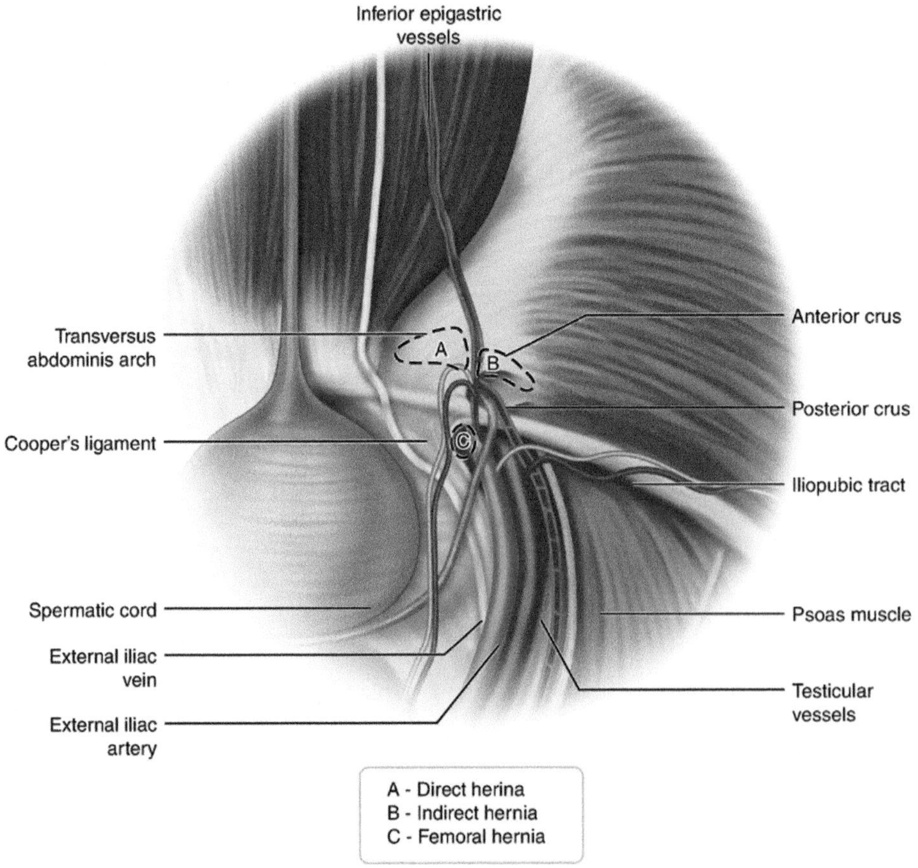

Figure 17.3 Myopectineal orifice from laparoscopic view.

genitofemoral nerve is a last resort and can be done in an open or laparoscopic fashion.
  - **Chronic neuropathic pain** is due to abnormal modulation of neural signals. These patients may demonstrate hyperesthesia or allodynia. Consultation with a pain management specialist is recommended.
- **Recurrence** is secondary to poor technique, an inherent connective tissue weakness, or poor healing secondary to smoking, obesity, coughing, malnutrition, or immune suppression.
  - Recurrence rates for Lichtenstein repair are about 2%, and laparoscopic repair under 5%. They generally occur medial to the mesh.
  - The diagnosis is made by physical exam with or without US or dynamic valsalva MRI. The latter has the highest sensitivity and specificity.
  - Observation is appropriate for asymptomatic recurrent inguinal hernia.
  - For symptomatic inguinal recurrences or any femoral recurrence, surgery is recommended.
  - If the initial repair was a tissue repair without mesh, then the recurrence can be approached in an open anterior, open posterior, or a laparoscopic fashion.
  - **If mesh was used initially, then the approach should be the opposite of the initial surgery.** A laparoscopic or open posterior approach is best for a recurrence after an open anterior mesh repair. An open anterior approach may be best for recurrence after a prior laparoscopic repair with mesh.
- **Ischemic orchitis** with testicular atrophy can result from complete dissection of a large sac and secondary injury of the testicular venous plexus with devascularization of the testicle.

FEMORAL HERNIA

- These account for <10% of all groin hernias, tend to occur in women, and are more likely to incarcerate than inguinal hernias.
- Contents herniate through the femoral space, which is medial to the femoral vein, posterior to the inguinal ligament, anterior to Cooper's ligament, and lateral to femoral lymphatics and the lacunar ligament. *See Figure 17.4.*
- **DIAGNOSIS**
  - Patients will have pain and a reducible bulge below the inguinal ligament with straining. Contents may incarcerate or strangulate.
- **TREATMENT**
  - Patients with easily reducible hernias can be scheduled for elective repair. There is **no role for observation** of minimally symptomatic femoral hernias.

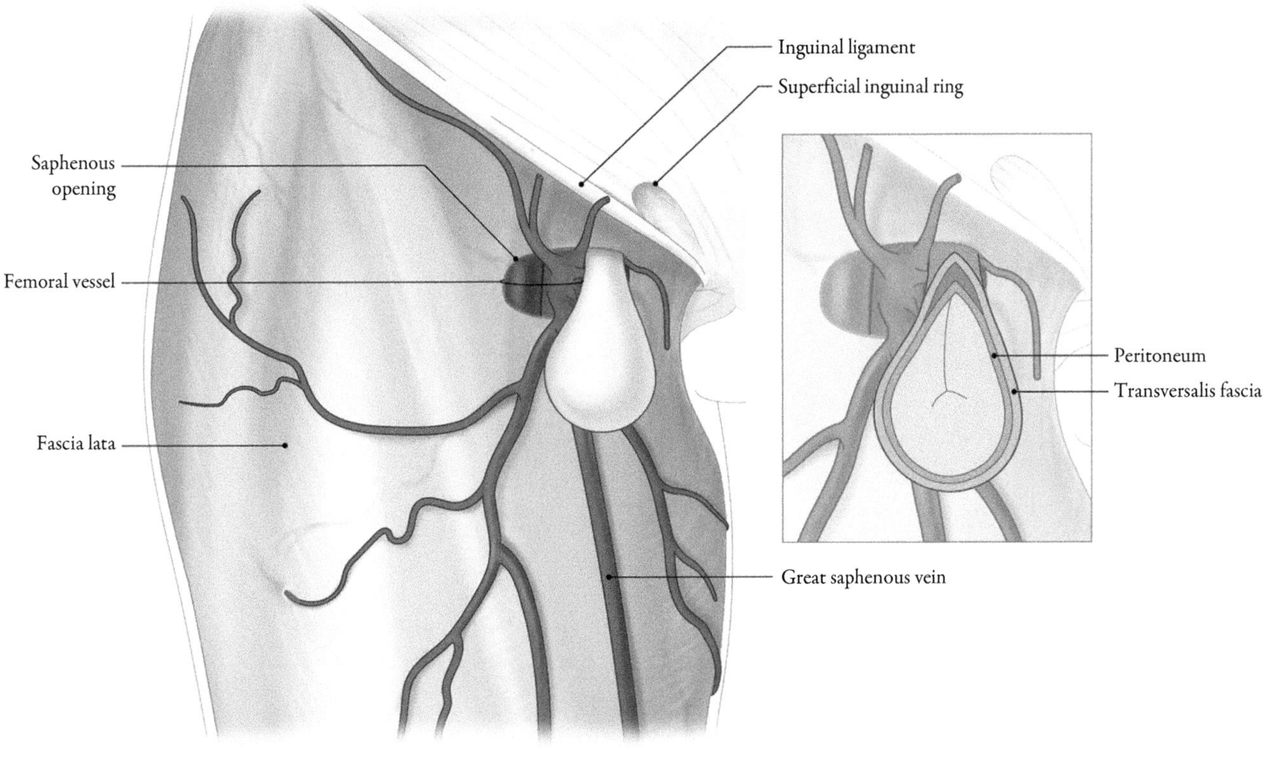

Figure 17.4 Femoral hernia.

- Manual reduction under appropriate analgesia or sedation should be attempted for incarcerated contents, provided that there is no concern for strangulation. This would be followed by operative repair at that same hospital visit.
- The operative approach is most often the open anterior fashion via an inguinal incision. The inguinal ligament can be transected for exposure and reduction of contents. Concomitant laparoscopy can aid in bowel reduction. Then a **mesh plug** can be placed into the femoral canal. The direct and indirect inguinal spaces can be addressed at the same time if needed. A **McVay tissue repair** is preferable if bowel contamination precludes the use of mesh.
- A **laparoscopic TEP or TAPP** can cover the entire myopectineal orifice, including the femoral canal. It is an option for elective repair (see inguinal hernia section above).

PUBIC INGUINAL PAIN SYNDROME

- This is also called athletic pubalgia and presents as chronic, recurrent inguinal pain related to exertion or position. There is no associated bulge. It is secondary to **unbalanced** rectus abdominis, hip adductor, and conjoined tendon pull on the superior pubic ramus and pubic tubercle.
- **DIAGNOSIS**
  - This is a clinical diagnosis only made after excluding a hernia with US, CT, or MRI.
- **TREATMENT**
  - Rest, ice, NSAIDs, physical therapy, or ultrasound therapy. Surgical tenotomy will offload forces on the tubercle and can be used if all of the above fail. If the pubalgia is due to weakness of the inguinal floor seen on a dynamic US, then mesh repair may be indicated.

VENTRAL HERNIA

- These occur on the anterior abdominal wall and can be epigastric, umbilical, or incisional. *See* Figure 17.5.
- They can present with a reducible bulge and intermittent pain. Incarceration or strangulation of hernia sac contents will produce constant pain. Some ventral hernias contain only non-reducible preperitoneal fat with no significant hernia sac.
- **EPIGASTRIC HERNIA**
  - These occur at an upper midline weakness of the linea alba, in areas of small vessel penetration known as vascular lacunae.
  - Many contain only preperitoneal fat. Sometimes there is a small hernia sac with contents.

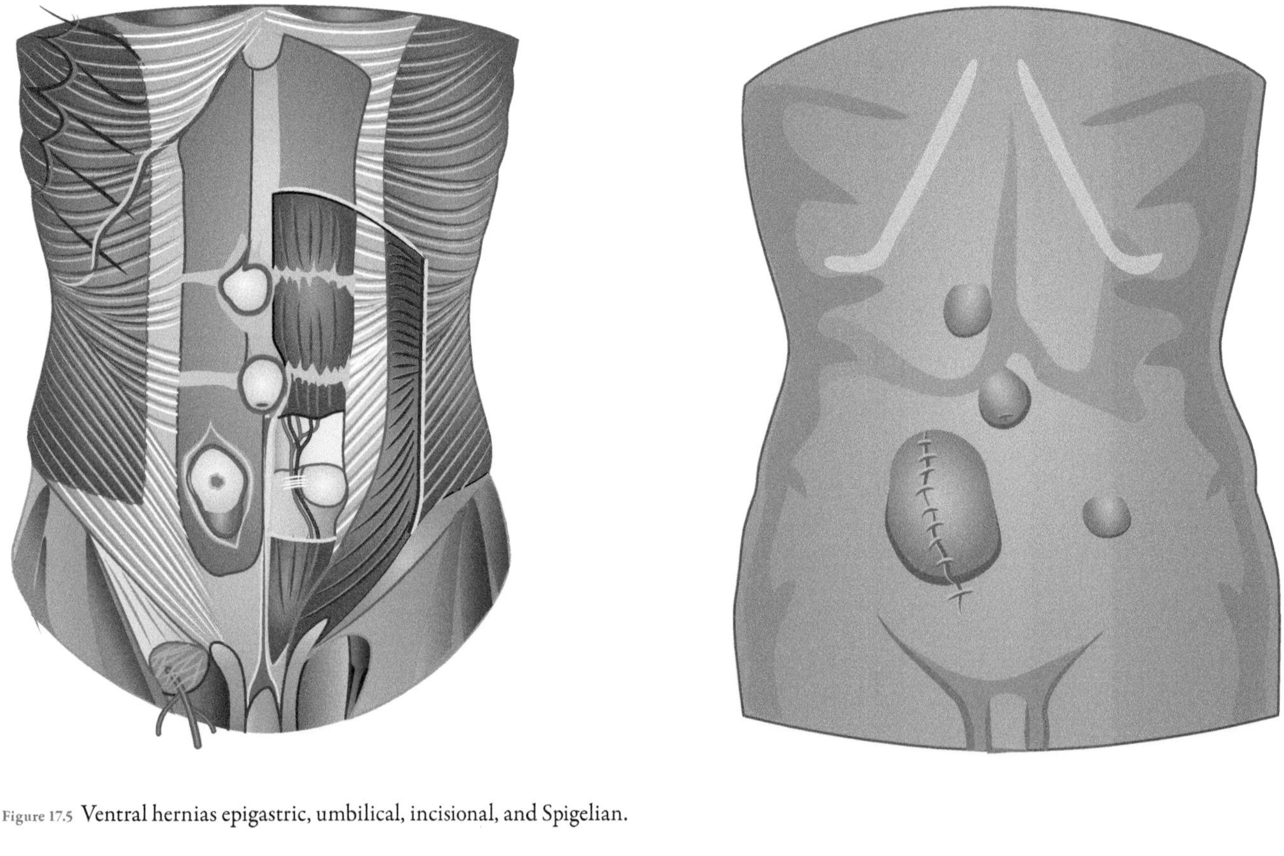

Figure 17.5 Ventral hernias epigastric, umbilical, incisional, and Spigelian.

- **DIAGNOSIS**
  - Patients can complain of an upper midline bulge and pain with straining. The bulge may not be reducible if it is primarily preperitoneal fat. An US or CT scan can be confirmatory if needed.
- **TREATMENT**
  - Elective repair is for reducible hernias or those containing only preperitoneal fat. Any incarcerated or strangulated hernia sac contents should undergo immediate repair at that same hospital presentation. Repair can be approached via a midline incision with dissection down to the defect, amputation or reduction of preperitoneal fat, transection or reduction of any hernia sac, and primary or mesh repair. A laparoscopic repair is described below.
- **UMBILICAL HERNIA**
  - This is a ventral hernia through a congenital defect in fascia at the umbilicus. Since it is a congenital defect, the fascial edges are generally healthier and stronger than incisional or epigastric hernias.
  - 80% of umbilical hernias close by age 5. Some are only detected later in adulthood and have a higher incarceration risk than in the pediatric population.
- **DIAGNOSIS**
  - Patients have a reducible bulge and discomfort at the umbilicus. Contents may become incarcerated or strangulated.
- **TREATMENT**
  - Patients < 5 years old should be observed unless there is incarceration, the diameter of the defect is >2 cm, or there are significant cosmetic concerns.
  - Cirrhotic patients should have their ascites medically optimized prior to repair.
  - Women of childbearing age can wait until they are done having children to undertake repair.
  - Elective repair is performed for reducible hernias or those containing only preperitoneal fat. Urgent repair is indicated for incarcerated or strangulated contents.
  - Open approach primary closure is adequate if the defect is <2 cm diameter. Mesh may be used if >2 cm. A laparoscopic repair is described below.
- **INCISIONAL HERNIA**
  - These are a common type of ventral hernia that complicate **10%–20% of laparotomy incisions**. Upper midline hernias are more common than lower midline hernias. Incisional hernias may also occur at 10 mm or larger laparoscopic port sites.
  - Risk factors include DM, **obesity, COPD,** connective tissue disorders (CTD), older age, immune suppression, or a prior infected surgical site.
  - Preventative measures at the time of initial laparotomy include:
    - Laparotomy closure with continuous 2-0 monofilament with bites taken 5–8 mm from the edge and 5 mm apart in a 4:1 suture to wound length ratio.
    - For very high-risk obese patients or those undergoing open abdominal aortic aneurysm (AAA) repair, a **prophylactic onlay synthetic mesh** at the time of laparotomy can decrease the postoperative hernia rate.
- **DIAGNOSIS**
  - Patients complain of a bulge and pain at the incision site. Contents may be reducible, incarcerated, or strangulated.
  - For patients who do not need immediate operative intervention, CT scan can delineate the exact extent of the incisional hernia, as well as other associated abdominal wall defects.
- **TREATMENT**
  - Elective or urgent repair depending on the presence or absence of incarcerated or strangulated hernia sac contents. Hernias containing only preperitoneal fat can be approached electively.
    - **Closure of the defect and placement of synthetic mesh** is ideal, provided there is minimal field contamination from compromised bowel.
    - Mesh should extend at least **4 cm beyond** the edges of the hernia defect.
      - **Synthetic non-absorbable mesh:**
        - **Polypropylene** mesh induces a strong inflammatory response and fibrosis, which allows it to incorporate into tissue. Elasticity of the surrounding native torso envelope may be compromised, and adhesions can develop as a result of fibrous scar.
        - **Expanded polytetrafluoroethylene (ePTFE)** does not induce as much inflammation and hence incorporation. It may be microporous and heavyweight or macroporous and lightweight. The latter can have enhanced bacterial clearance.
        - Both types of synthetic mesh have been used for ventral hernia repair in minimally contaminated fields.
      - **Synthetic absorbable mesh** is made from Vicryl.
      - **Biologic mesh** is an acellular collagen matrix derived from human, bovine, or porcine tissue. It allows neovascular ingrowth and is favored in a more heavily contaminated wound, or if excising an enterocutaneous fistula (ECF) at the same time as a hernia repair. The material will eventually be degraded, so should not be used in the bridging, inlay position as this guarantees hernia recurrence.
        - Collagen may be **non–crosslinked or crosslinked.** The latter is believed to have increased durability, tensile strength, and resistance to bacterial contamination, but decreased neovascular ingrowth.
        - **Bio-absorbable synthetic mesh** is a hybrid of the two types.
      - **Mesh Position** *See Figure 17.6.*

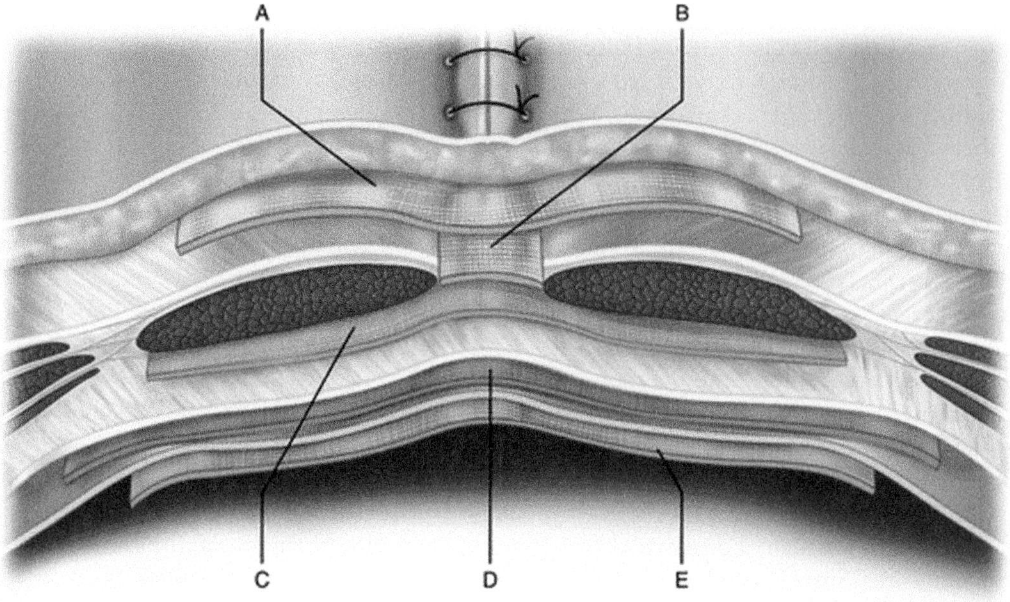

Figure 17.6 Positions of abdominal wall mesh- onlay (a), inlay (b), retrorectus sublay (c), underlay (d), intraperitoneal onlay (e).

- **Overlay**—superficial to the fascial closure, under the subcutaneous space.
- **Sublay**—posterior to the rectus muscles and anterior to a closed posterior rectus sheath. This has the lowest recurrence rate.
- **Underlay**—posterior to the posterior sheath and transversalis, but superficial to the peritoneal lining.
- **Intraperitoneal onlay**—under the peritoneal lining and in direct contact with intraperitoneal organs.
- A **bridging, inlay** position of the mesh can be used when the fascial edges cannot be brought together in spite of a release technique (see below). In this case, the mesh is the only layer covering the hernia defect.
- A sandwich technique can include a biologic extending 2 cm beyond the defect edges and a synthetic overlay.

- A **component release** can be done for fascial edges that cannot be apposed, or are under significant tension when doing so.
  - **Anterior component separation** involves incising the external oblique aponeurosis from cephalad to caudal, just lateral to the linea semilunaris.
    - It allows for unilateral **5 cm upper abdominal, 10 cm mid-abdominal, and 3 cm lower abdominal** medial advancement of the rectus/internal/ transversus complex on each side. The midline defect can then be closed without tension.
    - **Lipocutaneous flaps** must be raised in order to reach the semilunaris on each side and perform the release. Periumbilical and lateral perforating vessels into the subcutaneous layer should be preserved whenever possible.
    - Mesh is then placed in an **overlay position** atop the closed midline defect.
    - This procedure can increase Spigelian recurrences unless the mesh extends laterally.
    - Mesh can also be placed in the retrorectus sublay position; however, its lateral extension will be limited to the rectus space unless the posterior rectus sheath is opened.
  - **Posterior Component Separation**
    - Adhesiolysis must first be done to free bowel from the posterior sheath.
    - Then, **open the posterior sheath** just medial to the linea semilunaris, and free the rectus muscle from it in a medial to lateral direction. This is **myofascial separation.**
    - Be careful to preserve the inferior epigastric and lateral perforating vessels. Entirely free the posterior sheath from costal margin to arcuate line. Below the arcuate line, free the transversalis from the rectus muscle.
    - Once at the linea semilunaris, **incise the internal oblique contribution to the posterior sheath** from cephalad (where the transversus abdominus muscle lies posterior to it) to caudad (where it has already fused with the transversus fascia).
    - Then **transect the transversus abdominis muscle** starting in the upper abdomen and proceeding inferiorly through its fascial attachments. *See Figure 17.7.*

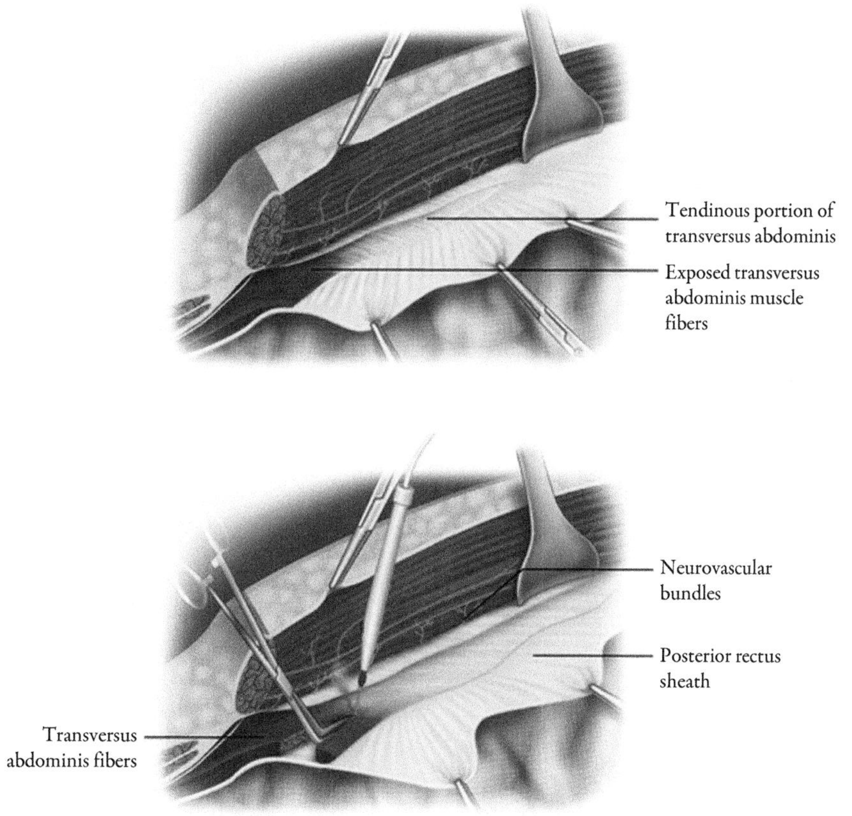

Figure 17.7 Posterior component separation by (a) opening posterior rectus sheath and then (b) transecting transversus abdominis muscle.

- The peritoneum is now exposed along the entire length from costal margin to superior pubic ramus, and can be dissected farther laterally and posteriorly until the central tendon of the diaphragm, lateral edge of the psoas muscles, and Cooper's ligaments are all exposed.
- Once the posterior components are sufficiently released in this manner, the posterior sheath will have enough mobility to be closed in the midline.
- Mesh is then placed in a **retrorectus sublay position.** Laterally, this will extend posterior to the transversus muscle and is technically in the underlay position.
- Transfix it laterally from the costal margin to the anterior superior iliac spine (ASIS) and medially at the xiphoid and symphysis. Then close the linea alba superficial to this.
- Skin and subcutaneous flap coverage may need to be raised overtop of the linea closure if the soft tissue does not meet on its own.

- With either technique, drains are left on both sides and removed when <30 ml/day.
- A longer period of intubation and possible paralysis may be needed for high plateau pressures postoperatively.

- **LAPAROSCOPIC VENTRAL HERNIA REPAIR WITH MESH**
  - This is associated with a **shorter length of stay, better pain control, and decreased wound infection risk.** Laparoscopy allows one to assess bowel integrity in chronic incarcerations, and also helps determine the presence of multiple small, clinically occult ventral hernias. However, it can theoretically miss hernias that contain only preperitoneal fat, since the peritoneal membrane is not herniating through the fascial defect in these cases.
  - Laparoscopy is not used if a hernia is **very large or very small**, the patient has systemic illness or a high suspicion for **compromised bowel**, there is a **hostile abdominal cavity**, or the patient has **skin breakdown** over the hernia site.
  - Use pre-op subcutaneous heparin venous thromboembolism (VTE) prophylaxis, epidural catheters, and a Foley if there is suspected adhesiolysis.
  - All ports are generally placed along one side of the abdominal cavity.
  - Mark the size of defect with spinal needles at edges and intraperitoneal measurement.
  - Primarily close the defect when possible, to prevent seroma formation.

- Mesh should extend >4 cm beyond fascial defect edges on each side. Even if the defect is closed with a suture passing device, the mesh size should be large enough to have 4 cm extension should the closure break down.
- The mesh is generally positioned as an **intraperitoneal onlay (IPOM)** with the anti-adhesive barrier facing viscera.
  - **Absorbable tacking** with a double crown technique around the perimeter helps hold the mesh in place.
  - **Transfascial sutures** can be used to secure position, but these may increase postoperative pain.
  - Alternatively, a peritoneal incision can be made and the parietal peritoneum dissected free of the abdominal wall in order to place an **underlay mesh.**
- **Overall recurrence rates are 5%–10%** whether open or laparoscopic, and depend on hernia width and contamination more so than comorbidities.
- A seroma anterior to the mesh and within a residual hernia sac is very common and managed conservatively.
- If an enterotomy is made at the time of adhesiolysis, one may need to convert to an open procedure for repair of the bowel and closure of the hernia defect primarily without mesh.
- If there is a mesh infection post-op, it likely represents an occult intestinal injury incurred during adhesiolysis.

SPIGELIAN HERNIA

- These hernias occur at the semilunaris and below the arcuate line.
- The external oblique aponeurosis often remains intact. If the external oblique aponeurosis is breached, the hernia sac can erupt into the subcutaneous (SQ) space and be more conspicuous. *See Figure 17.5.*
- **DIAGNOSIS**
  - Patients will complain of pain lateral to the rectus with straining. The pain may be constant if incarcerated. They may or may not complain of a bulge. In cases of an intact external oblique aponeurosis, a bulge may not be palpable. Confirmatory imaging with **CT or US** is often necessary unless the patient is in extremis or has peritonitis mandating immediate operative intervention.
- **TREATMENT**
  - An **open anterior approach** is best for urgent situations or large defects.
  - This often requires an incision through the external oblique aponeurosis.
  - Repair the defect in the internal aponeurosis and place an **onlay synthetic mesh** extending 4 cm beyond fascial defect edges, provided there is no bowel compromise.
  - One may use a biologic mesh in the onlay position in certain situations of minimal field contamination.
  - For heavy contamination, a tissue repair without mesh should be attempted.
  - For elective scenarios with a small hernia, a **laparoscopic TEP or TAPP with underlay mesh or IPOM** can be attempted. The defect should be closed and mesh extended 4 cm beyond edges in all cases.

LUMBAR HERNIA

- **Inferior** lumbar hernias occur at the triangle of Petit, bordered by the external oblique muscle laterally, latissimus dorsi medially, and the iliac crest inferiorly.
- **Superior** lumbar hernias are more common. They occur at the triangle of Grynfeltt bordered by the 12th rib superiorly, internal oblique muscle laterally, and erector spinae and quadratus lumborum medially. The external oblique and latissimus form the roof of the hernia defect.
- **DIAGNOSIS**
  - Patients can complain of pain with straining at these sites. If incarcerated, the pain will be constant. A bulge may or may not be palpable. Confirmatory imaging with **CT or US** is often necessary unless the patient is in critical condition mandating immediate exploration.
- **TREATMENT**
  - **Open repair** if the defect is >15 cm. The patient should be in the prone position, and the defect closed with a 3–5 cm extension of **synthetic overlay mesh**. Significant contamination from compromised bowel will preclude the use of mesh. A local tissue rearrangement rotational flap may be needed to cover very large defects.
  - **Laparoscopic TAPP repair** can be attempted if the defect is <10 cm and there is no suspicion for bowel compromise. The patient should be in a 45-degree contralateral decubitus position with the symptomatic side up. Ports will be placed on this side and the fascial edges closed. An underlay or IPOM mesh should extend 3–5 cm past the fascial edges.
  - Recurrence rates are as high as 15%.

PELVIC HERNIA

- **OBTURATOR HERNIA**
  - The obturator canal lies between the obturator internus and externus muscles within the obturator

foramen. The boundaries of the foramen are the superior, inferior, and ischial bony rami. The neurovascular bundle runs superiorly through the anteromedial portion of the foramen. Herniated contents run alongside this bundle. *See Figure 17.8.*

- **DIAGNOSIS**
  - Patients are often elderly, emaciated, women presenting with an acute incarceration on the right side. There is no palpable bulge.
  - **Howship-Romberg sign** is pain along medial thigh with extension, abduction, and internal rotation of hip, and is due to hernia contents causing increased tension within the canal irritating the obturator nerve.
  - Confirmation with an **abdominopelvic CT** is necessary, unless the patient clearly has peritonitis at which point an operation should ensue expeditiously.
- **TREATMENT**
  - Repair via an **open approach** if urgent or if bowel compromise is suspected.
  - The patient should be in a Trendelenburg position and the obturator membrane incised if needed for adequate bowel reduction.
  - The **defect can be closed using the obturator internus muscle.** Synthetic mesh can be placed if there was no bowel compromise or contamination.
  - Biologic mesh or tissue repair alone would be best in the setting of minimal or heavy contamination respectively.
  - A laparoscopic TAPP with underlay or IPOM mesh position can be attempted for elective cases.
- **SCIATIC HERNIA**
  - Occurs posteriorly through the greater sciatic foramen bound by the ischial spine, sacrum, and sacrospinous ligament. Diagnosis usually requires a confirmatory CT scan, and an open repair is recommended.

PARASTOMAL HERNIA

- These complicate nearly 50% of enteric or colonic stomas, mostly presenting within 2 years of construction.
- Risk factors include diabetes, **obesity, COPD,** connective tissue disorders, advanced age, immune suppression, or prior infected surgical site. A **stoma fascial defect >2.5 cm** is an independent risk factor.
- **DIAGNOSIS**
  - A bulge will be noticeable adjacent to the stoma. Patients may have pain or tenderness. Contents may be reducible or incarcerated. The hernia may impair stoma appliance fitting and cause leakage, leading to skin breakdown.
- **TREATMENT**
  - 80% of patients have minimal symptoms or are asymptomatic. No treatment is needed, and a **support belt** can be recommended.
  - Patients with non-reducible incarcerated or strangulated contents should undergo immediate open repair.

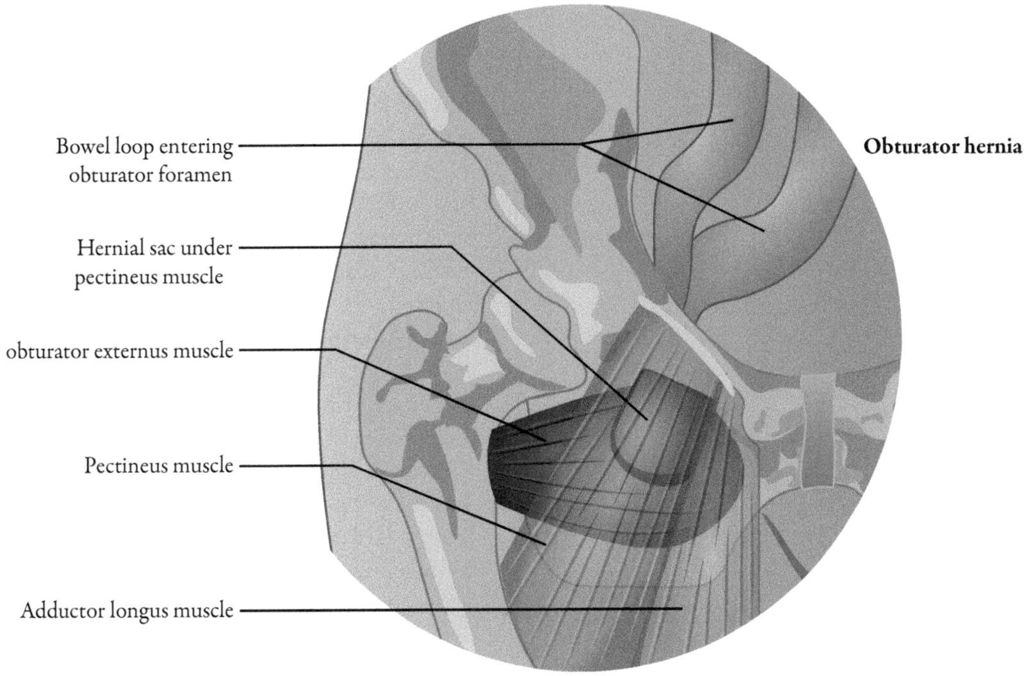

Figure 17.8 Obturator hernia.

- Patients with a reducible bulge accompanied by pain or stomal leakage due to poor appliance fit should undergo elective open or laparoscopic repair.
  - **Open repair** can simply involve adhesiolysis around the peristomal fascial defect and re-approximation of fascial edges with or without mesh.
  - **Laparoscopic Sugarbaker mesh repair** is the more commonly used technique in elective situations. *See Figure 17.9.*
    - The bowel leading into the stoma is mobilized so there is enough laxity that it can be pinned up against the abdominal wall laterally.
    - Measure the defect with spinal needles and an intraperitoneal ruler. Mesh should **extend 4 cm beyond the fascial defect edges, juxtaposing the afferent limb of stoma bowel against the parietal peritoneum laterally for a short distance**. Also the mesh should include any midline laparotomy hernia that may be present from the initial operation.
    - The non-adhesive side of the mesh will face the peritoneal cavity. This leaves the rough side of the mesh against the stoma bowel, and an unavoidable risk for future erosion.
    - **Transfascial sutures** can be used to secure the mesh. These should be spaced more closely to one another on the side of the lateralized stoma bowel. Secure lateral sutures first, and then medial. Tacks may also be placed.

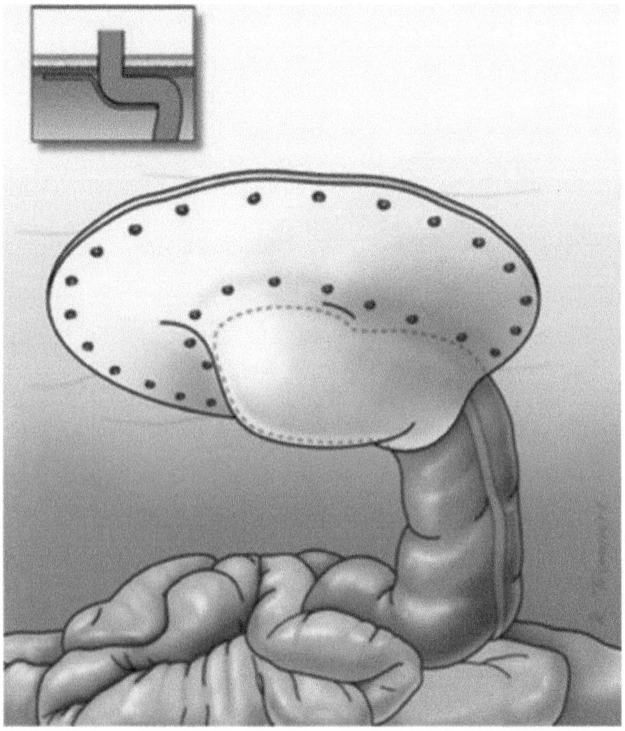

Figure 17.9 Intraperitoneal onlay mesh repair of parastomal hernia.

    - Complications include a 10% recurrence rate, 3% surgical site or mesh infection rate, and a 2% rate of postoperative bowel obstruction.
  - **Relocation of the stoma** may be needed if there is significant peristomal dermal pathology or mucosal ischemia of the stoma bowel extending below the fascial level, indicating compromised bowel.

OTHER ABDOMINAL WALL AND PERITONEAL CAVITY PATHOLOGY

- **Rectus Sheath Hematoma** occurs in anticoagulated patients with COPD who tear rectus muscle veins while coughing. The anterior and posterior rectus sheaths usually limit the size of the hematoma. The mass will be palpable and tender, without significant overlying skin changes. If below the arcuate line, the hematoma can be large. The mass will not change in size with movement **(Fothergill's sign).** If the hematoma is enlargening, **CT angiogram (CTA)** can determine if the inferior epigastric artery (IEA) is actively extravasating. Most cases can be managed with observation and reversal of coagulopathy with vitamin K, FFP, or PCC, Angioembolization for IEA extravasation or surgical evacuation are utilized for patients who have failed all of the above.
- **Omental Cysts and Tumors.** Cysts may be asymptomatic or symptomatic. Metastatic tumors are more common than primary omental malignancy. The omentum can torse independent of whether or not there is a mass. Symptomatic cysts or torsion should undergo resection. Metastases require biopsy. Further testing and treatment depend on the primary etiology.
- **Mesenteric Cysts and Tumors.** Cysts can be asymptomatic or symptomatic and are due to a disruption in lymphatics during development. They are generally separate from bowel, and so can be enucleated. The same is true for peripherally situated, benign tumors. Centrally located malignant tumors residing near the root of the mesentery often cannot be resected.
- **Encapsulating Peritoneal Sclerosis** can occur in patients on peritoneal dialysis. This is an immune mediated, fibrotic reaction that ultimately impairs the peritoneal lining's homeostatic capabilities. Treatment is with glucocorticoids, tamoxifen, and immunosuppression.
- **Retroperitoneal Fibrosis** can be due to beta-blockers, hydralazine, alpha-methyldopa, methysergide, or idiopathic. It is an immune mediated fibrotic reaction that can obstruct lymphatics and entrap and medially deviate both ureters.

18.

BREAST

ANATOMY AND PHYSIOLOGY

- The mammary ridges extend from the axillae to the inguinal region during the 5th–6th week as embryo. This may result in polythelia (accessory nipples and breast tissue) along this line in 1% of individuals.
- The breast is organized into multiple lobules arranged into **15–20 larger lobes**, with the greatest volume of tissue in the upper outer quadrants. **Cooper's ligament** is the suspensory connective tissue dividing the breast into segments, each containing multiple lobes.
- **Prolactin** stimulates lobular production of lipid and protein components of breast milk.
- **Minor ducts** drain lobules and contain a single epithelial layer. They merge into **major ducts** which have two epithelial layers. **Lactiferous sinus** sits below the nipple with major ducts draining into it. *See Figure 18.1*a.
- Below the ductal epithelium is the longitudinal smooth muscle **myoepithelial cells** containing myofibrils. **Oxytocin** causes contraction of myoepithelial cells.
- Deep to the myoepithelium is the basal lamina.
- Lymphocytes reside in the connective tissue below the basal lamina, where they produce antibodies that are excreted across duct epithelium into the duct lumen.
- **Estrogen** promotes duct development and growth of glandular tissue.
- **Progesterone** causes lobule development and maturation of glandular tissue.
- The **long thoracic nerve and lateral thoracic artery** sit over the serratus anterior.
- **Thoracodorsal nerve and artery** are posteriorly located and serve the latissimus dorsi.
- **Medial pectoral nerve** innervates the major and minor pectoralis muscles.
- **Lateral pectoral nerve** innervates the pectoralis major only.
- Intercostobrachial nerve transmits sensation from medial upper arm to the 2nd intercostal nerves.
- **Batson's valveless venous plexus** drains to the spine and may be the route for spread of bony metastases of breast cancer.
- **Sappey lymphatic network** lies below the nipple. 25% of lymphatic drainage goes to the mammary nodes and 75% to axillary nodes. Axillary node levels include **1**: inferior to pectoralis minor, **2**: anterior or posterior to pectorals minor. **3**: subclavicular.

SCREENING

- Women have a **12% lifetime risk** of invasive breast cancer (IBC).
- 5% of patients with IBC will have contralateral breast cancer within 10 years.
- Screening for IBC **decreases mortality by 20%–25%.**
- 10% of IBC have a negative mammogram and negative ultrasound (US).
- Self- and clinical breast exams are deemphasized as screening measures in comparison to imaging.
- **Screening mammogram** includes craniocaudal and medial-lateral oblique views every 2–3 years after age 40 and every year after age 50. Any finding should prompt a **diagnostic mammogram.**
  - Otherwise, screening may begin 10 years prior to the youngest age at which a first-degree relative received a breast cancer diagnosis.
  - May stop screening if a patient's life expectancy is <10 years.
  - Screening mammogram is reliable even if the patient has implants.
  - Tomosynthesis is mammography over a 3D arc. It increases detection of distorted architecture and possibly microcalcifications for those with dense breasts and high risk.
  - Contrast injection enhancement can increase cancer detection rates but is not standard of care.
- In addition to annual screening mammography, **magnetic resonance imaging (MRI)** yearly can be used to screen those patients with a >20% lifetime risk of IBC based on a risk model. MRI is also used for screening if a patient has a history of breast **radiation**, a known exposure that increases IBC risk, is **BRCA 1 or 2** positive, or has a relative who is BRCA 1 or 2 positive. The screening mammogram and MRI can be spaced 6 months apart.
- **Genetic panel testing** should be considered for those individuals with a cancer diagnosis at a young age,

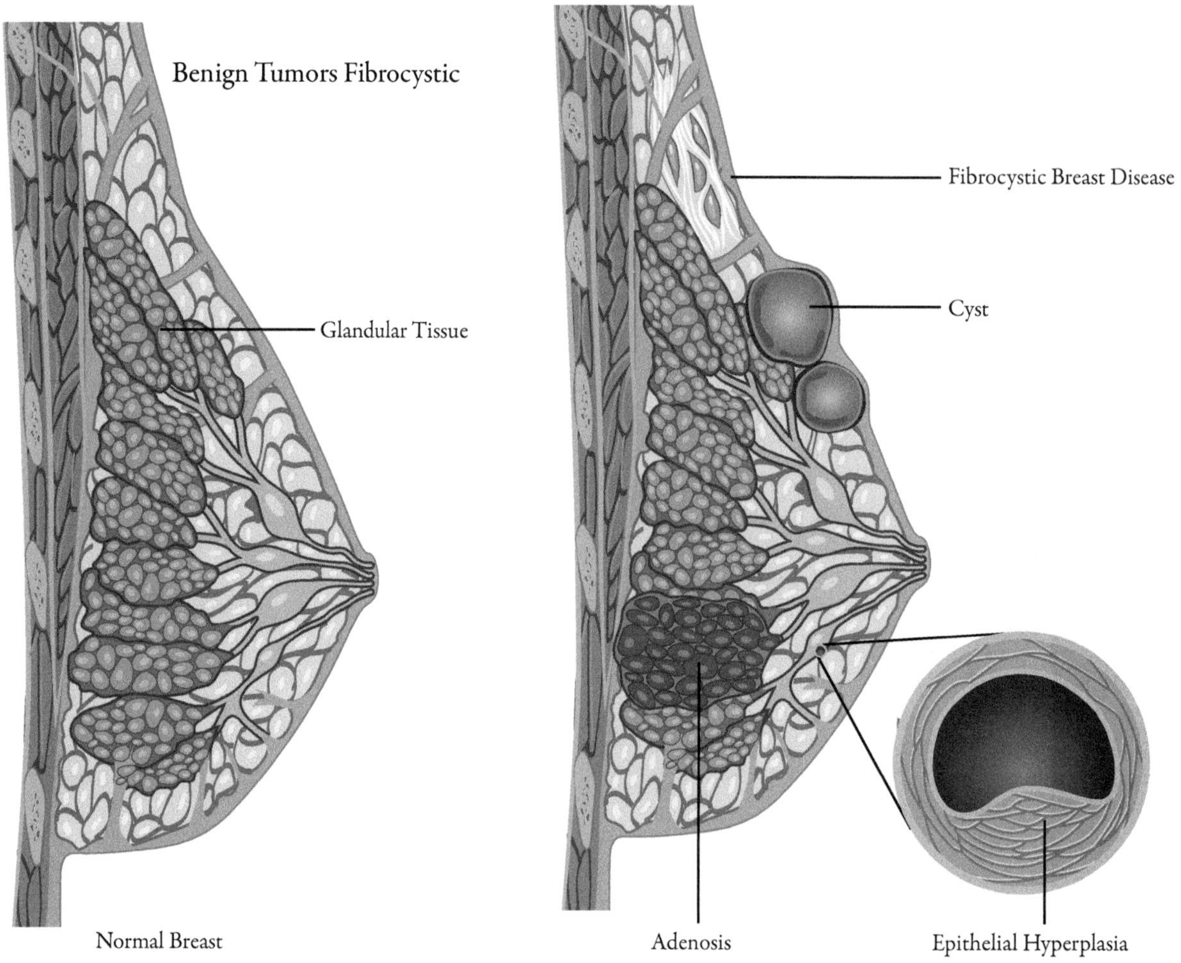

Figure 18.1 (a) breast architecture, and (b) benign glandular and ductal based disease.

multiple types of cancer, family members with cancer complexes, or those who belong to a high-risk patient population.

- Single site gene testing can be done in the case of a known familial mutation you are investigating. Otherwise, perform panel testing.
- Within any positive genetic result, there are benign, likely benign, uncertain significance, likely pathogenic, and pathogenic variants of mutations.
- If a patient tests positive for pathogenicity, all family members should be encouraged to be tested as well.
- If a patient tests positive with "uncertain significance," the import of this depends on the patient's history of cancer and family members' cancer status.

- Factors that are considered **high risk** include the following:
  - **Personal history of atypical hyperplasia, DCIS, or LCIS.**
  - BRCA 1 or 2 with a family history including **multiple** 1st- or 2nd-degree relatives, >2 relatives with premenopausal IBC, 1st-degree relative with **bilateral** premenopausal IBC, combination **breast and ovary** syndrome in the family, 1st- or 2nd-degree relatives with ovarian cancer, or any male relative with IBC.
- Factors that are considered **moderate risk** include the following:
  - Prior radiation, prior IBC, single 1st-degree relative with IBC, and first birth at >35 years old.
- Factors that are considered **low risk** include obesity, alcohol, postmenopausal hormonal therapy with progesterones, and Gail model risks.
  - **Gail model** can be used to assess risk so long as the patient has no personal history of IBC or in situ disease. Components include ethnicity, age at menarche, age at first birth, nulliparity, number of 1st-degree relatives with breast cancer, and any previous breast biopsies showing benign or atypical cells.

BRCA

- Individuals should be tested for BRCA 1/2 if they have triple negative cancer at <60 years of age, a new diagnosis of breast cancer with family members who have breast or ovarian cancer, or a close male family member with breast cancer.
- **BRCA1 is a mutated mismatch repair gene.** Tumors are poorly differentiated and ER/PR negative.
  - One may consider **bilateral prophylactic skin sparing or nipple sparing mastectomy** to decrease IBC risk by 90%. Otherwise, patients should receive mammograms alternating with breast MRI every 6 months.
  - Patients will also have a 40% ovarian cancer risk. One may consider **prophylactic total abdominal hysterectomy and bilateral salpingo-oophorectomy (TAH-BSO) after childbearing years** because there is no good screening for ovarian cancer. This will decrease ovarian cancer risk by 90% and IBC risk by 50%. It also improves survival.
- **BRCA2 is also a mutated mismatch repair gene.** These tumors are well differentiated and ER/PR positive.
  - One may consider **bilateral prophylactic skin sparing or nipple sparing mastectomies** to decrease IBC risk by 90%. Otherwise, mammograms alternating with breast MRI every 6 months.
  - **Tamoxifen** can decrease IBC risk by 60% since tumors are ER/PR+.
  - Patients have a 10% risk of ovarian cancer.
  - Male patients who are BRCA-2 have a **10% risk of breast cancer.**
  - Patients also have an increased risk of melanoma, gastric, biliary, and pancreatic cancer.

WORKUP OF A SYMPTOMATIC LESION

- **Diagnostic mammography** includes additional views beyond the craniocaudal and mediolateral oblique screening mammography views.
  - This modality has a 90% sensitivity and specificity for masses >5 mm. It is better in older patients with less dense parenchyma.
  - Irregular borders, a spiculated appearance, calcifications (linear, cluster, or branching), ductal asymmetry, or architecture distortion are all worrisome features.
- An **ultrasound (US)** is a better imaging modality for young patients with dense breast tissue.
- **BI-RADS system** categorizes mammographic and ultrasound imaging findings and determines further workup.
  - 0=non-diagnostic. In this case, the patient needs further imaging.
  - 1=no finding.
  - 2=benign.
  - 3=probably benign. The patient should receive a 6-month follow-up mammogram.
  - 4=suspicious lesion. This requires core needle biopsy (CNBx), and if an adequate sample is obtained and negative, then the lesion could potentially be followed with imaging.
  - 5=likely a malignancy. One must perform CNBx and expect a diagnosis of malignancy. If the CNBx is non-diagnostic or shows only a benign lesion, a patient in this category would still need an excisional biopsy (EXBx) to exclude sampling error, since the imaging is highly suspicious for cancer.
  - 6=biopsy-proven definitive malignancy.
- **Core needle biopsy (CNBx)** is guided by palpation, US, or stereotactic imaging.
  - In the prone or upright position, breast tissue is compressed between plates, and a 14g Trucut needle is used to obtain 4 to 12 vacuum-assisted samples.
  - Since the needle fires forward, it may be more difficult to obtain an adequate sample without injuring neighboring structures in situations where the lesion is near the pectoralis muscle, an implant, the dermis, or the nipple.
  - A radiopaque marker or wire is left in place for a possible future excisional biopsy.
- A **wire localized EXBx** can be done if suspicion for malignancy remains high after a non-diagnostic or benign CNBx.
- A palpable breast mass in a patient >40 years old should receive:
  - Physical exam, bilateral diagnostic mammograms, and a US followed by **CNBx**.
- A palpable breast mass in a patient <40 years old should receive:
  - A physical exam and **US**. A mammogram may not be as sensitive at this age because of dense breasts. Imaging may be followed by **CNBx** guided by palpation, US, or stereotactic imaging. If benign and BI-RADS 4 or less, one may avoid a wire localized EXBx and perform serial imaging.
- **Clinical staging** of a biopsy-proven malignancy includes:
  - 1. History directed at symptomatic metastases, personal risk factors, and familial cancer.
  - 2. An exam for palpable axillary, mammary, and clavicular lymphatic involvement.
  - 3. An **axillary US** at the time of diagnosis of a primary breast mass. After a breast mass CNBx, nodes will become reactive and may appear positive on an axillary US. Therefore it is best to perform the axillary US before any biopsy.

- Concerning lymph node features include a **cortex >3 mm thick, asymmetric bulging, transverse AP widening** more so than longitudinal craniocaudal lengthening, and the **absence of a fatty hypoechoic hilum**, indicating replacement of that hilum by tumor.
  - **Fine needle aspiration (FNA)** can evaluate any suspicious nodes on Ax US.
- 4. **Positron emission tomography (PET) CT** can be added for stage 3 disease or a history suggestive of metastases.

BENIGN LESIONS

- **FAT NECROSIS**
- Occurs after trauma or surgery and presents as a tender, discrete breast mass with well circumscribed eggshell calcifications on mammography. Patients may not recall the inciting trauma. CNBx is diagnostic.
- **CYCLICAL MASTALGIA OR MASTODYNIA**
- Usually bilateral and fluctuating with hormonal levels. Treatment includes reassurance and support bras. It is unclear if decreasing fat, nicotine, or caffeine intake helps. **Tamoxifen** can be prescribed for severe symptoms.
  - If a biopsy is done, **fibrocystic changes** are often seen, and the patient may have creamy nipple discharge and heterogeneous induration. *See Figure 18.1b.* Histology can show lobular or ductal epithelial hyperplasia (without atypia), apocrine metaplasia, or adenosis.
- **MASTITIS**
  - Generally a bacterial breast infection characterized by non-cyclical mastodynia or mastalgia and signs of inflammation. Staphylococcal species are common if the patient is lactating. Treatment is with oral antibiotics and continuation of breastfeeding.
    - US guided needle aspiration is the first-line therapy for any **abscess** resulting from duct obstruction. Incision and drainage is second-line therapy.
    - Consider inflammatory breast cancer in the non-lactating patient with signs of mastitis who is not responding to a trial of antibiotics.
    - **Granulomatous mastitis** is a chronic inflammatory process with or without fistulae that requires CNBx or EXBx for a diagnosis. It is not primarily an infectious process. Treatment is with excision. Some cases respond to glucocorticoids. Incisional drainage and antibiotics are used for a secondary infection.
    - **Plasma cell, periductal mastitis** also has non-cyclical mastodynia, with inflammation and a creamy nipple discharge. Treatment is with antibiotics.
- **PERIAREOLAR MAMMARY DUCT FISTULA**
  - Results from a chronically obstructed duct or subareolar lactiferous sinus that produces contained extraductal secretions which eventually become infected and evolve into an abscess. Over time, this necessitates to the edge of the areola, or a papillae within the areola, and develops into a chronic fistula. Treatment is with resection of the obstructed duct, abscess, and fistula.
- **GALACTORRHEA**
  - Bilateral nipple discharge secondary to an endocrine imbalance, increased prolactin, or anti-dopaminergic agents.
- **GALACTOCELE**
  - Acystic collection of non-infected milk in a lactating woman. Treatment is with aspiration if symptomatic. Observation is adequate if asymptomatic.
- **GYNECOMASTIA**
  - An overgrowth of subareolar ductal elements**.** Gynecomastia is not a risk factor for malignancy. It is secondary to decreased testosterone, increased estrogen, thyroid, renal, or hepatic disease, or medication side effects.
- **DIAGNOSIS**
  - Presents as a tender subareolar mass that can be unilateral in adolescence or bilateral in senescence. Clinical exam alone is sufficient in a young patient with bilateral involvement.
  - If unilateral in an elderly patient where the diagnosis is in doubt, mammography and biopsy will be needed to rule out carcinoma.
- **TREATMENT**
  - Includes **observation, tamoxifen,** or **surgical resection** of the subareolar ducts, plus liposuction of adjacent fat for cosmetic concerns, lesions >4 cm, or those not resolving by the end of puberty.
- **POLAND SYNDROME**
  - Unilateral hypoplasia of the shoulder, chest wall, muscles, and breast tissue in men. Most cases are minimally symptomatic.
- **CYSTS**
  - They are best evaluated by US.
    - If simple and asymptomatic, one may leave the cyst alone.
    - If simple and **symptomatic** with complete collapse after **aspiration** of serous fluid, there is no need for a biopsy.
    - Otherwise, **core needle biopsy the wall** of any non-serous, non-collapsing, recurrent, or heterogeneous cyst. Thereafter may perform EXBx if CNBx is concerning or the cyst is not resolving.
- **NIPPLE DISCHARGE**
  - Most commonly creamy, bilateral, and provoked. If the discharge is unilateral, spontaneous, bloody, or clear, this requires further investigation.
    - The differential diagnosis should include duct ectasia, fibrocystic disease, malignancy, or papilloma. **Intraductal papillomas** reside near the nipple and are

the most common cause of *bloody* discharge. They range from 5 mm to 5 cm in size. *See Figure 18.2.*

- **DIAGNOSIS**
  - Start with **mammography and US** for any suspicious features including spontaneous unilateral discharge that is bloody or serous.
    - **MRI** should follow a negative mammogram or US in patients whose discharge is not localizable to a quadrant by physical exam. Ductogram and ductoscopy are rarely used alternatives to MRI.
    - Stereotactic or US guided **CNBx** is indicated for a high BI-RADS lesion discovered on the above imaging to exclude malignancy.
- **TREATMENT**
  - If the lesion appears intraductal, then this is the cause of discharge. **Wire localization or lacrimal probe cannulation localization for an excisional biopsy** is the next step if the CNBx was benign.
    - If the CNBx suggests malignancy or in situ disease, treatment should follow as in any other scenario devoid of nipple discharge.
    - If lesion is not intraductal, then it is incidental and the workup of that lesion should follow the BI-RADS category like any other breast mass. However, the etiology of ductal discharge still needs to be determined.

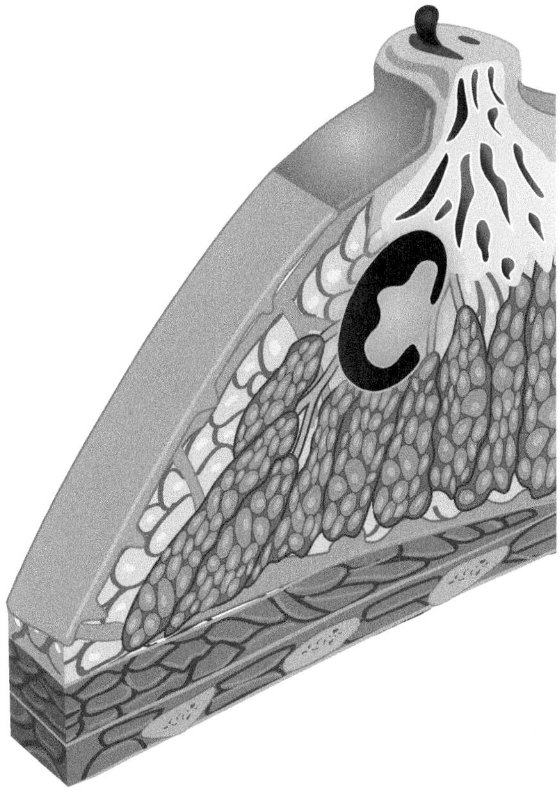

Figure 18.2 Nipple discharge from subareolar intraductal lesion.

    - If no lesion is seen on imaging, operative **lacrimal probe cannulation** may still be attempted, and a duct sector eliciting discharge can be excised.
    - **Subareolar duct exploration with resection** is an option if no localizable intraductal lesion can be found despite all of the above. This removes any intraductal lesion in proximity to the nipple areola complex. It can result in nipple hypoesthesia or anesthesia.
- **MONDOR DISASE**
  - A superficial thrombophlebitis of the lateral thoracic, thoracoepigastric, or superior epigastric veins. Workup includes mammography and US to exclude malignancy. Treatment is with NSAIDs and warm compresses.
- **ADENOMA**
  - Presents as a nontender breast mass with well circumscribed benign hyperplastic epithelium and sparse stroma on biopsy. It is often located near the nipple and is similar to an intraductal papilloma pathologically. It has no malignant potential, but generally should be excised for symptoms.
- **FIBROADENOMA**
  - Presents as a nontender mass in a patient 15-30 years old. Lesions may fluctuate in size and symptomatology with hormonal changes.
- **DIAGNOSIS**
  - **US guided CNBx** provides a definitive diagnosis. Biopsy shows a well circumscribed overgrowth of stroma compressing epithelium. An alternative would be just to watch without biopsy if the lesion is definitively BI-RADS 2 on imaging.
    - An **excisional biopsy (EXBx)** to exclude phyllodes can be considered when the patient is >40 years old or the mass is >3 cm and persistently growing. Other indications for an EXBx are increased stromal cellularity and mitoses on histology, or if imaging and biopsy results are not definitive for a fibroadenoma.
- **TREATMENT**
  - **Enucleation** if the lesion is symptomatic. **Cryoablation** with US monitoring is another option.
- **CYSTOSARCOMA PHYLLOIDES**
  - Most often rapidly growing and benign. They are stromal in origin and behave as aggressive variants of a fibroadenoma. Although rapidly growing, the majority are **benign** or at most have borderline features. 10% of cases are actually malignant and demonstrate a High Ki-67, necrosis, >5–10 mitosis per HPF, or spindle nuclei on pathology. There is no nodal spread and only rare hematogenous spread. Management includes **WLE and 1 cm margins**. Mastectomy is used if the lesion is large or the breast small.
- **HAMARTOMA**
  - Benign disorganized growths of tissue from multiple origins.

- **HYPERPLASIA**
  - Epithelial overgrowth categorized as mild, moderate, or advanced.
  - If >70% of the lumen is occupied or if there is atypia, then the patient will need an EXBx to exclude malignancy.
  - **Atypical ductal hyperplasia (ADH)** is a term used for up to 3 mm of hyperplasia with atypia. ADH quadruples the risk of IBC, and up to 30%–60% will have a nearby DCIS or invasive cancer on the EXBx specimen. DCIS is a term applied when there is more than 3 mm of ductal hyperplasia with atypia. *See Figure 18.3.*
  - **Atypical lobular hyperplasia (ALH)** is a term used for a non-distended lobule with hyperplasia and atypia. Generally it has no radiographic features, but quadruples the risk of IBC. Lobular carcinoma in situ (LCIS) is a term used for hyperplasia and atypia that have resulted in lobular distension.
  - **Radial scar (RS)** is a lesion less than 1 cm that demonstrates central sclerosis at a terminal duct site, secondarily pulling and distorting ducts and lobules on pathology. It can have a spiculated appearance on mammography and requires an EXBx for definitive diagnosis. Its presence doubles the risk of IBC. **Complex sclerosing lesion (CSL)** is the term applied to a radial scar that is >1 cm in size.
  - **Sclerosing adenosis** is a proliferation at terminal ducts and lobules with myoepithelial and stromal fibrosis. This may increase the risk of breast cancer, and is distinct from a CSL or radial scar.
  - ADH, ALH, LCIS, RS, CSL and sclerosing adenosis all need wire localized **EXBx** for a definitive diagnosis, and to exclude malignancy missed by biopsy sampling error. Positive margins are acceptable so long as all of the suspicious radiographic area is included in the sample. ADH, ALH, and LCIS can be treated with **anti-estrogens** regardless of ER status.

INVASIVE BREAST CANCER (IBC)

INVASIVE DUCTAL CARCINOMA (IDC)

- 10% of cases have an identifiable genetic mutation, and 15% have a family history of breast cancer.
- Malignancy originates from ductal epithelium that has progressed from atypia to carcinoma in situ to invasive disease.
- **DIAGNOSIS**
- IDC is discovered on screening mammography or presents as a nontender breast mass with or without skin changes, nipple and areolar changes, or palpable axillary, internal mammary, or clavicular nodes.
  - Initial imaging includes **bilateral diagnostic mammography and US.** Spiculated, branching calcifications, irregular borders, architectural distortion, and ductal asymmetry are characteristic of IDC.

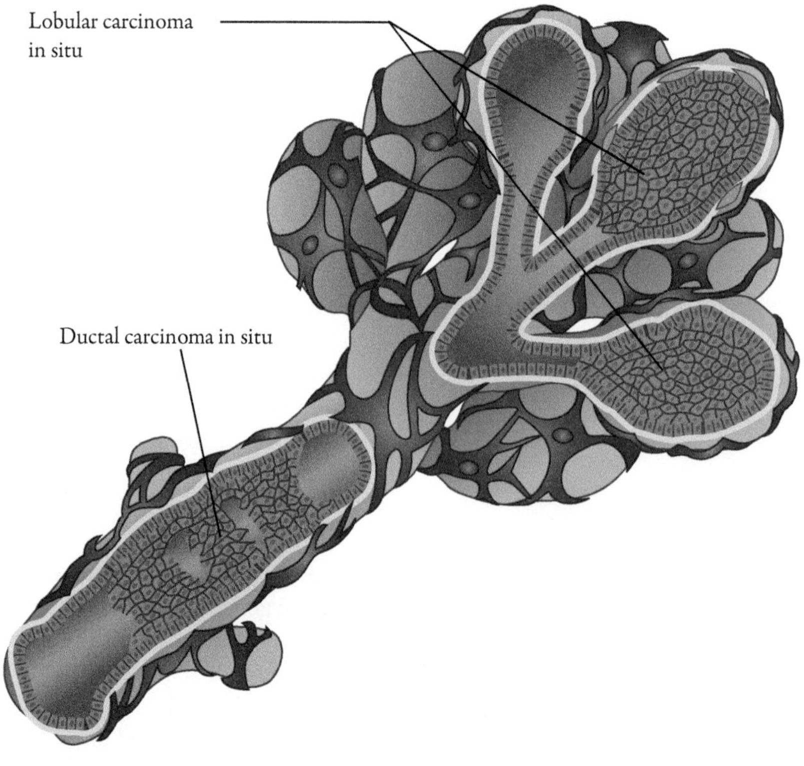

Figure 18.3 Lobular and ductal carcinoma in situ.

- **MRI** may be needed for patients with dense breasts, a lobular cancer diagnosis, a family history of young patients with breast cancer, or if the patient is BRCA positive. This will help determine disease burden and inform treatment options when taken together with pathologic results.
- **CNBx or EXBx** should follow imaging.
  - Aggressive features include VEGF, EGFR, HER2/neu, PCNA, Ki-67, p53, BCL-2, high mitotic index, and high S-phase fraction.
  - Less aggressive features include:
    - 1. ER/PR positive (75% of tumors). PR positive status is better than just ER positive.
    - 2. Low oncotype DX score.
    - 3. Tubular architecture, which has the best prognosis. Other types include medullary, papillary, mucinous, cirrhotic, and comedo. The latter has the worst prognosis.
- The primary lesion is staged as **T1:** <2 cm, **T2**: 2–5 cm, **T3** >5 cm, or **T4:** invasion into chest wall, skin edema, or skin ulceration. A T4 lesion may be seen in neglected, slowly progressive, invasive breast cancer or in rapidly progressive inflammatory carcinoma.
- Any palpable nodes or nodes visualized on **axillary US** should be sampled by **FNA.**
- Most tumors have a single focus, but some are **multicentric** in different quadrants, or **multifocal** sites within the same quadrant.
- Consider **CT chest, abdomen, and pelvis** imaging if the patient has a bulky, locally advanced T3 or T4 primary, or if they have >3 clinically positive axillary nodes.

- **TREATMENT**
  - **Partial mastectomy with whole breast radiation (WBRT) or a total mastectomy.**
    - Both treatments are considered equivalent and have a **<10% local recurrence at 10 years.**
    - **No ink on tumor constitutes adequate margins.** If frozen section or final pathology does not meet this standard, any or all margins can be re-excised, or the patient can receive a total mastectomy.
  - **Sentinel lymph node biopsy (SLNB)** is needed for accurate staging, unless clinically FNA positive nodes are already known to exist. In that case, an **axillary lymph node dissection (AxLND)** should be done. See below.
  - Young women with triple negative disease may elect for bilateral mastectomy.

INVASIVE LOBULAR CARCINOMA (ILC)

- Comprises 10% of all IBCs and originates from lobular epithelium.
- Classically the tumors are infiltrative, multifocal, and multicentric with no radiographic features or calcifications.
- Diagnosis and treatment are similar to IDC. Signet ring cells on pathology indicate a poor prognosis.

INFLAMMATORY BREAST CANCER

See Figure 18.4.

- This presents as a **rapid onset** of breast edema, erythema, and induration over weeks to months. These findings prove refractory to any empiric antibiotics given.
- **DIAGNOSIS**
  - **Start with US, diagnostic mammography, or MRI to find a parent mass that can undergo CNBx.** Biopsy can determine histology, ER/PR, and HER2/neu status.
    - **SLNB is not accurate** due to lymphatic invasion, and so should not be performed.
    - Dermal lymphatic invasion of adenocarcinoma can be seen on **skin punch biopsy.** This is most useful when no primary mass can be identified as a target for biopsy.
  - **CT chest and abdomen/pelvis evaluates for metastases.**
    - Brain MRI is reserved for patients with central nervous system symptoms.
- **TREATMENT**
- **Start with neoadjuvant chemotherapy** and, if a response is noted, one can proceed to a **modified radical mastectomy (MRM) 3–5 weeks later.**
  - MRM includes AxLND of level 1 and 2 and any palpable level 3 nodes.
  - This is followed by **adjuvant radiation** to include the chest wall, axillary, supraclavicular, and internal mammary basins 4–6 weeks after surgery.
  - **Endocrine adjuvant therapy** for 5 or more years if ER/PR positive.
  - **Trastuzumab adjuvant therapy** for 1 year if HER2/neu positive.
  - Either can begin immediately after radiation therapy.

MALE BREAST CANCER

- Risk factors include age >60 years old, history of steroid use, radiation, family history, Klinefelter syndrome (XXY genotype), decreased testicular function, cirrhosis, obesity, alcohol use, or **BRCA2 mutation.**
- 90% of cases are **IDC** since there are no lobules in the male breast.
- >80% are **ER/PR positive**, and most are HER2/neu negative.
- All require genetic testing, and 5%–13% will be BRCA2. Very few are BRCA1.
- **DIAGNOSIS**
  - Patients usually note a nontender, enlarging mass. This prompts a workup meant to distinguish the mass from gynecomastia.

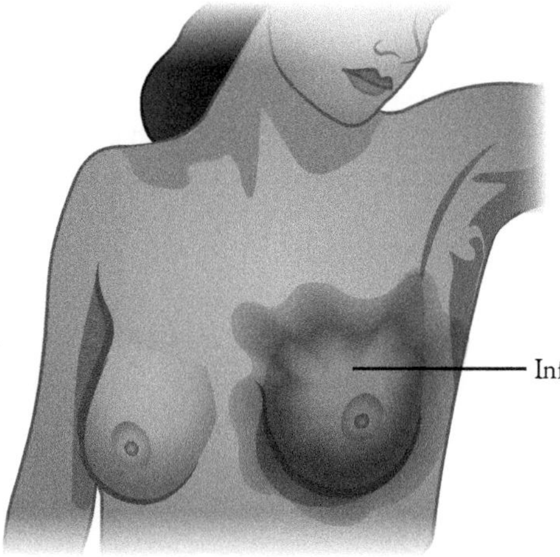

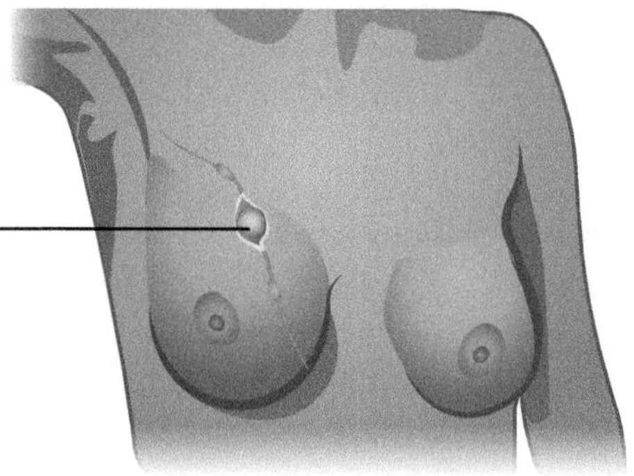

Figure 18.4 Inflammatory breast cancer.

- Testing includes diagnostic mammography, US with FNA of any suspicious nodes, CNBx of the mass, and possibly a skin or nipple punch biopsy of any concerning dermal features.
- CT chest and abdomen should be done to evaluate for metastases in patients who have a bulky primary tumor or positive nodes.

- **TREATMENT**
  - Neoadjuvant hormonal or chemotherapy can be considered if the primary is T3 or the patient is node positive.
  - Most patients require a **total mastectomy with SLNB**, since the primary lesion is usually subareolar and men have minimal breast tissue.
  - A modified radical mastectomy taking level 1 and 2 axillary nodes may be needed if the patient has a bulky primary or clinically positive nodes.
  - **Tamoxifen** is given as adjuvant therapy in all hormone receptor positive cases.
  - Any clinically positive nodes or SLNB positive nodes requires **AxLND**, since the Z-11 trial cannot be extrapolated to men, and they are getting a mastectomy without radiation.
  - Physical exam and mammography are adequate for postoperative screening.

DUCTAL CARCINOMA IN SITU (DCIS)

- Atypical ductal hyperplasia (ADH) that is >3 mm is considered DCIS. *See Figure 18.3.*
- **DIAGNOSIS**
  - It presents as a linear or segmental cluster of calcifications or microcalcifications (<5 mm) on screening or diagnostic mammography. Presentation

as a palpable mass or Paget's disease of the nipple are both much less common.
- Confirmation is by stereotactic **CNBx or wire localized EXBx** of the calcifications.
- There is a 15% risk of finding microinvasion on stereotactic CNBx or EXBx. These patients are upstaged to invasive carcinoma.

- **TREATMENT**
  - **Wire localized partial mastectomy with 2 mm margins** followed by **whole breast radiation (WBRT).** Any or all margins can be re-excised if not at least 2 mm on frozen section or final pathology. If IBC is noted on the specimen, "no ink on tumor" constitutes an adequate distance. Re-excision may be required to achieve this.
  - **Nipple- or skin sparing mastectomy with SLNB** is the best option for comedo-type DCIS, multicentric, extensive, or multifocal sites, large DCIS volume compared to the patient's breast size, positive margins after lumpectomy and serial excisions, or a recurrence after lumpectomy and WBRT.
  - SLNB is only used if:
    - 1. A mastectomy is chosen as treatment, since SLNB cannot be performed afterward. If a mastectomy were to be performed without SLNB, and the specimen ultimately revealed invasive cancer, then the patient would have to receive an axillary lymph node dissection of levels 1 and 2.
    - 2. The lumpectomy specimen shows microinvasion. There is a very low chance of finding a positive sentinel node in this situation, but SLNB should still be performed.
    - 3. DCIS presented as a palpable mass, in which case there is a higher likelihood of finding invasion on the final pathology of the lumpectomy specimen.
  - **Tamoxifen or anastrazole** should be added if the cells are ER+. This reduces ipsilateral and contralateral recurrence.
  - Oncologists may consider omitting radiation if the Oncotype DX DCIS score is low risk.

LOBULAR CARCINOMA IN SITU (LCIS)

- This is the term used for atypical lobular hyperplasia (ALH) that distends a lobule. *See Figure 18.3.*
- LCIS is a risk factor for invasive ductal cancer on either side, and some suggest may be a precursor to invasive lobular cancer. Often it is multifocal or multicentric.
- **DIAGNOSIS**
  - Generally noted incidentally on stereotactic or US guided CNBx done to evaluate neighboring calcifications. LCIS itself does not produce any imaging finding.
- **TREATMENT**
- **Requires wire localized excisional biopsy (EXBx) to exclude adjacent DCIS or invasive cancer**, which may be present in up to 20% of cases.
  - If the wire localized EXBx shows classic LCIS alone, **positive margins are acceptable.**
  - Re-excision to negative margins is suggested if pleomorphic LCIS is present on the biopsy.
- Prevention and surveillance of future carcinoma is the most important aspect of the diagnosis of LCIS because there is a **30%–40%** lifetime risk of invasive cancer on either side.
  - Perform **annual mammography with or without an annual MRI** staggered every 6 months.
  - Chemoprevention is with tamoxifen if premenopausal, or either **tamoxifen or aromatase inhibitors** if postmenopausal. These agents afford a 50%–60% risk reduction in IBC.
  - **Bilateral prophylactic skin or nipple sparing mastectomies** afford a risk reduction of >90% and can be considered for those who are BRCA1/2 or have a strong family history of invasive breast cancer.

PAGET'S OF NIPPLE

- This refers to **in situ adenocarcinoma of the nipple.** *See Figure 18.5.*
- **DIAGNOSIS**
  - Pathology shows pale, vacuolated, malignant cells in rete epidermal pegs on full thickness nipple skin biopsy.
    - The process is generally erupting from **underlying, subareolar, DCIS, or invasive ductal cancer (IDC),** so breast imaging needs to be performed as part of the diagnostic workup.
- **TREATMENT**
- **Skin sparing or simple mastectomy to include the tumor cells at the nipple areolar complex (NAC) as well as any primary underlying mass. SLNB** is also done in case the mastectomy specimen ends up showing invasive cancer.
- If the underlying mass is very close to the NAC, one may perform a central lumpectomy (including NAC) with radiation instead of a mastectomy.

PARTIAL MASTECTOMY

- Also known as a "lumpectomy," it is adequate treatment for smaller IBC or DCIS provided the patient is able to receive adjuvant radiation therapy. Partial mastectomy **accompanied by 50 Gy whole breast radiation (WBRT) over multiple weeks** is equivalent to mastectomy as treatment for IBC, both associated with a **5%–10% recurrence rate in 10 years.**

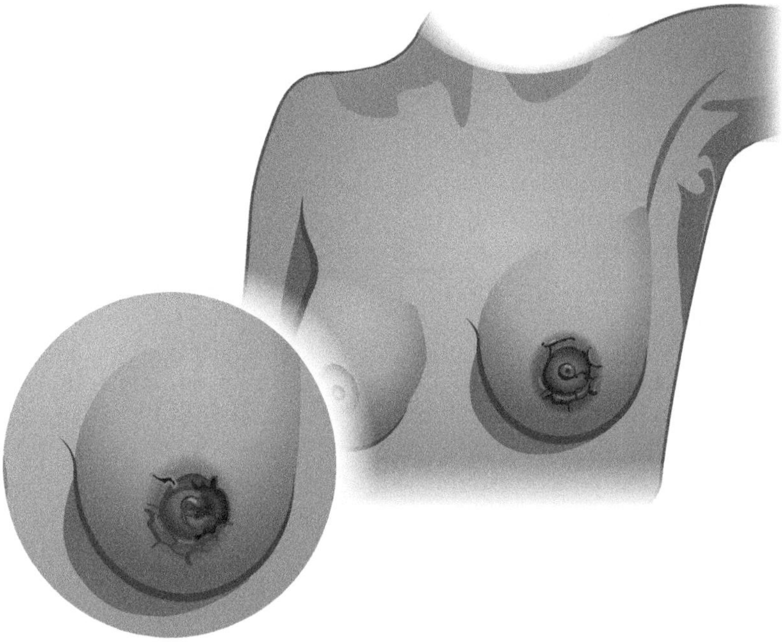

Figure 18.5 Paget's disease; carcinoma in situ of the nipple.

- Partial mastectomy can be used for multifocal tumors in proximity. It should not be used if the patient's tumor is multicentric or imaging displays diffuse calcifications, if the patient is pregnant, had prior thoracic radiation, has a history of scleroderma or SLE (precluding breast radiation), or has small breasts.
- Physical exam or US guided localization is adequate for a palpable mass.
- Preoperative wire localization is needed if the pathology is non-palpable. One may add bracketing devices that mark the radial margins.
- A curved incision is used for masses superior to the nipple, and a radial incision is used if inferior. 1 cm thickness subcutaneous flaps should be raised in all directions and the suspicious area excised.
- Once the specimen is removed, a specimen radiograph can be taken to ensure any calcifications or radiologic markers have been procured in the specimen.
- **No ink on tumor** is adequate margin for IBC, and **2 mm** is adequate margin for DCIS.
  - Inadequate margins require re-excision or total mastectomy.
  - A cavity shave can be performed for additional margins of 3–10 mm in areas that are positive or close to the gross margin of the lesion.
  - Some practitioners use radiofrequency ablation (RFA) at lumpectomy site to extend the margin or to treat the margins of a residual cavity after local resection.
- The residual cavity, should not be obliterated at closure, but rather it should be allowed to fill as a seroma.

MASTECTOMY

- There are a number of versions. Mastectomy was the original treatment modality for IBC and is equivalent to a partial mastectomy accompanied by whole breast radiation.
  - A **modified radical mastectomy (MRM)** takes the pectoralis fascia and level 1 and 2 axillary lymph nodes as well as all breast tissue, nipple complex, and overlying skin.
  - A **total mastectomy** takes breast tissue, skin, and the nipple via an oblique or transverse incision.
  - A **skin sparing mastectomy (SSM)** takes the nipple areolar complex, excisional biopsy scar, and breast tissue, leaving the remainder of the skin.
  - **Nipple sparing mastectomy (NSM)** uses a lateral inframammary incision and thin subareolar flaps to preserve the nipple. This can be done for primary

tumors that are peripheral, small, and without any skin involvement of the nipple or nipple discharge. A **subareolar tissue frozen section** should be negative, and there should be no findings of Paget's disease of the nipple. NSM may also be used as a prophylactic mastectomy. Recurrence rates are equal to an SSM in the appropriately selected patient.

- **Adjuvant radiation can be added to a mastectomy** in certain scenarios:
  - The primary is **T3 or greater**, or the margins are positive.
  - **N2 or greater** nodal status, including >4 axillary nodes positive, any mammary nodes positive, or any extracapsular nodal spread.
- A **contralateral prophylactic mastectomy** may be recommended for patients with a **genetic syndrome** (BRCA 1/2, PTEN-Cowden, p53-Li-Fraumeni), a strong **family history** of breast cancer without a known mutation, or a history of Hodgkin's lymphoma with **radiation** to the chest.
  - This reduces breast CA risk by 90%–95%, but does not affect overall survival.
  - Contralateral prophylactic mastectomy has a higher wound and overall complication rate than unilateral mastectomy.
  - In lieu of prophylactic contralateral mastectomy, one may consider additional surveillance imaging of the contralateral breast beyond standard mammography if there is:
    - Lobular histology of the index lesion.
    - Multi-centricity on the symptomatic side.
    - High-risk features of the primary lesion.
    - Tumor was ER/PR positive or the patient has a high Gail score.

RECONSTRUCTION AFTER MASTECTOMY

- This can be done with autologous tissue or an implant.
- **Transverse rectus abdominis flap** is the most commonly used. *See Figure 18.6.* It may be:
  - **1.** A **pedicled** flap from the contralateral superior epigastric artery.
  - **2.** A **free flap** from an inferior epigastric artery. Procurement of the flap may be preceded by **angiosome conditioning** after ligating the superior epigastric vessel. Occasionally, the superficial (rather than deep) inferior epigastric artery can be used to base a free flap.
  - 3. A **muscle sparing or perforator free flap** is where the donor site vasculature is skeletonized to retain muscle locally, and take only the vessel and a subcutaneous, dermal paddle.
- **Latissimus dorsi flap** is the second most common reconstruction, and includes a pedicle of skin and subcutaneous tissue based on the thoracodorsal vessel. *See Figure 18.7.*
- Latissimus flaps are used as coverage for an associated implant-based reconstruction, especially if radiation is part of the treatment regimen. This flap can serve as salvage coverage for a prior failed reconstruction. It can be performed as a free perforator flap sparing the latissimus muscle at its native site.
- Other options for small volume reconstruction include the following:
  - A **transverse upper gracilis** free flap based on the medial circumflex femoral vessels.
  - **Gluteal free flaps** based on the superior or inferior gluteal artery.

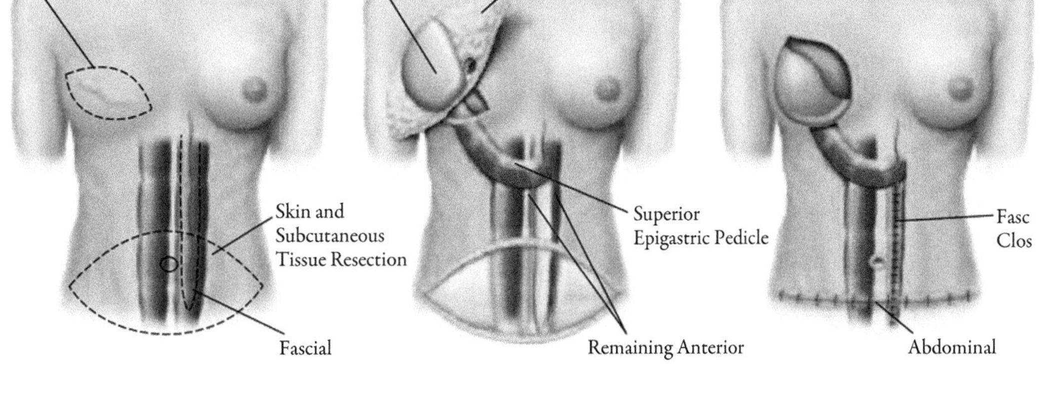

Figure 18.6 Pedicled contralateral transverse rectus abdominis muscle flap.

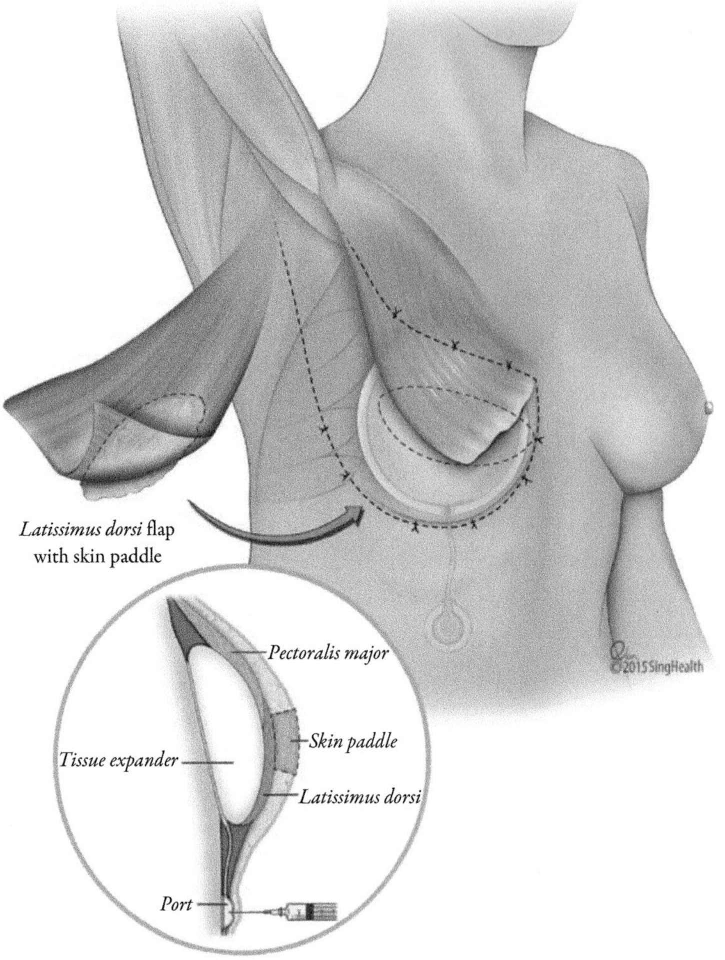

Figure 18.7 Latissimus dorsi flap with implant.

- A **trapezius or pectoralis major flap** based on the thoracoacromial artery, lateral thoracic artery, and intercostal perforators.
- **Implant-based reconstruction** can be immediate or delayed after a submuscular tissue expander is placed at the time of mastectomy. If an expander is used, the patient returns for weekly filling via a subcutaneous port until it is the desired size, which is generally at 2–3 months. At that point, the expander is left at this size for a few more months, and then exchanged for the implant. This can be augmented by an autologous latissimus flap. Biologic tissue matrices, like an acellular dermal matrix, can be used to help recreate the sling effect of the inframammary crease.
- One may **delay reconstruction** by 6–12 months after mastectomy in patients receiving radiation or who have yet to quit smoking**.** A submuscular tissue expander can be left in place post-op, and deflated for radiation sessions. Radiation can result in implant capsule contracture, so an autologous reconstruction is generally a better option when the time comes.
- All reconstructions are **complicated by smoking, obesity, and thin skin flaps** encroaching on the subdermal vascular plexus. The latter issue is especially important in immediate, implant-only reconstructions without autologous tissue.

AXILLA

- Nodal status is the most important prognostic factor for IBC.
  - **N1**: 1–3 axillary nodes. This equates with a 60% 5 years survival.
  - **N2:** 4–9 axillary nodes, any matted axillary nodes, or any isolated mammary nodes.
  - **N3**: supraclavicular nodes, infraclavicular nodes, mammary in combination with axillary nodes, or >9 axillary nodes.

- Distant metastases correlates with a 24% 5 years survival.
- **US of the axilla and FNA** of suspicious nodes can be done in any instance where clinical exam of the axilla is not definitively negative for lymphadenopathy. If any nodes are positive on FNA, then the patient is classified as having clinically positive nodes and will need an axillary node dissection. There is no role for SLNB unless one is planning neoadjuvant therapy and there was only one clinically positive node. In that case, SLNB can be performed after the neoadjuvant therapy in the hope of avoiding AxLND (see below).
- **Sentinel lymph node biopsy (SLNB)** is performed for patients with a **clinically negative axilla** (by exam and/or US) to determine if there has been occult tumor spread. *See Figure 18.8.*
  - It is utilized for patients with a T1 or T2 primary and no clinically detectable nodes, but can also be applied to T3 or multifocal or multicentric tumors.
  - A **tracer is injected into the peritumoral area, intradermal, or subareolar space drained by Sappey lymphatic network.** The purpose is to determine which nodes are the first to uptake that tracer. These nodes will be retrieved and examined for tumor presence. There are two types of tracers used:
    - **Preoperative 99Tc sulfur colloid** injection and imaging done to confirm the main nodal drainage basin within which to pursue the sentinel node. Generally this is axillary rather than mammary or supraclavicular. Intraoperatively, the nodes containing the highest level of radiotracer are identified and excised. Also, one should retrieve all nodes that have >10% of the highest node radioactivity. Most practitioners rarely take more than 4 nodes.
    - **Intraoperative isosulfan blue** injection can also be used, with nodal color guiding retrieval.
- **SLNB is not used if there are clinically positive nodes** by exam or US with FNA. Also, it is avoided in patients who have **inflammatory breast cancer** or another process believed to disturb the integrity of the lymphatic drainage system.
- **Positive sentinel nodes *classically* required an axillary lymph node dissection (AxLND)** to retrieve all nodes within the basin, except for the particular circumstance below.
  - If isolated tumor cells (ITC), micrometastases (<2mm), or 1–2 macrometastases (N1) are noted, but the primary tumor is T2 or less, and the patients will be receiving a partial mastectomy with WBRT and systemic chemotherapy or hormonal therapy, then one **may forego a dedicated axillary node dissection.** The belief is that the WBRT field captures the axillary nodes and assists in sterilizing this site when the tumor burden in this area is low.
  - Since mastectomy patients are not obligated to receive WBRT, similar SLNB findings cannot be extrapolated to this group.
- If all sentinel nodes are **negative**, there is a **1% nodal basin recurrence risk**
- SLNB is complicated by a **7% lymphedema risk.**
- **SLNB may be performed after neoadjuvant chemotherapy** for both clinical N0 and N1 patients, but one should use both tracers and gather at least 3 sentinel nodes, including the original node that

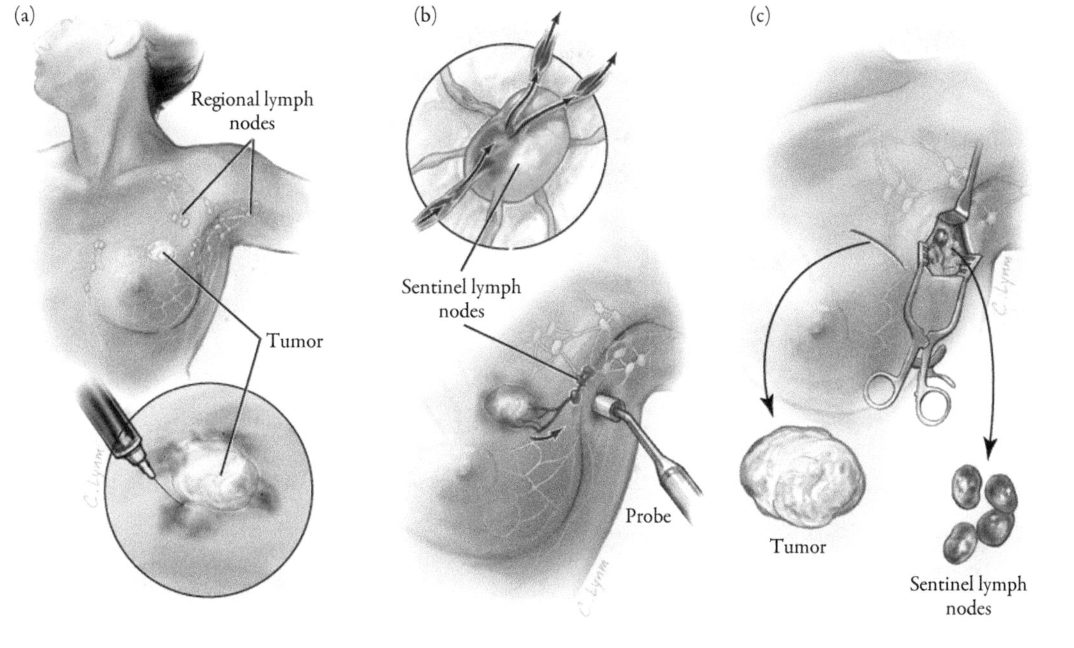

Figure 18.8 Sentinel lymph node biopsy.

qualified them as N1. This process is done to decrease the otherwise high false negative rate of SLNB in this setting. Any post-neoadjuvant chemotherapy positive node on SLNB (ITC, micros, macros) must receive an AxLND.

- AxLND **of level 1 and 2** is used for the following:
  - 1. **Clinically apparent nodes** by exam or US and FNA.
  - **2.** SLNB with **3 or more positive nodes.**
  - **3.** SLNB with **1-2 positive nodes but no plans for WBRT or systemic therapy.**
  - **4.** SLNB with **1-2 nodes with extracapsular extension or lymphovascular invasion.**
  - **5. Failure to identify a sentinel node**; or
  - 6. Any ITC or micrometastases on SLNB done *after* neoadjuvant therapy. Neoadjuvant chemotherapy can be given for breast cancer with clinically positive fixed or matted nodes (N2 status). Thereafter, AxLND can be done. If the patient just had N1 disease before neoadjuvant therapy, then dual tracer SLNB can be performed and possibly avoid AxLND.
  - Axillary node dissection **decreases local recurrence** in the above settings, but does not confer a survival advantage.
- The procedure is performed by opening the clavicopectoral fascia and defining boundaries. *See Figure 18.9.*
  - Pectoralis major and minor form the anterior and anteromedial border.
  - Subscapularis is the posterior boundary.
  - The axillary vein marks the superior border. The thoracoepigastric vein comes off of this, and posterior to the venous structures lies the **thoracodorsal nerve** innervating the latissimus dorsi.
  - The latissimus dorsi sits laterally, and the serratus anterior sits medially.
  - Medially run the **intercostobrachial nerves** heading to the medial surface of the upper arm.
  - Posterior to the venous structures in this medial area lies the **long thoracic nerve** innervating the serrates anterior. It descended from the brachial plexus medially in the neck, running over the middle scalene muscle, and behind the subclavian vessels to the axilla.
  - **Level 1** includes all tissue between the thoracodorsal and long thoracic nerves.
  - **Level 2** is posterior to pectoralis minor. Rotter's nodes are between the pectoralis major and minor, and are technically within level 2.
  - **Level 3** nodes are superior to the pectoralis minor.

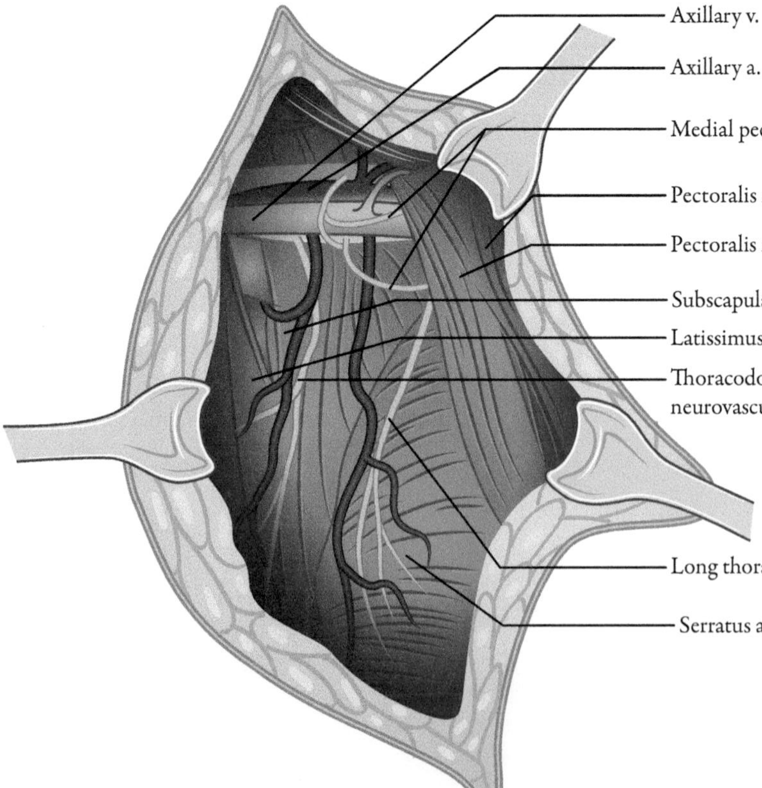

Figure 18.9 Boundaries of axillary lymph node dissection.

    - Retrieve a minimum of 10 nodes.
    - Leave drains until <40 ml/day output, at which time they can be removed.
  - **Additional dedicated nodal basin irradiation** after lymph node dissection can be considered if 4 or greater nodes are positive (N2 status), there is extracapsular nodal extension of tumor, high-grade features, lymphovascular invasion, or if genomic assays suggest aggressive biology. Nodal basin irradiation cannot currently be used in place of AxLND.
  - **Complications of AxLND** include:
    - **Lymphedema in 20%,** especially if the patient is obese and has been irradiated.
    - **Intercostobrachial cutaneous nerve injury** is more common than long thoracic or thoracodorsal nerve injuries.
    - Axillary vein thrombosis.
    - Lymphangiosarcoma 5-10 years later. This is known as Stewart-Treves syndrome and presents with a dark purple nodule on the affected, edematous, extremity. The tumor arises from dermal vascular endothelium. It may occasionally occur after axillary radiation alone. Treatment is with WLE or amputation.
- For cases of **recurrent breast malignancy in the axilla,** first confirm the diagnosis by exam or **US guided CNBx or FNA**. This should rule out lymphoma or melanoma. Then perform **breast imaging** for evaluation of recurrence at the primary site or chest wall. as well as **PET-CT** to evaluate for distant metastases.
  - Then consider AxLND into level 3 if not previously done and disease appears operable.
  - The other option would be basin irradiation if not previously done, or if there is a non-operable supraclavicular nodal burden. Both AxLND and basin irradiation can be combined.
  - Adjuvant chemotherapy and hormonal therapy will be needed.
- For the **initial axillary presentation of an occult breast cancer,** the diagnosis of breast origin must be confirmed by **CNBx or FNA** of the clinically apparent node.
  - Melanoma and lymphoma must be excluded on the biopsy specimen. Whole body imaging may be needed to assist in this evaluation.
  - Both **breast imaging with MRI and US** as well as systemic imaging with **PET-CT** should be done if occult cancer is suggested by a CNBx proven clinically positive node in the absence of a primary site in breast.
    - Any breast lesion identified by the above imaging should be biopsied to determine if it is the primary invasive carcinoma.
    - **Ipsilateral mastectomy** is considered equivalent to **ipsilateral WBRT** as treatment if no primary is found on MRI, and no other source is identified on whole body imaging
  - After the above, AxLND can be performed.
- For a **recurrent primary site breast tumor in a patient who had a previous SLNB**, a repeat SLNB cannot be performed. So the patient will need management of the recurrent primary as well as **AxLND to manage the axilla.**

RADIATION

- Standard treatment is **whole breast irradiation (WBRT) after partial mastectomy for 3–6 weeks daily, to a total of 50 Gy** (5000 CGy = 5000 RAD). The radiation may be given before or after chemotherapy. Higher doses have a skin sparing effect.
- WBRT may also be used **after total mastectomy for a T3 or greater primary or N2 or greater status.**
- Nodal basin targeted irradiation can be added to AxLND.
- Alternatives to WBRT at the primary site, such as intraoperative radiation or postoperative brachytherapy via indwelling catheters, are less often employed.

CHEMOTHERAPY

- **Adjuvant chemotherapy** has traditionally been indicated for primary tumors **>1 cm in size or node positive.** Tumors may be hormone receptor positive or negative and HER2/neu positive or negative. It is given in 3-week cycles consisting of a combination of anthracycline (doxorubicin, epirubicin), taxane (docetaxel, paclitaxel), and an alkylating agent (cyclophosphamide, carboplatin).
- **Hormonal blockade** is used for tumors that are ER/PR positive. It decreases recurrence by 30%–40%, and the risk of a second primary ipsi- or contralaterally by 50%. This amounts to an approximately 5% risk of contralateral breast CA in 10 years. Hormonal blockade also decreases metastases.
  - If ER/PR positive and premenopausal or male, consider **tamoxifen.** This is a breast estrogen antagonist that elsewhere acts as a weak estrogen agonist. It may cause hot flashes or edema and has a 1% risk of thrombosis and a 0.1% risk of endometrial cancer. It may be used as single adjunct therapy (without chemotherapy) if the oncotype DX score is <18.
  - If ER/PR positive and postmenopausal, consider an **aromatase inhibitor.** This decreases circulating estrogen levels, but can cause osteopenia and arthralgias. It may be used as single adjunct therapy (without chemotherapy) if the oncotype DX score is <18.
  - Hormonal therapy is given for a minimum of 5 years, but can be extended.

    - The **Oncotype DX score** is used to determine any benefit of chemotherapy over hormonal therapy alone for **ER/PR+, HER2/neu negative, >1 cm tumors that are low nodal burden**.
    - Hormonal therapy is also used for ADH, ALH, DCIS, LCIS, patients with a high Gail risk, or those that are BRCA2 positive.
    - Ovarian suppression with **oophorectomy, irradiation, or GnRH-LH agonists** is another option in the premenopausal patient.
- **Trastuzumab** is given monthly for one year in HER2/neu positive tumors.
    - These tumors have tyrosine kinase receptor overexpression and account for 20% of breast cancers.
    - Trastuzumab improves overall survival by 40%, but is known for its **cardiac toxicity**.
    - It may be part of combination therapy to retard the development of resistance. Pertuzumab also binds HER2.
- Combination molecular therapy with non-hormonal targets can be considered if resistance has developed in stage 4 disease.
- **Neoadjuvant chemotherapy** should be considered for any patient who would otherwise qualify for adjuvant chemotherapy based on clinical staging alone.
    - Further consideration should be given if the patient is **<40 years old**, the primary is **T3 or T4, nodal status is N1 or greater** (including matted nodes), or the pathology demonstrates a **high Ki-67**, or **triple negative receptor status** or hormone receptor negative but **HER2 positive**.
    - One must obtain multiple CNBx of the primary, and any multifocal or multicentric sites, in order to have enough tissue to best organize the neoadjuvant regimen.
    - The regimen can be given for 12 weeks, with surgery scheduled 2–3 weeks after completion.
    - Post-treatment evaluation of pathologic response includes diagnostic mammography or MRI. Alteration of the regimen may be needed if there is a lack of response on serial imaging or biopsies.
    - Neoadjuvant chemotherapy can be **given in the 2nd or 3rd trimester of pregnancy** and thereby allow time for the patient to deliver her fetus and later undergo breast conserving therapy and WBRT.
    - Patients are **more likely to qualify for breast conserving surgery** if chemotherapy is given in neoadjuvant fashion.
    - After completion of neoadjuvant therapy, patients may undergo **SLNB,** whether they had a clinically negative axilla or a single positive node on initial presentation.
        - If SLNB is to be undertaken, both isosulfan and T99 colloid should be used to ensure an adequate number of sentinel nodes identified, since the false negative rate of a single agent will be higher post-chemotherapy.
        - If a patient had a clinically positive node originally, this node must be recovered in the post-neoadjuvant SLNB.
        - Any positive nodes on a post-neoadjuvant SLNB, including isolated tumor cells (ITC) or microscopic deposits, should be treated by AxLND with or without nodal irradiation.
        - If all sentinels are negative, then AxLND can be avoided.
- Hormone therapy alone has been used neoadjuvantly for weeks or months, with surgery scheduled thereafter.

RECURRENCE

- **Local recurrence** includes skin, nipple, fascia, muscle, bone, or residual breast tissue.
    - If within 3 cm of the original tumor or if it has the same receptor status, then this is a true recurrence.
    - If >3 cm or it has a different receptor profile, then it may be a new primary.
- **DIAGNOSIS**
    - All patients with suspected local recurrence need a **biopsy** of the site for confirmation, **US with FNA** to evaluate regional recurrence at ipsilateral or contralateral axillary, clavicular, or mammary nodal basins, as well as **systemic imaging** to evaluate distant recurrences.
- **TREATMENT**
    - If the patient has metastatic or regional disease, they will need **chemotherapy** before any consideration is given to resection.
    - If the workup is negative for regional or distant recurrence and the patient had not previously had an SLNB, then **dual tracer SLNB** can be employed to avoid an AxLND. However, SLNB in this scenario can have a higher false negative rate.
    - Consider **resection of the primary site and a limited number of bone, hepatic, or pulmonary metastases** if the patient is younger, had a long disease-free interval before recurrence, is ER/PR positive or HER2/neu positive.
        - **If the local recurrence is after a previous lumpectomy and WBRT,** then resection will need to be a mastectomy. Autologous reconstruction should be favored, since the patient will have incurred microvascular changes and soft tissue fibrosis from the prior WBRT. These increase the risk of skin flap necrosis and seroma formation.
        - **If the local recurrence is after a previous mastectomy,** then treatment is wide local excision and plastic surgery assistance for closure. **Adjuvant radiation** may be used provided the patient had not received it previously.

- **If a post-mastectomy recurrence is extensive**, perform chemotherapy and then WLE 4–6 weeks later, including ribs and chest wall if needed. This can be followed by adjuvant radiation provided the patient had not previously received it.

CONSIDERATIONS IN PREGNANCY

- **SLNB with T99 sulfur colloid** is acceptable, but isosulfan blue has an anaphylaxis risk.
- **No radiation** can be given during pregnancy, so patients generally need a total, rather than partial, mastectomy. Surgery is safest in the second trimester.
- If a diagnosis of breast cancer is made in the late 3rd trimester, partial mastectomy and WBRT can be done after delivery.
- Patients **cannot receive hormone blockade or trastuzumab**.
- **Chemotherapy** can be administered in the **2nd or 3rd trimesters** as adjuvant or neoadjuvant therapy.

OTHER BREAST MALIGNANCIES

- Lymphoma is usually B cell origin. Excision is an option only for select low-grade tumors. CHOP chemotherapy is the main therapy otherwise.
- Squamous cell carcinoma can result from metaplasia of ductal epithelium.
- Adenoid cystic carcinoma is similar to the salivary gland type, and can be treated in the same manner with excision and radiation.
- Breast sarcoma is similar to malignant fibrous histiocytoma (aka, pleomorphic unclassified type).

19.

THYROID

ANATOMY AND PHYSIOLOGY

- The thyroid gland derives from the first and second pharyngeal arches. The average weight is 20 grams, and the organ contains 90% of iodine in the body. Parathyroid glands derive from pouches.
- The **superior thyroid artery** branches off the external carotid and descends to perfuse the gland. Ligating this artery close to the thyroid parenchyma helps spare the **external branch of the superior laryngeal nerve (EBSLN)** running superior and medial to the artery.
- The EBSLN is motor to the cricothyroid muscle. Injury will result in loss of voice projection and easy fatigability. *See Figure 19.1.*
- The **internal branch of the superior laryngeal nerve (IBSLN)** enters the thyrohyoid membrane cephalad of the superior thyroid artery pedicle. The nerve provides sensation to the larynx.
- The **inferior thyroid artery** branches off the thyrocervical trunk and ascends to perfuse the thyroid and both parathyroid glands. Ligating this artery close to the thyroid parenchyma can help spare its anterior branch that provides parathyroid perfusion
- The **recurrent laryngeal nerve** (RLN) can also be spared with careful inferior thyroid artery ligation. *See Figure 19.2.* The RLN is motor to all of the laryngeal muscles except the cricothyroid muscle. Injury can result in a paretic cord with a baseline adducted position. Patients will have hoarseness or, if both RLN are transected, airway occlusion.
  - The RLN may be non-recurrent in 0.5%–2% of cases, more often on the right side and associated with an ipsilateral arterial abnormality.
- The **superior parathyroid gland** lies posterior and lateral to the RLN.
- The **inferior parathyroid gland** lies anterior and medial to the RLN.
- The pyramidal lobe is a superior extension of the central thyroid isthmus. It runs along the thyroglossal duct tract. An IMA artery, derived from the aortic arch, perfuses the isthmus in 1% of patients.
- The **ligament of Berry** is a posteromedial fibrous tissue extension anchoring the gland.
- **Tubercles of Zuckerkandl** are posterolateral extensions of the gland residing near each RLN.
- Superior and middle **thyroid veins** drain into the internal jugular veins. An inferior vein drains into the brachiocephalic veins on each side.
- **Parafollicular C cells** reside within the superior and posterior region of the thyroid gland and make **calcitonin,** which inhibits osteoclasts and decreases serum calcium levels.
- **Functional Regulation**
  - **Thyrotropin releasing hormone (TRH)** is produced by the hypothalamus and directs the anterior pituitary to release **Thyroid stimulating hormone (TSH)**.
  - TSH binds to the TSH receptor on thyroid epithelium and moves iodine through the cuboidal epithelium into a central colloid matrix. Stimulating antibodies to the TSH receptor are a feature of Grave's disease.
  - **Thyroid peroxidase (TPO)** facilitates binding of iodine to tyrosine residues on thyroglobulin (TG) molecules within the colloid matrix. This produces **mono- and di-iodotyrosines (MIT, DIT)**. Antibodies to TPO are seen in Hashimoto's disease.
    - DIT and MIT combine to produce **T3,** which is the active thyroid hormone in serum with a half-life of 1 day.
    - Two DITs can combine to produce T4. This is 15 times more common than T3, and has a half-life of 7 days.
  - T3 and T4 are taken back up by cuboidal epithelium via pinocytosis and transferred into the serum, where they bind to thyroid binding globulin (TBG) and circulate. T4 is converted to T3 in the periphery by deiodinases.
  - T3 enters cells and binds to **nuclear thyroid hormone receptors.** Alpha receptors are in the CNS, and beta receptors are in the liver. T3 also increases sympathetic beta receptors.
  - **Primary hyperthyroidism** is the presence of excessive T3 and free T4, with feedback inhibition of TSH production.
    - **Thyroid storm** is unregulated thyrotoxicosis marked by fever, tachycardia, altered mentation, and gastrointestinal symptoms. It usually occurs in the setting of underlying Graves' disease (see below). Treatment is with **B blockade, glucocorticoids (to**

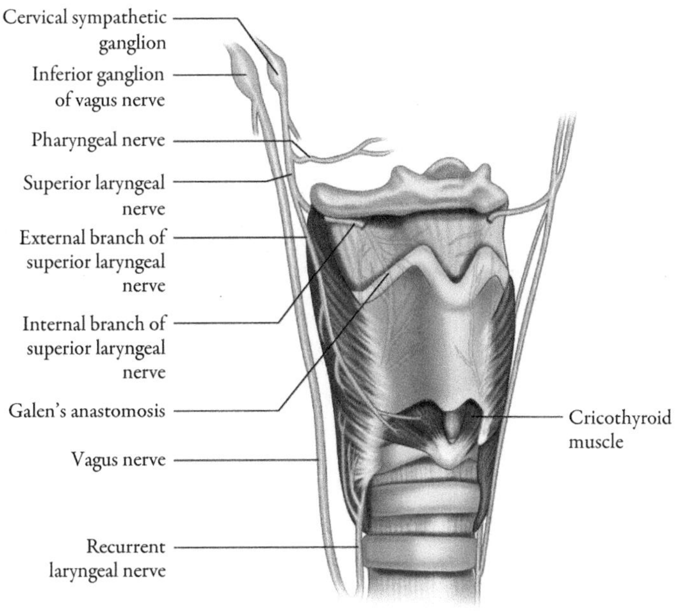

Figure 19.1 Laryngeal anatomy and innervation.

inhibit peripheral deiodinase activity), and Lugol's solution. The latter is a high-dose potassium iodide solution that paradoxically inhibits TSH activity. Less iodine is taken up by the thyroid epithelium, and so less T3 and T4 are made (Wolff-Chaikoff effect).
    - **T3 toxicosis** is a hyperthyroid patient with a normal free T4 but **increased T3**, possibly due to increased production or enhanced peripheral deiodinase conversion of T4.
- **Primary hypothyroidism** from glandular dysfunction is characterized by low circulating T3 and free T4, and high TSH representative of the hypothalamic and pituitary axis attempting to increase hormone production.
- **Secondary hypothyroidism** is much less common and results from hypothalamic or pituitary dysfunction. Serum evaluation will show a low TSH in addition to low T3 and low free T4.
- **Subclinical hypothyroidism** is a term used to describe a patient who appears clinically euthyroid with a normal free T4, but has a **high TSH.** Antibodies to TG, TPO, or TSH can be present. Patients may need levothyroxine supplementation prior to development of clinical hypothyroidism.
- **Sick Euthyroid** refers to a clinically hypothyroid patient who has low total and free T3 levels with **near normal T4 and TSH.** This could be due to decreased gland responsiveness to TSH or poor conversion of T4 to T3 in the setting of critical illness, starvation, trauma, or the postoperative state.

SINGLE THYROID NODULE

- Nodules are more common in females. 90% of thyroid nodules in adults will be **benign**. Children have a higher rate of malignancy.
- A sudden change in size or symptoms may be due to intra-nodular hemorrhage.
- A **family history** of thyroid cancer or a history of **6.5–2000 cGy radiation** both increase papillary cancer risk substantially. If the patient was exposed to >2000 cGy radiation, this level actually decreases the risk of papillary cancer due to epithelial destruction.
- **DIAGNOSIS**
    - Based on thyroid function tests (TFT), ultrasound (US), and fine needle aspiration (FNA).
    - **US** features that suggest malignancy include a heterogeneous, irregular, **hypoechoic mass,** with **internal calcifications. Growth in the anteroposterior diameter** increases the likelihood that the nodule is malignant. *See Figure 19.2.* Peripheral calcifications are indicative of a benign nodule.
        - If the lesion is symptomatic and cystic, drain the fluid. If the **cyst fluid is bloody, then a lobectomy** should be performed to exclude malignancy. If it was **serous, then one may follow clinically** and perform a lobectomy only if the serous fluid re-accumulates.
        - If the fluid aspirated is **colloid,** then the diagnosis is colloid goiter. Treatment includes supplemental **iodine and T4.** One can follow thyroglobulin (TG) levels and serial US.

BTA U-classification	Thyroid ultrasound & description	Fine needle aspiration cytology (FNAC)
U2 (a) – Benign: halo, iso-echoic/ mildly hyper-echoic	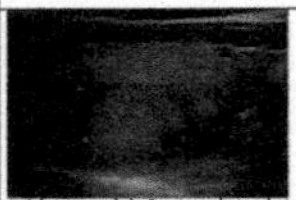 a. A benign nodule. It is iso-echoic relative to the thyroid, and surrounded by a hypoechogenic halo.	Not required*
U2 (b) – Benign: cystic change +/– ring down sign (colloid)	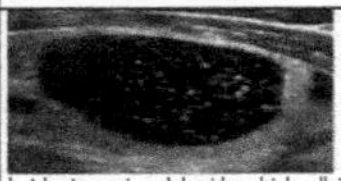 b. A benign cystic nodule with multiple colloids, which are seen as hyper-echoic spots with comet -tail.	Not required*
U2 (c) – Benign: Micro-cystic/ spongiform	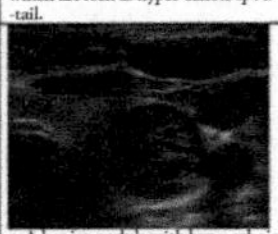 c. A benign nodule with hypo-echoic cystic spaces resulting in a spongiform or honeycomb appearance.	Not required*
U2 (d & e) – Benign: Peripheral egg shell calcification	 d. A benign nodule with eggshell calcification. Note the acoustic shadowing produced by the calcific ring. 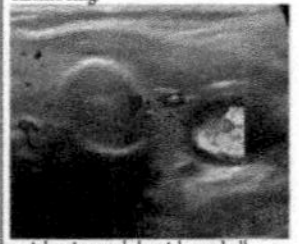 e. A benign nodule with eggshell calcification.	Not required*
U2 (f) – Benign: Peripheral vascularity	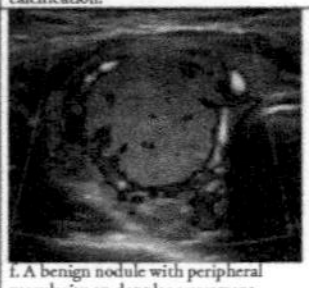 f. A benign nodule with peripheral vascularity on doppler assessment.	Not required*

BTA U-classification	Thyroid ultrasound & description	Fine needle aspiration cytology (FNAC)
U3 (a) – Indeterminate: Homogenous, hyper-echoic (markedly), solid, halo (follicular lesion)	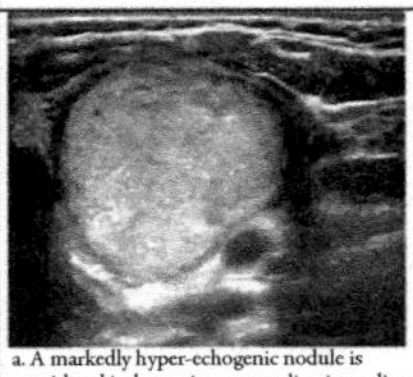 a. A markedly hyper-echogenic nodule is considered indeterminate regarding its malignant risk.	Required
U3 (b) – Indeterminate: ?Hypo-echoic, equivocal echogenic foci, cystic change	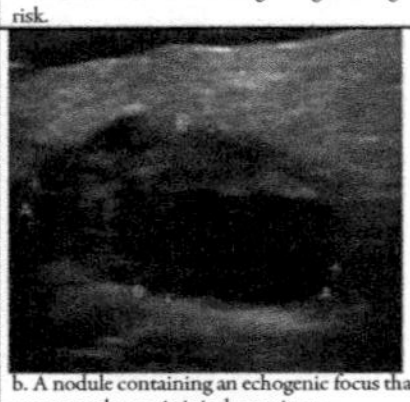 b. A nodule containing an echogenic focus that appears to be cystic is indeterminate.	Required
U3 (c) – Indeterminate: Mixed vascularity	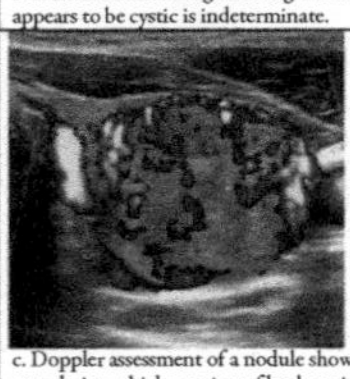 c. Doppler assessment of a nodule showing mixed vascularity, which consists of both peripheral and intra-nodular vasculature.	Required

BTA U-classification	Thyroid ultrasound & description	Fine needle aspiration cytology (FNAC)
U4 (a) – Suspicious: Solid, hypo-echoic (cf thyroid)	 a. A suspicious hypo-echoic nodule with signal lower than the surrounding thyroid tissue but higher than the strap muscle above.	Required
U4 (b) – Suspicious: Solid, very hypo-echoic (cf strap muscle)	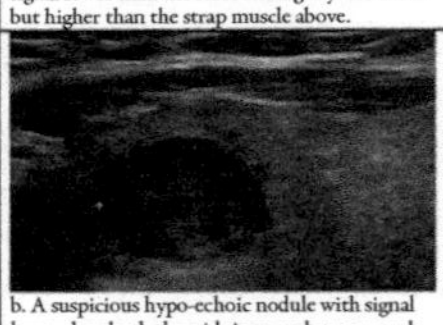 b. A suspicious hypo-echoic nodule with signal lower than both thyroid tissue and strap muscle.	Required
U4 (c) – Suspicious: Disrupted peripheral calcification, hypo-echoic	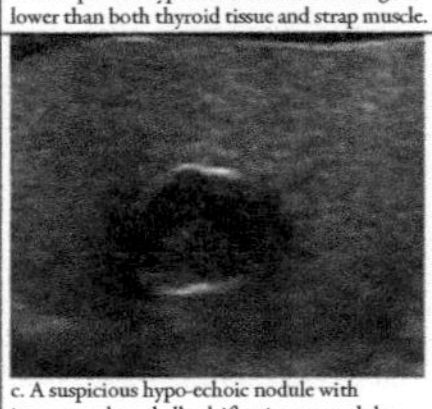 c. A suspicious hypo-echoic nodule with interrupted eggshell calcification around the edges.	Required
U4 (d) – Suspicious: Lobulated outline	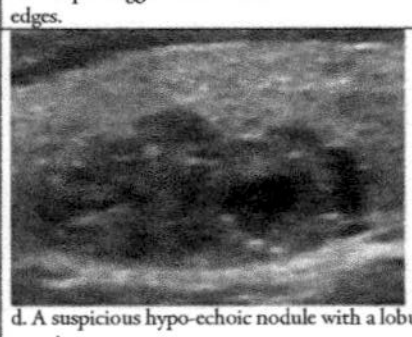 d. A suspicious hypo-echoic nodule with a lobular margin.	Required

BTA U-classification	Thyroid ultrasound & description	Fine needle aspiration cytology (FNAC)
U5 (a) – Malignant: Solid, hypo-echoic, lobulated/ irregular outline, micro-calcification (?Papillary carcinoma)	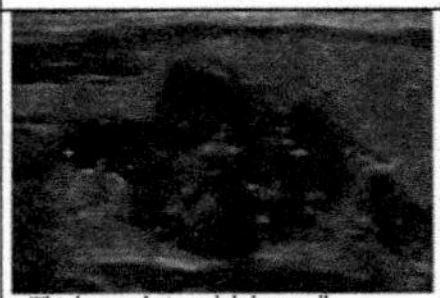 a. This hypo-echoic nodule has small hyper-echoic foci of calcification and an irregular lobulated contour. FNAC confirmed papillary thyroid cancer.	Required
U5 (b) – Malignant: Solid, hypo-echoic, lobulated/ irregular outline, globular calcification (?Medullary carcinoma)	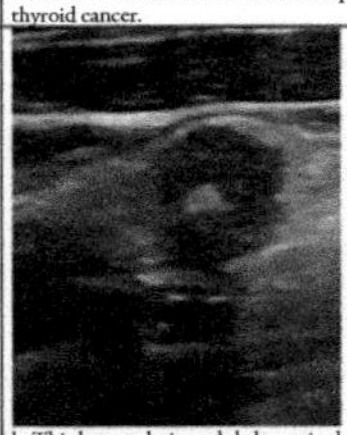 b. This hypo-echoic nodule has a single coarse globular calcification and an irregular contour. FNAC confirmed medullary thyroid cancer.	Required
U5 (c) – Malignant: Intra-nodular vascularity	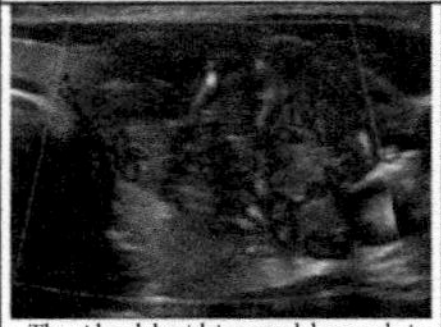 c. Thyroid nodule with intra-nodular vascularity. Later confirmed to be papillary thyroid cancer.	Required
U5 (d) – Malignant: Shape (taller > wide)	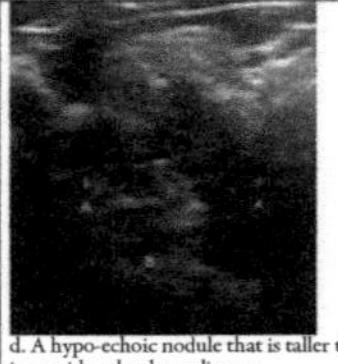 d. A hypo-echoic nodule that is taller than wide is considered to be malignant.	Required

Figure 19.2 Sonographic appearance of thyroid nodules.

- **FNA** is indicated for any nodule **>1 cm or any nodule <1 cm with atypical features, family history of thyroid cancer, or history of radiation**. Nodules <1 cm that do not fit these criteria may be followed with US. Bethesda categories of FNA results include the following:
  - **1. Nondiagnostic:** Repeat FNA and US should be done at 6–18 months.
  - **2. Benign:** If normal thyroid tissue is seen on FNA, it may be a solitary hot adenoma producing T4 and T3. TFTs should confirm this.
  - **3. Atypical cells: Repeat FNA with molecular testing.** There is a 5%–15% CA risk.
  - **4. Follicular cells or Hurthle cells:** This finding may indicate hyperplasia, a benign adenoma, or a malignant tumor in 15%–30% of cases. **Thyroid lobectomy or molecular testing** of the sample are indicated. Evidence of invasion on the lobectomy specimen confirms malignancy.
  - **5: Suspicious for cancer.** Lobectomy or thyroidectomy are indicated.
  - **6: Definite cancer.** Proceed with definitive surgical or medical therapy.
- **Core needle biopsy (CNBx)** should be considered if there is concern for unresectable, anaplastic, or non-thyroid malignancy (lymphoma or metastases). In these cases, thyroid surgery will not be curative, so the core needle specimen is done to provide more tissue for planning medical management with chemotherapy or biologic agents.
- **TFTs** help determine if the nodule is hormonally active.
  - Decreased TSH secondary to increased T4 and T3 production would indicate the patient is hyperthyroid.
  - The next step is to perform a **radionuclide I-123** to determine if the nodule itself is the source of hormones, or if the nodule is actually cold and the remainder of the gland is overactive.
- BRAF testing of a specimen can refine diagnostic accuracy for papillary thyroid cancer.

- **TREATMENT**
  - **Benign hot nodules** may either be **monitored** serially if the patient is asymptomatic, or treated medically with **methimazole, propylthiouracil (PTU), or I-131**. Lobectomy is reserved for failure of medical therapy to control hormone production and symptoms. Follow-up serial US and TFTs are indicated regardless of whether the nodule is treated.
  - **Cold nodules** may need an empiric lobectomy if there is concern for malignancy based on the US features, FNA results, family history or personal history of radiation, or growth rate. Benign large cold nodules may also undergo lobectomy for cosmetic reasons.
    - **Malignant nodules** are treated based on the type of malignancy (see individual sections below). Options range from a simple lobectomy for isolated papillary tumors <4 cm or follicular cancers <1 cm, to a subtotal or total thyroidectomy with or without selective or modified radical lymph node dissection at levels 2–6. Lymph node harvest depends on the type of malignancy, size of the primary, and the presence or absence of biopsy proven positive nodes (see the malignancy section below).
  - Postoperative calcium supplementation (1000–3000 mg) is indicated. Postoperative Vitamin D supplementation can be considered for any intra-op decrease in PTH level by >70%.

GOITER

- **DIFFUSE NONTOXIC (COLLOID) GOITER**
  - Due to iodine deficiency. This results in a buildup of colloid with the inability to construct hormone. An FNA will show colloid. Treatment is with **iodine and T4.** Follow-up TG levels and serial US for any nodule(s) noted.
- **MULTINODULAR NONTOXIC GOITER.**
  - Nodules should be evaluated based on US characteristics and may require FNA if >1 cm, suspicious features, or if the patient has risk factors for malignancy.
- **DIFFUSE TOXIC GOITER (GRAVES' DISEASE)**
  - **TSH receptor activating IgG antibodies** result in hyperthyroidism. Histology depicts columnar thyroid epithelium with decreased colloid.
- **DIAGNOSIS**
  - Patients are generally female and suffer from atrial fibrillation, heat intolerance, weight loss, palpitations, diaphoresis, gastrointestinal symptoms, pre-tibial edema from glycosaminoglycan deposition, and skin pigmentation.
  - They may have an orbitopathy and exophthalmos due to antigenic similarities between the TSH receptors and orbital fibroblasts and ocular muscles.
  - **TSH is low, and T3 and T4 are elevated.** There is **diffuse uptake** on an I-123 scan.
- **TREATMENT**
  - Starts with medical management. **Propranolol** 20–40 mg QID helps control sympathomimetic symptoms. **Methimazole** will inhibit TPO and new hormone production. **Propylthiouracil (PTU)** is preferred for pregnant patients. Medical management can take weeks or months for maximal inhibition, and is associated with an increased recurrence risk compared to other treatment modalities.
    - **Radioactive I-131** can destroy overactive thyroid epithelium. This is very effective, but contraindicated in children, pregnant patients, and those with severe

orbitopathy. Permanent hypothyroidism is a long-term side effect.
    - **Total thyroidectomy** can be undertaken after preoperative methimazole, beta-blockade, and potassium iodide for 2 weeks. This is reserved for patients who have failed medical management or are noncompliant, or those who cannot receive radioactive I-131 therapy (above). Also, patients with concomitant suspicious nodules or an extensive goiter with mass effect, or airway compression, can be considered for thyroidectomy.
- **TOXIC MULTINODULAR GOITER**
    - Seen in elderly women with hyperplasia secondary to **chronic low-grade TSH stimulation.** Multiple nodules will show uptake on a radioactive thyroid scan.
    - Management is **thyroidectomy**, since the patient is hyperthyroid *and* has potential for nodular neoplasia. **Radioactive I-131** is an option for the elderly or frail patient who is not a good candidate for thyroid resection.
- **SUBSTERNAL GOITER**
    - The majority are extensions of a normal cervical thyroid and are perfused by the superior and inferior thyroid arteries. They can be left alone if asymptomatic. In the rare case of a primary substernal goiter perfused by vessels from the brachiocephalic artery, a thyroidectomy should be performed.

THYROIDITIS

- **DRUG INDUCED THYROIDITIS**
    - **Amiodarone** can cause thyrotoxicosis from follicular cell destruction and secondary increased hormone release. Subsequent transient or permanent hypothyroidism can result.
        - Management is with **glucocorticoids.** A total thyroidectomy is indicated for persistent thyrotoxicosis in a patient who cannot be weaned from amiodarone and fails glucocorticoid treatment.
        - Hyperthyroidism can occasionally be from increased production due to the iodine content in amiodarone. This is less common, Treatment is with methimazole,
    - Other causative drugs include lithium, interferon-alpha, IL-2, or tyrosine kinase inhibitors.
- **ACUTE SUPPURATIVE BACTERIAL THYROIDITIS**
    - Due to **hematogenous inoculation or direct spread** of staph or strep into the thyroid gland. Some patients have an anatomic persistent connection of the thyroid with the pyriform sinus that can be seen by **endoscopy**.
    - Patients present with fever, tenderness, and dysphagia. Management is with **antibiotics, aspiration, or incision and drainage.**
- **SUBACUTE GRANULOMATOUS THYROIDITIS**
    - A cellular inflammatory response following a viral upper respiratory tract infection (**URI**).
    - Patients are generally female and present with a **tender gland, malaise, and an elevated erythrocyte sedimentation rate (ESR).** Initial hyperthyroid symptoms can last for 4–6 wks and are due to release of preformed hormone from inflamed thyroid epithelium. Up to 6 months of hypothyroidism can follow.
    - Management includes with **nonsteroidal anti-inflammatory medication (NSAIDs), aspirin (ASA), and glucocorticoids** if refractory. **Beta-blockade** can help with symptoms. Since no new hormones are being made, therapy with anti-TPO medications or I-131 is not effective.
- **SUBACUTE LYMPHOCYTIC THYROIDITIS**
    - Associated with anti-TPO and anti-TG antibodies in a **postpartum female.**
    - Patients have an initial **asymptomatic enlargement of the gland** with subsequent hyperthyroidism due to increased release of preformed thyroid hormone from inflamed epithelium for 2–4 months. This is followed by up to 6–12 months of hypothyroidism. 80% return to the euthyroid state eventually.
    - Management includes **beta blockers f**or symptoms. Hormone replacement may be needed for those with persistent hypothyroidism.
- **CHRONIC LYMPHOCYTIC THYROIDITIS (HASHIMOTO'S THYROIDITIS)**
    - The most common cause of hypothyroidism and goiter in the United States. It is characterized by **chronic lymphocytic** inflammation with **anti-TPO antibodies.** This results in decreased hormone production and increased pituitary TSH in response. There is an increased risk of lymphoma and possibly papillary thyroid cancer.
    - **DIAGNOSIS**
        - Patients are generally 30- to 50-year-old females, and may have a history of **childhood neck irradiation.** Confirmation is by anti-TPO, anti-microsomal, anti-TG, or anti-TSH antibodies.
    - **TREATMENT**
        - Start with **T4 replacement.** Symptoms may improve after 3–6 months. Rarely a partial thyroidectomy is employed for compressive symptoms or a suspicious nodule.
- **CHRONIC FIBROSING THYROIDITIS (REIDEL'S THYROIDITIS)**
    - Diffuse glandular fibrosis possibly accompanied by airway compression, involvement of strap muscles,

and carotid artery encasement. It is associated with primary sclerosis cholangitis (PSC), retroperitoneal fibrosis (RPF), and methysergide.

- **DIAGNOSIS**
  - Patients may complain of a firm, enlarging neck mass. **Open surgical biopsy** is required, since FNA and CNBx are often nondiagnostic.
- **TREATMENT**
  - Start with **glucocorticoids, tamoxifen, immunosuppression, and T4 supplementation.** An isthmectomy with or without a tracheostomy may be needed for compressive symptoms and airway control. The prognosis is poor.

ABNORMALITIES OF DESCENT

- **LINGUAL THYROID OR ECTOPIC THYROID**
  - Represents persistent tissue at the base of the tongue (foramen cecum). There is a 2% malignancy risk.
  - **DIAGNOSIS**
    - Symptoms include dysphagia, dyspnea, dysphonia, or an asymptomatic midline mass in a child.
    - In some patients this may be the only thyroid tissue, so if it is not clear by physical exam, a neck US and I-123 can help determine if a normal thyroid is present.
  - **TREATMENT**
    - **T4 suppression and I-131.** Resection is for patients in whom medical therapy fails, or there is a concern for malignancy.
- **THYROGLOSSAL DUCT CYST**
  - A remnant fluid-filled cavity along the pathway of thyroid gland descent. 1% of cysts harbor papillary cancer.
  - **DIAGNOSIS**
    - Patients may note that the cyst moves up and down with swallowing. Consider US or thyroid scan to confirm that a normal thyroid gland is present.
  - **TREATMENT**
    - Sistrunk procedure, which removes the cyst plus the midpoint, or all, of the hyoid bone. There is a 10% recurrence rate.

MALIGNANCY

- Thyroid cancer is the most common endocrine malignancy.
- Risk factors include male sex, age >50, history of low-dose childhood radiation, and a family history or known MEN2a or 2b syndrome.
- Patients present with a solid, solitary, cold, enlargening nodule. Accompanying dysphonia or dysphagia can be present. Workup includes a **neck US with FNA of a nodule >1 cm or nodules <1 cm with suspicious features, a family history of thyroid malignancy, or a personal history of radiation.**
  - FNA results may fall into one of five categories (see single thyroid nodule section above).
  - Suspicious neck lymph nodes may also undergo FNA, suggesting the diagnosis of an underlying thyroid cancer.
- **PAPILLARY CANCER**
  - Accounts for 85% of all thyroid cancer. Tumors are often multi-centric, especially if the patient has had radiation exposure or a family history of thyroid cancer. Papillary cancer commonly spreads to lymph nodes.
  - **DIAGNOSIS**
    - FNA of a thyroid nodule showing **Psammoma bodies** (calcified clumps of cells) with **Orphan Annie nuclei.**
    - FNA of a lateral neck lymph node showing ectopic thyroid tissue probably represents papillary cancer spread from the ipsilateral thyroid lobe.
    - Systemic imaging may be considered for locally advanced disease.
  - **TREATMENT**
    - **Lobectomy without I-131** is adequate for tumors <4 cm in size provided the patient is clinically node negative, without a family history of thyroid malignancy, and no history of radiation.
    - If the patient does not meet these criteria, then a **total thyroidectomy with or without a prophylactic *central* neck LN dissection** (CND) will be required.
    - Add a **radical neck LN dissection** if the primary is T3 or T4, there are clinically positive nodes on FNA, or there is extrathyroidal or extracapsular tumor extension.
    - **Radioactive iodine (I-131)** can be given 4–6 weeks later.
      - One must stop T4 supplementation for 6 weeks and start T3 instead.
      - T3 can be stopped 2weeks prior, and rTSH stimulation may be added.
      - I-131 can then be given, with the above interventions enhancing TSH release and hence increasing radioactive iodine uptake by thyroid cells.
      - Side effects include sialadenitis, bone marrow changes, leukemia, gastrointestinal symptoms, infertility, or parathyroid dysfunction.
      - One may give T4 to suppress future TSH production after I-131 therapy is complete.
    - **Radiation** is an option for unresectable disease, or cases non-responsive to I-131
    - There is a **10%–30% recurrence rate in the central or lateral neck.**
      - Follow **TG and anti-TG levels** to diagnose a recurrence. One may withdraw T4 and stimulate TSH prior to drawing these levels to increase sensitivity.

- One may also follow TSH and **US q6 months.** Radioactive iodine 123 scan or PET scan can localize the source of TG if an US is not diagnostic.
- A suspicious recurrent lesion can undergo **FNA and TG testing.**
- Prognosis is based on age and local invasion. For example, patients who are <45 years of age with metastases are at most stage 2. Those >45 years old with a tumor 2–4 cm in size are also stage 2.

- **FOLLICULAR CANCER**
  - The second most common thyroid malignancy. It is often unifocal, and spreads via a hematogenous rather than lymphatic route. Bone metastases can be seen.
- **DIAGNOSIS**
  - FNA of a thyroid nodule will show follicular cells (category 4). However these may represent hyperplasia, adenoma, or a true follicular malignancy (10%).
  - Distinction can be made by either **molecular testing** of sample cells to determine neoplasia or a **lobectomy** with frozen section for the definitive diagnosis.
  - For locally advanced cases, systemic imaging can be considered to exclude metastases
- **TREATMENT**
  - If a lobectomy specimen shows only hyperplasia, adenoma, or a **follicular malignancy in a nodule < 1cm, no further treatment beyond lobectomy is necessary.**
  - If a follicular cancer is **>1cm, treatment is with total thyroidectomy and i131** 4–6 wks later
  - Lymph node spread is uncommon, but a radical neck LN dissection would be done for clinically positive nodes or extrathyroidal or extracapsular tumor extension
  - Follow TG levels to screen for a recurrence and US to evaluate thereafter. There are **10–30% recurrence rates** in the central or lateral neck.
- **MEDULLARY THYROID CANCER**
  - Originates from parafollicular C cells that produce calcitonin. Excess calcitonin can lead to **diarrhea,** and occasionally flushing. Sporadic cases are more common than genetic syndromes like MEN 2a, 2b or familial medullary thyroid cancer. Medullary cancers are more aggressive than papillary or follicular ones, so early lymph node involvement is common, and metastases to lung, liver, & bone can be seen.
- **DIAGNOSIS**
  - FNA of a thyroid nodule showing medullary cancer and amyloid deposits. This is usually accompanied by **high calcitonin levels.**
  - A lateral neck US and FNA should be performed for any lateral compartment nodes
  - A very high calcitonin level or the presence of many cervical nodes should prompt **CT chest and abdomen to evaluate for metastases.**
  - Screening calcitonin levels can be obtained in patients genetically predisposed to medullary cancer
  - All confirmed cases of medullary thyroid cancer should be screened for MEN2a and 2b, Patients with either MEN 2a or 2b may have multiple, bilateral medullary tumors. Pheochromocytoma, if found, must be resected before addressing thyroid or parathyroid pathology
    - 2A: **Pheochromocytoma** can be present in up to 40%. Test for the **RET proto-oncogene** or check 24 hour urine normetanephrine levels. **Hyperparathyroidism** may be present in up to 30%. If MEN 2a is discovered early in life, then a prophylactic thyroidectomy and central neck dissection (CND) should be completed by age 5
    - 2B: **Pheochromocytoma** may be present in up to 40%. If diagnosed early in life, prophylactic thyroidectomy with CND should be completed by age 1
- **TREATMENT**
  - **Total thyroidectomy and central neck LN dissection (CND).**
  - Lateral neck dissection is reserved for FNA positive nodes in these compartments, or in some instances of a large primary or extensive central nodal disease burden
  - **Parathyroidectomy** should be done as well if the patient is hypercalcemic. One may remove and **autotransplant parathyroid glands to the forearm** in patients with **MEN2a.** Any subsequent hyperparathyroidism can be addressed by re-excising the transplanted tissue without having to go back into the neck.
    - One may **autotransplant inadvertently devascularized parathyroids into the sternocleidomastoid (SCM)** if the patient does not have MEN2a. These patients are less likely to develop hyperparathyroidism and need explantation.
  - **Radiation** therapy can be used for unresectable disease. Radioactive i131 will not affect medullary cancer.
  - Follow **calcitonin and CEA** serially for recurrence surveillance.
- **HURTHLE CELL CANCER**
  - **These are** multifocal and spread to lymph nodes as well as hematogenously to lung and bone. Diagnosis begins with FNA of a thyroid nodule showing Hurthle cells, then a lobectomy is needed for a definitive diagnosis of cancer. Hyperplasia and adenoma on the lobectomy specimen require no further treatment. A **total thyroidectomy** is necessary for confirmed Hurthle cell malignancy. A LN dissection is added for any clinically positive nodes. Radiation is employed for unresectable lesions.
- **ANAPLASTIC CARCINOMA**
  - Presents in elderly patients with a long-standing goiter. FNA can show only necrosis. **CNBx** or isthmectomy may be needed for a definitive diagnosis. **Management** is with a palliative partial thyroidectomy or tracheostomy for compressive

symptoms. Otherwise chemotherapy and radiation are the main therapies.

- **NON-HODGKIN'S B CELL LYMPHOMA**
  - May be secondary to Hashimoto's thyroiditis. Diagnosis is by FNA with flow cytometry, CNBx, or open biopsy and isthmectomy. Treatment is with chemotherapy and radiation.

THYROIDECTOMY PRINCIPLES

- Indications include papillary carcinoma > 4cm, any other thyroid malignancy >1 cm, a substernal goiter, goiter non responsive to medical therapy or with compressive symptoms, multi nodular goiter or a single large compressing nodule, or concerns over cosmesis.
- **Pre-op vocal cord visualization** with a fiberoptic scope is recommended to document a baseline asymptomatic cord paresis before re-operative surgery or undertaking a resection of a large mass.
- A standard incision is 6cm transversely at a cervical skin crease. A minimally invasive incision is 3–5cm and a video-assisted minimally invasive incision is < 3cm. A robotic assisted face-lift thyroidectomy is via a posterior auricular incision.
- The inferior and superior vascular pedicles should be ligated close to the capsule to avoid parathyroid ischemia or EBSLN injury, respectively.
- **Intraoperative recurrent laryngeal nerve (RLN) monitoring** is by ETT or surface electrodes. *See Figure 19.3.*
  - The **vagus nerve should be stimulated periodically within the surgical field** to note activation of the ipsilateral cord via the ETT or cutaneous monitoring. This ensures the entire RLN is intact. If one only stimulates distally along the RLN, then a more proximal injury of the RLN or vagus could be missed.
  - RLN monitoring is proven beneficial for **reoperations** and has a high negative predictive value for an uninjured nerve. However it has a **low positive predictive value** for injury since lack of stimulation of the cord may just be traction paresis.
- Postoperative hematoma usually develops in the first 6 hours after surgery. The strap muscles should be approximated loosely so any hematoma can decompress into the soft tissue layer, rather than compress the trachea. Hematomas with airway compromise should be evacuated immediately.
- **Transient hypocalcemia** is the most common complication. Paresthesias are the earliest symptoms of hypocalcemia. An intraoperative decrease in the PTH level by >70% or a postop PTH< 30 should prompt

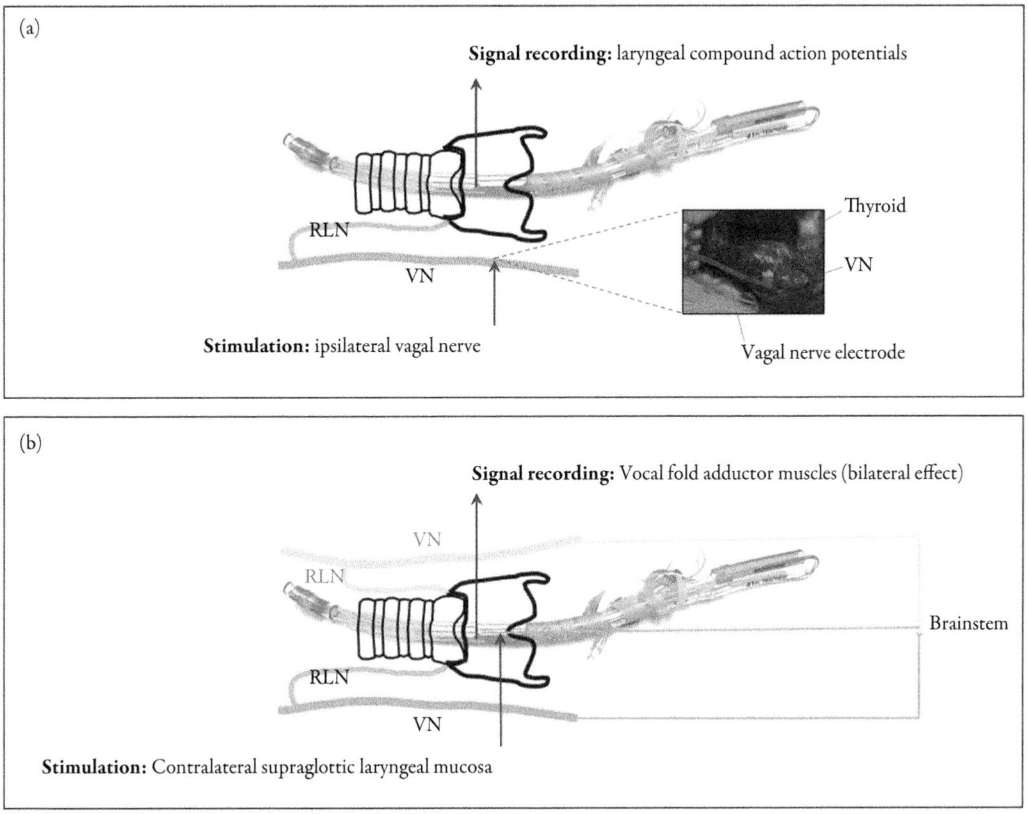

Figure 19.3 Intraoperative recurrent laryngeal nerve monitoring.

supplementation with calcium (1000 - 3000mg) and occasionally vitamin D.

- Inadvertently devascularized parathyroid glands can be autoimplanted if intra-op PTH and lack of adequate other glands suggest the patient may have postoperative hypoparathyroidism. The implant will take weeks to function, so the patient will need to remain on calcium and vitamin D supplementation until then.
- RLN injury is seen in 2–5% of cases, generally leaving the vocal cord in a **baseline Adducted position** without the ability for any further adduction during phonation or abduction during inspiration. Management options include observation, cord injection, voice therapy, or medialization of the cord. Nerve repair can be undertaken at 3–6 months to maintain bulk to the cord.

20.

PARATHYROID

ANATOMY AND PHYSIOLOGY

- **Superior parathyroids** derive from the 4th pharyngeal pouch, like the thyroid gland itself. They sit on the undersurface of the thyroid gland **posterolateral to the recurrent laryngeal nerve (RLN),** above or below the inferior thyroid artery. Ectopic locations include the tracheoesophageal (TE) groove, para- or retroesophageal space, or the carotid sheath.
- **Inferior parathyroids** derive from the 3rd pharyngeal pouch, like the thymus. They sit on the undersurface of the thyroid gland **anteromedial to the RLN** and below the inferior thyroid artery. They are often ectopic and can be located in the **tail of the thymus, within the anterior mediastinum, or reside intrathyroidal.**
- Both inferior and superior parathyroid glands are perfused by the inferior parathyroid arteries. *See Figure 20.1.*
- **Chief cells make parathyroid hormone (PTH)** with a half-life of 2–4 minutes. The normal level is 5–40. PTH increases calcium (Ca) reabsorption and phosphate loss in the distal convoluted tubule (DCT) of the nephron, increases osteoclast activity in bone, and enhances 1 hydroxylation of 25 OH vitamin D_3 (calcidiol) in the kidney. 1, 25 OH vitamin D_3 (calcitriol) in turn promotes intestinal Ca absorption. The net effect is to elevate serum calcium.
- **Calcitonin** is produced by parafollicular C cells of the thyroid gland, inhibits osteoclasts, and increases excretion of Ca and phosphates in the DCT.
- **Pseudohypoparathyroidism** describes high circulating PTH with nonresponsive end organ receptors.

HYPERCALCEMIA

- 99% of total body calcium resides within bone. 1% is in serum. Half of serum calcium is ionized, and the other half is bound to albumin, citrate, or phosphate.
- **Malignancy** is the primary diagnosis when hypercalcemia is discovered in an inpatient setting. It results from bone lysis secondary to metastases or from increased PTH related peptide (PTHrP) production in squamous cell carcinoma (SCC) or breast cancer.
- **Hyperparathyroidism** is the primary diagnosis when hypercalcemia is discovered in an outpatient setting. This includes adenoma and four gland hyperplasia.
- Other causes of hypercalcemia include hyperthyroidism, familial hypocalciuric hypercalcemia (FHH), immobilization, excess vitamin D_2 or D_3, milk-alkali syndrome, thiazide diuretics, or granulomatous disease with increased 1'-OH activity upon 25 OH vitamin D_3.
- Many patients are asymptomatic. If they do occur, symptoms include fatigue, thirst, nausea; abdominal, bone, and muscle pain; fractures, constipation, peptic ulcer disease, cholelithiasis, nephrolithiasis, pancreatitis, confusion, and depression.
- Management of hypercalcemia includes rehydration with intravenous fluids followed by bisphosphonates and calcitonin. Furosemide may be added once euvolemic. Mithramycin inhibits osteoclasts and can also lower serum calcium.
- **Renal leak hypercalciuria** results in low serum Ca levels and increased PTH with its secondary effects. Treatment is with thiazide diuretics to increase Ca reabsorption along the nephron, maintain serum calcium levels, and thereby decrease PTH production.

PRIMARY HYPERPARATHYROIDISM

- It is most commonly seen in older women or patients who received neck radiation 30–40 years prior. Etiologies include:
  - A nonpalpable **adenoma (85%)**. This risk is increased years after **neck radiation.**
  - **Four gland hyperplasia (10%)**. Some patients have **underlying MEN 1 or 2a**, another **familial syndrome**, or prior **lithium therapy**.
  - Double adenoma (4%).
  - Malignancy (1%).
- **DIAGNOSIS**
  - Often there is increased calcium as well as PTH levels in asymptomatic or symptomatic patients.
  - PTH causes a **hyperchloremic metabolic acidosis** by enhancing Cl absorption across the nephron while promoting bicarbonate loss. Urine phosphate and urine cAMP are also increased.

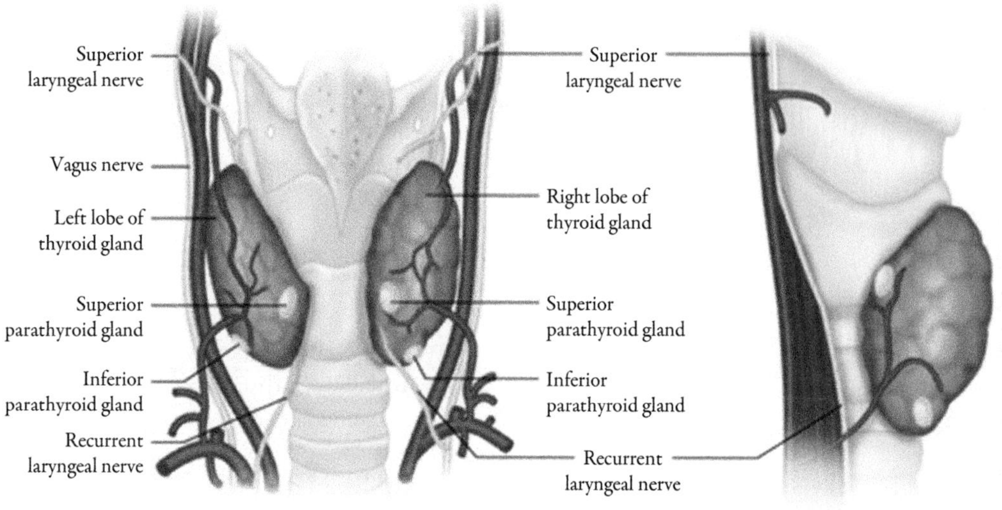

Figure 20.1 Parathyroid anatomy and perfusion - posterior and lateral views.

- Osteitis fibrosa cystica is a term describing increased osteoclast activity secondary to hyperparathyroidism. Subperiosteal resorption at middle phalanges can be seen on x-ray.
- **Secondary hyperparathyroidism** is when the parathyroid glands are reacting to low serum calcium from another etiology like vitamin D_3 deficiency. This can be excluded by checking 25-OH vitamin D_3 levels and renal function.
- **Familial hypocalciuric hypercalcemia (FHH)** should be excluded with a 24 hours urine Ca. Levels >400 are seen in primary hyperparathyroidism, while they are <100 in FHH due to a hyperactive Ca receptor in the DCT of the nephron enhancing reabsorption.
- Other alternative diagnoses to consider include a thoracic tumor, renal tumor, and multiple myeloma.
- The next step for confirmation is imaging.
  - **Tc99m sestamibi scan.** The tracer is taken up by thyroid and parathyroid tissue, but remains in the parathyroid glands longer. This can show if one side harbors an adenoma that retains the tracer longer or more avidly.
  - **Neck ultrasound (US)** allows visualization of parathyroid glands as oval, hypoechoic structures on the posterior surface of the thyroid. It is useful for localization of an adenoma or suggesting four gland hyperplasia.
  - CT or MRI can be added if the sestamibi scan and US are not congruent.
  - Venous sampling from each internal jugular vein can help with lateralization of the source of parathyroid hormone, should the above studies remain inconclusive.
  - Fine needle aspiration (FNA) of a suspected adenoma can prove that PTH is being produced.
- **TREATMENT**
  - Indications for surgery include: **age <50, glomerular filtration rate (GFR) <50, nephrolithiasis, or decreased bone mineral density on radius, spine, or hip x-rays.** Surgery will slow the loss of bone density.
  - If pregnant, **2nd trimester surgery is ideal** and can prevent neonatal hypocalcemia.
  - **Unilateral exploration** is preferred if preoperative imaging or venous sampling have identified an adenoma.
    - The patient's head should be elevated and neck extended. A transverse lower cervical incision is made, the platysma is divided, and the sub-platysmal plane dissected toward the suprasternal notch inferiorly and thyroid cartilage superiorly. Blunt retraction of the sternohyoid and sternothyroid off of the thyroid gland will allow for medial rotation of the gland pole.
    - Any adenoma should be resected. The intraoperative PTH baseline level should be noted, and a post-excision level should **drop by 50% in 10 or 20 minutes after resection.** Frozen section of a specimen can also confirm it is parathyroid tissue.
    - Alternatively, one may use a lateral incision between the anterior border of the sternocleidomastoid (SCM) and strap muscles to enter the tracheoesophageal (TE) groove and directly access a superior, posteriorly positioned gland.
  - **Bilateral exploration** to exclude hyperplasia or multiple adenomas is recommended if:
    - 1. Preoperative imaging fails to localize an adenoma.
    - 2. The intraoperative PTH level does not drop after a suspected adenoma is resected; or
    - 3. The patient has MEN 1, 2a, lithium exposure, or familial hyperparathyroidism.
    - Resection options for hyperplasia include:

- **Resect 3.5 glands** and clip or tag the 1/2 remnant after frozen section proves that the 1/2 gland specimen is parathyroid tissue.
- **Resect 4 glands and mince and auto-implant the most normal appearing gland** either below the sternocleidomastoid (SCM) or in 3–4 pockets in the non-dominant forearm brachioradialis muscle. The latter is used if the patient has MEN2a, since there is a high likelihood of postoperative hyperparathyroidism recurring. In such a situation, part of the implant can be easily accessed for re-excision without entering the neck.
  - Autotransplantation is recommended if the post-resection **intraoperative PTH level is <10.**
  - **The implants will take weeks to function.** So if an early postoperative PTH level is elevated, there is still functioning pathologic parathyroid tissue in the neck.
- A bilateral **cervical thymectomy** can be added in either case.

- **Remove the ipsilateral thyroid lobe** if there is suspicion for an **intrathyroidal adenoma** by intraoperative US, intraoperative gamma probe, persistently elevated PTH assays from the ipsilateral internal jugular vein, or a concern exists for **parathyroid cancer.** In the latter scenario, also perform an ipsilateral central neck dissection.
- If there is a missing gland or failure of the intraoperative PTH to drop after removal of "all four glands," then:
  - Check the cervical **thymus or intrathyroidal location** if the missing gland is an inferior one.
  - Check the **TE groove, carotid sheath, and paraesophageal or retroesophageal space** if the missing gland is a superior one.
  - If one is unable to find the adenoma despite a search of ectopic sites, intraoperative thyroid US, and empiric cervical thymectomy, recommendations are to leave the other glands alone and close the wound. *See Figure 20.2.*

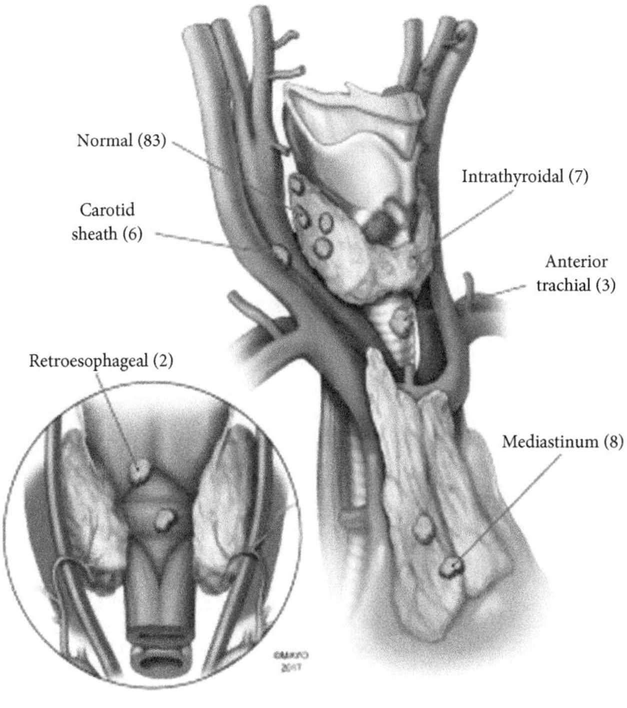

Figure 20.2 Ectopic parathyroid gland locations.

- **POSTOPERATIVE ISSUES**
  - **Postoperative transient hypocalcemia** presents as paresthesias *immediately* after the case. This is due to inferior thyroid artery insufficiency to the remaining parathyroid gland(s) or delayed activation of previously suppressed normal parathyroid glands after a dominant adenoma was resected. The hypocalcemia should resolve in days.
  - **Postoperative bone hunger** presents *days after* surgery. There is decreased serum calcium and HCO_3 since both are incorporated into bone. This can be prevented with **calcium supplementation, and occasionally dietary supplementation with vitamin D_2 (ergocalciferol) or D_3 (cholecalciferol) for 4–5 days.** Check the calcium level at 2 weeks to establish a new baseline, and then 6 months, and yearly.
  - **Postoperative hyperparathyroidism within 6 months** is due to a missed adenoma, an ectopic gland, baseline supernumerary glands (the last of which was abnormal), incomplete resection, malignancy, or hyperplasia misdiagnosed as an adenoma.
  - **Postoperative recurrent hyperparathyroidism after 6 months** is due to parathyromatosis, recurrent malignancy, or a recurrent or new adenoma.
  - The risk of permanent hypoparathyroidism after surgery is 10%–15%.
- **RE-OPERATIVE PRINCIPLES FOR RECURRENCE**
  - It is best to have **two imaging studies** (US, sestamibi, CT, or MRI) in concordance for localization.
  - **FNA cytology and PTH levels** can be taken from lesions, and selective venous sampling can help confirm the side needing intervention.
  - Re-operation is best performed as a limited unilateral exploration after **preoperative vocal cord evaluation,** since the risk of recurrent laryngeal nerve (RLN) injury is increased from 1% to **2%–15%.** Consider **intraoperative RLN monitoring** as well.
    - Consider a lateral approach between the lateral edge of strap muscles and the medial edge of the SCM, for unilateral exploration of an ectopic superior parathyroid.
  - The most common location for a "missing" gland is in the normal anatomic position.
  - The most common location for an ectopic gland depends on the glands' perceived origin.

- Inferior ectopic glands can reside intrathyroidal or within the thymus.
- Superior ectopic glands can reside in the TE groove, carotid sheath, or para- or retroesophageal spaces.
- One may still perform autotransplantation, **cryopreservation,** or both, for resected parathyroid glands in re-operative surgery.

SECONDARY HYPERPARATHYROIDISM

- This is most often from decreased 1, 25-OH vitamin D_3 in the setting of end stage renal disease (ESRD). The result is decreased calcium binding protein and decreased GI reabsorption, leading to a reflexive increase in PTH as it tries to maintain a low–normal calcium level.
- **DIAGNOSIS**
  - Elevated PTH produces osteodystrophy, nephrocalcinosis, ectopic (periarticular) calcification, and pruritus from calcium deposits in the dermis.
  - Unlike primary hyperparathyroidism, secondary hyperparathyroidism, in the setting of ESRD, has high phosphate levels.
  - The most serious manifestation of secondary hyperparathyroidism is **calciphylaxis,** which is characterized by arteriolar media deposition of calcium and phosphate, with secondary dermal ischemia. It is diagnosed on dermal biopsy, and treatment requires urgent parathyroidectomy (below).
  - Other etiologies of secondary hyperparathyroidism include Vitamin D_3 deficiency from dietary or sunlight limitation. **25-OH vitamin D_3 levels** should be checked, since it has a longer half-life than 1,25-OH vitamin D_3.
- **TREATMENT**
  - Calcium supplementation, vitamin D supplementation, decreased protein intake, and phosphate binding formulas for patients with ESRD.
  - **Cinacalcet** can be given to decrease the parathyroid's sensitivity to low circulating calcium and thereby mitigate secondary hyperparathyroidism and its deleterious tissue effects.
  - **Surgery** is reserved only for bone disease refractory to medical management or the development of calciphylaxis.
  - **Tertiary hyperparathyroidism** is the term used to describe the persistence of the hyperparathyroid state after successful renal transplantation. Treatment is with 3.5 gland parathyroidectomy or 4 gland parathyroidectomy with autoimplantation. A cervical thymectomy may be included in either case.

PARATHYROID CANCER

- This can present as elevated serum calcium and PTH levels with associated hypercalcemic symptoms, much like other causes of hyperparathyroidism.
- **DIAGNOSIS**
  - If a mass is palpable or the calcium levels are markedly high (>14) it almost certainly implies parathyroid malignancy rather than just an adenoma.
  - Workup includes an US and sestamibi scan.
  - Local lymph node involvement is common, and metastases can be present in the lung.
- **TREATMENT**
  - **Radical parathyroidectomy** includes the ipsilateral thyroid lobe and **LN dissection.**
  - **Radiation** can be used to prevent local recurrence and palliate any bony metastases. There is a 50% recurrence rate, with mortality being due to persistent hypercalcemia.

MULTIPLE ENDOCRINE NEOPLASIA (MEN)

- **MEN 1** (MENIN GENE).
  - **Parathyroid hyperplasia** is the most common symptomatic component. Treatment is 4 gland resection with forearm autoimplantation (see above).
  - **Pancreatic neuroendocrine tumor (pNET)** is the second most common component. Gastrinomas more so than insulinomas or nonfunctional pNETs. Half of tumors are malignant or multifocal.
  - **Pituitary prolactinoma** is the third component. It can present with galactorrhea in a female or bitemporal hemianopsia in a male patient.
- **MEN 2A** (RET PROTO ONCOGENE).
  - **Medullary thyroid cancer** is the hallmark of the syndrome. Diarrhea can be a common symptom. Tumors are often bilateral. This is the main cause of death in patients with MEN. Treatment is with a prophylactic thyroidectomy by age 5.
  - **Pheochromocytomas** are present in up to 40% of patients with MEN and can be bilateral. They must be removed before attempting thyroidectomy.

- **Parathyroid hyperplasia** is in up to 30% of patients. Treatment is 4 gland resection with forearm autoimplantation (see above).
- **MEN 2B** (RET PROTO ONCOGENE)
  - **Medullary thyroid cancer** is more aggressive in 2b. Diarrhea can be present. Tumors are often bilateral. This is the main cause of death in patients with MEN. Treatment is with a prophylactic thyroidectomy by age 1.
  - Mucosal neuromas and Marfanoid body habitus can be clues to the diagnosis.
  - **Pheochromocytomas** are present in up to 40% of patients with MEN and can be bilateral. They must be removed before attempting thyroidectomy.

21.

ADRENAL

ANATOMY AND PHYSIOLOGY

- Arterial supply to the adrenal gland:
  - Superior adrenal artery from the inferior phrenic artery.
  - Middle adrenal artery directly off the aorta.
  - Inferior adrenal from the renal artery.
- Venous drainage:
  - The left adrenal vein merges with the inferior phrenic vein, and together they drain into the left renal vein and to the inferior vena cava (IVC).
  - The right adrenal vein drains directly into the IVC. *See Figure 21.1.*
  - The adrenal cortex is the outer rim of the organ and composed of the glomerulosa, fasciculata, and reticularis, each with their own metabolic pathway. *See Figure 21.2.* An ectopic adrenal cortex may reside in the ovaries, testes, or spermatic cord.
- **The mineralocorticoid pathway is in the zona glomerulosa.** It derives from cholesterol to pregnenolone, progesterone, deoxycorticosterone, corticosterone, and aldosterone.
  - Aldosterone has a half-life of 20 minutes. It assists in reabsorption of sodium (Na) from sweat glands, salivary glands, gastrointestinal mucosa, and nephron tubules, thereby retaining intravascular volume. In the process, K is secreted.
  - Production is stimulated by angiotensin 2, hyperkalemia, and sometimes ACTH.
  - Production is inhibited by a high sodium diet, 2 L saline challenge, somatostatin, atrial natriuretic peptide, heparin, and dopamine. Aldosterone's effects can be blocked by spironolactone.
- **The glucocorticoid pathway is in the zona fasciculata** and derives from cholesterol to pregnenolone, 17 alpha pregnenolone, 17 alpha progesterone, deoxycortisol, and cortisol. The pathway is inhibited by metyrapone and etomidate.
  - Cortisol peaks at 4am–6am. It is bound to cortisol binding globulin (CBG, 85%) or albumin (15%). Once intracellular, it binds to cytosolic receptors and is taken into the nucleus. There it has inotropic, chronotropic, and vasoconstrictive effects. It is also proteolytic, gluconeogenic, and anti-inflammatory.
  - Corticotropin releasing hormone (CRH) is from the hypothalamus. It generates ACTH from pro-opiomelanocortin, and subsequent release from the anterior pituitary. ACTH promotes cortisol release.
- **The sex hormone pathway is in the zona reticularis** and derives from cholesterol to pregnenolone, 17 alpha pregnenolone, dehydroepiandrostenedione (DHEA), androstenedione, testosterone, and estrogen.
- The **adrenal medulla** is a neuroendocrine extension of the autonomic nervous system, receiving sympathetic stimulation and producing catecholamines. The metabolic pathway derives from:
  - Tyrosine converted to DOPA by **tyrosine hydroxylase.** This is inhibited by metyrosine.
  - DOPA converted to dopamine by **DOPA decarboxylase.**
  - Dopamine converted to norepinephrine via **dopamine beta hydroxylase.**
  - Norepinephrine converted to epinephrine by **phenylethanolamine N-methyltransferase** (PNMT). This enzyme is specific to the adrenal medulla and organ of Zuckerkandl.
  - Catecholamine are bound to albumin in the serum, and metabolized to metanephrine, normetanephrine, and VMA by catechol-O-methyltransferase (COMT) and monoamine oxidase (MAO).
  - Ectopic sites for the medulla are the organ of Zuckerkandl along the aorta (most commonly near the inferior mesenteric artery), around the carotid artery, in the para-aortic space, within the sympathetic thoracic chain, or in the bladder.

ASYMPTOMATIC INCIDENTAL ADRENAL MASS

- All visualized lesions first require a **hormonal screening evaluation** (see below).
- Radiographic features can help determine the lesions' etiology.
  - **Heterogeneous appearance or contrast enhancement to 80–90 Hounsfield units (HU) with slow washout (<50%) at 15 minutes,** are both consistent with a primary malignant tumor.

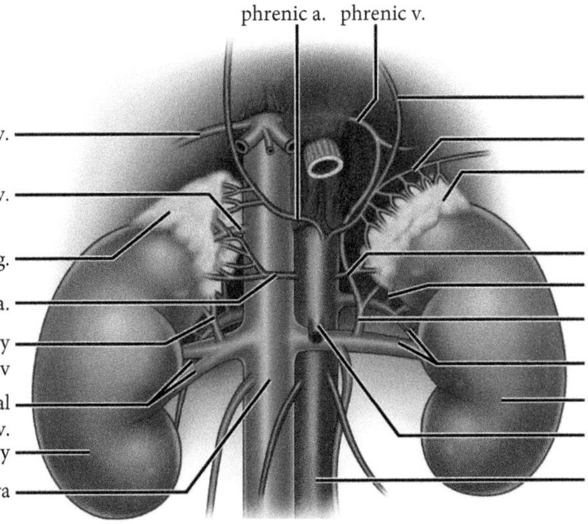

Figure 21.1 Adrenal anatomy and perfusion.

- A lesion that is 10 HU on non-contrast imaging and increases to 80–90 HU when contrast enhanced, but has **rapid washout** of >50% of contrast within15 minutes, is still consistent with a primary benign nonfunctioning tumor. A cortical adenoma is the most common benign lesion.
- -10 to -20 HU is consistent with a fat-predominant myolipoma. This is the second most common benign lesion. If imaging is definitive, this may be followed rather than resected, even if it is large.
- **Metastases** to the adrenal gland can be from lung, breast, melanoma, or renal cell carcinoma. These appear high attenuation on CT and hypermetabolic on PET.
  - First step is a **hormonal evaluation**, even if metastasis is highly likely and the patient has a known primary.
  - If the hormonal evaluation is negative and the single metastasis is believed to be *resectable*, then *no* FNA is needed, and **metastasectomy** can be attempted.
  - Fine needle aspiration (**FNA**) can be done after a negative hormonal evaluation when it is deemed that metastasectomy is not appropriate.
- Indications for resection of an adrenal lesion include the following:
  - **Heterogeneous appearance** on CT.
  - **Slow washout of contrast** at <50% at 15 minutes.
  - Benign appearance but **>4–6 cm or enlarging on serial imaging.**
  - **Hormonally active** lesion. Pheochromocytoma is more common than subclinical cortisol producing tumors. Aldosteronomas are the least common.
  - A **solitary metastasis** that is deemed resectable.
  - See below for adrenalectomy options (laparoscopic, endoscopic retroperitoneal, or open- anterior, lateral, or posterior).
  - If resection is not indicated by any of the above, perform **follow-up imaging** in 6 months and then yearly. Also add **yearly biochemical screening** to detect any subclinical Cushing's syndrome over the next 5 years or longer.

METABOLIC WORKUP FOR ADRENAL MASS

- These should be performed before any resection, core needle biopsy (CNBx) or fine FNA of a visualized adrenal mass, whether incidentally discovered or not.
- To evaluate for hyperaldosteronism, especially in patients with hypokalemia and hypertension, start with an **oral salt load or 2 liter saline infusion** to see if either suppresses renin and aldosterone production. **Aldosterone to renin ratios** can be checked, and if >20, they suggest primary aldosterone production independent of renin.
- To evaluate hypercortisolism:
  - First check a **24 hours urine cortisol or 11pm salivary cortisol** level.
  - If elevated, perform a **1 mg dexamethasone suppression** at 11pm. Morning cortisol production would be suppressed if the hypercortisolism was due to depression, alcohol, chronic kidney disease, or pregnancy.
  - If not suppressed, check an **ACTH level.**

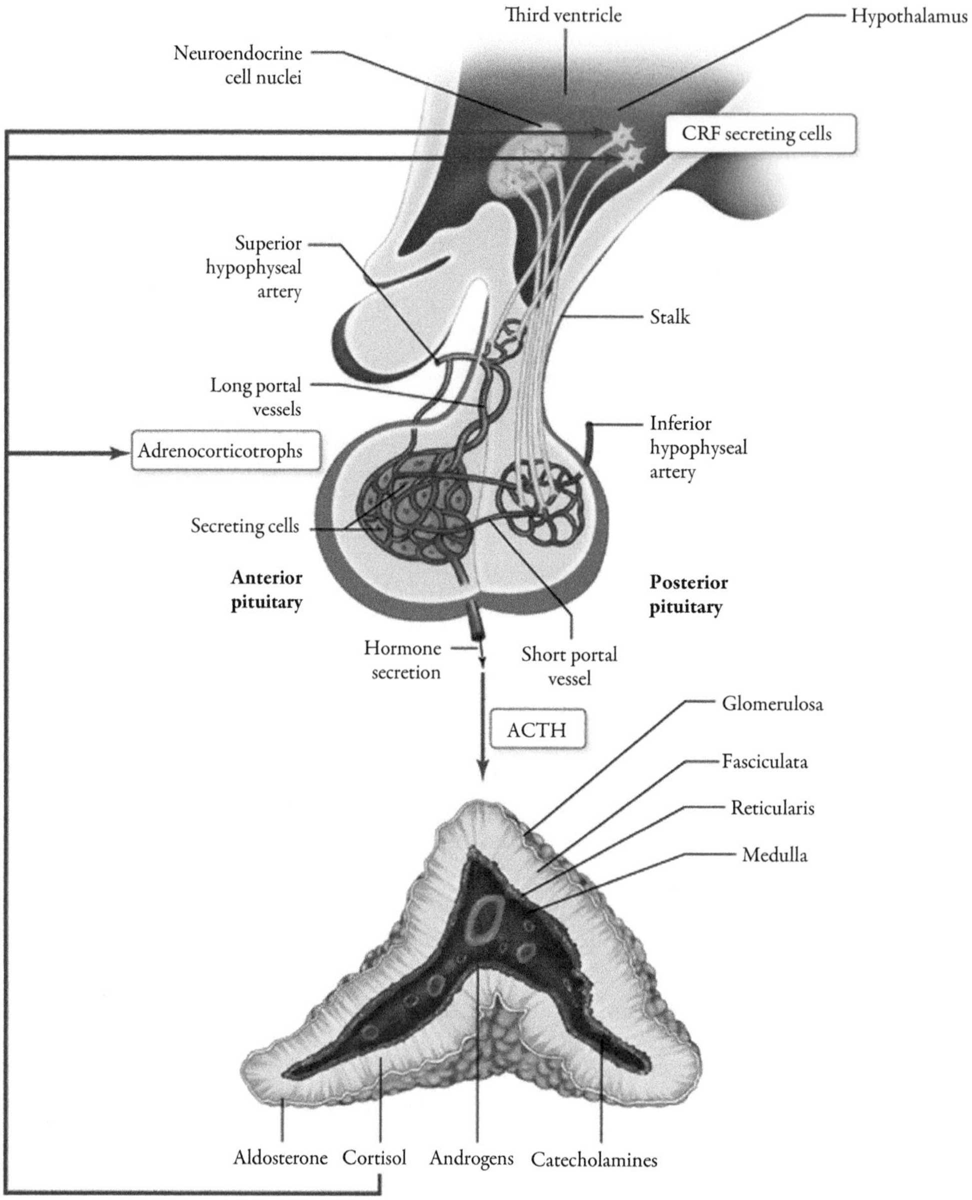

Figure 21.2 Adrenal architecture.

  - If this is low, obtain **CT adrenal imaging** for a hyperfunctioning adenoma.
  - If ACTH is high, perform a **high-dose dexamethasone suppression test**.
    - **Magnetic resonance image (MRI) the pituitary gland** if the morning cortisol is suppressed.
    - **CT the chest** for a small cell lung cancer if the morning cortisol is not suppressed.
- To evaluate for elevated sex hormone production, check 17-hydroxyprogesterone, 17-ketosteroids, DHEA, estrogen, and androgen levels. If elevated, any associated adrenal mass is almost always malignant.
- To evaluate for pheochromocytoma, check a **spot plasma metanephrine.** It has a high sensitivity, but a low specificity. So if the spot serum metanephrine is negative, then the diagnosis is ruled out. Add a **24 hours urine catecholamine** or metanephrine collection if the spot level is positive. This can confirm the diagnosis.

GLOMERULOSA

- **Hyperplasia** is the cause of hyperaldosteronism in 70% of cases. A hyperfunctioning **adenoma is** the cause in the remaining 30% of cases (Conn's syndrome).
- **DIAGNOSIS**
  - Patients can present with refractory hypertension and hypokalemia. **Salt load or 2 liter saline suppression test** will not alter aldosterone levels. **Aldosterone**

to renin ratios >20 are suggestive of unregulated production.

- **CT** is first-line imaging for a suspected adenoma, and MRI is a second choice. NP59 scintigraphy is less commonly used.
- **Venous sampling** of each renal vein should be done to confirm that any visualized mass is actually the source of the aldosterone. This is important, since hyperplasia is more common than Conn's syndrome, and true aldosteronomas are small and may mimic benign adrenal nodularity or a nonfunctional adenoma—the latter two being more common as a patient ages.
  - In venous sampling, aldosterone levels are compared to cortisol levels at each adrenal vein. The cortisol component corrects for catheter position and venous tributary dilutional differences between the right and left adrenal veins.
- **TREATMENT**
  - **Spironolactone for hyperplasia** or an **adrenalectomy for adenoma**. Bilateral adrenalectomy can be done for refractory hyperplasia. Fludrocortisone and endocrine supplementation should be given postoperatively.

FASCICULATA

- **HYPOCORTISOLISM**
  - **Secondary adrenal insufficiency from exogenous glucocorticoid suppression** is the most common cause of hypocortisolism. Patients are usually taking an equivalent of hydrocortisone 80 mg per day.
  - Primary hypocortisolism from **autoimmune destruction** is the second most common cause.
  - Other etiologies include pituitary dysfunction (secondary hypocortisolism), adrenal infection, infarction, hemorrhage, or metastases crowding the organ.
  - **DIAGNOSIS**
    - Patients can present acutely with fever, gastrointestinal symptoms, hypotension unresponsive to volume and vasopressor use, hypoglycemia, and hyperkalemia or acidemia.
    - Chronic presentations include hyperpigmentation (from high ACTH levels in primary insufficiency), weight loss, fatigue, weakness, or gastrointestinal symptoms.
    - A **high ACTH level and low cortisol and aldosterone levels** can be seen in primary gland dysfunction.
    - A random cortisol level is <15, and a subpar **response to 25 mcg of ACTH** injection is noted at 30 or 60 minutes post-injection. The change in measured cortisol is usually <9.
  - **TREATMENT**
    - Intravenous fluid resuscitation and dexamethasone 4 mg/day or hydrocortisone 50–100 mg q8 for 3 days. Fludrocortisone 0.05–0.1 mg/day can also be given.
    - Patients may need prophylactic stress dose glucocorticoids before elective surgery.
- **HYPERCORTISOLISM (CUSHING'S SYNDROME)**
  - Many cases will be subclinical, and based on laboratory results performed during the workup of an incidentaloma.
  - When present, symptoms can include headaches, fatigue, weakness, depression, insomnia, easy bruising, truncal obesity, moon facies, hypertension, osteoporosis, and diabetes. Hyperpigmentation can be present in cases of secondary hypercortisolism from high ACTH.
  - **1. IATROGENIC HYPERCORTISOLISM**
    - It is the most common etiology, resulting from daily glucocorticoid use
  - **2. PITUITARY ADENOMA PRODUCING ACTH**
    - This is termed Cushing's *disease,* and is the second most common cause of hypercortisolism. Production can be suppressed by a high-dose dexamethasone test (see metabolic workup above).
    - **MRI** can visualize a macroadenoma. **Petrosal venous sinus ACTH sampling** can localize a microadenoma.
    - **Management** is with **transsphenoidal surgery or stereotactic surgery or ablation**. Radiation can be employed for any residual disease. **Hormonal blockade** with metyrapone, aminoglutethimide, or ketoconazole is an alternative.
  - 3. **THORCIC ECTOPIC PRODUCTION OF ACTH**
    - Third leading cause of hypercortisolism. The paraneoplastic origin of ACTH can be small cell carcinoma, other lung carcinoids, or thymic tumors. These patients are not suppressed by a high-dose dexamethasone test (see metabolic workup above).
    - CT or MRI of the chest can help localize a culprit mass
    - Management includes resection of the primary when possible, bilateral adrenalectomy, or medical suppression of cortisol.
  - **4. ADRENAL ADENOMA**
    - Fourth leading cause of hypercortisolism. These patients have a low ACTH level (see metabolic workup above).
    - CT with IV contrast shows a homogeneous mass with rapid washout of contrast. Other options are abdominal MRI or NP-59 scintigraphy.
    - Management is with adrenalectomy. Postoperative hydrocortisone 10–20 mg every morning and 5–10 mg every evening may be needed, since the contralateral normal adrenal gland could remain suppressed. Fludrocortisone 0.05–0.1 daily is for mineralocorticoid replacement. Hormonal blockade can be attempted with ketoconazole metyrapone, or mitotane.

- **5. ADRENAL HYPERPLASIA**
  - A fifth cause for hypercortisolism. One must ensure there is no pituitary source of Cushing's disease. Then, **metyrapone or aminoglutethimide medical suppression** can be attempted. Bilateral adrenalectomy is reserved for refractory cases.
- **6. ADRENOCORTICAL CARCINOMA**
  - 50% of tumors are functional and produce mostly cortisol, or occasionally sex hormones with resultant virilization or feminization.
  - Patients can have symptoms from the mass effect or hypercortisolism. Alternatively, many are asymptomatic with an incidentally noted mass on CT.
    - CT contrast imaging shows a **heterogeneous contrast enhancing lesion up to 80–90 HU with slow contrast washout of <50% at 15 minutes.**
    - Stage 1 is a <5 cm primary, and stage 2 is >5 cm. Stage 3 are node positive. Stage 4 have metastases to the lung, liver, bone, or peritoneum.
    - Definitive diagnosis is only possible after pathologic examination of the resected specimen demonstrates local invasion, or a biopsy of metastases proves adrenocortical origin.
  - Management is with an **open radical adrenalectomy**. This should be attempted en bloc, even if the tumor has invaded the liver or if tumor thrombus is seen in the renal vein.
    - For non-resectable lesions, **debulking** may be beneficial to control symptoms from hormone overproduction. Mitotane or etomidate can help.
    - Chemotherapy options include etoposide, doxorubicin, or cisplatin. Radiation is an adjunct as well.
    - 30%–50% 5-year survival in the cases of complete resection.

RETICULARIS

- **CONGENITAL HYPERPLASIA**
  - This is from a defect in 21-hydroxylase or 11-beta-hydroxylase and leads to excess pregnenolone and progesterone, with conversion to the sex hormone pathway. The result is virilization in females or precocious puberty in males.
  - The 21-hydroxylase defect may sometimes result in sodium loss in nephrons, with secondary hyponatremia and hyperkalemia.
  - The 11-beta-hydroxylase defect does not result in salt wasting because deoxycorticosterone can still be produced, providing mineralocorticoid.
  - Suspicion can be confirmed with DHEA sulfate or testosterone levels.
  - Managment includes mineralocorticoid supplementation, genital surgery, and occasional consideration for bilateral adrenalectomy.
- **ESTROGEN OR ANDROGEN SECRETING TUMORS**
  - These can also arise from the reticularis and are often malignant and need resection.

MEDULLA

- **PHEOCHROMOCYTOMA**
  - Derived from neural crest cells of ectodermal origin. Most are located in the adrenal medulla, and more commonly are right-sided. *See Figure 21.3.*
  - Other locations include the organ of Zuckerkandl, sympathetic thoracic chain, carotid artery, or near the bladder.
  - Related entities include a **paraganglioma** (malignant extra-adrenal version of a pheochromocytoma) and a **ganglioneuroma** (a rare, benign extra-adrenal tumor of ganglion cells in the sympathetic chain). These may be seen as incidentalomas adjacent to the adrenal gland or within the area of the sympathetic chain. Definitive distinction may only occur at the time of excision.
  - 10%–40% of pheochromocytomas are malignant, 10% are bilateral, 10% occur in children, 10% are extra-adrenal, and 10% are familial.
  - Syndromes include multiple endocrine neoplasia (MEN) 2a or 2b, neurofibromatosis-1 (NF1), Tuberous sclerosis, Sturge-Weber, Carney's, or Von Hippel-Lindau (VHL) The latter includes renal cell carcinoma, cysts, and cerebellar or retinal hemangioblastomas.
- **DIAGNOSIS**
  - Patients may have symptoms of episodic hypertension, diaphoresis, or palpitations. They may have a decreased cardiac index and an increased systemic vascular resistance. Asymptomatic patients are usually diagnosed after chemical workup of an incidental adrenal mass seen on imaging.
  - A spot **plasma (nor)-metanephrine** level is the most sensitive test. If negative, it rules out the diagnosis.
  - If positive, this should prompt a **24-hour urine collection for catecholamines.**
    - **False positive** VMA levels can be seen with coffee, tea, fruits, vanilla, iodine, tca, or alpha and beta blockers. So there are dietary restrictions implemented prior to beginning the collection.

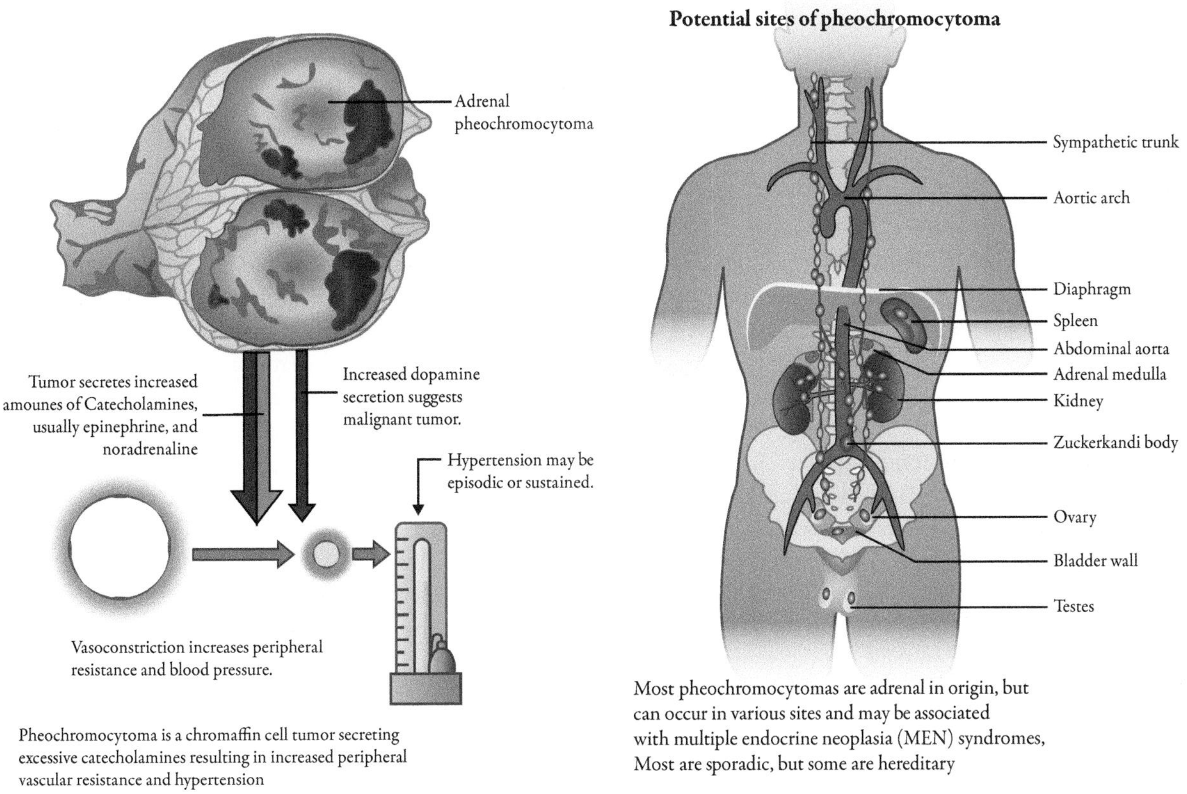

Figure 21.3 Pheochromocytoma.

- A 0.3 mg clonidine suppression test with persistence of catecholamines after 2–3 hours suggests their independent, pathologic production. This can be employed in equivocal cases.
- Chromogranin A levels are elevated, but are not specific for pheochromocytoma.
- **CT scan with IV contrast** is the imaging modality of choice. It will show a very bright lesion (Hounsfield units >80) with slow washout of contrast (<50% washout at 15 minutes).
  - Other imaging modalities include MRI with T2 imaging or a MIBG scan with I-123 for localization if the mass cannot otherwise be found, or practitioners are planning I-131 radioactive therapy in the presence of metastatic disease.
- **TREATMENT**
  - Adrenalectomy is the treatment of choice.
  - **Phenoxybenzamine** should be started at 10 mg BID, 2–4 weeks prior to the planned operation. The dosage may be increased by 10 mg every other day until the patient is orthostatic. This helps control arteriolar tone and prevent a hypertensive crisis intra-operatively.
    - Other alpha blockers, peripherally acting calcium channel blockers, or metyrosine can be used if the patient is refractory to a phenoxybenzamine load.
    - Beta blockers may be added for any tachycardia that ensues after alpha blockade.
  - Preoperative **intravenous crystalloid** replacement is mandatory.
  - Resection may be via a laparoscopic, endoscopic retroperitoneal, or an open approach. See adrenalectomy principles below.
    - A **cortical sparing adrenalectomy** can be done in certain cases of bilateral tumors. It involves preservation of the outer cortex along with the inferior phrenic artery and the adrenal vein on one side, and a complete adrenalectomy on the other side.
    - **Debulking** non-resectable lesions, metastases, or recurrences will help mitigate cardiovascular morbidity even if an R0 resection cannot be accomplished.
  - Postoperative rebound hypoglycemia and hypotension can result from loss of catecholamines and cortisol.
  - Local invasion on the specimen does not correlate with the presence of metastases.
  - Recurrence rates for all approaches are 5%–15%.
  - **Alternative treatments** for non-resectable lesions include arterial embolization, cryoablation, radiation, RFA, MIBG (I-131), octreotide, or chemotherapy.

ADRENALECTOMY PRINCIPLES

- A **laparoscopic approach** is reasonable for **benign appearing tumors <5 cm.** This may be via lateral or anterior ports.
  - Place 4 ports between the midclavicular and anterior axillary lines.
  - Mobilize the triangular ligament for exposure of the right adrenal gland, or the splenic flexure of the colon for exposure of the left adrenal gland.
  - Open Gerota's fascia and dissect laterally first and then superomedially. Ligate the vein early to control the efflux of vasoactive substances before further manipulation of the gland if operating for pheochromocytoma.
- An **endoscopic retroperitoneal approach** in the prone position can be used for **small, benign appearing bilateral tumors.** It is a preferred approach if a patient has prior peritoneal adhesions, and both adrenal glands can be addressed without repositioning.
  - 3 ports are placed subcostally below the 12th rib medially and 11th rib laterally. A mirror image set of ports can be placed on the other side if needed.
  - Insufflation may need to be up to 20 mmHg for adequate visualization in the retroperitoneal space.
- Suspected **malignancy or bilateral large tumors** should be addressed in an **open fashion.** This can be anterior, lateral, or posterior.
  - Anterior is via a single midline or bilateral subcostal incisions.
  - Lateral is via the decubitus position.
  - The posterior approach utilizes a hockey-stick incision that runs para-spinal and subcostal.
  - Inferior vena cava (IVC) and renal vessel control should be obtained before dissection of the medial tumor plane to help control hormonal efflux.

PITUITARY AND HYPOTHALAMUS.

See Figure 21.2.

- The **supraoptic nucleus within the hypothalamus controls ADH production.** When excessive and inappropriate (SIADH), treatment is with free water restriction, hypertonic saline, and occasionally demeclocycline.
- **Prolactinomas** can be micro- or macroadenomas. Prolactin levels will be >150. Females can have bilateral galactorrhea, while males may only present once the mass effect produces bitemporal hemianopsia. Treatment is with the dopamine agonist **bromocriptine.** Resection is pursued for macroadenomas, hemorrhage, a desire to be pregnant, or an associated CSF leak.
- **Acromegaly** is from growth hormone (GH) production. GH levels are usually >10, and IGF-1 levels are also elevated. Complications include hypertension, diabetes mellitus, valvulopathy, or cardiomyopathy. Treatment is with octreotide or resection and radiation.
- **Nelson syndrome** describes pituitary hyperplasia, hyperpigmentation, and amenorrhea in the setting of elevated corticotropin releasing hormone (CRH) after a bilateral adrenalectomy. Treatment is with pituitary adenoma resection and glucocorticoid replacement to provide feedback to the hypothalamus and decrease CRH production.
- **Craniopharyngiomas** are pituitary in origin and present with a visual field defect, headaches, or diabetes insipidus. They are not metabolically active. Treatment is resection.
- **Sheehan's syndrome** describes pituitary ischemia in the setting of shock. Patients can subsequently have poor lactation, abnormal menses, or any other secondary endocrine organ insufficiency. Treatment is with hormone replacement covering all aspects of pituitary function.

SURGICAL SUBSPECIALTIES

22

CRITICAL CARE

CARDIOVASCULAR PRINCIPLES

- Normal **cardiac output (CO)** is 4-8 L/minute. CO equals heart rate (HR) times stroke volume (SV).
  - **SV** is left ventricular end diastolic volume (LVEDV, preload) - left ventricular end systolic volume (LVESV). Ejection fraction (EF) is SV/LVEDV.
  - **HRs** up to 150 lead to increased CO. This is known as the Bowditch effect. Thereafter, further increases in HR drop CO due to decreased diastolic filling time and short diastolic subendocardial perfusion time. This impairs LVEDV and contractile strength respectively. The result is diminution of SV and hence CO.
  - 25% of CO goes to the kidneys, 15% to the brain, and 5% to myocardium.
- **Oxygen delivery (DO_2)** = CO × arterial oxygen content (CaO_2). CaO_2 is 1.34 × hemoglobin (Hgb) × arterial saturation of hemoglobin (SaO_2) + 0.003 × partial pressure of oxygen in arterial blood (PaO_2). **Oxygen consumption (VO_2)** = CO × (CaO_2 − CVO_2)
  - The DO_2 to VO_2 ratio is generally 5:1, and normal **central venous oxygen saturation ($ScVO_2$)** returning to the right atrium is 75%.
  - Most organs can increase extraction when DO_2 drops. Myocardium, however, is at maximum extraction at baseline, and therefore is more reliant on DO_2.
  - Normal **mixed venous oxygen saturation (MVO_2)** is 65%, generally lower than $ScVO_2$. This is because it includes coronary venous return whose SVO_2 is 30% (representative of myocardial oxygen consumption) mixed into the systemic venous return.
  - Myocardial O_2 consumption increases with increasing HR or increasing wall tension from chamber dilation or higher vascular resistance.
  - Decreased $ScVO_2$ is from increased peripheral tissue extraction or decreased delivery.
  - Increased $ScVO_2$ is from decreased extraction due to:
    - Distributive shock and shunting of blood past tissue beds (sepsis, hepatic failure, anaphylaxis, adrenal insufficiency). Shunts can also be post-traumatic or surgically constructed.
    - Impaired mitochondrial function (cyanide, sepsis); or
    - Decreased metabolic activity (hypothermia, paralysis, sedation).
  - Left ventricle saturation is typically lower than post-alveolar pulmonary capillary saturation due to addition of bronchial venous blood in pulmonary veins.
  - p50 is the PaO_2 at which arterial hemoglobin saturation (SaO_2) is 50%. Under physiologic conditions, the p50 is at a PaO_2 of 27 mmHg. Right shift of the **oxy-hemoglobin curve** will offload more oxygen to tissue and increase p50. Causes of a right shift include heat, increased CO_2, acidemia, increased 2,3 DPG, or increased ADP. A left shift of the oxy-hemoglobin curve decreases p50.
  - **Alveolar pressure of oxygen (PAO_2)** = 0.21 (760 −47 mm Hg water pressure) - pCO_2/0.8. **Arterial pressure of oxygen (PaO_2)** is generally 10–15 mmHg lower than PAO_2 in a normal physiologic state. Pathologic processes that obstruct diffusion of oxygen across the alveolar-capillary membrane will result in a larger PAO_2 - PaO_2 gradient.
- **Systemic vascular resistance (SVR)** = 80 x (MAP - CVP)/cardiac output (CO).
  - Normal SVR is 800–1400. Increased afterload (high SVR or aortic stenosis) reflexively increases CO in a healthy heart. This is known as the Anrep effect.
  - Decreased peripheral vascular compliance tends to exaggerated pulse pressure peripherally, so peripheral arteries demonstrate higher SBP and lower diastolic blood pressure (DBP) (wider pulse pressure). The mean arterial pressure (MAP) remains the same, when compared to central arteries.
- Normal values are as follows: for **central venous pressure (CVP)**, 5–10 mmHg; **pulmonary artery pressure (PAP)**, 25/15 mmHg; and **pulmonary capillary wedge pressure (PCWP)**, is 7–15 mmHg.
- **PCWP** is obtained using a pulmonary artery (PA) catheter with the balloon inflated in West Zone 3 and measured during end expiration. The wedge represents LA pressure and is dependent on LV compliance and aortic and mitral valve function. PCWP may be higher

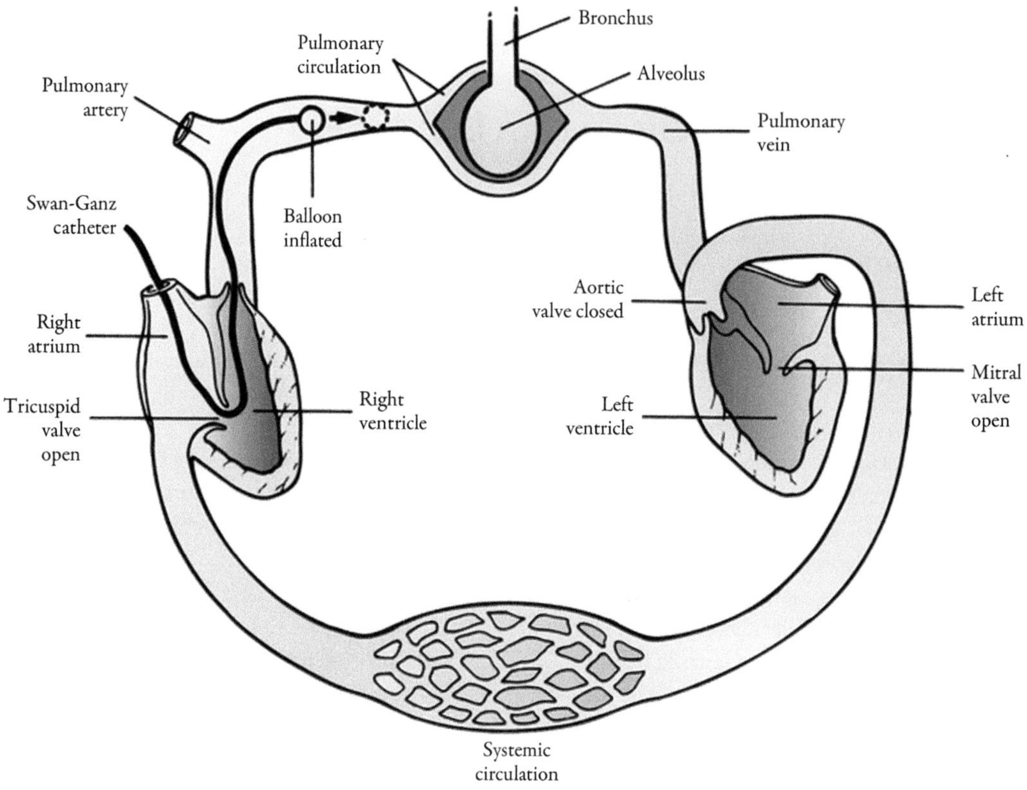

Figure 22.1 Pulmonary capillary wedge measurement.

and not representative of LAP if the distal pulmonary vasculature is occluded or there is severe alveolar membrane disease or elevated intra-alveolar pressures. *See Figure 22.1.*

- Hemoptysis may represent pulmonary artery rupture from the PA catheter. Management is by taking down the balloon, increasing positive end expiratory pressure (PEEP), and intubating the non-affected side with single lung ventilation and balloon inflation in the mainstream bronchus for protection. In the stable patient, bronchoscopic attempts at topical control and interventional angioembolization are options prior to resorting to thoracotomy and lobectomy.

CARDIOVASCULAR PHARMACOLOGY

- Cardiac pacemaker cells are regulated by vagal parasympathetic neural input, sympathetic neural input, circulating catecholamines via beta receptors, and intracellular calcium. Dromotropy refers to the rate of conduction through the sinoatrial and atrioventricular nodes. Chronotropy refers to the overall heart rate.
- Cardiac myocytes are regulated by circulating catecholamines via beta receptors, intracellular calcium, and mechanical stretch. Inotropy refers to contractile strength.
- Arteriolar and vascular wall muscle tone is regulated by the following: *See Figure 22.2.*
    - 1. Sympathetic stimulation of alpha 1 receptors.
    - 2. Endocrine contributions via circulating angiotensin, aldosterone, kininogen.
    - 3. Endothelial production of prostaglandins, nitric oxide (NO), thromboxane, and endothelin.
    - 4. Vascular smooth muscle intrinsic autoregulation.
    - 5. Neighboring tissue production of CO_2, H+, K, heat, histamine, and serotonin.
- **Catecholamine Production Pathway**
    - **Tyrosine** is metabolized by tyrosine hydroxylase to DOPA. Tyrosine hydroxylase is inhibited by metyrosine.
    - **DOPA** is metabolized by DOPA decarboxylase to dopamine.
    - **Dopamine** is metabolized by dopamine B hydroxylase to norepinephrine.
    - **Norepinephrine** is metabolized by PNMT to epinephrine in the adrenal medulla and organ of Zuckerkandl.

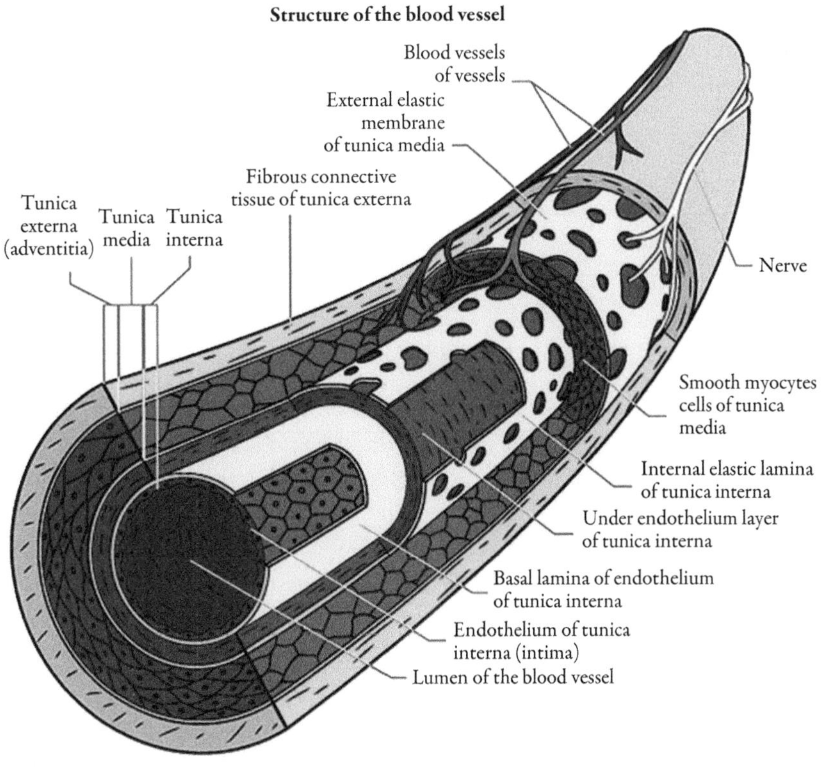

Figure 22.2 Arteriolar tunica muscular wall.

- Catecholamines are metabolized by **COMT and MAO** to normetanephrine, metanephrine, and vanillylmandelic acid (VMA).
- Alpha 1 receptors mediate vascular smooth muscle contraction, gluconeogenesis, glycogenolysis. **Phenylephrine** is a pure alpha 1 agonist. Dopamine, norepinephrine, and epinephrine all have alpha 1 effects. Hydralazine is an alpha 1 blocker causing vasodilation.
- Alpha 2 receptors mediate venoconstriction and negative feedback to central sympathetic nerve fibers. **Clonidine and dexmedetomidine** are alpha 2 agonists.
- Beta 1 receptors increase chronotropy and inotropy. **Dobutamine, dopamine, norepinephrine, and epinephrine** all have B1 effects.
- Beta 2 receptors relax bronchial and vascular smooth muscle. **Dobutamine and epinephrine have** B2 effects.
- **Cyclic AMP phosphodiesterase (cAMP PDE) inhibitors**, like milrinone, act on the intracellular targets of the B1 receptor in myocytes and result in increased cytosolic calcium, leading to increased inotropy. They do not affect pacemaker cells or conduction tissue, so there is no change in chronotropy or dromotropy. Pulmonary and peripheral vascular smooth muscle dilate under the effects of cAMP PDE inhibitors because of intracellular shunting of calcium into the sarcoplasmic reticulum.
- **Vasopressin** 1 receptors cause peripheral vasoconstriction. Vasopressin 2 receptors produce water reabsorption in the renal collecting ducts. Vasopressin 3 receptors increase factor VIII and VWF release from endothelial cells, improving platelet adherence. Desmopressin has receptor 2 and 3 effects but no vasoconstriction.
- **Nitroprusside** is a direct arteriolar vasodilator via NO production.
- **Nitroglycerin** causes venodilation and coronary artery dilation, also via NO production.
- **Digoxin** inhibits the sodium potassium ATPase pump, secondarily increasing myocyte cytosolic calcium and inotropy. It also slows the AV node by enhancing vagal neural tone. Myocytes may have increased sensitivity to hypokalemia. Digoxin has also been associated with mesenteric ischemia.

SHOCK

- Shock is defined as inadequate tissue oxygenation and can be distributive, hypovolemic/hemorrhagic, cardiogenic, or obstructive.
- **DISTRIBUTIVE SHOCK**
  - Can be neurogenic, adrenal, hepatic, or septic in etiology. Mitochondrial toxins like cyanide or hydrogen sulfide can also produce distributive shock.

System	Score				
	0	1	2	3	4
Respiration PaO_2/FIO_2, mmHg (kPa)	≥400 (53.3)	<400 (53.3)	<300 (40)	<200 (26.7) with respiratory support	<100 (13.3) with respiratory support
Coagulation Platelets, X10^3 μL^{-1}	≥150	<150	<100	<50	<20
Liver Bilirubin, mg dL^{-1} (μmol L^{-1})	<1.2 (20)	1.2–1.9 (20–32)	2.0–5.9 (33–101)	6.0–11.9 (102–204)	>12.0 (204)
Cardiovascular	MAP ≥70 mmHg	MAP <70 mmHg	Dopamine <5 or dobutamine (any dose)[2]	Dopamine 5.1–15 or epinephrine ≤0.1 or norepinephrine ≤0.1[a]	Dopamine >15 or epinephrine >0.1 or norepinephrine >0.1[a]
Central Nervous System (CNS) Glasgow Coma Scale score[b]	15	13–14	10–12	6–9	<6
Renal Creatinine, mg dL^{-1} (μmol L^{-1})	<1.2 (110)	1.2–1.9 (110–170)	2.0–3.4 (171–299)	3.5–4.9 (300–440)	>5.0 (440)
Urine output, mL per day				<500	<200

FIO_2: fraction of inspired oxygen; MAP: mean arterial pressure; PaO_2: partial pressure of oxygen.

[a]Catecholamine doses are given as$\mu g kg^{-1} min^{-1}$ for at least 1 h.

[b]Glasgow Coma Scale scores range from 3 to 15; higher score indicates better neurological function.

Figure 22.3 SOFA score.

- **Systemic inflammatory response syndrome (SIRS)** was a term used to describe systemic vital sign abnormalities due to a non-infectious insult.
- **Sepsis** is defined as an infectious source with a sequential organ failure assessment (SOFA) score >2. *See Figure 22.3.*
  - **Quick SOFA (qSOFA) includes RR >22, SBP <100, or altered mentation.** This allows up-front identification of patients likely to have sepsis and a prolonged ICU course.
- **Septic shock** describes patients with sepsis (infectious source and SOFA score of >2) plus a lactate >2 and hypotension requiring vasopressors despite adequate volume resuscitation.
  - Endotoxin lipid A from Gram negative rods (GNR) leads to increased IL1 and TNF alpha production, vasodilation, capillary leakage, and microvascular thrombosis.
  - Neurohormonal responses include increased endogenous epinephrine, norepinephrine, renin, ADH, ACTH, and hyperglycemia secondary to insulin resistance.
  - Treatment is with 30 ml/kg crystalloid in the first hour, broad antibiotics, and then norepinephrine with or without vasopressin.
  - In patients who cannot be stabilized on vasopressors despite adequate volume resuscitation and cardiac function, one may consider adjunct glucocorticoid therapy for a presumed adrenal insufficiency relative to the infectious stress.
- **Neurogenic shock** is due to loss of sympathetic systemic vascular tone from a high cervical spine injury that disrupts sympathetic fibers cephalad of their exit points in the thoracic spine. *See Figure 22.4.* Patients may be bradycardic as well, if the injury is above the origin of the cardiac accelerator fibers. This is generally diagnosed after exclusion of hemorrhagic or septic shock in a patient with new-onset paraplegia. Treatment is with crystalloid volume resuscitation to counterbalance the increased capacitance from systemic venous dilation. Thereafter, if they are still hypotensive, patients may require a pure alpha agonist (if they are tachycardic) or norepinephrine (if they are bradycardic).
- **HEMORRHAGIC SHOCK**
  - Class 1 (<15% blood volume loss) is associated with tachycardia or a normal pulse.
  - Class 2 (15%–30% blood volume loss) produces tachycardia and a narrowed pulse pressure.

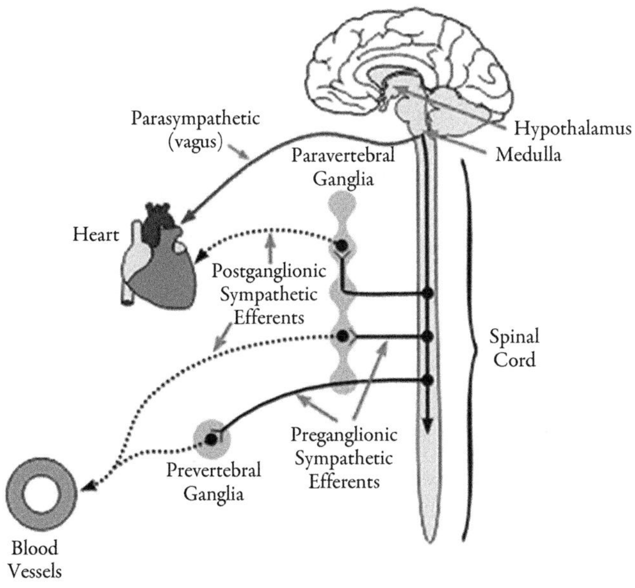

Figure 22.4 Autonomic innervation of the cardiovascular system.

- Class 3 (>30% blood volume loss) decreases BP and urine output.
- Class 4 (>40% blood volume loss) results in end organ dysfunction.
- Treatment is with 1:1:1 PRBC, FFP, and platelet resuscitation and addressing the site(s) of hemorrhage.

- **CARDIOGENIC SHOCK**
  - Due to a massive myocardial infarction, acute valvular insufficiency, or tamponade with impaired atrial diastolic filling.
  - Treatment depends on the etiology and includes revascularization, valve replacement or repair, IVF, and pericardial decompression.
  - Intra-aortic balloon pump (IABP) can be placed to improve coronary perfusion and systolic flow, as long as there is no aortic regurgitation. The tip of the device should lie just distal to the left subclavian artery (SCA). Inflation should be at the T wave (diastole) to increase DBP and coronary perfusion. Then deflation should be at the P wave (just before systole) to decrease afterload. *See Figure 22.5.*
  - Percutaneous left ventricular assist device (LVAD) and extracorporeal membrane oxygenation (ECMO) can serve as temporary mechanical support devices in place of, or in addition to, an IABP.

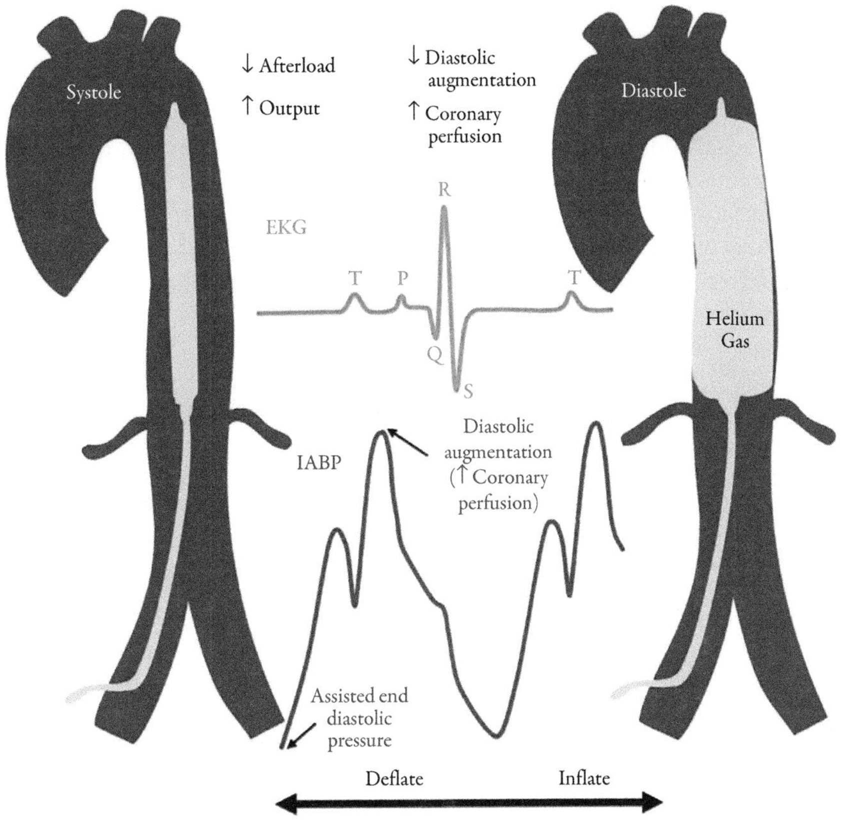

Figure 22.5 Intra-aortic balloon pump during systole and diastole.

- **OBSTRUCTIVE SHOCK**
  - Includes tension hemithorax, alveolar overinflation, and massive pulmonary embolus.
  - Tension hemithorax is usually from air (tension pneumothorax) and is treated with emergent thoracostomy.
  - Alveolar overinflation requires treatment of bronchospasm, bronchiolar edema, and control of ventilatory parameters to permit prolonged, slow expiration and fewer inspiratory cycles. This will decrease intra-alveolar pressure and relieve pulmonary capillary compression.
  - Pulmonary thromboembolism is treated with systemic thrombolytics if the patient has arrested. If the patient has not arrested but remains unstable despite inotropes and vasopressors, consideration should be given to right heart catheterization and suction thromboembolectomy, or open surgical thromboembolectomy. Otherwise, systemic heparin alone is sufficient for a stable patient. An IVC filter can be inserted for patients who are intolerant of anticoagulation, cannot receive any anticoagulation due to concomitant bleeding risk, or have had recurrent emboli despite therapeutic anticoagulation. A pulmonary artery (PA) catheter inserted while the patient is in the left lateral decubitus position can aspirate a pulmonary air embolus.

NEUROLOGIC

- **DELIRIUM**
  - Diagnosis is based on the confusion assessment method (CAM) noting acute onset of inattention with a fluctuating course, plus either (a) disorganized thinking or (b) altered level of consciousness.
  - It typically occurs in elderly patients ASA class 3 or 4 designated as frail (see below) who are receiving opioids or anticholinergics. Delirium increases hospital length of stay (LOS), postoperative nonsurgical morbidity and mortality, and the need for nursing home (NH) transfer. Longer duration of delirium correlates with an increased chance for long-term depressed cognition.
  - Disrupted white matter can be seen on MRI. Initial non-pharmacologic treatment is preferable to medication.
- **FRAILTY**
  - Patients designated as frail have abnormalities in the components below:
    - **Cognition** evaluated by a mini-mental status exam (MMSE) or mini-cog exam.
    - **Depressed mood** associated with exhaustion or social isolation.
    - **Comorbid diseases** as measured by the Charlson index.
    - **Weight loss and malnutrition,** including a low albumin.
    - **Disability** and limited mobility as measured by the capacity for activities of daily living (ADLs) and a timed up-and-go test of >12 seconds to stand, walk 3 meters, and sit back down.
    - Frailty is associated with longer LOS, readmission, need for discharge to NH, and postoperative morbidity, delirium, and mortality.
- **BRAIN DEATH**
  - Diagnosed by **two neurologic exams separated by 6–12 hours** documenting the absence of:
  - 1. Any **motor function** (with the exception of spinal reflexes); and
  - 2. Any **brainstem activity.**
  - Plus the patient should have a **positive apnea test.**
    - 100% FiO_2 is administered, the ventilator is turned off, and a baseline ABG is obtained. Then, observation for any respiratory effort. The test is positive if at **10 minutes** the patient has made no respiratory effort and has a repeat ABG showing a pCO_2 of 60 or an increase of >20 mmHg from the baseline pCO_2 value.
    - The test is indeterminate if the patient develops hypoxia or cardiovascular instability during the 10-minute period.
  - Hypothermia, hypotension, circulating intoxicants, drug overdose, and severe metabolic derangements preclude the diagnosis of brain death by the described examinations.
  - **Confirmatory tests** including cerebral angiography, nuclear medicine brain perfusion, transcranial Doppler (TCD), or electroencephalogram (EEG) may be needed in the event the diagnosis cannot be made by the above measures.
- **CEREBRAL PERFUSION PRESSURE (CPP)**
  - This is measured as mean arterial pressure (MAP) minus intracranial pressure (ICP), and should ideally be maintained at >60. Management of intracranial hypertension is aimed at preventing further neurologic injury.
  - 1. **Drainage of any focal collection** of fluid or blood and removal of any tumor-producing adjacent compressive effects.
  - 2. **Maintenance of MAP** so that CPP remains >60 mmHg. An intraventricular or intraparenchymal monitor may be used to follow ICP.
  - 3. **Head of bed elevation.**
  - 4. Control of **pCO_2 in the mid-30 range.**
  - 5. **Sedation,** anticonvulsants, temperature control, and glucose control.
  - 6. **Hypertonic saline or mannitol (1g/kg)** to decrease cerebral edema. The latter will decrease ICP within 20 minutes by drawing out intracellular fluid and decreasing blood viscosity, which can improve perfusion. One must keep serum osmolarity <300.

- 7. **Neuromuscular junction (NMJ) blockade** may be necessary if the above measures are inadequate.
- 8. **Decompressive craniotomy** is a last resort for diffusely edematous cerebral tissue in the absence of a focal mass effect.

- **CRITICAL ILLNESS POLYNEUROPATHY**
  - Axonal degeneration of motor and sensory nerves.

PULMONARY

- Total lung capacity (TLC) is composed of inspiratory reserve volume (IRV), tidal volume (TV), expiratory reserve volume (ERV), and residual volume (RV). *See Figure 22.6.*
  - **Functional residual capacity (FRC)** is ERV + RV.
  - **Forced vital capacity (FVC)** is TLC minus RV, and is generally >15 ml/kg. It is measured by having the patient take a maximum inhalation, followed by maximum exhalation down to RV.
    - **Forced expiratory volume in one second (FEV1)** is the amount of air expired in one second during testing for FVC.
      - FEV1 >2 liters allows one to tolerate pneumonectomy.
      - FEV1 >1.5 liters allows one to tolerate a lobectomy.
    - Restrictive lung diseases will decrease all capacities, but may have an increased FEV1.
    - Obstructive lung diseases will increase TLC (due to an increased RV). FVC can be normal or low. FEV1 is characteristically decreased to a greater extent than other parameters.
  - **Inspiratory capacity (IC)** is the volume that can be inspired at the end of normal expiration (IRV + TV).
- **Pulmonary vascular resistance (PVR)** is determined by a number of factors. Hypoxia or acidemia cause pulmonary vasoconstriction. Alkalemia, prostaglandins, and nitric oxide cause pulmonary vasodilation. Long-standing alveolar-capillary membrane destruction or fibrosis can elevate PVR.
- Ventilation (v) and perfusion (Q) are both higher in West Zone 3 (the dependent area of the lung). But the v/Q ratio is lowest in this region, since the increase in perfusion is relatively greater than the increase in ventilation.

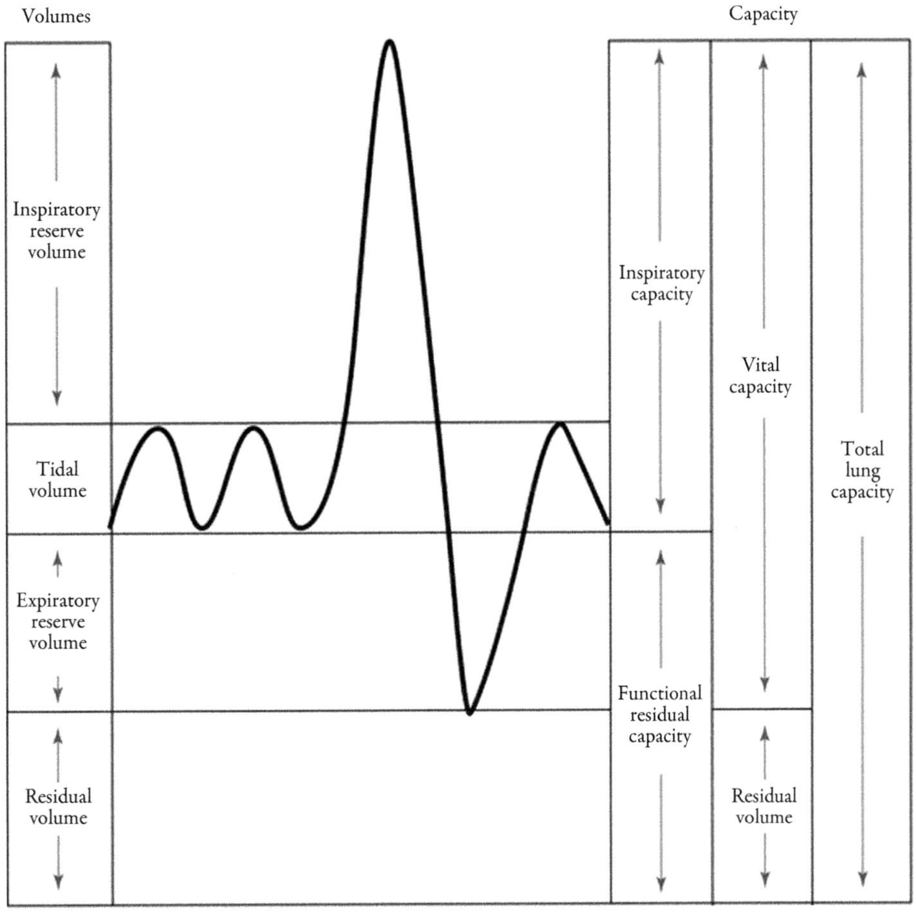

Figure 22.6 Lung volumes and capacities.

- **POSITIVE PRESSURE VENTILATION**
  - **Pressure support** simply supplies an expiratory pressure (PEEP) and an additional inspiratory pressure. Patient effort determines the number of inspiratory cycles as well as the duration of each cycle.
  - **Assist control mode ventilation** involves setting either a tidal volume or a peak pressure in order to deliver a given amount of air during a minimum set number of inspiratory cycles. Additional attempts at inspiration are supported with the same tidal volume or peak pressure. The inspiratory time, flow rate, and expiratory time can all be controlled. PEEP is usually supplied during exhalation.
  - **Synchronous intermittent mandatory ventilation** is a combination of volume or pressure assist control (utilized for a set number of breaths per minute) plus pressure support (utilized for any additional breaths attempted).
  - **Positive end expiratory pressure (PEEP)** can improve recruitment, FRC, and oxygenation. Excessive PEEP may decrease LA filling from alveolar compression of neighboring pulmonary capillaries. This decreases the volume of blood that can transit through the pulmonary circulation and hence LV preload. In this manner, excessive PEEP can decrease cardiac output.
  - If a low level of positive pressure is used throughout the inspiratory as well as expiratory cycles, it is termed **continuous positive airway pressure (CPAP).** This can be delivered without endotracheal intubation and is known as **noninvasive positive pressure ventilation (NPPV)**. Additional positive pressure can be given during each inspiratory cycle. NPPV can be useful for congestive heart failure (CHF) or chronic obstructive pulmonary disease (COPD) exacerbations.
  - Barotrauma to alveoli occurs at peak pressures >50 or plateau pressures >30. This can result in pneumothorax, pneumomediastinum, or long-term damage to the alveolar-capillary unit.
  - **Weaning from positive pressure ventilation and extubation.** Initial considerations include the absence of tachypnea or tachycardia, an adequate pH and pCO_2, normal mentation, the ability to follow commands when sedation is paused, presence of a cuff leak, and minimal secretions. In addition, the parenchymal criteria below should be met:
    - Negative inspiratory force (NIF) >20–30.
    - Rapid shallow breathing index (RSBI) <105 and ideally <80.
    - Minute ventilation (MV) <10 L/minute.
    - FiO_2 requirement <40%.
    - PEEP requirement of 5 or less.
- **ACUTE RESPIRATORY DISTRESS SYNDROME (ARDS)**
  - Marked by a capillary leak with neutrophilic and proteinaceous alveolar exudates.
  - There is increased TNF alpha and IL- during the **exudative phase at 24–96 hours.** This is followed by the **proliferative phase at 3–10 days** and the **fibrotic phase beyond 10 days.**
  - ARDS is often secondary to pneumonia (PNA) or sepsis, aspiration, multi-trauma, large surface area burns, or pancreatitis.
  - A loss of type 2 pneumocytes leads to a loss of surfactant and alveolar collapse.
- **DIAGNOSIS**
  - Bilateral acute-onset infiltrates without suspicion for CHF or volume overload by echocardiography or wedge pressure.
  - PaO_2/FiO_2 of <300 while receiving a PEEP of 5. Mild ARDS has ratios of 200–300, moderate 100–200, and severe <100.
  - **Shunt fraction >20% or dead space >60%** of TV are other indicators of severity in ARDS. Pathophysiologic dead space is within terminal bronchioles and alveolar units that have decreased perfusion. Anatomic dead space is along the trachea through to the proximal bronchiole, since there is no gas exchange interface.
- **TREATMENT**
  - **Low TV ventilation** of 6 ml/kg, **prone position** to change West zones, and **neuromuscular junction (NMJ) blockade.** Other, less common modes of ventilatory support are listed below.
  - Airway pressure release ventilation (APRV) includes settings for high pressure and time at this pressure, as well as low pressure and time at this low pressure. The patient spontaneously breathes through this assistance.
    - The difference between the high and low pressures determines volume of air and hence ventilation and pCO_2 levels. The duration of the low pressure time can assist with expiration in the case of CO_2 retention.
    - The duration of the high pressure time and increasing the value of the low pressure setting (PEEP) can determine oxygenation.
  - High-frequency oscillating ventilation (HFOV) is for pediatric patients and includes a low TV and high respiratory rate.
  - **Extracorporeal membrane oxygenation (ECMO)** can be considered if PaO_2/FiO_2 <100, shunt fraction >30%, plateau pressures >45, and hypercarbic acidemia with a pH <7.15 despite standard ventilatory care. This can be venovenous if cardiac function is adequate. Otherwise it should be venoarterial.
- **HOSPITAL ASSOCIATED PNEUMONIA (HAP)**
  - Pulmonary symptoms start **>48 hours after admission.** *H. fluenza,* MSSA, Strep pneumo, and anaerobes are the main organisms. GNR and MRSA predominate if >4 days after admission.

- **VENTILATOR ASSOCIATED PNEUMONIA (VAP)**
  - Symptoms or signs occur **>48 hours** after being placed on a ventilator. VAP has the same microbiology as HAP. Diagnosis is confirmed by a protected specimen brush or **bronchoalveolar lavage.** Prevention of VAP includes head of bed elevation to > 45 degrees, aspiration of secretions above the patient's tracheal cuff, and chlorhexidine (CHX) oral swabs

RENAL

- Renal arteries bifurcate into interlobar and arcuate vessels which give rise to the afferent arteriole that directly flows to Bowman's capsule and thereby filters plasma into the nephron lumen.
- The nephron has proximal and distal convoluted tubules (PCT and DCT) separated by the loop of Henle. The loop dips deep into the medullary interstitium. The solute concentration difference between the nephron lumen and the cortical and medullary interstitium determines the net movement of ions and water across tubular epithelium into peritubular capillaries, This is how plasma volume and composition are regulated. After passing through the loop and DCT, the collecting ducts carry the remaining filtrate as water is further reabsorbed. CDs drain into renal calyces, renal pelvis, and ureter.
- **Renin** is produced during periods of low perfusion pressure at the juxtaglomerular apparatus (JGA) or in response to low sodium concentration at the macula densa or catecholamine beta agonism. *See Figure 22.7.*
- Renin stimulates the liver to convert angiotensinogen to angiotensin 1. (AG1). This is converted to angiotensin 2 (AG2) via angiotensin converting enzyme (ACE) in pulmonary vasculature.
- **AG2** mediates peripheral vascular smooth muscle constriction, tachycardia, and myocardial contractility. It also directs the adrenal zona glomerulosa to release aldosterone.
- **Aldosterone** promotes sodium reabsorption along the distal convoluted tubule (DCT) of the nephron.
- Na and H_2O reabsorption are inhibited by **atrial natriuretic peptide (ANP)** released from the atrial wall when it is distended.
- **Antidiuretic hormone (ADH)** is released by the posterior pituitary when a patient is hyperosmolar or when peripheral venous tone is low. It causes vasoconstriction via V1 receptors, and water reabsorption at collecting ducts via V2 receptors. V3 receptors promote platelet adhesion.
- **POSTOPERATIVE AKI**
  - Most often due to intraoperative hypotension. Other causes include postoperative blood loss anemia, hypovolemia, dehydration, impaired cardiac function, or nephrotoxic medications damaging tubular cells or renal interstitium.
  - Evaluation of urine and serum can help definitively determine the etiology.
  - **Fractional excretion of sodium (FENa)** is most useful for patients with oliguric AKI. It is calculated as urine Na × serum Cr/urine Cr × serum Na.
  - An increased urine OSM >500 and a urine Na <20 with FENa <1% all suggest a pre-renal cause to AKI.

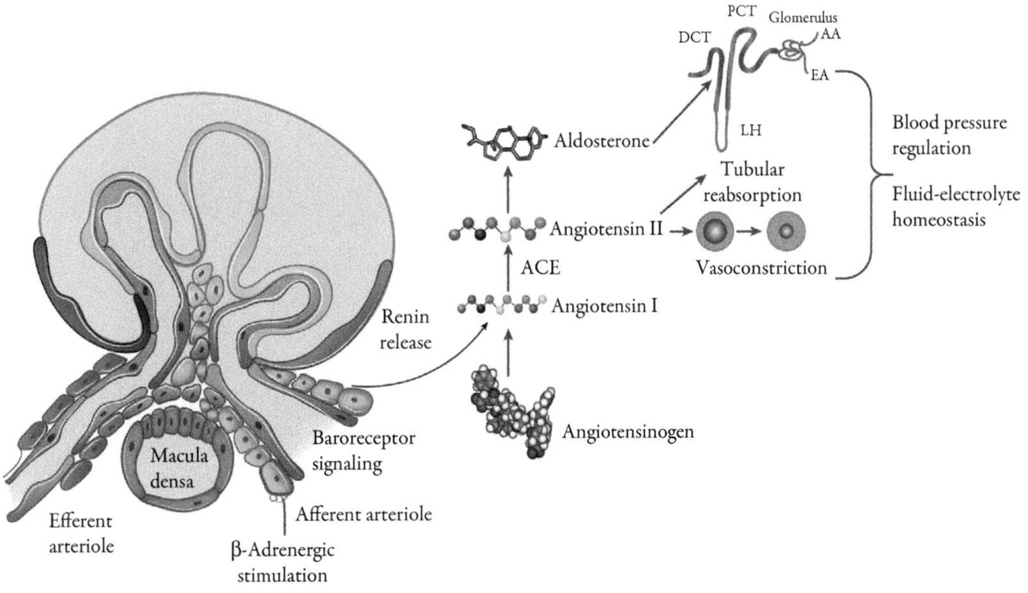

Figure 22.7 Renal glomerular and tubular endocrine regulation.

This could be from hypovolemia, anemia, or low perfusion pressure due to cardiogenic or thoracic obstructive pathology. Widespread renal arteriolar constriction can produce a similar pre-renal picture. The diminished glomerular filtration rate (GFR) incites functioning tubular cells to avidly reabsorb sodium along the nephron in order to preserve intravascular volume.

- Urine Na >40 and FENa >3% suggest that AKI is due to tubular cell dysfunction impairing sodium resorption. Tubular necrosis could have been secondary to a prolonged ischemic pre-renal state, or have developed independent of perfusion. There may be granular casts in a urinalysis.
- The **fractional excretion of urea (FEUrea)** may be needed if the patient is on diuretics inhibiting Na reabsorption, and thus falsely elevating the FENa.
- **Cerebral salt wasting** occurs after a brain injury, and inhibits sodium reabsorption along the nephron. This produces increased urine Na, urine OSM, and urine volume with resultant hypovolemia.

- Indications for urgent **hemodialysis (HD)** or **continuous renal replacement therapy (CRRT)** include hyperkalemia, uremia, acidemia, hypervolemia, bleeding secondary to thrombocytopathy, and end organ damage from exogenous toxins normally reliant on renal excretion.
  - CRRT has slower flow and is more physiologic than HD, since it provides continuous modification of plasma composition. It is safer for patients with hemodynamic instability.
  - Critically ill patients with multi-organ dysfunction and new renal failure may benefit from early institution of CRRT even in the absence of the above urgent criteria.
  - **Ultrafiltration** is when volume is removed but the plasma composition is not altered.
- Nonsteroidal anti-inflammatory medications (NSAIDs) decrease prostaglandin (PG) production resulting in renal arteriolar vasoconstriction and pre-renal AKI that can progress to tubular dysfunction or necrosis.
- Serum myoglobin and antibiotics like aminoglycosides and penicillins can cause direct tubular injury.
- **Contrast nephropathy** is non-oliguric and appears to have components that are pre-renal (FENa <1%) as well as intrarenal, with direct tubular damage and tubular casts seen in urine.
  - Creatinine increases by at least 25% from baseline within 24–48 hours.
  - Contrast injury most often does not require HD unless the patient had baseline CKD with a creatinine >4.
  - It can be prevented by nonionic iso-osmolar contrast and prehydration. Some practitioners choose to give N-acetylcysteine (NAC) 600 mg twice on the day prior to contrast administration.

23.

TRAUMA

PRESENTING PRINCIPLES AND IMMEDIATE EMERGENCY DEPARTMENT INTERVENTIONS

- Field deaths are due to cardiac or major vascular lacerations or devastating brain, brainstem, or upper cord injuries. ED and early hospital course deaths are due to brain injury or hemorrhage. Days to weeks later in a patient's hospital course, death is due to multi-organ failure or sepsis.
- Hemorrhagic categories and their physiologic findings include the following:
  - **Class 1 (<15% blood volume lost): tachycardia or normal pulse.**
  - **Class 2 (15%–30%): tachycardia, narrowed pulse pressure.**
  - **Class 3 (>30%): decreased BP and urine output.**
  - **Class 4 (>40%): end organ dysfunction.**
- **ED resuscitative thoracotomy** is for:
  - Neck or extremity trauma and CPR <5 minutes.
  - Blunt trauma and CPR <10 minutes (or loss of pulse in ED after blunt trauma).
  - Penetrating torso trauma and CPR <15 minutes.
  - T4–T5 incision along rib space from sternocostal junction to the posterior axillary line.
  - Open pericardium, repair myocardium, clamp aorta, and clamp pulmonary hilum if needed.
  - Extend a transverse sternal incision to clamp the R hilum if an R thoracostomy tube demonstrates hemothorax suspicious for major lung injury on that side, and there was no need for an L hilar clamp.
  - The right pleural space can also be accessed from trans-anterior mediastinal blunt dissection via a left thoracotomy if needed. If a hemothorax is encountered in this manner, then transverse sternotomy to clamp the R hilum should ensue.
  - Survival is 1%–3% for gunshot wounds and 8%–10% for stab wounds.
- **Retrograde aortic balloon occlusion** can be performed from femoral artery percutaneous access for those patients without a cardiac injury who are in extremis due to suspected infradiaphragmatic, intraperitoneal, or retroperitoneal hemorrhage.
  - Zone 1 (descending thoracic aorta) balloon inflation is at 43 cm–48 cm. This site is used empirically in a moribund patient expected to have life-threatening sub-diaphragmatic hemorrhage. *See Figure 23.1.*
  - Zone 3 (infrarenal aorta) balloon inflation is at 21 cm–31 cm and can be used when imaging or laparotomy suggest arterial hemorrhage is originating below the renal arteries.
- **External temporary pelvic binder** should be placed for suspected hemorrhagic shock and a physical exam or pelvic x-ray suggesting an unstable pelvis, particularly in anteroposterior compression injuries.
- **Massive transfusion protocol (MTP)** should be instituted for any patient with active hemorrhage and vital sign abnormalities. It allows the patient to receive pRBC:FFP:PLTs in a 1:1:1 ratio.
  - Universal donor blood is used most often in MTP. It is O+ for males and O- for females.
  - Type specific blood is screened for ABO compatibility, but not screened for antibodies or all minor antigens. So patients can have minor, delayed hemolytic or non-hemolytic reactions.
  - Type and cross-matched blood is identical in minor antigens and antibodies, but takes longer to obtain.
  - **Thromboelastography** can guide product resuscitation and is associated with increased FFP use and a mortality reduction.
  - **Tranexamic acid (TXA)** prevents plasmin formation and decreases hemorrhage, MI, and mortality if given within 3 hours of the initiation of hemorrhage. It is most effective if given <1 hour after the traumatic event. An 8-hour infusion should follow the bolus.

DIAGNOSTIC MODALITIES

- **Chest x-ray** to evaluate for blunt or penetrating thoracic injuries, pneumothorax (PTX), hemothorax (HTX), or major diaphragm disruption.
- **Pelvis x-ray** evaluates for anteroposterior, vertical shear, or bucket handle bony injuries. Major disruptions of the pelvic ring posteriorly can be associated with retroperitoneal, pelvic, and intraperitoneal hemorrhage.

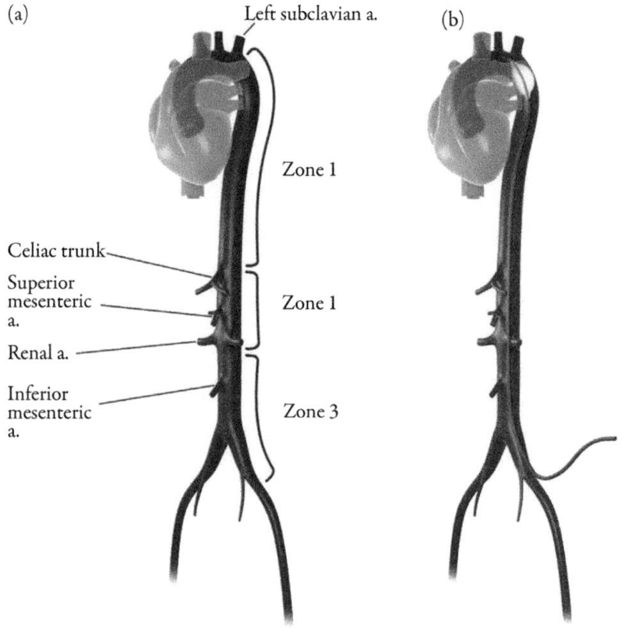

Figure 23.1 Retrograde aortic balloon occlusion in Zone 1 or 3.

- **Focused abdominal sonogram for trauma (FAST)** is for suspected hemorrhage in a patient presenting with tachycardia, narrow pulse pressure, or hypotension. FAST may miss retroperitoneal blood, small volumes of intraperitoneal fluid (<80 ml), and hollow viscus injuries.
  - **Diagnostic peritoneal lavage (DPL)** is used in rare instances when one needs to determine the type of fluid seen on FAST, or if no FAST is available. DPL should be supraumbilical in pregnancy or patients with pelvic fractures. A study is considered positive if it demonstrates >100k RBC in blunt trauma, >10k in penetrating trauma, >500 WBC, food particles, bile, urine, feces, or an elevated amylase. A negative DPL may miss retroperitoneal or contained mesenteric injuries.
- **CT abdomen and pelvis (CTAP)** is the most commonly used modality to diagnose intraperitoneal fluid, solid organ injury, and retroperitoneal injury in a stable patient.
  - CT may miss hollow viscus injury. In some instances it demonstrates secondary signs of bowel wall thickening, small amounts of free fluid without a source, small amounts of free air, or changes in the appearance of the mesentery.
  - CT may miss a diaphragm injury in left thoracoabdominal stab wounds. This is best diagnosed by laparoscopy.
- **Local wound exploration** can be employed to determine anterior fascial penetration for stab wounds in a patient with an otherwise benign exam.
  - If penetrated, then the patient should be observed with serial abdominal exams to determine if there has been visceral injury. If exams at any time thereafter suggest a visceral injury, then a diagnostic laparoscopy or laparotomy can be performed.
  - If no fascial penetration is noted on local wound exploration, then the patient may be discharged home.
- **Diagnostic laparoscopy** is for a stable patient with an indeterminate CT scan with or without a clinically worsening exam to evaluate bowel. It is also used for concerns of a diaphragm injury in a penetrating thoracoabdominal wound.

LAPAROTOMY INDICATIONS

- Free fluid on FAST in an unstable patient or a transient responder to the initial resuscitation.
- Large amounts of intraperitoneal fluid with the appropriate exam or vital signs, or an unreliable exam due to concomitant neurologic injuries.
- CT scan suggestive of a hollow viscus injury with the appropriate exam or vital sign changes, or inability to follow the abdominal exam due to mental status.
- Specific solid organ or adjunctive abdominal organ injuries may also need laparotomy if they produce hemodynamic instability, in spite of transfusion, with or without a preceding interventional radiology procedure.
- A **seatbelt sign** may indicate small bowel (SB) injury, sternal fractures, or lumbar fractures. Its presence may decrease the threshold to perform laparoscopy or

laparotomy if other factors (exam, vitals, imaging) are indeterminate.

- At laparotomy, if hemoperitoneum is encountered, **all 4 quadrants should be packed** with lap pads before beginning inspection of each quadrant independently.
  - Once **solid organ hemorrhage** is addressed, retroperitoneal hematoma should be evaluated. Open all **penetrating trauma hematomas** encountered, unless retrohepatic caval in location. Open all **blunt trauma hematomas** encountered unless they are stable and non-expanding in Zone 2, 3, or a retrohepatic caval hematoma.
  - Once intraperitoneal and retroperitoneal hemorrhage has been addressed, then one can **run the bowel** from ligament of Treitz through ileocecal junction and the entire colon. Also check the **lesser sac** for pancreatic and duodenal injuries.
  - See individual organ trauma sections below for management.
- **Damage control surgery** is recommended if temp <35, pH <7.2, base deficit >15, and multi-unit transfusions have been required intraoperatively, with or without vasopressors.
  - Damage control includes packing of venous bleeding sites, stapling off bowel injuries in discontinuity, and leaving the abdomen open with a temporary abdominal closure wound vacuum device.
  - Damage control will not work for arterial bleeding; this must be ligated or repaired. Use of a shunt or polytetrafluoroethylene (PTFE) is acceptable despite contamination in these cases.
  - If the laparotomy is negative but the patient remains unstable, then a transdiaphragmatic pericardial window is indicated, and the pleural spaces should also be decompressed empirically.
- **Abdominal compartment syndrome** is suspected when a patient has a tense abdominal wall in the setting of low urine output or high peak pulmonary pressures due to upward diaphragm displacement.
  - Patients may have multiple solid or hollow viscus organ injuries, or may have generalized visceral organ edema and ascites due to a recent large-volume resuscitation after hemorrhage or extensive burns.
  - **Intra-abdominal pressure** is measured via the bladder at end expiration on a ventilator in a sedated and paralyzed patient.
  - Diagnosis is confirmed with a **pressure >25 or a pressure >20 with any end organ dysfunction** (most commonly elevated pulmonary pressures and low urine output).
  - Management is by laparotomy to relieve increased SVR, restrictions to functional residual capacity (FRC) and total lung capacity (TLC) from upward diaphragm displacement, and impairments in preload from IVC compression. At that time, associated organ injuries should be addressed. Occasionally, a paracentesis alone can be therapeutic if the compartment syndrome is due to ascites.

BRAIN AND CRANIAL VAULT

See Figure 23.2.

- **Glasgow Coma Scale (GCS)** cumulative score determines traumatic brain injury (TBI) severity.
  - Mild (13–15 points) is often a concussion, and in the absence of any other concerning findings or coagulopathy does not necessarily need imaging. Concussions can produce brief confusion (Grade 1), amnesia (Grade 2), or transient loss of consciousness (Grade 3).

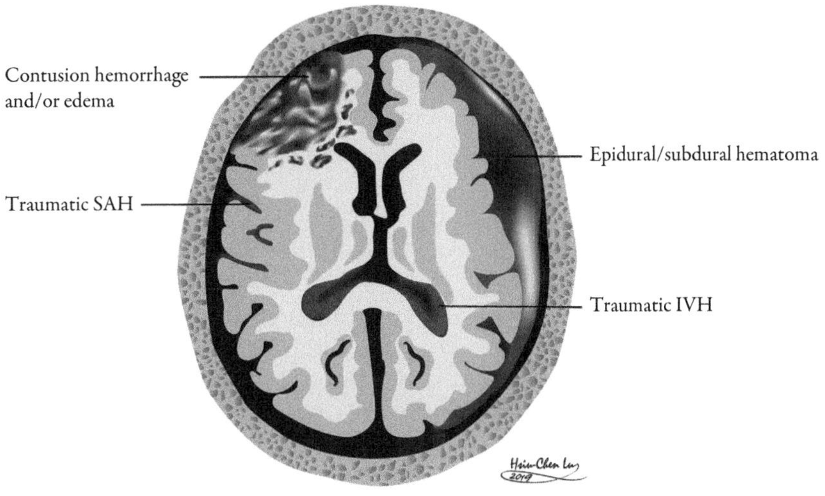

Figure 23.2 Types of traumatic intracranial hemorrhage.

- Moderate (9–12 points), and severe (3–8 points) TBI require brain imaging. Also, any patient with focal neurologic findings will need imaging and possibly CTA of the neck and brain, unless hemodynamic instability prompts operative care first.
- **Cerebral contusions** occur on the undersurface of the frontal and anterior temporal lobes as well as the upper midbrain. A hematoma may exist within it, especially in elderly or coagulopathic patients. Edema peaks at 24–48 hours.
- **Intraparenchymal hematoma (IPH)**, if post-traumatic, is located in the temporal lobe and may be associated with a background cerebral contusion. If secondary to hypertension, it will be in the deep grey matter. And, if associated with amyloid angiopathy, IPHs are peripheral and superficial. Treatment is often conservative, with medical management of systolic pressure, intracranial pressure (ICP), and coagulation status. Evacuation can be considered if <1 cm from surface, >20–50 cm3, or the patient has a decreasing GCS, midline shift, or effacement of basilar cisterns on imaging.
  - A less common cause of IPH is an arteriovenous malformation (AVM) rupturing intraparenchymally. These are treated by coil embolization, microsurgical resection, or stereotactic radiosurgery if small. Cavernous or venous hemangiomas or capillary telangiectasias have low intraparenchymal rupture risk.
- **Subarachnoid hemorrhage (SAH)** caused by trauma is associated with neighboring contusion and possibly IPH. They are treated conservatively, with BP and symptom management, serial imaging, and correction of coagulopathy.
  - A spontaneous rupture of a congenital aneurysm at the circle of Willis can produce a non-traumatic SAH. Blood in these cases generally pools around the brainstem and cisterns, and patients require further imaging to find an underlying aneurysm or AVM. Headache and stiff neck are seen with spontaneous ruptures. Treatment is with coil or clip of the aneurysm and hypertension hypervolemia hemodilution therapy to prevent vasospasm that can occur within 1–3 weeks thereafter. If it occurs, vasospasm is managed with calcium channel blockers.
  - Some SAH may be chronic with membranes that lead to recurrent acute hemorrhage episodes.
- **Subdural hematoma (SDH)** is due to torn bridging veins. The collection of blood does not cross midline, as it is restricted by the dural reflection. Treatment is to evacuate any SDH causing >1 cm shift or neurologic deterioration. This is done via craniectomy.
  - Chronic SDH may be evacuated by a burr hole. A subdural to peritoneal shunt can be used for long-term drainage.
  - A secondarily infected SDH can result in a subdural empyema producing thrombosis of crossing cortical arteries and subsequent cerebrovascular accident (CVA).
- **Epidural hematoma (EDH)** is an extradural hemorrhage due to a skull fracture lacerating the middle meningeal artery or another large cranial vessel. There need not be any underlying brain injury, so it has the best prognosis of any intracranial hemorrhage. Neurosurgery should generally operate for >0.5 cm thickness since the rate of expansion can be rapid and ensuing shift will compromise neuronal perfusion. Nonoperative management can be considered in select cases if GCS >8, hematoma is thin, and <30 cm2 volume.
- **Intracranial pressure (ICP) monitors** can be intraventricular or intraparenchymal. They are used in patients with a GCS <8 or suspected elevated ICP with inability to follow the neurologic exam and no indication for immediate operative intervention. They assist in calculating cerebral perfusion pressure (CPP) using mean arterial pressure (MAP) - ICP.
  - Goal is to keep the ICP <20 and CPP >60 with sedation, paralysis, raising the head of the bed, hyperventilation to pCO_2 mid-30s, ventriculostomy catheter drainage, and a craniectomy if needed.
  - Mannitol or hypertonic saline can be given for a transient reduction in ICP in patients exhibiting signs of impending herniation. Sodium level should be kept <150 and osmolarity <300.
  - Complications of ICP monitor placement include hematoma (1%) and infection (5%). Patients with ICP monitors generally receive anti-epileptic medications prophylactically for one week.
- **Skull fractures** are most commonly managed non-operatively unless the fragments are displaced a distance greater than the width of bone, or the fracture is open, contaminated, or associated with a persistent cerebrospinal fluid (CSF) leak despite conservative therapy.
  - **Frontal sinus fracture** that involves the posterior wall is in communication with the cerebral space. The sinus may need obliteration (cranialization) with sacrifice of the nasofrontal duct. The nasolacrimal duct must be preserved.
  - **Basilar skull fractures** can produce raccoon's eyes, Battle's sign, hemotympanum, or CSF otorrhea or rhinorrhea. CSF leaks are managed conservatively without antibiotics. A lumbar CSF drain can be placed to decompress the leak.
  - **Temporal bone fractures** are one component of the basilar skull. 20% are transverse in orientation and from frontal trauma. There may be associated cranial nerve VII or VIII dysfunction. 80% are longitudinal from a lateral blow with associated conductive

hearing loss. Cranial nerve VII injury is less likely in these cases.

- Cranial nerve VII paralysis may require urgent decompression and primary repair, interposition graft with greater auricular, cervical, or sural nerves, or use of cervical plexus nerve or ipsilateral XI or XII transferred to the distal VII stump for re-innervation.
- If the facial nerve dysfunction is delayed, then it can be due to edema alone and evaluated by electromyography or nerve conduction studies. If repair is needed later, temporalis, masseter, or digastric **muscle grafts** are used if there is fascial muscle atrophy.
  - Otherwise a **composite graft** of gracilis, latissimus, pectorals minor, serratus (and their corresponding nerves anastomosed to the proximal VII stump) are needed in long-standing cases where all facial muscle neuromuscular junction (NMJ) endplates have been lost.

SPINE

- **Spinal shock** is transient areflexia and flaccidity after a cord contusion or neuropraxic injury.
- **Complete cord** injury results in loss of all motor or sensory function below the lesion. Spinal reflexes may remain intact. **Neurogenic shock** is from high thoracic or cervical cord injury with loss of sympathetic fibers. Treatment is with fluid resuscitation and vasopressors to achieve mean arterial pressure (MAP) >85 ensuring cord perfusion.
- **Anterior cord syndrome** involves loss of motor and pain and temperature sensation below the level of the lesion, due to corticospinal tract (CST) and lateral spinothalamic tract (LST) dysfunction.
- **BrownSequard syndrome** is a hemicord injury with loss of contralateral pain and temperature, and ipsilateral motor and proprioception below the level of the lesion.
- **Central cord syndrome** is distal upper, more so than lower, extremity weakness due to hyperflexion or cord contusion.
- **Conus medullaris and caudal equine syndromes** involve compression near the level of cord termination or below. Symptoms include lower extremity weakness with alterations in bowel and bladder function.
- **American spinal injury impairment (ASIA)** categorizes the severity of the above types of cord syndromes.
  - A: Plegia below the injury, with no motor or sensory function in the sacral segments.
  - B: Sensory function preserved, but no motor function.
  - C: Paresis with less than half of muscle groups having at least 3/5 strength below lesion.
  - D: Paresis with more than half of muscle groups having at least 3/5 strength below lesion.
  - E: Patient is at baseline, pre-injury status.
- **CT cervical spine** is adequate to clear collar in a patient who is obtunded. Otherwise, clearance involves physical examination to determine that the patient is alert, not intoxicated, has no midline tenderness, no focal neurologic deficits, and no distracting injury.
- **Atlanto-occipital dissociation** is treated with plate fixation from occiput to cervical spine.
- **Jefferson fractures** of the C1 ring are due to axial load. Treatment is with immobilization alone, unless there is >7 mm of combined displacement of lateral masses. In that case, surgery can be considered.
- **Odontoid fractures** include type 1 through the tip of the dens, type 2 through the base of the dens, and type 3 through the vertebral body at the base of the dens. Type 2 fractures need surgical stabilization. The others can be immobilized.
- **Hangman fracture** is a bilateral C2 pedicle fracture. Treatment is with external fixation.
- **Unilateral (rotational force) or bilateral (distraction force) jumped facets** are treated with immobilization in traction.
- Philadelphia or Miami J collars are used for longer term stabilization of C3–C7 injuries.
- Vertebral body fractures are defined by the column involved:
  - **Anterior column** includes injury to the anterior longitudinal ligament and the anterior half of the vertebral body. These are compression fractures that can be treated non-operatively.
  - **Anterior/middle column** includes the posterior half of the vertebral body and posterior longitudinal ligament. These are burst fractures treated with plate and screw fixation with or without a bone graft.
  - **Middle/posterior columns** include the facet joints, lamina, and spinous processes. This type of burst fracture is a chance fracture. Treatment is with plate and screw fixation with or without a bone graft.
  - Fall from a height onto feet is associated with lumbar spine and calcaneus fractures.

FACE AND NECK

- A nasal **septal hematoma** needs drainage to prevent loss of the cartilaginous septum and collapse of the mid-vault. **Anterior epistaxis** can be packed. **Posterior epistaxis** needs posterior packing and occasionally endoscopic control of bleeding or maxillary artery embolization.
- **Facial fractures** can demonstrate periorbital ecchymoses, instability to the midface, anesthesia in the V2 distribution, malocclusion, and nasal or

oropharyngeal bleeding. They are best imaged with a facial CT scan.

- **Orbital bone fracture** is usually the inferior wall. Surgery is for enophthalmos, entrapment, impaired upward vision not corrected by forced adduction, or if a patient has diplopia >2 weeks.
  - Supraorbital fissure or orbital apex syndrome results in a deficit to cranial nerve 2. Treatment is with corticosteroids or surgery.
  - Carotid-cavernous sinus fistula is treated with angioembolization.
- **Naso-orbital ethmoidal fractures** can have telecanthus on exam. Two-thirds have a CSF leak, but this can be managed conservatively or with an occasional lumbar CSF drain.
- **Le Forte fractures** extend through the pterygoid plate and imply a higher risk of concomitant cervical spine injury. *See Figure 23.3.* They often need operative repair.
  - Type 1 fracture is across the lower maxilla.
  - Type 2 is through the nasal bridge and across the maxilla.
  - Type 3 extends laterally through the orbit.
- **Zygomaticomaxillary complex (ZMC) fractures** include breaks in the zygomaticomaxillary, zygomaticofrontal, and/or zygomatic arch. The arch should be repaired first to establish symmetry of the cheek, then the maxilla can be approached through a gingivobuccal incision and the frontal through an upper eyelid incision. Finally, maxillomandibular fixation (MMF) can be performed if needed.
- **Mandibular fractures** are diagnosed by symptoms of malocclusion and the tongue blade test. A panorex x-ray or CT scan can delineate the extent of injury. Treatment is with rigid or semi-rigid MMF or open reduction and internal fixation (ORIF).

- **Penetrating neck trauma** secondary to stab wounds have traditionally been divided into zones
  - Zone 1 is from the clavicle to cricoid and includes the trachea and upper esophagus. These patients may have a pneumothorax by clinical exam or x-ray.
  - Zone 2 is from the cricoid to angle of mandible and includes the hypopharynx and accessible neck vasculature.
  - Zone 3 is above the angle of the mandible and associated with cerebrovascular injuries that are not easily accessible.
  - **Immediate exploration** is recommended for any patient presenting with **hard signs of vascular injury.** These include pulsatile bleeding, an expanding hematoma, a pulse deficit at the cephalad carotid or upper extremity, signs of a CVA, Horner's syndrome, or hypotension due to field blood loss.
    - Large amounts of air bubbling from the wound or dissecting into the subcutaneous space are also considered hard signs of injury with regard to the aerodigestive tract, and should prompt exploration as well.
    - Carotid artery injuries can be repaired primarily with prolene. If a segment is missing or the repair would be under tension, the injury can be temporized with an intraluminal **shunt** while harvesting the greater saphenous or jugular veins as a graft.
    - For devastating internal carotid artery injuries, **transposition** of the proximal external carotid artery to the distal internal carotid artery and ligation of the internal carotid artery proximal to its injury can obviate the need for reconstruction of the internal carotid.
    - Unilateral internal jugular vein injuries may be ligated.
    - **Primary repair of the trachea or esophagus** with absorbable suture via a left neck approach is appropriate. The esophagus can be repaired in two layers. Drains should be left in place.
  - **CT angiogram (CTA)** is appropriate for patients without hard signs of vascular or aero digestive injury who are not being immediately transported to the OR. Some of these patients will have soft signs of vascular injury, including a non-expanding hematoma, non-pulsatile compressible bleeding, adjacent nerve paresis (other than Horner's syndrome), or symptoms of aerodigestive injury without air bubbling from the wound.

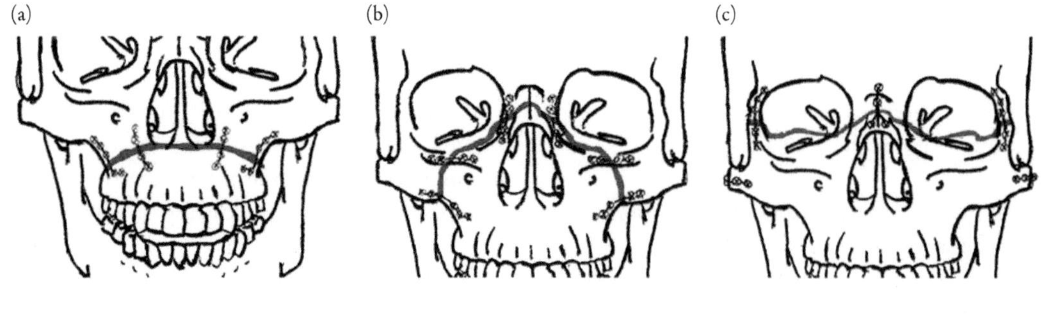

Figure 23.3 Le Forte fractures.

- Although most practitioners would obtain a CTA, some select stab wounds to Zone 2, without any signs or symptoms at all, may be followed with serial exams without imaging. Gunshot wounds (GSW), however, will still need a CTA even in the absence of symptoms.
- After a negative CT or a negative aerodigestive exploration, concerns of an esophageal injury should prompt an **esophagram** done with water soluble contrast. Since this can miss up to 15% of injuries, it is generally repeated with dilute barium. If still negative, an **esophagoscopy** should follow. This combination has the best sensitivity for cervical esophageal injury. A contained cervical esophageal injury seen on endoscopy or imaging may be followed clinically without intervention. A diet may be gradually reintroduced.
- If not contained, **primary esophageal repair** should be via the left neck using absorbable suture in two layers. If there is significant contamination and the injury is subacute or chronic, one may just leave cervical drains. See the Chapter 7, Esophagus, for diagnosis and management of intra-thoracic esophageal injuries.
- **Primary tracheal repair** should be performed whenever possible. A laryngeal release can be used if there is greater than 4 cm defect missing. See the trachea section of Chapter 29, Thoracic, for more details.

- **Blunt Neck Trauma** is more common than penetrating injury.
  - **CTA neck** evaluating for a carotid or vertebral artery injury is indicated for an abrasion across the area of carotid, Le Forte or major maxillary or mandibular fractures, C1–C3 fracture, transverse process fractures through the vertebral foramina, facet dislocation, posterior epistaxis, cerebral ischemic signs or symptoms, suspicion for diffuse axonal injury (DAI), or presence of depressed GCS precluding a complete neurological exam.
  - Blunt carotid injuries are generally 2–3 cm distal to the carotid bifurcation.
    - Carotid injury Grade 1 is <25% luminal occlusion from intimal injury.
    - Grade 2 is > 25% luminal occlusion from intimal injury.
    - Grade 3 is a pseudoaneurysm.
    - Grade 4 is complete thrombotic occlusion.
    - Grade 5 is a transection.
    - Carotid injury is managed with **heparin** (goal PTT 40–50) or aspirin to prevent thromboembolic CVA for grades 1,2, and 4 provided the patient remains asymptomatic.
      - **Aspirin** may be used in place of heparin if the patient is otherwise dischargeable or will remain an inpatient with intrameningeal extension of the vascular injury or other concomitant injuries that are at high risk for bleeding.
    - **Stent or surgical repair,** with or without a GSV vein patch, is reserved for symptomatic patients or those with Grade 3 pseudoaneurysmal injuries.
      - Grade 4 injuries can be operated on if <2 hours old and the patient has a preserved level of consciousness. Grade 5 injuries may receive an operation under similar circumstances, or if the associated hemorrhage is not controlled.
  - **Vertebral artery injuries are graded like carotid injuries. Intimal disruption may be treated with heparin or antiplatelets** to prevent thromboembolic CVA in asymptomatic patients. One may embolize or ligate and exclude the vertebral artery and eliminate the embolic risk in symptomatic patients.
  - **Esophageal or tracheal injuries** are less common than in penetrating trauma, but are approached similarly. If there is a laryngeal fracture, one may still secure the airway with endotracheal intubation or cricothyroidotomy in the ED.

CARDIAC AND AORTIC

- A **stable patient with a penetrating box injury** and hemopericardium by FAST exam generally is taken to the OR for **pericardial window** performed via a 5–8 cm vertical incision directly over the xyphoid. The xyphoid is excised and the retrosternal space is entered bluntly. The anterior diaphragmatic reflection of the pericardium is grasped and opened, and the pericardial space drained externally. This may be sufficient if only old, dark blood is found and the patient remains hemodynamically stable.
  - If fresh blood is noted at the time of the window, then a sternotomy should be done in spite of hemodynamic stability at that moment.
  - In some instances, practitioners may choose to observe a stable patient with a penetrating box injury and small volume pericardial fluid, reserving pericardial window and sternotomy only for hemodynamic changes. This is a less common course of action.
- An **unstable patient with a penetrating box injury** and hemopericardium needs immediate median sternotomy. An emergency department resuscitative thoracotomy is utilized if the patient is moribund or pulseless and cannot make it to the operating room. Median sternotomy is the ideal exposure for myocardial, ascending aorta, BCA, L CCA, or BCV injuries.
- **Blunt cardiac injury.** Patients with a sternal fracture are considered higher risk. The diagnosis can be ruled out by a normal ECG.
  - If the ECG is abnormal, including sinus tachycardia, then one may consider **inpatient telemetry**

monitoring and possibly serial troponins. The latter are not mandatory.
- For persistent or recurrent arrhythmia in a stable patient, consider a **transthoracic echocardiogram**.
- For persistent or recurrent arrhythmia causing hemodynamic instability, exclude hemorrhage, obstructive shock, cardiac tamponade, or neurogenic etiologies to hypotension. Then obtain an echocardiogram and ECG to exclude ischemia. Monitor the patient in the ICU for inotropic or vasopressor support, and have cardiology involved.
  - The range of injuries include small or large myocardial muscle contusions, valvular injury, traumatic coronary artery thrombosis, or papillary muscle, septal, or free wall rupture requiring cardiac surgery and cardiopulmonary bypass. *See Figure 23.4.*

- **Aortic injury** is suggested by a rapid deceleration injury in a patient complaining of chest pain.
  - CXR can demonstrate a widened mediastinum, apical capping, or loss of the aortic contour. These findings can be due to distortion of the aortic wall itself or, more often, mediastinal hematoma from associated venous injuries. Historically, upper rib fractures on CXR have implied a significant enough force to also disrupt the aorta. Up to 20% of patients will have a normal CXR.
  - Definitive diagnosis is by CTA. The injury is most commonly located at the ligamentum arteriosum. Less commonly it is at the diaphragmatic hiatus.
  - Management first involves addressing other life-threatening injuries. Then the aorta can be approached with thoracic endograft or open repair with or without partial left heart or full cardiopulmonary bypass during the repair. Aortic injuries have a 1% mortality per hour and a 10% paraplegia rate postoperatively.

LUNG AND THORACIC CAGE

- **Non-hilar lung injuries** are generally the result of peripheral penetrating trauma or blunt trauma with rib fractures lacerating pulmonary parenchyma. They are best managed with tube thoracostomy after CXR demonstrates a pneumothorax or hemothorax.
  - An unstable patient probably has a **tension pneumothorax** due to one-way entry of air into the pleural space and resultant shifting of the mediastinum and kinking of great veins. Treatment is with needle or finger decompression at T4–T5 space in the anterior or mid-axillary line. Then place a chest tube through the decompression site.
  - A **sucking chest wound (open pneumothorax)** results from a chest wall injury that is greater than two-thirds diameter of the trachea. This allows environmental air to enter the pleural space during inspiration. *See Figure 23.5.* The pleural air causes lung collapse every inspiratory cycle. There may or may not be any significant pulmonary parenchymal injury. Treatment is with coverage of the wound on three sides to prevent entraining air and also allow for expiratory release. Then a chest tube can be placed at another site along the hemithorax.
- An **operating room thoracotomy** is indicated for more than 1500 ml of initial bloody output from a chest tube placed for a hemothorax. Thoracotomy is also indicated for output greater than 250 ml/hour for 3 hours, or any continued hemorrhage from a chest tube with concurrent hemodynamic instability. These patients may have cardiac, aortic, or **hilar, or major pulmonary vessel injuries.**
  - **Left anterior thoracotomy** is best for access to the descending aorta, proximal left subclavian artery injuries, and left hilar lung injuries.
  - **Right posterolateral thoracotomy** is best for access to the distal trachea, carina, and proximal left bronchus, right bronchus, right hilar injuries, or esophageal injuries.
  - **Supraclavicular incisions** allow intervention on distal R or L subclavian artery injuries.
  - **Infraclavicular incisions** are used for axillary artery injuries. The clavicle can be resected for complete exposure.
  - **Tracheobronchial injury** may require contralateral mainstem bronchus intubation prior to operative repair. Patients may benefit to have their chest tube on water seal. Repair if PTX with air leak >3 days.
  - A pulmonary tractotomy can be performed using a stapler device to lay open a penetrating parenchymal injury and address hemostasis and pneumostasis of the exposed surfaces. Then the open wound bed and staple lines can be re-approximated with a running locked stitch.
- **Video assisted thoracic surgery (VATS)** is utilized for the evacuation of a retained hemothorax not resolved after one chest tube, persistent air leak >3 days, or non-expansion of the lung from any pleural cause.
- Primary **blast injury** pressurization causes alveolar rupture and pulmonary contusion. Blast effect is enhanced in an enclosed space. Hollow viscus gastrointestinal (GI) structures and the tympanic membrane can also rupture from blast injury. Secondary blast injury is from shrapnel, tertiary injury is from the body being thrown, and quarternary associated injuries are from burn, crush, or toxic inhalation.
- **Rib fractures** are generally managed conservatively.
  - An **anesthetic rib block, thoracic wall catheter, or epidural anesthesia catheter** can **decrease mortality** after multiple rib fractures by improving pain control and ventilation.
  - **Rib plating** for internal fixation with or without sternal plating can **decrease ventilation time** in

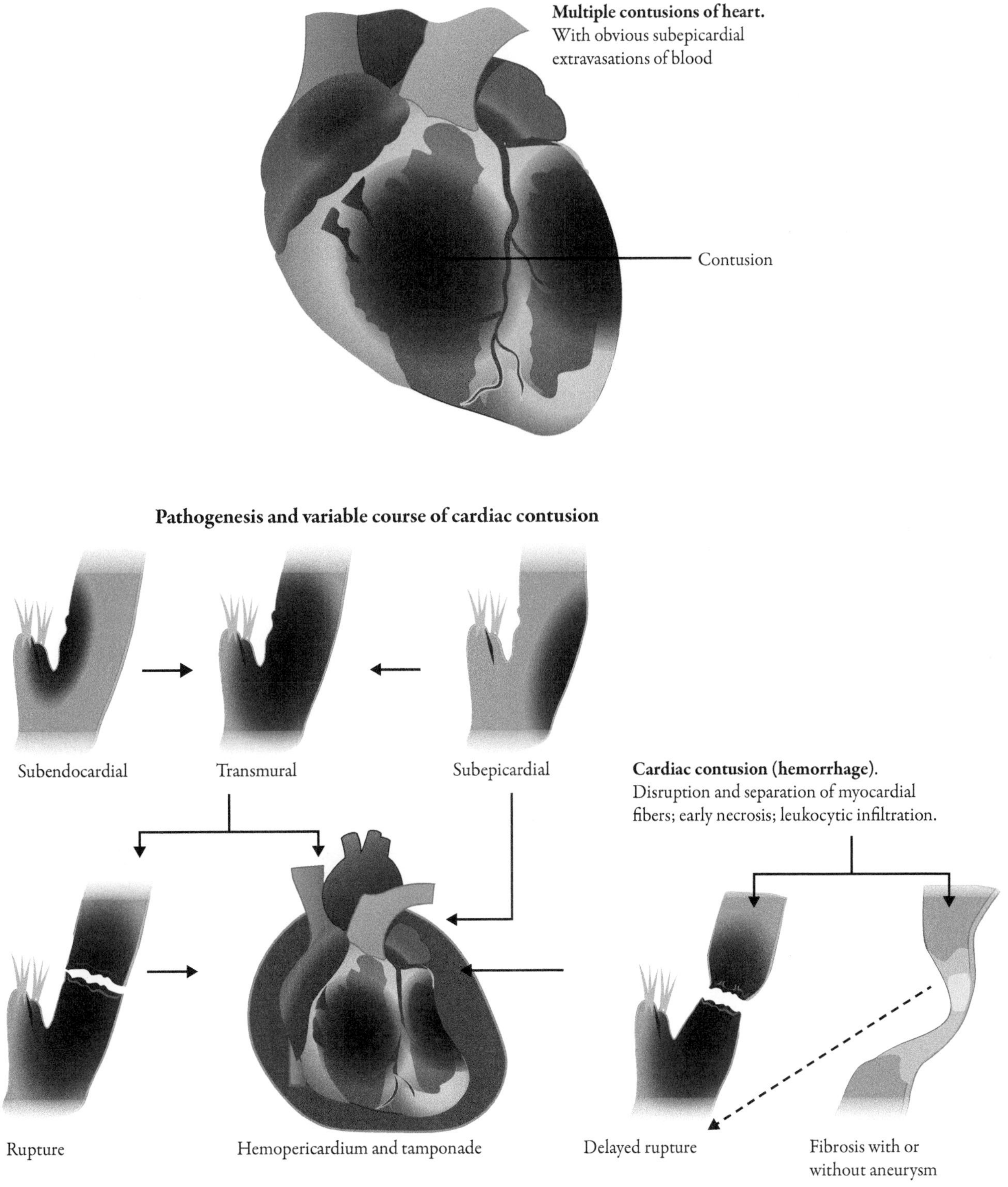

Figure 23.4 Blunt cardiac injury.

patients with flail segments and no significant underlying pulmonary contusion (or parenchymal disease) as the primary reason for their ventilator dependence.

- **Diaphragm Injuries** can be from a blunt trauma blowout (occurs on the left more commonly than the right) or penetrating trauma.
- Penetrating trauma can have missed injuries and delayed incarceration. Therefore, consider **diagnostic laparoscopy** for any thoracoabdominal penetrating wound, especially if on the left, and even if CT imaging is negative.
  - Primary repair is with non-absorbable suture with or without synthetic mesh via an abdominal approach,

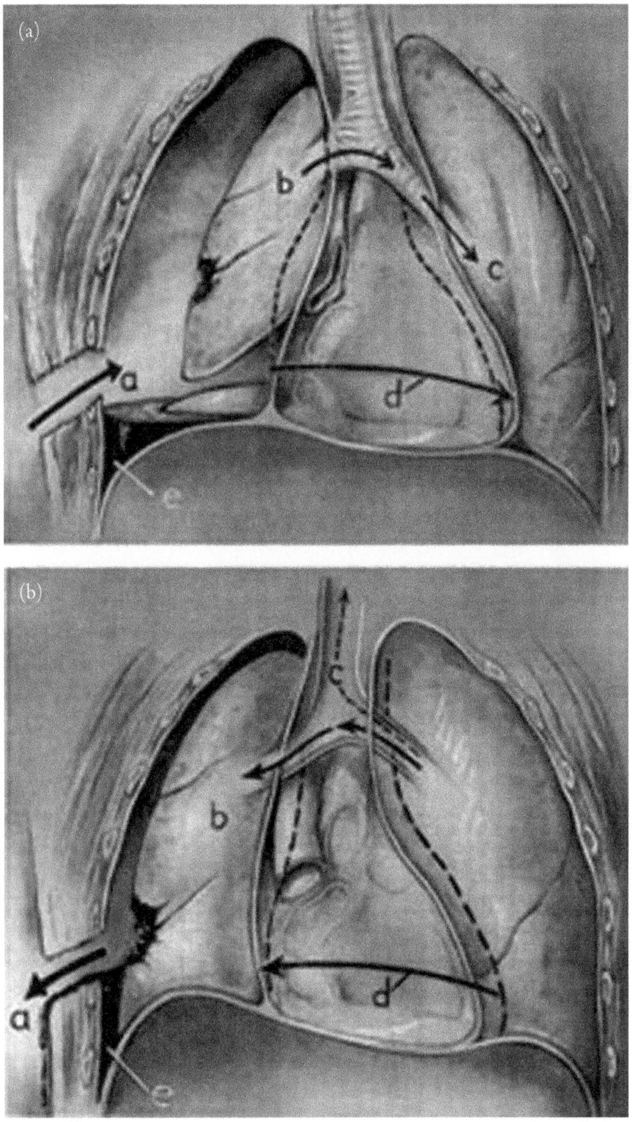

Figure 23.5 A sucking chest wound as a cause to tension pneumothorax.

provided it has been <1 week since the traumatic event. This allows one to address other intra-abdominal injuries as well.
- May transect the diaphragm from its attachments laterally and transpose it superiorly in order to reduce tension on the repair of a large defect.
- Repair via a thoracic approach if >1 week since injury, in order to take down adhesions.

VASCULAR

- After blunt or penetrating injury, **hard signs** requiring immediate exploration for a vascular injury include:
  - Active hemorrhage.
  - A pulse deficit.
  - An expanding pulsatile hematoma.
  - Distal limb ischemia.
  - A bruit or thrill.
- **Soft signs** requiring further imaging with CT angiogram include:
  - A non-pulsatile hematoma.
  - Neural deficit adjacent to a vascular structure.
  - A measured ankle-brachial index (ABI) or brachio-brachial index (BBI) < 0.9.
  - A history of hemorrhage in the field prior to arrival.
- Retroperitoneal zones and major vessel exposure strategies. *See Figure 23.6.*
  - Incising the left triangular ligament, opening the hepatogastric ligament, and incising the right crus of the diaphragm to access the cephalad supramesocolic aorta and gain vascular control.
  - Opening the lesser sac from an infragastric route or hepatogastric ligament to access the peripancreatic vessels and supramesocolic aorta.
  - Lifting the transverse colon and releasing the ligament of Treitz with retraction of the duodenum

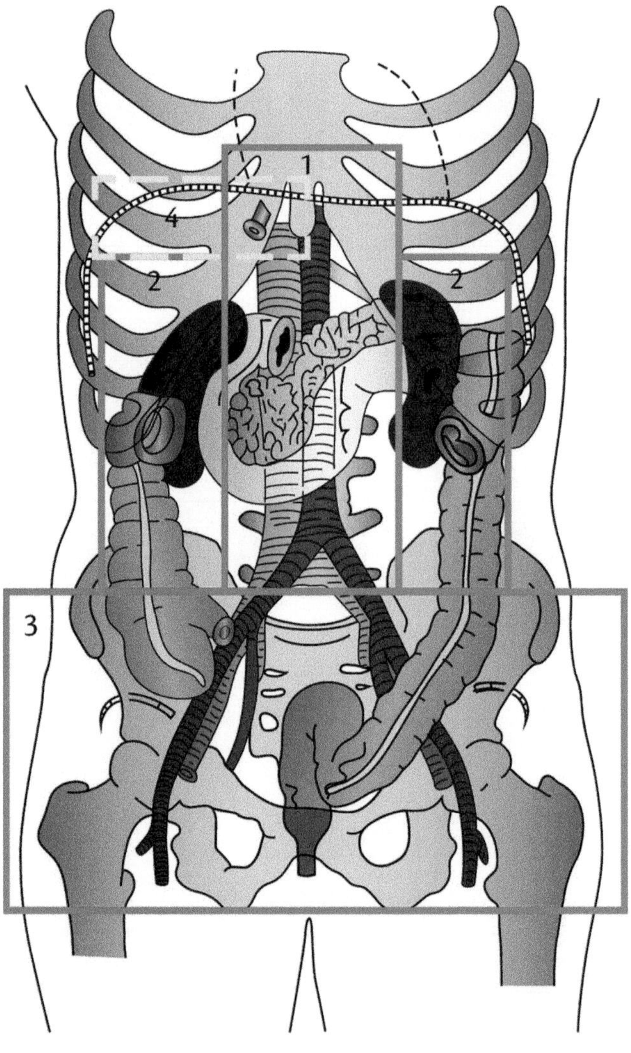

Figure 23.6 Retroperitoneal zones 1–4.

to the patient's right to access the proximal inframesocolic aorta.
  - **Cattell** right medial visceral rotation of the ascending colon *See Figure 23.7.*
  - **Kocher** maneuver medially rotating the C loop of the duodenum.
  - **Mattox** left medial visceral rotation of the descending colon, spleen, distal pancreas, kidney. *See Figure 23.8.*
- For the unstable extremis patient with intraperitoneal bleeding not controlled with packing, one must manage arterial inflow to mitigate hemorrhage. Options include:
  - A supraceliac aortic clamp.
  - Descending thoracic aortic clamp.
  - Retrograde aortic balloon occlusion in descending thoracic aorta Zone 1.
- Arterial injuries may be **temporarily shunted** in line, to control hemorrhage and provide distal perfusion while the patient is being stabilized. *See Figure 23.9.*
- Arterial re-anastomosis should be end–end, and performed after branch ligation to improve mobility and reduce tension. A **Fogarty embolectomy** distally and proximally will help clear inflow and outflow prior to the anastomosis.
  - **SVG interposition** grafts taken from the contralateral greater saphenous vein (GSV), will be needed for missing segments in vessels with diameters <6 mm.
  - A **PTFE interposition** graft will be needed for missing segments in vessels with >6 mm diameter. It is acceptable to reconstruct with PTFE even in the setting of bowel contamination.

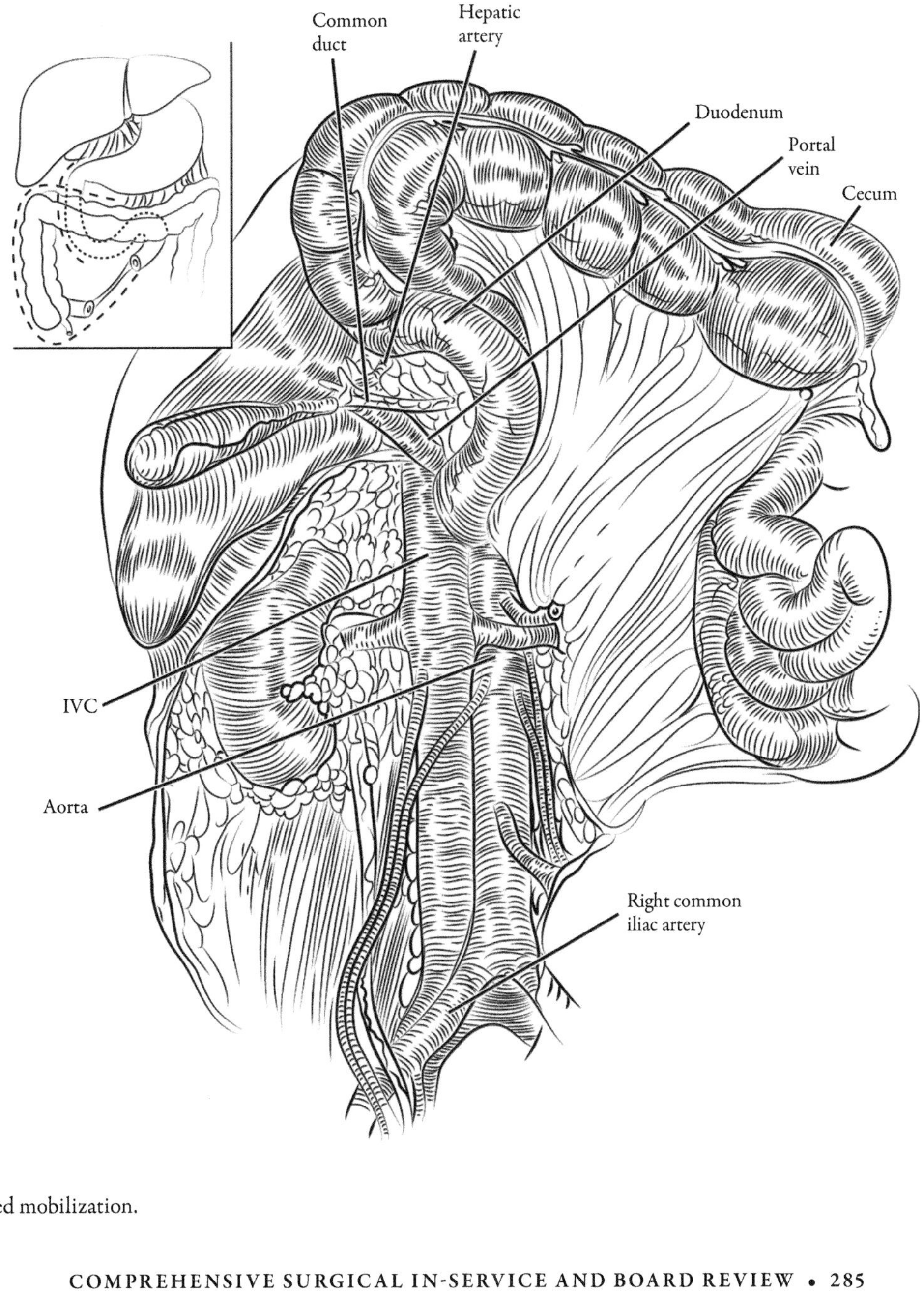

Figure 23.7 Right-sided mobilization.

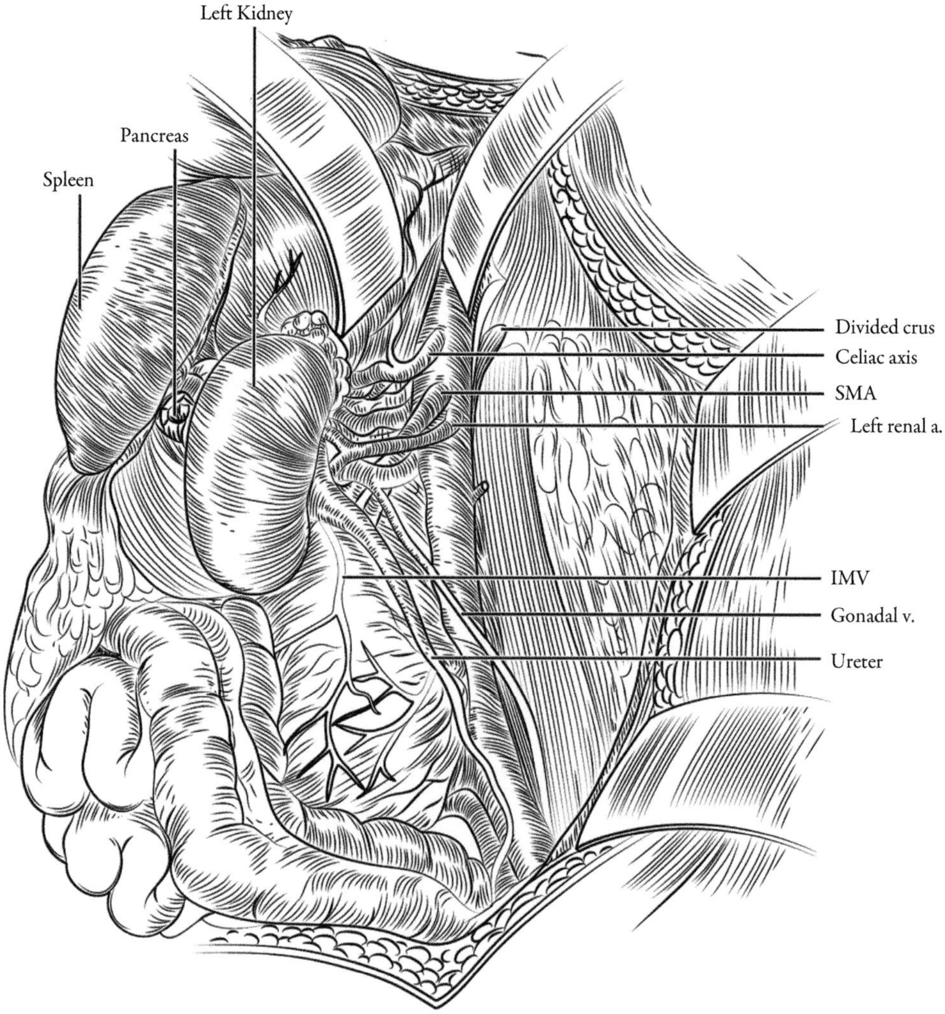

Figure 23.8 Left-sided mobilization.

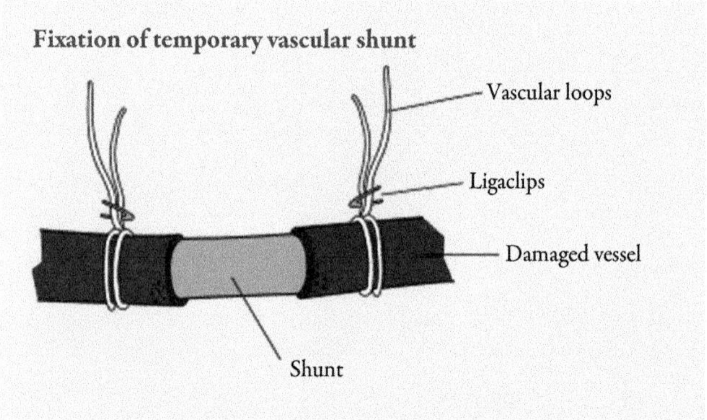

Figure 23.9 Temporizing shunt for vascular injury.

- Before using an interposition graft, consider **arterial transposition** for injuries not reparable primarily, but having a parallel arterial conduit in close proximity. Examples include anastomosis of the proximal ECA to the distal ICA and ligation of a proximal ICA injury. Or anastomosis of the proximal IIA to the distal EIA and ligation of a proximal EIA injury.
  - An injured subclavian artery may be **stented or ligated** because collaterals from the vertebral artery, thyrocervical trunk (which includes the supracervical, transverse cervical, ascending pharyngeal, and inferior thyroid branches), and the dorsal scapular, circumflex scapular, subscapular arterial complex can all feed the distal subclavian or axillary artery.
- **Venous repair** should be attempted at the following injury sites: SVC, suprarenal IVC, brachiocephalic, subclavian, axillary, femoral, or popliteal veins.
  - For the larger veins with full-thickness injuries, posterior wall repairs can be accomplished via the anterior venotomy injury.
  - For missing segments not amenable to primary repair, one may use interposition PTFE to keep vein patency for 1–2 weeks while venous collaterals form.

    - At venous sites other than those listed above, including the infrarenal IVC, ligation is acceptable for patients in extremis.
- There is a higher mortality with penetrating proximal vascular injuries.
- There is a higher amputation rate with blunt distal vascular injuries.
- **Compartment syndrome** is due to restoration of flow after a prolonged limb ischemia, usually more than 4–6 hours. Injury is mediated by **PMNs during the reperfusion phase**, resulting in muscle edema within a fixed fascial compartment. When compartmental pressure exceeds capillary pressure, ischemia ensues once again.
    - Common etiologies include a crush injury of a limb, supracondylar fractures, proximal tibial fractures, or arterial injury or occlusion followed by reperfusion. Patients having concomitant venous and arterial injuries in the same limb are at higher risk.
    - Other cases may be chronic and exertional, leading to elevated pressures in the anterior and lateral lower leg compartments.
    - The diagnosis is clinical. Symptoms and signs start with pain with passive range of motion progressing to paresthesia, anesthesia, paralysis, poikilothermia, and pulselessness last. Associated acute kidney injury is due to myoglobin casts. If needed for confirmation, compartment pressures can be measured and are usually >20, or a calculated delta pressure is <30 (DBP - compartment pressure).
        - Management is prompt fasciotomy. For the lower extremity, a four-compartment fasciotomy is via a lateral incision over the fibular area to access both the anterior and lateral compartments, and a medial incision running just posterior to the tibial edge to access the superficial and deep posterior compartments. *See Figure 23.10.*
        - The anterior compartment contains the tibias anterior, extensors, and deep peroneal nerve. It is essential to recognize the fascial interface separating the anterior and lateral compartments, so as to ensure that both compartments are independently opened to their full extent in a cranial-caudal orientation.
        - The lateral compartment contains the peroneus muscles and neighbors the superficial peroneal nerve, which can be injured.
        - The deep posterior compartment has the tibialis posterior, flexors, and the posterior tibial nerve. It is accessed by detaching the soleus and entering the area just posterior to the tibial bone.
        - The superficial posterior compartment has the soleus, gastrocnemius, and the sural nerve nearby.
    - An upper extremity volar fasciotomy starts at the ulnar side of the upper arm, crosses the antecubital fossa to the middle or radial side of the forearm, and heads down to the middle of wrist. This helps release the median nerve at both sites proximally and distally. An upper extremity dorsal fasciotomy is on the radial side of the upper and forearm.

STOMACH

- Injuries are diagnosed by CT scan, or at laparoscopy or laparotomy for another indication.
- **Grade I–III** injuries can be repaired primarily in two layers.
- **Grade IV** injuries involve significant destruction and devitalization of a portion of the stomach and generally require a partial resection (antrectomy or partial gastrectomy) with Billroth I or II reconstruction.
- The lesser sac should be opened to evaluate any posterior gastric wall, pancreatic, or vascular injury.

DUODENUM

- The mechanism of injury is most often blunt and has four grades:
    - **Grade I**: Hematoma or laceration without perforation seen on CT.

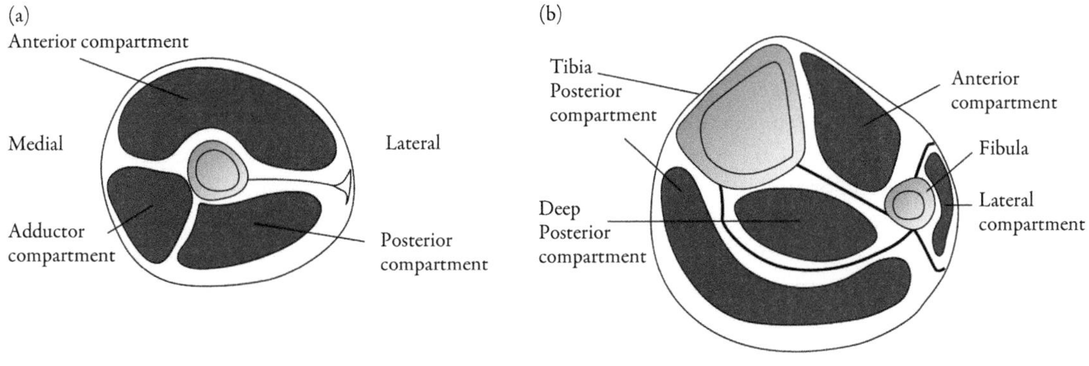

Figure 23.10 Leg compartments - upper (a) and lower (b).

- **Grade II**: Disruption of less than 50% of circumference.
- **Grade III**: Disruption of 50% to 75% circumference of D2 or 50% to 100% of D1, 3, 4.
- **Grade IV**: Disruption of more than 75% circumference of D2 involving the ampulla or distal common duct.
- **Grade V** injury involves massive disruption of the duodenopancreatic complex or duodenal devascularization.

- Injuries are diagnosed by CT scan or during laparoscopy or laparotomy for another indication.
- Management depends on the injury grade.
  - Grade 1 can be managed with **observation and nasogastric tube (NGT) decompression** for 2–3 weeks. Some patients may need supplemental total parenteral nutrition (TPN). If a Grade I injury is discovered intraoperatively, the serosa of the hematoma may simply be **incised and decompressed.**
  - Grades II–IV are managed with either **primary repair** with or without a jejunal serosal patch or **resection and anastomosis.**
    - Another option is pyloric exclusion and gastrojejunostomy with a proximal intraluminal drain of the injured duodenal stump and a distal feeding jejunostomy.
    - Duodenojejunostomy involves stapling proximally and distally to the duodenal injury and taking proximal duodenum and anastomosing it to the side of the distal jejunum.
  - Grade V injuries should be managed with drain placement in the acute period, and consideration for a delayed pancreaticoduodenectomy.

JEJUNUM AND ILEUM

- The small bowel is most commonly injured after penetrating trauma.
- After blunt trauma, loops of bowel may have been transiently caught between the object striking the abdomen and the patient's spine. This can result in blowout of the bowel secondary to a **temporary closed loop physiology under high pressure**.
- Free air on CT would be diagnostic of a hollow viscus injury. Secondary signs of injury on CT include free fluid without a source (no solid organ injury to explain the fluid), bowel wall thickening, mesenteric stranding, or hematoma.
- Management includes a mandatory exploration if peritonitis is present or the patient has evidence of hollow viscus injury on imaging. If they have an unreliable abdominal exam accompanied by secondary signs of bowel injury by CT, a seatbelt sign, or chance fracture of lumbar vertebrae, they may also need exploration. Otherwise, if the abdominal exam, vitals, and CT are benign, one can perform serial abdominal exams and operate for any worsening findings.
  - Transverse **primary repair** is the treatment of choice if <50% of the wall is injured
  - **Resection and anastomosis** if >50% of the wall is injured or there are multiple smaller lacerations in close proximity.
  - Also evacuate large mesenteric hematomas. If bowel perfusion is compromised by any mesenteric injury, then resect that segment of bowel and its mesentery.

COLON

- Injuries are diagnosed by CT scan or during laparoscopy or laparotomy for another indication.
- Management is based on the degree of injury.
  - Transverse **primary repair** is adequate if <50% of the wall is damaged and there is minimal fecal contamination. **Resect and anastomose** injuries of >50% colonic wall. Consider proximal diversion if a low colorectal or coloanal anastomosis is constructed.
  - An **end stoma** rather than anastomosis can be employed if there is heavy, prolonged contamination preoperatively or if the patient is systemically ill.
  - Damage control surgery and colonic discontinuity is best for the critically ill, hypothermic, acidemic, or hemorrhaging patient. The colon can be stapled off and the abdomen left open.
  - Open any paracolonic hematoma and resect any colon with questionable perfusion.

RECTUM

- Injuries are diagnosed by physical examination for low extraperitoneal canal injuries, a CT scan, or during laparoscopy or laparotomy for another indication.
- Management depends on the degree of rectal injury and whether it is above or below the peritoneal reflection. *See Figure 23.11.*
  - **Intraperitoneal injuries** are treated with a primary repair or resection and anastomosis based on colonic principles above.
  - **Extraperitoneal injuries** are treated with proximal colonic diversion alone if not reachable from a transanal approach. If low anorectal within 5 cm of the verge, one may repair the injury trans-anally, with or without proximal diversion.
    - If involving the sphincter mechanism, patients may need a wraparound **sphincteroplasty**, post-anal intersphincteric levatorplasty, and a future gracilis transfer for neosphincter construction. An artificial

balloon sphincter or sacral nerve stimulator are other options to try and reestablish continence in the future.

HEPATIC

- Blunt injuries are more common than penetrating, with **90% successfully managed non-operatively.** *See Figure* 23.12. Delayed bleeding can occur at >48 hours after injury.
- Grade 1: <1 cm laceration, <10% subcapsular hematoma.
- Grade 2: 1–3 cm laceration, 10%–50% subcapsular hematoma, 1 cm–10 cm intraparenchymal hematoma (IPH).
- Grade 3: >3cm laceration, >50% subcapsular hematoma, >10cm IPH.
- Grade 4: 25%–75% of one lobe injured; 1–3 segments involved.
- Grade 5: >75% of one lobe, >3 segments involved, central hepatic vein or inferior vena cava (IVC) injury.
- Grade 6: Shattered and avulsed liver.
- Injuries are diagnosed by CT scan with IV contrast or during laparotomy for another indication.
- Management depends on hemodynamic stability and imaging findings. Most injuries can simply be observed.
  - **Angioembolization** is for a hemodynamically stable patient with arterial blush on CT angiogram.
  - **Exploratory laparotomy** is reserved for hemodynamic instability despite initial resuscitation.
- Pack all four quadrants upon entry, and then focus on the liver. First take down falciform, triangular, and coronary ligaments for mobility, and control hemorrhage with perihepatic packing, manual compression, topical hemostatics, argon coagulation, and blunt needle repair with or without omental graft insertion.
- **Pringle maneuver** can be repeated for up to 15 minutes each time to control hepatic arterial or portal venous tributary bleeding. It will not control hepatic venous bleeding.
- Explore any **portal triad hematoma.**
  - Primary repair of the portal vein or SMV may require pancreatic transection or partial pancreatectomy.
  - Primary repair of the common bile duct is acceptable for injuries <50%. Choledochoenterostomy is needed if >50% of the circumference of the duct is involved.
  - The proper hepatic artery can be primarily repaired or reconstructed with saphenous vein graft.
- For a **penetrating injury through the liver,** consider placing a Penrose drain over a balloon catheter into the tract. This can be inflated for intraparenchymal tamponade.
- **Retrohepatic caval** control is by infrahepatic IVC clamp and a suprahepatic IVC clamp, with diaphragmotomy if needed for intrapericardial control of the suprahepatic IVC.
- For patients being managed non-operatively who have CT evidence of a **biloma,** a percutaneous drain may be placed.

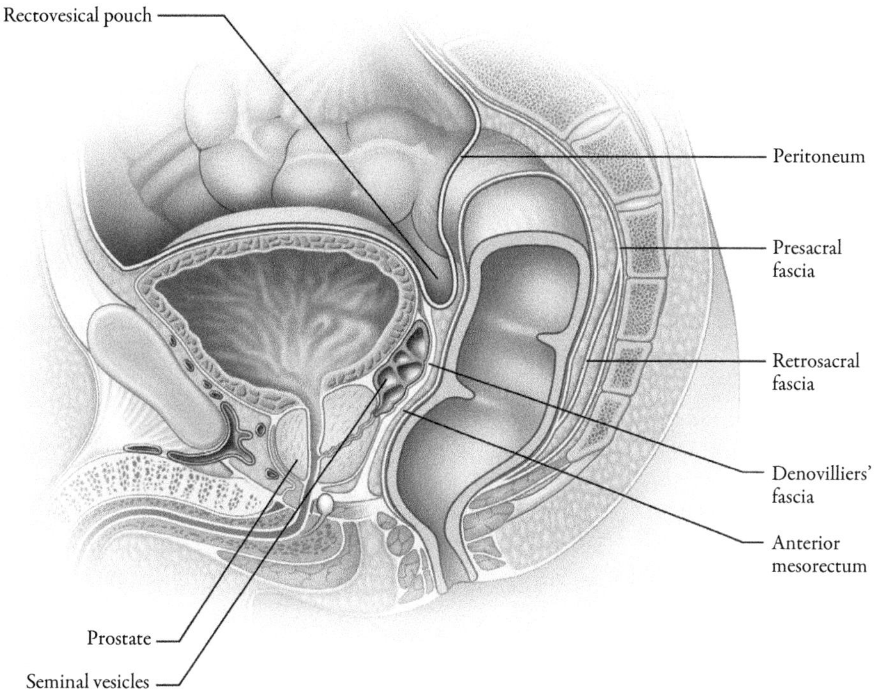

Figure 23.11 Intraperitoneal and retroperitoneal rectum.

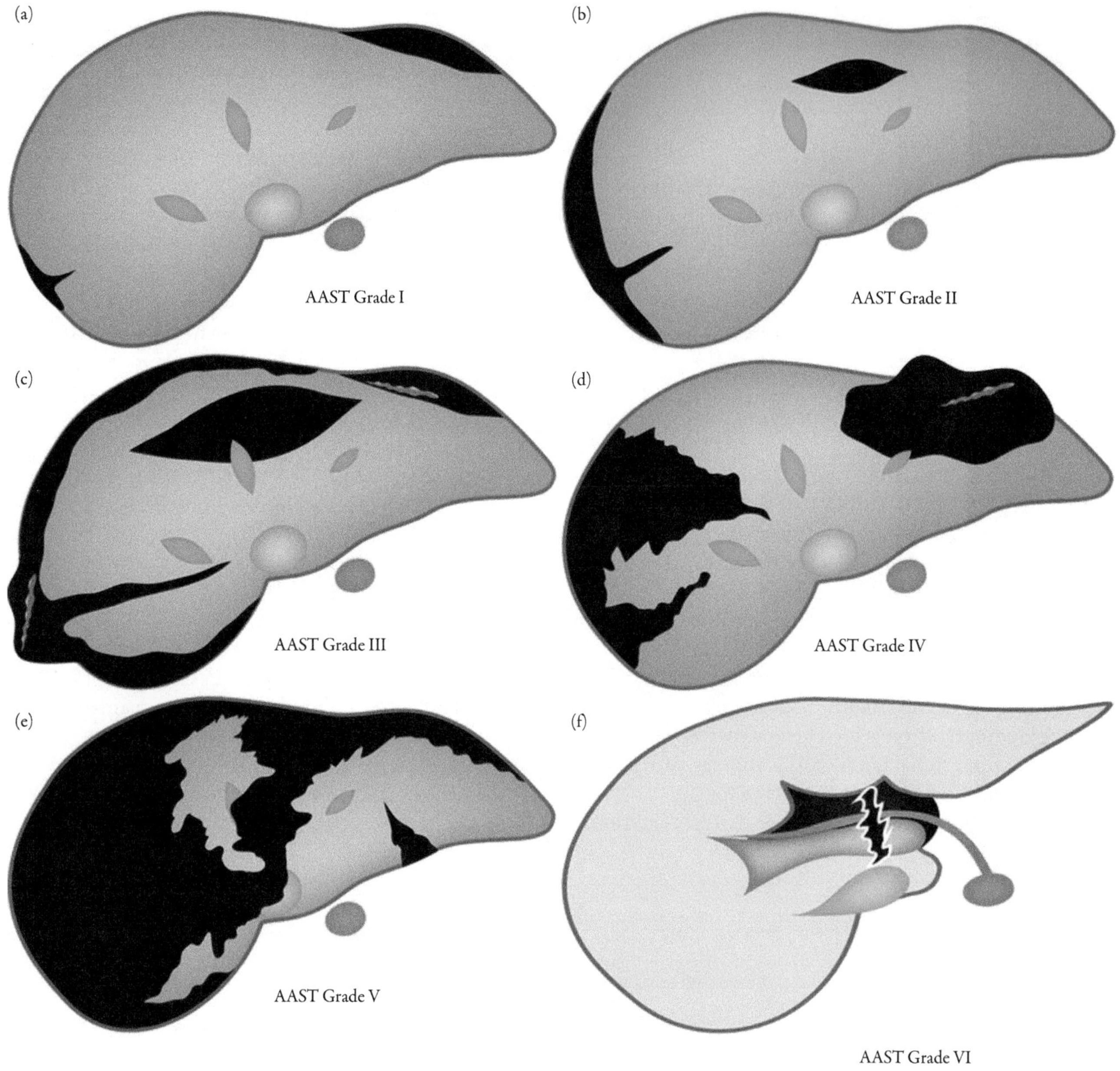

Figure 23.12 Grades of liver injury.

    - A parenchymal bile leak <300–500 ml/day may continue to be managed with drain alone for up to 14 days.
    - For higher volume or prolonged leaks, a bilogram via the drain can note the site of injury. Endoscopic retrograde cholangiopancreatography (ERCP) sphincterotomy and stent can encourage internal drainage and divert bile from the injured parenchyma.
    - Cholangiogram and delayed operative intervention are indicated for patients not improving with the above.
- **Delayed upper gastrointestinal (UGI) bleeding** after liver parenchymal injury should be evaluated with esophagogastroduodenoscopy (EGD). If **hemobilia** is noted extruding from the ampulla, then an intraparenchymal arterial to biliary communication exists. Treatment is with angioembolization.

SPLENIC

- Splenic injury or rupture is usually in the setting of major trauma. Occasionally, the traumatic event was not noted or considered significant by the patient, and they present with delayed rupture.
- Atraumatic splenic rupture can occur secondary to a hematologic malignancy or infection that resulted in organ engorgement and fragility.

- Patients can have left upper quadrant abdominal pain with referral to their left shoulder, pleuritic left lower chest pain, or more generalized abdominal pain from hemoperitoneum. Stable patients with suspicion for intra-abdominal trauma or splenic injury can receive CTAP with IV contrast, and the grade of injury can be defined. *See Figure 23.13.*

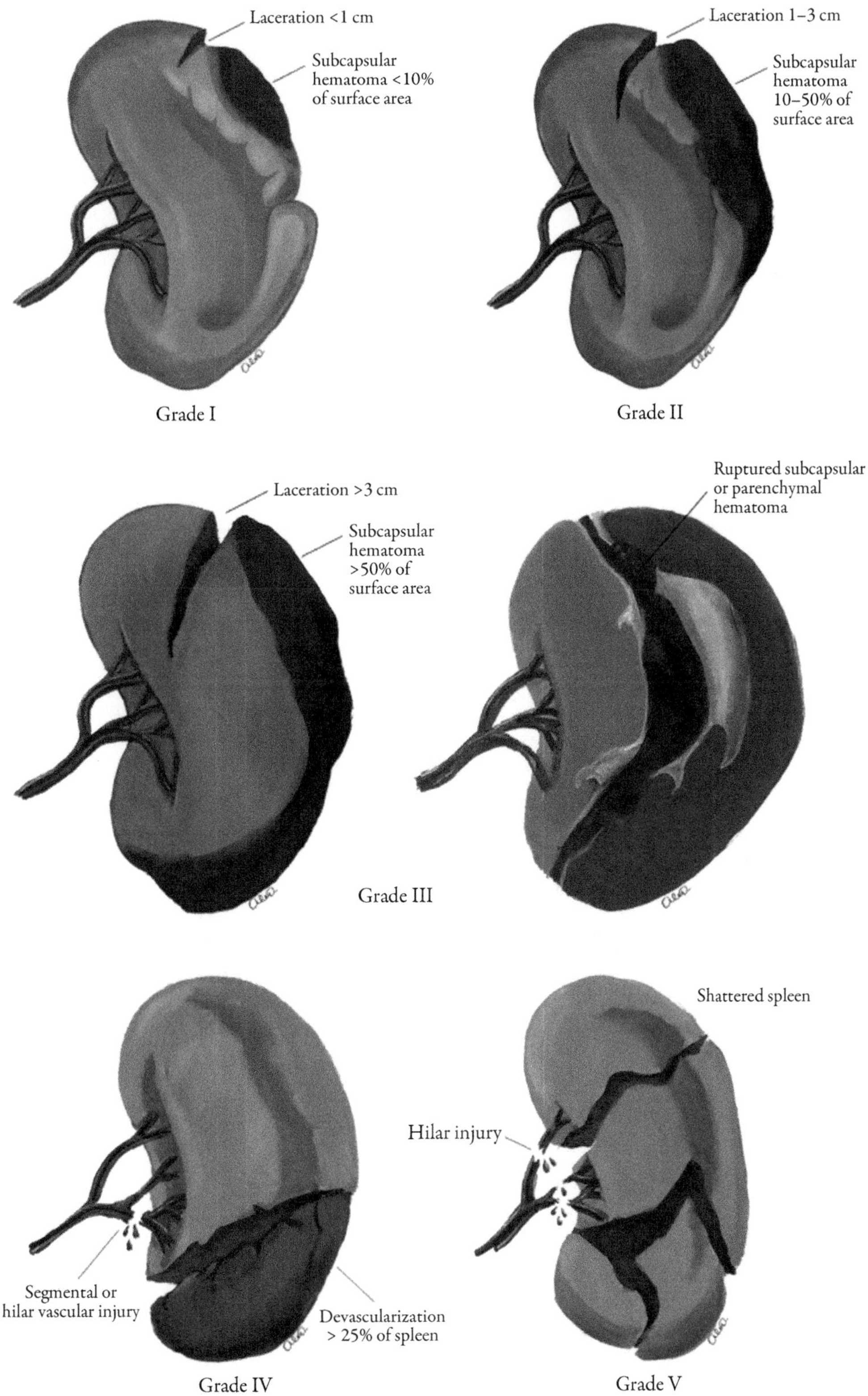

Figure 23.13 Grades of splenic injury.

- **Grade 1:** <10% subcapsular hematoma or <1 cm laceration.
- **Grade 2**: 10%–50% subcapsular hematoma or 1–3 cm parenchymal laceration, or <5 cm intraparenchymal hematoma (IPH).
- **Grade 3**: Subcapsular hematoma >50%, laceration >3 cm, IPH >5cm, or any extravasation or pseudoaneurysm associated with a splenic injury.
- **Grade 4:** >25% of the spleen devascularized.
- **Grade 5**: A shattered spleen with hilar devascularization.

- Management depends on hemodynamic stability and imaging characteristics. A significant number of injuries can simply be observed.
  - An unstable patient with a positive FAST US for intraperitoneal blood may need **exploratory laparotomy,** four quadrant packing, and possibly emergent splenectomy if this is the source.
  - A **nonoperative trial** can be continued despite any of the above grades of injury, provided the patient remains hemodynamically stable after initial fluid resuscitation or up to 2 units pRBC transfusion, and the patient has no other major injuries. However, one must recognize that 10% of low-grade injuries, and 75% of high-grade injuries, fail a nonoperative trial.
  - **Selective angioembolization** is for a stable patient with Grade 3 injury, discrete arterial extravasation on CTA. **Nonselective angioembolization** is an option for the stable patient in whom a nonoperative course is chosen.
  - **Splenectomy** is reserved for the following:
    - Hemodynamic instability despite initial resuscitation.
    - Continued product transfusion requirements in a nonoperative trial, with or without a preceding selective or nonselective angioembolization.
    - A high-grade injury with other confounding injuries needing laparotomy, or accompanying CNS injuries that would make the patient intolerant of any decrease in their cerebral perfusion pressure.
  - **Splenorrhaphy** is a consideration in a stable patient, often a young child, in whom one-third of the organ can be saved.
    - Topical cautery, argon coagulation, microfibrillar collagen, gelatin sponge, or methylcellulose can be used for hemostasis of the cut edge of spleen.
    - Horizontal mattress suture plication can be used with organic pledgets, and possibly a tongue of omentum sewn in place for hemostasis.
    - Mesh wrap of the entire spleen can be done with a purse-string suture.
    - **Partial splenectomy** is an option for deeper lacerations, provided one-third of the organ can be preserved. Techniques include:
      - Ligation of the inferior or superior pole arterial and venous tributaries.
      - Stapled or cautery excision of the devitalized portion of the spleen.
      - Oversewing the cut margin with pledgets, with or without an omental inclusion.
  - **Delayed hemorrhage** can occur up to 6 weeks post-injury. After that time, the parenchyma has usually healed.
  - **Immunizations** should be given prior to discharge. The highest risk for post-splenectomy sepsis would be within 2 years of removal in a child whose hematologic disorder prompted the splenectomy (see below).

PANCREATIC

- Penetrating injuries are most common. Blunt injuries are associated with vertebral chance fractures or a seatbelt sign.
- Injuries are diagnosed by CT scan or at time of laparoscopy or laparotomy for another indication.
- Management depends on the grade of injury.
  - **Grade 1** (minor contusion) and **Grade 2** (major contusion) are both managed non-operatively or, if in the OR for another reason, closed suction drains can be placed. Open any peripancreatic hematoma if you are already in the OR addressing other injuries.
  - **Grade 3** injuries include a distal pancreatic duct transection with or without a peripancreatic fluid collection. Management includes a stapled distal pancreatectomy. Some patients may also need a splenectomy. Ligation of the duct proximal to the staple line can decrease postoperative leak rates.
  - **Grade 4** injury is a proximal duct transection with or without a pancreatic head injury and fluid collection. Complex pancreatic head injuries may be managed with drains in the acute setting and Whipple in a delayed fashion.
- Major central disruptions can benefit from a central pancreatectomy with Roux-en-Y pancreatojejunostomy to drain the remnant tail, if the patient is stable. The head can be stapled off on its distal end, with the proximal end simply left to drain into the duodenum via an uninjured proximal pancreatic duct.

UROLOGIC

- **Renal Trauma**
  - The renal vein lies anterior to the artery, which is anterior to the renal calyces draining into the renal pelvis and proximal ureter.
  - All **penetrating trauma** in proximity to the kidney in a stable patient needs CT imaging with IV contrast.

Patients may need rectal contrast as well, to evaluate the neighboring ascending or descending colon. **Blunt trauma** with gross hematuria in adults, or >50 RBC/HPF in pediatrics, needs renal imaging with CT IV contrast provided the patient is stable. Microscopic hematuria on an initial urinalysis in an adult who was hypotensive at the time of collection may also be imaged.

- **Renal Injuries**. *See Figure 23.14.*
  - **Grade 1**: Subcapsular hematoma.
  - **Grade 2:** Perirenal hematoma or <1 cm depth cortical laceration.
  - **Grade 3**: >1.cm depth medulla laceration or extraparenchymal bleeding confined. Gerota's fascia.
  - **Grade 4:** Laceration into the collecting system, renal artery, or vein.

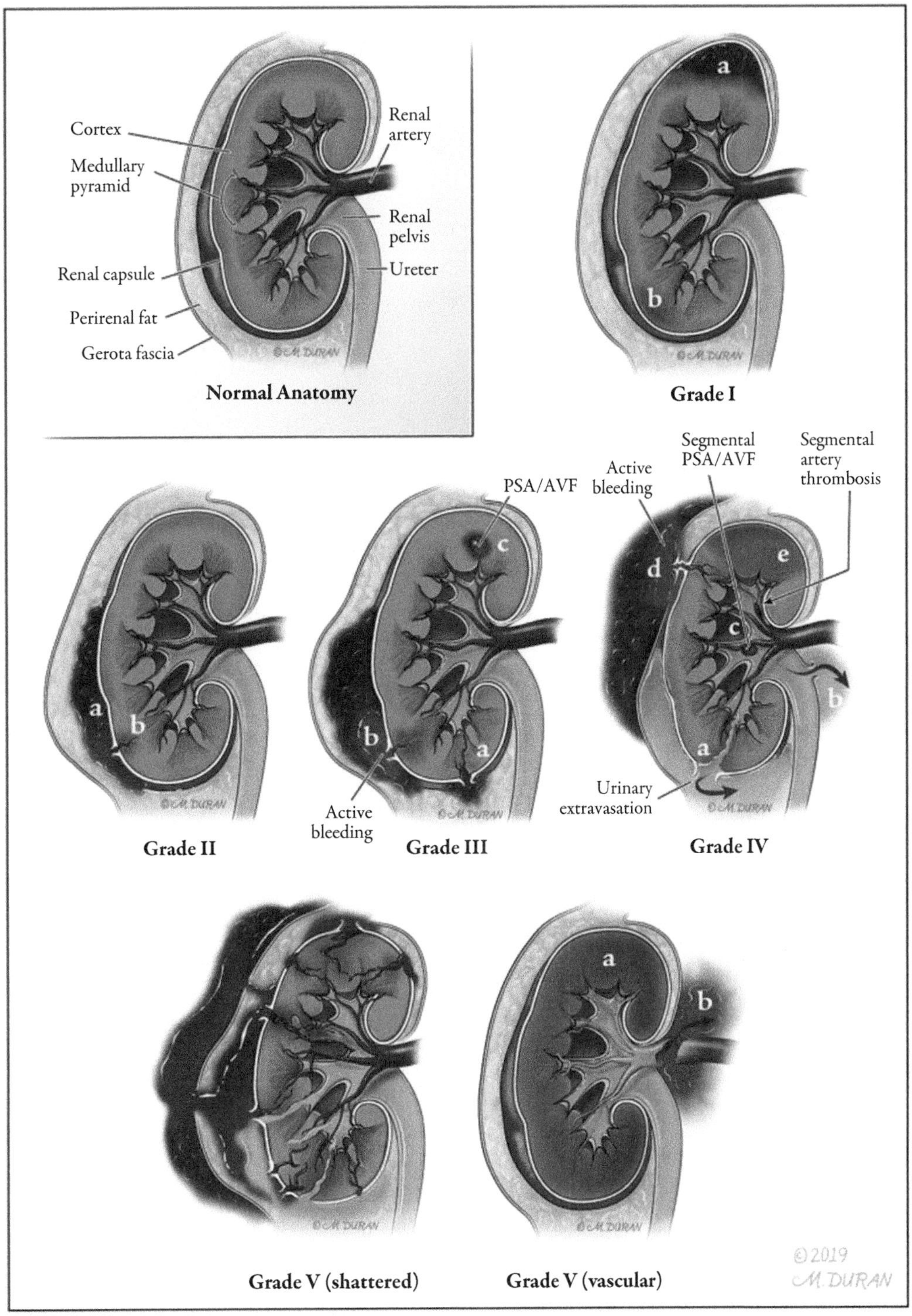

Figure 23.14 Grades of renal injury.

- **Grade 5**: Shattered parenchyma or avulsed hilum.
- **Operative exploration is needed for any unstable patient.** Most often a nephrectomy will be required in these cases.
  - One may ligate the left renal vein near the IVC, and still perform renal repair in an attempt to spare the kidney, since adrenal and gonadal vein collaterals exist that can continue to drain the organ. But ligation of the right renal vein necessitates a nephrectomy.
- **Blunt trauma Zone 2 perirenal hematomas** found when operating on other injuries can be left alone, unless they are expanding.
- **Penetrating trauma Zone 2 perirenal hematomas** need exploration unless a preoperative CT was available and showed normal perfusion and no major arterial or venous injury.
- Perirenal drains should be left within any urinoma encountered intraoperatively. A ureteral stent can be placed in the subacute period of a collecting system injury.
- **Angioembolization** is used for stable patients with arterial blush. At times it can be considered if a patient becomes unstable after CTA, provided that the interventional suite can be accessed rapidly, operative renal access would be otherwise difficult, and the extravasation represents the sole injury.
- A **complete renal artery thrombosis** can be the result of an isolated deceleration intimal injury. Most of the time the patient has multiple other injuries to attend to, and it has been more than 4 hours since injury, so ischemic damage has already set in. In this common setting, **observation** is the appropriate management provided there is a normal functioning contralateral kidney.
  - **Salvage** may be attempted at any time if the patient is uninephric with a renal artery thrombosis.
  - On rare occasion, a complete renal artery thrombosis is encountered at less than 4 hours since the traumatic event, and the patient has no other pressing injuries—in that case, treatment with **endovascular thrombectomy and stent** may be attempted.
- **Ureteral trauma**
  - Blood supply enters medially in the superior two-thirds and laterally in inferior one-third of the ureter. The periureteral soft tissue carries perfusion longitudinally along the ureter and should be left intact. All injuries require intervention.
  - Cystourethroscopy with stent can be sufficient for a **partial injury or a ureteral contusion.**
  - A **proximal injury** is primarily repaired with absorbable sutures in a spatulated fashion over a stent, provided the two ends can meet without tension. If missing a segment of more than 2 cm and the patient is critically ill, then ligation and percutaneous nephrostomy should be done. Transureteroureterostomy, ileal substitution, and renal auto-transplantation with ureteroneocystostomy are options for the stable patient.
  - A **distal injury** is at or below the common iliac vascular bifurcation. Management is generally re-implantation as ureteroneocystostomy, with a Boari flap bladder cuff or psoas hitch as needed.
  - Place drains and maintain a Foley catheter in all patients. The latter prevents vesicoureteral reflux in the setting of a stent keeping the ureteral orifice open.
- **Bladder Injury**
  - Patients present with gross hematuria, blood at the urethral meatus, and a perineal or scrotal hematoma. There may be accompanying pelvic rami fractures. Confirmation is with CT cystogram.
  - Management depends on whether the injury is contained by the peritoneal reflection superiorly.
    - **Extraperitoneal** injuries are from pelvic bone fragments. Treatment is with Foley drainage for 1–2 weeks provided there are no other pelvic injuries indicating a need for operative intervention. In the latter case, primary bladder repair can be undertaken at the same time.
    - **Intraperitoneal** injuries are from blunt trauma with a blowout of the dome of the bladder through the peritoneal reflection and into the abdominal cavity. Treatment is with exploration and repair.
    - Bilateral nephrostomies can divert complex ureteral orifice injuries in an unstable patient.
- **Urethral trauma**
  - **Disruption** may occur at the prostatic, membranous, or penile urethra.
  - **An injury is suspected by a straddle mechanism, pubic diastasis, rami fractures**, blood at the urethral meatus, or a high-riding prostate on digital rectal exam. The patient may have urinary retention. Confirmation is with a **retrograde urethrogram** x-ray at bedside.
  - Urology may choose to stent with a Foley catheter if it is a contusion, stretch injury, or only partial thickness. Healing may then ensue around the catheter.
    - **Suprapubic cystostomy diversion and delayed repair** should be pursued if the injury is from a blunt mechanism and there is complete transection.
    - Urology may consider **primary repair**, in addition to suprapubic diversion, if it is an isolated penetrating injury to the urethra.
- **Injuries to Buck's fascia and fractured tunica albuginea**, whether penile or testicular, should be repaired by a urologist.

PELVIC

- **Anterior** pelvic fractures are associated with venous oozing and **posterior** pelvic fractures with internal iliac arterial tributary or sacral venous plexus bleeding. *See Figure 23.15.*

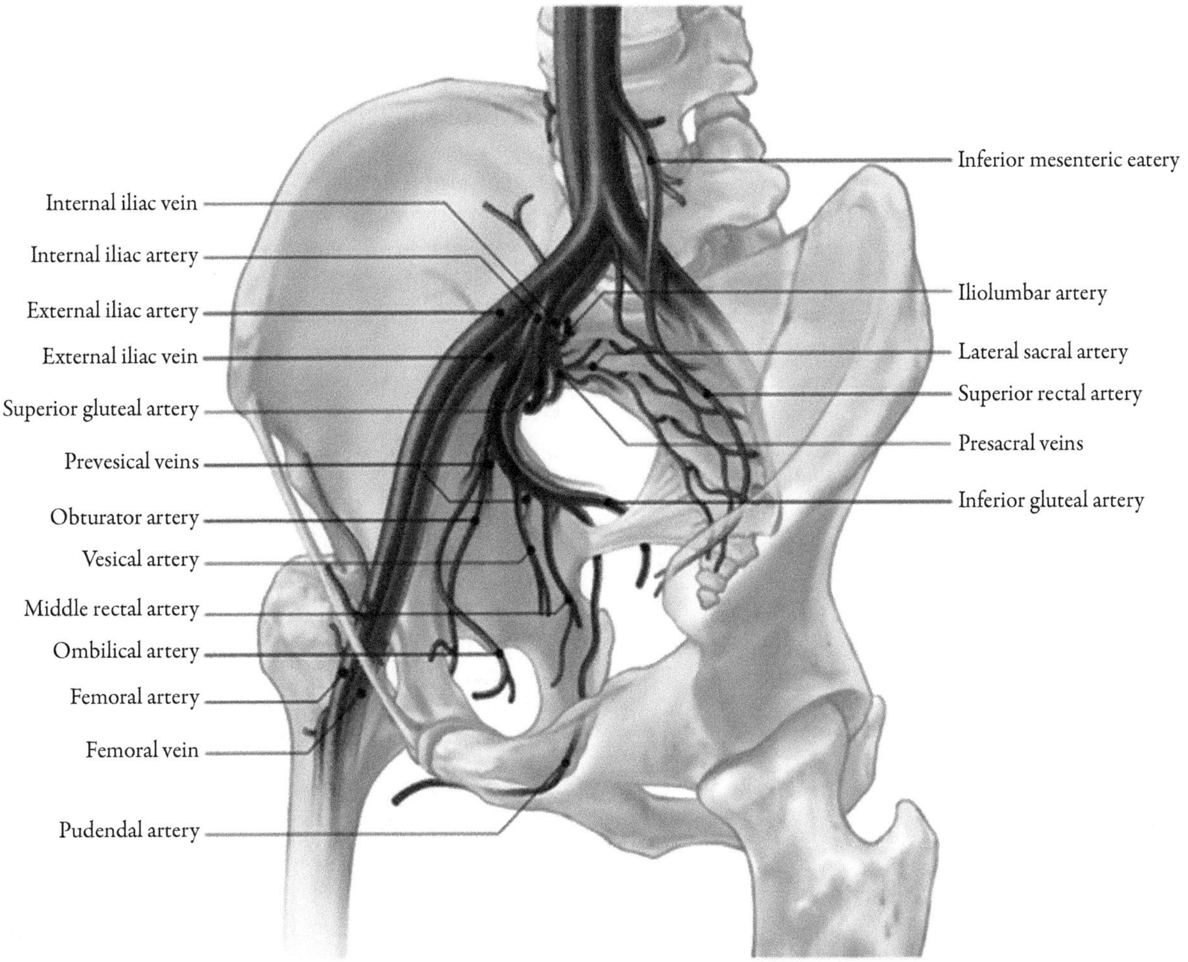

Figure 23.15 Internal iliac arterial and venous tributaries as source to hemorrhage in unstable pelvic fractures.

- A patient with an unstable pelvis by physical exam needs a **temporary binder.** This may be a specific device or just sheets and clamps. Binding works best for anteroposterior injuries rather than unilateral fractures from shear or lateral force.
  - If there is free peritoneal blood on FAST and the patient has unstable vital signs, proceed to **laparotomy.**
  - If the FAST is negative for intraperitoneal blood, but the patient remains unstable despite initial resuscitation, then an assessment for cardiac tamponade, tension pneumothorax or hemothorax, and repeat FAST exam should ensue in the next few minutes.
    - If still no etiology is found for the altered hemodynamics, then it should be presumed to be from pelvic and retroperitoneal blood associated with the unstable pelvic fractures.
    - **Pelvic packing** in the operating suite is the next best step. A midline incision can be taken down to the transversals fascia, just outside the peritoneal cavity, and three rolled-up lap pads are placed in the preperitoneal space on each side of the bladder.
  - If the FAST is negative and the patient stabilizes after initial resuscitation, imaging including a CTA of the abdomen and pelvis can be performed. If there is arterial extravasation, then endovascular **arterial embolization** by interventional radiology is the next step.
  - Formal **orthopedic external or internal fixation** should only be done if the patient has remained stable and no intraperitoneal intervention, preperitoneal packs, or angioembolization are needed in the immediate period.
- Any expanding Zone 3 blunt trauma hematoma can be explored or undergo angioembolization. If not expanding, then the hematoma should be left alone.
- Penetrating trauma Zone 3 hematomas noted in the OR should be explored even if they are non-expansile.
- Consider colonic diversion in cases of major pelvic injury.

SOFT TISSUE AND BONE

- **Displaced limb fractures in patients lacking a distal pulse** should undergo immediate reduction. If the pulse returns, then one may proceed with trauma imaging, including a CTA of the limb.
  - If the pulse does not return, then the patient should be taken to the OR for angiogram, vascular repair with or without a temporary shunt, possible saphenous vein harvest, and bone stabilization with an external fixator. If obtained prior to the OR, CT imaging should be prompt and meant to exclude major torso injuries that would take precedence over a pulseless limb.
- **Orthopedic injuries in proximity to major vascular structures but without a hard sign of vascular injury** (for example femoral-tibial dislocation), may receive an ABI or BBI. No further imaging of the limb is needed if the ABI is >0.9. A CTA of the limb should be obtained for an ABI <0.9.
- **Open fractures with a soft tissue defect <1 cm or >1 cm with minimal tissue damage** should be treated with antibiotics against Gram-positive cocci (GPC) and washout.
  - **Extensive tissue damage** may need antibiotics against GPC and Gram-negative rods (GNR) and washout.
  - Actual **wound contamination** requires antibiotics against GPC, GNR, and anaerobes as well as a washout.
  - **External fixation** may be best for open fractures needing washout, or closed fractures with extensive tissue damage.
  - Consider **amputation** for patients with major vascular injuries, extensive soft tissue devitalization, and nerve transection.
- **Morel-Lavallée lesion** is shearing of the dermis and subcutaneous layer off of fascia without an actual break in the dermis. *See Figure 23.16.* The dermis may necrose due to ischemia from loss of perforating vessels. Fluid collections in the subcutaneous space should be aspirated. If >50 ml; then incise, decompress, and place a closed suction drain.
- See the Chapter 30, Vascular, for diagnosis and treatment of compartment syndrome,
- See the Chapter 27, Neurosurgery and Orthopedics, for other specific injuries.

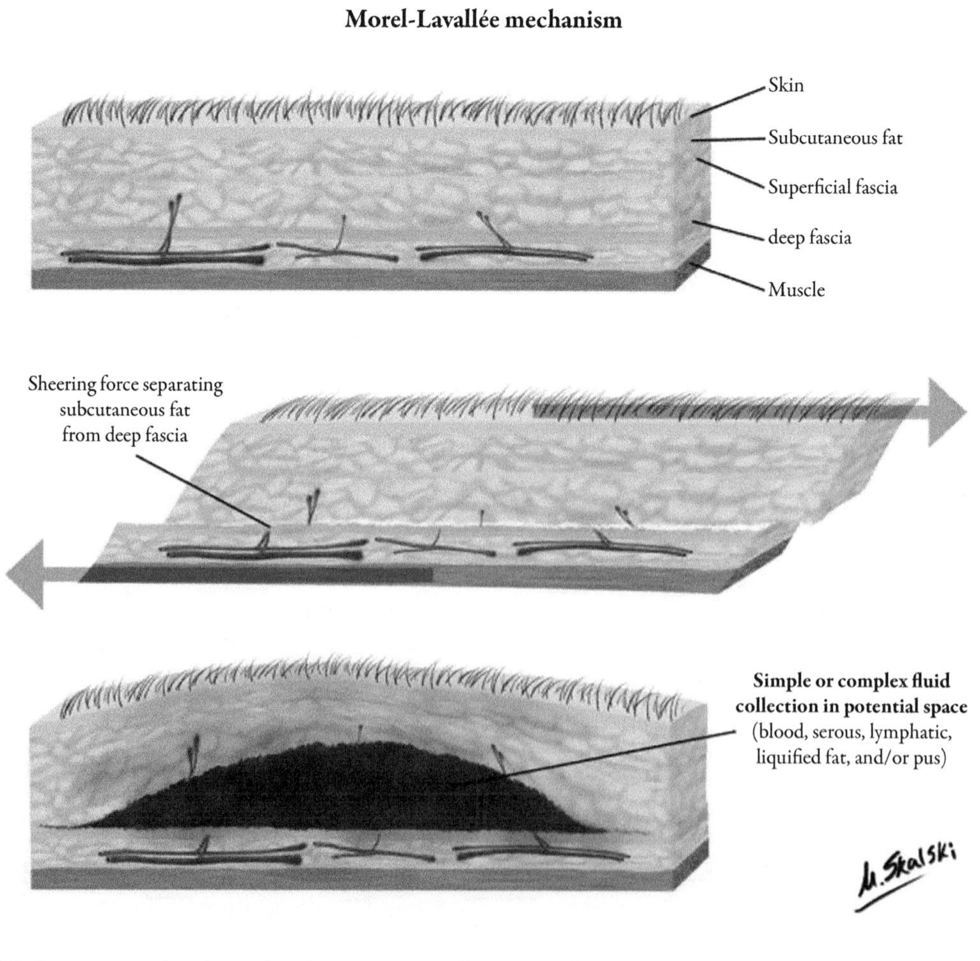

Figure 23.16 Internal degloving injury that shears the subcutaneous layer from the underlying fascia.

PREGNANCY CONSIDERATIONS

- A mother's blood volume is increased by 50% during pregnancy. Cardiac output, glomerular filtration rate (GFR), pelvic vein size, and minute ventilation are all also increased.
- Systemic vascular resistance (SVR), functional residual capacity (FRC), and gastrointestinal and biliary motility are all decreased.
- Thrombotic risk is increased due to venous stasis and hypercoagulability.
- Sensory fibers and rectus muscles are stretched across a thin, tense abdominal wall, so the abdominal exam can be less sensitive for detecting intraperitoneal injury.
- During the first trimester, the uterus is protected by pelvic bones. At 12 weeks, the uterus is at the symphysis pubis. At 20 weeks, fundal height is approximately 20 cm and at the umbilicus.
- Priority in trauma to the pregnant patient goes to the mother. ATLS principles and ABC based resuscitation should be followed similar to a non-pregnant patient. Placing a **bump under the right side** can help relieve caval compression by a gravid uterus.
- **Pelvic fractures** during pregnancy are associated with greater than 35% fetal mortality.
- Traumatic **placental abruption** is diagnosed by uterine tenderness, uterine contractions, and fetal tachycardia due to fetal hypoperfusion. A KB test quantifies the volume of fetal to maternal blood mixture.
- **Uterine rupture** can present similarly to placental abruption, but uterine contractions are not evident, and fetal death is more common. The uterus can be primarily repaired or may require a hysterectomy. *See Figure 23.17.*
- A **trauma C-section** may be needed in the setting of a perimortem shock state secondary to hemorrhage or tamponade, provided the uterus appears gravid. The procedure can still be performed up to 5 minutes after cardiac arrest. A gravid uterus limiting visualization during emergent laparotomy is another reason for performing a C-section in a trauma scenario.
- Radiation exposure via CT is 3–5 RADS and by x-ray it is 0.1-1 RADS. Either are permissible in any trimester of pregnancy. However, US and MRI are often preferred initial abdominal imaging in a stable, non-trauma pregnant patient, provided time permits.

RADIATION

- High dose is considered greater than 1 Gray (Gy) or 100 centigray (cGy or RAD) over a short period of time.
- **Hematologic signs** of radiation toxicity include pancytopenia within 48 hours at 1–4 Gy exposure. Lymphocyte count nadirs at 48 hours and can be prognostic of the degree of toxicity.
- **GI symptoms** occur within 1 hour if the patient has been exposed to more than 6 Gy. This portends a higher risks of death in 1–2 weeks.
- Higher radiation dose exposures lead to **dermal and neurologic symptoms**.
- Treatment starts with removing, irrigating, and disposing of clothes as well as the irrigant solution.

HYPOTHERMIA

- Primary hypothermia is from the environment, and secondary is from medical disease or ingestion. Mild hypothermia is 35°–32° C, moderate 32°–28° C, and severe is less than 28° C.
  - **Passive rewarming** starts with removal of wet clothing and use of blankets. This is adequate for

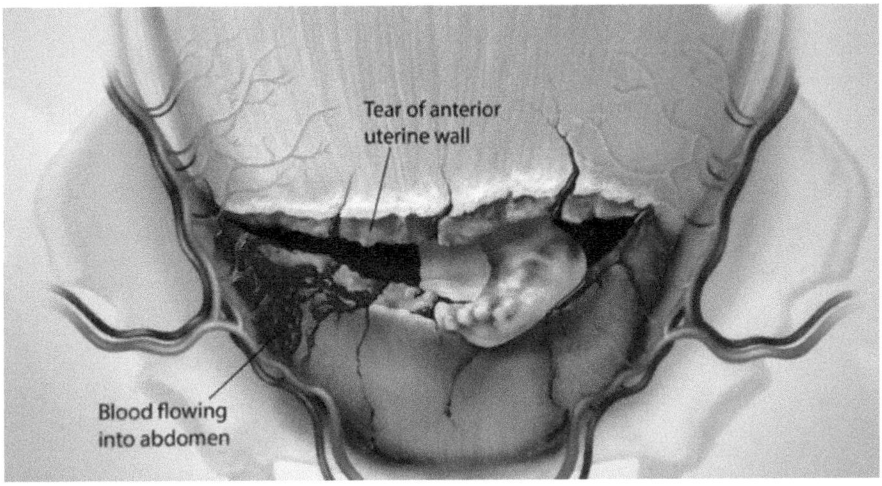

Figure 23.17 Uterine rupture from trauma or labor.

stable patients with mild or moderate hypothermia. It can be augmented by **active external rewarming** with a Bair Hugger and warm packs.

- **Active internal rewarming** is most commonly done with humidified warm oxygen and warm intravenous fluid.
- **Invasive active internal rewarming** is via mediastinal, pleural, and peritoneal lavage for patients with severe hypothermia and hemodynamic or rhythm instability.
- **Extracorporeal membrane oxygenation (ECMO)** can transfer heat by conduction to the patient's bloodstream as it flows through the circuit. It is the fastest way to normalize temperature for patients with severe hypothermia who have arrested or have persistent rhythm instability despite other measures.

- **Frostbite**
  - Frostnip is first-degree frostbite defined by superficial epidermal ice crystal formation with no permanent sequelae.
  - Second-degree frostbite has clear bullae, and third-degree has hemorrhagic bullae secondary to full-thickness dermal freezing and subcutaneous vascular injury.
  - Fourth-degree frostbite presents as necrosis down to bone.
  - Managment is with rapid rewarming in a 40°C bath. Tissue plasminogen activator (TPA) can be considered in patients with poor response to rewarming who incurred injury within 24 hours, and have a post-warmth ischemic time of less than 6 hours. Prostacyclins can also help improve microcirculatory flow. The true necrotic demarcation of digits may not be declared until 2 weeks or later.

24.

BURN

ANATOMY AND PHYSIOLOGY

- The **epidermis** is composed of keratinocytes, lymphocytes, and Langerhans antigen presenting cells. Melanocytes reside in the basal keratinocyte layer and transfer melanin to neighboring keratinocytes, which acts as an umbrella cap protecting their nuclei from UV light.
  - **Mechanoreceptors** (Meissner's corpuscles) are in the lower epidermis and superficial dermis.
  - **Eccrine and apocrine glands** produce sweat in order to regulate serum composition. They are innervated by the sympathetic nervous system and densely populate the palms, soles, axillae, anogenital region, and nipple.
  - **Hair follicles** are associated with a sebaceous gland. Apocrine glands can drain into the hair follicle as well.
  - The lower part of the epidermis consists of rete pegs interdigitating with the dermal papillae. *See Figure 24.1.*
- The **dermis** has a superficial layer (papillae) composed of loose structural proteins and its own vascular network. The deeper reticular dermis contains thick horizontal proteins and a separate vascular network.
  - **Pacinian corpuscles** monitor pressure.
  - **Ruffini endings** perceive warmth.
  - **Krause bulbs** signal cold.
  - Lymphatics reside in the deeper dermis and do not have valves. The subcutaneous lymphatic network within adipose tissue below does contain valves.

DIAGNOSTIC ASSESSMENT

- Burns have a bimodal age presentation at <4 years old and >65 years old, and generally occur at home.
- **Etiology**
  - **Scald burns are the most common.** They are due to water, hot oil, grease, or other chemicals. The latter may need to be inactivated or diluted for adequate treatment. Scalds are worse with clothes on, as that prolongs the contact time between skin and a hot interface.
  - **Flame burns** are more likely than other types to present to a hospital and need admission.
  - **Flash burns** are from exposure to a brief, intense burst of heat. Clothing is protective.
  - **Contact burns** are from conduction of heat via touch of a stove or electrical source.
- **Zones and Degree** *See Figure 24.2.*
  - The **central zone** is the deepest area of a burn where there is coagulative necrosis. It is surrounded by a **stasis zone** which can remain partial or progress to full thickness. The perimeter of the burn is a **hyperemic zone.** Each zone can have a different depth, and hence degree, of burn injury.
  - **1st-degree** burns are limited to the epidermis. They heal by re-epithelialization from the edges or from hair follicles.
  - **2nd-degree** burns are partial thickness.
    - **Superficial partial thickness burns** blanch with pressure and contain vesicles or bullae. They heal by re-epithelialization from the wound edges, hair follicles, or other appendages.
    - **Deep partial thickness burns** may be pale, with white and pink areas. Sensation is decreased, and hair follicles may be lost or easily fall off with pressure. These injuries would take weeks to heal by re-epithelialization, since minimal appendages remain as a source for epithelial migration within the wound. Epithelial re-lining must occur from the edges only. Often, deep partial thickness burns are treated with excision and a skin graft.
  - **3rd-degree** burns are full thickness, white, insensate, leathery, and often charred. They will require excision and a skin graft.
  - **4th-degree** burns are through the entire dermis and part of adipose, muscle, or bone.
- **Surface Area**
  - Rule of 9s allows a clinician to rapidly estimate the percentage surface area burned. 1st-degree burns are not included. The percentages applied to each body site are as follows:
    - Head 9%, each arm 9%, each leg 18%, torso 36%, genitourinary area 1%, and palm 1%.
  - **Burn center transfer** should be considered for the following:

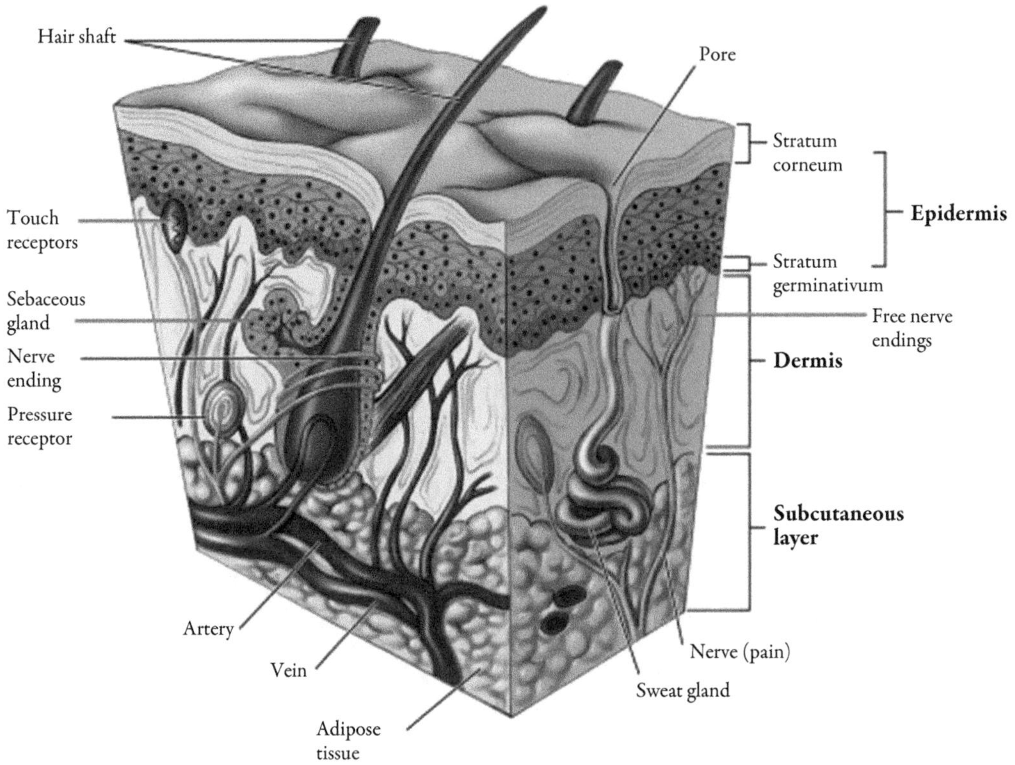

Figure 24.1 Skin and subcutaneous architecture.

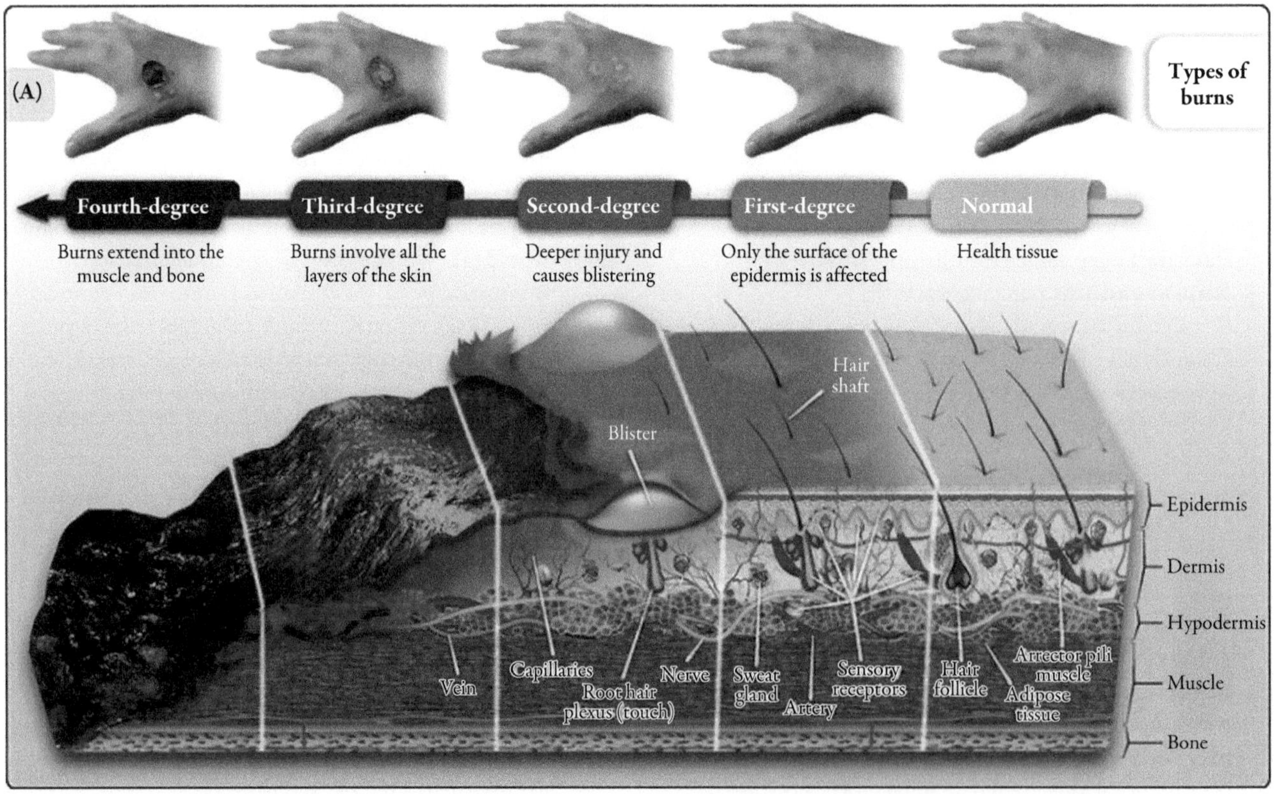

Figure 24.2 Burn depths.

- >20% total body surface area (TBSA) 2nd-degree burn. The threshold is lowered to **10%** for patients <10 or >50 years old.
- > 5% 3rd-degree burn in any age group.
- Burns over hands, feet, joints, face, perineum.
- Inhalation, electrical, or chemical burns.
- Patients with multiple comorbidities, psychological, or social issues.

- **Suspected abuse or neglect** based on sharply demarcated scald burns of uniform depth, or stocking glove distribution and without splash marks. Suspect cases can present in a delayed fashion with conflicting histories. There may be physical exam or radiographic evidence for old injuries.

TREATMENT

- An urgent **escharotomy** is done for any full thickness circumferential dermal burns to extremities compromising distal vascular or neural function. Escharotomies can also be done over the torso for full thickness circumferential burns that impair ventilation.
- **Fluid resuscitation** should be with lactated ringers (LR).
  - If >30% TBSA burned, use a volume between the Parkland formula (4 × % TBSA × weight in kg) and the modified Brooke formula (2 × TBSA % burn × kg).
  - Half of the volume is given in the first 8 hours following the burn, and the second half of volume over the next 16 hours. The rate may be titrated to ensure a urine output >0.5 ml–1 ml/kg/hour in the adult, or 2–4 ml/kg/hour in a child or infant. Patients with inhalation injury may require more fluid to reach urine output goals.
  - After 24 hours one may switch to colloid if further resuscitation is needed. This may help avert abdominal compartment syndrome or worsening pulmonary edema.
  - Parkland or Brooke formulas are not used in electrical or chemical burns that have induced deeper tissue injury, since requirements cannot be calculated simply by the dermal area involved.
- **Vitamin C** may scavenge free radicals and may decrease IV fluid requirements.
- **Nutritional Supplementation**
  - There is increased catabolism and protein loss across wounds. As such, the urine urea nitrogen is not an accurate measure of nitrogen balance.
  - Calculation of caloric requirements.
    - 35–40 kcal/kg/day and 2.0–2.5 g protein/kg/day for severe burn patients.
    - Alternatively, one may calculate individualized caloric needs at 20–25 kcal/kg/day + **30** kcal/kg/day x %burn and 0.8–1.0 g protein/kg/day + **3** g protein/day x %burn.
    - Early gastric or duodenal feeding prevents an ileus. Patients should be on a strict insulin regimen.
  - **Propranolol** can decrease hypermetabolism, and **oxandrolone** can promote anabolism and preserve lean body mass.
- **1st-degree and superficial partial thickness 2nd-degree** burns can just receive soap, water, and topical antibiotics like bacitracin or Neosporin.
- **Deep partial thickness 2nd-degree burns or any deeper burn need excision** after crystalloid resuscitation. This functions as source control for the inflammatory response, driver of catabolism, and potential infection.
  - Ideally, the excision should be within 72 hours of presentation, unless the burns involve the face, hands, feet, or genitalia. In these cases, surgery may be deferred for 1 week in order to minimize the extent of resection.
  - **Early autograft** decreases the risk of infection, evaporative loss, and protein loss.
    - **Split thickness skin grafts (STSGs)** are harvested and meshed to allow them to cover a larger area. They have decreased primary graft contraction, but increased recipient site contraction. Donor sites include the back, buttocks, thighs, and scalp.
    - **Full thickness skin grafts (FTSGs)** have increased primary graft contraction, but decreased recipient site contraction. FTSGs allow for the full benefit of dermal elasticity and range of motion. They are best suited for burns over mobile areas and concerns of cosmesis. Donor sites include retro- and preauricular, supraclavicular, or inguinal sites. The subdermal fat should be removed.
    - Delayed grafting is okay if the patient is systemically ill, culture of the wound excision has >100k bacteria, or there is a lack of donor sites.
    - Grafts may be placed directly over muscle, granulation tissue, peritendineum, or periosteum.
    - Once grafted, **imbibition** of nutrients begins at <48 hours, **inosculation** and capillary alignment starts at 2–4 days, and **revascularization** matures by 5 to 7 days. Sensation returns at 10 weeks to 2 years.
    - Thinner grafts vascularize and adhere more easily than thick ones. The main cause of failure is sub-graft fluid collection and hematoma, so the grafted area should be immobilized.
  - **Other Temporary Grafts**
    - **Homografts (cadaveric allografts) function** as temporary coverage for 2–4 weeks to promote revascularization and adherence, conditioning the

wound bed for permanent coverage. After this they are removed and replaced with an autograft.

- **Xenograft (porcine or bovine)** last for 2 weeks but do not vascularize. Then they are removed and replaced with an autograft.
- Other **dermal substitutes** include AlloDerm or Integra, which has an outer silicone that is later removed, and an inner collagen and chondroitin remaining. Biodegradable temporizing matrix (BTM) can be used as synthetic coverage to promote the development of a neodermis. All synthetic dermal substitutes are followed by placement of an autograft.
- **Cultured epidermal autograft keratinocytes** have been used as coverage.

- **Facial burns** should receive topical antibiotics for the first week, then an FTSG.
- **Hand burns** should receive FTSG with a period of immobilization, and then range of motion to prevent contracture.

INFECTION

- Secondary infection of burn wounds originates from bacteria surviving in the depths of epidermal appendages and then erupting to the surface. This results in microvascular thrombosis and conversion of the stasis zone to a coagulative zone.
  - Impaired granulocyte chemotaxis and cellular immunity predispose to infection.
  - Pseudomonas and *S. aureus* are the most common organisms.
  - Signs include discoloration, apparent deepening of the burn wound, edema, hemorrhage, necrosis, loosening of an eschar, and purulence. Formal diagnosis requires a wound culture demonstrating >100,000 bacteria.
  - **Topical antimicrobial therapy**
    - Topical **bacitracin** is for wounds that are nearly healed and not overtly infected.
    - Topical **mupirocin (Bactroban)** is preferred if there is MRSA in the wound culture.
    - **Silver sulfadiazine (SS)** has limited eschar penetration and may inhibit epithelialization. It is effective for candida, but not pseudomonas. Complications include neutropenia and thrombocytopenia, and it should be avoided in patients who have a sulfa allergy. SS is not commonly used on newly placed grafts.
    - **Silver nitrate (SN)** has limited eschar penetration. It is also ineffective against pseudomonas. Complications include electrolyte changes and methemoglobinemia. An alternative agent is 0.5% sodium hypochlorite (Dakin's solution).
    - **Silver embedded dressings** are moistened with sterile water before application, and left on for up to 1 week. Silver inhibits bacterial cellular respiration. These can be used on clean superficial partial thickness wounds that do not need to be evaluated daily.
    - **Mafenide acetate (MA)** penetrates eschar and is active against pseudomonas and GPCs. Complications include metabolic acidosis and pain from application.

OTHER TYPES OF BURNS

- **Hydrofluoric acid** dermal burns are treated with topical, intravenous, or intra-arterial calcium.
- **Powder** burns, such as aluminum hydroxide in concrete powder or lye, should be wiped away before beginning irrigation.
- **Hot tar** burns need lipophilic solvent or adhesive remover.
- **Chemical or intravenous medication extravasation** into the subcutaneous adipose layer can be managed expectantly or with liposuction to enhance clearance.
- **Electrical Burns.** Alternating current (AC) causes muscle tetany, and direct current (DC) causes a patient to be thrown from the source. The degree of dermal involvement does not correlate with underlying visceral injury, which can include demyelinating polyneuritis, myelitis, cataracts, visceral solid or hollow organ necrosis or perforation, or rhabdomyolysis. Treatment is supportive. Cardiac monitoring is recommended for any electrical source >1000 volts. Fasciotomy may be required with extensive muscle injury.
- **Lightning** may lead to cardiac arrest from transient brainstem dysfunction.
- **Corneal burns or abrasions** are diagnosed by fluorescein and treated with topical antibiotics.
- **Alkali** chemical burns cause liquefactive necrosis of pharyngeal and esophageal mucosa with deeper injury than **acid**-induced coagulative necrosis of mucosa. Treatment is with irrigation and endoscopy to quantify injury.
- **Frostbite** *See Figure 24.3.*
  - **Frostnip** is first-degree frostbite defined by superficial epidermal ice crystal formation with no permanent sequelae.
  - **2nd-degree** frostbite has clear bullae and **3rd-degree** has hemorrhagic bullae from full thickness dermal freezing and subcutaneous vascular leakage.
  - **4th-degree** frostbite presents as necrosis down to the bone.
  - Management is with rapid rewarming in a 40° C bath. Tissue plasminogen activator (TPA) can be considered in patients with poor response to rewarming who incurred injury within 24 hours, and

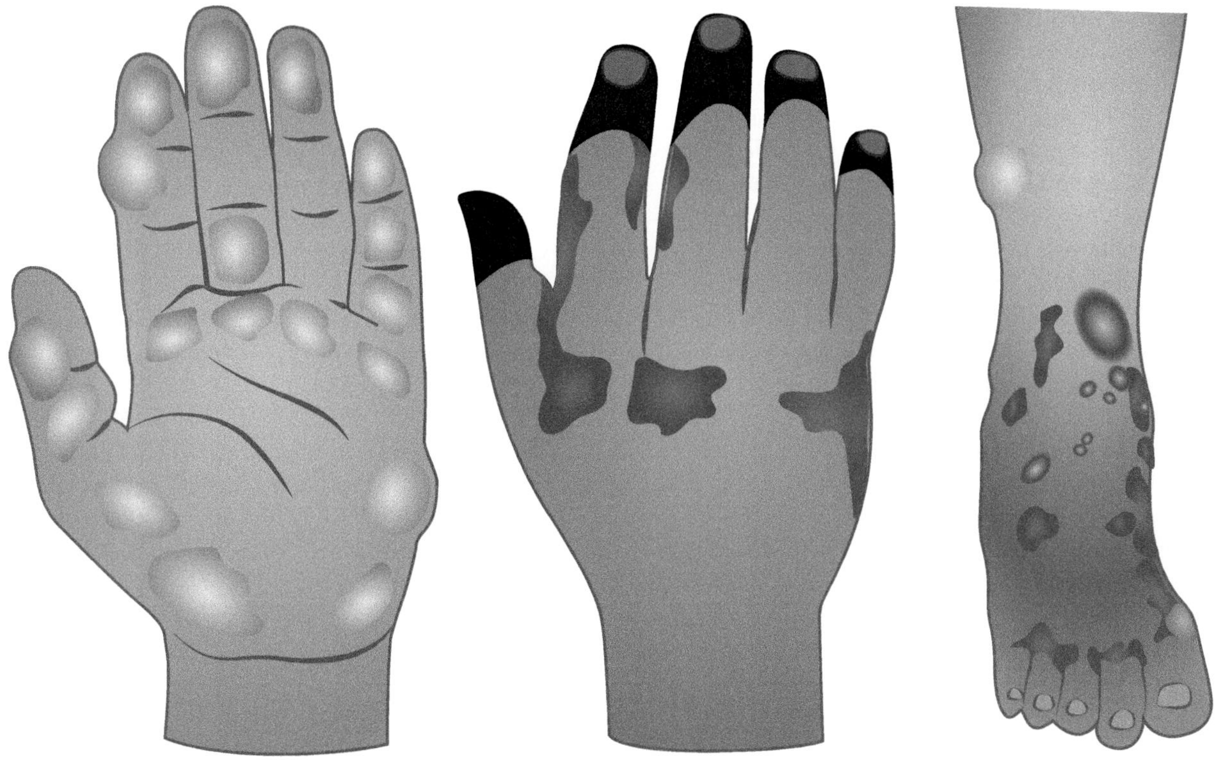

Figure 24.3 Frostbite.

have a post-warmth ischemic time of less than 6 hours. Prostacyclins can also help improve microcirculatory flow. The true necrotic demarcation of digits may not be declared until 2 weeks or later.

- Primary **hypothermia** is from the environment, and secondary hypothermia is from medical disease or an ingestion. Mild is 35°–32° C, moderate 32°–28° C, and severe is <28° C. Treatment options include:
  - **Passive rewarming** includes removal of clothing and use of blankets for a hemodynamically stable patient with mild or moderate hypothermia.
  - **Active external rewarming** with a Bair Hugger and warm packs is for the stable patient with mild or moderate hypothermia.
  - **Active internal rewarming** involves humidified warm oxygen, warm IV fluids, and warm gastric, bladder, mediastinal, pleural, and peritoneal lavage. The latter invasive procedures would only be used for hemodynamic or rhythm instability in severe hypothermia.
  - **Extracorporeal membrane oxygenation (ECMO)** allows heat transfer by conduction and is the fastest way to normalize the temperature. This is for severe hypothermia patients who have arrested or have persistent hemodynamic instability after the above measures.
- **Immune or infection mediated dermal injury** can have similar physiologic effects as thermal injury.
  - **Scalded skin syndrome** is a superficial epidermal injury from staphylococcal toxin.
  - **Erythema multiforme** presents as target lesions secondary to herpes or mycoplasma pneumoniae.
  - **Stevens-Johnson syndrome** is <10% TBSA epidermal injury, and **toxic epidermal necrolysis (TEN)** is >50% TBSA epidermal injury. Nikolsky sign is positive. Both are secondary to antibiotics, other medications, or a viral infection. Treatment is to remove the offending agent, resuscitate, and temporarily cover wounds with a homograft or xenograft. Other measures include intravenous immune globulin (IVIG), plasma exchange, and non-glucocorticoid immunosuppression.

INHALATION INJURY

- **Thermal damage** mostly involves the oropharynx and upper airway. It is maximal at 12 to 24 hours.
  - Patients may have facial burns, singed eyebrows, singed nasal hairs, a hoarse voice, wheezing, or carbonaceous sputum or soot in the posterior pharynx.
- **Toxic irritants** and carbonaceous material injure the lower airways, and may take a few days to manifest. This can be diagnosed and cleared by early bronchoscopy, unless severe hypoxemia precludes this procedure.

- **Carbon monoxide** poisoning creates carboxyhemoglobin (COHb), which left-shifts the oxyhemoglobin curve. The result is an inability to offload O_2 to tissue, and "100%" arterial saturation with high venous saturations as well. **Cyanide toxicity** impairs mitochondrial oxygen utilization, leading to high venous oxygen levels and distributive shock.
- Carbon monoxide or cyanide poisoning can be suspected by noting high venous oxygen saturation due to impaired offloading and utilization of O_2.
- COHb levels of >30% result in confusion, >40% lethargy and coma, and >60% death.
- Treatment starts with. intubation for any upper airway compromise that threatens patency, or for significant lower airway injury producing a PaO_2:FiO_2 ratio <250.
  - One may start by using **low tidal volume (TV) lung protective strategies** and **prone positioning.** However, liberalizing this strategy may be necessary to achieve adequate oxygenation and ventilation in severe inhalation injury.
- Some patients may need high-frequency oscillating ventilation (HFOV) or ECMO.
- Bronchoscopy can help clear airways and delineate injury.
- Aerosolized heparin may prevent fibrin airway casts.
- Aerosolized N-acetylcysteine (NAC) can act as a free radical scavenger.
- Carbon monoxide poisoning is managed with 100% FiO_2, which decreases the half-life of COHb to less than 60 minutes. Positive pressure ventilation can enhance clearance. Hyperbaric oxygen is only occasionally utilized.
- Cyanide toxicity is managed with **hydroxycobalamin** to create cyanocobalamin, or **sodium thiosulfate** to create thiocyanate, both of which siphon off cyanide. **Nitrites** create methemoglobin which can bind cyanide and keep it from mitochondria.
- Inhalation injury is complicated by pneumonia (PNA) in up to 50% of cases.

25.

PEDIATRICS

VOLUME AND COMPOSITION

- An infant has 80 ml/kg blood volume and 65 ml/kg when over 7 years old.
- Glomerular filtration rate (GFR) is about 20 ml/minute for children and approximately 70 ml/minute as an adult. Pediatric urine OSM can rise up to only 600 mOsm/kg, while adults can concentrate their urine up to 1200 mOsm/kg and hence have better volume retention.
- Pediatric sodium (Na) requirement is 2–5 mEq/kg and potassium (K) is 1–3 mEq/kg.
- Dextrose infusions should be at 4–6 mg/kg/minunte. Lipid requirements are 0.5–3.0 g/kg/day. and protein requirements are 2–3.5 g/kg/day.
- **Product transfusion** should proceed at 10 ml/kg for packed red blood cells (pRBC), fresh frozen plasma (FFP), and platelets.
- **Bolus of crystalloid** is 20 ml/kg each and can be repeated 2–3 times as needed for resuscitation of hypovolemic and septic patients.
- **Maintenance intravenous fluid (mIVF)** is calculated as 4 ml/kg/hour for the first 10 kg, 2 ml/kg/hour for the second 10 kg, then 1 ml/kg/hour for remaining weight.
  - Use normal saline (NS) for all patients except **neonates, who should receive 1/4 NS.**
  - Grams of weight loss is equivalent to the mls of dehydration. One may add 1/2 of the total mls of dehydration to the calculated total mIVF throughout each day.

NECK MASS

- **Branchial cleft cysts** are diagnosed by physical exam and may present acutely enlarged due to infection or as a noninfected draining sinus from a dermal pit.
  - **1st cleft cysts** are at the angle of the mandible and have a sinus that connects to the external auditory canal (EAC).
  - **2nd cleft cysts** are at the anterior border of the sternocleidomastoid (SCM) and have a sinus tract that passes between the carotid bifurcation to connect to the pharyngeal tonsillar pillar.
  - **3rd cleft cysts** are medial to the SCM in the lower neck and have a sinus that runs posterior to the common carotid and drains into the pyriform recess.
  - Treatment of all types is by resection of the cyst, sinus tract, and any external fistula. Acutely infected cysts can be temporized with antibiotics or incision when necessary.
- **Thyroglossal duct cyst (TGD)** resides along the line of descent of the thyroid gland from the foramen cecum. It presents as a cystic midline mass at 2 to 4 years of age, when neck fat diminishes. The mass moves up and down with swallowing. If solid, then it is an ectopic thyroid and may be the sole thyroid tissue of the patient.
  - Check thyroid function tests (TFT) and ultrasound (US) or thyroid scan I-123 to confirm a native thyroid is present, if not palpable in the normal position.
  - Management is with the **Sistrunk procedure**, which includes resection of the TGD, the hyoid, and any tract to the base of the tongue.
  - Incise and drain any secondarily infected TGD, deferring resection until the infection is resolved. Some cases will develop a postoperative salivary fistula.
- **Persistent enlarged lymphadenopathy** should undergo FNA. The differential diagnosis includes bacterial lymphadenitis, cat scratch disease, scrofula, lymphoma, or cystic hygroma.
  - **Lymphadenitis** is the most common, and can be treated with antibiotics for 2 weeks. Drainage or excision are reserved for failure of antibiotics.
  - **Cystic hygroma** resides lateral to the SCM and is the result of obstructed lymphatics in utero. Diagnosis is made on prenatal US. Polyhydramnios from impaired swallowing can also be present. Treatment is with percutaneous aspiration, serial excisions, and sclerosing agents.
- **Rhabdomyosarcoma** may present in the head and neck in childhood. Nodal involvement and metastases are evident in up to one-quarter of cases. The genitourinary area and extremities are other sites of presentation. Treatment is with resection, neoadjuvant or adjuvant chemotherapy, and radiation.
- **Hemangiomas** grow during the first year of life, then they involute over the next decade. Subtypes include superficial strawberry hemangiomas or deep cavernous

hemangiomas. Treatment is with **beta blockade** if the lesion is causing a functional limitation. Other options include glucocorticoids, interferon alpha, and vincristine. Surgical resection is a last resort.

- **Vascular malformations** do not recede with time. Capillary and venular malformations are known as port wine stains. Venous malformations are larger and treated with **sclerosis or laser therapy,** sometimes over multiple sessions. Arteriovenous malformations are treated with preoperative **embolization and then surgical resection.** Lymphatic malformations, like cystic hygroma, can be treated with **aspiration, serial excisions, or sclerotherapy**.
- **Torticollis** in an infant can present with an associated fibrotic mass in the sternocleidomastoid. Treatment is with physical therapy; no further imaging or biopsy is needed.

RESPIRATORY TRACT

- **Choanal atresia** is a unilateral nasal passageway obstruction by bone or mucous membrane. It presents with poor sucking due to inability for nasal breath. Treatment is surgical recanalization.
- **Laryngomalacia** presents in infancy with intermittent airway obstruction and stridor. It is due to immature epiglottic cartilage. Patients may not tolerate feeding, since the act of swallowing covers the laryngeal inlet and results in respiratory difficulty. Symptoms are worse in the supine position. Diagnosis is confirmed with fiberoptic laryngoscopy showing an omega shaped epiglottis, short aryepiglottic folds, and prolapsing arytenoid mucosa. The condition usually resolves by 1 year of age. A supraglottoplasty may be needed in severe cases affecting growth.
- **Laryngeal papillomatosis** may be related to maternal HPV. It usually improves by puberty. Otherwise, endoscopic or laser resection may be needed.
- **Tracheomalacia** has fragmented tracheal rings. It presents with recurrent upper respiratory obstructive symptoms and usually **improves by 1–2 years of age**. Some patients may need aortopexy anteriorly to the sternum to prevent it from compressing a pliable trachea posteriorly. Select patients may need tracheoplasty in adulthood.
- **Tracheoesophageal fistula (TEF)** usually include some degree of esophageal atresia. Aspiration occurs overtop a blind esophageal pouch or via gastric refluxate transiting the fistula into the trachea. There are multiple types. *See Figure 25.1.*
  - A is a blind esophageal pouch and gasless abdomen on x-ray. It is the second most common type.
  - B is a TEF between the proximal blind esophageal pouch and trachea.
  - **C is a TEF to the distal esophagus and is the most common** type. Air will be present in the stomach.
  - D is a TEF to the proximal pouch <u>and</u> the distal esophagus.
  - H is a TEF without a blind pouch. In the absence of esophageal atresia, it presents later in life as recurrent aspiration pneumonia.
  - One-fifth of cases have associated congenital **cardiac abnormalities or VACTERL association defects.** Initial workup should include **echocardiogram and spinal US screening** before contemplating esophageal surgery. Prognosis is based on the severity of cardiac defects, and is better for patients weighing more than 1.5 kg.
  - Management is via a right extrapleural thoracotomy, mobilization of the esophagus, division of azygous vein, ligation of fistula, and closure of the tracheotomy. An end–end esophageal anastomosis

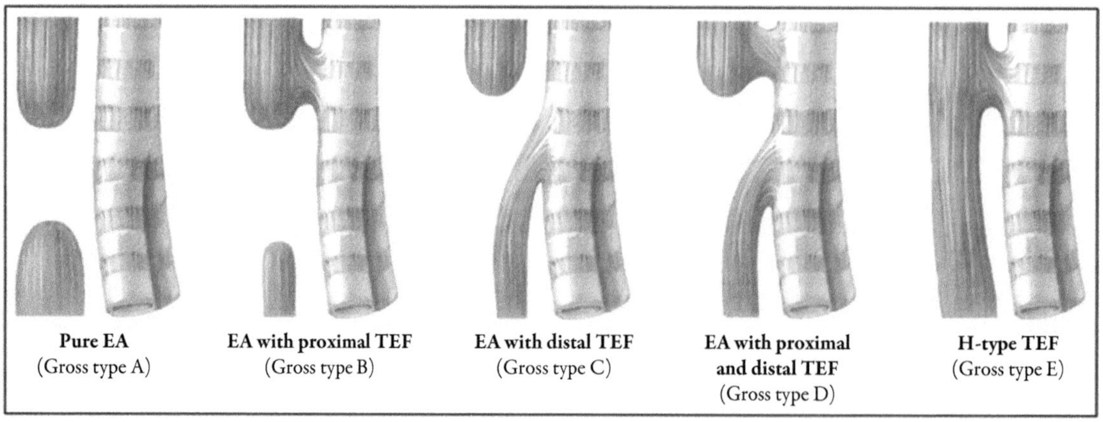

Figure 25.1 Types of congenital tracheoesophageal fistula.

is performed with or without a gastrostomy tube (G tube).

- The most common long term complication of treatment is **gastroesophageal reflux disease (GERD).**
- **Strictures** can present early or late, and are treated with dilation via nasal and direct gastric routes.
- A **salivary leak** may require re-operation if large and presenting early. If smaller and hence presenting late, drainage, antibiotics, and pulmonary toilet may be sufficient. Most late presentations resolve with this conservative management.
- Empyema, recurrent fistula, and tracheomalacia at the repair site are other less common complications.
- For patients less than 2.5 kg, in respiratory distress, or systemically ill, definitive repair can be delayed. Temporizing maneuvers include a drainage tube into the proximal blind pouch and a G tube inserted for gastric decompression while supporting ventilation. One may need to clamp the G tube if ventilation worsens due to stealing tracheal air by way of a type C, D, or H TEF. These patients may benefit from urgent fistula ligation.

- **Pulmonary sequestration** describes lung parenchyma with no communication to the parent tracheobronchial tree. It is most common in the left lower lobe and can present as pneumonia.
  - Arterial supply can be anomalous and from the thoracic or abdominal aorta via the inferior pulmonic ligament.
  - **Extralobar sequestrations** present early in life and have other congenital anomalies. Venous drainage is systemic via the azygous vein. Treatment is with **excision.**
  - **Intralobar sequestrations** present later in life. Venous drainage is pulmonary, and arterial inflow is from the abdominal aorta. Since this is within a lobe, treatment is **lobectomy.**
- **Congenital overinflation** is failure to develop cartilaginous support of bronchioles, most often in the left upper or right middle lobes. It results in alveolar hyperinflation and obstruction of pulmonary capillaries with hemodynamic compromise. **Emergent lobectomy** is indicated. Bronchoscopy is not useful unless it is an older child with the suspicion of a bronchial foreign body or mucous plug.
- **Cystic adenoid malformation** (congenital pulmonary airway malformation) is a contained area of mucous-secreting epithelium and elastin within a multiloculated cyst that originated from terminal airways. There is no cartilage or alveolar structure within the malformation. There is a high risk of malignant transformation. Presentation is with recurrent or non-resolving infection, and treatment is with lobectomy.
- **Bronchogenic cysts** are unilocular cysts containing cartilage and originating from a proximal bronchus posterior to the carina. Presentation is usually after the neonatal period, and treatment is with resection or marsupialization and extirpation of the lining.
- **Bronchiectasis** is pathologic dilation of bronchi due to cartilaginous weakening from repeated infections, foreign body impaction, or cystic fibrosis (CF). Some patients are immune compromised. Treatment is with medical management. Resection or lung transplantation are reserved for severe, recalcitrant cases.
- **Meconium aspiration** is the most common reason for respiratory difficulties in a neonate. Other etiologies include respiratory distress syndrome (RDS), persistent pulmonary hypertension (PPHN), sepsis, or congenital diaphragmatic hernia (CDH). Patients not manageable with positive pressure ventilation may need extracorporeal membrane oxygenation (ECMO).
- **Mediastinal masses** are categorized by location.
  - Posterior masses are generally of **neurogenic** origin and include neurofibroma, ganglioneuroma, and neuroblastoma.
  - Middle mediastinal masses are often cysts, lymphoma, or teratoma.
  - Anterior masses can be **lymphoma, germ cell tumors (teratoma or seminoma), thyroid enlargement, or thymoma.**

DIAPHRAGM, CHEST, AND ABDOMINAL WALL

- In a **Bochdalek diaphragmatic hernia** there is a left posterolateral diaphragm defect and no actual hernia sac. 80% of patients have associated cardiac or neural abnormalities or malrotation. *See Figure 25.2.* Respiratory distress is due to native pulmonary hypertension, lack of surfactant, and pulmonary hypoplasia. Occupation of a hemithorax by gastrointestinal (GI) contents can cause mediastinal shift and obstructive shock.
  - Prenatal ultrasound (US) or MRI can make the diagnosis. Otherwise, after birth neonates have respiratory difficulties, an absence of breath sounds, and a CXR showing abdominal contents residing in the hemithorax. A contralateral lung area to head circumference ratio of less than 1 is a poor prognostic sign.
  - Management begins with nasogastric (NG) tube placement to decompress the proximal GI tract in the hemithorax, followed by mechanical ventilation. High-frequency oscillating ventilation (HFOV) and inhaled nitric oxide (iNO) are options if needed.
    - Venoarterial (VA) ECMO can be used if the neonate is >2 kg and >34 weeks gestational age without high-grade intraventricular hemorrhage.
    - Supportive measures can be continued for 24 to 72 hours, followed by repair.

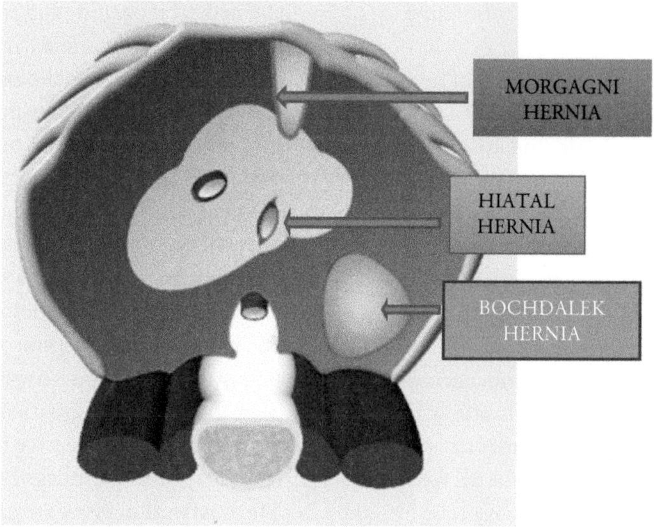

Figure 25.2 Types of non-traumatic diaphragmatic hernia.

- At the time of repair, three-fourths of cases require **mesh** to reconstruct the diaphragm, and sometimes the abdominal wall as well. Its use implies an increased risk of recurrence.
- Up to half of patients have **chronic lung disease** after repair.
- A **Morgagni diaphragmatic hernia** is a defect located anteriorly and to the right of midline. They are found incidentally later in life, but still need repair.

- **Pectus excavatum** is a posteriorly displaced lower sternum. Repair is with a Nuss bar if the Haller index is >3. Otherwise a Ravitch repair may be indicated in older patients.
- **Pectus carinatum** is most often treated with an orthotic brace. Repair is for cosmetic reasons.
- **Gastroschisis** is a failure of the small bowel herniation, 270-degree counterclockwise rotation, and reduction process in fetal life. All patients therefore have **malrotation** and many have **jejunoileal atresia** due to mesenteric vascular or umbilical vein insufficiency. *See Figure 25.3.*
  - Defects reside to the right of midline, usually are less than 4 cm, and have no peritoneal sac. The herniated small bowel is inflamed, stiff, and often dysmotile.
  - Management starts with NPO status, total parenteral nutrition (TPN), and **silo reduction** for 1 to 2 weeks. This is followed by **traditional closure or sutureless closure** using the umbilical stalk. Enteral feeding is slowly reintroduced.
- **Omphalocele** is a failure of lateral abdominal wall fold migration to midline, so the peritoneal sac and umbilical vasculature are retained in a midline defect. *See Figure 25.3.*
  - It is seen in premature neonates and those with intrauterine growth retardation. Patients may have an elevated AFP. It is associated with **other, non-intestinal congenital abnormalities** and therefore has a generally worse prognosis than gastroschisis.
    - Cantrell Pentalogy describes an omphalocele plus cardiac and inferior pericardial defects as well as lower sternal and central diaphragm defects.
  - Management is with **topical iodine or Silvadene for re-epithelialization** of the peritoneal lining over the course of 2 to 3 months. The defect can be closed primarily either early or late in the course. There is no need for a silo, because the peritoneal cover is intact.
- **Prune-belly syndrome** is absence or hypoplasia of abdominal wall muscles plus pectus excavatum, respiratory insufficiency, hypotonic lower urinary tract with reflux, bilateral cryptorchidism, and hip dysplasia. Treatment is with abdominal wall reconstruction and bilateral orchidopexy between 6 to 12 months of age.
- An **umbilical hernia** is a failure of closure at the linea. It is more common in African American infants. Most defects will close by 5 years of age, so observation can continue until then. If symptomatic at any time, the defect is greater than 2 cm in diameter, or it persists beyond the age of 5, then primary repair with absorbable suture should be performed.
- **Patent urachus** may communicate with the bladder lumen and extrude urine from the umbilicus. Otherwise, the remnant may simply be a blind-ended urachal sinus or a urachal cyst. Treatment is by resection.
- An **omphalomesenteric duct** communicating with an underlying Meckel's diverticulum would extrude mucous from the umbilicus.
- An **umbilical granuloma** has no internal organ communication, and can be treated with topical silver nitrate.

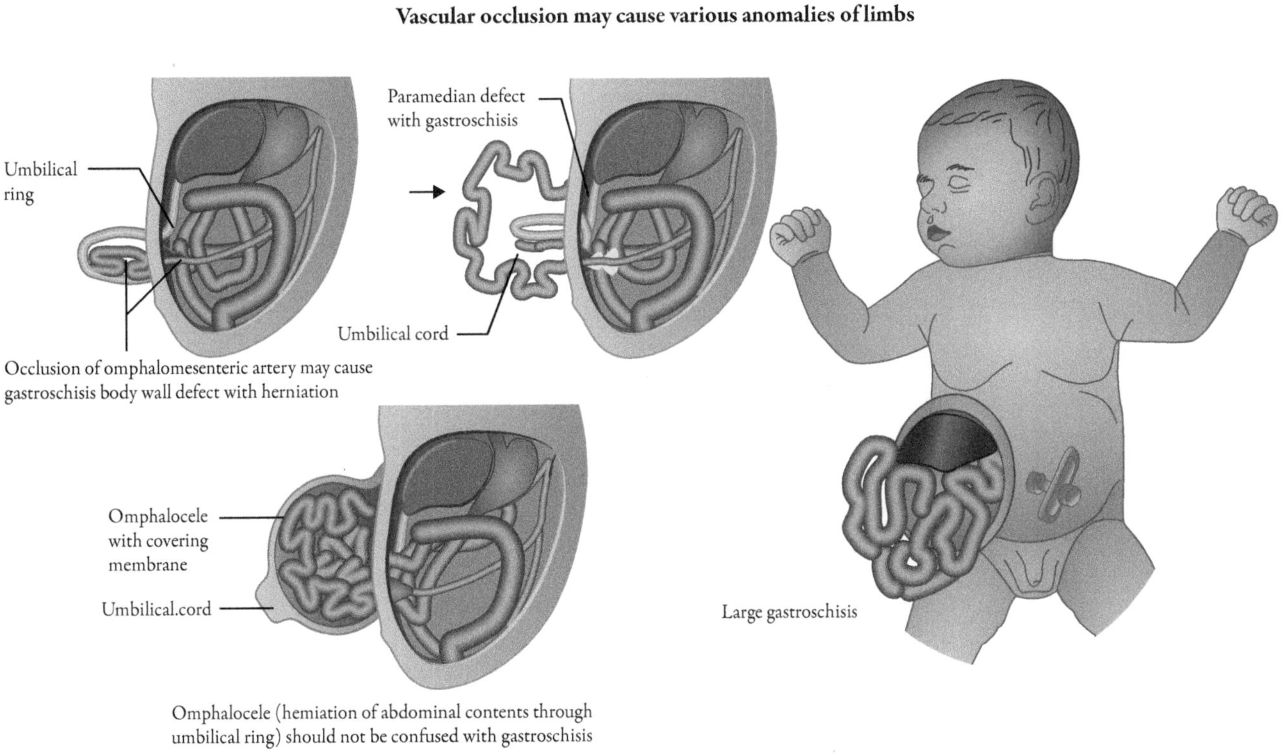

Figure 25.3 Omphalocele and gastroschisis.

- **Inguinal hernias** are indirect and classically present in a premature infant with a patent processus vaginalis (PPV). They are more common in male patients and on the right side. Inguinal hernias **incarcerate in 20%** of patients during childhood. They may have a sliding component containing the fallopian tube, ovary, colon, or bladder. Treatment is with **high ligation of the sac.** No need for repair of the inguinal floor, and no need to explore the contralateral side.
- **Non-communicating hydroceles** do not fluctuate in size and resolve at less than 1 year of age. **Communicating hydroceles** will fluctuate because of a PPV. Treatment is with opening and eversion of the hydrocele sac plus ligation of PPV if the hydrocele is communicating, or if it is non-communicating but present beyond 1 year of age.
- **Undescended testicles** reside above the inguinal ring. **Ectopic testicles** are below the ring but not within a sac. Patients may have a hypoplastic scrotum without rugal folds, or they can have underlying Prune-belly syndrome
  - US can make the diagnosis, followed by MRI if needed.
  - Management is with **orchidopexy** before 2 years of age via an inguinal incision. If not possible, try again in 6 months. At that point, one may consider testicular artery transection to liberate the testicle and give time for cremasteric and vas deferens vascular collaterals to form. This can be followed by another attempt at delayed orchidopexy 6 months later. Consider **hormonal treatment** initially if the patient has bilateral undescended testicles.
- There is a higher seminomatous cancer risk inherent to the undescended testicle. This does not change with treatment, although fertility will improve with treatment.

LIVER AND GALLBLADDER

- **Physiologic jaundice** starts at 2–3 days of life and improves by one week. If jaundice presents earlier or fails to clear, consider congenital infection, choledochal cyst, biliary atresia, or inspissated bile from TPN, hemolysis, or cystic fibrosis (CF).
- **Choledochal cysts** have a fibrotic cyst wall without a mucosal lining. They are due to in utero reflux of pancreatic enzymes via the common channel and, if not addressed, carry a risk of obstructive cholangitis as well as metaplasia and eventual cholangiocarcinoma.
  - Diagnosis is by prenatal ultrasound (US) or as workup for abdominal pain in the toddler years. There are 5 different types. *See Figure 25.4.*
    - **Type 1 is diffuse extrahepatic dilation.** This is the most common.
    - Type 2 is focal saccular extrahepatic dilation (diverticulum of duct).

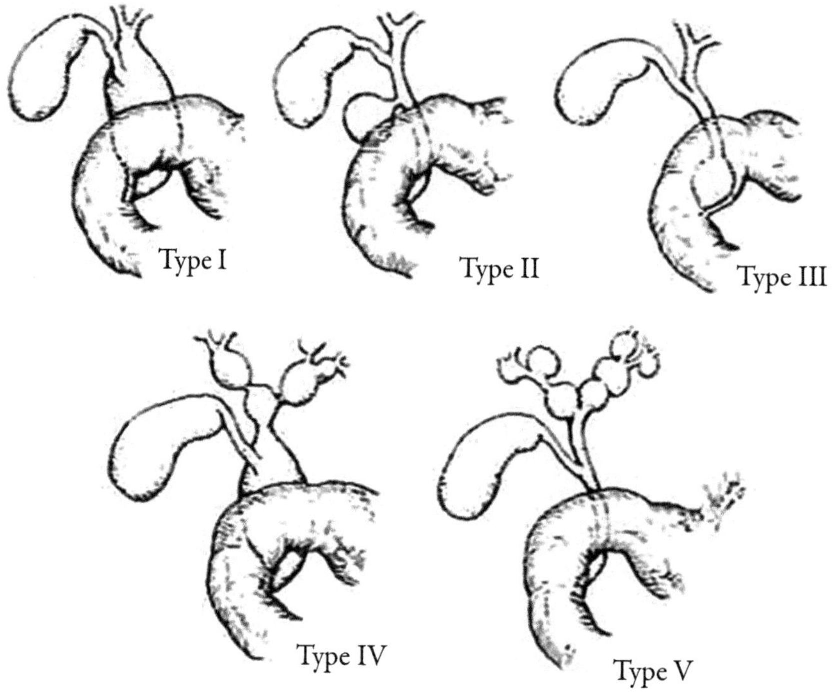

Figure 25.4 Types of choledochal cysts.

- Type 3 is a dilated distal, intramural portion of duct (choledochocele).
- Type 4a is multiple intra- **and** extrahepatic cysts. 4b is extrahepatic cysts only.
- Type 5 is intrahepatic cysts only.
- Management is with **hepaticojejunostomy** for type 1, **focal resection** in certain cases of type 2, **endoscopic sphincterotomy** or resection and choledochojejunostomy for type 3, and **lobectomy with hepaticojejunostomy** versus **liver transplantation** for types 4 or 5.

- **Biliary atresia** is suspected if jaundice has persisted beyond 2 weeks after birth and there is a negative metabolic liver screen. Cholangitis is one of the more common complications of atresia and obstruction.
  - **US and cholangiograph**y showing atretic ducts. Biopsy will show periportal fibrosis and cirrhosis. A nuclear scan shows liver uptake, but no radiotracer ever makes it into the intestinal lumen.
    - Subtypes include OHi 1 (CBD atresia), OHi 2 (hepatic duct atresia), and **OHi 3 (intrahepatic duct atresia).** OHI3 is the most common.
  - Management is with a **Kasai hepatoportoenterostomy** performed within 3 months of life to prevent cirrhosis. One-third improve, probably because ductules are patent at a portal plate level. One-third will expire. One-third fail to improve and need a **liver transplant**. This is the most common indication for a liver transplantation in the pediatric population.
- **Hepatoblastomas** are primitive hepatic tumors and associated with hemihypertrophy, Beckwith-Wiedemann, and familial adenomatous polyposis (FAP).
  - Lesions are usually unifocal and present before 4 years of age. An increased AFP and sometimes BHCG are seen. Patients may have precocious puberty.
  - Stage 1 is defined by a localized tumor that underwent an R0 resection. Stage 2 tumors underwent an R1 resection. Stage 3 tumors are R2 resections with gross residual disease.
  - Additional treatment includes cisplatin-based chemotherapy in a neoadjuvant or adjuvant fashion. Prognosis is based on histology and achieving an R0 resection.

NEURAL-NEUROBLASTOMA

- This is the most common solid tumor outside of the central nervous system (CNS) in children. It arises from neural crest cells and presents as an asymptomatic mass before 2 years of age.
- Tumors originate at the adrenal gland or anywhere along the sympathetic chain. Ganglioneuroblastomas and ganglioneuromas are well-differentiated variants.
- Patients most often present with a painless mass in infancy or toddler years. Some individuals will have raccoon's eyes due to orbital invasion, opsoclonus-myoclonus as a paraneoplastic manifestation, or

blueberry muffin spots as sites of extramedullary hematopoiesis. Initial imaging is usually **US followed by IV contrast CT scan or MRI** to determine the stage of disease.

- **Stage 1** is localized tumor that undergoes an **R0 resection.**
- **Stage 2** is localized tumor that undergoes an **R1 or R2 resection.**
- **Stage 3** is a tumor that **crosses midline** with or without lymphadenopathy.
- **Stage 4** is an advanced tumor with distant metastases. Stage **4S** describes a **localized tumor with distant metastases to bone marrow and skin.** This may have a better prognosis than other Stage 4 patients.
- Some tumors release catecholamines including **VMA, HVA,** and **metanephrines,** producing hypertension. VIP production can result in diarrhea.
- Lactate dehydrogenase (LDH) may be elevated and high neuron-specific enolase (NSE) levels can suggest metastases.

- Management is with resection of the adrenal and kidney with or without doxorubicin chemotherapy in a neoadjuvant or adjuvant manner. Diploid tumors have a better prognosis, and N-Myc amplification confers a worse prognosis.

RENAL-NEPHROBLASTOMA

- This is the second most common solid, non-CNS, tumor in children less than 2 years of age. If over 2 years, it is the most common.
- Patients often will have an asymptomatic mass of renal origin. Some can have hemihypertrophy; Beckwith-Wiedemann syndrome (macroglossia, gigantism, visceromegaly, hypoglycemia), or WAGR syndrome (Wilms, aniridia, genitourinary abnormalities, mental retardation). Imaging is with **US** followed by **IV contrast CT or MRI.**
  - **Stage 1** is limited to the kidney and underwent an R0 resection.
  - **Stage 2** extends beyond the kidney but still undergoes an R0 resection.
  - **Stage 3** is an extensive tumor removed by R1 or R2 resection.
  - **Stage 4** includes metastases to lung, bones, and liver.
  - **Stage 5** are bilateral nephroblastomas.
- Management is by resection of the mass and kidney. Neoadjuvant chemotherapy is used for Stages 3, 4, or 5. Thereafter, surgery can be performed and the tumor extracted from the renal vein if needed. Survival is 85%, and prognosis is based on grade and the completeness of resection.

TERATOMA

- This is a tumor containing multiple tissue types foreign to the location where it is discovered. Patients can have increased AFP and bHCG.
- **Sacrococcygeal teratomas** are present at birth and are benign in 90%. There is increased malignant potential if present at over 2 months of age. Treatment is with resection and coccygectomy. Some patients can benefit from fetal intrauterine surgery if the lesion is causing high output heart failure.
- **Ovarian and mediastinal teratomas** present in adolescence and adulthood, and are treated by resection.

GASTRIC AND ESOPHAGEAL

- **Gastroesophageal reflux (GER)** is due to transient lower esophageal sphincter (LES) relaxation as a neonate or infant
  - Patients may have trouble with feeding, spitting up, or regurgitation. The diagnosis is clinical in most scenarios. **Upper gastrointestinal contrast imaging (UGI)** can be done to exclude gastric obstruction, duodenal atresia, or malrotation. A **24 hours pH probe** will confirm the diagnosis.
  - Treatment starts with propping up the baby during feeds, thickening feeds with rice or cereal, and adding a histamine blocker (H2B) or proton pump inhibitor (PPI). Most cases resolve in infancy and do not need surgery.
    - For severe, persistent symptoms that qualify as gastroesophageal reflux disease (GERD) and failure of medical management, patients will need a nasojejunal (NJ) or gastrojejunal (GJ) feeding tube or a Nissen fundoplication.
- **Pyloric stenosis** classically presents in a 3- to 6-week-old firstborn male.
  - Patients will have projectile vomiting and sometimes a palpable olive mass on exam. Laboratory workup reveals low NaCl and K, but high HCO_3. Patients may have paradoxical aciduria with K in it. This is due to NaHCO3 reabsorption along the proximal convoluted tubule (PCT) in response to hypovolemia and low glomerular filtration rate (GFR). Confirmation is by an **US** demonstrating over 4 mm thickness and over 16 mm length to the pyloric musculature.
  - Start management with **resuscitation** and K repletion in maintenance fluids after resuscitation (D5 1/2NS with KCl). Then an open or laparoscopic **pyloromyotomy** can be performed from the antrum across the pylorus. The incision should extend down to the submucosal layer. Both edges of muscle

should be independently mobile. There is a 1%–3% perforation rate.

INTESTINE

- Small and large bowel are not distinguishable by x-ray in neonates, as no plica circularis or haustra are visible.
- **Duodenal atresia (DA)** is from **failure of recanalization** of the duodenum during development. 20% of patients have trisomy 21. It can be associated with cardiac, renal, or other gastrointestinal (GI) abnormalities.
  - DA presents soon after birth as bilious emesis and failure to pass meconium in a **well-appearing neonate. Abdominal x-ray (AXR)** will show gastric and duodenal air (double bubble) without any distal air in the GI tract.
    - An **upper gastrointestinal contrast study (UGI)** will show accumulation up to the first part of the duodenum but no further. The contrast should be suctioned out to prevent aspiration. This test is diagnostic but not always necessary.
    - If the patient actually has a duodenal web, annular pancreas, or malrotation with volvulus, then there may be some distal air on AXR and passage of some contrast on UGI.
  - Evaluation for associated abnormalities should proceed after DA is diagnosed.
  - Surgical treatment is a **proximal transverse to distal longitudinal duodenoduodenestomy**. A Foley can be passed distally in case there are any, as yet undiagnosed, downstream atresias.
- **Malrotation** is failure of the normal 270-degree counterclockwise rotation that usually takes place at 10–12 weeks gestational age, prior to the return of enteric contents into the abdominal cavity. When malrotated, the duodenum and all small bowel remain on the patient's right. The **cecum resides at approximately midline, from which Ladd's bands extend to the right upper quadrant (RUQ)** anterior to the duodenum.
  - An **UGI** contrast study will show the duodenal-jejunal junction on the patient's right side in malrotated anatomy.
  - **Midgut volvulus** can complicate malrotation as the small bowel and cecum turn in a clockwise fashion, with Ladd's bands then compressing and obstructing the duodenum. The distal small bowel is also compromised as it suffers from a closed loop obstruction and occlusion of its twisted mesenteric vasculature. *See Figure 25.5,*
    - If midgut volvulus occurs, it is generally in the first year of life. Presentation includes **bilious vomiting** from the duodenal obstruction distal to the ampulla of Vater, and **systemic illness** from small bowel ischemia in the closed loop.
    - The duodenum and stomach will be distended on **AXR.** There may be central air-fluid levels seen within the distal, closed loop of occluded small bowel. In the correct clinical picture, these AXR findings can be diagnostic and need not be accompanied by a UGI contrast study.
  - Management is with laparotomy, **counterclockwise de-torsion** of the small bowel and cecum, **lysis of Ladd's bands,** resection of any frankly necrotic small bowel, placement of the cecum in the left lower quadrant (LLQ), performing an appendectomy, and broadening the small bowel (SB) mesentery. Re-anastomosis can be attempted provided the patient's clinical condition allows for it. Otherwise, this should be reserved for a future operative visit and the patient left in discontinuity or with double-barrel enterostomies in the interim.
- **Jejunoileal atresia** is due to an **intrauterine vascular accident** resulting in an atretic segment or segments of intestine. It can be associated with gastroschisis.
  - Patients present early in the neonatal period with bilious emesis and failure to pass meconium. However, since there is generally no intestinal ischemia or compromised bowel, they **appear well.**
  - **Abdominal x-ray or UGI** with follow-through may demonstrate a transition point with proximal air-fluid levels. Barium enema may be needed as part of the diagnostic process. It can suggest alternative pathology like Hirschsprung disease, treat meconium ileus or meconium plug, and visualize any accompanying microcolon in the setting of small bowel atresia.
    - There are **four types** of atresia. Type 1 is **mucosal** atresia only, with the remainder of the bowel wall intact as well as its mesentery. Type 2 is atresia of the **entire bowel wall** leaving a fibrous band. The mesentery is still uninvolved, however. Type 3a has a wedge shaped **mesenteric atretic defect** in addition to the intestinal wall. Type 3b has a larger mesenteric defect and bowel with a distal Christmas tree pattern. Type 4 has multiple atretic areas.
  - Management is with **small bowel resection and anastomosis.** The proximal dilated bowl may need to be tapered. The distal bowel may need tailoring to increase luminal diameter. A Foley catheter balloon can be passed distally to exclude additional mucosal atretic segments prior to anastomosis.
- **Necrotizing enterocolitis (NEC)** presents in premature neonates after the first week of life, when their intestines have become colonized with bacteria. They subsequently can develop bacterial multiplication, impaired motility, and bacterial translocation into their intestinal wall. Impaired mesenteric perfusion due to

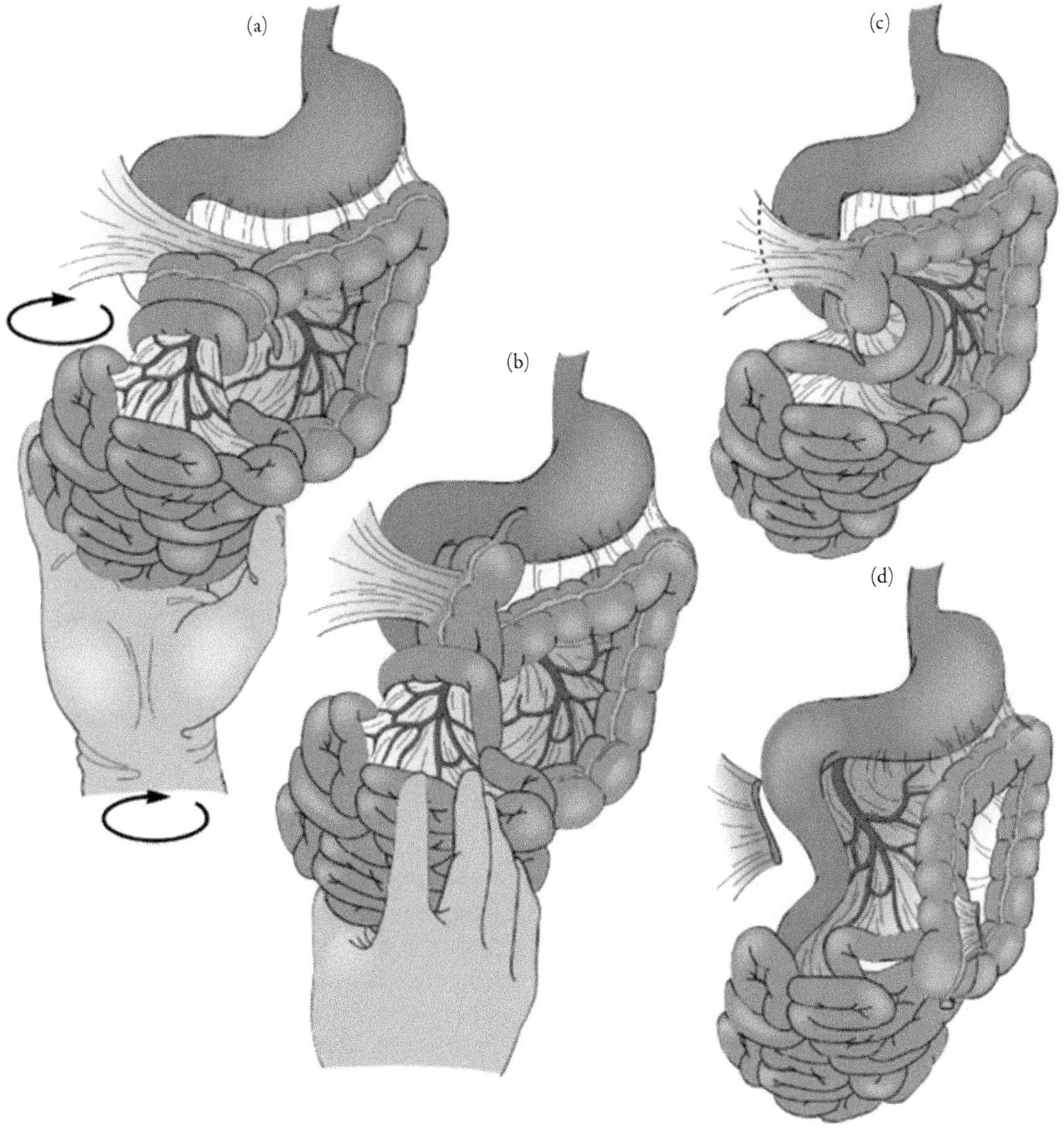

Figure 25.5 Malrotation with midgut volvulus. Occlusive Ladd's bands and manner of de-torsion.

congenital heart disease, intrauterine growth retardation (IUGR), or perinatal hypoxia can contribute to NEC.

- Signs include irritability or lethargy, trouble with feeding, and abdominal distension. **Type 1 NEC** has the above clinical findings and increased residuals after feeds. AXR may show an **ileus. Type 2 NEC** has bilious vomiting, bloody stool, abdominal tenderness, and an AXR with **pneumatosis intestinalis** in addition to ileus. **Type 3 NEC** has all of the above plus **pneumoperitoneum** on AXR or lateral decubitus view.
- Initial **conservative** management with nil per os (NPO) nasogastric tube (NGT) decompression, IVF, and TPN. This will be sufficient for the majority of patients.
  - **Surgery** is reserved for pneumoperitoneum or clinical deterioration and development of peritonitis despite conservative management. One must resect all infarcted bowel and bring double-barrel proximal and distal end stomas to the skin surface. Once the patient has recovered and is of adequate weight, microcolon or small bowel atresias can be excluded by contrast studies before embarking on an anastomosis.
  - Patients can suffer from short bowel syndrome after resection and anastomosis, or strictures in the colon or elsewhere after recovery from medically managed NEC.
- **Meckel's diverticulum** is the most common congenital GI malformation. It is a remnant of the vitelline duct on the antimesenteric border of the distal small bowel. It is a true diverticulum containing all layers of the bowel wall in addition to its own mesentery, the mesodiverticulum.
  - The rule of 2's generally applies: The diverticulum resides 2 feet from the ileocecal (IC) valve, affects 2% of the population, is twice as common in boys than in

girls, is symptomatic in 2%, and contains 2 kinds of mucosa: pancreatic more commonly than gastric.
  - Meckel's diverticulum presents as **painless lower gastrointestinal bleeding (LGIB)** from ulceration of the small bowel mucosa on the mesenteric border adjacent to the diverticulum. The **ulceration** occurs secondary to gastric mucosa producing acid.
    - A less common presentation is **obstruction** from a fibrous band, a mesodiverticular band, or because the diverticulum is acting as a lead point for intussusception.
    - An **omphalomesenteric fistula** may also be seen with extrusion of mucous from the umbilicus. *See Figure 25.6.*
    - Confirmation is by a **pertechnetate scan** localizing to gastric mucosa. Provided the indication for ordering the test is painless LGIB in a pediatric patient, the test has a high sensitivity, since these are the patients likely to have gastric mucosa at their diverticulum.
  - **Diverticulectomy** is the best option for an incidentally discovered Meckel's diverticulum in a child, or if any patient is suffering from diverticulitis at any age.
  - **Segmental resection,** including the diverticulum and neighboring small bowel distally and proximally, is the best option if this bowel is inflamed or ulcerated.
  - An asymptomatic Meckel's diverticulum incidentally discovered in an adult should be left alone.
- **Meconium ileus** is an ileocecal valve obstruction due to thick meconium. 10% of these patients may have cystic fibrosis (CF) suggested by high levels of NaCl in sweat, or corroborative genetic testing for a mutation in the CF transmembrane receptor in the patient or parents.
  - AXR can show soap suds and no air-fluid levels in a distended neonate who hasn't passed meconium. Occasionally, elevated intraluminal small bowel pressure can result in perforation that is **contained as a meconium pseudocyst, or free** with peritonitis.
  - Treatment for non-perforated patients is by **gastrografin or N-acetylcysteine (NAC) enemas** q12 hours until all the meconium has passed. If this fails, one may perform laparotomy, **enterotomy, and manually remove meconium or use antegrade enemas.** A stoma may be constructed to facilitate postoperative antegrade irrigation. Less commonly, involved SB that is compromised or perforated may need to be resected.
  - **Distal intestinal obstruction syndrome** is the meconium ileus counterpart in adults. It is treated similarly with gastrografin enemas.
- **Intussusception** is the most common etiology of obstruction for infants and children aged 3months to 3 years. Typically there is no lead point. In some cases, the lead point can be hypertrophied ileal Peyer's patches, lymphoma, or a Meckel's diverticulum. The **intussuscipiens** is the distal bowel into which the **intusussusceptum** has invaginated. The resulting high intraluminal and intramural pressures can exceed capillary perfusion pressure and give rise to mucosal sloughing, seen as currant jelly stools. *See Figure 25.7.*
  - Abdominal US shows a **target sign.** This study is superior to an x-ray and avoids the radiation of a CT scan.
  - Management starts with air-contrast enema (at up to120 mmHg pressure) or a barium enema (maximum height of the barium column being 3–5 feet). One may attempt pneumatic or hydrostatic reduction twice. Pediatric patients can be discharged home after a brief observation period following successful reduction.
    - **Surgery** is reserved for those **failing two attempts** at the pressure or height limits, or patients with multiple **recurrences (5%–10%), peritonitis, or free air.** One may remove the appendix at the same time. Isolated bloody or currant jelly stools are not, on their own, an indication for surgical exploration. These patients can still undergo pneumatic or hydrostatic reduction.
    - **In adult patients, symptomatic intussusceptio**n is an indication for surgical resection, as this is likely associated with a malignant lead point. **Asymptomatic, incidental intussusception** seen on CT scan of an adult patient can be managed with observation alone.
- **Mesenteric cyst** is a structural abnormality of the lymphatic system. Symptoms are from bowel obstruction or ulceration that leads to mucosal bleeding. The cyst contains serous and proteinaceous fluid and remains separate from the small bowel (SB) mucosa. Treatment is with **resection of the cyst alone,** preserving the mesenteric vessels and adjacent SB.
- **Mesenteric duplication** is a mucosal lined cavity that shares a wall with the SB, lying within leaves of ileal mesentery. Symptoms are from bowel obstruction or ulceration that leads to mucosal bleeding. Treatment is **resection of the cavity, including the neighboring confluent SB when necessary.** If resection is not possible, perform enterotomies to access the duplication lumen, and strip the mucosa so the duplication cavity collapses.

COLORECTAL AND ANUS

- For **appendicitis**, see the section in Chapter 10, Colon.
- **Imperforate anus** describes the failure of the urorectal septum to descend to the sphincter muscles. It is more common in males. A **lateral radiograph** should be done to note the lowest point of rectal gas relative to the coccyx. Check for associated VACTERL with an **echocardiogram, and spinal US** to look for spinal cord abnormalities. These tend to be directly proportional to the severity of atresia. MRI is better if the patient's bones have already ossified.

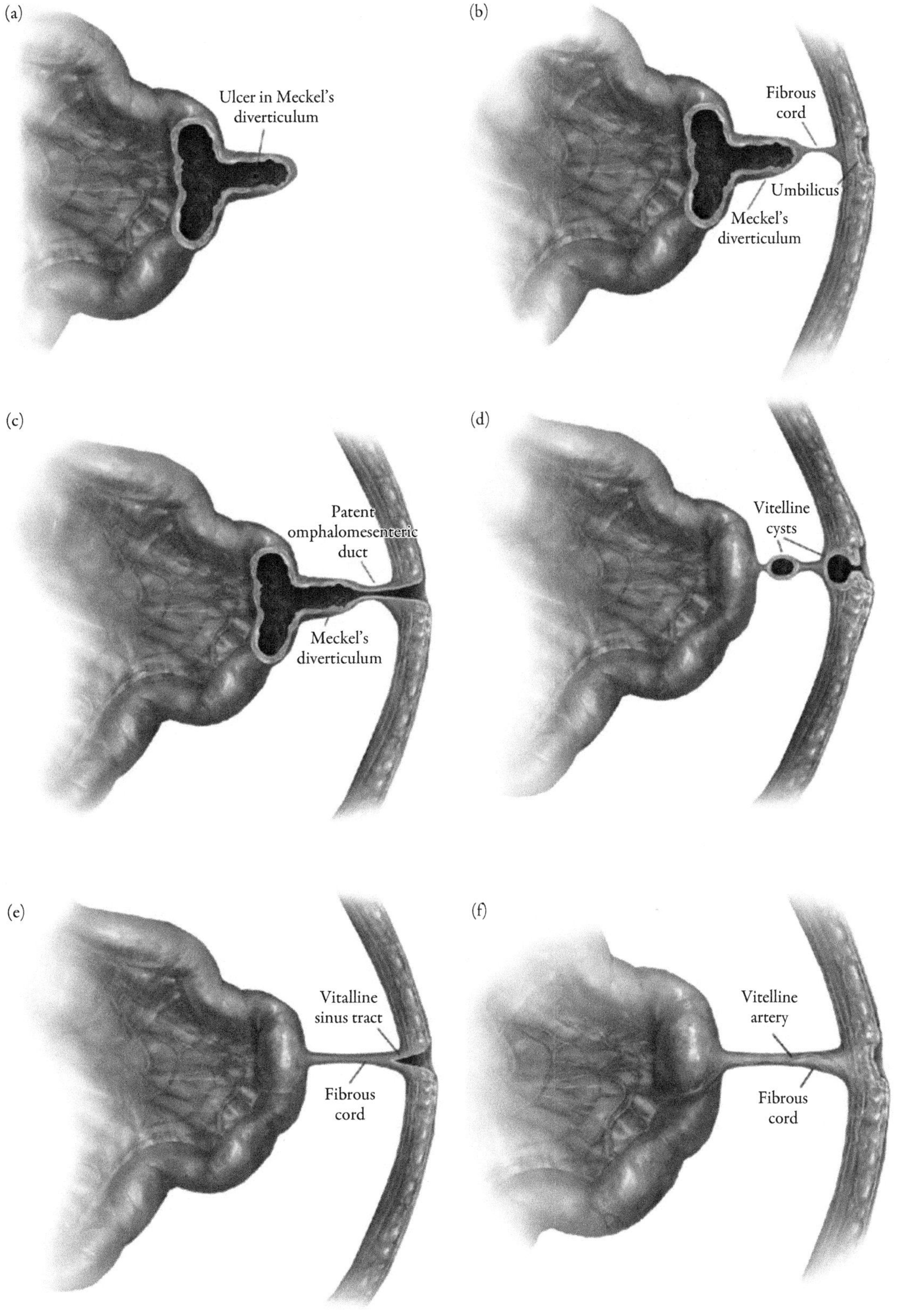

Figure 25.6 Defects of vitelline, omphalomesenteric anatomy.

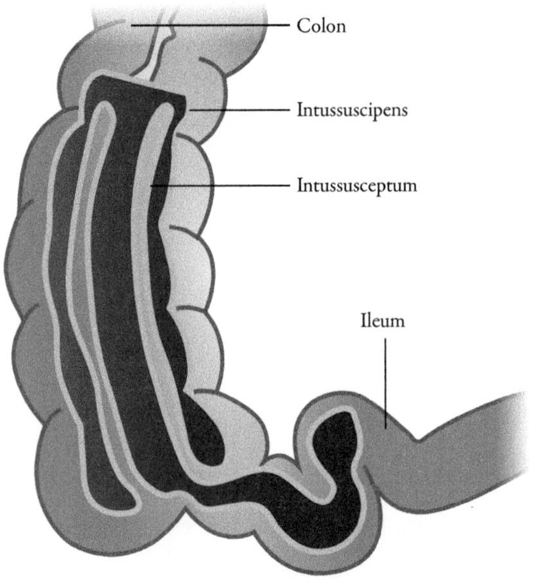

Figure 25.7 Ileocolic intussusception.

- **A high imperforate anus is above the levators.** It generally has a fistula to the vagina or to a persistent common channel (cloaca) in females. Meconium is expelled at the vagina. Males may have a fistula to the bladder, prostatic, or bulbar urethra, with meconium in urine.
  - Treatment is with a **double-barrel, dual ended colostomy** at the time of presentation in order to completely defunctionalize the rectum and keep stool out of the vagina or bladder. A subsequent distal colostogram can precisely define the length of rectum and location of the fistula. This is followed by a **posterior sagittal anorectoplasty (PSARP)** with laparoscopic assisted mobilization of the rectum, ligation of the fistula, and anal reconstruction at **3–6 months of age**. The procedure can be complicated by postoperative incontinence.
- A **low imperforate anus is below the levators.** It generally has a fistula to the vestibule or perineal skin with meconium extruding from this skin opening. Treatment is with a **PSARP at the time of diagnosis.** No stoma or laparoscopic assistance is needed. The anorectum is pulled down into the area of sphincter muscles. It can be complicated by constipation that may require postoperative stricture dilation.

- **Hirschsprung disease.** The characteristic finding is rectal aganglionosis at submucosal (Meissner's) and intermuscular (Aurbach's) neural plexi. Aganglionosis results from failure of neural crest cells to migrate caudally by 12 weeks gestational age, and is associated with GDNF and RET genes. Hirschsprung disease is more common in males.
- Neonates have failure to pass meconium at 24 hours and they may have explosive diarrhea with anorectal digital exam. Children and adults may present with chronic constipation and a very short segment of aganglionosis.
  - Confirmation is with **biopsy at 1, 2, and 3 cm proximal to the dentate line** to determine the level of aganglionosis. This is usually near the **rectosigmoid junction.** Suction biopsy is a mucosal partial thickness sample, and is usually adequate. Full thickness biopsies are the gold standard. Neural hypertrophy may be noted, and there can be increased acetylcholinesterase.
  - **Barium enema** can exclude an upstream microcolon secondary to a proximal small bowel atresia or another process. Note that a barium enema cannula may bypass a very short, low anorectal aganglionotic segment and hence be falsely negative.
  - **Manometry** can demonstrate a lack of receptive rectal wall relaxation.
  - Hirschsprung colitis is the leading cause of death and is a sequelae of chronic partial obstruction. Some patients will have diarrhea.
- After the extent of aganglionosis is determined by serial biopsies, a **colostomy** can be constructed above this level in order to decompress the patient in the acute period if they are ill. Otherwise, in a stable patient one can proceed to definitive management without a colostomy. This includes resection or bypass of the aganglionic segment and coloanal anastomosis.
  - The **Suave** procedure is a submucosal endorectal dissection with pull through of the proximal colon into a cuff of aganglionic muscle.
  - The **Duhamel** procedure bypasses an aganglionic segment in the retrorectal space with a proximal and distal end-side stapled anastomosis between the colon and anorectum. The anterior wall of this new neorectum is aganglionic, and the posterior wall is composed of normal, innervated colon. This procedure is good for a long-segment aganglionosis in an older child. There is a decreased stricture rate and a less extensive pelvic dissection.
  - The **Swenson** procedure is a complete aganglionic segment resection and an end to end anastomosis.
  - Adults presenting with chronic constipation and a short-segment distal aganglionosis can be treated with a myotomy in the anorectal area.
  - Treatment of Hirschsprung colitis is by rectal irrigation, emergency colostomy, or colectomy.

MISCELLANEOUS CNS, FACE, GENITOURINARY

- **Craniosynostosis** can present as early fusion at the sagittal, coronal, or lambdoidal sutures before 2 years of age.

- Subependymal **intraventricular hemorrhage** is due to systemic illness or cardiopulmonary disease in a premature neonate. Treatment is with intraventricular catheter drainage.
- **Spina bifida occulta** is a defect in the vertebral column, most often asymptomatic.
- **Meningocele** is when the meningeal layer herniates through the defect.
- **Myelomeningocele** is when the spinal cord along with the meninges herniates through the defect. Patients may have hydrocephalus, a weak upward gaze, aqueductal stenosis, or Arnold Chiari 2 malformation.
- An **encephalocele** is brain and its meningeal covering herniated through a skull defect.
- **Cleft lip** should be repaired after 10 weeks of age so long as the patient's weight is >10 lbs, and Hgb >10 mg/dl. Patients may need orthotics or nasoalveolar molding prior to repair.
- **Cleft palate.** The primary palate is anterior to the incisive foramina. The secondary palate is posterior and includes the soft palate. Repair should be done sometime in the first year of life, with mucoperiosteal flaps for hard palate coaptation and possibly a second procedure for the soft palate. This is to prevent misarticulations, velopharyngeal insufficiency (VPI), and nasal speech. VPI can also occur postoperatively. Patients must be at least 6 months post-op and have an adequate vocabulary in order to appropriately assess VPI.
- **Cranial or facial atrophy or microsomia** is treated with distraction osteogenesis of the mandible.
- **Vaginal atresia** can present along the spectrum of imperforate hymen to more severe atresia and hydrocolpos. This should be repaired in the post-puberty adolescent years.
- Simple ovarian cysts may be observed.
- **Ambiguous genitalia** can result from:
  - An XY patient with absence of SRY or Müllerian inhibitory substance (MIS), or low testosterone, dihydrotestosterone, or androgen receptors.
  - An XX patient with ovotestis or congenital adrenal hyperplasia (CAH).
- Transuterine fetal surgery includes vesicoamniotic shunting for lower urinary obstruction, myelomeningocele repair to prevent amniotic fluid damage to the spinal cord, and the EXIT procedure for airway obstruction.

MISCELLANEOUS SOFT TISSUE, VASCULAR

- **Congenital melanocytic nevus** can be small <1.5 cm (0.5% melanoma risk), large >10 cm, or giant >20 cm (5%–10% melanoma risk). Treatment is with chemical peel, laser dermabrasion, or excision after tissue expansion. A biopsy should be obtained first if melanoma transformation is suspected.
- **Vascular occlusion** in pediatric patients should prompt consideration of fibromuscular dysplasia (FMD). Treatment is with heparin and observation if post-angiographic, even in the presence of motor deficits. When necessary, surgical reconstruction should employ an arterial conduit. Veins reinforced with Dacron or PTFE can be used if no arterial conduit is available.
- **Limb abnormalities** include digit duplication (most often the thumb), digital fusion (syndactyly) of soft tissue or bone and cartilage, and constrictive band syndrome limiting limb length growth in utero.

26.

NEUROSURGERY AND ORTHOPEDICS

CRANIUM PRINCIPLES

- The envelope of the cranium, from superficial to deep, follows the mnemonic SCALP and includes the following: **Skin, Cutaneous soft tissue, Aponeurotica of galea (extension of the frontalis muscle), Loose connective tissue, and Periosteum.**
  - Soft tissue grafts can be placed to any layer. Burr holes drilled into the diploe may be useful if grafting directly onto bone. Adjuncts include tissue expanders and composite grafts including the rib, outer table of another portion of the skull, or a prosthesis.
  - Any osteomyelitis or osteitis of a bone flap requires debridement due to poor perfusion and hence poor antibiotic penetration.
- Circle of Willis. *See Figure 26.1.*
  - Two **vertebral arteries** ascend and merge into one basilar, which bifurcates into two **posterior cerebral arteries.** These can be compressed by uncal brain herniation. The posterior cerebrals are connected to middle cerebral arteries via two posterior communicating arteries.
  - **Anterior cerebral arteries** branch off middles and connect via one anterior communicating artery. The anterior cerebral arteries can be compressed by falcine brain herniation.
- **Cranial Nerve (CN)**
  - 1 is olfactory, 2 optic, 3 oculomotor (controls the majority of globe movements), 4 trochlear (controls the superior oblique muscle), and 6 abducens (globe abduction).
  - 5 trigeminal contains three branches that are sensory to the entire face. **Trigeminal neuralgia** results in shooting pain along the facial distribution. Treatment is with nerve transection, stereotactic radio surgery (SRS), or microvascular decompression via a posterior fossa craniectomy.
  - 7 facial (motor to the face and anterior tongue taste), 8 vestibulocochlear, 9 glossopharyngeal, 10 vagal, 11 accessory, and 12 hypoglossal.
- The **supraoptic nucleus** of the hypothalamus releases ADH in response to high serum osmolarity or low venous tone. **Diabetes insipidus (DI)** is from lack of ADH. It results in loss of water in urine and euvolemic hypernatremia. DI is treated with free water and Desmopressin (DDAVP). **Syndrome of inappropriate antidiuretic hormone (SIADH)** results in concentrated urine and euvolemic hyponatremia. Treatment is with free water restriction and diuresis. **Cerebral salt wasting** is when a cerebral insult causes loss of sodium in urine and hypovolemic hyponatremia.
- Grey matter represents mostly neuronal cell bodies on the cerebral and cerebellar cortical surface, White matter is deep to this, and composed of neuronal axons. Deep grey matter consists of neuronal cell bodies nestled within the center of the cerebrum, and includes the caudate, putamen, globes pallidus, and thalamus. Deep brain surgery is by stereotactic lead placement into the deep grey matter to treat motor regulation disturbances like tremor, Parkinson's disease (PD), or dystonia.
- **Cerebrospinal fluid (CSF) volume** approximates 150 ml at any one time, with 500 ml per day made by the choroid plexus. Leaks from a cribriform plate fracture result in rhinorrhea, and those from temporal bone injuries produce CSF otorrhea. Diagnosis of a **CSF leak** is by checking B-transferrin levels of the fluid. Most are managed conservatively without antibiotics. Occasionally, a lumbar CSF drain or surgical repair with coverage will be necessary.
- Normal **cerebral perfusion pressure (CPP)** should be >60 mmHg. It is calculated as mean arterial pressure (MAP) minus intracranial pressure (ICP). Normal ICP should be <20 mmHg, and is determined by the composite of brain parenchyma, CSF volume, and any other space-occupying lesion residing within the limited capacity of a rigid skull.
  - Intraventricular pressure monitors are ideal for following patients with a GCS <8 in whom the neurologic exam cannot be utilized. Intraparenchymal pressure monitors (with or without oxygen and temperature sensors) are an alternative to intraventricular monitors.
  - Management of intracranial hypertension is aimed at preventing further neurologic injury.
    - 1. Drainage of any focal collection of blood or tumor producing adjacent compressive effects.
    - 2. Maintenance of MAP so that CPP remains >60 mmHg. An intraventricular or intraparenchymal monitor may be used to follow ICP.

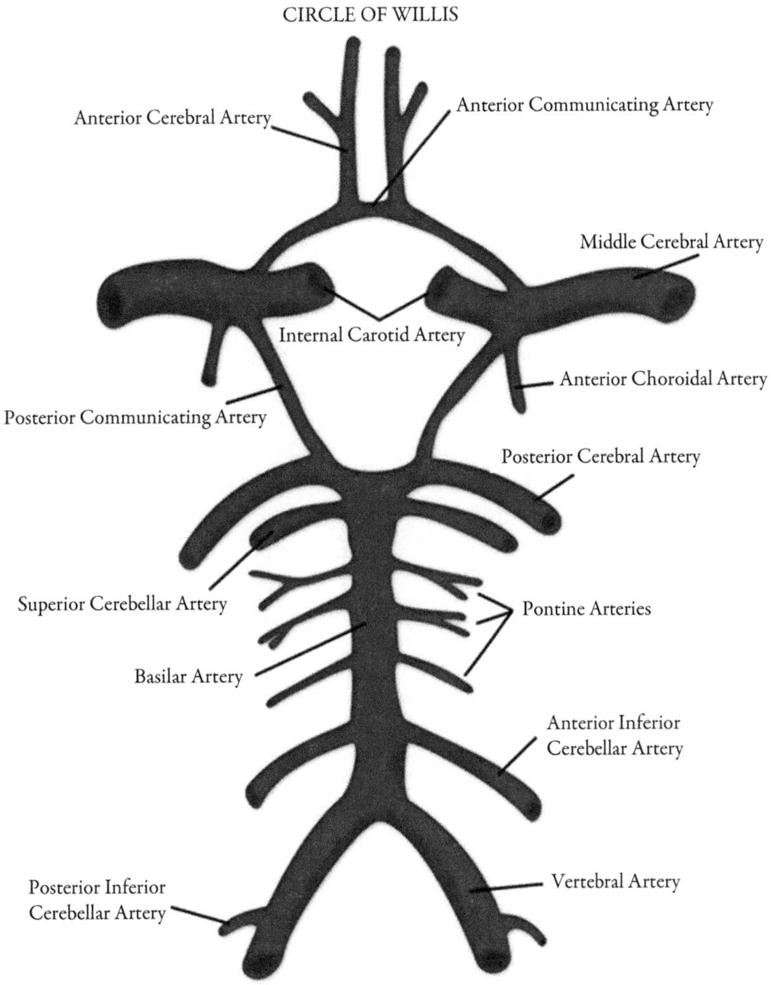

Figure 26.1 Circle of Willis.

- 3. Head of bed elevation.
- 4. Control of pCO_2 in the mid-30 range.
- 5. Sedation, anticonvulsants, temperature control, and glucose control.
- 6. Hypertonic saline or mannitol (1g/kg) to decrease cerebral edema. The latter will decrease ICP within 20 minutes by drawing out intracellular fluid and decreasing blood viscosity to improve perfusion. One must keep serum osmolarity <300.
- 7. Neuromuscular junction (NMJ) blockade if the above measures are inadequate.
- 8. Decompressive craniotomy to accommodate diffusely edematous cerebral tissue in the absence of a focal mass effect.

BRAIN MASS

- **Metastases** are the most common brain mass and primarily come from the lung, breast, kidney, gastrointestinal tract, or melanoma.
- Primary brain tumors most often originate from glial cells and include **astrocytoma** (2/3rds of which are high-grade glioblastoma multiforme), **oligodendrocytoma**, or **microglial** origin (brain macrophages).
- A medulloblastoma is a neuronal origin tumor in the infratentorial position and is more common in children. Ganglioglioma is a mixture of neural and glial origin cells. Ependymomas come from the lining of the ventricles and may obstruct CSF flow or produce drop metastases to the spinal cord. Tumors from the choroid plexus are mostly papillomas, and not malignant.
- A **meningioma** comes from the arachnoid layer.
- **Pituitary tumors** include microadenomas that are <1 cm, or macroadenomas >1 cm. Prolactin may be overproduced. Dopamine can inhibit this. Craniopharyngiomas are pituitary origin and not usually hormonally active.
- Hemangioblastomas can be seen in patients with Von Hippel-Lindau syndrome (VHL), who also may have renal cell carcinoma and pheochromocytoma.

- Other tumors include epidermoid, dermoid, and teratoma.
  - **Abscesses** can be distinguished from masses by CT or MRI. They can be the sequelae of preceding cerebritis or hematogenous spread
- **DIAGNOSIS**
  - The primary imaging modality for cranial pathology is **CT for hemorrhage and MRI for mass lesions.** T2 weighted images show edema. Diffusion weighted images are for ischemic areas, and gadolinium contrast can help define neoplasia or abscess.
- **TREATMENT**
  - Eligible masses may be **resected** via craniectomy or undergo **stereotactic radiosurgery (SRS).** This uses a linear accelerator or gamma knife for lesions under 3 cm that are not adjacent to the medulla or spinal cord. Proton beam or cyber knife are alternatives.
  - **Abscess** treatment is with 6–8 weeks of antibiotics along with aspiration or glucocorticoids for neighboring edema.

NEURAL AND NEUROVASCULAR INJURIES

See Figure 26.2.

- **CONCUSSION**
  - This is from a jarring of the brain within CSF and the cranial vault. Subcellular-level injuries occur, and they are not appreciable on CT or MRI. A concussion is graded clinically. Grade 1 is confusion, 2 amnesia, and 3 is transient loss of consciousness. Treatment is supportive with brain rest and avoidance of further trauma.
- **DIFFUSE AXONAL INJURY (DAI)**
  - Results from a sudden deceleration of brain parenchyma producing sheared axons that retract and curl into a ball. Patients can have symptoms that range from mild cognitive deficits to coma. An MRI can grade severity. Treatment is supportive.
- **INTRAVENTRICULAR HEMORRHAGE (IVH)**
  - Can occur in the subependymal layer in premature infants with systemic or cardiopulmonary illnesses. Hemodynamic changes can result from the amount of hemorrhage.
  - Adults may have IVH from extension of intraparenchymal bleeding that decompresses through cerebral tissue into the open space of the ventricle. Treatment of IVH is with a ventricular catheter.
- **CEREBRAL CONTUSIONS**
  - Occur on the undersurface of the frontal and anterior temporal lobes, or within the upper midbrain. Edema is maximal at 24–48 hours, and a hematoma within the contused parenchyma can develop, more commonly in elderly individuals.
  - Patients may have headache, loss of consciousness, or altered mentation. Treatment is conservative. Rarely patients will need a decompressive craniectomy.
- **INTRACEREBRAL HEMATOMAS**
  - Secondary to trauma (temporal lobes), spontaneous from hypertension (deep grey matter), or amyloid angiopathy (peripheral and superficial within the parenchyma). A less common cause is intraparenchymal rupture of an arteriovenous malformation (AVM). Other vascular malformations such as cavernous or venous hemangiomas, or capillary telangiectasias, have low rupture risk.

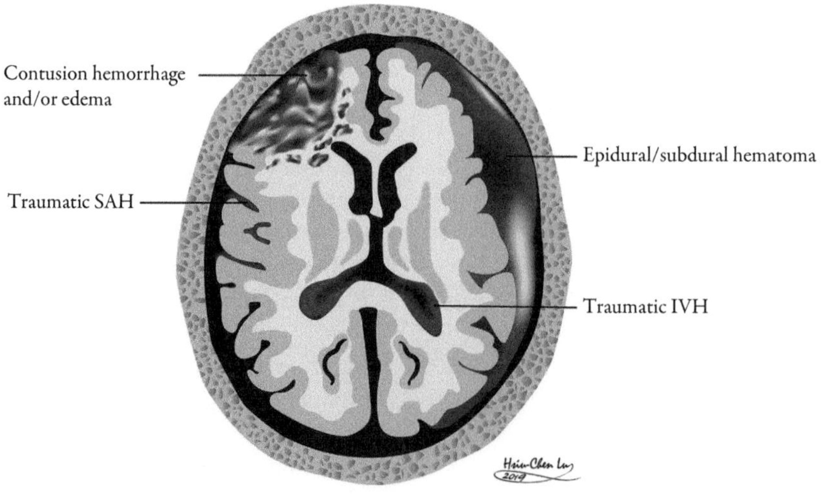

Figure 26.2 Types of traumatic intracranial hemorrhage.

- **DIAGNOSIS**
  - Patients may have a headache, loss of consciousness, focal neurologic signs, or altered mentation prompting CT imaging.
- **TREATMENT**
  - Conservative management will suffice for most patients, with control of blood pressure and any coagulopathy or platelet dysfunction.
  - Evacuation can be considered for hematomas that are <1 cm from the cortical surface, or those causing a decreased GCS, midline shift, or effacement of a basilar cistern. Ruptured AVM can be treated by coil embolization, microsurgical resection, or SRS if under 3 cm in size.
- **SUBARACHNOID HEMORRHAGE (SAH)**
  - Caused either by trauma or spontaneous rupture of a congenital aneurysm at the circle of Willis branch points. Some SAH may be chronic, with vascularized membranes that lead to recurrent acute hemorrhage episodes.
- **DIAGNOSIS**
  - Patients complain of a headache and, if aneurysmal in origin, this may be accompanied by a stiff neck. Loss of consciousness, focal neurologic signs, or altered mentation can all be present. CT imaging is usually confirmatory. Occasionally, a nontraumatic SAH requires lumbar puncture demonstrating xanthochromia to secure the diagnosis.
- **TREATMENT**
  - Spontaneous aneurysmal SAH is treated with coiling or clipping the aneurysm, followed by hypertension hypervolemia hemodilution therapy to prevent vasospasm in the early period. If it occurs, vasospasm can be managed with calcium channel blockers.
  - Posttraumatic SAHs are managed conservatively.
- **SUBDURAL HEMATOMA (SDH)** results from torn bridging veins. The hematoma does not cross midline and has a higher mortality than epidural hematoma (EDH) because it is usually associated with underlying parenchymal injury.
- **DIAGNOSIS**
  - Symptoms include headache with or without focal neurologic signs and alteration of mentation. Those with a midline shift may have a depression in their GCS. CT imaging is confirmatory.
- **TREATMENT**
  - Evacuation by craniectomy is indicated for >1 cm midline shift or any neurologic deterioration.
  - Chronic SDH may be evacuated by a burr hole with membranes stripped. Placement of a subdural to peritoneal shunt is occasionally required.
  - Untreated, a secondary subdural empyema may develop and result in sepsis, infectious cerebritis, or even traversing cortical artery thrombosis leading to a cerebrovascular accident (CVA).
- **EPIDURAL HEMATOMA (EDH)**
  - This is the result of a skull fracture that lacerates the middle meningeal artery or venous sinuses. Blood then accumulates under high pressure outside the dura. This dissects the dura off the undersurface of the skull and the hematoma takes on a lentiform shape, compressing neighboring brain parenchyma and resulting in secondary neuronal ischemia. Almost all cases are post-traumatic.
- **DIAGNOSIS**
  - Since brain parenchyma is not primarily injured by the impact, patients who have loss of consciousness on impact quickly regain normal mentation. They may subsequently deteriorate if the EDH grows to compress parenchyma and causes secondary brain ischemia. CT imaging is confirmatory.
- **TREATMENT**
  - Generally evacuation by craniectomy is indicated for >0.5 cm thickness EDH.
  - Nonoperative management can be considered on occasion for a small, localized EDH and normal GCS.
  - EDHs have the best prognosis after treatment because the underlying brain did not undergo any primary injury at the time of impact.
- **BRAIN DEATH**
  - Can be declared if:
    - There is **no motor activity** other than spinal reflexes.
    - There is **no brainstem activity** on two exams separated by 6–12 hours; and
    - The patient has a **positive apnea test** where they breathe 100% FiO_2, a baseline arterial blood gas (ABG) is obtained, then they are observed. The test is positive if there is increased pCO_2 and no respiratory effort at 10 minutes. It is indeterminate if cardiovascular instability or hypoxia ensues during the observation period.
    - **Confirmatory testing** includes cerebral angiogram, nuclear medicine brain perfusion, transcranial Doppler (TCD), or electroencephalogram (EEG) if the two motor exams and apnea test are not diagnostic.
    - Hypothermia, hypotension, intoxicant or drug overdose, and metabolic derangements preclude the diagnosis of brain death.

SPINAL CORD

See Figure 26.3.

- **Proprioception** is carried in dorsal columns that synapse at the medulla, cross the brainstem, and synapse at the thalamus before going to the post-central gyrus of the cerebral cortex.
- **Sensory fibers for pain and temperature** enter the cord and synapse at the dorsal horn, cross the cord, and ascend as the lateral spinothalamic tract (LST).

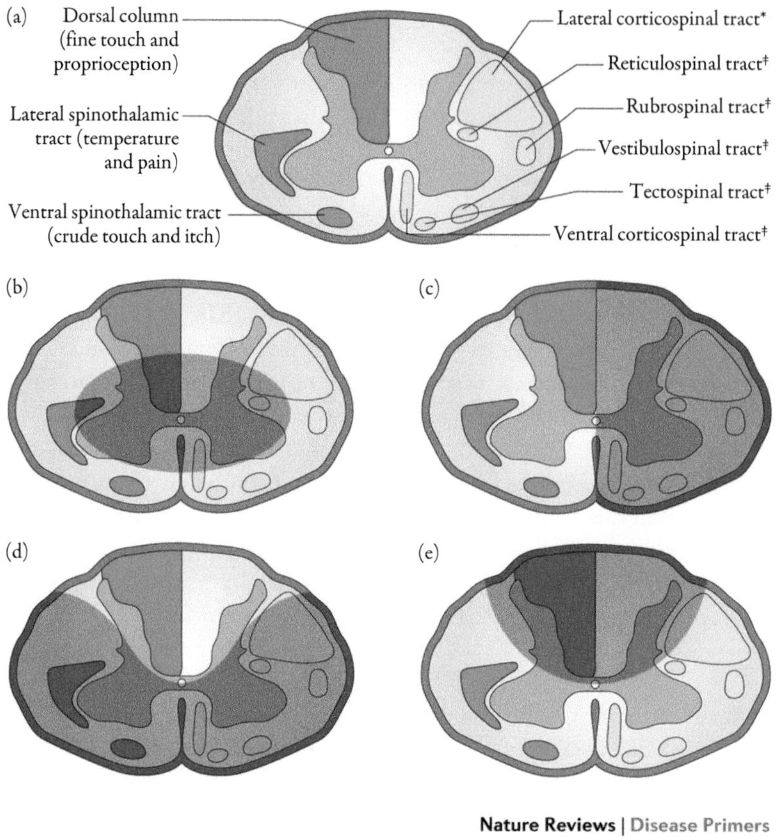

Figure 26.3 Partial cord syndromes central (b). hemicord (c). posterior (d). anterior (e).

Fibers pass the medulla and synapse at the thalamus before going to the post-central gyrus of the cerebral cortex.

- **Upper motor neuron fibers** travel in the lateral corticospinal tract (CST) and carry impulses to the ventral horn to initiate and regulate peripheral nerve motor function.
- Spinal shock is transient areflexia and flaccidity after trauma to the spinal cord. If it has no permanent injury, the motor deficits will resolve.
- **Neurogenic shock** is from a high thoracic or cervical cord injury with loss of sympathetic fibers. Treatment is with intravenous fluid resuscitation, inotrope, and vasopressor support for a mean arterial pressure (MAP) goal of >85 to improve cord perfusion.
- **Anterior cord syndrome** is loss of motor and pain sensation below the level of injury, and results from injury or ischemia to the CST and LST bilaterally. Proprioception and general tactile sensation can be preserved.
- **Posterior cord syndrome** is loss of proprioception and tactile sensation from injury or ischemia to the dorsal columns bilaterally. Motor function and pain and temperature sensation can be preserved. Posterior cord syndrome is rare.
- **Brown-Sequard hemicord** syndrome presents as loss of contralateral pain and temperature and ipsilateral motor function and proprioception. It is due to penetrating injury to one side of the spinal cord.
- **Central cord syndrome** results in upper more than lower extremity weakness, and is due to cord hyperflexion.
- **ASIA** classification for cord injuries:
  - A: Plegia and lack of sensation below the level of injury.
  - B: Plegia but preserved sensation below the level of injury.
  - C: Paresis, <50% of muscle groups at 3/5 strength below the injury.
  - D: Paresis, >50% of muscle groups at 3/5 strength below the injury.
  - E: Baseline, pre-injury status.
- **CT cervical spine** is the initial imaging modality to evaluate the cervical cord, and is sensitive enough that a negative exam is adequate to clear a rigid collar in an obtunded patient.
- **Atlanto-occipital dissociation** requires plate fixation from occiput to cervical spine.
- **Jefferson fracture** of the C1 ring is due to axial load. Treatment is with immobilization. If there is more than

7 mm of combined displacement of lateral masses, then surgical stabilization may be necessary.

- **Odontoid fracture** include type 1 through the tip (stable), type 2 at the transverse alar ligament (unstable, needs surgical fixation), and type 3 into vertebral body (often stable).
- **Hangman fractures** involve both pedicles of C2 and are treated with external fixation.
- **Unilateral or bilateral jumped facets** are from rotation or distraction injury forces respectively. They are treated in traction.
- **Vertebral body fractures** can involve the anterior column (compression), anterior and middle columns (burst), or middle and posterior columns ("chance" fracture). Compression fractures are managed conservatively, and burst or chance fractures require plate and screw fixation with or without a bone graft.
- Spinal cord tumors are classified by their location within the canal.
  - **Extradural tumors** are outside of the meningeal covering and are more likely malignant and metastatic from the prostate or breast. Treatment is generally chemotherapy targeted to the primary cancer, and any palliative vertebral column treatment to mitigate cord compression.
  - Intradural tumors are more likely benign. **Intradural extramedullary tumors** are within the meninges but outside of the spinal cord matter, and include meningiomas, neurofibromas, and schwannomas. The latter may extend from the intradural to extradural space in a dumbbell fashion, and can be associated with neurofibromatosis 2 syndrome. Diagnosis is by imaging and biopsy if needed. Surgery is generally for those lesions that are symptomatic, growing, or appear malignant. **Intradural intramedullary tumors** include ependymoma and astrocytoma.

PERIPHERAL NERVE

See Figure 26.4.

- **Neuropraxia** is demyelination from blunt injury or pressure. Re-myelination occurs spontaneously, and full recovery is expected by 6 weeks.
- **Axonotmesis** is when the axon and its myelin covering are damaged. However, the **endoneurium and perineurium are still intact.** So after distal **Wallerian degeneration** of myelin and axon, the proximal bud will grow at 1 mm/day within the confines of the endo/perineurial covering, until it reaches a distal target neuromuscular junction (NMJ) or sensory receptor.
- **Neurotmesis** is from transection of the axon, myelin, endoneurium, perineurium, and often the epineurium. This leads to a neuroma as the proximal bud attempts to sprout without guidance from the endoperineurium. A neuroma-in-continuity may develop if there is partial neurotmesis. Nerve conduction studies can be utilized to see if the proximal nerve can conduct past the neuroma to the distal nerve. If so, there is no need for resection and repair. Neurolysis can release a constricting scar from a neuroma-in-continuity.
- **Early repair within 3 days** is recommended for penetrating injuries where a nerve is sharply transected, or for an expanding hematoma with compression of an adjacent nerve, or for any worsening of peripheral nerve function after an injury.
- **Subacute repair within 3–4 weeks** is recommended for blunt transection, provided there is no progressive deterioration in nerve function requiring more urgent intervention. Delaying repair helps define the extent of injury after Wallerian degeneration is complete. This can guide debridement and reconstruction.

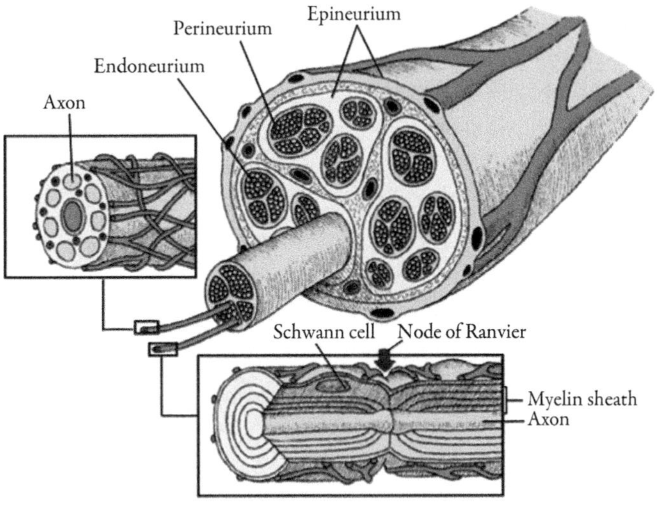

Figure 26.4 Peripheral nerve architecture.

- **Electromyography (EMG) and nerve conduction studies (NCS)** are done at 3-4 weeks post-injury to determine the extent of Wallerian degeneration. If there is no conduction, then one can resect back to healthy, conducting fascicles with flush transection of the nerve at that point. The same can be done distally at the site of healthy, conducting fibers.
- Proximal and distal mobilization allows anastomosis of the epineurium of the proximal and distal nerve bundle in the appropriate orientation using epineural vessels, appearance of fascicles, and stimulation of different fascicles to align corresponding motor and sensory units. Histochemical stains can assist with appropriate alignment.
- The goal is for the proximal bud to have epineural and perineural guidance for sprouting at 1 mm/day toward the distal target.
- Autogenous nerve grafts are most often taken from the sural nerve. The antebrachial, radial nerve sensory branch, superficial peroneal, distal interosseous, greater auricular nerve, or cervical plexus are other options.
- Autogenous or synthetic nerve conduits may be used to house proximal and distal nerve endings (each placed within the conduit) and provide a framework for regrowth in small diameter, short gaps of distal sensory nerves.
- **Neurotization** is nerve transfer from an adjacent normal nerve to an area distally along an injured nerve. It is best if only a few fascicles from the distal portion of a donor nerve are utilized, as they are more likely to be purely motor or sensory fibers. The majority of the donor fascicles will remain in place servicing their target tissue.

- **Delayed repair** at 3–6 months is for lesions that were believed to be neuropraxia or axonotmesis, but failed to recover over the preceding months.
- **Late repair** at over 1 year post-injury is for salvage of a partially functional nerve, with or without a prior repair attempt. If an injury is distant from the NMJ, then the distal NMJ and muscle will atrophy by 18–24 months. This may preclude functional recovery even if a sprout reaches the target after that point. Such patients may need muscle or tendon transfer. Tinel's percussion can note the leading edge of recovering fibers.
- Best results are seen in young patients who undergo nerve repair within 6 months of a distal, clean, sharp, transection to a pure motor or sensory nerve.
- **Causalgia** is when sympathetic neural sprouts regrow into a partially injured, mixed motor and sensory nerve.

ORTHOPEDIC PRINCIPLES

- Within **cortical bone,** osteoblasts synthesize non-mineralized bone, and osteoclasts resorb bone. The Haversian canal system houses the vascular supply and brings nutrients to the cortex. **Trabecular bone** lies within the cortex and is porous and filled with red or yellow marrow. *See Figure 26.5.*
- Cartilage receives nutrients via diffusion from synovial fluid.
- A **torus** fracture is bending of one side of cortex, usually in a pediatric patient.
- A **greenstick** fracture is through one side of the cortex, while the other side remains bent but not broken.
- A **transverse or oblique** fracture is when the cortex is broken on both sides.
- **Salter-Harris** fractures are through growth plates (physis). *See Figure 26.6.*
  - 1: Fracture line is directly through growth plate.
  - 2: Proximal metaphysical fracture extending into the growth plate.
  - 3: Fracture extends from the plate into the distal epiphysis. It is treated with closed reduction and pins.
  - 4: Is a combination of 2 and 3.
  - 5: Is a crush injury of the physis.
- **Avascular necrosis (AVN)** results from fractures that disrupt the blood supply to the scaphoid, humeral head, femoral head, or talus.
- **Nonunion** can occur when the two ends of bone are in motion, poorly vascularized, poorly aligned, or the patient is malnourished or immune compromised. Smoking is a risk factor for nonunion.
- **Osteomyelitis** may be hematogenous spread of staphylococcus into the metaphysis of long bones, or secondary to septic arthritis, traumatic inoculation, or soft tissue destruction and infection eroding down to a bony prominence. In the latter case, multiple organisms are involved. Diagnosis is by MRI and bone biopsy. Treatment is with IV antibiotics with or without debridement or resection. Long-term oral antibiotics can be adequate if there is an appropriate response to initial therapy.
- **Complex regional pain syndromes** may be sequelae of bone and soft tissue injury, with or without a concomitant nerve injury.

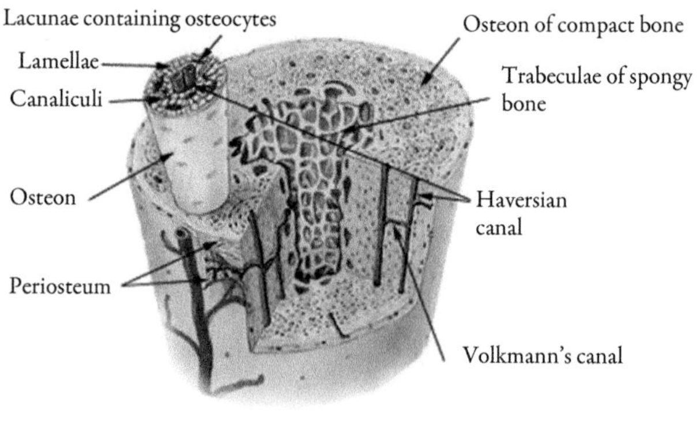

Figure 26.5 Cortical and trabecular bone architecture.

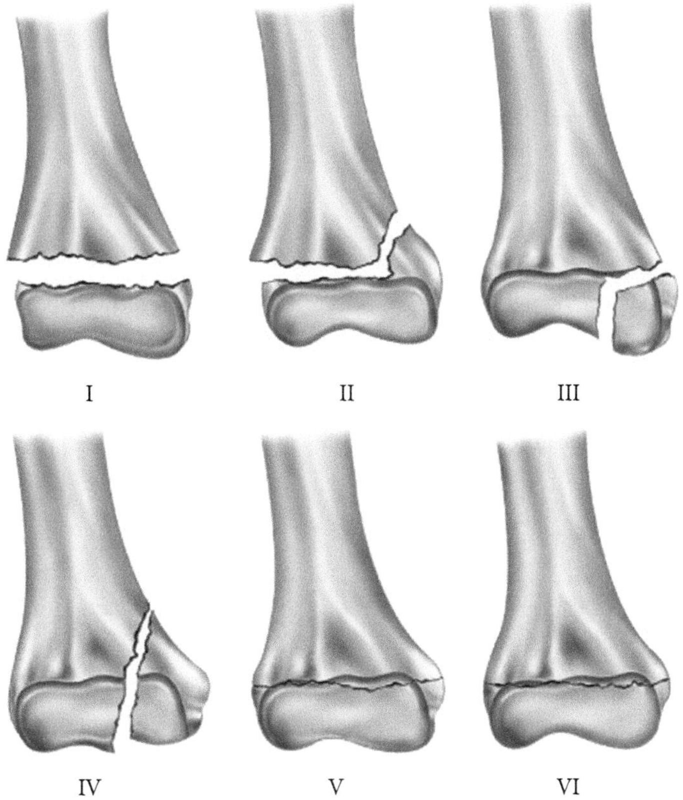

Figure 26.6 Types of Salter-Harris fractures through the physis.

CERVICAL SPINE AND UPPER EXTREMITY

- **Cervical stenosis or disc herniation** with secondary myelopathy or motor radiculopathy is treated by foraminotomy or anterior cervical discectomy and fusion. See the spinal cord section above for more on cervical spine injuries.
  - Cervical root 1 is motor to the strap muscles via the ansa cervicalis. C2–C4 innervates the neck and scalp. C3–C5 also innervates the diaphragm.
  - The cervical roots corresponding to upper extremity weakness are as follows:
    - C5 root (from a C4–C5 disc protrusion): weak biceps.
    - C6: weak biceps and brachioradialis.
    - C7: weak triceps and wrist extension.
    - C8 (C7–T1 disc protrusion): wrist flexion and intrinsic hand muscle weakness.
  - The axillary nerve comes from C5–C6 and innervates the deltoid for sensation and shoulder abduction.
  - The musculocutaneous nerve is from C5–C7 and provides sensation to the lateral part of the forearm and motor control over the brachialis, coracobrachialis, and biceps.
- **Clavicle fracture** treatment is mostly with a sling. If the distal clavicle is fractured and the acromioclavicular or coracoclavicular ligament is involved, then surgery may be needed.
- **Impingement** can be secondary to subacromial bursitis, biceps tendonitis, or supraspinatus tendonitis. Treatment is most often conservative. A subacromial bursectomy or acromioplasty is occasionally needed.
- **Rotator cuff tear** involves the supraspinatus, infraspinatus, teres minor, or subscapularis tendons, and is treated conservatively.
- **Acromioclavicular separation** is initially managed with a sling. Surgery is for type 3 injuries where the ligament and the neighboring coracoclavicular ligament are both completely transected.
- **Scapular fracture** is managed conservatively, with surgery reserved for involvement of the glenoid fossa.
- **Humeral head fractures** may be up to 4 parts involving the head, greater and lesser trochanters, and shaft.
- **Glenohumeral dislocation (shoulder dislocation)** are often anterior, with axillary nerve stretch. Posterior dislocations are less common and can result in axillary artery injury. Associated shoulder injury can include a labrum tear (Bankart lesion) or humeral head fracture (Hill-Sachs deformity), or rotator cuff tear. Dislocations should be reduced promptly, and distal neuromuscular function confirmed. Associated shoulder injuries are

managed conservatively. Recurrent dislocations may warrant surgery.

- **Midshaft humerus fracture** can result in a radial nerve neuropraxia that improves over months. Management is with a sling or open reduction internal fixation (ORIF) for significant displacement.
- **Supracondylar humerus fractures** are treated with ORIF in all adults and in any pediatric patient with displaced fragments. Non-displaced or minimally displaced pediatric fractures can be treated with closed reduction. An associated **brachial artery impingement** usually improves after reduction. If not, the brachial artery may need repair. Median nerve injury, compartment syndrome, and an ischemic (Volkmann's) contracture can result if the brachial artery is not addressed in a timely fashion.
- **Olecranon fractures** are treated with a sling, or ORIF for displaced fragments from the pull of the triceps muscle tendon.
- **Monteggia fracture** is a proximal ulnar fracture and radial head dislocation treated by ORIF.
- **Nightstick fractures** are of the ulnar shaft and managed by ORIF.
- **Radial head fracture** is treated by sling and early range of motion. Surgery is for patients with limited pronation and supination or an associated ulnar coronoid process fracture.
- **Colles distal radius fracture** can be managed with closed reduction.
- **Galeazzi distal radius fracture** has an associated radio-ulnar joint dislocation, and ulnar malalignment and is treated by ORIF.
- **Forearm fasciotomy** compartment release runs from the ulnar side of the volar upper arm, obliquely across the antecubital fossa, down the radial side of the volar forearm, and then curving back across the middle of the wrist into the palm. This helps release the median nerve from the transverse carpal ligament. Dorsal release is along the radial side of the upper arm and forearm.

HAND

See Figure 26.7.

- A deep palmar arterial arch originates primarily from the radial artery and perfuses digits 1 and 2.
- A superficial palmar arterial arch originates primarily from the ulnar artery and perfuses digits 3 through 5.
- The **neurovascular digital bundle** runs laterally, just on the volar side of each finger crease.
- The **transverse carpal ligament** connects the scaphoid to the pisiform, where it forms the base of Guyon's canal. The volar carpal ligament is superficial to this and forms the roof of Guyon's canal.
- **Median nerve** (C6–T1 roots) gives rise to the anterior interosseous nerve, innervating the forearm muscles, and then proceeds to the 1st and 2nd lumbricals and the thenar compartment, with the exception of the adductor pollicis. It also provides sensation to the volar and distal dorsal digits 1, 2, 3, and half of 4. Carpal tunnel syndrome results in median nerve compression from the carpal ligaments. Treatment is with a volar splint, glucocorticoid injection, and endoscopic or open carpal ligament release.
- **Ulnar nerve** (C8–T1 roots) innervates the hypothenar compartment and dorsal and volar interossei. as well as lumbricals and adductor pollicis. It also provides sensation to the volar and dorsal sides of digits 5 and half of 4. Cubital tunnel syndrome is from ulnar nerve compression at the medial condyle. Patients will have a weak grip with positive Tinel's and Froment's signs. Treatment is with a splint or cubital tunnel release.
- **Radial nerve** (C5–C8 roots) has a motor branch to forearm extensor muscles and a sensory branch to the dorsal hand thenar web space.
- **Hand Infections**
  - ***Staphylococcus aureus*** **(SA)** is the most common organism. *Eikenella corrodens* is in human bites, *Pasteurella* in dog and cat bites, and *Bartonella* in cat scratch disease. Marine infections can transmit *Mycobacterium marinum* or *Vibrio* species.
  - **Collar button** abscesses occur at interdigital web spaces.
  - **Paronychia** is from SA or oral flora. Treatment is with hot soaks, antibiotics, incision and drainage, and nail removal if needed. Fungal infections can be chronic.
  - **Felon** is an abscess residing between the septae of the volar distal finger pulp, sometimes after minor trauma. Treatment is with a lateral incision for drainage.
    - Herpetic whitlow causes a vesicular eruption over the distal phalanx. It can also initiate a felon. Diagnosis can be made by Tzanck smear of a vesicle scraping.
  - **Flexor tenosynovitis** is often from penetrating trauma. Diagnosis is clinical and based on **Kanavel signs**, which include fusiform digit swelling, tenderness over the volar digital sheath, a baseline flexed position of the distal (DIP) and proximal interphalangeal joints (PIP), and pain with passive extension of the finger. Treatment is with **sheath irrigation** via an open incision over the volar aspect of the digit, or small transverse proximal and distal incisions with an irrigating catheter.

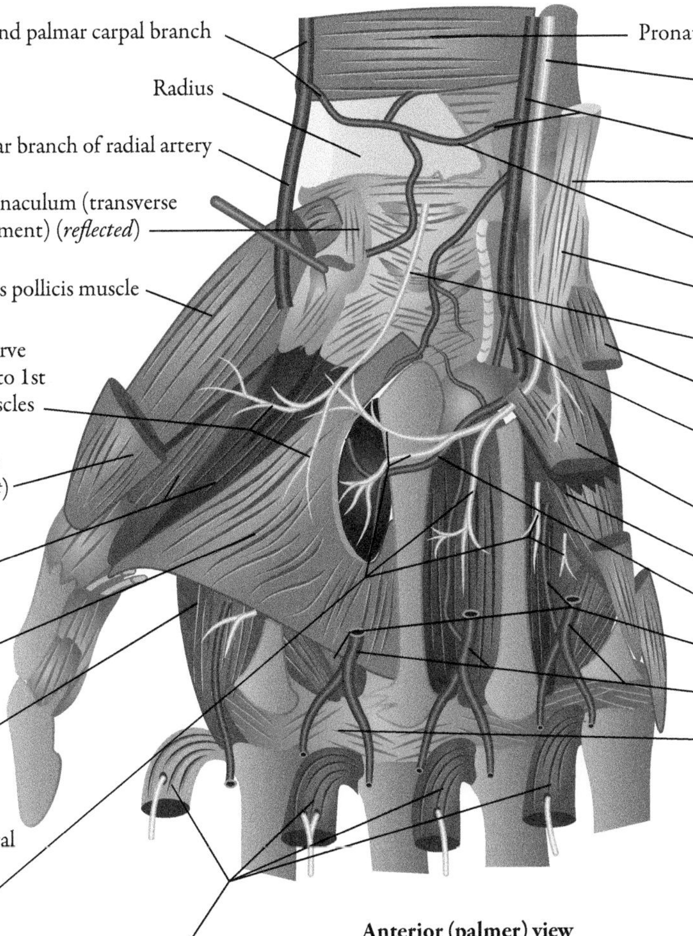

Figure 26.7 Neurovasculature and intrinsic muscles of the hand.

  - The thumb flexor sheath leads into the **radial bursa** proximally, which extends along the flexor pollicis longus (FPL) into the forearm via the space of Parona.
  - The fifth digit flexor sheath leads into the **ulnar bursa and palmar bursa** proximally, and the forearm via the space of Parona. For these bursae, drainage may require multiple separate incisions over each site.
- **Dorsal** hand infections can be drained at bedside if subcutaneous, or in the operating room if subaponeurotic.
- **Thenar and midpalmar space** infections are deep to the flexor tendons and bursa in the palm, but superficial to the interossei.
- **Hypothenar space** infections are lateral to the 5th flexor and superficial to the 5th metacarpal.

- Noninfectious tenosynovitis (trigger finger) is when the tendon catches within a flexor sheath near the metacarpal phalangeal joint. Treatment is with a splint, sheath glucocorticoid injections, or surgical release.

- **Dupuytren's contracture** is secondary to diabetes and alcohol. Collagenous nodules and cords on the palmar fascia result in a flexion claw hand contracture with limited passive range of motion. Treatment is with injection, fasciotomy or fasciectomy.
- **Scaphoid fracture** has a risk of AVN. Treatment is by immobilization with thumb spica or volar splint. Occasionally, surgical realignment and fixation is needed.
- **Ganglion cyst** is a dorsal wrist joint capsule herniation. Spontaneous resolution by 2 years is commonplace. Otherwise, resection of the cyst and pedicle may be needed.
- **Traumatic amputation** should be treated by re-implantation of any pediatric digit, an adult thumb, or any adult digit amputated proximal to the PIP.
- Soft tissue grafts can be used to cover loss of digital tissue when reimplantation is not indicated. Donor sites include the adjacent finger, base of thumb, split thickness grafts from any standard site, full thickness

grafts from the hypothenar area, a radial axial flap, or any free graft. Another option is to integrate digits into the abdominal or groin envelope until the soft tissue covering has matured.

- **High-pressure injection injuries** from a paint or grease gun may require decompression. Thenar, midpalmar, and hypothenar spaces can be decompressed by incisions directly over the area. Interossei compartments can be released of high pressure from dorsal incisions between metacarpals 2–3 and 3–4.
- **Tendon ruptures** of a superficial or deep flexor will result in impaired flexion of digits and should be repaired in the operating room. Rupture of an extensor tendon mechanism can be repaired in the operating room or emergency room.

LUMBAR SPINE

- The obturator nerve mediates hip adduction via the adductor muscle group. The superior gluteal nerve is for hip abduction and the inferior gluteal for hip extension via the gluteal muscles. The femoral nerve mediates knee extension via the quadriceps.
- **Spinal stenosis** is due to loss of disc height, bulging of an annular ligament, or osteophytes. Treatment is with epidural injection, decompression, and stabilization.
- **Scoliosis** can be congenital, degenerative, metabolic, neurogenic, or myogenic in origin. A brace is adequate if 20–40 degrees of scoliosis is present. Spinal fusion with pedicle screws, wires, hooks, and connecting rods may be needed if more than 40 degrees angulation exists.
- **Spondylolisthesis** is subluxation of a vertebral body, generally in an adolescent gymnast complaining of back pain. Spondylolisthesis may be secondary to traumatic injury as well.
- **Herniated nucleus pulposus (herniated or bulging disc)** is more common in the lumbar than the thoracic spine. Patients may have pain localized to their back, or signs of nerve root impingement.
  - L3 nerve root corresponds to the L2–L3 disc space. and compression produces weak hip flexion. The L4 root is responsible for knee extension, L5 for toe and ankle dorsiflexion. S1 root mediates plantar flexion at the ankle.
  - Diagnosis is by MRI, and treatment is with NSAIDs and heat. Surgery with a posterior midline discectomy or laminectomy is reserved for patients with motor deficits or refractory pain.
- See the Chapter 23, Trauma, or the spinal cord section above for more on spine fractures.

PELVIS, HIP, AND FEMUR

See Figure 26.8.

- **Pelvis fractures** can be due to anterior–posterior forces, lateral force, or vertical shear injuries. Trauma management principles should be the first priority. See Chapter 23, Trauma, for details.
  - Mechanically unstable pelvic fractures are treated with plate and screws through the ileum, sacrum, and symphysis once the patient's hemodynamics are controlled.
  - Isolated, minimally displaced anterior ring fractures can be treated with weight bearing as tolerated (WBAT).
  - Elderly patients with pelvic fractures have increased blood transfusion requirements and increased mortality, especially if repair is delayed.

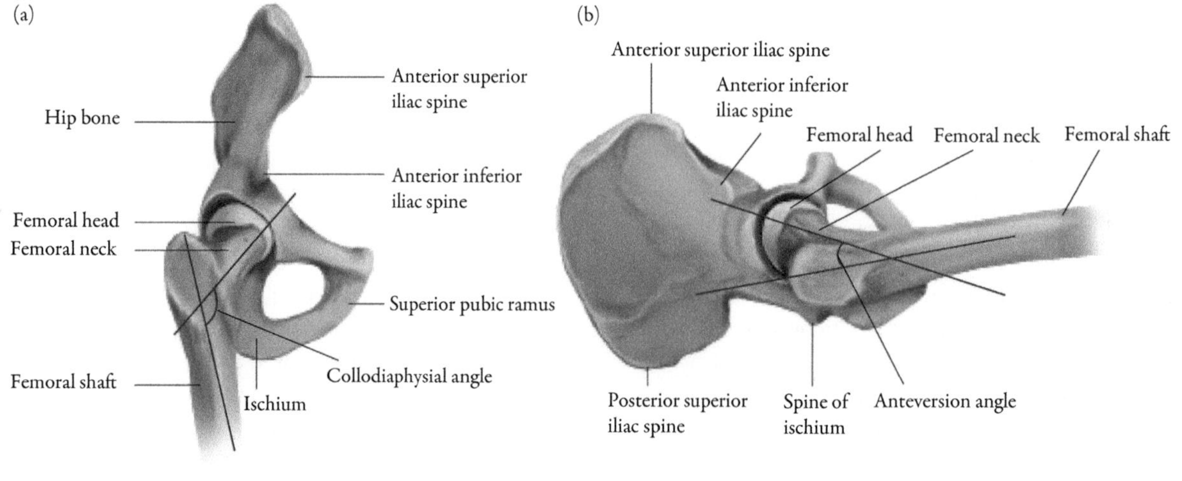

Figure 26.8 Femoral-acetabular articulation.

- **Impingement** of the femoral head against the anterior superior labrum results in impaired flexion, internal rotation, and adduction. There may be associated labrum tears or loss of cartilage. Treatment is with acetabuloplasty.
- **Posterior femoral head dislocation** presents with adduction, internal rotation, and sciatic nerve compression.
- **Anterior femoral head dislocation** presents in abduction, external rotation, and occasionally a femoral artery injury.
- **Femoral neck fractures** have a risk of AVN of the femoral head. Prevention is by early ORIF with nails through the greater trochanter into the neck and head. Other treatment options include a hemiarthroplasty replacing the head and neck or a total arthroplasty replacing or relining the acetabulum as well.
- **Intertrochanteric fracture** treatment is with a sliding hip screw and plate along the lateral cortex, or an intramedullary nail with lateral screw fixation.
- **Subtrochanteric femur fracture** fragments are often displaced due to the pull of opposing muscle groups. Treatment is with an intramedullary nail.
- **Arthritis** treatment is NSAIDs or injection with corticosteroid, lidocaine, or hyaluronic acid. Surgery involves osteotomy for realignment, arthrodesis with auto or allograft, or arthroplasty with biologic or cement fixation. Hemi- and total arthroplasty are other options. Complications of arthroplasty include **periprosthetic loosening**, nerve injury, **periprosthetic fractures**, limb length discrepancy, iliopsoas impingement, tendonitis, or dislocation of prostheses.
- **Pediatric Femoral Head and Shaft Pathology**
  - **Hip fracture (acetabulum, femoral head, neck, or proximal shaft)** is classically treated with a hip spica if under 6 years, or a flexible nail if 6–14 years old. Closed reduction is for diaphyseal fractures.
  - **Legg-Calve-Perthes** is AVN of the femoral head epiphysis in a toddler. Treatment is with range of motion exercises.
  - **Slipped capital femoral epiphysis** occurs in obese adolescents when the femoral head epiphysis moves in relation to the physis and metaphysis. Treatment is closed reduction.
  - **Congenital hip dislocation** presents as a neonate with a positive Ortolani and Barlow maneuver. Diagnosis is by ultrasound or x-ray. Treatment is with immobilization in abduction.

FEMOROTIBIAL

See Figure 26.9.

- **Anterior cruciate ligament (ACL)** injuries are from lateral trauma. Diagnosis is with the anterior drawer test and MRI. Treatment is with physical therapy or surgery utilizing an allograft or an autograft from the patellar or quadriceps tendon,
- **Posterior cruciate ligament injuries (PCL)** are diagnosed by a posterior drawer test and MRI, and managed conservatively.
- **Medial meniscus injuries** may be treated by arthroscopic debridement or repair
- **Collateral ligament injuries** are managed conservatively if the tear is small. Surgery is reserved for large tears or other associated injuries.
- **Posterior knee dislocation,** without hard signs of vascular injury, requires an ankle-brachial blood pressure index (ABI) and a CT angiogram if the ABI is < 0.9. If there are hard signs of popliteal artery injury, then emergent operative revascularization is needed. See Chapter 23, Trauma, for details.
- **Patellar fractures** are managed conservatively unless displaced fragments or comminution mandates ORIF.
- **Patellar tendon (PT) or quadriceps tendon (QT) injuries** should be repaired with immobilization in extension and gradual introduction of range of motion exercises.
- Tibial fractures should first be evaluated for distal perfusion and neural function, then splinted, and have repeat exams to exclude compartment syndrome. **Tibial plateau fracture** treatment is conservative for minimally displaced simple fractures, all others need ORIF. **Tibial shaft fractures** are treated with an intramedullary (IM) nail. Any open tibial plateau or shaft fracture can receive external fixation.
- **Plantaris muscle rupture** produces edema and ecchymoses to the popliteal fossa. It is managed conservatively.

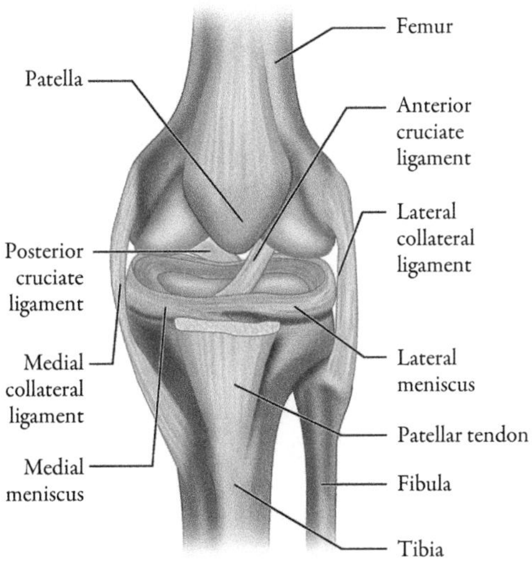

Figure 26.9 Femoral-tibial articulation.

- **Osgood-Schlatter** is a tibial tubercle apophysitis from quadriceps traction in adolescents. Treatment is with activity limitation and immobilization if severe.
- **Arthritis** treatment is NSAIDs or injection of corticosteroids, lidocaine, or hyaluronic acid. Surgery includes uni-compartmental or total arthroplasty resurfacing of the distal femur, tibia, and patella, with a metal or polymer liner plus cement fixation. Complications include periprosthetic loosening, nerve injury, periprosthetic fracture, limb length discrepancy, tendonitis, or dislocation.
- **Common peroneal nerve injuries** are from fibular head fractures, lithotomy surgical positioning, or prolonged compression. The result is impaired ankle dorsiflexion (like L5 nerve root impingement) but also impaired ankle eversion.
- Soft tissue loss over the distal third of lower leg may need free flap coverage, since there is usually no sufficient local random or axial pedicled flap available.
- See Chapter 30, Vascular, for diagnosis and treatment of compartment syndrome.

TIBIOTALAR AND FOOT

See Figure 26.10.

- **Malleolar fractures** are managed with immobilization. Bi- or trimalleolar fractures or syndesmotic tibiofibular injuries need ORIF.
- **Plafond (pilon) tibial fracture** is managed with ORIF or external fixation if the soft tissue envelope is very edematous.
- **Talar and calcaneal fractures** are managed with closed reduction, or ORIF if severely displaced. If the patient has fallen from a height onto both heels, check for an associated lumbar spine fracture.
- **Ankle sprains** involve tears in the talofibular or calcaneofibular ligaments, usually due to foot inversion.
- **Metatarsal fractures** should be immobilized. Surgery is reserved for Jones fractures and those injuries refractory to attempts at immobilization and non–weight bearing.
- **Equinovarus** (clubfoot) in infants requires application of a cast.

DIABETIC FOOT AND GANGRENE

- Microvasculopathy and neutrophil dysfunction are contributory to non-healing diabetic foot infections. 80% of patients have a neuropathy. Acute, superficial infections often harbor Gram-positive cocci (GPC), while chronic, deep infections harbor Gram-negative rods (GNRs) and anaerobes.
- **DIAGNOSIS**
  - Patients often have an **insensate chronic wound at a bony prominence** of their foot. There may be only a small area of dry tissue loss, ulceration, or necrosis; a chronic draining sinus; or a more extensive wet wound with or without necrosis.
  - Ulcers over 2 cm diameter, over 3 mm depth; those that probe freely to bone; or those associated with a sausage digit or erythrocyte sedimentation rate (ESR) >70 mm/hour probably have osteomyelitis.
  - **Perfusion studies** should be obtained, including ABI and imaging with duplex ultrasound (US) or CT angiogram (CTA). An ABI >1.3 represents a non-compressible vessel, and a follow-up toe-brachial blood pressure index (TBI), toe pressure, or pulse volume recording (PVR) waveforms will help determine the need for revascularization.
- **TREATMENT**
  - Management depends on whether the gangrene is wet or dry, and the degree of systemic illness.
    - **Wet gangrene** is partially perfused and has poorly defined borders. This should be **amputated and debrided immediately**. See Chapter 30, Vascular, for principles of amputation.
    - **Dry gangrene** has sharp borders, and **revascularization should be done prior to amputation.** This may be treated with surgical debridement, and intravenous transitioned to oral antibiotics for the long term, depending on the bone margin bacterial load.
    - Nutritional maximization, control of diabetes (HGB A1C goal <7.5), orthotics that prevent future pressure ulcerations, and negative pressure wound therapy (NPWT) can help preserve viable neighboring tissue. Important adjuncts are non-adherent occlusive dressings and pressure offloading
    - Revascularization is indicated for failure to heal or an initial ABI <0.4. Bypass to the anterior or posterior tibial arteries should be done prior to amputation, provided there is no infected tissue necessitating urgent amputation. See Chapter 30, Vascular, for details of tibioperoneal revascularization.

BONE TUMORS

- **Metastases** are the most common. Breast and prostate cancer produce blastic metastases. Lung, rectal, renal, and thyroid cancers produce lytic metastases. Treatment is chemotherapy aimed at the primary tumor type. Internal fixation or radiation can be employed for impending fracture or spinal cord compression.
- **Multiple myeloma** is a primary malignancy of bone marrow. Treatment is by chemotherapy with or without internal fixation to prevent fracture displacement. Solitary plasmacytoma can be managed with radiation.

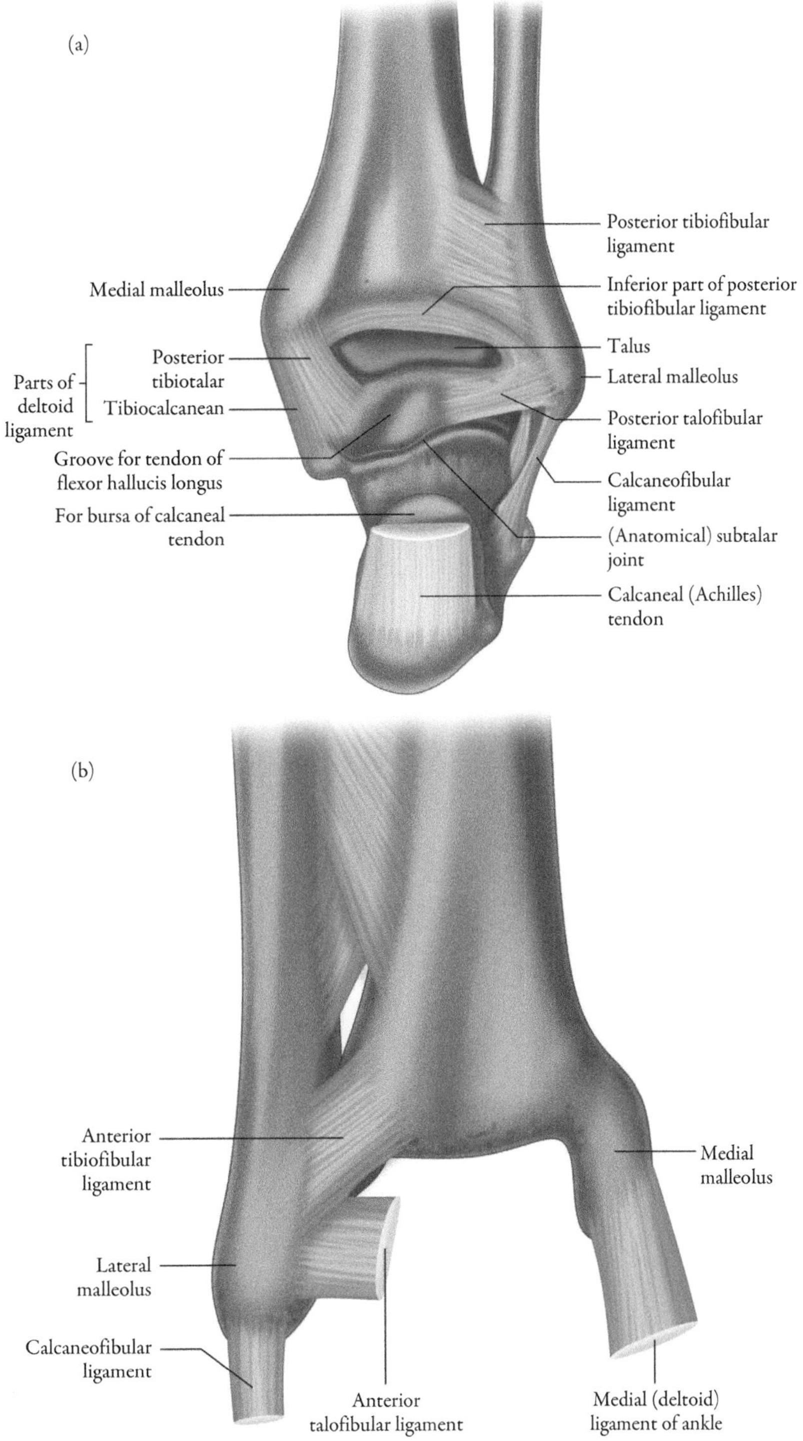

Figure 26.10 Tibiotalar articulation.

- **Osteosarcoma** is the most common malignant sarcoma. Patients are often under 20 years of age, and present with pain or a mass at the knee. A periosteal sunburst reaction can be seen on imaging. Treatment is resection with a focus on limb preservation. Neoadjuvant chemoradiation may be necessary. Metastasectomy is an option in select circumstances.
- **Ewing's sarcoma** is the second most common primary malignant sarcoma. Patients are under 30 years of age and present with a mass at the diaphysis of the femur and a periosteal reaction. Diagnosis is by marrow biopsy showing small blue cells. Chemoradiation may be utilized in addition to resection.

- **Chondrosarcoma** is the third most common primary malignant sarcoma. Patients are over 40 years of age and present with lesions commonly at the pelvis, shoulder, and ribs.
- Angiosarcoma, hemangioendothelioma, hemangiopericytoma, and malignant fibrous histiocytoma are other, less common, sarcomas within bone.
- **Giant cell tumor of bone** presents in females under 40 years of age with an epiphyseal lytic lesion at the knee, distal radius, or proximal humerus. It may occur as a nodule in a volar wrist or digit tendon sheath.
- **Osteochondroma** is the most common benign tumor of bone. Patients are under 20 years old. **Osteoid osteoma** appears as a 1 cm lucency with a sclerotic rim. Patients have nocturnal pain. Treatment is symptomatic with NSAIDs. **Enchondroma** is a lytic lesion at the physeal cartilage. Fibrodysplasia, periosteal chondroma, chondroblastoma, desmoplastic fibroma, ossifying fibroma, and adamantinoma are other benign tumors. Treatment for benign tumors is curettage or resection if necessary.

27.

HEAD AND NECK

ANATOMY

See Figure 27.1.

- The soft tissue envelope for the neck, from deep to superficial, includes skin, platysma, **superficial** fascia over the sternocleidomastoid (SCM), **pretracheal** fascia over trachea, esophagus, and carotid, and **prevertebral** fascia covering bone.
- The submandibular triangle lies between the digastric bellies and the submental triangle between the anterior digastrics.
- Carotid triangle is between the SCM, posterior digastric, and omohyoid, and the muscular triangle is between the omohyoid strap muscles, SCM.
- Supraclavicular triangle lies between the inferior omohyoid, posterior border of SCM, and clavicle. Lymph nodes here can be from breast, lung, or a gastrointestinal malignancy.
- Posterior triangle is between the posterior border of SCM, trapezius, and clavicle.
- The **accessory nerve (XI)** exits through jugular foramen and emerges high on the posterior border of SCM, above Erb's point and the greater auricular nerve. It is most commonly injured during surgery within the posterior triangle, such as a modified radical neck dissection (MRND).
- **Auriculotemporal** provides sensation anterior to the auricle. If injured, it can regrow into adjacent sympathetic nerves, leading to **gustatory sweating (Frey's syndrome)**.
- **Greater auricular nerve (GAN)** is from cervical root 2 and 3 and emerges at the mid-portion of the posterior border of SCM (Erb's point). It provides sensation to the pre- and post-auricular space, lobe of the ear, and inferior and posterior auricle. GAN may be injured during parotid surgery or a facelift.
- **Lesser occipital nerve** also emerges from the bundle at Erb's point.
- **Long thoracic nerve** traverses over the middle scalene muscle.
- The trigeminal nerve is composed of V1, V2, and V3 branches plus the corda tympani, which carries taste from the anterior tongue.
- **Facial nerve (VII)** has 5 branches: temporal, zygomatic, buccal, marginal mandibular, and cervical.
  - If lacerated at a point lateral to the lateral canthus, the nerve will need **primary repair** with 8-0 or 10-0 suture. A GAN or sural cable graft or a gracilis muscle graft may be needed.
  - For a temporal bone fracture with VII palsy and no recovery at 1 week, **temporal bone decompression** can be considered.
- **Glossopharyngeal nerve** is motor to the stylopharyngeus and palatoglossus and assists in swallowing. It also carries taste from the posterior tongue.
- **Hypoglossal nerve** is motor to the tongue musculature (except palatoglossus). It can be injured during submandibular gland resection or dissection near the carotid bifurcation.
- **Vagus nerve** gives rise to the recurrent laryngeal, which pierces the cricothyroid membrane and provides motor innervation to all of the intrinsic muscles of the larynx.
- **Superior laryngeal** external branch goes to the cricothyroid muscle (controls voice endurance), and internal branch through the thyrohyoid membrane is sensory to the larynx.
- **Thyrocervical trunk** is composed of the suprascapular, transverse cervical, ascending pharyngeal, and inferior thyroid arteries.
- **Thoracoacromial trunk** can be used for a pectoralis muscle flap. Perfusion is also augmented by branches from the internal thoracic artery.
- **Cervical nodes** are classified by their location *See Figure 27.2.*
  - Level 1a submental, **Level** 1b submandibular.
  - Level 2a and 2b suprahyoid jugular separated by CN XI.
  - Level 3 over the jugular between hyoid and cricoid.
  - Level 4 over the jugular below cricoid.

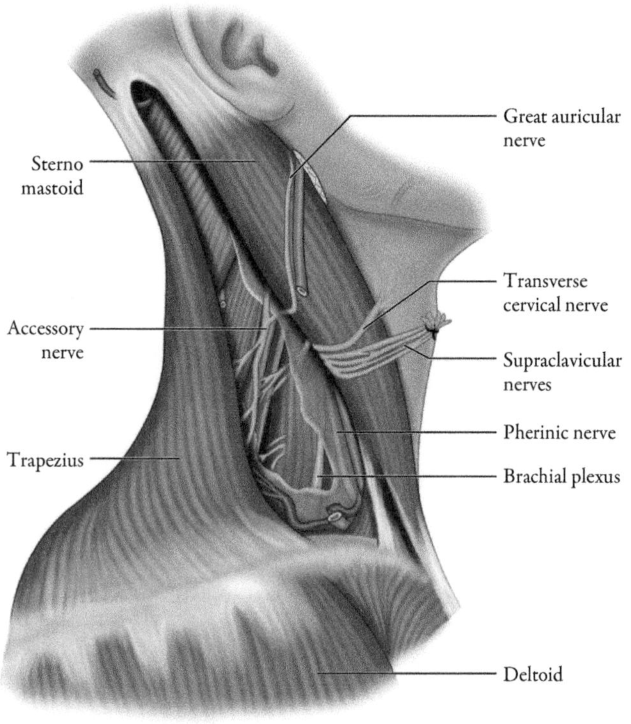

Figure 27.1 Superficial neck anatomy and neural innervation.

- Level 5a and 5b posterior to SCM separated at level of cricoid.
- Level 6 central.
- Level 7 mediastinal.

OCULAR

- **Lacerations** may be closed primarily if <25% of the lid margin is involved. Other options are canthotomy and advancement flap or a full thickness skin graft (FTSG) from the contralateral upper lid.
- For mild **ptosis,** the upper lid tarsal plate, conjunctiva, and levator aponeurosis can be excised. For severe ptosis, a frontal muscle flap can be employed.

EAR

- **Pinna lacerations** simply need skin approximation, there is no need to suture cartilage. Other options include a wedge resection of devitalized tissue with primary closure, posterior auricular flap, or a chondrocutaneous advancement flap.
- **Auricle hematomas** should be drained to prevent a long-standing deformity (cauliflower ear).
- Cerumen in the external auditory canal is the most common cause of hearing loss.
- **Otitis media (OM)** can be acute at less than 3 weeks or chronic with an effusion at over 8 weeks. The tympanic membrane (TM) may be retracted. Treatment is with oral antibiotics. Some patients will need myringotomy, tympanoplasty, or ossicular reconstruction.
- **Mastoiditis** is from extension of suppurative OM and is treated with antibiotics with or without mastoidectomy.
- **Cholesteatoma** is an inclusion cyst from the inner TM lining, treated with excision.
- Herpes simplex virus (HSV) of cranial nerve (CN) 7 can produce **Bell's palsy** without any visible lesions. Varicella zoster virus (VZV) of CN 7 can produce **Ramsay Hunt syndrome** with lesions in the external canal
- **Acoustic neuroma** of CN 8 may grow into the cerebellopontine angle. Treatment is with resection or radiation.
- **SCC of the ear** can have metastases to the parotid in 20% of cases. Treatment is with Mohs surgery with or without radiation, or MRND. More extensive resection can involve subtotal temporal bone removal.

NASAL CAVITY

- The nasal cavity is lined with **respiratory and olfactory epithelium.**
- **Polypoid or nonpolypoid sinusitis** can be treated with antihistamines, humidified inhalation, glucocorticoids, and endoscopic surgery to relieve obstruction at the sinus ostia.
- **Fungal sinusitis** can be allergic, invasive, or secondary to a fungal debris occupying the sinus. Allergic disease is treated with glucocorticoids and irrigation, and invasive disease requires resuscitation and debridement. Fungal debris must be removed from the sinus cavity to relieve the obstruction.
- **Nasal cavity tumor**s are most often primary **squamous cell carcinoma (SCC)** and present with obstruction or epistaxis. Treatment includes endoscopic resection for small lesions, open **wide local excision (WLE)**, palatectomy, maxillotomy, or orbital exenteration. **Neoadjuvant or adjuvant radiation** may be employed to fortify margins. **Adjunctive or definitive chemotherapy** may be used for advanced or unresectable patients respectively. Post-resection coverage of soft tissue defects can require split thickness skin graft (STSG), full thickness skin graft (FTSG), or random, axial pedicled, or free grafts. Composite grafts can include iliac bone, radius, or ribs.
- **Nasal fractures** should be reduced after edema subsides. **Septal hematoma** should be drained to prevent deformity.

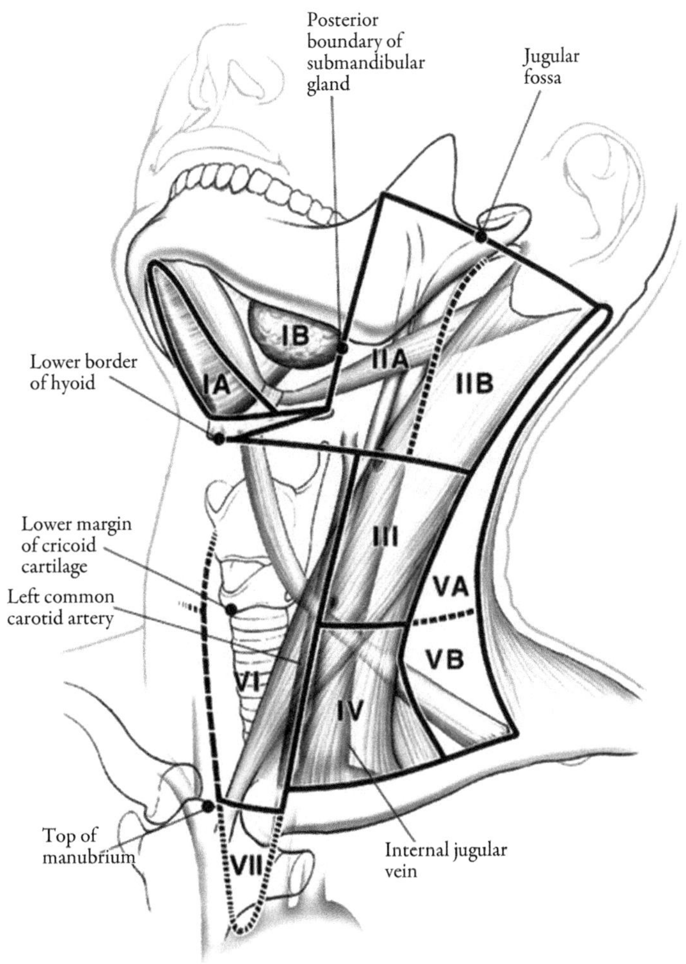

Figure 27.2 Cervical lymph node stations.

- **Cerebrospinal fluid (CSF) rhinorrhea** is managed conservatively for up to 3 weeks. A lumbar catheter can decrease CSF pressure. Repair of refractory leaks can be transethmoidal. Fixation of displaced facial fractures may help decrease the leak.
- **Epistaxis** is most often from the anterior nasal cavity and responds to anterior packing. 10% of epistaxis is from the posterior nasopharynx tributaries of the maxillary artery. Endovascular maxillary or ethmoidal artery embolization may be needed if bleeding does not respond to a posterior pack or endoscopic intervention.
- **Nasopharyngeal tonsils (adenoids)** can obstruct eustachian drainage and result in recurrent otitis media (OM). Adenoid resection may be needed for prevention of OM, and can be complicated by velopharyngeal reflux or post-obstruction pulmonary edema.

ORAL CAVITY

- The oral cavity is composed of the lips, mouth, floor, anterior tongue, hard palate, and anterior tonsillar pillars. The primary palate is anterior to the incisive foramina. The secondary palate is posterior to these foramina, including the remainder of the hard palate and all of soft palate.
- **SQUAMOUS CELL CARCINOMA (SCC)**
  - **Tobacco and alcohol** have historically caused oral cavity and larynx SCC, now both are less common. **HPV**-associated oral squamous cell cancer is now the most common form of head and neck SCC.
  - A typical presentation is a young male with a large cervical nodes demonstrating SCC and a small primary tumor in the palatine or lingual tonsils.
  - **Lower lip SCC** is from **ultraviolet (UV) light exposure.** This can also cause minor salivary gland tumors and melanoma. Upper lip basal cell carcinoma (BCC) is less common.
  - **Tongue** stratified epithelium SCC may invade into the underlying glossal muscles or into the mandible, which may need composite resection glossectomy and partial mandibulectomy.
  - Verrucous lesions from the **buccal mucosa** may be well differentiated SCC and less aggressive. **Floor of**

mouth SCC may invade the tongue and spread along salivary ducts. **Hard palate** SCC can be secondary to irritation from dentures, or arise from minor glands at the hard–soft palate junction.

- **DIAGNOSIS**
  - Generally this is made by **endoscopy and CT of the head and neck.** Further imaging may be needed to exclude metastases. A T2 lesion is 2–4 cm in size.
  - Oral cavity SCC spreads to nodal stations 1, 2, and 3. N stage is determined by the size and number of nodes involved.
- **TREATMENT**
  - Start with **WLE** and add a modified radical neck dissection (**MRND) or selective lymph node dissection (SLND)** if the primary is >4 mm depth, >4 cm size (T3), has high-risk pathologic features, close margins, perineural or bone invasion, or clinically positive nodes are identified in stations 1–4. CN XI is one of the most common nerves injured during MRND in the posterior triangle. Intraoperative stimulation to note trapezius muscle contraction can help protect the nerve.
  - **Postoperative radiation** may be added for the above features or any R1 resection. It can be complicated by osteoradionecrosis of the mandible. **Adjunctive or definitive chemotherapy** can be added for advanced or unresectable cases respectively.
  - **Coverage** options for residual defects include STSG or FTSG, random, axial pedicled, or free grafts, or possibly composite grafts including iliac bone, radius, or ribs.
  - For lip lesions, primary closure is possible after resection of less than one-third of the lip. Otherwise a variety of flap coverage techniques can be employed; transferring from the upper lip to a central or lateral lower lip defect, or performing a central or lateral lower lip advancement flap.

- **OTHER ORAL TUMORS**
  - Includes ameloblastoma (odontogenic epithelial malignancy) and osteogenic sarcoma from maxillary or mandibular bone. A radicular (inflammatory) cyst develops at the root of teeth and is benign.
- **OTHER LIP AND JAW PATHOLOGY**
  - **Maxillary or mandibular fracture** treatment is with maxillomandibular fixation (MMF) for 4–6 weeks with or without open reduction and internal fixation (ORIF).
  - **Temporal-mandibular joint (TMJ) dislocations** require prompt reduction.
  - **Cleft lip** can be repaired after 10 weeks of age provided the patient weighs >10 lbs. and has an Hgb >10.
  - **Cleft palate** should be repaired at 12 months to prevent velopharyngeal reflux and nasal speech.

PHARYNX

See Figure 27.3.

- **NASOPHARYNGEAL TUMORS**
  - SCC is from **Epstein Barr Virus (EBV).** Patients present with epistaxis or middle ear effusion secondary to eustachian tube obstruction. Nodes may be present in station 5. Treatment is **radiation with or without chemotherapy.** EBV titers can be followed for response to therapy.
  - Other tumors include benign papilloma or lymphoma; the latter is more common in children.
  - Angiofibroma is a vascular tumor producing unilateral epistaxis in a young male. CT should exclude sinus involvement. Treatment is by preoperative angioembolization and then resection.
- **OROPHARYNGEAL TUMORS**
  - **SCC is from HPV, tobacco, or alcohol.** Betel nut, Plummer-Vinson syndrome, and immune suppression are other causes. Half of patients have a p53 mutation and one-eighth have a second primary in the head and neck, lung, or esophagus.
  - Patients may have a globus sensation, persistent sore throat, or dysphagia.
  - **Minor salivary gland tumors** can be mucoepidermoid, adenoid cystic, or adenocarcinoma.
  - Nodes in stations 2–4 may be positive.
  - Treatment of SCC or salivary gland tumors in the oropharynx is with **minimally invasive resection when feasible, otherwise WLE** (including glossectomy or mandibulectomy if needed) with or without MRND, or **radiation and/or chemotherapy.** See the oral cavity section above. Lymphoma can also present in the oropharynx of young adults and is managed with chemotherapy.
- **HYPOPHARYNGEAL TUMORS**
  - SCC is similar in etiology to the oropharyngeal site. Nodes in stations 3, 4, or 6 may be positive. Treatment is **WLE including partial or total laryngectomy for advanced lesions as needed. MRND or radiation** may be added in cases of excessive tumor burden, aggressive features, or compromised margins. Patients should be evaluated for synchronous or metachronous esophageal, bronchial, or lung cancer.
  - Reconstruction of residual defects can be by primary closure, split or full thickness skin graft, local random flaps, or pedicled flaps including deltopectoral, trapezius (based on the transverse cervical artery), scapular (on the circumflex scapular artery), pectoralis (from the thoracoacromial trunk and lateral thoracic arteries), or latissimus (fed by the thoracodorsal artery). A free flap can be composite and taken from

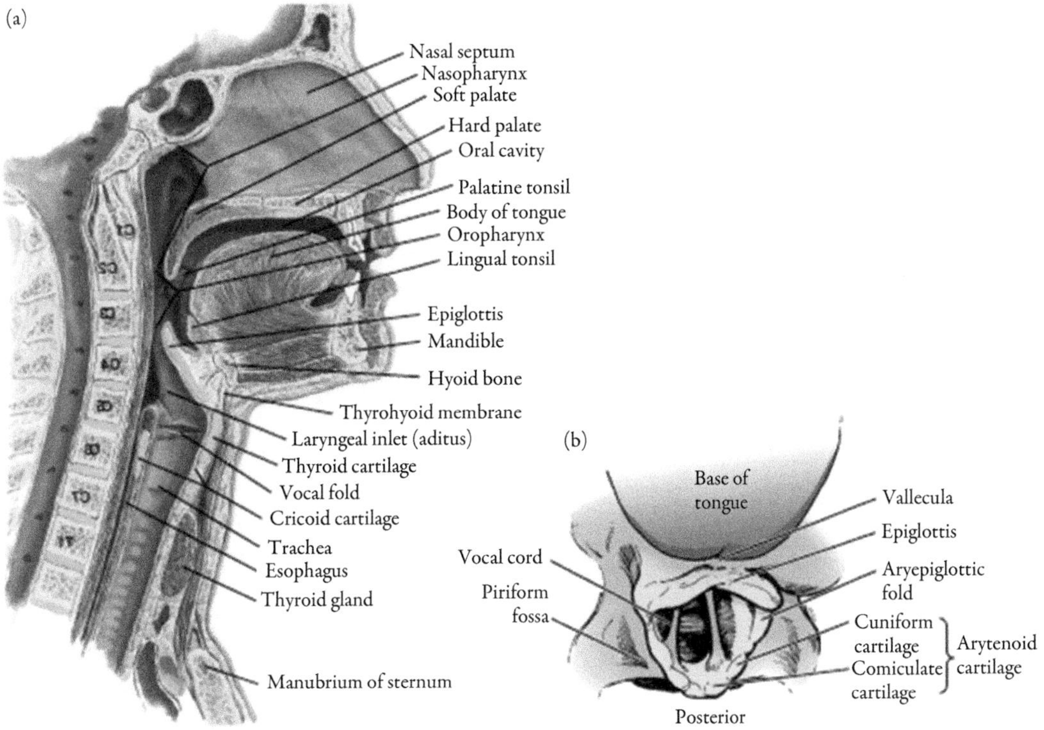

Figure 27.3 Nasal and oral cavities in relation to nasopharynx and oropharynx.

the radial forearm, anterolateral thigh, iliac crest, fibula, or jejunal, or gastric.

- **PHARYNGEAL INFECTIONS**
  - **Peritonsillar abscesses** should be aspirated and incised if necessary.
  - **Retropharyngeal abscesses** are most often in children and should be drained via a posterior pharyngeal incision in the operating room.
  - **Parapharyngeal abscesses** can be drained via a lateral neck or submental incision.
  - **Ludwig's angina** is a phlegmon of the submental space without a true, drainable abscess. Management is with airway control via intubation or tracheostomy, and dental source control as well as parenteral antibiotics.
- **OBSTRUCTIVE SLEEP APNEA (OSA)**
  - OSA can be associated with excessive pharyngeal tissue, obesity, micrognathia, or retrognathia. Symptoms include snoring, daytime somnolence, or more than 10 apneic or hypopneic episodes per hour. Treatment options include an oral appliance to ameliorate posterior pharyngeal obstruction, continuous positive airway pressure (CPAP), tonsillectomy or adenoidectomy, uvulopalatopharyngoplasty (UPPP), base of tongue resection, mandible advancement, hyoid suspension, or tracheostomy.

LARYNX

See Figure 27.4.

- **Supraglottic larynx** has respiratory epithelium lining the epiglottis, aryepiglottic folds, and false vocal folds. Pathology here drains via lymphatics piercing the thyrohyoid membrane to nodal stations 2, 3, and 4.
- **Subglottic larynx** extends down to the inferior cricoid and is also lined with respiratory epithelium.
- The **glottis** is that portion of the larynx including the true vocal cords that is lined by stratified squamous epithelium. Pathology here drains to nodal stations 3, 4, and 6.
- **Malignant tumors** include SCC, neuroendocrine, and salivary gland tumor. Patients with laryngeal malignancies may have voice alterations and should be evaluated for **synchronous or metachronous lesions or lung cancer.** Treatment of laryngeal malignancy is with **minimally invasive endoscopic surgery when feasible,** chemotherapy, or radiation for larynx preservation whenever possible. WLE includes total or partial laryngectomy.
- **Benign laryngeal pathology** includes **papilloma** from HPV 6 or 11, cord edema from smoking or reflux, cord cyst, leukoplakia, or **granuloma** secondary to intubation, reflux, or voice abuse.

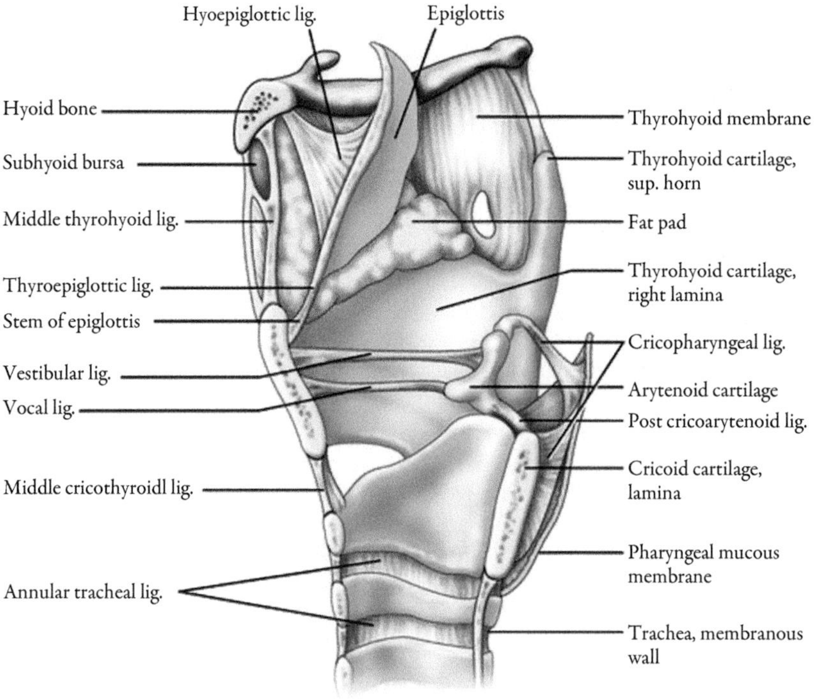

Figure 27.4 Laryngeal anatomy.

- **Epiglottitis** is managed with airway control in the OR by nasal or orotracheal intubation or tracheostomy. Parenteral antibiotics directed at GPC and GNR are also required.
- **Cord paresis or paralysis** may cause a hoarse voice and aspiration. It results from tumor invasion or recurrent laryngeal nerve dysfunction. Treatment is vocal cord injection, speech therapy, surgical cord medialization, or reinnervation by nerve transfer from the ansa cervicalis. If bilateral cord dysfunction, the patient may need cord lateralization, arytenoidotomy, or tracheostomy.

SALIVARY GLAND

See Figure 27.5.

- **Parotid gland** has serous output. **Stenson's duct** runs near the transverse facial artery and buccal branches of the facial nerve. Lacerations in this area should be repaired over a stent to prevent sialocele formation.
  - **Suppurative parotitis** is due to dehydration causing stasis in the duct and secondary infection with *S. aureus*. Treatment is with rehydration, salivation, antibiotics, and drainage if needed.
  - 80% of **parotid tumors** are benign. However, given how common parotid neoplasia is, it remains the most frequent site for malignant salivary tumors as well.
- **Sublingual glands** have mucinous output, and **submandibular glands** have half serous and half mucin output. 25% of tumors in either of these glands are benign. Sialadenitis is mostly submandibular or sublingual. Treatment is with sialogogues or incision to remove stones. Gland excision can be done if the inflammation is recurrent.
- **Minor glands** reside at the junction between the hard and soft palate.
- **Malignant salivary gland tumors** include mucoepidermoid carcinoma and adenoid cystic carcinoma. These tumors invade nerve roots and grow indolently. Treatment is **WLE, total parotidectomy, radiation, and lymph node dissection if necessary**. Adenoid cystic carcinoma is radiosensitive. Some tumors are similar to breast ductal tissue and respond to hormonal or Her2 blockade.
- **Benign salivary gland tumors** present as unilateral or bilateral painless parotid masses. They include pleomorphic adenoma in a middle-aged female, and Warthin tumor in a middle-aged male smoker. Treatment is with **superficial parotidectomy**.
- Hemangioma, neural sheath tumors, and lipomas are benign non-epithelial salivary tumors.

UNDIFFERENTIATED NECK MASS

- A parapharyngeal mass anterior to the styloid is likely salivary in origin. If posterior to the styloid, it is likely neural, vascular, or lymphoid in origin.

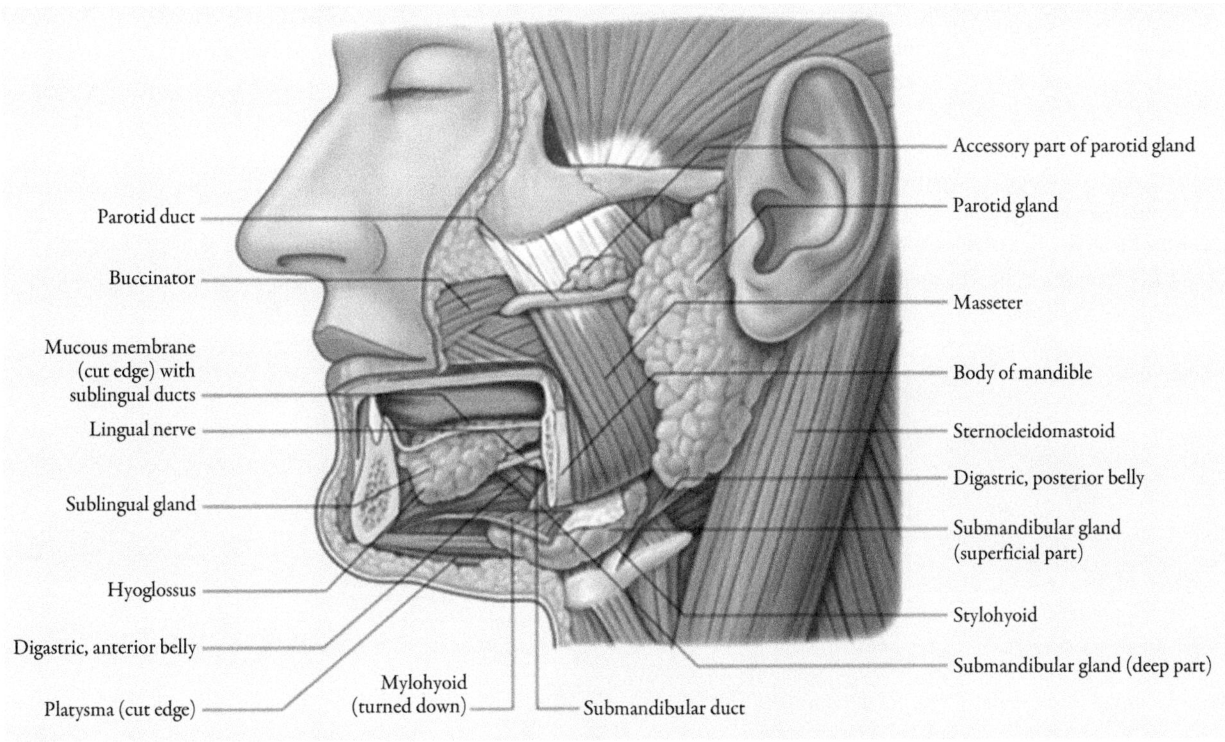

Figure 27.5 Sublingual, submandibular, and parotid glands.

- A **hemangioma** is the most common benign mass in adults.
- Lateral neck masses (branchial cleft cysts and cystic hygroma) and midline masses (thyroglossal duct cysts and ectopic thyroid) are discussed in Chapter 25, Pediatrics.
- **Lymphadenopathy** can be due to bacterial infection of the node (lymphadenitis), cat scratch disease, tuberculous infection of the node (scrofula), lymphoma, or invasion by squamous cell or adenocarcinoma.
  - **Lymphadenitis** is the most common, and can be treated with antibiotics for 2 weeks. Incision and drainage, or excision, is reserved for failure of antibiotics.
  - **Scrofula** is cervical tuberculous lymphadenitis producing a nontender lateral neck mass. Diagnosis is by fine needle aspiration (FNA) or excisional biopsy with acid fast bacilli testing. Patients may be asymptomatic but should be considered contagious. There need not be any pulmonary involvement. Treatment is with a multidrug regimen for 6 months.
  - **Lymphoma and carcinomas** are generally diagnosed and treated after biopsy confirmation. See below.
- **MALIGNANT NECK MASS**
  - These include primary or metastatic SCC, metastatic adenocarcinoma, metastatic melanoma, lymphoma, sarcoma, or neural origin tumors.
- **DIAGNOSIS**
  - Start with CT, MRI, and US neck imaging for the primary and any nodes. Irregular nodes >1.5 cm with central necrosis may be malignant. PET CT imaging is useful if an infraclavicular primary is suspected.
  - **Fine needle aspiration (FNA)** should be performed via exam, US, or CT guidance with a 22g needle on a 20 ml syringe. The needle is passed through the lesion multiple times at different angles with continued aspiration. The suction should be released before the needle is removed and the sample placed on a slide.
    - Most often the specimen will reveal **SCC** and should be tested for **HPV** (suggests an underlying oropharyngeal primary) and **EBV** (suggests an underlying nasopharyngeal primary).
    - If **adenocarcinoma** is seen, then it is likely the patient has breast, lung, or GI metastasis to a head and neck node.
    - **Melanoma or lymphoma** are other possible diagnoses.
    - If the FNA appears **infectious or inflammatory**, then trial 2 weeks of antibiotics and re-evaluate. TB should be treated with a multidrug regimen (above).
    - If the FNA is **non-diagnostic**, then it may be repeated, or a core needle biopsy (CNBx), or excisional biopsy (EXBx) may need to be performed. The latter is reasonable if the mass is small and easily accessible.
    - If sampling a node remains non-diagnostic or it proves to be metastatic SCC, and a primary could not be

located by CT imaging, then perform a **panendoscopy** including nasal, oral, hypopharyngeal, laryngeal, esophageal, and bronchial sites.
- The pyriform and base of tongue can be biopsied, an empiric bilateral tonsillectomy can be performed, and an ipsilateral lingual tonsillectomy is also an option in search for a primary.
- If all of the above are negative, then a neck exploration and **EXBx** should be pursued, even for a mass that is difficult to access. **Modified radical neck dissection (MRND)** can be done at the same time if the frozen section of the EXBx specimen is positive for SCC. *See Figure 27.6.* The patient may need an empiric tonsillectomy as well.

- **TREATMENT**
  - Any primary identified via any part of the above process is treated with **resection** if it resides in the

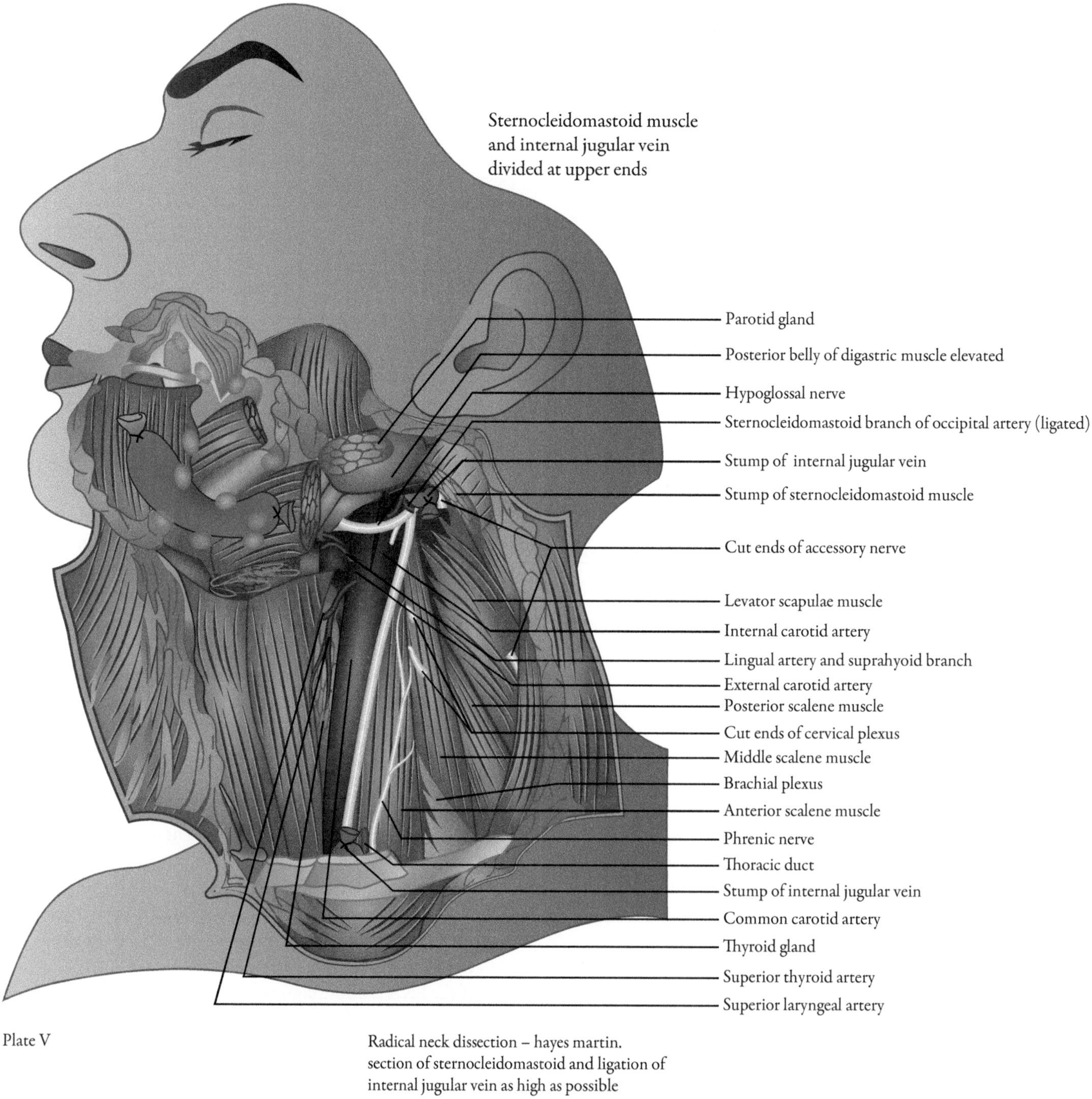

Figure 27.6 Modified radical neck dissection.

nasal, oral, or select oropharyngeal, hypopharyngeal, or laryngeal sites (see corresponding sections above). Nasopharyngeal tumors are treated with radiation and chemotherapy, without surgery.

- **Adjuvant neck radiation after resection** is indicated if the patient has >1 metastatic node, large nodes, extracapsular nodal spread, positive margins, or in the setting of an unknown primary believed to be in the neck The nasopharynx and oropharynx may be included in the radiation field, unless the patient's specimen is HPV-positive (then radiation can be directed at the oropharynx alone).
- If the entire workup remains negative for a primary, but cellular evaluation of the original node was consistent with carcinoma, then **empiric chemotherapy and radiation** to the nasal, oral, and hypopharynx with or without MRND is an option.
- **Definitive radiation** is utilized for nasopharyngeal and non-resectable oropharyngeal, hypopharyngeal, and laryngeal primaries.

28.

CARDIAC

CORONARY ARTERIES

- The right coronary artery (RCA) gives off the Sino-atrial (SA) nodal branch and a posterior descending artery 85% of the time (right dominant coronary system). *See Figure 28.1.*
- **Myocardial infarction (MI)** is most often due to thrombosis atop a ruptured atherosclerotic plaque or severe, chronic narrowing from a mature plaque in the setting of increased metabolic demand. Most patients with acute MI will go to the cardiac catheterization laboratory for prompt recanalization of the culprit vessel. Complications of MI include immediate dysrhythmias, rupture of the septal, free wall, or papillary muscles within one week, or long term aneurysmal dilation of the ventricular wall. Septal or free wall rupture can be closed with a pericardial patch. A papillary muscle tear can rarely be repaired and more commonly the mitral valve will need replacement.
- **Coronary artery bypass graft surgery (CABG)** is indicated for the following. *See Figure 28.2.*
  - Left main coronary artery stenosis.
  - Dual vessel coronary artery stenoses with diabetes.
  - Proximal left anterior descending and circumflex stenoses; or
  - Three vessel coronary artery stenoses in a non-diabetic.
  - Low ejection fraction, an inability to stent, or recurrent stent occlusions are other indications for CABG.
  - Mortality is <2%. Preoperative cardiogenic shock is the most important indicator for increased mortality. Tight postoperative glucose control decreases sternal wound infection but does not affect mortality.
  - There is an 80% patency of saphenous vein grafts (SVG) at 5 years and 95% patency of left internal mammary (LIMA) to left anterior descending (LAD) at 20 years.

VALVES

- **Aortic stenosis** is the most common adult valve lesion. Senile calcific degeneration occurs with age. Bicuspid valves have a higher stenosis rate earlier in life. The high resistance of a stenotic aortic valve increases left ventricular (LV) workload during the isovolumetric contraction phase of the cardiac cycle. Symptoms include angina, syncope, and heart failure. The latter implies a 1- to 2-year life expectancy.
  - Aortic valve replacement is needed when symptomatic, the gradient is >50 mmHg, or valve aperture area is <1 cm^2.
  - Bioprosthetic valves are for patients desiring pregnancy, those with a high risk for bleeding on anticoagulation, or elderly patients. Mechanical valves are used for younger patients who plan to take anticoagulation. Stentless homografts have decreased gradients and a lower risk for endocarditis, but may have less longevity.
- **Aortic regurgitation** requires surgery when severe and symptomatic, LV end diastolic diameter is increased, or ejection fraction is decreased.
- **Mitral regurgitation** results in left atrial (LA) and left ventricular (LV) distension with increased workload. It may be from myxomatous change with prolapse, leaflet perforation, chordae rupture, or papillary muscle dysfunction, and can be complicated by atrial fibrillation. Valve repair or replacement is recommended for severe MR that is symptomatic, associated with LV distension or a reduced ejection fraction (EF), or complicated by elevated pulmonary artery pressure or atrial fibrillation.
- **Mitral stenosis** is a sequelae to rheumatic fever. Symptoms are from pulmonary edema or distal emboli. Catheter based balloon valvuloplasty is the main treatment modality for patients with commissural fusion. Replacement is recommended when the stenosis is categorized as severe, and the patient is symptomatic and the valve not amenable to balloon valvuloplasty.
- For patients with **paroxysmal or persistent atrial fibrillation,** a Cox-Maze procedure and left atrial appendage ligation can be added to any cardiac valvular surgery.
- **Endocarditis** is a bacterial or fungal infection of platelet-fibrin aggregates on damaged valve leaflets. It can occur in native or prosthetic valves. Common organisms include *strep viridans, staph aureus, enterococcus,* or Gram-negative rods (GNR). Tricuspid

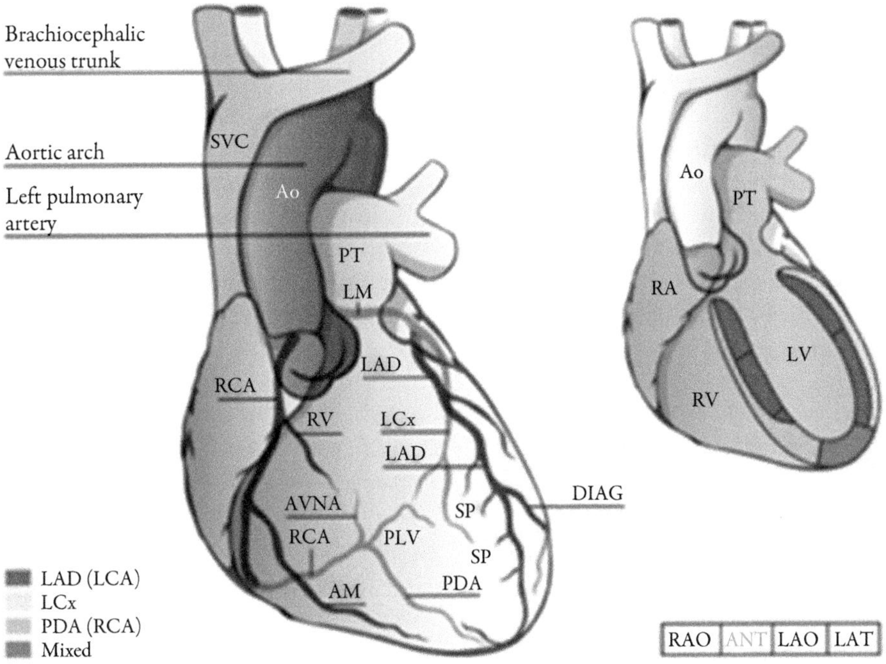

Figure 28.1 Coronary artery anatomy.

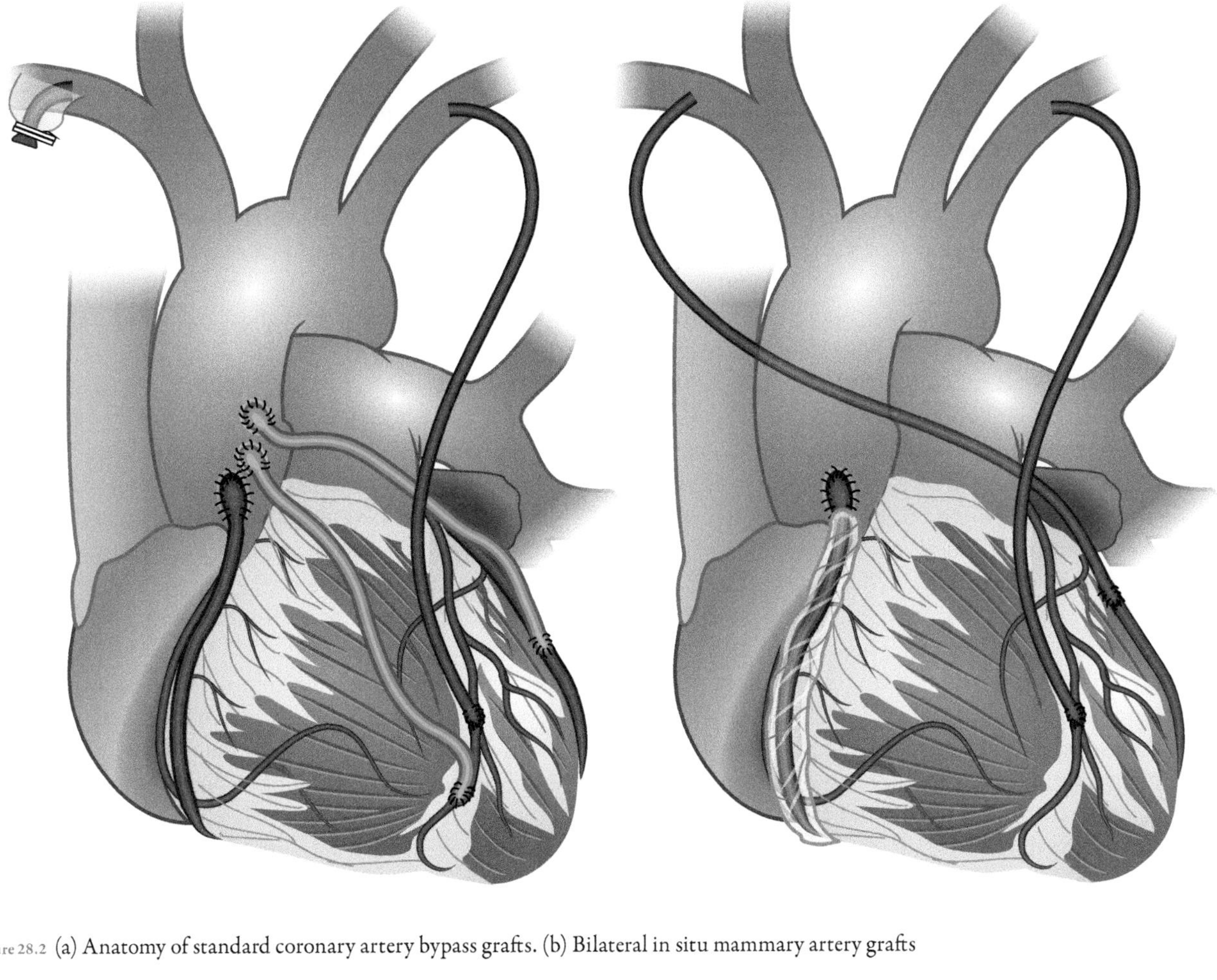

Figure 28.2 (a) Anatomy of standard coronary artery bypass grafts. (b) Bilateral in situ mammary artery grafts

or pulmonary valve endocarditis is in IV drug users. *S. aureus*, GNR, and pseudomonas are common.

- The diagnosis is based on a combination of Dukes criteria, including positive blood cultures, characteristic echocardiogram findings, fever, peripheral emboli, and immunologic phenomenon.
- **Management** starts with IV antibiotics. Surgery is reserved for cardiogenic shock or heart failure secondary to valve incompetence, occlusion or fistulization between cardiac chambers, loss of structural integrity to a valvular apparatus, or the fibrous trigone of the heart, perivalvular abscess, or septal abscess with heart block, purulent pericarditis, recurrent emboli or certain large vegetations. In the absence of these findings, antibiotics may be utilized over a prolonged period. If however there is persistent fever or positive blood cultures despite appropriate antibiotics (and no other undrained infectious focus in the body), then surgery should be re-considered.
- **Procedural prophylaxis** is recommended for dental interventions in patients with a prosthetic valve, prior endocarditis, cardiac transplant recipients, or congenital heart disease that is un-repaired, recently repaired, or those repaired with a residual defect.

CONGENITAL CARDIAC

- **Fetal Circulation.** Oxygenated umbilical vein blood returns through the left portal vein, ductus venosus into the left hepatic vein, and then into the inferior vena cava (IVC). It is funneled through the right atrium (RA) and into the patent foramen oval (PFO) guided by the eustachian valve and tendon of Todaro. The blood enters the left atrium and can go to left ventricle and cerebral circuit. Deoxygenated cerebral venous return is funneled from the superior vena cava (SVC) into RA and right ventricle (RV), through pulmonary artery (PA) and ductus arteriosum to descending aorta and umbilical artery. It is oxygenated in the placenta and returns via the umbilical vein. *See Figure 28.3.*
- **Right to left shunts** produce cyanosis and can eventually cause polycythemia, hyperviscosity and thrombosis, cerebrovascular accident (CVA) from paradoxical emboli, and brain abscesses or endocarditis from bacteria bypassing pulmonary macrophages.
  - **Tetralogy of Fallot** is the most common cyanotic lesion and describes a ventricular septal defect (VSD), overriding aorta, pulmonic stenosis, and RV hypertrophy. A Blalock-Taussig (BT) shunt can be done if an early intervention is needed to preserve pulmonary blood flow. Otherwise most patients should receive a single stage repair of the stenotic pulmonary outflow, widening of the RV outflow tract with a patch or an RV to PA conduit, and closing the VSD. Medical management prior to repair includes hydration, sedation, opiates, and oxygen to decrease pulmonary vascular resistance (PVR), squatting to increase systemic vascular resistance (SVR), and beta blockade to prevent a tet spell.
  - **Transposition** of the great arteries, **tricuspid atresia, and truncus arteriosus** are other common cyanotic lesions.
- **Left to right shunts** include ventricular septal defect (VSD), atrial septal defect (ASD), patent ductus arteriosus (PDA), and anomalous pulmonary venous return (TAPVR) where one or more pulmonary veins (PV) may drain into the innominate vein, coronary sinus, or portal veins. Pulmonary venous return could be partially or completely obstructed in these cases.
  - A pulmonary to systemic blood flow ratio (Qp/Qs) <1.5 is tolerated well. A ratio >2 may require repair of the underlying lesion if symptomatic or persistent.
  - Eventually, increased pulmonary blood flow and pressure may result in a rise in PVR, due to thickening and loss of compliance of pulmonary arterioles and microvasculature. This can progress to the point that the shunt is reversed and now flowing right to left (Eisenmenger syndrome).
  - **Ventricular septal defect (VSD)** is the most common congenital cardiac abnormality. Perimembranous location is more common than muscular, inlet, or outlet positions. The magnitude of the shunt depends on the ratio of PVR to SVR. 80% close spontaneously. Most patients are **symptomatic at 4–6 weeks** of life when PVR has dropped and the shunt fraction increases. Symptoms of heart failure or failure to thrive warrant closure. Otherwise, a VSD may be repaired in infancy or childhood if the shunt fraction is clinically tolerable, and the child is still feeding and growing.
  - **Atrial septal defect (ASD)** produces a left to right shunt determined by the ratio of **RV to LV compliance.** 80% are ostium secundum defects in the area of the fossa ovalis (FO). Ostium primum defects are from failure of septum primum fusion and have associated MV and TV pathology or Trisomy 21. ASDs should be repaired if they are symptomatic, enabling paradoxical emboli, or persistent with a Qp/Qs >2.
  - **Patent ductus arteriosus (PDA)** generally closes in infancy with or without indomethacin assistance. If a duct has persisted into childhood, then it likely has a structural abnormality that will not respond to indomethacin. Treatment is with endovascular closure or surgical ligation.
- **Pulmonic stenosis** results from fused cusps and produces post-stenotic dilation of the main pulmonary artery as well as right ventricular hypertrophy. Treatment is with balloon valvuloplasty or surgical repair.

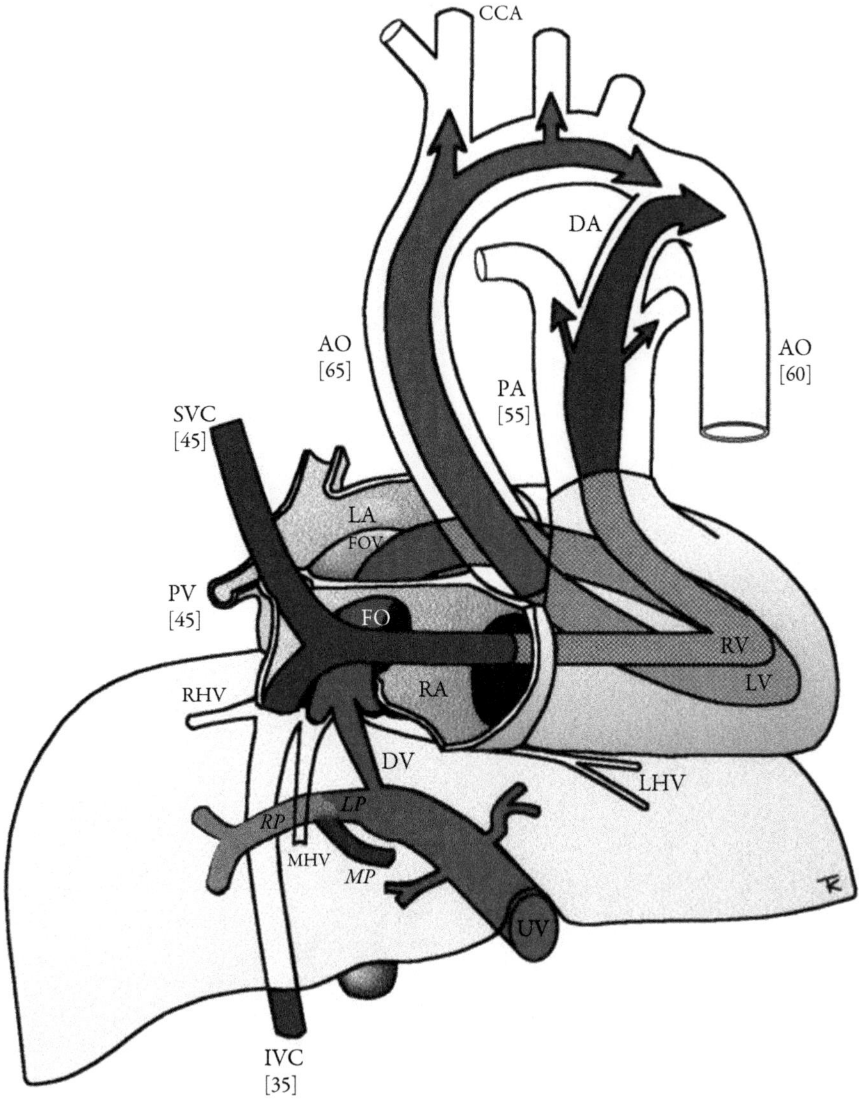

Figure 28.3 Fetal circulation shunting oxygenated blood to the left atrium and ventricle, and deoxygenated blood to the right ventricle.

- **Aortic coarctation** generally occurs at the isthmus near the PDA and can result in cardiogenic shock and pulmonary edema in the neonatal period. If less severe, patients can present with upper extremity hypertension in childhood. Treatment includes resection and anastomosis or patch aortoplasty.
- **Aortic stenosis** should be treated if the gradient is more than 30 mmHg at rest, or more than 50 mmHg with exertion. Interventions include endovascular balloon valvuloplasty, surgical valvuloplasty, use of a pulmonary autograft, or bioprosthetic aortic valve replacement.
- **Aortic regurgitation** can be related to subaortic stenosis or a VSD jet. Symptoms or a high LV end systolic diameter warrant attempts at repair or replacement.
- **Vascular rings** are associated with double aortic arch, right-sided arch, or aberrant origin and course of a subclavian vessel and tend to encircle both the trachea and esophagus. Patients have respiratory symptoms and sometimes dysphagia.
    - A **PA sling** is when the left PA arises from the right PA and courses between the trachea and esophagus. It may be associated with complete tracheal rings that independently are causing tracheal narrowing.

OTHER PATHOLOGY

- **Cardiopulmonary bypass (CPB) sequelae** include decreased blood viscosity due to hemodilution from priming crystalloid, fluid shifts from the vascular to the interstitial space, hemolysis, coagulopathy, fibrinolysis, and an inflammatory cascade. Moderate hypothermia is maintained during bypass to protect end organs. Aortic, left atrial (LA), and left ventricular (LV) vents can help de-air when CPB is being weaned. Post-CPB

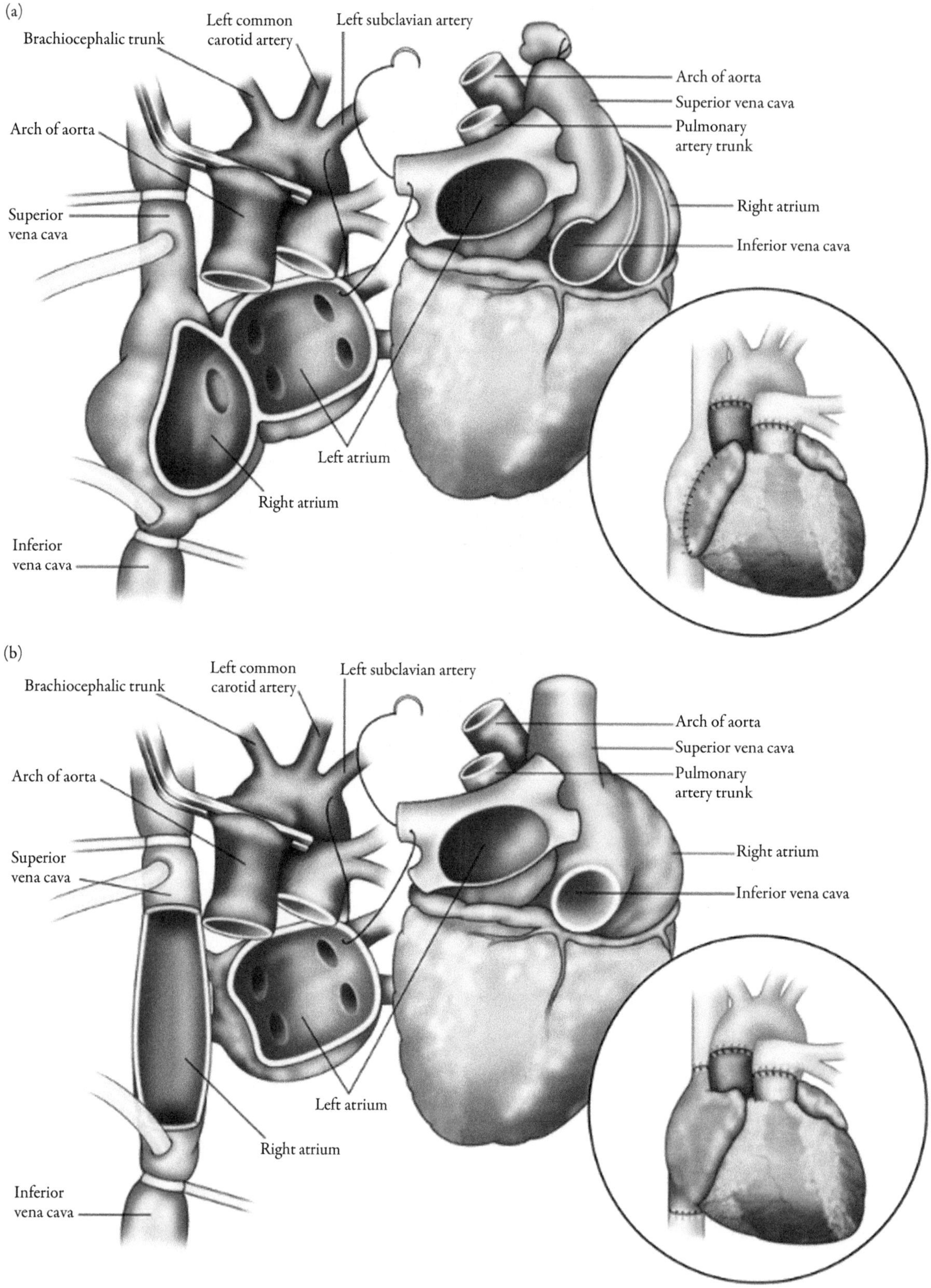

Figure 28.4 (a) Recipient anatomy after cardiectomy. (b) Donor heart after implantation.

encephalopathy or organ dysfunction may be due to microcirculatory sludge.

- **Post-pericardiotomy syndrome** presents with fever, pleuritic, positional chest pain, and diffuse ST elevation. Treatment is with NSAIDs and sometimes colchicine. If occurring at 2–4 weeks postoperative, it may be autoimmune mediated (Dressler syndrome).
- **Constrictive pericarditis** can be idiopathic or secondary to radiation, tuberculosis, viral infections, or malignancy. Treatment is with pericardiectomy of all tissue lying between phrenic nerves.
- **Cardiac tamponade** can present with Beck's triad of jugular venous distension, muffled heart sounds, and hypotension. Patients may have exaggerated pulsus paradoxus, Kussmaul's sign, and equalization of chamber pressures on invasive monitoring. Diagnosis is by echocardiogram, and treatment is with a pericardial drain, pericardial window, or sternotomy depending on the type and origin of the pericardial fluid.
- **Neoplasia.** A benign myxoma in the left atrium is the most common tumor. Angiosarcoma is the most common malignant intracardiac tumor. Metastases are most commonly from lung cancer.

CARDIAC TRANSPLANTATION

- Candidates should have a life expectancy <1 year due to **cardiogenic shock, ejection fraction (EF) <20%, heart failure class III or IV, and reversible end-organ dysfunction.** They should not have a recent diagnosis of malignancy or another terminal condition. They should not have irreversible organ dysfunction or fixed, elevated pulmonary vascular resistance.
- Patients may be on inotropic and chronotropic chemical support as an inpatient or outpatient.
- Some candidates will be on percutaneous or implanted left ventricular, right ventricular, or biventricular support. Including extracorporeal mechanical circulation.
- Heart transplantation is most often an orthotropic procedure where the native heart is removed and the donor organ is implanted in its place.
- The graft can withstand **4–6 hours of cold ischemic time** after procurement from a donor. A length of donor aorta, pulmonary artery, and both vena cava are taken with the heart. Also the left atrium is transected near the pulmonary vein confluence.
- For implantation, the left atrium is anastomosed first, followed by inferior vena cava, pulmonary artery, aorta, and then cross clamp removal and superior vena cava (SVC) completion.
- **Complications**
  - **Primary graft dysfunction** requires prolonged use of postoperative inotrope or mechanical support. It is generally related to a long ischemic time or poor status of the donor organ.
  - Elevated pulmonary vascular resistance or right heart dysfunction can be managed with milrinone, nitric oxide, and prostaglandins.
  - **Acute rejection** has a lymphocytic myocyte infiltration. Empiric serial myocardial biopsies are performed via right heart catheterization in order to screen for this. Infection and acute rejection are the main causes of death in the early postoperative period.
  - **Chronic rejection** with diffuse coronary vasculopathy and progressive stenosis is the main cause of death after one year.
  - **Skin cancer, post-transplant lymphoproliferative disorder (PTLD), and other malignancies** can occur over the long term. Malignancy is the main cause of death after 5 years.
  - One-year survival after heart transplantation is 80% and 5 years survival is 60%–70%.

29.

THORACIC

TRACHEA

- Percutaneous **tracheostomies** are considered equivalent to an open procedure, but with a lower wound infection rate. Bronchoscopic assistance is useful but not mandatory, and percutaneous needle access can be augmented with ultrasound (US) guidance. Tracheostomy can decrease the overall time of mechanical ventilation.
- A **tracheoinnominate artery fistula** can result from a tracheostomy placed below the 3rd cartilage ring. *See Figure 29.1.* Patients may have a small herald bleed before the onset of large volume hemorrhage. Treatment is with **overinflation** of the tracheostomy cuff to tamponade the sidewall of the bleeding innominate artery. This is followed by removal of the tracheostomy and **transtracheal digital pressure** on the innominate artery, pinning it against the undersurface of the sternum while the patient is **endotracheally intubated** and the balloon cuff inflated distally to protect from bloody aspiration. *See Figure 29.2.* The patient can then undergo **manubriotomy or sternotomy** with ligation of the proximal innominate artery and direct tracheal repair.
- A **tracheoesophageal fistula** can be congenital or acquired. The latter is from the endotracheal tube (ETT) cuff pressing against the esophagus, with or without a concomitant nasogastric tube. *See Figure 29.3.* Treatment is to replace the tracheostomy with a **large volume cuff ETT terminating below the fistula**. The balloon cuff is inflated to prevent air from rising and entering the fistula and esophagus. Natural, delayed healing of the TEF may occur when it is isolated in this manner. Also, esophageal secretions transgressing the fistula into the trachea will collect above the balloon cuff and not disseminate down to alveoli. Some patients will need a decompressive gastrostomy and feeding jejunostomy. If delayed healing does not occur, **treatment is with tracheal resection and re-anastomosis and direct esophageal repair** with a sternohyoid flap. This is ideally done after the patient has weaned from the ventilator.
- Small, well-apposed **tracheal lacerations** involving less than one-third of the circumference of the trachea may be treated **non-operatively.** Otherwise, repair is via a **cervical collar incision** for the upper trachea, **median sternotomy** for the proximal intrathoracic trachea, or **right posterolateral thoracotomy** for the distal intrathoracic trachea, carina, or takeoff of the left mainstem bronchus. A left posterolateral thoracotomy may be needed for injuries to the mid- or distal left bronchus.
- **Tracheal stenosis** is secondary to mechanical trauma from intubation, scarring at a tracheotomy entry site, or prolonged ETT or tracheostomy cuff pressure against the tracheal mucosa. Presentation is usually 1–6 weeks after the above incidents.
  - The ideal treatment is **resection and re-anastomosis** after rigid bronchoscopic assessment of the stricture in the OR. An absolute contraindication to resection is a long-segment stenosis or prolonged need for positive pressure. Relative contraindications include prior neck radiation with its associated microvascular disease. Other treatment options for tracheal strictures are palliative and temporizing, and include **bronchoscopic dilation, laser ablation, or a tracheal stent.**
  - Intravenous anesthetics are preferred over inhaled agents, and a small ET tube passed beyond the stenotic segment is placed for airway control prior to resection. Graduated sized dilators may be needed to pre-dilate the stenotic segment if <6 mm diameter.
  - A collar incision, incorporating any tracheostomy site, should have superior and inferior flaps raised to the cricoid and sternal notch respectively. One may divide the manubrium to expose the carina. Circumferential tracheal dissection should proceed **immediately atop the trachea** to prevent injury to the recurrent laryngeal nerve (RLN), and should extend for no more than 1cm proximal and 1cm distal to the stenotic segment. Resection of the stenotic segment is followed by a temporary separate distal ET tube placement for ventilation during repair via interrupted anastomotic non-absorbable sutures. **Hyoid cartilage laryngeal release** can be done from a separate incision when a segment of trachea longer than 4 cm is resected. When the repair is complete and the distal tube has been removed, the orotracheal tube is

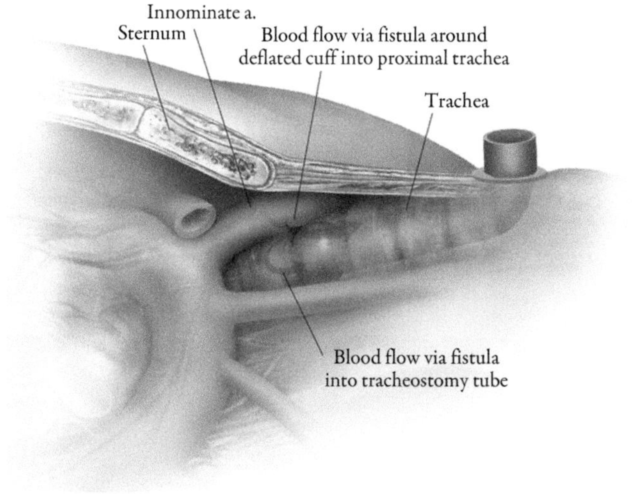

Figure 29.1 Anatomy of tracheal to innominate artery fistulisation.

advanced past the anastomosis. One may place a **strap muscle flap** between the tracheal reconstruction and the innominate artery, over the anastomosis itself. A **submersion leak test** is performed with ventilation while the endotracheal cuff is down. **Submental sutures** to the pre-sternal skin will keep the neck in flexion for one week post-op and reduce tension on the anastomosis.

- Ideally the patient should be extubated in the OR or within 1or 2 days. Racemic epinephrine and glucocorticoids can prevent laryngeal edema. If prolonged ventilation is needed post-op, a low tracheostomy should be performed below the anastomosis.
- **Dehiscence** is the most feared complication. Risk factors include diabetes mellitus (DM), resection of a segment longer than 4 cm, prior tracheostomy, or re-operative repair. Diagnosis is by bronchoscopy prompted by a post-op air leak with subcutaneous emphysema. Treatment is by repair buttressed with strap muscles or resection and re-anastomosis. A permanent tracheostomy is a last resort. Other complications include wound and lung infection, laryngeal edema, temporary or permanent swallowing dysfunction, or granulation tissue development requiring bronchoscopic debridement.

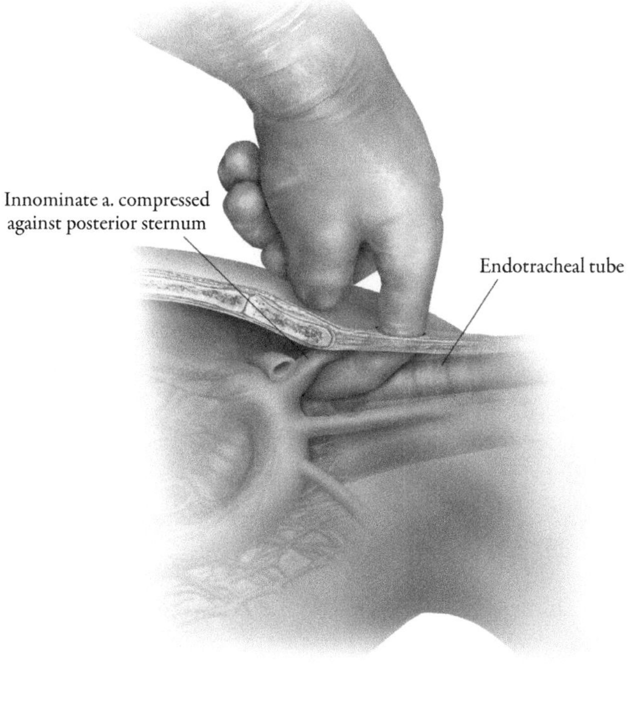

Figure 29.2 Extra-tracheal digital compression of innominate artery against the sternum.

- **Tracheal masses** may be benign (papilloma or hemangioma) or malignant, including squamous cell carcinoma (SCC) or adenoid cystic salivary carcinoma. Management can include bronchoscopic resection for discrete lesions or formal tracheal resection after staging workup is complete, with or without radiation and adjunctive therapy.

LUNG

- Trachea and bronchi are supported by cartilage around their anterior and lateral circumference. The posterior wall is membranous. Bronchi divide into segmental bronchi.
- Each upper lobe has apical, anterior, and posterior segments. The right middle lobe has a medial and lateral segment, and the left lingual has a superior and inferior segment. Each lower lobe has a superior segment and anterior, posterior, medial, and lateral basal segments. Within each segment, sub-segmental bronchi lead into distal bronchioles that do not have cartilage but instead have smooth muscle with contracts to control lumen diameter. Terminal bronchioles lead into alveolar ducts and sacs, all of which contribute to gas exchange.
- The right lung comprises 55% of total volume and the left lung 45%. Tidal volumes are generated by the diaphragm (80%) and intercostals (20%). Accessory muscles of respiration include the sternocleidomastoid (SCM), scalenes, levator scapulae, and serratus posterior. The thoracic cavity is lined by parietal pleura.
- **Type 1 pneumocytes** line alveoli and are involved in gas exchange. **Type 2 pneumocytes** make surfactant. They comprise 60% of alveolar cells, but only occupy 3% of the alveolar lining. Pores of Kohn allow direct air exchange between alveoli. Lung parenchyma is covered by visceral pleura.
- **Lymph node stations** may be hilar at the right or left mainstream bronchus (station 10), interlobar at

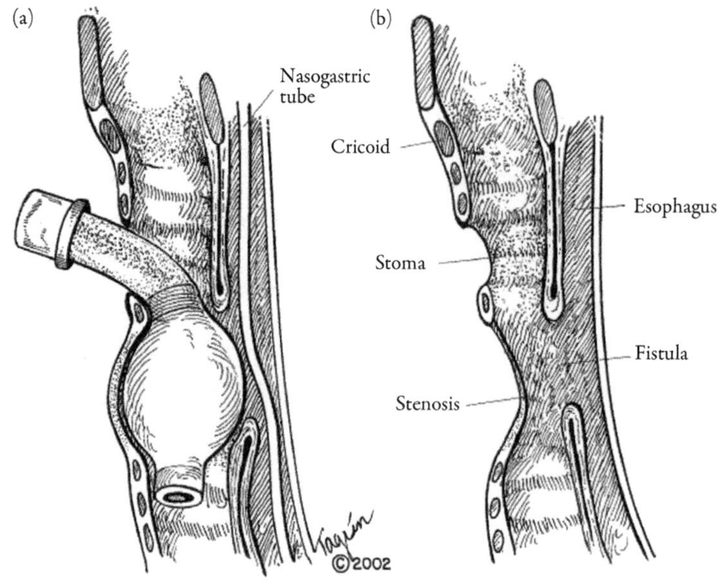

Figure 29.3 Tracheostomy cuff pressure creating a tracheoesophageal fistula.

the branch points between bronchi (station 11), lobar along an upper, lower, or middle bronchus (station 12), or segmental at bronchial branch points within a lobe (station 13). Lung pathology can also spread to mediastinal and aortopulmonary window nodes. *See Figure 29.4.*

- Prior to any surgical resection, **pulmonary function tests** can evaluate the patient's tolerance of the proposed procedure. In order for reliable interpretation, the patient must be able to cooperate with full effort during testing.
  - **Forced expiratory volume in 1 second (FEV1)** is that portion of the vital capacity that a patient can actively exhale in one second. The **preoperative FEV1** should be >60% of the predicted value of a normal subject with similar age, height, and gender. Preoperative FEV1 >2000 ml implies the patient can safely receive a pneumonectomy. Preoperative FEV1 >1500 ml is sufficient for a lobectomy. A **postoperative predicted FEV1** >800 ml (after resection of the given lung segments) or >40% of the postoperative predicted value for the given lung resection in a patient of similar age, height, and gender, would be sufficient.
  - **Diffusion capacity for carbon monoxide (DLCO)** is dependent on alveolar architecture, capillary surface area, hemoglobin concentration, and lung perfusion. It is a good predictor of pulmonary complications post-op and quality of life after a major lung resection. **Preoperative DLCO** should be >10 ml/min/mmHg or >60% of the value of a normal subject of similar age, height, and gender. A **postoperative predicted DLCO** >40% of the postoperative predicted value for the given lung resection in a patient of similar age, height, and gender would be sufficient.
- A **VQ scan or lung perfusion scan** can be used to determine the contribution of the segment of lung to be removed. If its bronchi are occluded or it is not being perfused, then it is noncontributory, and its removal need not be included in the predicted postoperative calculations above.
- Other disqualifiers for lung resection include **preoperative hypercarbia** (pCO_2 >50) or **hypoxia** (pO_2 <60 at rest) or **impaired maximum oxygen consumption** during cardiopulmonary exercise testing (VO_2 max <10 ml/kg/min or <50% of that predicted in a similar age, height, weight normal subject).

CARCINOMA

- **Squamous cell carcinoma (SCC)** is central in location, either in situ or invasive, and strongly related to smoking. Parathyroid hormone related protein (PTHrP) may occasionally be overproduced.
- **Adenocarcinoma** is more common than SCC. It is peripheral in location and occurs in both smokers and nonsmokers. **In situ** disease is lepidic only, and does not invade the basement membrane **Micro-invasive** disease is lepidic with less than 5 mm invasion. **Lepidic predominant** has more than 5 mm invasion but still is predominantly within the alveolar airspace. **Adenocarcinoma** proper has more than 5 mm invasion and a minimal lepidic component.

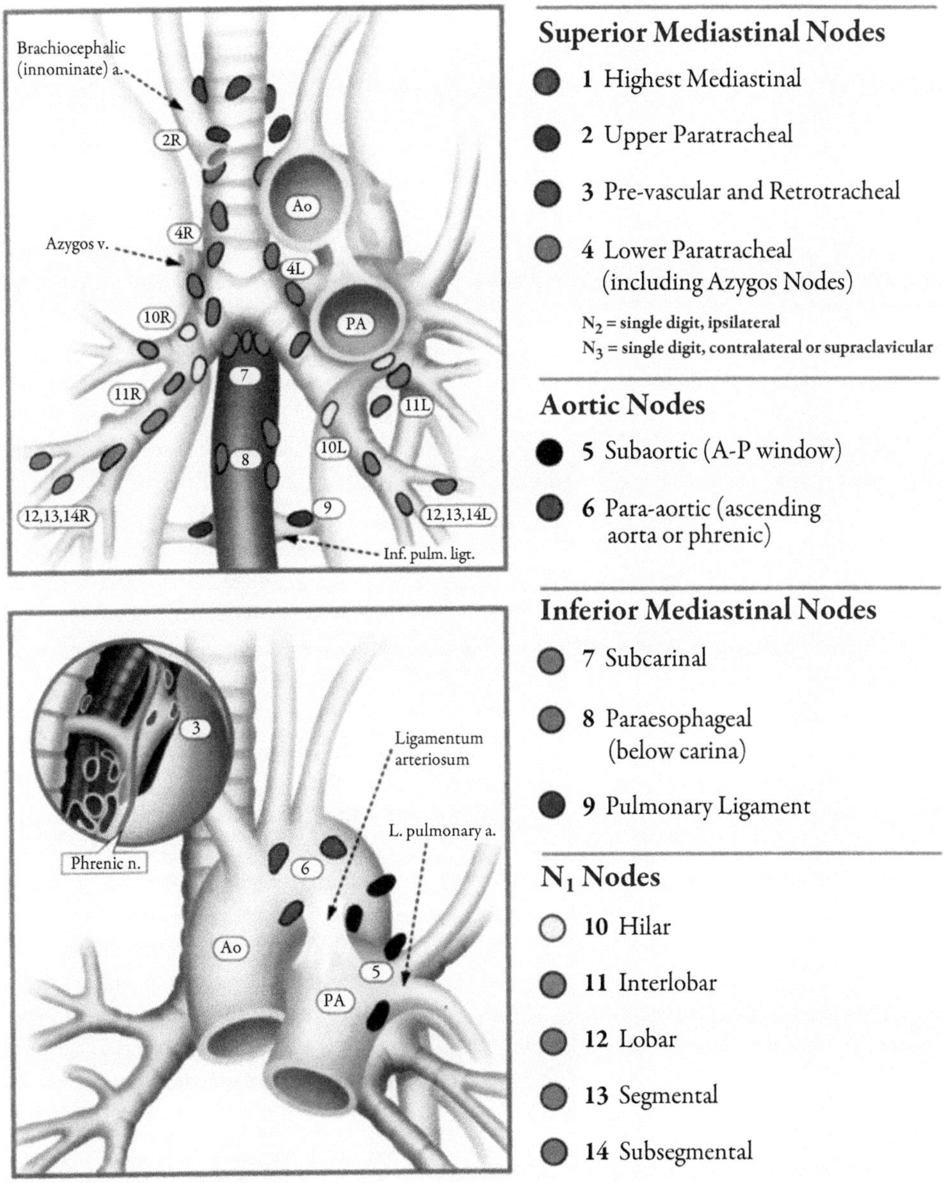

Figure 29.4 Mediastinal and hilar lymph node stations.

- **DIAGNOSIS**
  - This may be made by screening studies, incidentally on imaging obtained for other purposes, or after presentation with cough, hemoptysis, chest pain, or shortness of breath.
  - Since lung cancer is often asymptomatic, low dose CT screening is offered to patients at risk. These include individuals 50–75 years of age who have a large smoking history and currently smoke or have quit less than 15 years ago.
    - Homogeneous, smooth solid nodules >8mm need serial imaging if the patient has a low pretest cancer risk, or biopsy if high pretest cancer risk.
    - Solid nodules < 8 mm need serial imaging only if the patient has a high cancer risk.
    - Non-solid ground glass nodules >5 mm need serial imaging or biopsy depending on the pretest cancer risk.
    - If an old CT is available, all such lesions should be compared to note stability or growth in the intervening time period.
    - Lesions >3 cm spiculated, stippled, with eccentric calcification in smokers are more likely primary malignancies.
    - Multiple smooth, round nodules or masses are likely metastases from colon, renal, sarcoma, melanoma, ovarian, or endometrial origin.
- **CT chest and abdomen pelvis** establishes a T & N stage. **CT-PET** establishes M stage.
  - **T2 lesions** are 3–5 cm in size or involve the visceral pleura, mainstem bronchus, or result in partial lobe collapse.

- **T3 lesions** are 5–7 cm in size or involve the chest wall, pericardium, or phrenic nerve. Any satellite nodule in the same lobe would also qualify as T3.
- **T4 lesions** are 7 cm or greater, invade cardiovasculature, diaphragm, or mediastinum (including trachea, esophagus, or vertebral body). The presence of a satellite lesion in a separate lobe on the same side also qualifies as T4.

- **Biopsy** of the primary can be accomplished in a number of ways.
  - Bronchoscopy is best for a central tumor or nodes. **Endobronchial ultrasound (EBUS)** is used to guide transbronchial biopsy via fine needle aspiration (FNA) or core needle biopsy (CNBx). Direct or transbronchial forceps biopsy is another option for larger tumors invading the airway lumen.
  - Bronchoscopic **brushings or washings** can be diagnostic at times.
  - **Transesophageal endoscopic ultrasound (EUS)** guided FNA may be the best approach for tumors adjacent to the esophagus.
  - **Transthoracic CT guided** FNA or CNBx is the preferred initial method of sampling peripheral tumors
  - **Thoracoscopic** excisional biopsy (EXBx) by wedge resection or forceps are utilized when other approaches are not feasible or have yielded inadequate tissue. CT guided localization with a wire or dye injection may be necessary prior to a thoracoscopic biopsy.
- **N1** status includes hilar nodes. **N2** is subcarinal or ipsilateral mediastinal nodes. **N3** includes contralateral mediastinal nodes.
  - **Cervical mediastinoscopic or EBUS guided biopsy** is indicated for any nodes in stations 2, 4, or 7 that are larger than 1 cm on CT, or for any nodes seen in the setting of a primary that is T2 or T3, since there is a greater chance of mediastinal nodes harboring malignancy. Any avid node on CT-PET should also be biopsied.
  - **Anterior mediastinotomy** (Chamberlain procedure) or **thoracoscopy** are options to biopsy station 5 or 6 nodes in the aortopulmonary window.
  - **Transesophageal EUS** FNA may be necessary to sample nodes in the peri-esophageal area that are not accessible by the methods above.
  - **M** status is positive if there is a malignant effusion, tumor in a contralateral lung, or distant metastases.
- **TREATMENT**
  - **Neoadjuvant chemotherapy** is considered for N2 disease, poorly differentiated tumors, or certain T2 or T3 primaries with somatic or visceral invasion (for example a Pancoast tumor). Surgery can be offered to N2 patients after neoadjuvant therapy if a low number of stations were positive and had minimal malignant burden each. Resection may include nerves or other involved somatic structures.
  - Surgical options can be performed thoracoscopically, robotically, or via thoracotomy.
    - A dual lumen tube, or single lumen tube with bronchial blocker, is placed, and the lung on the operative side is allowed to passively collapse as the nonoperative side is isolated and ventilated.
    - For thoracoscopy, a camera port is placed in the mid-axillary line at the 7th–8th interspace. Working ports are placed posteriorly and anteriorly in upper rib spaces, with the anterior port serving as the extraction site. CO_2 insufflation of the pleural space can facilitate decompression of the lung on the operative side.
    - **Wedge resection or segmentectomy** are reserved for excisional biopsies, peripheral in situ disease, peripheral low-grade malignancies, or in certain cases for patients who cannot tolerate a lobar resection.
    - **Lobectomy** is the most common treatment for lung cancer and involves controlling the lobar arterial and venous channels as well as stapling across the lobar bronchus and the neighboring fissure. The most common complication is atelectasis or an air leak from either the portion of stapled parenchyma (generally resolves) or the lobar bronchus stump (may require re-intervention).
    - **Pneumonectomy** requires a thoracotomy and controlling the pulmonary artery and veins before stapling across the mainstream bronchus. This can be complicated by arrhythmias or a bronchial stump leak.
    - **Persistent air leak** after lung resection lasts longer than 3 to 5 days. It is treated by placing the chest tube to water seal, using a one-way valve, instituting chemical pleurodesis or an autologous blood patch into the pleural space, bronchoscopic plugging of the airway corresponding to the segment of leaking lung parenchyma, and surgical revision of parenchymal or bronchus staple line with or without a pleural tent or buttress.
  - **Stereotactic radiation** is an alternative for patients who are not surgical candidates and have a lesion more than 2 cm from the proximal bronchus.
  - **Radiofrequency ablation (RFA)** is an alternative for poor operative candidates who have lesions less than 3.5 cm and no hilar spread, parietal spread, or nearby major vessel.
  - **Adjuvant chemotherapy** is for N1 or greater and also can be considered for certain T3 lesions.
  - Recurrence most commonly presents as metastases to the brain, opposite lung, liver, or adrenals within 3 years of treatment.

OTHER LUNG MALIGNANCIES

- **Neuroendocrine** tumors comprise 20% of all lung cancer.
  - **Carcinoid** tumors are central and vascular. They may protrude into the bronchial lumen and produce serotonin. They can be localized by an octreotide scan.
  - **Atypical carcinoid** tumors are peripheral and large-cell. Both can be treated by resection and lymphadenectomy.
  - **Small-cell** are aggressive neuroendocrine tumors, generally central in location, and can cause superior vena cava (SVC) syndrome, hypertrophic pulmonary osteoarthropathy (HPO), Lambert-Eaton myasthenia, neuropathy, or inappropriate release of ACTH or ADH. Treatment is with cisplatin and etoposide. Rarely, radiation is used if the disease is limited to one hemithorax; Five-year survival is less than 5%.
- **Salivary gland tumors** are central, slow-growing, and/ rarely metastasize. Adenoid cystic cancer is from a submucosal gland and spreads along perineurial lymphatics. Treatment is with resection and radiation. Mucoepidermoid and mucous gland adenomas are treated by resection.

BENIGN LUNG PATHOLOGY

- **Hamartoma** is the most common benign tumor and presents as asymptomatic popcorn calcifications in a fat density lesion on CT. Serial CT scans can ensure no significant growth.
- An **abscess** is secondary to aspiration of mixed oral flora most commonly in the superior segment of the right lower lobe. It may take 1–2 weeks to cavitate; Treatment is with antibiotics alone. In time, the abscess will spontaneously drain into its associated bronchus. Lung resection is reserved for large, unresolving abscesses, or if there is a concern for malignancy. See the Chapter 3, Micro-organisms and Antimicrobials, for other lung infections including tuberculosis and fungal pathogens.
- **Massive hemoptysis** is defined as more than 600 ml of blood per day, or bleeding that results in hemodynamic changes. Etiologies include bronchiectasis, infection, and trauma.
  - The source is usually bronchial artery bleeding, and death is due to bloody aspirates disseminated across multiple lung segments resulting in impairment of alveolar gas exchange. Death is generally not via hemorrhagic shock.
  - Treatment is by **placing the pathologic side down,** and isolated intubation of the opposite side. Flexible bronchoscopy with **epinephrine injection or saline iced lavage** can be initial measures. A **bronchial occluder** can isolate the bleeding segmental bronchus so that other parts of the lung are not flooded with blood. **Rigid bronchoscopy** may allow for topical control of hemorrhage. Catheter-based **embolization** of the culprit bronchial artery is the next step. Some patients need a **lobectomy of pneumonectomy** for definitive control of hemorrhage.
  - Other less common causes of hemoptysis include:
    - 1. Rasmussen aneurysms of the pulmonary artery in association with a neighboring tuberculous cavity.
    - 2. Arteriovenous malformations (AVM). Other presentations for AVM include paradoxical venous origin emboli to the brain, and brain abscesses from transpulmonary bacterial transit bypassing alveolar macrophages.
- **Broncholiths** are secondary to external granulomatous invasion into the airway lumen and require surgical removal or lung resection.
- **Bronchostenosis** can be secondary to chronic fibrosis, with narrowed bronchial airways obstructing airflow and leading to downstream alveolar insufficiency.
- **Bronchiectasis** is pathologic dilation of bronchi due to cartilage damage from infection, tumor, cystic fibrosis (CF), poor ciliary function, chronic granulomatous disease, or alpha-1 antitrypsin deficiency. Treatment is generally conservative, with parenteral or inhaled antibiotics and mucolytics.
- See the Chapter 25, Pediatrics, for intralobar and extralobar **sequestration.**

PLEURA

- The pleural space contains 10–15 ml of physiologic fluid at any one time. Fluid is produced by the parietal layer and absorbed by the visceral layer.
- **Simple transudative effusions or non-infected exudates** can be treated with observation, diuresis, or thoracentesis posteriorly between the scapular tip and posterior axillary line.
- **A parapneumonic effusion** resides adjacent to infected lung tissue. If <10 mm thick on a decubitus film, it can be managed with antibiotics alone. If >10 mm thick, it needs pleural drainage with a small catheter in addition to antibiotics for pneumonia. These effusions generally have a normal pH and no organisms early on. **Complicated** parapneumonic effusions have signs of infection including low pH, low glucose, and organisms on Gram stain or culture.
- **Empyemas** are complicated parapneumonic effusions with purulence and >500 WBC.
  - **Exudative phase** is during the first week and may have a thickened pleura and minimal loculations. The pH of fluid will be low, and organisms will be present. It is treated with antibiotics and a chest tube. If the effusion is not completely drained by a chest tube,

chemical deloculation can be attempted with **10 mg TPA and 5 mg dornase** into the chest tube twice daily for 3 days. The tube should be clamped for 1–2 hrs after each administration.

- **Fibroproliferative phase** is during the second week. The pleural space becomes more loculated, generally necessitating **video assisted thoracoscopic (VATS) deloculation.**
- **Organized phase** is during the third week and generally needs **open decortication** to manually remove loculations and a thickened pleural rind. Ideally, this will allow the underlying lung parenchyma to expand and fill the thoracic cavity. If the lung is chronically entrapped and non-re-expanding, open drainage into the environment via a surgically constructed wound may be necessary to allow clearance of the infection (Eloesser flap). Since the lung is frozen in capacity and the visceral perimeters of the pleural cavity fixed, the lung will not collapse on exposure to atmospheric pressure.

- A **chylothorax** is characterized by lymphocytes and long-chain triglycerides within a "milky" pleural effusion. The cisterna chyli collects lymph from mesenteric and visceral abdominal channels. It sits just below the aortic hiatus (L2) and passes through the aortic diaphragmatic hiatus at T12 to become the thoracic duct along the patient's right posterior mediastinum. A right-sided chylothorax results from pathology below thoracic level 5 or 6, and a left-sided chylothorax is from pathology above that level. Note that if a patient is NPO or taking only clear liquids, the effusion may not appear chylous despite a lymphatic origin. Half are secondary to trauma or surgery, and half are secondary to malignancy, commonly lymphoma.
  - The diagnosis is suggested by noting a "milky" appearance of a pleural effusion, measuring a fluid triglyceride level above 110, noting an excess of lymphocytes, and a positive Sudan red stain.
  - Management depends on the daily output.
    - Less than 500 ml/day chylous output from a pleural drain can be treated with a **medium chain triglyceride diet or nonfat diet.**
    - 500–1000 ml/day chylous output can be treated by making the patient **NPO and starting total parenteral nutrition (TPN) with or without octreotide.**
    - If >1000 ml/day despite the above measures, then treatment should include:
      - 1. If secondary to trauma or surgery, **direct ligation** of the thoracic duct or the culprit leaking tributary within the surgical field if possible. This would include cervical re-exploration if that was the initial operative site.
      - 2. Otherwise, **embolization** or **ligation of the thoracic duct in the right lower chest** if not accessible at the site of leakage.
      - 3. **Talc pleurodesis** if the leak is secondary to a malignancy.
- **Recurrent pleural effusions** are treated with the following:
  - 1. **Mechanical or talc VATS pleurodesis or bedside doxycycline pleurodesis**, provided that the lung can re-expand.
  - 2. A **chronic drainage tube** for lungs that cannot expand.
  - 3. An **open drainage procedure** if the lung is trapped under a visceral pleural surface scarred in place. In this situation, the lung cannot collapse further upon exposure to atmospheric air.
  - In recurrent **malignant** effusions, cytology is often negative for SCC, lymphoma, or mesothelioma. It can be positive in up to 50% of cases of adenocarcinoma. Systemic imaging and bronchoscopy should be pursued to identify a primary tumor in the setting of recurrent effusions that are suspected to be malignant. VATS biopsy of suspicious pleural-based lesions can be diagnostic of primary site or metastatic malignancy.
- **Pneumothorax (PTX)** is a collection of environmental air within the pleural space. It generally arises from the alveolar lumen via a tear in the visceral pleura. However, occasionally the air is introduced during the inspiratory phase directly through a breach in the tracheobronchial tree or the chest wall.
  - **Primary, spontaneous PTX** occurs without any underlying lung pathology. Patients are generally young and tall. They are more commonly right-sided and originate from a ruptured apical visceral pleura atop a bleb. There is a 20% recurrence rate, and a 60% chance of having a third PTX after the first recurrence. The chance of having a 4th PTX is 80% after that point.
  - **Secondary spontaneous PTXs** are from bronchiolar obstruction or expiratory collapse causing alveolar air-trapping and resultant rupture of the visceral pleural lining. Secondary spontaneous PTX occurs in the setting of asthma, COPD, or airway foreign bodies. Treatment includes medical management of the primary bronchiolar problem in addition to pleural drainage.
  - Other causes of **secondary *non-spontaneous* PTX** are penetrating or blunt trauma, ventilator-induced barotrauma, or postsurgical resection. Catamenial PTXs are secondary to visceral pleural endometrial implants and pleuroperitoneal foramina.
  - Patients may present with pleuritic unilateral chest pain, decreased breath sounds, and hyper-resonance to percussion. Pleural air can be confirmed on chest x-ray (CXR) or CT scan.

- Management depends on hemodynamic stability, imaging characteristics, and the perceived likelihood for spontaneous healing or future recurrence.
  - Hemodynamic instability implies that enough air has built up to cause mediastinal shift and impair caval drainage into the right atrium. These patients should receive **immediate tube thoracostomy without imaging.**
  - If the patient is stable and the PTX is small and <3 cm distance from the cupola, one may **observe and repeat a CXR** in 6 hours to ensure stability. Patient then can be discharged if stable. Otherwise, if the PTX is large or growing on serial CXRs, then perform **tube thoracostomy** drainage to allow the lung to re-expand and the visceral pleural lining to heal as it scars against the parietal pleura. The chest tube can be transitioned from suction to water seal as this is accomplished.
  - **Thoracotomy or VATS blebectomy, or wedge resection and mechanical pleurodesis** should be considered if a patient has a primary spontaneous PTX that is recurrent, bilateral, or developed a tension phenomenon at any point in the past. Those patients who do not meet the above criteria but have an unresponsive air leak for >3 days, or non-re-expansion of lung after tube thoracostomy, also qualify for surgical intervention aimed at removal of the culprit lung and pleurodesis. Additionally, surgery can be considered for patients in a high-risk profession or those living in a remote area.
  - A PTX associated with a **postoperative prolonged air leak** beyond day 3 can be treated with water seal, Heimlich valve, chemical pleurodesis, autologous blood patch into the pleural space, bronchoscopic plugging of the culprit airway segment, or surgical revision of a parenchymal or bronchus staple line.
- **Hemothorax** is a collection of blood in the pleural space after trauma or surgery. A 28F chest tube should be placed in most instances, especially if imaging suggests >200 ml of blood. If not completely resolved after one chest tube, a VATS evacuation should be performed within 1 week, before the fibrinous phase sets in.
- **Mesothelioma** is a pleural-based malignancy seen 20–50 years after asbestos exposure. It is believed to start as parietal pleural plaques that coalesce. Most cases are locally advanced and metastatic at the time of presentation. Pleural fluid cytology is diagnostic in only 10% of patients, so incisional or excisional biopsy of a lesion is usually necessary. Treatment is with extrapleural pneumonectomy, extrapleural decortication, intraplerual chemotherapy, or talc palliation.
  - Less commonly, a single benign fibrous visceral pleural nodule may be present. This can be diagnosed by VATS pleural biopsy and treated with local resection.

MEDIASTINUM

- The **superior mediastinum** is above the pericardial reflection. The **inferior mediastinum** is divided into three sections. **Anterior** mediastinum lies in front of the pericardium and contains the thymus and lymph nodes. The **middle** mediastinum contains the pericardium and heart. The **posterior** mediastinum lies behind the pericardium and includes the descending aorta, esophagus, and spinal canal with neural structures. The thoracic duct crosses from right to left in the posterior mediastinum at the T4–T5 level. *See Figure 29.5.*
- **Median sternotomy** provides the best exposure for the brachiocephalic vein, right brachiocephalic artery (BCA), the proximal right or left common carotid, and ascending aorta and arch. A high left anterior thoracotomy is the best exposure for the proximal left subclavian artery
- **Left posterior thoracotomy** is for exposure of the descending aorta.
- **Right posterior thoracotomy** provides access to the esophagus, distal trachea, carina, right bronchus, or proximal left bronchus.
- **Mediastinal masses** may reside in the anterior, middle, or posterior sections.
  - Half of mediastinal masses are metastatic from the lungs, gonads, or lymphoma.
  - The other half of mediastinal masses are primary, and mostly thymic in adults and neurogenic in pediatrics. Half of primary tumors are malignant, often producing local symptoms.
  - Patients may have central chest pressure or a sensation of fullness. Others will be asymptomatic. CXR can show abnormalities to the mediastinal contour. CT scan is diagnostic of a mass.
  - Biopsy options include:
    - **CT guided transthoracic needle biopsy** as an initial step.
    - **Cervical mediastinoscopy** to assess lymph nodes in stations 2, 4, and 7.
    - **Anterior mediastinoscopy** (Chamberlain procedure) for access to stations 5 and 6.
    - Less commonly, a thoracotomy, thoracosternotomy, or thoracoscopy are required for a **surgical biopsy**.
  - **Thymoma** is the most common anterior mediastinal tumor. Over 50% are symptomatic, and 50% have myasthenia gravis (MG), while only 10%–20% of

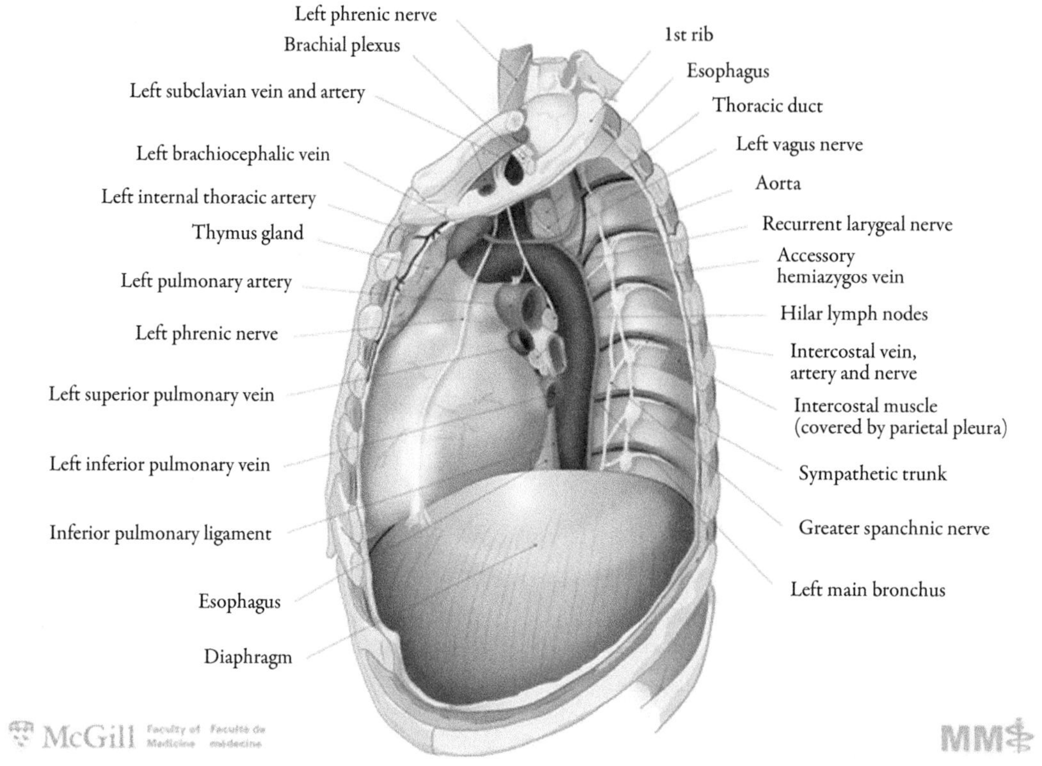

Figure 29.5 Anatomy of the mediastinum.

MG patients have a thymoma. MG is not associated with thymic carcinoma or neuroendocrine thymic tumors.

- Thymic lobules are divided by inward extension of the fibrous capsule. Each has outer packed **thymocytes (T lymphocytes)** and **epithelial cells** from which thymoma, thymic carcinoma, or neuroendocrine tumors are derived. The central medulla has glandular cells pending differentiation. All cells are replaced by lipid-laden macrophages with aging.
- Phrenic neurovascular bundles run laterally. Perfusion is from bilateral internal thoracic arteries and sometimes thyrocervical trunk tributaries. Drainage is to the innominate vein superiorly.
- Diagnosis is by CT scan. Stage 1 and 2 tumors are localized. Stage 3 is invasion into adjacent lung, pericardium, or great vessels. Stage 4 have distant pericardial and pleural implants or hematogenous or lymphatic spread. PET-CT can determine if distant masses are likely to be metastases.
  - No biopsy is needed for tumors that are <5 cm or cystic, or those masses that are definitively diagnosed by CT as an encapsulated thymoma. Otherwise, **CT guided transthoracic biopsy** is the first step to determine histology and architecture.
- Tumors may be **resected transsternally or thoracoscopically** without a biopsy if the CT is diagnostic of a thymoma Stage 1 or 2, and the mass is small, well encapsulated, and involving only intrathymic tissue or parathymic fat.
- **Neoadjuvant radiotherapy or chemoradiotherapy** should precede resection or debulking for Stage 3 and locally advanced Stage 4 cases. Patients with hematogenous or lymphatic spread are also Stage 4; however, they are treated with definitive chemoradiotherapy.
- Bilateral thoracoscopic port placement can assist with resection.
- The entirety of thymic tissue and anterior mediastinal fat should be resected caudally to cranially. If it is thymic carcinoma, the perithymic, intrathoracic, and cervical nodes should be removed. One phrenic nerve may be resected along with the specimen. Both pleural spaces should be inspected for drop metastases.
- Any myasthenia patient refractory to medical management can qualify to have their residual thymic tissue, hyperplasia, or thymoma removed. Anesthesiologists should avoid neuromuscular junction (NMJ) blockers. Preoperatively, anti-cholinesterases may be stopped, to decrease secretions and cholinergic crisis.
- Prognosis for thymoma is based on **T stage and achieving an R0 resection. Adjuvant radiotherapy** is used for R1 resection.
- Local recurrences can be managed surgically. Medical management is best for distant recurrence.

- **Germ cell tumors** are the second most common type of anterior mediastinal tumor.
  - Half are **mature teratomas**, which are benign. They are more common in pediatrics and contain ecto, meso, and endodermal tissue with a heterogeneous and calcified appearance on CT. AFP and BHCG are negative. Treatment is with resection.
  - **Seminomatous tumors** are the most common malignant germ cell tumor. Patients are generally asymptomatic young males. Diagnosis is by CT imaging and biopsy. One must exclude a gonadal primary. Treatment is with **chemotherapy (bleomycin, etoposide, cisplatin)** with or without radiation. Most regress, and there is an 85% survival at 5 years. Response to therapy can be assessed with PET-CT and patients can have remnant teratoma, fibrosis, or residual cancer. A second round of medical therapy may be indicated for residual disease or recurrence. Surgery is no longer the mainstay of treatment.
  - **Non-seminomatous germ cell tumors** are generally symptomatic and present in young males. AFP and BHCG are elevated. Diagnosis is by CT and image guided biopsy. Treatment is with **cisplatin-based chemotherapy and surgical resection for any residual mass.** Adjuvant radiation is sometimes included. Follow AFP and BHCG to normalization postoperatively.
- **Lymphoma** represents 20% of anterior mediastinal tumors. Non-Hodgkin's lymphoma can be large B cell, lymphoblastic, or other types. Hodgkin's lymphoma has a better prognosis. Ann Arbor staging system and the presence or absence of B symptoms help evaluate the extent of disease. Diagnosis is by CT and PET-CT needle biopsy of FDG avid areas. Treatment is with chemotherapy and radiation. Any residual FDG avid mass should be biopsied to ensure it is negative for tumor.
- **Thyroid goiter** represents 15% of anterior mediastinal tumors. It may be primarily substernal or an extension of a cervical goiter. Both should be resected via a cervical incision with manubriotomy if needed.
- **Middle Mediastinal Masses**
  - **Lymphadenopathy** may be lymphoma, metastatic lung cancer, sarcoidosis, or fungal infection. Diagnosis is by biopsy via cervical mediastinoscopy, anterior mediastinoscopy, or endoscopy. Treatment varies by etiology.
  - **Mediastinal cysts** can be **bronchogenic, esophageal** (present distally in the middle mediastinum), or neuroenteric. **Pericardial cysts** are at the right anterior cardiophrenic angle and can be left alone if asymptomatic and not enlargening.
  - **Angiofollicular lymphoproliferative disorder** (Castleman disease) is in HIV patients. It is sometimes associated with HHV8. Treatment is resection if unicentric, or multimodal chemoradiotherapy and immunomodulation if multicentric.
- **Posterior mediastinal masses** are neurogenic and more common in pediatrics. Diagnosis is by CT guided biopsy if not definitive by imaging. MRI is best to delineate any neural foraminal involvement.
  - **Schwannoma (neurilemoma)** should be surgically resected. An intraspinal dissection may be necessary. **Neurofibromas** can be managed conservatively. 40% are associated with neurofibromatosis (NF1). Imaging characteristics and an aggressive growth rate may suggest malignant sschwannoma or neurofibrosarcoma.
  - **Ganglioneuromas** are benign and derived from cell bodies of the sympathetic chain. **Ganglioneuroblastomas** are less well differentiated. **Neuroblastomas** are undifferentiated, malignant, hormonally active, and more common in children less than 2 years of age. See Chapter 25, Pediatrics.
  - **Paragangliomas** include mediastinal pheochromocytomas and chemodectomas. They are malignant and hormonally active. Localization is based on CT imaging or nuclear medicine studies.
  - Other tumors in the posterior mediastinum include lipoma, liposarcoma; cavernous, capillary, or venous hemangiomas; lymphangioma, fibroma, or fibrosarcoma.
- **Acute mediastinitis** is generally secondary to sternotomy, spontaneous or iatrogenic esophageal perforation, or extension from a cervical or oropharyngeal infection. Post-sternotomy risk factors for wound infection and subsequent mediastinitis include poorly controlled diabetes, smoking, elderly, obese, bilateral internal thoracic artery harvest, and a long cardiopulmonary bypass time.
  - Patients present with chest pain and a systemic inflammatory response. There may be drainage from a sternal wound or evidence of sternal dehiscence. CT scan will show mediastinal fluid collections and air. When suspected as the source, extravasation from the esophagus can be seen if enteric contrast is given.
  - Management is with broad antibiotics and a cervical approach for debridement if the process is limited to the superior mediastinum above the carina or T4. Otherwise, the patient will need a thoracotomy for adequate drainage of infected tissue, and addressing an esophageal origin if one exists. Muscle flap coverage may eventually be needed.
- Spontaneous **pneumomediastinum** or minimal post-traumatic pneumomediastinum can be observed if believed to be alveolar in origin. Otherwise, if vomiting, fever, or mediastinal or pleural fluid is present, then perform an esophagram, as the patient may need management per mediastinitis (above).
- **Chronic, fibrosing mediastinitis** is secondary to tuberculosis, histoplasmosis, radiation, or autoimmune disease. This can encase and impair mediastinal organ

function, including the superior vena cava (SVC), trachea, or esophagus. Treatment is aimed at the infectious or immune etiology.

- **Superior vena cava (SVC) syndrome** is secondary to small-cell lung cancer invasion, luminal stenosis from central lines, or fibrosing mediastinitis. Treatment is with a stent and radiation therapy if malignant.

CHEST WALL

- **Chondromas** are benign cartilaginous tumors at the anterior costochondral junction.
- **Chondrosarcoma** is the most common malignant tumor in the chest wall, and generally painful.
- **Fibrous dysplasia and fibrosarcoma** are at the posterolateral rib, and sometimes present with pain.
- **Osteochondroma** is a benign tumor originating from the rib cortex in adolescents.
- **Osteosarcoma** is the most common bone malignancy overall, and presents with pain and a sunburst appearance on x-ray.
- **Ewing's sarcoma** emanates from a rib and appears as an onion peel lesion on imaging.
- Other sarcomas include malignant fibrous histiocytoma (MFH), liposarcoma, fibrosarcoma, leiomyosarcoma, rhabdomyosarcoma, and synovial sarcoma.
- **Desmoids** arise from fascia and fibroblasts, and may evolve within a prior scar.
- Lipoma, hemangioma, schwannoma, and neurofibromas can also present in the chest wall.
- Treatment is with **wide local excision (WLE)** of the mass, oftentimes including a normal rib above and below. When necessary, the diaphragm may be transected along lateral margins to preserve phrenic neurovasculature, and the diaphragm reconstructed. **Neoadjuvant chemotherapy** can be used for osteosarcoma and Ewing's sarcoma. In desmoids, positive margins are acceptable in order to preserve function. **Radiation** may be an adjunct. Isolated metastases from osteosarcoma may be resected.
    - Defects >5 cm or those that require removal of 4 or more ribs need some form of **chest wall stabilization**. This can include **synthetic mesh** (polypropylene mesh, PTFE, or methyl methacrylate) provided there is low infection risk. Otherwise bioprosthetic mesh may be used. The mesh is then covered by a **myocutaneous or myofascial pedicled or free flap**. Options include pectoralis major, latissimus dorsi, rectus abdominis, or external oblique.
    - Defects <5 cm or consisting of 3 or fewer ribs can be reconstructed with only autologous tissue including **iliac crest, another rib, or fibula** if stabilization is needed. This is covered by a fascial graft or myocutaneous or myofascial pedicled or free flap. If no stabilization is required, then these smaller defects can simply be covered with any of the above pedicled or free flaps.
    - Biologic mesh is used for young children.
    - Prognosis of chest wall tumors is based on **grade and chemo responsiveness.**

LUNG TRANSPLANT

- Indications for lung transplant include advanced COPD, interstitial pulmonary fibrosis, pulmonary arteriopathy, and CF.
- Qualified recipients should have a life expectancy <2 years, low FEV1 and DLCO, limited exercise tolerance, dependence on home O_2, or pulmonary hypertension refractory to medical management.
- **Single lung** transplant is most often performed. **Double lung** transplants, however, are more common for young patients and those with CF, alpha-1 antitrypsin deficiency, and primary pulmonary hypertension.
- Donor organs can be stored for 6 hours in cold solution. The donor bronchus should be cut just proximal to its lobar bifurcation. Vessel loop occlusion of the pulmonary artery (PA) can determine if cardiopulmonary bypass (CPB) or extracorporeal oxygenation (ECMO) are needed. The bronchial anastamosis is performed before the PA, and pulmonary vein (PV) anastomoses. The implanted lung will have no bronchial arterial flow or major lymphatic drainage.
- Postoperatively patients should be maintained on **low FiO_2, diuresed,** and have **regular bronchoscope evaluations.** Bronchial strictures may be dilated.
- **Primary graft failure** is defined as infiltrates without another etiology and a low pAO_2:FiO_2 ratio. Donor aspiration, contusion, infection, severe hypoxemia, and prolonged organ ischemic time could have contributed.
- **Reperfusion injury** can lead to acute respiratory distress syndrome (ARDS) and is the main cause of early death.
- **Acute rejection** has perivascular lymphocytosis and is managed with enhanced immune suppression.
- **Chronic rejection** has bronchiolitis obliterans and is the main cause of late death.

30.

VASCULAR

ANATOMY AND PHYSIOLOGY

- Arterial and venous walls are composed of endothelium, intima, internal elastic lamina (IEL), media, external elastic lamina (EEL), adventitia, and vaso vasorum atop this. *See Figure 30.1.* Arteries have a more developed media. Veins are thinner, with less connective tissue, and an endothelial lining that forms valves.
- **Arterial intimal atherosclerotic plaque** is associated with hypertension (HTN), diabetes (DM), and hyperlipidemia (HLD). Macrophages deep to the endothelium absorb fat and lipids to become foam cells, and smooth muscle proliferation is initiated. Intimal plaque disruption results in thrombus formation, fibrosis and further luminal narrowing. A 50% reduction in lumen diameter corresponds to a 75% reduction in cross-sectional area.
- **Hypercoagulable states** can result in venous or arterial thromboses even in the absence of plaque formation. Acquired conditions are most commonly from **smoking, heparin-induced thrombocytopenia (HIT), and antiphospholipid syndrome (APS).** Inherited hypercoagulability is most commonly from **factor V Leiden resistance to protein C.** There are also **prothrombin defects** that result in enhanced coagulation. Hypercoagulability has a stronger association with bypass graft occlusion then does a history of diabetes, coronary artery disease (CAD), or end stage renal disease (ESRD).
- Endovascular Principles
  - Wires are categorized by their diameter, coating, stiffness, and tip angle.
  - Balloons are categorized by diameter and length and are used for dilation. They can be low profile for pre-dilation to bypass a tight lesion. The shoulder length of a balloon may injure adjacent normal endothelium.
  - **Self-expanding stents (SES)** are for long, tortuous lesions. They are generally oversized by 1–2 mm. Examples include nitinol bare metal stents (BMS) or drug-eluting stents (DES).
  - **Balloon expandable stents (BES)** are for short, rigid, straight lesions near an arterial orifice or bifurcation. PTFE may line the stent for its entire length, or the bare stent edges may be exposed for traction.
  - Stents may be placed in a subintimal tract for a chronic total occlusion (CTO) of the native lumen.
  - Stents need **Plavix for up to 6 weeks after deployment, and aspirin indefinitely.**
- Vascular injuries are described in Chapter 23, Trauma.

CAROTID ARTERY

- 85% of **cerebrovascular accidents (CVAs)** are ischemic. Most are associated with hypertension. If embolic, one-third are cardiac in origin, and about one-third are atheroembolic from plaque in the ipsilateral carotid artery. 10% of CVAs are small vessel, lacunar.
- **CAROTID ATHEROSCLEROTIC PLAQUE**
- This develops at low velocity, low shear stress areas on the lateral walls of the common carotid artery (CCA) as it bifurcates into the internal carotid artery (ICA) and external carotid artery (ECA). This leads to stenosis.
- **DIAGNOSIS**
  - Patients are discovered in one of four ways
    - 1) Symptomatic patients have an unstable plaque that embolizes distally into the cerebral circuit and produces either a transient ischemic attack (TIA) or CVA. The ischemic symptoms are usually secondary to emboli lodged in a distal cerebral artery, not from an absolute reduction in flow at the site of the stenosis. **TIA** are due to small quantities of embolic material and resolve within 2 hours. Most last in the range of 5–15 minutes. 25%–45% of these patients will also have a CVA within 5 years. Larger quantities of embolic material produce a CVA. 50% of patients will have another CVA within 5 years, and half of these will be fatal.
    - 2) Asymptomatic patients noted to have a **bruit** may receive a duplex ultrasound (US). A unilateral bruit could be from plaque narrowing the lumen of the ipsilateral carotid, or it could be from increased flow in the ipsilateral carotid due to a contralateral, flow-limiting, plaque.
    - 3) Asymptomatic patients may be noted to have carotid stenoses **incidentally on a neck imaging study** done for other purposes. This would prompt further evaluation by duplex US.

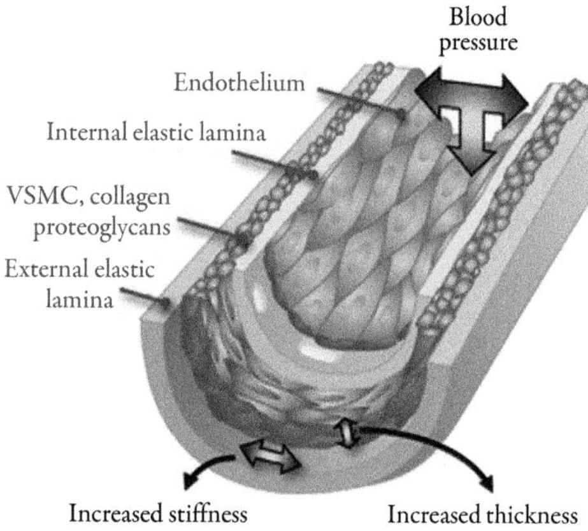

Figure 30.1 Arterial wall and adaptation.

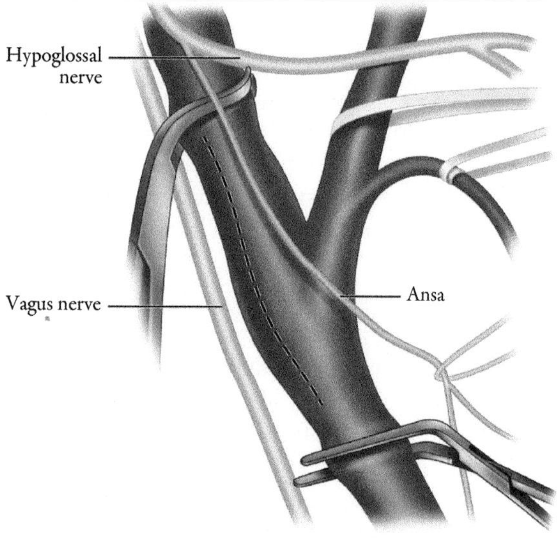

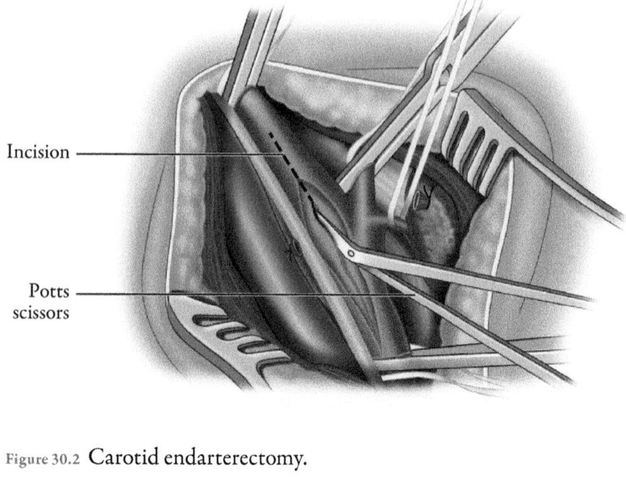

Figure 30.2 Carotid endarterectomy.

- 4) Asymptomatic patients in certain high-risk populations receive **screening carotid ultrasounds**, at which time the diagnosis is made.
- **Duplex US** can reveal both the degree of stenosis and the morphology and longitudinal extent of the plaque.
  - The **ICA has biphasic flow and a broad systolic peak** due to low distal resistance.
  - The **ECA has triphasic flow, and a narrow systolic peak** and dicrotic notch.
  - Velocities >230 cm/s during systole, >140 cm/s during diastole, or an ICA:CCA velocity ratio >4 signifies a >70% luminal reduction in that area of the carotid artery.
  - Echolucent, irregular, ulcerated, heterogeneous plaque may have higher lipid content or intra-plaque hemorrhage and therefore be less stable. This corresponds with a higher CVA rate at any given degree of stenosis compared to an echogenic, homogeneous plaque.
- Magnetic resonance imaging (**MRI**) of the neck and brain is another diagnostic option, but it may overcall stenosis. 10% of asymptomatic patients will have a CVA on MRI or CT. This then qualifies any ipsilateral carotid lesion as "symptomatic."

- **TREATMENT**
  - **NASCET (North American Symptomatic Carotid Endarterectomy trial)** recommends a CEA for symptomatic (hemispheric weakness or amaurosis fugax) carotid artery stenosis **>70%,** or a hemispheric TIA with stenosis **>50%** in an otherwise healthy candidate.
    - When symptomatic, the endarterectomy should be performed within 2 weeks unless major CVA changes are seen on the brain CT. CEA may be done more urgently for crescendo TIAs.
  - **ACAS (Asymptomatic Carotid Surgery trial)** recommends a CEA be done for **>80%** asymptomatic stenosis. An otherwise healthy asymptomatic candidate may receive a CEA at **>60%** stenosis if thel risk of stroke or death is estimated at <3%.
  - **Principles of Endarterectomy** *See Figure 30.2.*
    - Local or regional anesthesia has an improved cardiac stability compared to general anesthesia, but not necessarily a decreased myocardial infarction (MI) rate.
    - In the setting of bilateral stenoses, repair the more stenotic and dominant side first.
    - An incision is made along the anterior border of the sternocleidomastoid (SCM) or transversely within a skin crease.
    - Avoid the **greater auricular nerve** to the earlobe, and divide the facial vein.
    - Avoid the **hypoglossal nerve** sitting above the carotid bifurcation. This is the most common cranial nerve injured.

- One may need to ligate the SCM arterial branch from the external carotid artery (ECA) for mobilization.
- Identify the ansa cervicalis and vessel loop the ECA, superior thyroid artery (STA), and internal carotid artery (ICA).
  - **Exposing a high carotid bifurcation** may require nasotracheal intubation so that the mandible ramus is closed (and shifted more anteriorly). Other options are division of the posterior belly of the digastric, keeping in mind that the hypoglossal nerve runs just under this. Resection of the styloid process, subluxation of the mandible anteriorly, or a vertical ramus osteotomy are other maneuvers to help expose a high bifurcation. Care should be taken not to injure CN 11 or 9 in this high location.
- Preserve the **vagus nerve** to prevent recurrent laryngeal nerve (RLN) hoarseness, and the **mandibular branch of the facial nerve** to prevent lip droop.
- 70–100 u/kg of heparin should be given and allowed to circulate for 3 minutes.
- **A shunt should be used in any symptomatic patient who had a CVA or TIA.**
- For asymptomatic patients, the ICA can be **test clamped** to exclude electroencephalogram (EEG), somatosensory evoked potential (SSEP), transcranial Doppler (TCD) changes, or neurologic signs (in patients receiving local-regional anesthesia). Any of the above would prompt the use of a shunt in an asymptomatic patient.
  - If there are no such changes, then one can measure the **stump pressure** by unclamping ICA and clamping CCA and ECA. Use a needle to measure the pressure at CCA distal to clamps. If >50 mmHg, then retrograde flow via the circle of Willis and down the ipsilateral ICA is adequate, and no shunt is needed for the asymptomatic patient.
  - If the contralateral side is severely stenosed, one may still consider a shunt in spite of a negative test clamp and adequate stump pressures.
- Re-clamp ICA, CCA, and ECA in that order, and perform a **longitudinal arteriotomy** with or without a shunt placed distally into the ICA and then back-bled before being tucked into the proximal CCA.
- An **endarterectomy** is then done via a longitudinal incision starting proximally at CCA and coursing distally over the ICA, with or without tacking sutures to keep residual plaque flush with the arterial wall at interfaces cephalad and inferior to the endarterectomy site. An eversion endarterectomy can be done for the ECA.
- The arteriotomy should be closed with a **patch** and back-bled by releasing and re-clamping the ICA before the final suture. Then release ECA and CCA, and finally ICA clamps in that order.
  - Alternatively, an eversion endarterectomy of the ICA can be done via a transverse oblique transection just distal to the bifurcation. Roll the adventitia back, remove plaque, and re-anastomose end to end without a patch or shunt. The clamps can be back-bled by releasing in reverse order to purge the surgical site lumen of air and debris.
  - A **completion duplex US** should be performed in all cases.
- **Perioperative Considerations**
  - **Antiplatelets** and a **statin** should be given perioperatively.
  - Mortality is <1%. A postoperative CVA occurs in 1%–2% of previously asymptomatic patients, and 2%–6% of symptomatic patients. MI occurs in 2%–4% of patients.
  - If a **postoperative neurologic deficit is noted in the recovery room**, the patient should be immediately taken back to the OR. If a pulse is present, then do an on-table angiogram and possible thrombolysis, since the occlusion is cephalad of the operative site. Do not use a balloon embolectomy device in the cerebral circulation. If no pulse is present, then open the wound, re-explore, remove the thrombus, and revise the endarterectomy and patch.
  - A **post-op neurologic deficit at 12–24 hours** should be evaluated by noncontrast CT and CT angiogram for hemorrhage or embolic CVA.
  - **Neurologic symptoms at 48 hours or more postoperatively** may be hyperperfusion due to increased flow into a maximally dilated distal cerebral bed without autoregulatory vasoconstrictive capabilities. This can result in a CVA, seizures, headache, or confusion. CT scan should still be done to exclude hemorrhage. Treatment is with beta blockade.
- **Recurrent stenosis** is defined as >50% narrowing of the lumen diameter and occurs in about 10% of patients after CEA for asymptomatic disease. This is decreased in patch closure as opposed to primary closure.
  - If it occurs **within 1 month, it was a technical error.**
  - If between **1 month and 2 years**, it is probably secondary to **myointimal hyperplasia** in the endarterectomy bed. This is smooth and rubbery. It is unlikely to embolize and can be watched if asymptomatic, as it may regress. In the *rare* instance where the hyperplasia causes an absolute luminal reduction that leads to a TIA or CVA, one may consider carotid artery stent (CAS) for intervention (see below).
  - If it occurs **more than 2 years after CEA, then the etiology is new atherosclerotic disease.** This can lead to atherothromboembolism to the distal cerebral circuit.

- **Surgical treatment** can be considered if there is >50% reduction in lumen from a technical or atherosclerotic re-stenosis and the patient is **symptomatic.** Repeat endarterectomy and patch angioplasty or resection and interposition saphenous vein, or polytetrafluoroethylene (PTFE) are options. Re-operation has increased rates of death, MI, CVA, and cranial nerve injury.
- Most **asymptomatic recurrences are managed medically.**
- **Carotid artery stenting (CAS)** simply tacks the plaque against the arterial wall to prevent atherothromboembolism. It is for **symptomatic patients only** and may be preferable in the setting of prior irradiation, a tracheostomy, or a hostile neck not amenable to surgery.
  - It is not used in the asymptomatic or in elderly patients with high embolic potential.
  - Periopeartive **dual antiplatelets** are given, and aspirin is continued indefinitely.
  - **Embolic protection devices (EPD)** include a **distal filter** or **reversal of flow** with inflation of balloons in the proximal CCA and proximal ECA while deploying the stent.
  - Despite EPD, there is an increased **periprocedural CVA risk compared to CEA.** This is due to brittle carotid and aortic plaque morphology.
  - A balloon-expanding stent is rigid, with increased radial force, and best suited for great vessel orifices. A self-expanding stent is better for the cervical region.
  - Given that the native plaque remains plastered against the arterial sidewall, there will appear to be residual "stenosis" after this procedure and higher "re-stenosis" rates. Myointimal hyperplasia, the effects of the stent impairing arterial wall elasticity (resulting in faster US velocities), and the stent causing artifact on CT angiogram (CTA) or MRI can all also create the impression of re-stenosis.
  - **CAS may be equivalent to CEA for the treatment of recurrent symptomatic stenosis.** In-stent recurrent stenosis (ISR) after CAS should be managed with repeat endovascular intervention. CEA with stent removal and patch, or resection with interposition graft, are utilized if the repeat endoscopic intervention failed.
- **CAROTID ANEURYSMS**
  - These are secondary to atherosclerotic degeneration, trauma, iatrogenic procedural injury, extension from a neighboring local neck or oropharyngeal infection, hematogenous seeding of microorganisms, or an infection of a prior patched repair.
- **DIAGNOSIS**
  - A pulsatile mass may be noted, with the local mass effect of the aneurysm producing cranial nerve dysfunction or Horner's syndrome. Thromboembolic complications include TIA and CVA.
  - Clinical suspicion is confirmed by US, and the exact anatomic dimensions and extent can be visualized by CTA or MRA.
- **TREATMENT**
  - **Resection and a GSV (or SFA) interposition graft.** This has a 1%–10% CVA risk.
  - **Ligation** may be required for a patient in extremis with adequate distal stump pressures.
  - The external carotid (EC) can be transposed to the distal internal carotid (IC) via a transcranial route if needed.
  - An endovascular stent (covered or bare metal) is a second-line treatment if the patient is a poor operative candidate or the lesion is inaccessibly high or low. Embolization is an option if the aneurysm has a narrow neck.
- **CAROTID DISSECTION**
  - Typically occurs in younger patients along the internal carotid artery, 2–3 cm cephalad of the common carotid bifurcation. They can be spontaneous or associated with blunt neck trauma or torsion. The patient may or may not have a history of a connective tissue disorder.
- **DIAGNOSIS**
  - Patients have a history of neck pain, TIA, or CVA. Patients may have had prior overt or occult neck trauma.
  - Injuries are graded based on US or CTA appearance.
    - Type 1: Subintimal with <25% of the lumen involved.
    - Type 2: Subintimal with >25% of the lumen involved.
    - Type 3: Subadventitial with pseudoaneurysmal dilation.
    - Type 4: A completely occluded lumen.
- **TREATMENT**
  - **Antiplatelets or anticoagulants if asymptomatic.**
  - **Endovascular or surgical repair is reserved for symptomatic patients or those with Grade 3 injuries** at risk of progressive enlargement and rupture.
  - Complete traumatic carotid occlusion is generally not treated with surgery unless the injury is less than 2 hours old, and the patient has a preserved level of consciousness. Revascularization of older injuries can result in subsequent intracerebral hemorrhage.
- **OTHER CAROTID PATHOLOGY**
  - **Carotid Bulb Tumors** arise from neural crest cells. Most are asymptomatic and demonstrate Lyre's sign (splaying of the ECA and ICA on imaging). The tumor may encase nerves or neighboring vessels. Treatment is with resection via a peri-adventitial dissection, with or without preoperative embolization.
  - **Carotid Coil** pathology results in kinking and ischemic symptoms secondary to the patient's neck position.

- **VERTEBRAL ANEURYSMS OR DISSECTION**
  - These are managed medically with antiplatelets or anticoagulants in order to mitigate distal thromboembolization. Endovascular coil occlusion or stent is a second-line treatment.
  - Ligation is reserved for patients who have thromboembolic complications despite the above measures, provided a proximal and distal site on the vertebral artery are surgically accessible.

UPPER EXTREMITY VASCULATURE

- **UPPER EXTREMITY ARTERIAL ISCHEMIA**
  - This may present as claudication, paresthesias, motor deficits, or evidence of tissue loss due to distal digital embolization.
  - **Subclavian artery (SCA) stenosis** is from atherosclerosis (more common at the origin of the left subclavian), arterial thoracic outlet syndrome (TOS, see below), Takayasu's arteritis, giant cell arteritis, Buerger's disease, mixed connective tissue disorder, scleroderma, azotemic arteriopathy secondary to end stage renal disease (ESRD), antibody-mediated vascular inflammation, or primary vasospasm.
    - **Collaterals** around the subclavian and axillary artery include the thyrocervical trunk (via the suprascapular and transverse cervical arteries) and the adjacent, distal dorsal scapular artery (via the circumflex and subscapular arteries). These tend to preserve flow in the setting of acute or chronic subclavian artery occlusions, and allow for ligation or sacrifice of the subclavian artery if a patient is in extremis.
    - If chronically occluded at the origin of the left subclavian or proximally along the right subclavian, the vertebral artery may provide retrograde flow to the limb, stealing perfusion from the circle of Willis. The resulting **subclavian steal** produces neurologic symptoms with limb exertion.
  - **Emboli** to the upper extremity vasculature is most often cardiac in origin, and less often arterioembolic from mural thrombus of the brachiocephalic or subclavian artery.
  - Like the lower extremities, **trauma-induced ischemia** may be blunt, penetrating, iatrogenic or due to repetitive activity, as in the hypothenar hammer syndrome (HHS).
- **DIAGNOSIS**
  - Once upper extremity ischemia is suspected, compare segmental blood pressures (BP) in each extremity. This can determine the level of occlusion.
  - A CTA of the extremity or a duplex US will delineate the occlusion.
  - Isolated digital pressures under 70 mmHg, a gradient of more than 30 mmHg from wrist to digit, or interdigital gradients of more than 15 mmHg can indicate small vessel occlusion or vasospastic disorders
  - For subclavian steal, a vertebral artery US can show reversal of flow.
- **TREATMENT**
  - This depends on the level of the steno-occlusive disease
    - **Endovascular stents** are for left subclavian artery origin stenosis or subclavian artery trauma. Subclavian artery transposition to the side of the carotid, or a carotid to subclavian or axillary artery bypass with greater saphenous vein (GSV) or PTFE are other options with better long-term patency.
    - The brachial artery can be exposed for **repair, interposition, or bypass graft** by incising and dissecting along the medial and posterior groove of the biceps muscle.
    - Isolated radial or ulnar thromboses with adequate collateral flow do not need intervention, and may be **observed.**
    - HHS requires ulnar artery exposure at the hand, with resection of the disease segment and interposition vein graft. Balloon angioplasty, endovascular thrombectomy, or thrombolytics are alternative treatments.
    - Most digital small vessel disorders are treated medically with **antiplatelets, cilostazol, calcium channel blockers, or digital sympathectomy.**
- **THORACIC OUTLET SYNDROME (TOS)**
  - Neurogenic, venous, or arterial are the three types
  - Abnormal upper rib anatomy, a hypertrophied anterior scalene, thickened subclavius muscle, or prominent sternocostal and costoclavicular ligament can result in compression of neuromuscular structures as they exit the thoracic space to head toward the upper extremity. *See Figure 30.3.*
  - **Neurogenic TOS** is from compression of the lower brachial plexus and is the most common form. Patients have recurrent pain radiating in the distal extremity and distribution of the lower brachial plexus. They may have associated weakness. **Diagnosis** is by an **elevated arm stress test (EAST or Roos test)** and scalene block test or intrascalene botulinum toxin injection. **Treatment** is with **physical therapy.** Trans-axillary rib, scalene, and subclavius resection is only used for refractory cases.
  - **Subclavian venous TOS** is from chronic impingement of venous outflow with effort-induced edema and eventual thrombosis of the subclavian vein. Patients may have a diffusely swollen arm, worsened in the hyper-abducted position. **Diagnosis** is by US of the axillary and subclavian veins. **Treatment** is with **thrombolysis, anticoagulation,** and later **trans-axillary** rib, scalene, and subclavius resection.
  - **Subclavian artery TOS** is the least common. **Diagnosis:** It can present with upper extremity claudication and paresthesias. **Adson's test** reveals a decreased peripheral pulse on ipsilateral neck turn,

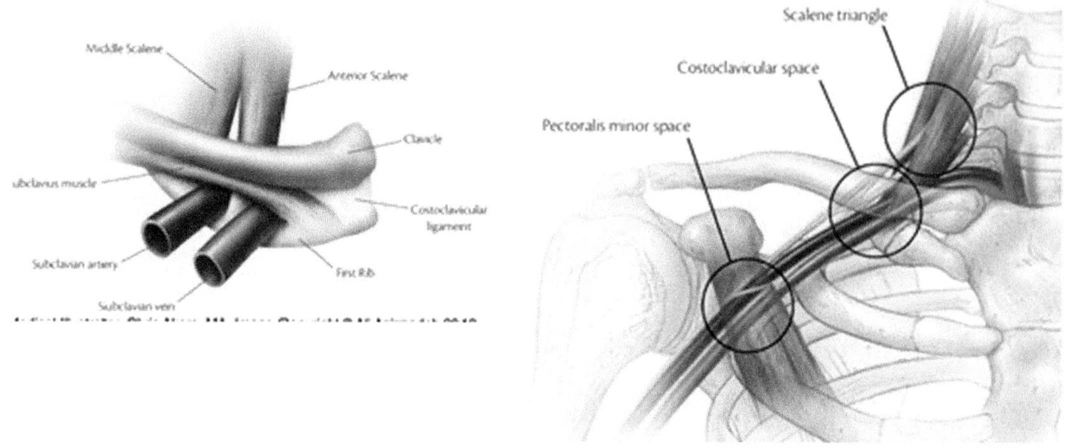

Figure 30.3 Sites of thoracic outlet neuromuscular impingement.

due to impingement of the artery by a hypertrophied anterior scalene muscle. Aneurysmal dilation of the subclavian artery can be seen distal to its impinged point. The aneurysm can be lined with mural thrombus and result in distal thromboemboli. **Treatment** is with a **supraclavicular approach for rib, scalene, and subclavius resection**. The subclavian artery may need to be repaired, and a separate lateral infraclavicular incision can help with distal axillary artery access.

- **ABNORMALITIES OF SCA ORIGIN**
  - **Right aortic arch with a retroesophageal left SCA** can lead to trouble swallowing solids (dysphagia lusoria) as the aberrant vessel courses posterior to the esophagus.
  - An **aberrant right subclavian artery** from the distal left (normal) aortic arch (0.5%) may travel retroesophageal, and is also associated with dysphagia lusoria, plus thoracic duct drainage into the R internal jugular (IJ) to superior vena cava (SVC) confluence, and a non-recurrent right laryngeal nerve.

HEMODIALYSIS ACCESS

- One should establish **autogenous distal access** when patients reach chronic kidney disease Stage 4 (CKD 4). Preoperative vein mapping may be needed if veins are not palpable with a tourniquet or the patient uses intravenous drugs, is obese, or had a prior peripherally inserted central line. A minimum vein diameter of **3 mm** and radial artery diameter of **2 mm** are preferred. Brachial artery diameter should be a minimum of **5 mm.**
- The order of preference for construction of an autogenous fistula are: radiocephalic, brachiocephalic, brachiobasilic, prosthetic radial-brachial vein straight graft, brachial artery-vein forearm prosthetic loop graft, brachial artery to axillary vein straight prosthetic graft, and lastly, axillary artery to vein upper arm loop prosthetic graft. *See Figure 30.4.*
- The procedure involves dissecting a neighboring artery and vein, ligating venous tributaries, and transecting the vein. Then the proximal open end of the vein is anastomosed to the side of the neighboring artery. The distal end of the vein is ligated. If a graft is used, it is usually sewn end to side at both the artery and vein sites. Aspirin and any anticoagulant can be continued throughout the operation, provided the INR is <2.5.

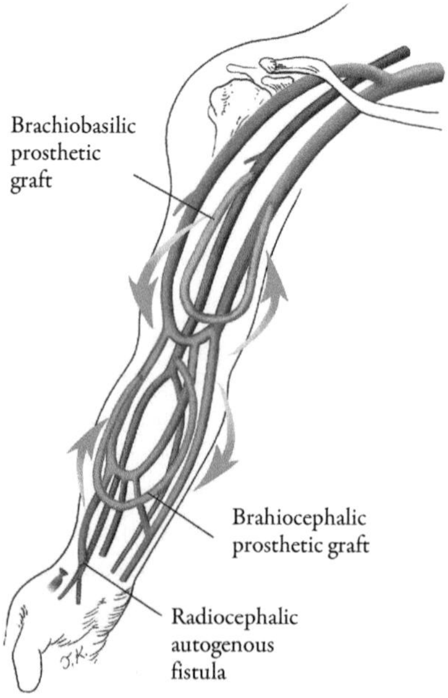

Figure 30.4 Hemodialysis access.

- An autogenous fistula is considered mature when a US demonstrates **>6 mm diameter** and flow **>400–600 ml/minute.** The fistula should ideally be located **within 6 mm of the dermal surface.**
- If **prosthetic graft** is needed because of lack of autogenous conduit, then the development of venous anastomotic intimal hyperplasia will limit the lifespan of the graft. So, only establish this access once the patient is actually hemodialysis dependent. Grafts are 6 mm diameter and may be used once fibrous scar forms at >10 days postoperatively.
- If there are no options for an upper extremity fistula or graft, then **lower extremity access** can be constructed as a looped prosthetic graft from femoral artery to vein, or femoral artery to greater saphenous vein.
- **Hemodialysis access with reliable outflow (HeRO graft)** is for patients with central venous stenosis and no other access options. Wire access is obtained across the central stenosis, and a tunneled catheter is placed into the right atrium. The tunneled proximal end of the catheter is attached to a PTFE graft. The other end of the PTFE is anastomosed to the side of the brachial artery or the venous end of a previously constructed fistula. This process bypasses the central stenosis that led to fistula failure in the first place, allowing the new graft to empty into the RA.
- **COMPLICATIONS OF VASCULAR ACCESS**
  - **Infection** is most frequent. Hemodialysis catheters and prosthetic grafts have the highest risk. Treatment is with removal of the infected line or infected portion of graft. A new graft would need to be tunneled in a different site, using the uninfected, incorporated old graft portions where available. Autogenous fistula can be treated with IV antibiotics alone, and surgery reserved for refractory cases.
  - **Thrombosis** of a graft or fistula is treated by percutaneous or open thrombectomy. Then a contrast study should be done to identify any sites of stenosis that may have contributed to the thrombosis.
  - **Stenosis** is most often on the venous anastomotic end of a **prosthetic graft** and produces enhanced pulsatility in the graft, with an eventual weak inflow, thrill, or bruit. If stenosis is on the arterial anastomotic end, pulsatility will be diminished first, along with the thrill or bruit. For **central venous stenosis**, the outflow obstruction results in pulsatility and an eventual weak thrill or bruit in either a graft or fistula dependent on the central vein for runoff. For **autogenous fistulae**, the anastomotic site can become stenotic, resulting in poor inflow with a lack of pulsatility and a weak thrill or bruit. The fistula may fail to mature. Treatment of all situations is **endovascular dilation with or without a stent.** Open surgical revision is a last resort.
  - **Steal syndrome** complicates 15% of cases. Risks factors include female gender, age greater than 60, diabetes, or brachial artery inflow disease.
    - Treatment is with **ligation** if steal is present immediately after surgery and accompanied by severe motor symptoms. For patients presenting sub-acutely without ischemic motor deficits, other options include:
    - **Distal revascularization with interval ligation (DRIL)** performed by ligating the native vessel below the fistula and placing a new autogenous vein conduit from the proximal artery to the distal artery. This proximalizes (and hence prioritizes) inflow for the distal limb.
    - **Reconstruction using distal inflow (RUDI)** performed by ligating the fistula and constructing a new conduit from the distal arterial tree into the outflow end of the fistula. This preserves distal limb flow via native arteries, but requires a healthy distal arterial donor for the new conduit to be attached to.
    - **Plication or banding** can be used if the accelerated venous runoff is resulting in mild distal sensory symptoms or congestive heart failure (CHF).
  - **Aneurysmal change** to the venous end of a fistula needs endovascular treatment directed at upstream venous stenosis. If there is skin breakdown over the aneurysmal area, surgical revision of that site is needed.
- **PERITONEAL DIALYSIS**
  - Requires a catheter residing in the lower peritoneal cavity and nightly exchanges of effluent to equilibrate plasma composition via molecular movement across the mesothelial peritoneal membrane.
    - Laparoscopic insertion using a **rectus sheath tunnel** approach and **omentopexy** helps decrease adhesions and prevent mechanical occlusion. Mechanical occlusion may necessitate laparoscopic repositioning, but only occasionally requires conversion to hemodialysis.
    - Removal and replacement of the catheter is only needed if there is fungal growth, multiple episodes of bacterial peritonitis, or a soft tissue infection along the tract that will not clear. Repeated episodes of peritonitis or catheter infection are the main reasons prompting conversion to hemodialysis.

THORACIC AORTA ANEURYSM

- The aneurysm may involve the aortic root, ascending aorta, arch, or descending aorta. *See Figure 30.5.* Thoracic aneurysms are less common than aneurysms at the infrarenal site. Etiologies are degenerative changes in the media, infection, or connective tissue disorders.
- **Ehlers-Danlos (type 4)** is a lack of type 3 collagen in the vessel wall.

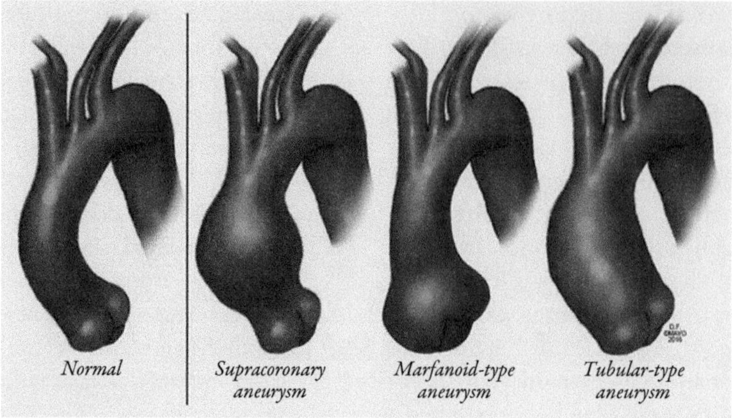

Figure 30.5 Ascending aorta aneurysm.

- **Loeys-Dietz syndrome** patients have a bifid uvula, cleft palate, hypertelorism, and a defect in the TGF-B pathway that results in aneurysm development.
- **Marfan syndrome** is a lack of the fibrillin backbone for elastin. There are increased TGF-B levels and increased mucopolysaccharides in the media layer, resulting in cystic medial degeneration and aneurysm development, ectasia, or dissection.
- **Treponema (syphilis)** causes an infectious endarteritis of the aortic wall vasa vorum. This compromises perfusion to the outer component of the vascular wall and secondary aneurysmal development.
- **Takayasu's arteritis** results in intimal stenoses of the arch and cephalic and brachial vessel branches.
- **Temporal arteritis** involves all layers of the aortic wall, leading to both stenoses and aneurysm.
- Patients with a **bicuspid aortic valve** have a x10 aortic dissection rate, increased matrix metalloproteinases (MMP), and decreased fibrillin in the media.
- A **bovine arch** refers to a brachiocephalic artery (BCA) giving rise to the right common carotid artery (CCA), right subclavian artery (SCA), and left CCA. It is associated with decreased fibrillin within the media, and increased incidence of aneurysms of the arch.

- **DIAGNOSIS**
  - Aneurysms are most often asymptomatic and incidental on CT imaging. Rarely a patient will present with a contained rupture or a secondary dissection of the aortic intima.
- **TREATMENT**
  - Surgery is indicated if the aneurysm is
    - Growing by **>1 cm per year** (a normal growth rate is < 0.1cm/year).
    - **>6 cm** greatest diameter.
    - **>5 cm greatest diameter with a concomitant luminal dissection or a baseline connective tissue disorder.**
    - **>4.5 cm diameter at the aortic root or ascending aorta in a patient needing aortic valve replacement for stenosis or regurgitation.** The ascending aorta can also be indexed against the patient's own descending aorta on imaging.
  - **Surgical replacement of the ascending aorta alone** can be done under cardiopulmonary bypass (CPB) via the right atrium and aortic arch cannulation. An aortic hemiarch or total arch both require an **alternative arterial cannulation site.** This is most often via a graft sewn to the right axillary artery, or a femoral artery cutdown.
    - **Deep hypothermia** at 18°–22°C with circulatory arrest is needed if the brachiocephalic artery (BCA), L CCA, and L SCA must be sewn to the graft in a total arch, or if there is no room for aortic clamp placement in a hemiarch repair. Clamping the BCA during this time allows deep hypothermia to be accompanied by **antegrade cerebral perfusion (ACP)** via the R axillary graft while the rest of the body remains without circulation.
    - The **distal/superior arch confluence anastomosis is performed first, often requiring hypothermic circulatory arrest with or without ACP.** A felt pledget strip can be used if the aortic media is believed to be a liability for anastomotic leak. If circulatory arrest was utilized, once the distal anastomosis is done, CPB can be re-established via the axillary graft and a clamp on the aortic graft just inferior to the distal anastamosis.
    - The patient can be rewarmed while a **proximal anastomosis** is completed. This can be at the sinotubular junction or at the annulus. If the latter, any concomitant valvulopathy with regurgitation is treated by adding **re-suspension of the aortic valve leaflets, or use of a composite graft that includes a prosthetic valve and aortic conduit.** Other options are homografts or stentless porcine grafts. Any proximal annular anastomosis will require **coronary reimplantation.**

- **Thoracic endovascular aortic repair (TEVAR)** can be performed for aneurysmal dilation distal to the takeoff of the L carotid artery.
  - Preoperative **L SCA to carotid transposition or carotid to L SCA bypass graft** can be done if one is planning to **cover the L SCA orifice** with an endograft, and this is preferred in most cases. It is mandatory if the patient has a dominant L vertebral artery, an incomplete circle of Willis, an LUE AV fistula, or a LIMA to LAD coronary graft. In all these instances, antegrade flow must be maintained in the proximal SCA.
  - Alternatively, one may use a **preconstructed fenestrated endograft** accommodating the L SCA orifice.
  - Preoperative **cerebrospinal fluid (CSF) drainage with a lumbar drain** will mitigate spinal cord ischemia, and is especially important when lining the distal descending thoracic aorta.
  - A cutdown to the common femoral artery (CFA) and placement of a large sheath allows for device deployment. Percutaneous CFA access on the contralateral side is done with a 5F sheath. Once deployed, balloon dilation of the proximal and distal seal will optimize apposition with the native aortic endothelium. Pre- and post-deployment angiography should be done.
  - TEVAR has a **lower mortality and complication profile** than open repair, including a 2%–3% paraplegia rate from covering intercostals or the L SCA. Endovascular repair necessitates **serial imaging and a higher rate of repeat interventions.**

THORACOABDOMINAL AORTIC ANEURYSMS

- **DIAGNOSIS**
  - These are generally asymptomatic and found incidentally on imaging. Contained rupture and secondary intimal dissection are less common presentations.
  - They are classified by the **Crawford system.** *See FigAure 30.6.*
    - Class 1: Descending thoracic aorta (DTA) aneurysm only.
    - Class 2: DTA aneurysm and entire abdominal aorta is aneurysmal.
    - Class 3: Inferior DTA aneurysm and entire abdominal aorta aneurysm.
    - Class 4: Entire abdominal aorta is aneurysmal without a thoracic component.
    - Class 5: A short DTA aneurysm extending to the suprarenal abdominal aorta.
- **TREATMENT**
  - Historically this has been surgical, with endovascular approaches reserved for elderly paticnts and thosc where major tributary orifices can be covered.
    - For elective cases, **preoperative testing** includes echocardiography, nuclear medicine stress testing, and

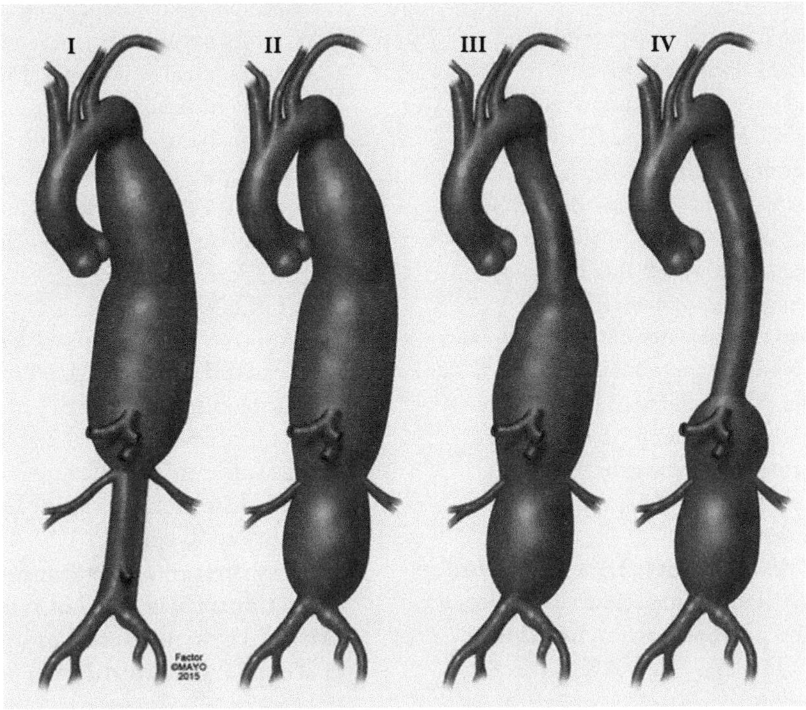

Figure 30.6 Thoracoabdominal aorta aneurysms—Crawford classification.

coronary angiography. Preoperative pulmonary function tests (PFTs) should be considered for a large descending thoracic aortic component that would require single lung ventilation. An Fev1>1L and a normal pCO_2 are ideal.

- **CSF drainage** via a lumbar catheter helps improve spinal cord perfusion pressure and decrease ischemia.
- Intraoperative **motor evoked potential (MEP) monitoring** can direct selective intercostal artery perfusion for spinal cord preservation. **Moderate hypothermia at 32°C–34°C is utilized.**
- A **thoracoabdominal incision** starts anywhere between the 5th to 8th interspace posteriorly, runs parallel to the rib, and extends anteriorly up to the lateral margin of the rectus abdominis muscle. The incision is directed inferiorly, and parallel to the rectus from that point on.
- A retroperitoneal exposure requires **medial and anterior retraction of the peritoneal membrane** and ureter. Ligation of left lumbar vein will liberate the left renal vein and allow for medial and anterior retraction of the left kidney.
- A **peripheral perimeter transection of the diaphragm** allows both cavities to be continuous.
- **Partial left heart bypass (LHB)** can offload the LV and lengthen tolerable end organ ischemic time when the aorta is clamped in a Crawford type 1–3 aneurysm repair. Partial LHB is via a 24F cannula in the inferior left pulmonary vein (or left atrium) for inflow, and 8mm Dacron to left common iliac artery (CIA) or CFA for outflow.
  - After LHB is initiated, cross-clamp proximal and distal to the aneurysm, open the aortic sac, and initiate renal perfusion and intercostal perfusion via **selective cannulation** with cold crystalloid plus systemic methylprednisolone or mannitol.
  - Complete the proximal aortic anastomosis, then advance the clamp inferiorly and clamp the proximal graft.
  - Patch the intercostals and then advance the clamp inferior to the intercostal anastomoses.
  - Patch the celiac, superior mesenteric artery (SMA) and right renal artery as one unit onto the aortic graft. A side graft can be used for a direct end to end left renal artery anastomosis, after which the clamp can be advanced.
  - Finally, the distal aortic anastomosis or B-CIA anastamoses can be completed. Trim and close the aneurysm sac over the graft.
- Open repair has a **7.5% mortality, 2.5% spinal cord ischemia** (despite MEP directing selective intercostal perfusion), and **>10% postoperative acute kidney injury (AKI)** rate. Only 2%–8% of AKI patients actually end up needing dialysis. A preoperative creatinine >1.8 predicts postoperative renal failure, and this in turn increases mortality. One may consider prophylactic renal artery stenting, revascularization, or endarterectomy prior to intervention on the thoracoabdominal aneurysm.

AORTIC DISSECTION

- **DIAGNOSIS**
  - Symptoms include sharp or tearing chest pain that may radiate into the back. Associated organ malperfusion can be seen as CVA, upper or lower limb ischemia, or visceral ischemia.
    - Patients may have an underlying connective tissue disorder or sympathomimetic drug use.
    - Classified as acute if <2weeks, subacute at 2–6 weeks, and chronic at >6 weeks. The latter have a thick flap on imaging.
    - **CTA** is the imaging modality of choice. **Transesophageal echocardiogram (TEE)** can be used in unstable patients. MRI requires prolonged imaging time.
    - **Stanford Type A** dissections involve the ascending aorta. DeBakey 1 includes the ascending and descending. DeBakey 2 is the ascending only. These can be complicated by aortic root dilation, regurgitation, occlusion of coronary ostia, or rupture of the false lumen into pericardial space. *See Figure 30.7.*
    - **Stanford type B** dissections involve the descending thoracic aorta. DeBakey 3a is the thoracic aorta only. DeBakey 3b includes the thoracoadbominal aorta. 20% are complicated by impending rupture, malperfusion, refractory hypertension, or refractory pain.
    - **An intramural hematoma (IMH)** is managed similarly to an aortic dissection. Its etiology may be a very small intimal break not visualizable on CT imaging, rupture of vasa vorum within the media of the aortic wall, or a **penetrating aortic ulcer (PAU)** allowing blood to accumulate in the media. *See Figure 30.8.*
- **TREATMENT**
  - Medical treatment begins with **beta blockade, then vasodilators** for a goal heart rate of 60–70 and systolic blood pressure (SBP) of 110. This decreases stress at the site of intimal tear. Medical therapy alone may be sufficient for uncomplicated Stanford type B aortic dissections or type B IMH. There is a 10% mortality at 30 days.
  - **Surgery is reserved for Stanford type A dissections/IMH/PAU or type B dissections/IMH/PAU with visceral or limb malperfusion or secondary aortic dilation portending rupture.** In acute type A dissections, mortality is 1%–2% per hour.

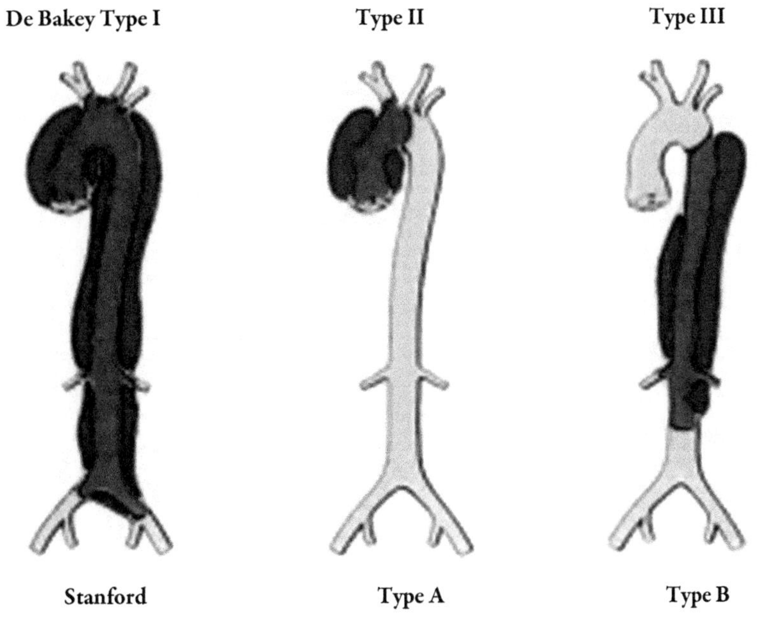

De Bakey	
Type I	Originates in the ascending aorta, propagates at least to the aortic arch and often beyond it distally
Type II	Originates in and is confined to the ascending aorta
Type III	Originates in the descending aorta and extends distally down the aorta or, rarely, retrograde into the aortic arch and ascending aorta
Stanford	
Type A	All dissections involving the ascending aorta, regardless of the site of origin
Type B	All dissections not involving the ascending aorta

Figure 30.7 Stanford and Debakey aortic dissection classification.

- Cardiopulmonary bypass (CPB) is via the following:
  - 1. Femoral venous or right atrial cannula; and
  - 2. Femoral arterial cannula or 8–10 mm Dacron graft sewn to the right axillary artery via an infraclavicular cutdown.
- Once exposure has been obtained and the ascending aorta and arch dissected, one can perform:
  1. **Isolated ascending aorta replacement** up to the takeoff of the brachiocephalic artery without hypothermic arrest, provided a clamp can be placed across the distal ascending aorta.
  2. **Hemiarch** keeping the common confluence of brachiocephalic vessels so the distal arch and descending aorta can still be perfused via the axillary cannula, provided the aorta can be cross-clamped obliquely. Otherwise, a hemiarch will require deep hypothermia to 18°C, and circulatory arrest if a clamp cannot be placed and hence the arch cannot be perfused (see thoracic aneurysm section above). A clamp on the BCA will allow for antegrade cerebral perfusion (ACP) via the axillary cannula, however.
  3. Traditional **total arch**, which leaves the BCA, L CCA and L SCA on an island and therefore requires deep hypothermic circulatory arrest with or without ACP. This is less commonly done. The BCA, L CCA, and L SCA may need to be re-implanted separately in instances where the dissection flap extends into one of their lumens.
- The **distal/superior arch confluence anastomosis is performed first, often requiring hypothermic circulatory arrest with or without ACP** (see thoracic aneurysm section above). A luminal felt pledget strip and bioglue can ensure apposition of the components of the distal aortic wall. A **frozen elephant trunk extension as an endoluminal covered stent** can be left in place distal to the left subclavian takeoff, and later extended into the descending thoracic aorta.

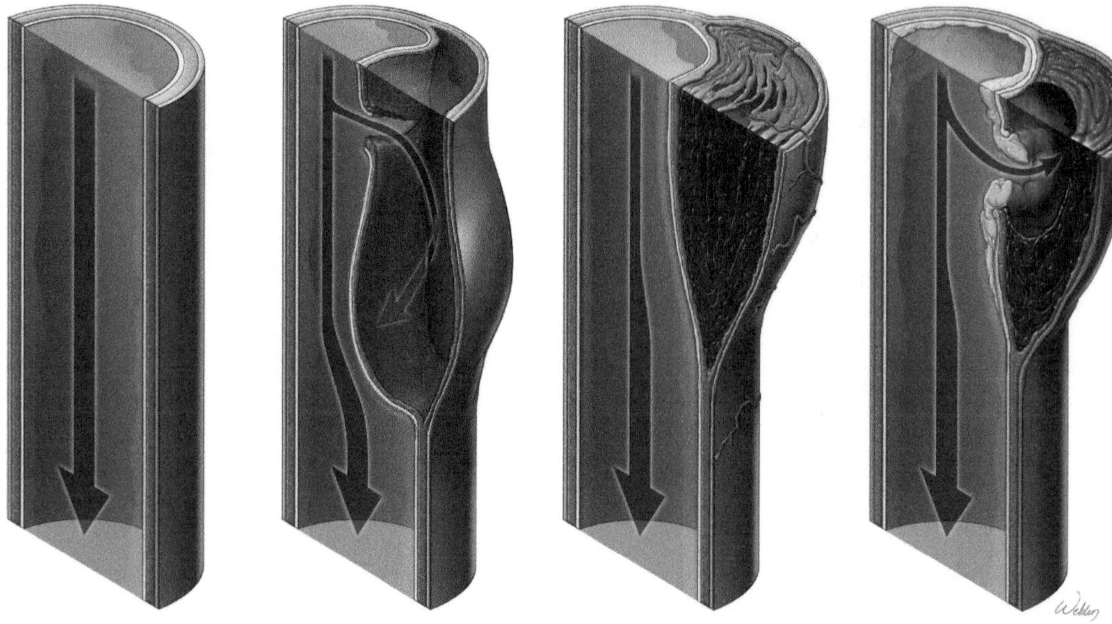

Figure 30.8 Acute aortic syndromes.

- **Once the distal anastomosis is done, CPB can be re-established via the axillary graft** and a clamp on the aortic graft just inferior to the distal anastomosis. The patient can be rewarmed while a **proximal anastomosis** is completed. This can be at the sinotubular junction or at the annulus. If the latter, any concomitant valvulopathy with regurgitation is treated by adding **re-suspension of the aortic valve leaflets or use of a composite graft that includes a prosthetic valve and aortic conduit.** Other options are homografts or stentless porcine grafts. Any proximal annular anastomosis will require **coronary reimplantation.**
- Rarely, a simple narrow neck saccular ascending aortic aneurysms can be resected and undergo **patch aortoplasty.**
- Surgery carries a 25%–30% mortality, a 5% CVA risk, and 10% myocardial infarction (MI), renal failure, and paraplegia risk.

- **Thoracic endovascular aortic repair (TEVAR)** can be used for **visceral or limb malperfusion** in a patient whose **site of intimal break is distal to the L carotid artery.** Coverage of the intimal break entry site allows the true lumen to be reclaimed distally. Also, **any site of aortic dilation should be covered** as well.
  - In some instances, **fenestration** of a thick chronic dissection flap serves to depressurize the false lumen, equilibrate it with the true lumen, and hence improve any malperfusion secondary to insufficiency at the ostia of true lumen branches.
  - TEVAR has a 2%–3% paraplegia rate. Serial surveillance CT scans should be performed at 1, 6, and 12 months.

AORTIC ARCH OCCLUSIVE LESIONS

- **DIAGNOSIS**
  - Patients present with insufficiency of carotid or upper extremity vasculature. There may be blood pressure discrepancies between each arm.
  - Most commonly the etiology is atherosclerosis, but connective tissue and autoimmune disorders are also common. Confirmation is with a CTA of the chest.
- **TREATMENT**
  - Indicated if >75% of the orifice of the carotid, subclavian, or brachiocephalic artery is involved. Intrathoracic and cervical approaches have high long-term patency.
    - **Median sternotomy or mini-inverted T sternotomy** down to T3 can be used for innominate artery or proximal left carotid artery reconstruction (endarterectomy, transposition, or bypass graft). Traditional total arch replacement would require median sternotomy and circulatory arrest (see above).
    - A **transverse supraclavicular incision** for carotid-subclavian artery bypass or subclavian-subclavian bypass using PTFE.
    - A **left posterior thoracotomy** for proximal subclavian artery access. This is more often managed endovascularly.

- **Endovascular stents** across ostial lesions are second-line therapy and may extend 1–2 mm into the aortic lumen. Buddy wires, balloons, or kissing stents can maintain patency of adjacent bifurcation vessel ostia when addressing the primary lesion.

ABDOMINAL AORTA

- Abdominal aortic aneurysms (AAA) are more common in elderly men. The aortic wall has **decreased elastin** and **increased matrix metalloproteinases (MMP)** in its media layer.
- **DIAGNOSIS**
  - 75% of AAAs are asymptomatic. A single screening ultrasound (US) is used for those >65 years of age with smoking history or family history of AAA. Other individuals >75 years of age may also qualify for a screening US.
  - Symptomatic AAA is due to rapid expansion, adjacent organ compression, mural occlusion and ischemia, or mural debris and embolization producing trash foot or blue toes.
  - Rupture generally occurs at the left posterolateral wall, 2–4 cm below the renal artery orifices. HTN and COPD are the main risks for rupture. There is a 30% mortality if they reach the hospital alive. CT may show continued or unconfined rupture, or just periaortic stranding implying impending rupture.
  - For a patient in extremis, a **bedside US** can confirm AAA and, in the appropriate clinical setting, the patient may need to proceed to the OR expeditiously. For stable, symptomatic patients, a **CT angiogram of the abdomen and pelvis** is confirmatory. Juxta- or pararenal AAA is defined as dilation starting within 1cm of the renal arteries.
    - **Inflammatory AAA** is a sterile, immune mediated, fibrotic process in the wall that adheres the aorta to the duodenum and ensnares ureters. It presents with back pain, weight loss, and an elevated erythrocyte sedimentation rate (ESR).
    - **Mycotic aneurysms** have an infection within the wall. Common organisms include salmonella or *S. aureus*. They infect an aortic plaque, resulting in aortitis and then aneurysmal dilation. Blood cultures are usually positive.
- **TREATMENT**
  - An asymptomatic AAA should be repaired at **>5 cm in women and >5.5 cm in men.** There is an increased rupture risk if expansion rate is **>1 cm/year** (normal expansion is 2–3 mm/year).
  - Also females, smokers, those with a family history of AAA, HTN, or COPD have a higher risk of rupture at any given diameter.
- **Every 6-month US surveillance** may be beneficial if the AAA has been noted to expand >3 mm/year or if the AAA is approaching 5.5 cm. Otherwise, **yearly US** is sufficient.
  - In certain outpatient situations, preoperative coronary revascularization can be deferred with no increase in cardiovascular complications. Exceptions would be those patients with left main disease, EF <20%, or severe valvulopathy.
  - For a ruptured AAA presenting to the emergency department, **permissive hypotension** should be employed until operative proximal control can be obtained via endoaortic balloon or transperitoneal supraceliac clamp (after taking down the left triangular ligament, opening the hepatogastric ligament, and dividing the right crus of the diaphragm). A thoracoabdominal approach can be used for supraceliac control in select patients). Wherever one starts, the clamp should be moved as inferiorly as possible once the neck of the aneurysm is defined. An inframesocolic clamp may be adequate if the rupture is below the superior mesenteric artery (SMA).
  - There is a reported increased risk of AAA rupture within 30 days after an unrelated laparotomy.
- **Open Repair**: A **transperitoneal approach** starts with inframesocolic aortic exposure obtained by mobilization of the left renal vein, ligating adrenal, and gonadal tributaries. *See Figure 30.9.*
  - For supraceliac aortic clamping, the left triangular ligament should be taken down, the gastrohepatic ligament opened, and the right crus of the diaphragm incised.
  - Right or left medial visceral rotation will expose the length of the aorta. Left medial rotation involves mobilizing the spleen, pancreas, descending colon, and left kidney.
  - A **lateral retroperitoneal approach** is below the 11th rib. It is not often used if there is right renal artery stenosis. This approach decreases post-op ileus, but does not change pulmonary complications.
  - A **lateral thoracoabdominal approach** starts above the 10th interspace with an incision parallel to the chosen rib from the posterior axillary line and extended anteriorly to the lateral edge of the rectus to expose a suprarenal aortic aneurysm. The left kidney can be reflected anteriorly.
  - Once accessed, **longitudinal opening of the aorta** is followed by endarterectomy, thrombectomy, and oversewing lumbar apertures.
  - Cold renal preservation fluid can be delivered into the renal artery ostia.
  - A **Dacron tube or bifurcated, short neck, graft** is the conduit of choice.

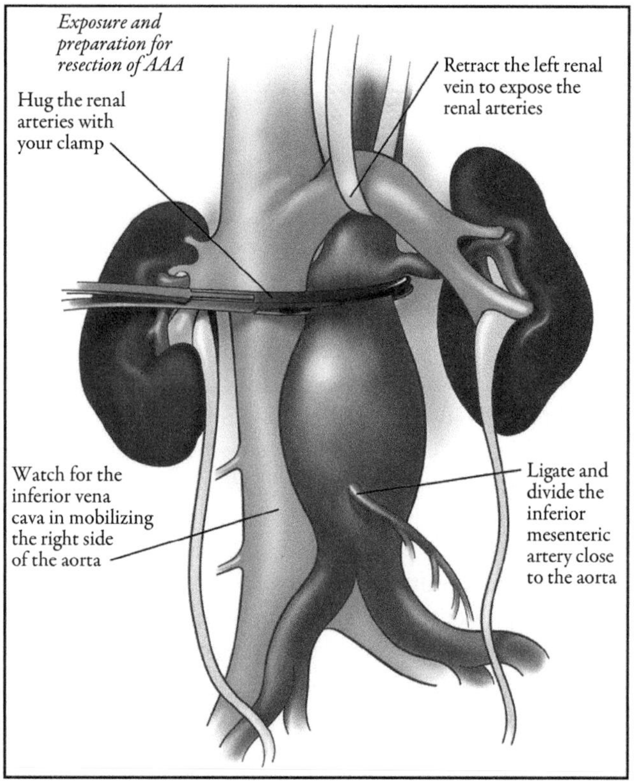

Figure 30.9 Infra-renal aortic aneurysm.

- One may **reimplant the inferior mesenteric artery (IMA)** if it demonstrates poor backflow or a back pressure <40 mmHg, or if the SMA is known to be very stenotic. IMA reimplantation is not needed if it has zero backflow or brisk backflow.
- Ideally, **one internal iliac artery should be preserved** to prevent vasculogenic impotence.
- Finally, **close the aorta sac** over the graft and close the retroperitoneal reflection (in a transperitoneal approach).
- For inflammatory aneurysms, preoperative ureteral stents may be used to assist in identifying a ureteral injury when performing resection and interposition graft.
- For mycotic abdominal aortic aneurysms, treatment is with resection and extranatomic bypass (axillo-bifemoral bypass or thoracofemoral bypass tunneling retroperitoneally posterior to the left kidney)

- **Outcomes and Complications**
  - **Neurogenic impotence** is from disruption of hypogastric nerves.
  - **Vasculogenic impotence** is from covering the orifice of the internal iliac artery.
  - **Ischemic colitis** is due to lack of antegrade flow down the IMA. It presents with postoperative abdominal pain and diarrhea with or without blood. Diagnosis is by sigmoidoscopy or colonoscopy showing discrete areas of ischemia or ulceration corresponding to the IMA distribution. The distal rectum and anal canal are spared, since they are in the internal iliac artery territory. Management is supportive. Full-thickness necrosis is managed with colectomy and colostomy.
  - Operations >5 hours in length are independently associated with increased mortality or cardiopulmonary morbidity.
  - Pulmonary, cardiac (2%–6%), and renal (<2%) complications are predicted by their respective preoperative comorbidities. Cardiac dysfunction is the most common cause of early death. Renal dysfunction is the most common cause of late death.
  - In certain patients, a retro-aortic renal vein can incur injury from the proximal clamp.
  - **Mortality is <5% when elective.** Supra- and juxtarenal repairs have a 5%–10% mortality rate. **Mortality is 30%–50% when emergent.**
  - Postoperative **aortic graft infections** are either early and acute from *S. aureus* or late and indolent from *S. epidermis*. Blood cultures are generally negative. Treatment is with an extra-anatomic bypass by axillo-bifemoral, or a thoracofemoral approach tunneling retroperitoneally and posterior to the left kidney. A transobturator bypass may be utilized for an infected femoral anastomosis. Once an alternate route of circulation is established, resection of the infected graft can be performed. There is a 10%–50% mortality and amputation rate.
  - **Aortoenteric fistulae** develop at >6 months post-op and result from a chronic infection at the proximal anastomotic site that erodes into the 3rd or 4th portions of the duodenum or the adjacent jejunum. *See Figure 30.10.*
    - Patients may have a minor herald upper gastrointestinal bleed (UGIB) prior to the onset of painless major hemorrhage, hematemesis, or melena. The fistulization site may be missed on upper endoscopy, so definitive diagnosis is by **abdominal pelvis CT** showing loss of the interface between the proximal aortic anastomosis and the duodenum or jejunum. There may be extraluminal peri-aortic air indicative of a chronic perigraft infection. A tagged WBC scan can also suggest a chronic infection.
    - Management includes a **temporary aortic stent,** followed by resection, intestinal repair or resection, and an **extranatomic vascular bypass.** An alternative is to place a new **homograft** within the infected field to reestablish aortic continuity and avoid an extranatomic bypass.
  - **Endovascular repair (EVAR)** indications are the same as open repair: 5 cm aneurysm for women and 5.5 cm for men. However there should be adequate caliber iliofemoral vessels for delivery of the device.

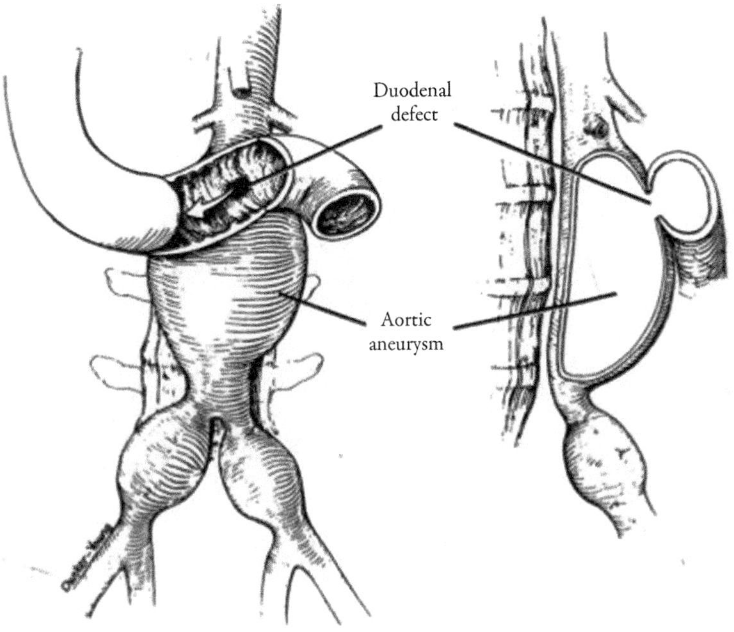

Figure 30.10 Aortoduodenal fistula.

- Hybrid procedures involving retroperitoneal exposure of the external iliac artery (EIA) or common iliac artery (CIA), with a conduit sewn in place for graft deployment, can bypass stenotic external iliac and femoral vessels.
- In order to accommodate the stent graft, a **10–15 mm length of aortic neck proximally and common or external iliac arteries distally** are needed. The aortic neck angle should be less than 60 degrees.
- **Diameter of the proximal neck should be 20–30 mm, and the distal iliac artery 10–15 mm** to allow for an adequate seal, prevent endoleak, and prevent shift. The internal iliac artery will need to be occluded with embolic material if the distal stent will cover its orifice. This prevents back-bleeding into the endosac causing a type 2 leak.
- To preserve major branch orifices, a **chimney or snorkel graft** can be placed into the celiac, superior mesenteric, or main renal arteries via a transbrachial or an axillary approach, while the aortic stent graft is deployed by the iliofemoral route. Alternatively, a **custom fenestrated graft** can also be constructed to accommodate such critical branches. The IMA or orifice to an *accessory* renal artery may be covered by the stent graft.
- In emergent ruptured situations with proximal aortic endovascular balloon control, one may deploy the endograft and then move the balloon within the main body of the stent before the contralateral gate is entered and contralateral limb deployed.
- EVAR is not suitable for patients with underlying connective tissue disorders or an infection of the aortic wall.
- EVAR has been used in patients with AAA who are pending a planned laparotomy for enterocolonic or genitourinary malignancy. Prior to the laparotomy, the AAA can be lined in a standard fashion provided one is sure that covering the IMA will not sacrifice colonic health. The EVAR is not believed to be at any great risk of bacterial inoculation during the subsequent surgery, since the peri-graft area is not breached.
- Alternatively, patients with an AAA who will be undergoing open sterile organ resections (liver, spleen, or adrenal) have received a concomitant open AAA repair without inoculation risk.
- **Outcomes and Complications**
  - In the emergent setting, patients who anatomically qualify for EVAR may have lower mortality, MI, renal, and respiratory failure compared to those who had to receive open surgery.
  - Mortality is **<2% when elective**, and **up to 25% when emergent.** These are about half of the corresponding 30-day mortalities for open repair. However, at 2 years this benefit is no longer seen, and mortality is equal between the two approaches.
  - 3% of patients have limb vascular occlusion or pelvic or buttock ischemia.
  - 1% have colorectal ischemia and 0.25% spinal cord ischemia.

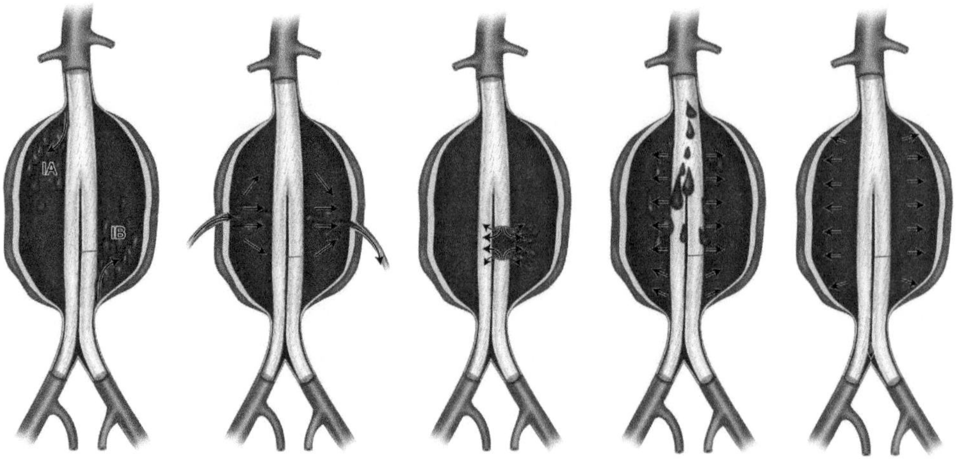

Figure 30.11 Types of endoleak.

- **Postoperative surveillance** with CT scans and US is needed at 1 month and then every 6–12 months to detect endoleaks. *See Figure 30.11.*
  - **Type 1 endoleak.** These are normally detected on completion angiogram as leakage from the aortic or iliac artery lumen around the proximal or distal end of the stent graft and into the endosac. Management includes placement of an aortic cuff or iliac extension limb. Oversizing the proximal and distal endograft by 10%–20% and performing balloon angioplasty at these sites will minimize type 1a (proximal) and 1b (distal) endoleaks.
  - **Type 2 endoleak.** Occurs in up to 30% of patients and is due to collateral vessel back-bleeding into the endosac. Two-thirds of cases resolve spontaneously, sometimes simply after cessation of procedural anticoagulation. Type 2 leaks are **treated only if they persist for >6 months or are symptomatic.** Iliolumbar artery embolization of the back-bleeding vessel, or laparoscopic or videoscopic ligation of a collateral vessel are two options.
  - **Type 3 endoleak.** This is leakage of blood in between components of the stent graft and into the endosac. They are managed with a relining stent graft. Multicomponent grafts should have their individual pieces overlapping by at least 3 cm, and balloon angioplasty at these areas of overlap can help prevent type 3 endoleaks.
  - **Type 4 endoleak.** Blood makes its way through an intact graft lining into the endosac. These are rare, and may resolve after cessation of procedural anticoagulation.
  - **Type 5 endoleak (endotension).** There is no visible leak of contrast on fluoroscopy; however, serial outpatient CT scans show an enlarging endosac with contrast enhancement or thrombus. The presumption is a slow, occult leak of blood at a small site along the endograft.

AORTOILIAC OCCLUSIVE DISEASE

- This is most often due to atherosclerotic plaque and insidious narrowing, sometimes compounded by an acute thrombosis atop a ruptured portion of the plaque.
- A less common presentation would be acute occlusion from a large embolus lodging at the aortoiliac bifurcation.
- In chronic situations, collaterals form to bypass the stenosis *See Figure 30.12.*
  - 1, **Inferior mesenteric artery (IMA)** to the hemorrhoidals and then retrograde flow up the internal iliac artery, thereby perfusing the common and external iliacs.
  - 2, **Lumbar arteries** to the gluteals collateralizing with the internal iliac or circumflex femoral arteries, thereby perfusing the external iliacs and distal extremities.
  - 3, **Subcostal or lumbar arteries** to the circumflex iliac, which meets the common femoral artery (CFA).
  - 4, **Superior epigastric artery** descending to the CFA.
- **DIAGNOSIS**
- Presentation of chronic disease is with Leriche syndrome, defined by buttock and thigh claudication, impotence, and diminished femoral pulses.
  - Acute occlusion of the aortoiliac bifurcation from an embolus has sudden onset of bilateral acute limb ischemia.

- CT angiogram (CTA) of the aorta with runoff to bilateral lower extremities is the diagnostic test of choice.

- **TREATMENT**
  - Medical management with smoking cessation, exercise, antiplatelets, and statins are first line for claudication, even in this setting of iliac disease.

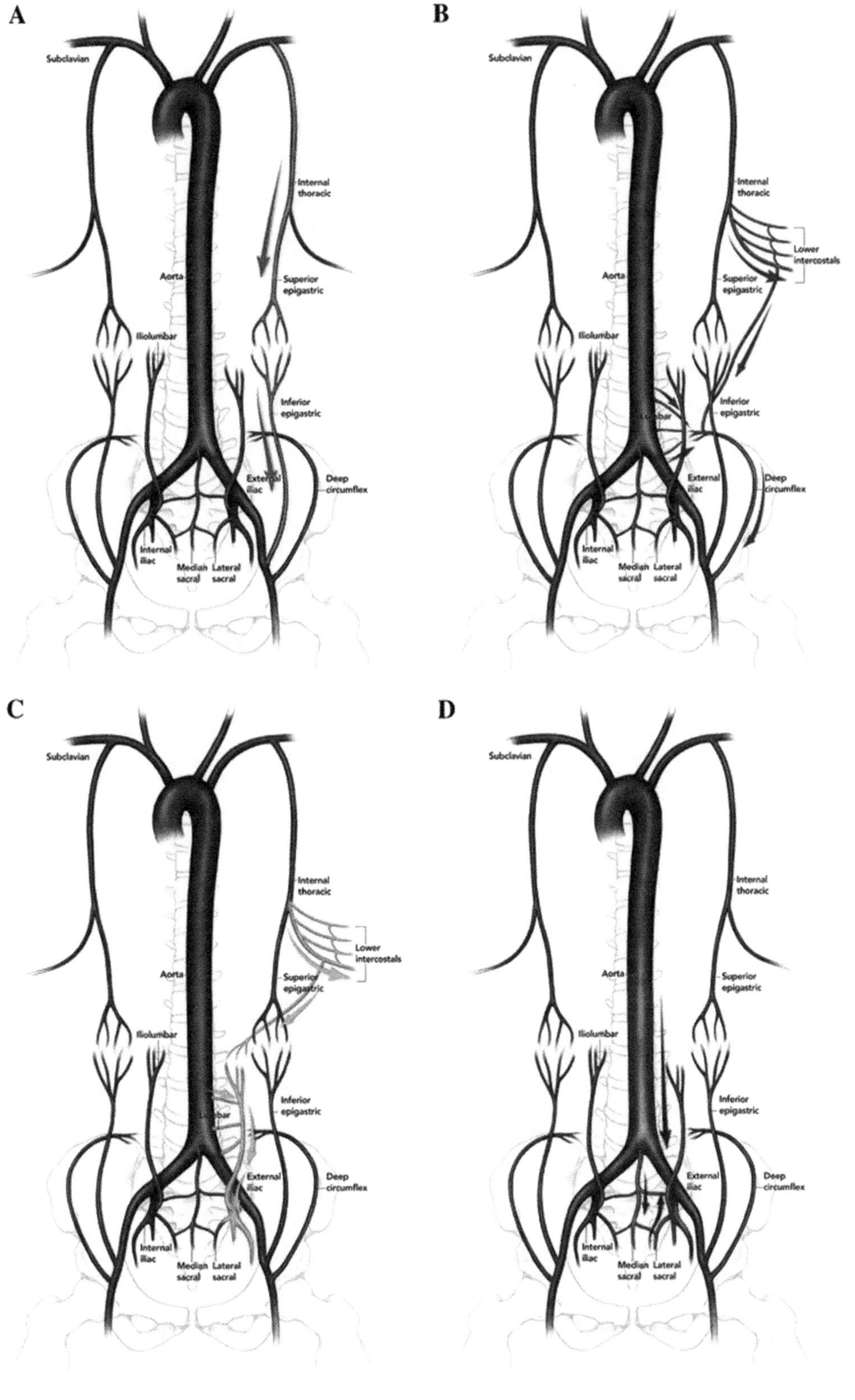

Figure 30.12 Collaterals bypassing aorto-iliac occlusion.

Cilostazol may be beneficial provided the patient does not have heart failure.

- If a patient has critical limb ischemia (CLI), defined as rest pain, tissue loss, or sensory or motor nerve dysfunction, then treatment is with:
  - 1. **Common femoral artery cutdowns and proximal Fogarty embolectomies** at the common iliac level. This is the fastest way to reestablish perfusion in the setting of an embolic occlusion resulting in acute limb ischemia. Subsequent workup should focus on finding the embolic source.
  - 2. **Aortobifemoral bypass** is for chronic, multiple, long segment lesions. The principles of the procedure are below:
    - Expose bilateral CFA and perform adequate CFA and profunda femoris artery (PFA) endarterectomies for outflow.
    - Transperitoneal exposure of the abdominal aorta.
    - If there is suprarenal atherosclerotic disease, one may need a cephalad clamp for suprarenal endarterectomy. Occasionally the supraceliac or distal thoracic aorta can be used for inflow.
    - Tunnel graft limbs directly anterior to EIA bilaterally and under the inguinal ligaments. Tunneling should be done before any proximal anastomosis.
    - A proximal **end–end aorta to Dacron proximal anastomosis** is then completed after resection of a small segment of aorta superior to the IMA, allowing the graft to lie flat in the retroperitoneum. This should be done as close to the renal arteries as possible, diminishing the chance of future infrarenal atherosclerotic disease compromising inflow of the graft. An end–end anastomosis eliminates all antegrade pelvic perfusion, and leaves the internal iliacs dependent on retrograde external iliac flow.
    - If significant EIA occlusions exist such that they cannot carry retrograde flow for pelvic perfusion, then an **end–side proximal aorta to Dacron anastomosis** should be done instead. This preserves some antegrade aortic to iliac flow for pelvic perfusion. Native antegrade emboli can still occur, and there may be an increased risk of aortoenteric fistula at the proximal anastomotic site.
    - The distal anastomoses of each Dacron limb are then performed to the anterior surface of the common femoral arteries. This completes the reconstruction.
    - **Patency at 5 years is greater than 85%,** as opposed to a 50%–75% patency rate for an axillo-bifemoral bypass.
    - Complications include the following:
      - Infrarenal aortic atherosclerosis can lead to proximal graft narrowing and occlusion.
      - Intimal hyperplasia and atherosclerosis of the femoral artery at the anastomotic site can lead to a unilateral limb occlusion. This is the most common late complication.
      - **Pseudoaneurysms** occur in about 1% of cases. They result from suture fracture, graft material fatigue, a thin wall after CFA endarterectomy, or a chronic infection.
      - **Femoral site infections** contiguous with graft material may require rerouting a new graft via the anteromedial obturator foramen to the SFA or popliteal artery, and then explanting the old limb graft, debriding, and ligating the infected vessel.
  - 3. **Axillo-bifemoral bypass (Ax-bifem)** is an axillary anastomosis placed medially along the vessel (to prevent avulsion during shoulder abduction) after the pectoralis major muscle is split along its fibers, and the clavipectoral fascia opened. One should use the axillary artery on the side with the higher BP for inflow.
    - The conduit is usually ringed PTFE to the ipsilateral CFA. Then a femoral PTFE graft jumps off from the hood of the distal ax-fem graft and is tunneled in the subcutaneous suprapubic area to the contralateral CFA.
    - This bypass should be used for patients with cardiopulmonary and hepatic comorbidities, a stoma, or a hostile abdomen from prior adhesions, peritoneal dialysis, or radiation. It also is a **rapid way of reestablishing perfusion to both legs in non-embolic aortoiliac occlusions** suffering from the beginnings of distal tissue loss.
    - The patient may need a future distal extremity bypass, which can be staged after the inflow is established.
    - 5 year patency rate for an Ax-bifem is **50%–75%.**
    - **Ax-bifem** has a similar wound complication rate but less pulmonary and renal complications than an aortobifemoral bypass.
    - Steal from the ipsilateral upper extremity (UE) may occur, especially if the patient has a proximal subclavian artery stenosis. This area may need to be ballooned.
    - Other **complications** include occlusion of the graft and thromboemboli to the ipsilateral brachial or axillary artery, or a pseudoaneurysm at any of the anastomotic sites.
  - 4. **Femoral-femoral bypass** can be done under local or regional anesthesia with a stent to the donor iliac. This will address a unilateral aortoiliac occlusion.
  - 5. **Endovascular repair** is for discrete short segment stenoses or occlusion.
    - Any lesion with more than 50% stenoses or >10 mmHg systolic pressure drop across it should be treated.
    - An **ipsilateral retrograde approach affords pushability through a CIA lesion.** Percutaneous transluminal balloon angioplasty (PTBA) can be followed by balloon-expanding stents if there is

residual stenosis, a residual post-balloon gradient >5 mmHg, or a dissection.
  - If stenotic at the CIA ostia, bilateral CFA access with balloons and stents simultaneously deployed at both CIAs can prevent shifting of plaque to the less diseased side.
  - **A contralateral antegrade approach is best for navigating an EIA lesion** without causing a dissection. Use PTBA and self-expanding stents if near the inguinal ligament.
  - A **transbrachial approach** offers the benefits of pushability, but requires a cutdown and traversing the aortic arch. This incurs descending aorta embolization risks.
  - A **hybrid approach** involves a common femoral and profunda endarterectomy, and a **retrograde ipsilateral PTBA and stent** for an iliac lesion.
- 5-year primary patency rate is **75%–85%**.
- Endovascular repair can also be used for CIA aneurysms >3.5 cm or EIA aneurysms >3 cm. Surgical repair is a second option at these sites.

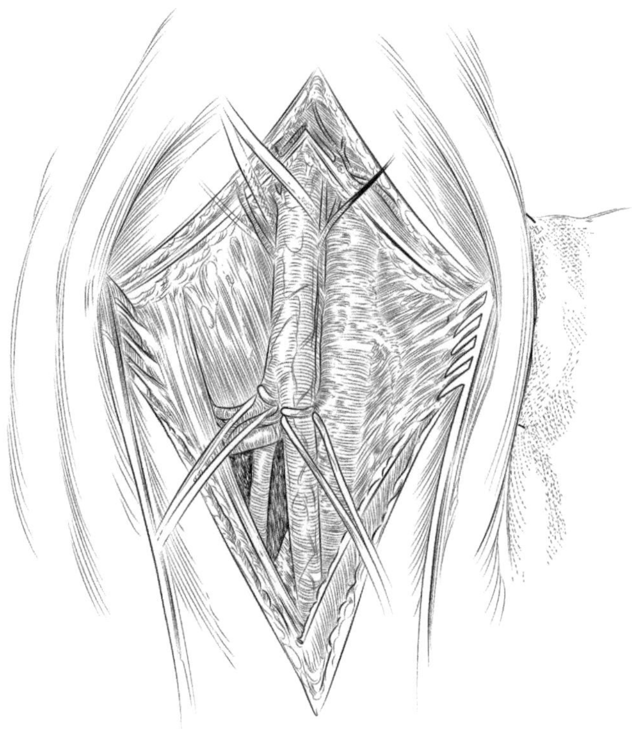

Figure 30.13 Common femoral and profunda femoral exposure.

PROFUNDA FEMORIS ARTERY (PFA)

- The PFA bifurcates posterolaterally from the common femoral artery (CFA) and the superficial femoral artery (SFA), and runs **anterior to the pectineus muscle and posterior to adductor longus.**
- Medial and lateral circumflex femoral vessels are the first branches, and they collateralize with superior and inferior gluteals, obturator vessels, and the pudendals. The main trunk of the PFA continues as a series of perforators through the adductor muscles.
- The **proximal PFA** is accessed via an inferior extension of an incision over the CFA. *See Figure 30.13.*
  - An **endarterectomy** can then be done to improve the collateralized flow below the knee in the setting of SFA disease. This can also act as an inflow source for a distal bypass, or alternatively serve as an outflow landing site for aortic, contralateral femoral, or axillary origin bypasses. The endarterectomy should be closed with bovine pericardium or Dacron. If the field is infected, an autogenous vein or SFA wall patch can be used instead.
  - When needed, an anterolateral or anteromedial incision (in relation to the Sartorius muscle), starting a few centimeters distally, can expose the **middle segment of the PFA.** This segment sits medial to adductor longus and lateral to the vastus medialis. Distally, a posteromedial incision carried posterior to the adductor longus and anterior to adductor brevis muscles can expose the **distal segment of the PFA** if necessary.

FEMORAL-POPLITEAL

- **CLAUDICATION**
  - The most common symptom of femoropopliteal artery atherosclerotic disease is claudication. This generally occurs at an ABI of <0.7–0.9 at rest, which may drop to <0.6 with exercise.
- **DIAGNOSIS**
  - **Patients should have a consistent history of claudication at a given distance.** Symptoms occur one level below the occlusive lesion.
    - The **Rutherford classification** helps categorize claudicants. Class 1–3 are mild, moderate, and severe claudication. **Class 4 is rest pain**, Class 5 designates ischemic tissue loss, and Class 6 designates necrosis (see below).
    - CIA occlusion or stenosis leads to buttock and thigh claudication.
    - EIA occlusion or stenosis leads to anterior thigh claudication.
    - Proximal SFA occlusion or stenosis causes calf claudication.
    - Distal SFA or popliteal occlusion or stenosis causes foot claudication.
    - Imaging can include **US or CTA.** The former is better at evaluating distal arterial patency, and the latter is better at evaluating the inferior aorta and iliac system.

- **TREATMENT**
  - Start with smoking cessation and exercise. The patient should walk until they feel pain, then rest and repeat this process for 30 minutes per day x 3 per week. At this stage, patients have a **1% per year amputation risk.**
  - Complementary medical management includes blood pressure control, a statin with or without ezetimibe, aspirin, or Plavix. The latter is based on coronary artery disease (CAD) and cerebrovascular accident (CVA) risks.
    - The **risk of cardiac death in a claudicant is 3%–5%/year.** This is higher than the risk of limb loss, so medical management is key to improving mortality.
    - **Cilostazol** is a phosphodiesterase inhibitor that causes smooth muscle relaxation and platelet inhibition. Its use should be avoided in patients with congestive heart failure (CHF), renal, or hepatic disease. After initiation, 75% of claudicants are stable or improve.
    - Glucose control for a goal hgba1c below 7% is also important.
- **CRITICAL LIMB ISCHEMIA (CLI)**
  - Characterized by rest pain (ABI <0.5), tissue loss (ABI < 0.4), or nerve dysfunction. Patients may have pallor, dependent rubor, and hair loss. They often have multiple arterial stenoses in series and are classified as Rutherford 4 through 6. Approximately **25% need amputation within a year, and 25% of patients die of a cardiovascular event in one year.**
- **Endovascular therapy** is an option for discrete short segment stenoses or occlusions, especially if the patient's predicted lifespan is less than 2 years or they have no adequate venous conduit.
  - Contralateral CFA micropuncture access is obtained, and a sheath is parked after traversing across to the diseased side. The patient is heparinized, the lesion traversed with a wire (a support catheter parked close by can improve pushability), and then the catheter advanced past the lesion to ensure adequate positioning in the distal arterial tree.
  - Balloon angioplasty can be performed and a self-expanding stent deployed for any residual stenosis, dissection, or vessel recoil. Stents may be covered or uncovered. They are not used as much in the popliteal area. Drug-coated balloons or drug-eluting self-expanding stents can be used in order to inhibit intimal hyperplasia.
  - Atherectomy by laser, directional, or rotational methods are adjuncts.
- **Surgical therapy** is for extensive lesions and a lifespan predicted to be more than 2 years.
  - An **endarterectomy of the CFA and PFA** via a longitudinal incision and patch angioplasty will improve inflow and symptoms of claudication. Distal tissue loss is dependent on adequate caliber infrapopliteal vessels, so this may not improve after proximal endarterectomy alone.
  - Long segment or nearly occlusive SFA lesions should receive surgical **bypass** rather than endovascular therapy.
  - Conduit options include:
    - 1. Autologous saphenous vein is ideal and should be >3 mm diameter and free of a sclerotic wall.
    - 2. Spliced upper or lower extremity vein grafts can be used if there is no adequate saphenous vein.
    - 3. Polytetrafluorethylene (PTFE) and cryopreserved cadaveric veins are less attractive conduit options. PTFE is inferior to autologous vein for above and below the knee landing zones. One may consider an autologous vein cuff for the distal anastomosis.
  - The donor site is usually CFA after undergoing an open endarterectomy.
  - The recipient site is either of the following:
    - **Above the knee popliteal artery.** This is accessed by an incision along the anterior border of Sartorius and deepened between Sartorius posteriorly and vastus medialis anteriorly. The adductor tendon is divided to expose the artery medial to the vein. *See Figure 30.14.*
    - **Below the knee popliteal artery.** This is accessed via an incision 1 cm posterior to the tibia and retraction of the gastrocnemius posteriorly. The pes anserinus is divided, and the artery lies lateral to veins. *See Figure 30.15.*
  - The ipsilateral greater saphenous vein (GSV) is harvested, and the origin-branch technique is used at the distal GSV transection point, in order to create a wider proximal anastomosis after reversal.
  - A tunnel is made in a sub-Sartorius position, or sometimes subcutaneous level.
  - 5-0 suture for proximal and 6-0 for distal anastomoses. Patency is assessed with palpation, US, or angiography at the end of the procedure. Consider anticoagulation if the patient has poor distal runoff.
  - **Outcomes**
    - 5 year patency rate is **75%.** Loss of patency at <30 days is technical; if from 30 days to 2 years it is due to myointimal hyperplasia, and at >2 years it is due to progression of atherosclerosis.
    - **Surveillance** is with ABI and US q3 months for one year, q6 months for 2 more years, and every year thereafter. This should be performed regardless of the presence or absence of symptoms.
    - Impending failure may be heralded by a peak systolic velocity >300 cm/s, a PSV ratio >3 across a lesion, diffuse low flow (<40 cm/s), or a drop in the ABI by 0.15. Any of these findings should prompt repeat angiography, regardless of the presence or absence of symptoms.
    - **Repeat interventions** may be via percutaneous transluminal balloon angioplasty. Surgical re-interventions are for stenoses >2 cm in length or within 3 months of the original operation.

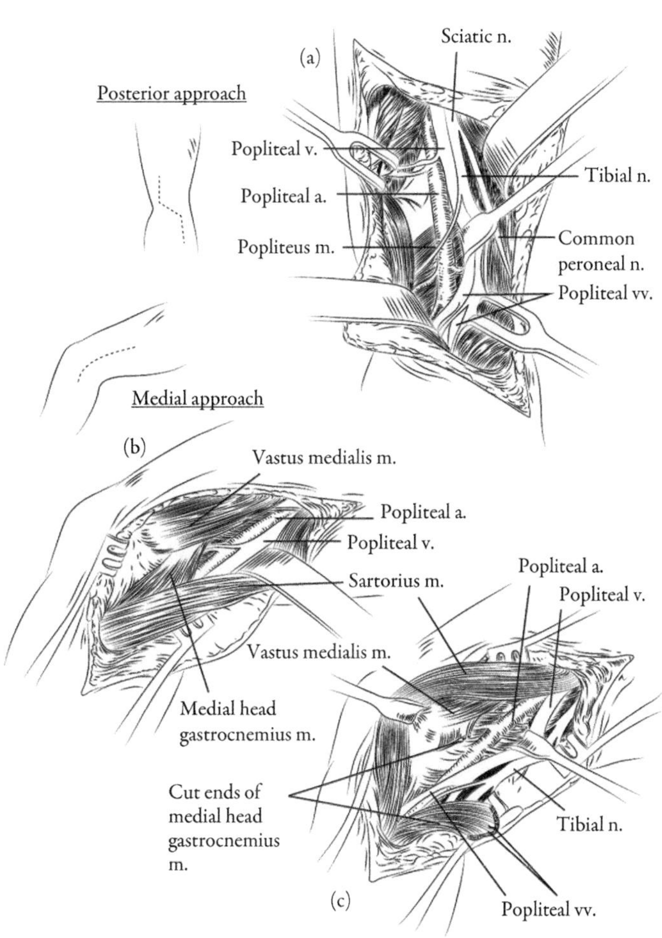

Figure 30.14 Proximal popliteal artery exposure.

- **FEMORAL ANEURYSM**
  - The CFA location is more common than PFA. Approximately 50% of patients have a concomitant AAA or popliteal aneurysm.
- **DIAGNOSIS**
  - Most patients are asymptomatic and diagnosed by physical exam with US or CTA imaging. When present, symptoms can be due to mass effect, thrombosis, thromboembolism, or claudication.
- **TREATMENT**
  - Indications for intervention include any **symptomatic** aneurysm, and asymptomatic aneurysms **>3 cm** or those harboring significant **intraluminal thrombus**.
    - The preferred approach is **aneurysm resection and interposition graft** from the distal EIA to distal CFA at its bifurcation point, or independent anastomoses to the PFA and SFA. The repair may include a distal jump graft. The aneurysm sac may be retained.
  - **Anastamotic pseudoaneurysms** are due to suture line disruption, stress on the graft material, or a low-virulence infection with *Staphylococcus epidermidis* producing exoslime. Surveillance with US is reasonable in asymptomatic patients, until the pseudoaneurysm is >2–3 cm. Symptomatic or enlargening anastomotic pseudoaneurysms should be resected and replaced with an interposition graft.
- **Infected pseudoaneurysms** due to intravenous drug use (IVDU) are treated with resection and ligation with extra-anatomic bypass using autogenous conduit, ideally the contralateral GSV. The obturator foramen is one option for the course of a bypass graft.
- **Iatrogenic post-puncture pseudoaneurysms** are treated with:
  - **Observation** and serial US if <2 cm.
  - **US guided thrombin injection** if they are 2–4 cm, narrow neck, and not infected. This is better than US guided compression alone.
  - Otherwise, **direct operative repair** of the puncture site is indicated if the pseudoaneurysm was refractory to thrombin injection, is >4 cm size, has a wide neck, is complicated by neural compression, or has overlying skin changes.
  - Collagen-based vascular closure devices at the time of the initial puncture can help prevent pseudoaneurysm formation.

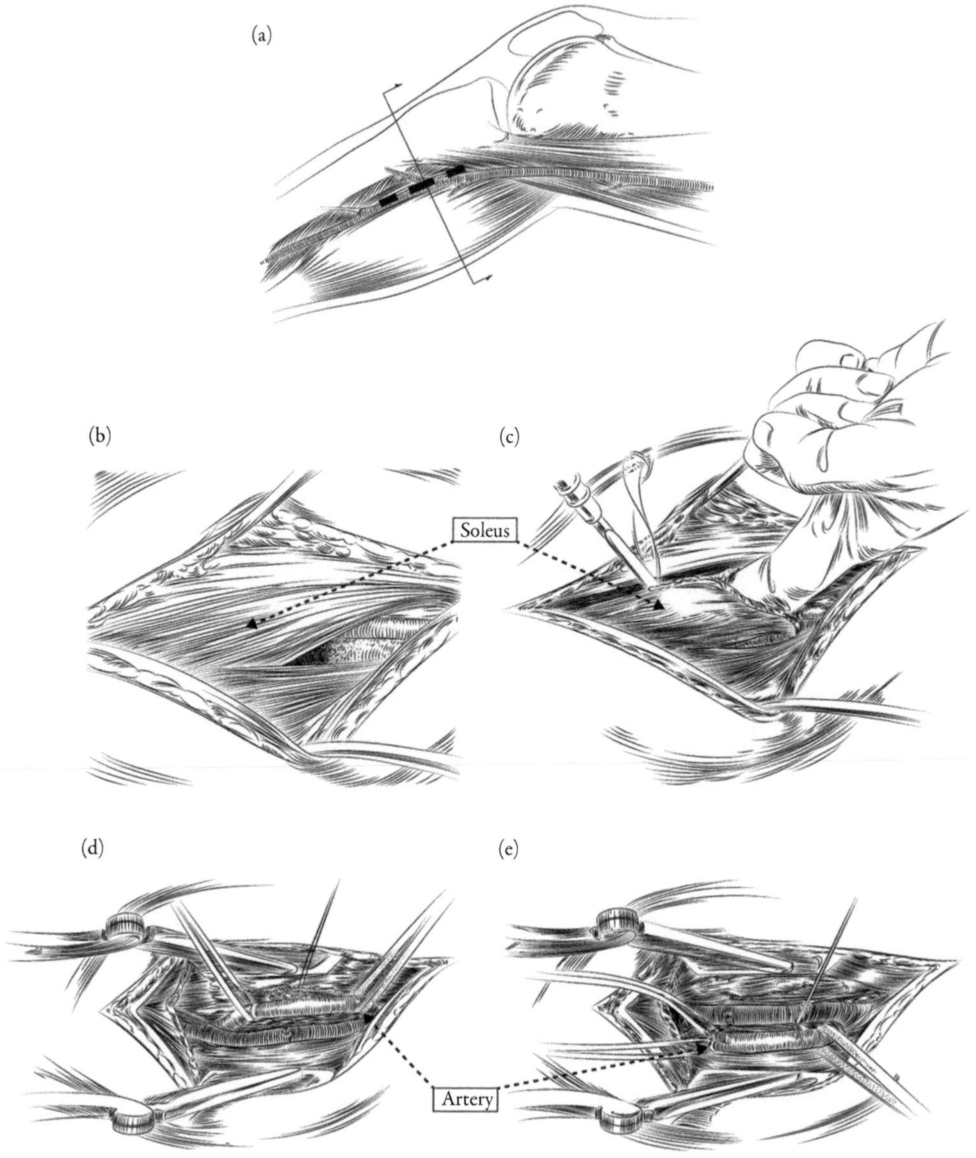

Figure 30.15 Distal popliteal artery exposure.

- **Femoral arteriovenous fistula** is secondary to trauma or a complication of prior surgery. Patients may have lower extremity steal symptoms or a thrill or bruit on exam. US is confirmatory. Treatment includes observation if asymptomatic and iatrogenic. Otherwise, surgical primary repair, patch angioplasty, or interposition graft should be undertaken.
- **POPLITEAL ANEURYSM**
  - These are due to atherosclerotic degeneration associated with tobacco. **50% have another aneurysm elsewhere, including the contralateral popliteal site.** 14% of AAA patients have popliteal aneurysms.
  - **DIAGNOSIS**
  - Most are asymptomatic and incidentally found by physical exam and US or CTA imaging. Other presentations include thromboembolism, blue toe syndrome, claudication, or adjacent compressive symptoms of the popliteal vein or tibioperoneal neural branches.
- **TREATMENT**
  - Resection and **interposition graft** is indicated for any symptomatic aneurysm, or an asymptomatic aneurysm **> 2 cm** in size.
    - The posterior approach is preferred, especially if the aneurysm has produced adjacent compression. A lazy S incision from **superomedial to inferolateral allows harvest of the distal GSV or proximal lesser saphenous vein (LSV), and ligation of all geniculate branches.** *See Figure 30.16.* The proximal extent of the aneurysm is between the biceps femoris and semimembranosus and semitendinosus muscles.

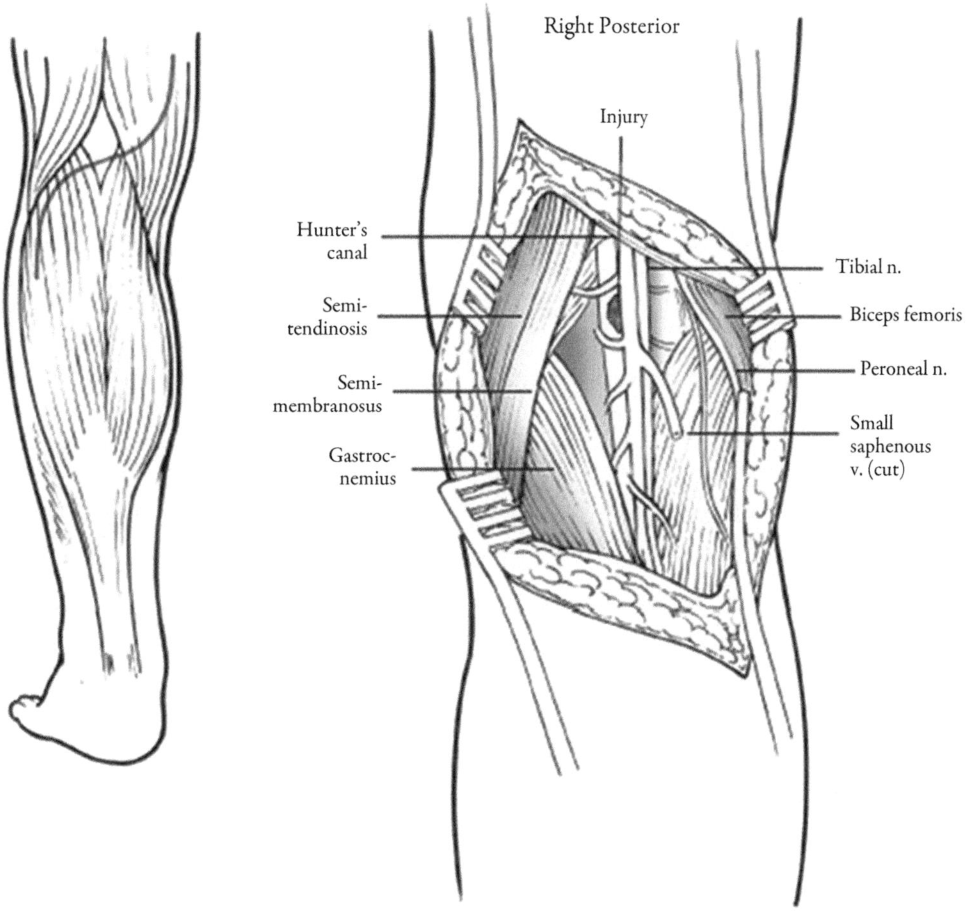

Figure 30.16 Posterior exposure of popliteal artery.

- Another treatment option is **surgical bypass via a medial approach** in an emergent situation, such as trauma or Rutherford 2B ischemia (see acute limb ischemia section below).
- Thrombolysis can be considered for Rutherford 2a acute ischemia.
- An endovascular stent graft can be used for asymptomatic elderly poor operative candidates, whose aneurysms have >1–2 cm length necks and adequate distal runoff.
- Asymptomatic aneurysms <2 cm may undergo US surveillance.

TIBIOPERONEAL

- Patients with tibioperoneal occlusive disease are less likely to have claudication and **more likely to present with critical limb ischemia (CLI)** symptoms. These include rest pain, tissue loss, or nerve dysfunction. Common comorbidities include renal disease, CAD, and DM.
- **DIAGNOSIS**
  - **ABI** <0.9 or TBI <0.7, or pressure drop across a lesion of >30 mmHg are abnormal.
  - ABI >1.4 is unreliable, since it implies non-compressibility of vessels. The lumen may be stenotic, but the vessel wall does not collapse on external pressure. The quality of the signal (triphasic, biphasic, or monophasic) may help assess luminal flow in the setting of non-compressible vessels. **Pulse volume recording (PVR)** can also evaluate lumen patency in these cases.
  - Evaluation of patients with CLI can show ABI <0.4, calf pressure <50 mmHg, toe brachial index (**TBI**) <0.5, or a toe pressure <30 mmHg. A minimum **$TcPO_2$** of 60 mmHg is needed for tissue healing.
  - **Duplex arterial US** is the next study if ABI or segmental pressures are abnormal.
  - **Formal angiography has better resolution than a CTA for tibioperoneal disease.**
- **TREATMENT**
  - **In the setting of dry, non-infected gangrene, including metatarsal head pressure necrosis (mal**

perforans ulcers), revascularization should be done before any amputation. This can salvage marginally perfused tissue, demarcate necrosis, and can augment healing of any future amputation site. Exceptions include gangrene limited to the distal digits, which may be allowed to auto-amputate regardless of the timing of an intervention.

- In the setting of **wet gangrene, systemic illness,** or a heel ulcer penetrating down to the calcaneus, debridement and **amputation should take place prior to attempts at revascularization.**
- **Endovascular interventions** are for discrete lesions <4 cm length stenosis, or <2 cm length occlusion. Endovascular therapy is also indicated if there are no autogenous conduit options.
  - Access is via the contralateral CFA, the brachial artery, or ipsilateral CFA. Retrograde access from the ipsilateral tibial artery is also an option.
  - Best visualization of tibial and peroneal anatomy is obtained with full strength contrast via a catheter in the popliteal artery. More proximal iliac and femoral arterial disease can be seen with half strength contrast or CO_2 angiography.
  - Percutaneous transluminal balloon angioplasty **(PTBA) with a drug-eluting stent (DES)** is the best intervention, since the caliber of these small vessels seems to benefit from stent placement and inhibition of myointimal growth. DES can be used even if there is no post-balloon stenosis, dissection, or recoil. An atherectomy device can be employed for recanalization prior to stent deployment.
  - **Dual antiplatelet therapy (DAPT)** is recommended for endovascular interventions. Surveillance US and ABI should be done at 1, 3, and 6 months post-intervention.
- **Surgical bypass** with autogenous vein is used for complex, multiple, longer segment stenoses or occlusions. Lesions resulting in significant tissue loss may also warrant surgical rather than end-vascular therapy. Ideally, patients should have a predicted lifespan >2 years.
  - Inflow can be taken from CFA, PFA, superficial femoral artery (SFA), or popliteal artery, depending on the length of available conduit.
  - Outflow can be to anterior tibial (AT) or posterior tibial (PT) arteries, whichever feeds the angiosome of the area of tissue loss. The peroneal artery is less desirable as a target.
  - The outflow target need not be circumferentially dissected. An Esmark bandage and tourniquet at the thigh, rather than clamps, can be sufficient for proximal vascular control during anastomosis.
    - Expose the **proximal or middle PT** with an incision 2 cm posterior to medial edge of tibia, retracting gastrocnemius and soleus posteriorly, and finding the artery lateral to the veins between the soleus and flexor digitorum compartment.
    - The **distal PT** is found posterior to the medial malleolus, between the flexor digitorum and hallucis tendons.
    - **Plantar artery tributaries** of the PT can be found by extending the incision from posterior to the medial malleolus over the medial edge of the foot.
    - The **middle segment of peroneal artery** is exposed from the same incision as that used for the middle segment of PT, but the soleus and PT neurovascular bundles are retracted posteriorly, and the dissection continued deeper (laterally).
    - **Distal peroneal artery** is exposed by lateral distal leg incision and fibulectomy.
    - **Distal AT artery** is exposed by an incision between the tibia and fibula anteriorly and retracting tibialis anterior and extensor digitorum muscles.
    - **Dorsalis pedis (DP) artery** is exposed by an incision between the 1st and 2nd metatarsals.
    - The ideal conduit is GSV. This can be in a non-reversed fashion for size match with a valvotomy to facilitate distal flow along the conduit. It could even be left in situ. Otherwise, the GSV conduit can be reversed and tunneled through the adductor hiatus, popliteal fossa, between the heads of the gastrocnemius, or via a subcutaneous course distant from the GSV harvest bed, and then anastomosed in standard fashion.
- If no adequate GSV conduit is available, consider an **endovascular intervention** as a second choice. Other options include the LSV, or basilic and cephalic veins in continuity or spliced together. **Cryopreserved vein or PTFE** (with a vein cuff for the distal anastomosis) have much lower patency.
  - Postoperatively, patients receive a **statin and single antiplatelet** with surveillance US and ABI at 1, 3, and 6 months. For GSV, there is a **50% 5-year patency,** which is not influenced by the level of the outflow target.
- **Anticoagulation can be considered for patients with poor runoff.** Construction of a distal arteriovenous fistula (AVF) can provide a low-resistance circuit for outflow, and improve runoff and possibly graft patency.

ACUTE LIMB ISCHEMIA

- This is the sudden onset of ischemic symptoms with a cool limb and impairments of sensory or motor function due to an acute complete arterial occlusion with inadequate collateral reserve. The etiology is either **thromboembolic or in situ thrombosis atop a ruptured plaque.**

- An embolus is organized thrombus from the atrial lumen, valve, left ventricle wall, aortic or iliac aneurysm wall, or paradoxically from a vein traversing through a right to left intracardiac or intrapulmonary shunt. It lands at bifurcation sites. The common femoral artery bifurcation is the most common, followed by aortoiliac bifurcation, and then the popliteal trifurcation. Emboli in the distal brachial artery at the radio-ulnar bifurcation are less common.
- **DIAGNOSIS**
  - The limb should be markedly different from their other limbs and the patient's baseline vascular status. Patients may have a water hammer pulse just proximal to the occluded site and a normal contralateral limb in the setting of an embolic etiology. Distally they often have nonpalpable and non-Dopplerable pulses with a cool extremity and variable limitations to sensory and motor function.
    - Patients may have evidence of baseline arterial insufficiency in their contralateral limb, or elsewhere, if the etiology is a thrombosis in situ.
    - Categories of acute limb ischemia are:
      - **Rutherford 1**: normal cap refill.
      - **Rutherford 2a**: sensory disturbance
      - **Rutherford 2b**: motor disturbance.
      - **Rutherford 3**: rigor. Veins may be thrombosed as well.
    - The source is evaluated with **echocardiogram and CTA of the entire aorta, chest, abdomen, and pelvis, with extremity runoff.**
- **TREATMENT**
  - For all patients, start **heparin** 80–100 u/kg and 15–20 u/kg/hr for a goal PTT 60–80, provided there is no contraindication. Argatroban is an alternative.
  - If **Rutherford 2a** and the occlusion is within 14 days, one may add **intra-arterial recombinant tissue plasminogen activator (rt-PA)** for 24–36 hours via contralateral CFA access, a sheath parked in the ipsilateral iliac artery, and a multi-sidehole catheter residing *within* the thrombus. A suction or mechanical endovascular **thrombectomy** can also be performed. Runoff may be assessed via angiography when across the lesion.
    - Thrombolysis infusion is halved if fibrinogen levels are <150, and halted if fibrinogen levels are <100. A heparin GTT with goal PTT 30–50 is used to prevent thrombosis locally, within the catheter.
    - Patients may need a subsequent, less aggressive, vascular reconstruction.
    - There is a **1% CVA risk but lower MI risk with thrombolysis.**
    - **25% of patients fail to recanalize** and require surgery, but the amputation and mortality rates are no different than if surgery was done first.
  - For **Rutherford 2b or 3,** patients need operative management tailored to either embolus or thrombosis in situ
    - Suspected thromboemboli require **arteriotomy and Fogarty thromboembolectomy** with rt-PA for up to 24–36 hours to address thrombi that have accrued distally.
      - The procedure is performed via femoral cutdown, transverse arteriotomy, and retrograde embolectomy with 5-6F Fogarty balloon catheter. Also, perform antegrade embolectomy down PFA and SFA to popliteal artery with 3–4F Fogarty balloon catheter. One may augment this with a longitudinal distal popliteal or tibial arteriotomy and embolectomy retrograde and antegrade from these sites with 2–3F Fogarty balloon catheters. For an acute aortoiliac embolus, the procedure can be performed bilaterally, simultaneously.
      - UE embolectomy is via longitudinal brachial arteriotomy using a 4F balloon proximal and 3F distal.
    - **Atheroemboli** are composed of cholesterol and platelets. They lead to inflammation when lodged at end-arterial sites. The result is renal ischemia, blue toes, or trash foot. Most often they are from an **aortoiliac mural plaque or aneurysm.** Treatment is with a stent graft or resection with an end–end proximal aortobifem to address the source of emboli.
    - For thrombosis in situ, patients need **open thrombectomy, endarterectomy with patch angioplasty, bypass, or resection and interposition graft.**
  - Acute limb ischemia carries a **10%–30% amputation rate at 30 days.**
  - Consider adding **multi-compartment fasciotomy** for any patient with greater than 4–6 hours of ischemia, especially if there is concomitant venous insufficiency.
  - For patients with a prior surgical intervention:
    - Thrombosis of a prior PTFE graft will produce symptoms that are generally worse than the patient's preoperative state.
    - Thrombosis of a prior saphenous vein graft results in a return to baseline symptomatology, but the saphenous vein wall may necrose.
    - If <30 days post-op or the patient exhibits Rutherford 2b or 3 symptomatology, the etiology is likely a technical error, and **open thrombectomy and graft revision** may be needed.
    - If >30 days post-op and Rutherford 2a, **catheter thrombolysis, mechanical thrombectomy, and percutaneous transluminal balloon angioplasty (PTBA) of an intimal hyperplastic lesion or an atherosclerotic plaque** may be attempted after the lumen has been reclaimed and imaged.

AMPUTATION

- **Trauma:** Amputation is more commonly needed in blunt distal extremity injuries as opposed to proximal, penetrating extremity injuries. The latter have increased mortality.
- **Ischemia and infection**: In the setting of **dry, non-infected gangrene, including metatarsal head pressure necrosis (malum perforans ulcers), revascularization should be done before any amputation.** This can salvage marginally perfused tissue, demarcate necrosis, and can augment healing of any future amputation site.
  - Exceptions include gangrene limited to the distal digits, which may be allowed to auto-amputate regardless of the timing of revascularization.
  - In the setting of **wet gangrene, systemic illness,** or a heel ulcer penetrating down to the calcaneus, **debridement and amputation should take place prior to attempts at revascularization.**
  - Persistent rest pain with failed revascularization is another indication for amputation.
- A palpable pulse one level superior to the amputation site predicts that the site will heal.
- **Above the knee amputation (AKA)** is appropriate for advanced pathology above the knee, or disease below the knee in a non-ambulatory or contracted patient. 90% heal, and 40% of patients who were previously ambulatory may walk again. A palpable common femoral pulse and intact profunda femoris tributaries suggest the amputation will heal. A fishmouth incision is most often used.
- **Below the knee amputation (BKA)** is appropriate for advanced pathology below the knee, provided there is no knee contracture, muscle wasting, or extensive soft tissue loss. 75% heal, and the majority walk. A palpable popliteal pulse, an ankle pressure >60 mmHg or $TcPO_2$ >40 mmHg predict a high likelihood of the stump healing.
  - An anterior transverse incision is made 10 cm distal to the tibial tuberosity around two-thirds of the limb circumference. A posterior flap should extend 10 cm distal to the anterior incision. The extension should be for a distance equal to about one-third of the limb circumference. All tibioperoneal arteries and veins should be controlled. Transect the fibula 1 cm proximal to the tibia. Fascial closure may take the form of myoplasty (suturing muscle overtop bone) or myodesis (suturing muscle to bone).
  - Use a **brace or a rigid dressing with plaster** to protect the stump staple line and prevent contracture. Serial stump-shrinker applications and rehabilitation are needed for the next 3 months. This is followed by initial prosthesis training for 6 months, and then a permanent prosthesis is fitted and used for 3–5 years with continued physical therapy.
- **Complications.** Mortality is primarily from cardiovascular events postoperatively—CHF, CVA, chronic kidney disease (CKD). Some patients suffer from prolonged postoperative pain and phantom limb pain. There is about a 12% risk of needing a contralateral major amputation. This is highest in patients with DM and CKD. Local soft tissue wound infections can be managed with debridement and antibiotics. Osteomyelitis of the stump will need stump revision.
- **Other amputations**
  - **Symes** amputation across tibial joint surface.
  - **Chopart** retains talus and calcaneus.
  - **Lisfranc** retains tarsal bones.
  - **Transmetatarsal amputation** is angled to produce shorter metatarsal lengths from medial to lateral, and is covered with a long plantar flap.
  - **Ray amputation** resects the digit as well as the metatarsal head.
  - **Digital amputation** may be at the metatarsophalangeal or interphalangeal joint.
  - **Upper extremity amputations** are generally aimed at retaining as much functional limb length as possible.

VENOUS DISEASE

- Proximal veins have more adventitia and fewer valves compared to distal veins. Distal veins recanalize better after a deep venous thrombosis (DVT).
- **CHRONIC VENOUS INSUFFICIENCY (CVI)**
  - This may be due to the post-thrombotic syndrome (PTS) after a DVT or from reflux of one or more deep or superficial veins.
- **DIAGNOSIS**
  - Symptoms and signs include aching, **edema**, nocturnal cramps, ulcers, epidermal **hyperpigmentation**, dermal and subcutaneous **sclerosis,** and destruction of lymphatics with eventual lymphedema. Insufficiency. An **ultrasound (US)** can evaluate for underlying DVT or venous dilation.
  - **Augmented venous pressure** is increased in chronic insufficiency. Patients with a pressure > 80 mmHg have ulcerations 80% of the time.
  - **Calf pump exercises** do not decrease the venous pressure as markedly if there is CVI. And pressure more quickly returns to baseline if there is reflux from upstream. The latter can be diagnosed by a US placed cephalad to a deflating BP cuff.
  - A **Trendelenburg test** is done to determine the site of insufficiency. A tourniquet is placid on the proximal thigh (compressing the GSV) and proximal lower leg (compressing the LSV), and then the patient stands up

to see if varicosities develop immediately. This would indicate incompetent perforators. If no varicosities fill initially, but develop after release of one of the tourniquets, then that site contains the incompetent saphenous vein.

- **CEAP Classification:**
  - C1: small varicosities and C2 significant varicosities.
  - C3: edema and C4 lipodermatosclerosis.
  - C5: healed venous ulcer and C6 active venous ulcer. Lateral ulcers are from an incompetent LSV, and medial ulcers are from a refluxing greater saphenous vein (GSV). If the saphenous system appears normal, refluxing perforating branches of Boyd or Cockett in the vicinity of an ulcer may be the etiology. Culprit refluxing veins are generally seen as dilated 3.5 mm on US.

- **TREATMENT**
  - Start with **compression stockings at 30–40 mmHg**. Rarely, valve reconstruction or valve auto- or cryotransplantation are done.
    - Venous ulcers are treated with an **Unna boot every week** (zinc oxide, glycerin, calamine, gelatin, Mg-aluminum). This heals 90% of ulcers in 5 months, especially if <3 cm. If unresponsive to the Unna boot, then a US evaluation for refluxing veins would be the next step.
    - **Ablation of a refluxing GSV or LSV can be with laser or radiofrequency (RFA)** for a vein <2 cm in diameter and cephalad of the ulcer, *See Figure 30.17.* **Ligation** is preferred if the refluxing GSV is >2 cm diameter. Both procedures **reduce ulcer recurrence** and require that the deep venous system is open. They do not accelerate healing of an active ulcer, however.
    - **Subfascial endoscopic perforator surgery** involves ligation of refluxing Boyd (around the knee) or Cockett (around the ankle) perforators to spare the dermis from high venous pressures and subsequent ulceration.

- **VARICOSE VEINS**
  - These are superficial enlargements and need not be associated with any underlying DVT, chronic venous insufficiency, saphenous reflux, or incompetence of perforator veins. Risk factors include smoking, obesity, and low activity.
    - Diagnosis by physical exam can be augmented by an US to exclude underlying DVT, but is not mandatory.
    - Management of isolated varicosities is with **sclerotherapy**. Treat proximal before distal varicosities. Foam sclerosant or **stab phlebectomy** is preferred if the varicosities are >3 mm in diameter.

- **SUPERFICIAL THROMBOPHLEBITIS**
  - Inflammation associated with a thrombosed superficial vein. It is treated with **NSAIDs and warm compresses.** If located <3 cm from the GSV–CFV junction, or there have been recurrent thromboses, or the patient is hypercoagulable, then **anticoagulate for 6 weeks** (consider fondaparinux), provided an initial DVT US shows no extension into the deep system. After 6 weeks, a repeat US of the deep veins can be performed.

- **SUPPURATIVE THROMBOPHLEBITIS**
  - An infected thrombosed vein. It is treated with antibiotics and resection of the vein if superficially located in soft tissue. If the vein is in the pelvis, then antibiotics and anticoagulation alone are utilized. Vein ligation or resection in this location is reserved for those who are nonresponsive.

- **DEEP VEIN THROMBOSIS (DVT)**
  - Pathophysiology is based on Virchow's triad of stasis, hypercoagulability, and endothelial injury.
    - Categories considered **highest risk** for venous thromboembolism (VTE) include spinal cord injury, long bone fractures, and patients undergoing general and

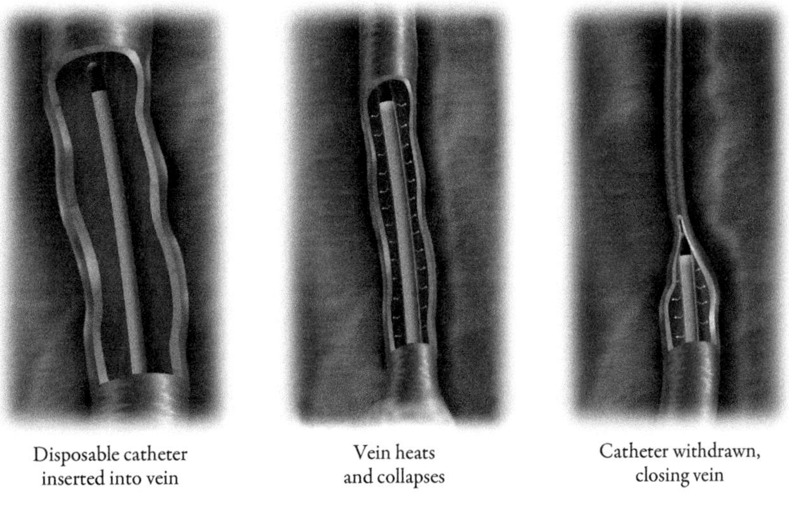

Figure 30.17 Venous ablation.

orthopedic surgery. Malignancy and other traumatic injuries are also at risk for venous thrombosis.

- **VTE prophylaxis for high and moderate risk patients is with Lovenox qd or BID** or with heparin TID. The latter has a higher bleeding risk, but is preferred if the patient has renal insufficiency. Prophylaxis should continue for **4 weeks post-op abdominopelvic surgery for weight loss, malignancy, or inflammatory bowel disease.**
- Categories considered **lower risk** for VTE include laparoscopy surgery or a single risk factor of smoking, obesity, or immobility alone. Prophylaxis is with ambulation and sequential compression devices to the lower extremities.
- Cardiac surgery patients do not receive pharmacologic VTE prophylaxis in the immediate post-op period.

- **DIAGNOSIS**
  - US from the CFA down through popliteal veins can demonstrate a hyper-echoic lumen that is non-compressible.
  - A CT venogram is used for locales that cannot be accessed by US, or for equivocal findings on US. Magnetic resonance venography (MRV) or a formal venogram are options if US or CT venogram are not diagnostic, but clinical suspicion remains high.
- **TREATMENT**
  - 30–40 mmHg knee-high **compression stockings.**
  - **Lovenox 1 mg/kg BID** has decreased bleeding compared to unfractionated heparin and is a better option if the patient has an underlying malignancy.
  - **Unfractionated heparin IV infusion** is used for patients who are at higher bleeding risk and may need to have anticoagulation abruptly held. It is also preferred over Lovenox if the patient has renal insufficiency.
  - Either Lovenox or heparin infusion should be transitioned to **warfarin,** except in the setting of underlying malignancy. In that case, continuing Lovenox is preferred.
  - **Anti-Xa agents like rivaroxaban or apixaban** can be used in place of warfarin for a proximal non-malignant iliofemoral DVT.
  - **Dabigatran** is an oral direct thrombin inhibitor that can also be used for DVT.
  - The **duration of anticoagulation** should be 3 months if the thrombosis is provoked, 3–6 months if there is no identified provoking factor, or 6 months or longer if it is perceived the patient has an ongoing thrombotic risk.
  - An isolated distal calf (soleal) DVT below the knee without other hypercoagulable risks may be **observed, and a repeat US performed in 1-2 weeks.** Anticoagulation would be indicated if the thrombosis progresses above the knee.
  - If the thrombosis is related to a central line, then the line may only stay in place if it is of high necessity and cannot be re-sited. Otherwise it should be removed.
  - **Catheter directed thrombectomy and thrombolysis** can be considered for the following:
    - 1. A proximal iliac DVT with **phlegmasia alba dolens** (intact internal iliac vein tributaries), **phlegmasia cerulea dolens** (occluded internal iliac vein tributaries), or for any limb threatened by poor perfusion secondary to the venous obstruction.
    - 2. Reduction in the severity of **post-thrombotic syndrome.** This syndrome occurs in up to 50% of patients within 2 years of the diagnosis of DVT.
    - 3. **May-Thurner Syndrome.** Left iliac vein thrombosis due to compression by an overlying right iliac artery. In this case, an iliac vein stent may also need to be deployed. *See Figure 30.18.*
    - 4. **Paget-von Schroeder** venous thrombosis of the axillo-subclavian veins that is effort-induced. In this case, a **trans-axillary** rib, scalene, and subclavius resection may need to be performed after the patient is re-canalized by thrombectomy, thrombolysis, and anticoagulation (see upper extremity section on thoracic outlet syndrome, above).
    - In order to augment any recanalization and intervention on the iliac vein, one may construct an AVF by transecting the GSV and sewing its proximal end to the side of the adjacent femoral artery. This will allow pulsatile arterial blood to flow through the GSV and up into the iliac vein, preventing stasis and thereby keeping it patent.
- **PULMONARY THROMBOEMBOLISM (PTE)**
  - PTE often originated from a lower extremity DVT. It can also occasionally come from an upper extremity deep vein, pelvic vein, or from a thrombus generated within the right atrium or ventricle.
- **DIAGNOSIS**
  - Symptoms include pleuritic, usually lateralized, chest pain (peripheral pulmonary emboli) or shortness of breath (central emboli). Confirmation is by **CT pulmonary angiogram** of the chest. **Echocardiography** can also suggest a large PTE via demonstration of a dilated pulmonary artery or right ventricular strain.
- **TREATMENT**
  - **Lovenox or unfractionated heparin (UFH)** should be given for the same duration of treatment as for a DVT.
  - **Systemic thrombolysis** is used if the patient is in obstructive shock and cannot be adequately resuscitated. **Catheter directed thrombectomy with or without thrombolysis** is an option for

patients with a sub-massive embolus, provided they can be resuscitated and temporized with inotropes, vasopressors, or extracorporeal support.

- Sternotomy and **surgical thromboembolectomy** is a last resort for the patient who has failed other forms of treatment or is not a candidate for catheter based thrombolysis. In a chronic or subacute setting with organized and incorporated thrombus in the pulmonary arterial tree, surgical retrieval and pulmonary endarterectomy under hypothermic circulatory arrest may be necessary.
- **Retrievable inferior vena cava filter** may be indicated as prevention of PTE in the following cases:
  - Patients who developed recurrent VTE or PTE despite therapeutic anticoagulation; or
  - Patients who have a contraindication to anticoagulation of VTE or PTE; or
  - Patients who have complications from anticoagulation.
  - IVC filters are placed below the renal veins.
    - Venogram should be done before placement to ensure adequate size of cava, position of renal veins, and absence of thrombus. A venogram is also done prior to retrieval to ensure that no clot is embedded below the filter, as this could embolize if removal is attempted.
    - The filter may be removed if no longer needed, since 20% will have a recurrent DVT, and 1%–3% will have filter migration, thrombosis, or occlusion at the filter site.
    - Other complications include mechanical failure, tines embedding through the vein wall, or secondary infection of the filter and subsequent bacteremia.
  - **Superior vena cava (SVC) filters** can prevent an UE DVT from resulting in a PE. This can occur in up to 9% of cases.

LYMPHATIC DISEASE

- Lymphatics are not present in bone, muscle, tendon, or cartilage.
- The channels have no basement membrane. Valves are present in subcutaneous lymphatic channels, but not dermal lymphatics.
- Lymphatic insufficiency results in interstitial accumulation of protein and cells, fibrosis of lymphatic walls with loss of valve function, inflammatory cell infiltrate with fibrinoid thrombi obstructing channels, and finally, fibrosis of soft tissue with findings of hyperkeratosis, acanthosis, adipose hypertrophy, and pitting edema. Eventually the edema becomes indurated and non-pitting.
- Areas of the body with poor lymphatic function have a baseline higher risk of secondary infection.
- **LYMPHATIC INSUFFICIENCY**
  - Categorized as primary or secondary.
    - Primary includes **congenital lymphedema** in female patients under 2 years old who have **bilateral lower extremity edema. Lymphedema praecox** is the #1 cause of primary lymphedema and occurs in females <35 years of age who have **unilateral LE** edema. **Lymphedema tarda** is the term used for primary insufficiency in females over 35 years of age.
    - Secondary causes include infection with *W. bancrofti*, cancer therapy (lymphadenectomy, radiation), trauma, bacterial infections, and obesity.

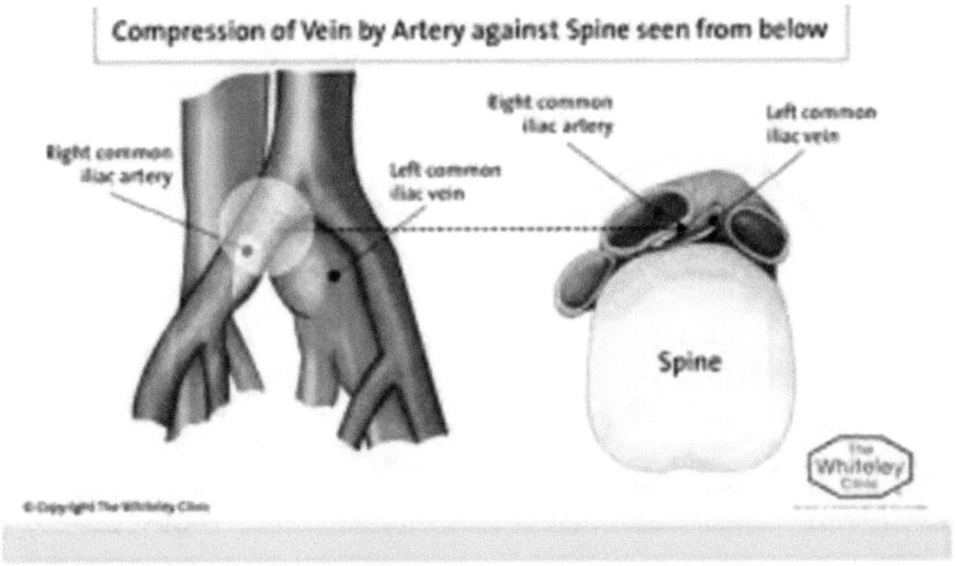

Figure 30.18 Left common iliac vein thrombosis due to overlying right iliac artery. May-Thurner syndrome.

- **DIAGNOSIS**
- Physical exam findings are the key, with additional imaging reserved for select cases.
  - **Lymphoscintigraphy** can be used as an adjunct and the radio tracer injected in the interdigital dermal or subcutaneous (SQ) space. Normal findings should be **inguinal node enhancement by 60 minutes and pelvic node enhancement by 180 minutes.** There would be no uptake if there is dermal or SQ lymphatic insufficiency.
  - **Indocyanine green (ICG) lymphangiography** injects dye directly into the lymphatic lumen and can map out lymphatic channels with greater detail.
- **TREATMENT**
  - Start with knee-high or thigh-high graduated compression garments at **40–60 mmHg, weight loss, and skin care.**
  - **Manual or automated pump massage** can be added for decongestion 4–6 hours a day at 30–60 mmHg.
  - Subdermal end–end or end–side lymphovenous anastomoses can be attempted after ICG lymphangiography demonstrates suitable lymphatic channel candidates. This can improve volume, symptoms, and infection risk, provided it is done early in the course before fibrosis sets in.
  - Vascularized lymph node transfer from the groin to an axillary post-lymphadenectomy site allows for lymphangiogenesis and enhanced lymphatic pumping.
  - **Debulking with liposuction or excision of SQ tissue**, with or without skin grafting, is used once fibrotic tissue changes have set in, and physiologic recanalization (above) would no longer be sufficient.
- **LYMPHOCELE**
  - Most commonly, this will be seen after an inguinal dissection. It may originally appear as a non-resolving seroma that re-accumulates rapidly after attempted aspiration. Lymphoceles are treated with percutaneous drainage, resection, or ligation of a feeding channel (identified by lymphangiography via isosulfan blue injection into the dermis of the distal limb).
- **LYMPHANGIOSARCOMA**
  - Presents as a red and blue plaque on an extremity and may have lung metastases. It occurs years after axillary lymph node dissection (AxLND) with or without radiation. Treatment is with wide local excision (WLE) or amputation.

OTHER LIMB PATHOLOGY

- **COMPARTMENT SYNDROME**
  - Due to restoration of flow after blunt muscle damage or prolonged limb ischemia of more than 4–6 hours. **Injury is mediated by polymorphonuclear cells (PMN) during the reperfusion phase**, resulting in muscle edema within a fixed fascial compartment. As compartmental pressure rises to exceed capillary pressure, secondary ischemia ensues.
  - Common etiologies include crush of a limb, supracondylar fracture, proximal tibial fracture, or arterial injury or occlusion followed by reperfusion. Patients having concomitant venous and arterial injuries in the same limb are at highest risk. Other cases may be chronic and exertional, leading to elevated pressures in the anterior and lateral lower leg compartments.
- **DIAGNOSIS**
  - Symptoms and signs start with **pain** with passive range of motion progressing to **paresthesia,** anesthesia, paralysis, poikilothermia, and pulselessness last.
  - If needed for confirmation, compartment pressures can be measured and are usually >20 mmHg. Or, a calculated delta pressure (diastolic blood pressure – compartment pressure) will be <30 mmHg.
  - Creatinine kinase may be elevated. Associated acute kidney injury is due to myoglobin casts.
- **TREATMENT**
  - Prompt four-compartment fasciotomy via a lateral incision over the fibular area to access both the anterior and lateral compartments, and a medial incision running just posterior to the tibial edge to access the superficial and deep posterior compartments. *See Figure 30.19.*
    - The **anterior compartment** contains the tibias anterior, extensors, and deep peroneal nerve. It is essential to recognize the fascial interface separating the anterior

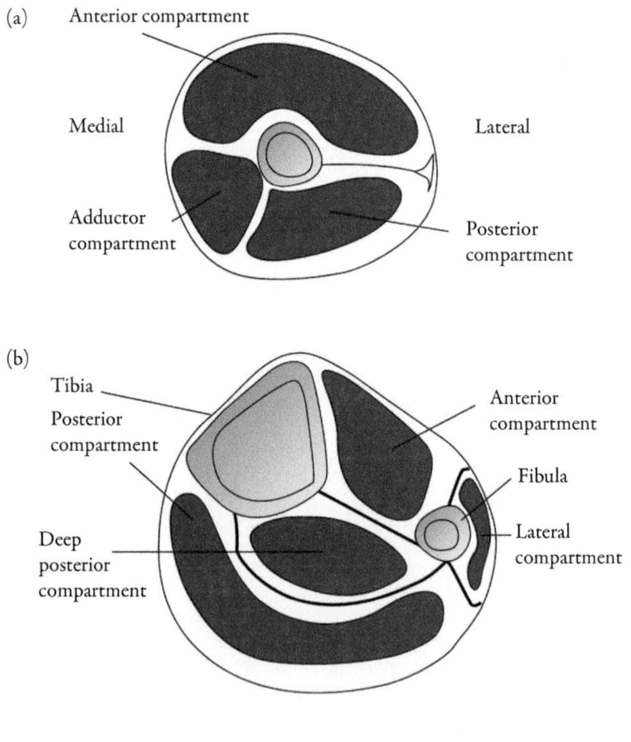

Figure 30.19 Compartments of leg- upper (a). lower (b).

and lateral compartments, so as to ensure that both compartments are independently opened to their full extent in a cranial-caudal orientation.

- The **lateral compartment** contains the peroneus muscles and neighboring superficial peroneal nerve, which can be injured during the incision.
- The **deep posterior compartment** has the tibialis posterior, flexors, and the posterior tibial nerve. It is accessed by detaching the soleus and entering the area just posterior to the tibial bone.
- The **superficial posterior compartment** has the soleus, gastrocnemius, and the sural nerve nearby.

- An **upper extremity fasciotomy** is done on the volar and dorsal sides.
  - Volar fasciotomy starts at the ulnar side of the upper arm, crosses the antecubital fossa to the middle or radial side of the forearm, and heads down to the middle of the wrist. This helps release the median nerve at both sites proximally and distally.
  - Dorsal fasciotomy is on the radial side of the upper arm and forearm.
- **POPLITEAL ENTRAPMENT SYNDROME**
  - Occurs in men in their 40s with intermittent claudication due to impingement of the popliteal artery by the medial head of the gastrocnemius.
  - Patients may have a loss of pulses or drop in ABI in plantar flexion. 30% have venous impingement, too.
  - Treatment involves **resection of the tendinous attachment of the medial head of gastrocnemius** as well as **arterial reconstruction** if the wall has incurred chronic injury. The contralateral side may need treatment even if asymptomatic.
- **ADVENTITIAL CYSTIC DISEASE**
- Occurs in men in their 40s with intermittent claudication due to a mucin-filled cyst originating from the joint space. The scimitar sign describes the appearance of the impinged popliteal artery on angiogram.
- Treatment is with **cyst resection or enucleation.** Treatment of the contralateral asymptomatic side may be necessary.

ACUTE MESENTERIC ISCHEMIA

See Figures 30.20 and 30.21.

- **PATHOPHYSIOLOGY**
  - Encompasses the following etiologies:
    - 1. Thrombosis atop a ruptured mesenteric vascular plaque. This is the **most common** cause of mesenteric ischemia, and the second highest mortality.
    - 2. Embolus to the superior mesenteric artery.
    - 3. Ischemia related to overall low flow in mesenteric vasculature due to low cardiac output (non-occlusive ischemia). This has the **highest mortality** of all types.
    - 4. Mesenteric venous thrombosis producing SB mucosal congestion and secondary inflow ischemia.

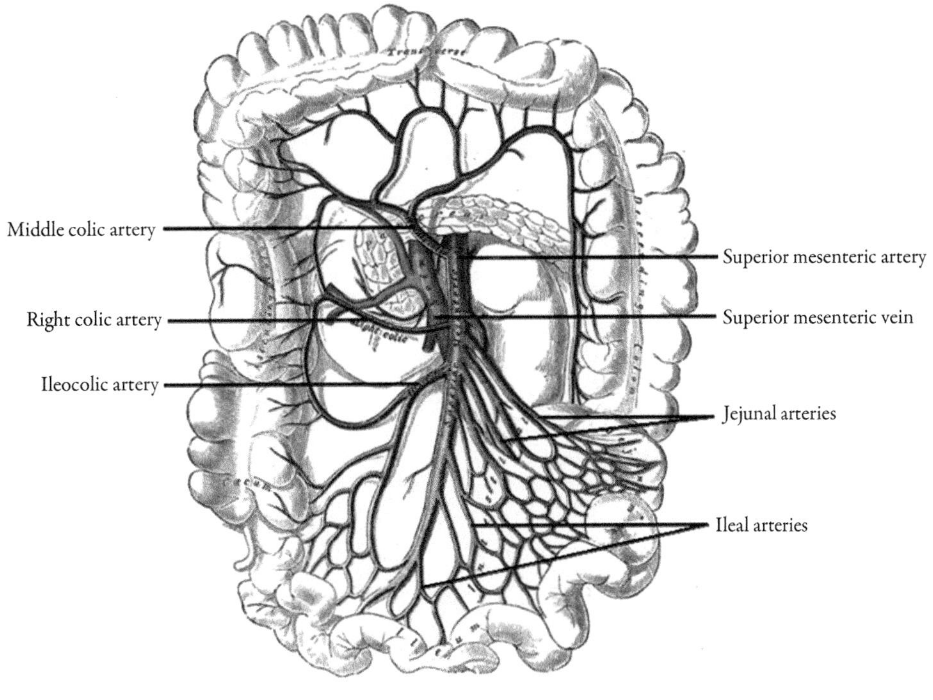

Figure 30.20 Mesenteric vascular anatomy.

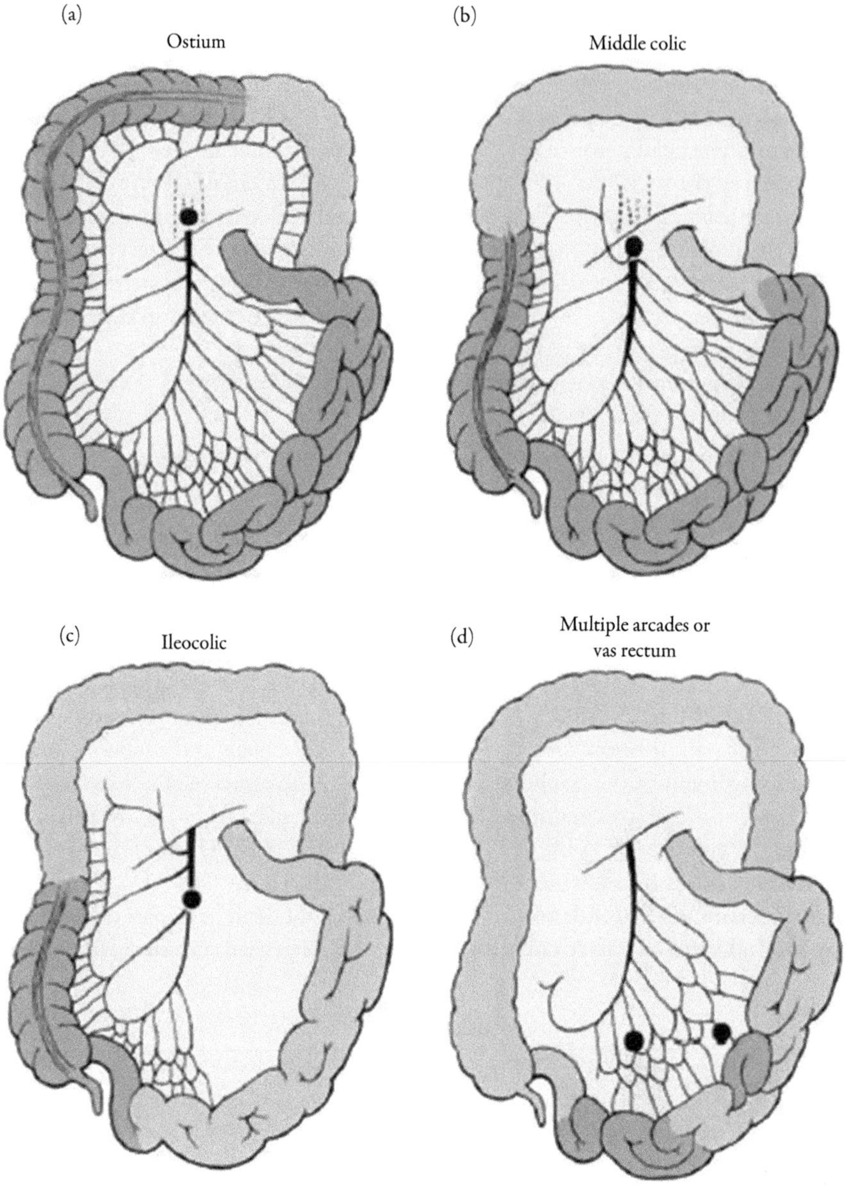

Figure 30.21 Distribution of vascular ischemia.

- **DIAGNOSIS**
  - Presents with upper gastrointestinal (UGI) symptoms of pain and nausea or vomiting.
  - Diarrhea or hematochezia are less common.
  - The abdominal exam may be soft and nontender early on, when the ischemia is limited to the mucosal lining. The patient will still have severe pain from mucosal oxygen deprivation, however.
  - Eventually the ischemia extends full thickness to the serosal covering. At that time, palpation of the abdominal wall will bring the parietal peritoneum up against ischemic serosa and result in involuntary guarding.
  - The diagnosis is confirmed by a CT angiogram (**CTA**) showing:
    - 1. **Tapering of superior mesenteric artery** within 1–2 cm of its origin if thrombosis atop plaque.
    - 2. **Abrupt meniscus cutoff of SMA** several centimeters from origin if embolic.
    - 3. Patent SMA and IMA if non-occlusive ischemia; or
    - 4. Thrombosis somewhere along portal venous system.
  - For patients who are unstable or who have an equivocal CTA, a **diagnostic laparoscopy** or exploratory laparotomy may be required for diagnosis and therapy.
- **TREATMENT**
  - Initially start a heparin drip for any suspicion of thrombosis or embolus, provided there are no contraindications.
  - **Embolus** is treated by laparotomy and exposure of the SMA.

- The SB is retracted medially, and the transverse colon cephalad.
- The ligament of Treitz is transected, and the 4th portion of duodenum retracted to the patient's right side.
- Then, the retroperitoneal reflection at the root of the mesentery is opened, and a transverse SMA arteriotomy proximal to the takeoff of the middle colic artery is performed.
- **Distal and proximal Fogarty balloon embolectomies** can retrieve clot.
- Finally, the transverse arteriotomy is closed primarily.

- **Thrombosis atop a plaque** is treated by laparotomy, SMA exposure as above, a longitudinal arteriotomy, and **thrombectomy**.
  - A retrograde wire and stent of the proximal SMA may be performed.
  - The longitudinal arteriotomy can be closed by patch angioplasty.
  - Some patients may need **autogenous vein bypass** from the supraceliac aorta, distal aorta, or right common iliac artery (CIA) to the proximal SMA, just inferior to the pancreas.
  - A few patients will need an additional bypass to the celiac artery, common hepatic artery, or proximal splenic artery.
  - Pure endovascular approaches for embolus or thrombosis in situ are not regularly used if there is need for assessment of bowel viability.
- **Nonocclusve ischemia** is treated by supportive measures, fluid resuscitation, optimization of cardiac output, antibiotics, or, less commonly, SMA catheterization and papaverine infusion (30–60 mg/hour for 24 hours).
  - A laparotomy may be needed to assess bowel viability, but is not routinely done.
  - **Ischemic colitis** is one type of non-occlusive ischemia. It manifests at Griffiths point, Sudek's point, and the distal ileocolic watershed areas. 80% of these cases are successfully managed medically, provided the ischemia is limited to superficial mucosa. See Chapter 10, Colon.
- **Mesenteric vein thrombosis** is treated with a heparin drip.
  - Intra-arterial thrombolytics via the SMA or splenic artery have been used if ischemia is refractory to anticoagulation.
  - There is no role for venous thrombectomy.
  - Patients may need a laparotomy and bowel resection if the clinical picture is consistent with full-thickness necrosis of the bowel wall. But this is not common. Strong consideration should be given to a mandatory second look laparotomy in all operative cases.

CHRONIC MESENTERIC ISCHEMIA

- **PATHOPHYSIOLOGY**
  - Long-standing, progressive celiac and SMA artery atherosclerotic stenoses is the hallmark of this condition.
  - The metabolic demands of food intake, luminal metabolization, and peristalsis overwhelm oxygen supply in the setting of impaired perfusion. This results in mucosal ischemia.
  - **Median arcuate ligament syndrome** is a less common cause of chronic ischemia due to an extrinsic impingement of the celiac artery and impaired foregut perfusion.
- **DIAGNOSIS**
  - Patients with chronic ischemia have food fear and postprandial pain with weight loss. Their abdominal exam is usually benign, however.
  - Suspicious cases who have concomitant renovascular disease can receive a **duplex US of mesenteric vasculature**. This can also be used as **surveillance** in known cases. Concerning features include:
    - Peak systolic velocity (PSV)>275 cm/s or end diastolic velocity (EDV) >45 cm/s.
    - PSV:AO ratio >3 is indicative of >70% diameter reduction.
    - PSV <40 is indicative of critical stenosis.
    - A test meal of 8 oz. protein drink can be given, and then a repeat US evaluation of the SMA to note any gradient change. Gradient changes in the celiac or IMA indicate this vessel is providing critical collateral blood flow.
  - **Diagnostic angiography or CTA** should follow for patients with a positive US. Alternatively, if there is a high clinical suspicion for the diagnosis and the patient does not have renovascular disease, CTA can be the initial modality.
- **TREATMENT**
  - An endovascular **balloon-expanding stent** can be placed across a 1–2 cm calcified, ostial, or proximal stenotic lesion.
  - A **self-expanding stent** may be better for lesions distal along the artery. Stents have a 30% failure rate at 1 year.
  - The definitive treatment is **revascularization** of the SMA with or without revascularization of the celiac or IMA as well.
    - **An antegrade PTFE** bypass goes from the supraceliac aorta (transcural exposure) to the end of the celiac artery beyond the diseased segment. A second graft can traverse from the supraceliac position in a retro-pancreatic course to the side of the SMA.
    - **A retrograde PTFE** bypass goes from the side of the infrarenal aorta or right iliac artery to the side of the SMA or side of the common hepatic artery.
    - An aortic endarterectomy extending 2 cm into the ostia of the celiac, SMA, and renal arteries requires medial visceral rotation and is less commonly done.

- Median arcuate ligament syndrome is treated with transection of the ligament and arterial reconstruction if needed.
- **Postoperative duplex surveillance** of an SMA stent or bypass graft is done every 6 months to detect myointimal hyperplasia. This can be followed by **angiography** if criteria indicate a >70% stenosis. **Endovascular intervention** is done to assist patency in these cases, even if the patient is asymptomatic.

OTHER VISCERAL ARTERIAL DISEASE

- **RENAL ARTERY (RA) STENOSIS**
  - Results in renin overproduction by a hypoperfused juxtaglomerular apparatus (JGA) distal to a narrowed arterial lesion. It is the etiology to hypertension in 5%–10% of patients.
  - Atherosclerotic lesions are most commonly in the proximal renal artery and ostia. The left side is more often involved than the right.
  - Fibromuscular dysplasia (FMD) can involve the middle or distal renal artery. The right side may be more commonly involved in FMD.
- **DIAGNOSIS**
  - Most patients are asymptomatic. Some will have hypertension refractory to medical management. Other presentations include acutely worsening renal function or flash pulmonary edema.
  - Ultrasound is the initial study. A peak systolic velocity **(PSV) >180 cm/s or PSV ratio of >3.5 across the lesion** are both indications for additional imaging. Ultrasound can also evaluate **intrarenal interlobular and arcuate arteriolar resistance indices.** The peak systolic intrarenal velocity minus the end diastolic velocity is divided by the peak systolic velocity. Higher indices indicate that diastolic flow is impaired due to *arteriolar* narrowing.
  - Confirmation of arterial disease is by **CTA or formal angiography.**
  - Other modalities that can suggest the need for angiography include a captopril renal scan leading to a decrease in GFR on the side of the stenotic lesion, renal vein renin level ratios >1.5 compared to the contralateral normal side, or renal vein to systemic renin ratio >0.24. All indicate JGA hypoperfusion.
- **TREATMENT**
  - **Asymptomatic patients are managed medically** with statin, antiplatelets, smoking cessation, glucose control, and an ACE inhibitor with or without additional anti-hypertensive therapy. ACE-inhibitors should be avoided if there is bilateral renal artery stenosis.
  - In **symptomatic patients, PTBA and stent with or without Plavix** is the first-line therapy. A covered stent may be needed if the patient has a renal artery aneurysm >1.5 cm. Surveillance should continue with US q6–12 months post-therapy.
  - Second-line surgical options include:
    - 1. Transverse endarterectomy across the aorta and bilateral RA.
    - 2. Vein graft bypass from the side of the aorta, side of common hepatic artery, or end splenic artery to the distal end of the ipsilateral RA after its transection just distal to the stenosis.
    - 3. Excision of the stenotic area and kidney reimplantation.
    - 4. Nephrectomy for persistently high renin levels in an atrophic kidney.
    - 5. Complex renal artery aneurysms can be repaired after nephrectomy in a hypothermic ex vivo fashion, after which the kidney can be auto-transplanted.
- **VISCERAL ARTERY ANEURYSMS**
  - May be found incidentally, and are treated with **liquid or coil embolization or a covered stent.** *See Figure 30.22.* Indications for treatment include:
    - Any symptomatic aneurysm.
    - Asymptomatic aneurysms >2–3 cm; or
    - Any splenic artery (SA) or visceral aneurysm in a female of childbearing age.
    - Open surgical alternatives are less commonly used, but include **ligation of a proximal splenic artery aneurysm, excision of a mid-segment aneurysm, or resection and splenectomy for distal SA aneurysms.**
- **VASCULITIS**
  - An autoimmune attack on all sizes of arteries, producing some combination of pain at the site of arterial involvement, distal ischemia, and systemic symptoms (deriving from the inflammatory response). *See Figure 30.23.*
  - **Takayasu's arteritis** involves the aorta, coronaries, carotids, and pulmonary arteries. It is an **intimal process** that results in occlusions at mid-portions of vessels, not branch points. The typical patient is an Asian female with ischemic symptomatology, pain over the artery site, and systemic symptoms. Treatment is with glucocorticoids. Balloon angioplasty is not recommended for stenotic inflammatory lesions.
  - **Temporal arteritis** is an aortic and carotid arteritis with **granulomas and giant cells in all layers of the vascular wall.** This leads to both aneurysm formation and luminal occlusion. Patients are typically older women with headaches, ischemic symptoms in the external or internal carotid artery distribution (especially the ophthalmic artery), and tenderness at the temporal artery. Treatment is with empiric glucocorticoids followed by a temporal artery biopsy for confirmation.

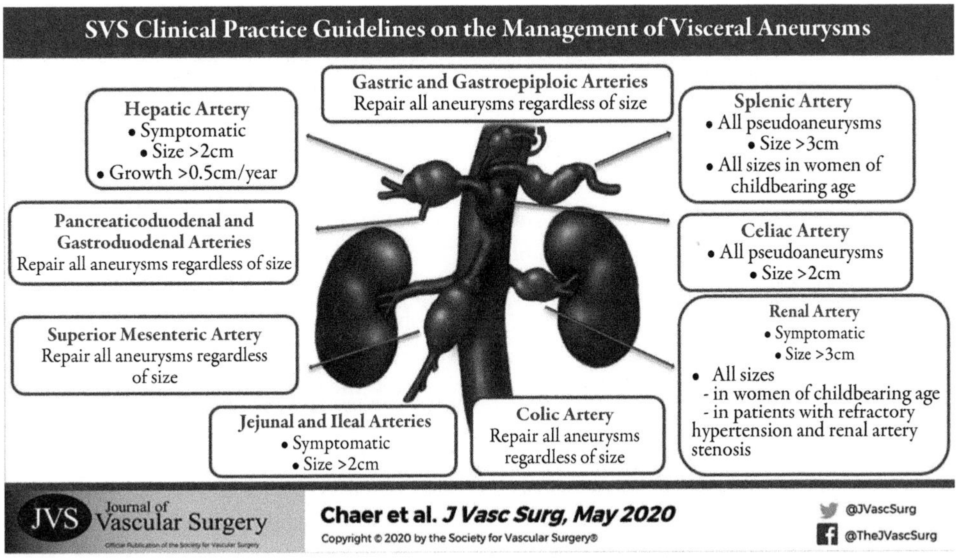

Figure 30.22 Visceral artery aneurysms.

- **Polyarteritis nodosa** is a necrotizing vasculitis of small and medium-sized arteries in viscera, renals, and skin, which produces local and systemic symptoms. Treatment is with glucocorticoids.
- **Kawasaki's** arteritis is fibrinoid necrosis of medium-sized arteries (coronaries) in an infant. Treatment is with glucocorticoids, intravenous immunoglobulin (IVIG), and low-dose aspirin
- **Hypersensitivity angiitis** is a type 2 or 3 allergic reaction of distal vasculature producing a skin rash and systemic symptoms. Treatment is with calcium channel blockers (CCB), pentoxifylline, and stopping the offending agent.
- **Bechet's** disease is a lymphocytic vasa vorum degeneration producing oral and genitourinary ulcers, DVTs, and superficial thrombophlebitis.

- **FIBROMUSCULAR DYSPLASIA (FMD)**
- This describes fibroplasia of the media of arterial walls, with dysplasia more common than hyperplasia. Only 5% of cases involve the intima.

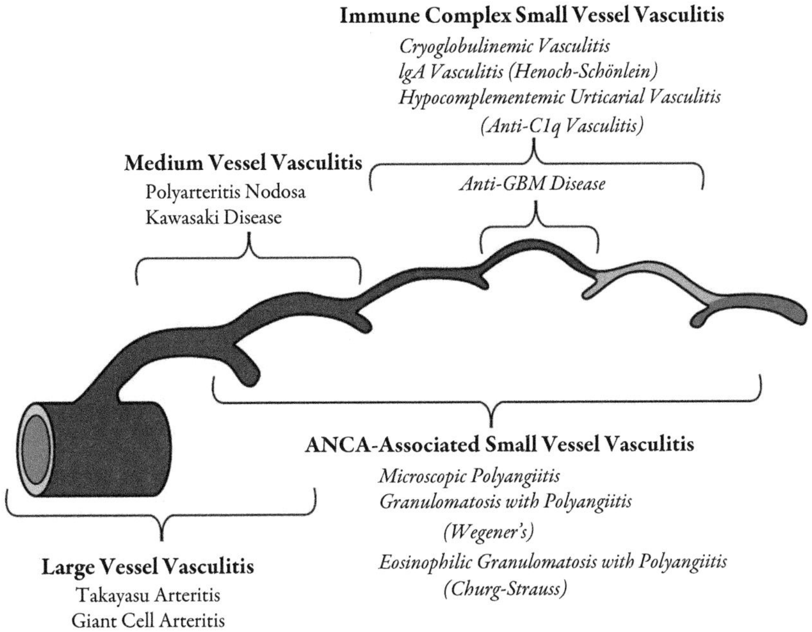

Figure 30.23 Types of vasculitis.

Figure 31.1

- Carotid arteries, renal arteries, and iliacs are the main vessels affected. Subclavian and mesenteric vessels are not normally involved.
- **DIAGNOSIS**
  - Patients are generally 30- to 40-year-old females presenting in one of three ways:
    - 1. TIA or CVA symptoms due to **CCA or ICA disease.** Bilateral distal ICA lesions are common. One-fourth of patients have CNS aneurysms.
    - 2. Hypertension due to **distal renal artery** involvement causing hypoperfusion and increased juxtaglomerular apparatus (JGA) renin production.
    - 3. **Iliac artery** involvement producing ischemic thigh or buttock symptoms. This is the least common presentation.
    - Confirmation is with CTA or formal angiogram showing a string of beads sign. MRA can also detect stenoses.
- **TREATMENT**
  - **Antiplatelet medication** if the patient is asymptomatic. **PTBA** without stenting is used for symptomatic lesions.
- **BUERGER'S THROMBOANGIITIS OBLITERANS**
  - Presents as distal extremity ischemia in a smoking male <45 years old with corkscrew collaterals, inflammatory cells confined to the intima, and impaired arteriolar relaxation with subsequent thrombosis.
  - Patients may have superficial thrombophlebitis of venules. This is a diagnosis of exclusion.
  - Treatment is with smoking cessation, prostaglandins, phosphodiesterase inhibitors, antiplatelets, and anti-endothelins.
- **PRIMARY RAYNAUD'S**
  - Patients have characteristic coolness, color change, and numbness of digits upon exposure to cold. It is generally bilateral and without tissue loss. There should be a negative capillary exam, ESR, and serology.
  - Treatment of primary Raynaud's is with avoidance of triggers and starting nifedipine, or sildenafil as second line. Fluoxetine has also been used.
    - Surgical options include video assisted thoracoscopic cervical sympathectomy or lumbar sympathectomy (surgical or chemical), or digital sympathectomy by removing the adventitia after a test dose of amide analgesic.
    - **Secondary Raynaud's** may have a positive capillary exam or other indices of an underlying autoimmune process or upstream arterial pathology. Treatment should be directed at the primary process.
- **RADIATION ARTERITIS**
  - This is secondary to injury of the most sensitive area in the vessel wall, the endothelial cell. Early sloughing and thrombosis is followed by late fibrosis at 1 to 10 years post-treatment. Advanced atherosclerosis can occur decades later.

31.

UROLOGY AND GYNECOLOGY

RENAL

- The renal vein lies anterior to the renal artery. The latter has 5 segmental branches. The renal pelvis collects all urine via calyces, and is the most posterior of the three structures.
- The right renal artery runs posterior to the inferior vena cava (IVC).
- The left renal vein crosses anterior to the aorta, and its branches include gonadal, ascending lumbar, and adrenal-inferior phrenic veins.
- **Renal cell carcinoma** (**RCC**) arises from proximal convoluted tubule (PCT) cells. It is secondary to smoking.
  - Presentation is most commonly as an incidental mass seen on imaging performed for another reason. Alternatively, large tumors may exert mass effect or present with hematuria. Paraneoplastic presenting features include erythrocytosis from erythropoietin production, hypoglycemia from insulin, hypertension from renin, hypercalcemia from parathyroid hormone-related protein (PTHrP), and Cushing's syndrome from excessive ACTH.
    - Tumors may be multifocal or recurrent in the setting of an underlying syndrome like **Von Hippel-Lindau (VHL).** These patients may also have cysts, hemangioblastomas of the cerebellum or retina, and pheochromocytoma.
    - **CTAP** defines the extent of local disease. A T2 lesion is >7 cm, T3 involves the renal veins or perinephric tissue, and T4 extends beyond Gerota's fascia. *See Figure 31.1.* **No biopsy** is needed; imaging characteristics are sufficient. Lymph nodes involvement qualifies as N1 status, or Stage 3.
    - One-third of patients have metastases at the time of diagnosis, most commonly to the lung. **PET imaging** is an important part of the workup.
  - Management is with **radical nephrectomy,** which includes the adrenal gland, Gerota's fascia, and adjacent lymph nodes. Tumor thrombus within the renal vein and IVC can be extracted and followed by venous reconstruction if needed. Adjunctive therapy includes **TK inhibitors,** IFN alpha, and IL-2. Partial nephrectomies can be considered if the patient would otherwise be hemodialysis (HD) dependent with a total nephrectomy.
- Other tumors include **transitional cell carcinoma of the renal pelvis** (which requires resection with a ureter and bladder cuff), **metastases** from breast cancer, and **angiomyolipomas** (benign hamartomatous growths associated with tuberous sclerosis).
- Renal inflammation can be from **pyelonephritis (WBC casts in urine), glomerulonephritis (RBC casts in urine), or interstitial nephritis (arthralgias and serum eosinophilia).** Granular casts are seen in acute tubular necrosis (ATN).
- **Nephroureterolithiasis** is most commonly calcium-oxalate. Steatorrhea binds calcium intra-intestinally and allows oxalate to be absorbed into the bloodstream. **Struvite stones** (magnesium ammonium phosphate) are staghorn in shape and occupy the renal calyces and pelvis. They are the result of urease-producing proteus infections in an alkaline urine. **Uric acid stones** can be seen in patients with an ileostomy or myeloproliferative disorder. Congenital **cysteine** stones are in acidic urine.
  - Alpha blockers or calcium channel blockers can facilitate passage of stones; however, those greater than 6 mm are less likely to pass.
  - If medical management fails, options include extracorporeal shock wave lithotripsy (ESWL), ureteroscopy, stent, percutaneous nephrostomy, or open nephrolithotomy.
- See the Chapter 23, Trauma, for renal injuries.

RENAL TRANSPLANTATION

- The kidney is one of the most commonly transplanted organs. Recipient diabetic and vascular nephropathy causing end-stage renal disease (ESRD) are the main etiologies to failure. It is a heterotopic transplant, since the native kidneys are left in situ, and the donor organ is placed at a separate site in the pelvis.
- Most transplants are deceased donor renal transplants (DDRT) with the minority being living donor renal transplants (LDRT). In the latter, the organ is procured as a laparoscopic left nephrectomy. Donor complications include wound infection and pulmonary embolus.

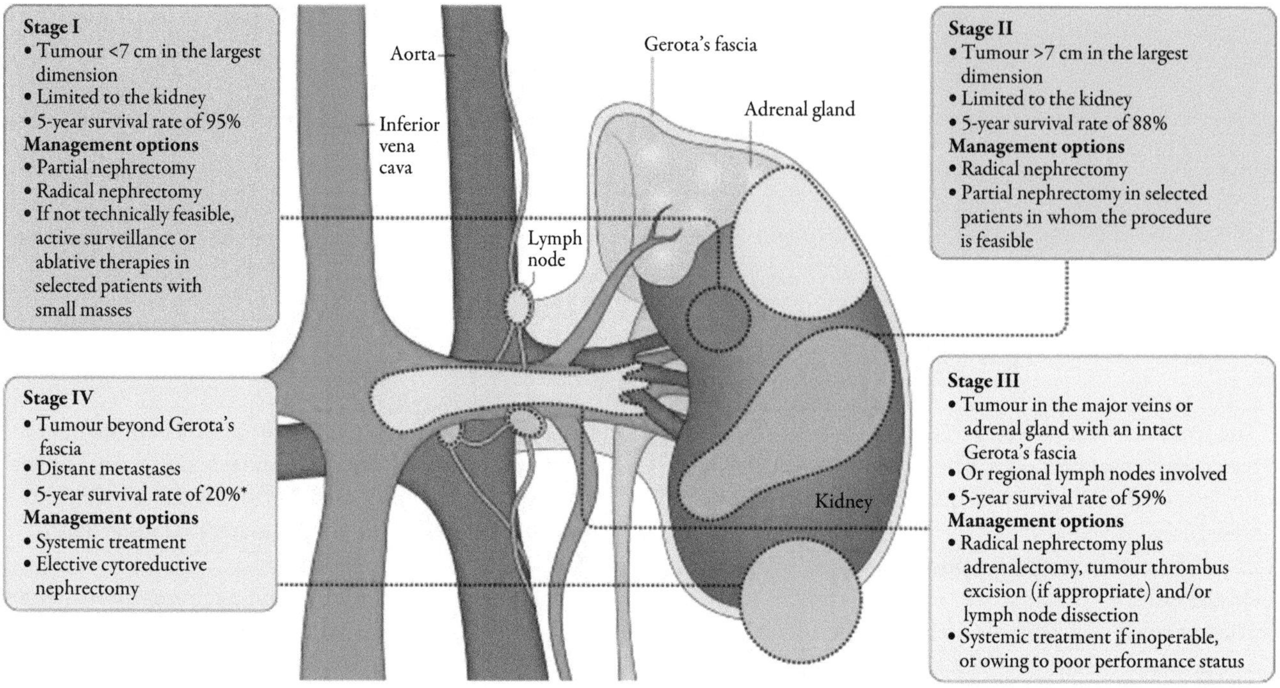

Figure 31.1 Stages of renal cell carcinoma.

- Transplants may be performed in spite of a donor UTI or a donor increase in creatinine up to 3. Viral infections like HBV, HCV, CMV, EBV are also not absolute contraindications. An HIV-positive organ can still be used provided the recipient is also positive, and the viral load is zero with a CD4 >200.
- The recipient must be in good cardiovascular health, since **postoperative mortality is from stroke and myocardial infarction (MI).** Recipients also should be free of low-grade cancer for 2 years and high-grade cancer for 5 years.
- Once explanted, the organ can be stored for up to **48 hours of cold ischemic time** in a lactobionate cold solution. The organ is prepared on back table, and periureteric tissue is preserved since this carries donor ureter wall perfusion from the hilum.
- Arterial and venous anastomoses are performed to the external iliac artery (EIA) and vein EIV. This is followed by a ureteroneocystostomy anastomosis with or without a stent in place. *See Figure 31.2.*
- **Complications** are generally detected by performing a **graft ultrasound** for the workup of a low urine output after excluding an obstructed Foley catheter and treating any hypovolemia.
  - **Urinoma** can be treated with a stent and drainage.
  - **Lymphocele** may compress the ureter. Treatment is with drainage with or without sclerosis. A peritoneal window can be performed if refractory to drainage.

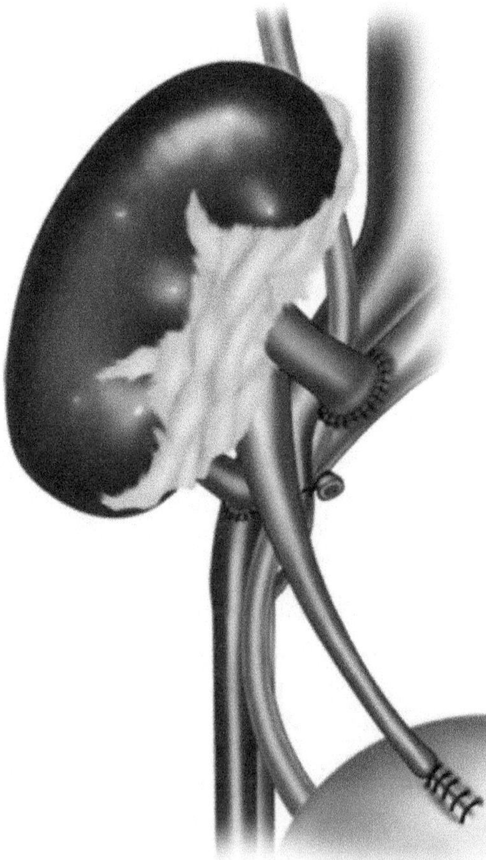

Figure 31.2 Anatomy of renal transplantation.

- **Renal artery anastomotic stenosis** is diagnosed with an ultrasound (US) and treated with a stent. Acute thrombosis occurs in <1%, and a chronic stricture can be secondary to ischemia.
- **Renal vein thrombosis** can present with graft congestion and proteinuria or hematuria. Treatment is with thrombectomy, anticoagulation, and re-transplantation if needed.
- **Post-op acute tubular necrosis (ATN)** can present as oliguria and an elevated creatinine. There are no diagnostic US findings. If the patient requires HD within the first week post-op, this is termed "delayed graft function" and may have been secondary to prolonged ischemic time.
- Post-op **hyperglycemia** is due to tacrolimus, cyclosporine, or glucocorticoids.
- Post-op viral infections can be treated with ganciclovir for CMV or acyclovir for HSV or VZV.
- **Outcomes**
  - **90% of deceased donor renal transplants (DDRT) and 95% of living donor renal transplants (LDRT) are functional at 1 year.**
  - A transplant recipient lifespan increases to 20 years from approximately 10 years when on hemodialysis (HD).
  - Some recipients need a native nephrectomy for chronic, recurrent pyelonephritis, vesicoureteral reflux (VUR), or refractory renovascular origin hypertension.
  - Patients transplanted for focal segmental glomerulosclerosis (FSGS) are prone to early post-op recurrences manifesting as proteinuria. Treatment is with plasmapheresis.

URETER

See Figure 31.3.

- **Proximal injuries** should be primarily repaired with **absorbable** sutures in a spatulated fashion over a stent. If missing more than a 2 cm segment, then ligation and percutaneous nephrostomy should be done if the patient is critically ill from multisystem trauma. Otherwise, a urologist can be involved for transureteroureterostomy, ileal substitution, or renal auto-transplantation with ureteroneocystostomy in stable patients.
- **Distal injuries** at or below the common iliac bifurcation are generally re-implanted as a ureteroneocystostomy with or without a Boari flap bladder cuff or psoas muscle hitch.
- Periureteral soft tissue should be preserved during dissection, so as not to limit perfusion.
- All repairs should be over a ureteral stent and receive drains and a Foley catheter to prevent urine reflux from the bladder.

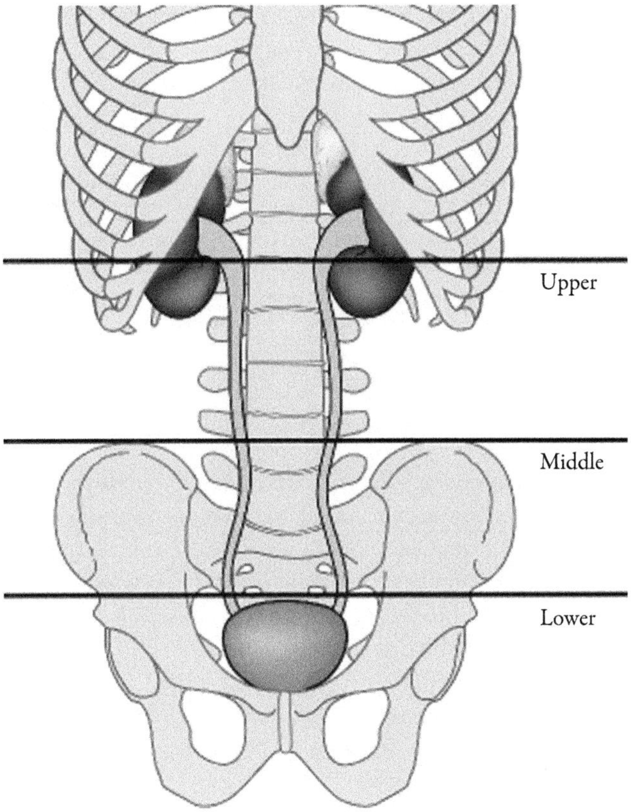

Figure 31.3 Sectors of potential ureteral injury.

- Ureteral duplications and ureteroceles are treated with resection and ureteroneocystostomy.

BLADDER

- **Transitional cell cancer** is the most common. It presents as painless hematuria in a male smoker. Diagnosis is by cystoscopy and CT imaging if believed to be extensive. Treatment is with intravesicular BCG and endoscopic resection if T1. Cystectomy with ileal conduit (or a continent reservoir) is needed for higher T stage lesions. Chemotherapy and radiation are also added if T2 or greater. Squamous cell carcinoma is less common than transitional.
- **Neurogenic upper motor neuron (UMN) bladder dysfunction** is from pathology above T12, leading to spastic incontinence. Partial bladder outlet obstruction with an active detrusor muscle can produce a similar type of incontinence.
- **Flaccid lower motor neuron (LMN) bladder dysfunction** is from pathology below L1 or direct involvement of the pelvic parasympathetic nerves. This leads to overflow incontinence. A similar pattern of incontinence can result from bladder outlet obstruction with a weakened detrusor muscle.

- **Stress incontinence** is from loss of the urethral angle and sphincter mechanism. Treatment is with Kegel exercise and urethral or pubovaginal suspension procedures.
- Patent urachus with or without a cyst is managed with resection of tract, closure of bladder dome, and relief of bladder outlet obstruction.

MALE GENITOURINARY

See Figure 31.4.

- **PROSTATE**
  - **Benign prostatic hypertrophy (BPH)** develops in the prostatic transition zone. Symptoms include nocturia, frequency, hesitancy, a weak urinary stream, retention, and overflow incontinence. Treatment is with **alpha blockers, 5 alpha reductase inhibitors, or transurethral resection (TURP).**
  - **Adenocarcinoma** is in the posterior lobe.
    - An elevated prostate specific antigen (PSA) raises suspicion of hte diagnosis. Confirmation is with transrectal biopsy followed by CT chest and abdomen/pelvis to evaluate the local extent of disease and any lung or osteoblastic bone metastases.
    - Localized disease is treated with **radiation or radical prostatectomy,** including the seminal vesicles and lymph nodes. Extracapsular or metastatic disease is treated with **radiation and androgen deprivation.** The latter includes leuprolide for LHRH blockade, flutamide for testosterone blockade, finasteride for 5 alpha reductase inhibition, or bilateral orchiectomy. Observation is an option for elderly patients with low-grade tumors.
      - Post-treatment **surveillance** is with PSA levels, which should be zero if a resection was done. An elevated PSA may represent residual primary site disease, metastases, or just BPH or prostatitis.
    - **Screening** with PSA can begin at 50 years of age or 40 years if high risk. Elevated alkaline phosphatase can be seen in tumors with capsular invasion.
- **TESTICLE**
  - **Germ cell cancer** is either seminomatous or nonseminomatous. An undescended testicle is at increased risk for a germ cell tumor.
    - Diagnosis is by ultrasound (US) of a mass noted on physical exam. This is followed by chest and abdomen/pelvis CT for metastases to the para-aortic area, retroperitoneum, or lungs.
    - Management is with **orchiectomy via an inguinal incision.**
      - **Seminoma** is the primary subtype. **Retroperitoneal radiation (XRT)** should be given after orchiectomy. **Chemotherapy** follows if there is bulky residual retroperitoneal, para-aortic nodal burden, or metastases. Retroperitoneal lymph node dissection (RPLND) can be used for persistent nodes despite the above.
      - **Nonseminomatous subtypes** include teratoma, embryonal carcinoma, choriocarcinoma (produces BHCG), and yolk sac tumor (produces AFP). **Retroperitoneal lymph node dissection (RPLND)** should follow orchiectomy. **Chemotherapy** may be added. Repeat surgical resection is reserved for residual disease after the above.
  - **Torsion** occurs toward the midline. Extravaginal torsion is in neonates. It occurs outside the tunica vaginalis, and includes the cord. Intravaginal torsion is in adolescents. Treatment is with bedside de-torsion and then operative exploration for full de-torsion and bilateral orchiopexy. Orchiectomy should be done if the testicle is necrotic at the time of exploration.
  - **Epididymitis** is believed to result from increased abdominal straining and urine refluxing from the bladder through the vas deferens. This causes chemical inflammation of the epididymis, which can result in secondary bacterial infection. Sexually transmitted organisms are common in a young adult and/or coliform bacteria in an older patient. Treatment is with antibiotics aimed at the perceived bacterial source.
  - **Spermatoceles** are cystic outpouchings of the epididymis and are treated with resection.
  - **Varicoceles** are more common on the left, and are due to left renal vein thrombosis extending into the gonadal vein. A varicocele on exam should prompt renal imaging for tumor or vein thrombus.
  - A **hydrocele** is a residual epithelial-lined encasement of the testicle.
    - If there is a **patent processus vaginalis,** there is communication with the peritoneal cavity. The hydrocele will fluctuate in size, and it will need **resection.**
    - If it is **non-communicating,** it will not fluctuate in size. Non-communicating hydroceles that persist past 1 year of age will need resection. Aspiration and sclerosis is an alternative treatment.
- **URETHRA AND PENIS**
  - A **urethral injury** should be suspected after a straddle mechanism, pubic diastasis or rami fracture, or a high-riding prostate on digital rectal exam. Blood may be seen at the urethral meatus, and the patient may have urinary retention.
    - A **retrograde urethrogram** x-ray at bedside can reveal a disruption at the prostatic, membranous, or penile urethra.
    - **Suprapubic cystostomy** for diversion with delayed repair is the most common treatment for blunt injuries causing complete transection.

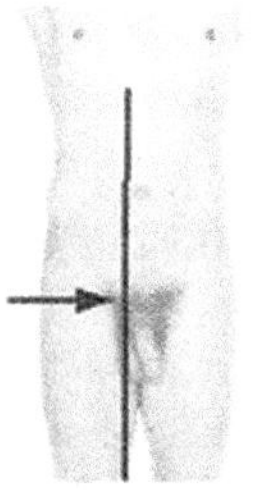

Prostate and seminal vesicles are examined rectally through palpation and through transrectal sonography (fig. 5.250). The rectal fascia (waldeyer's fascia) and the rectoprostatic fascia (denonvilliers' fascia) are important in staging the spread of rectal and prostate cancer. A surgical approach to the prostate can be done by suprapubic access extraperitoneally, retropubically through retzius' space, transversically (through the bladder), or transperoneally.

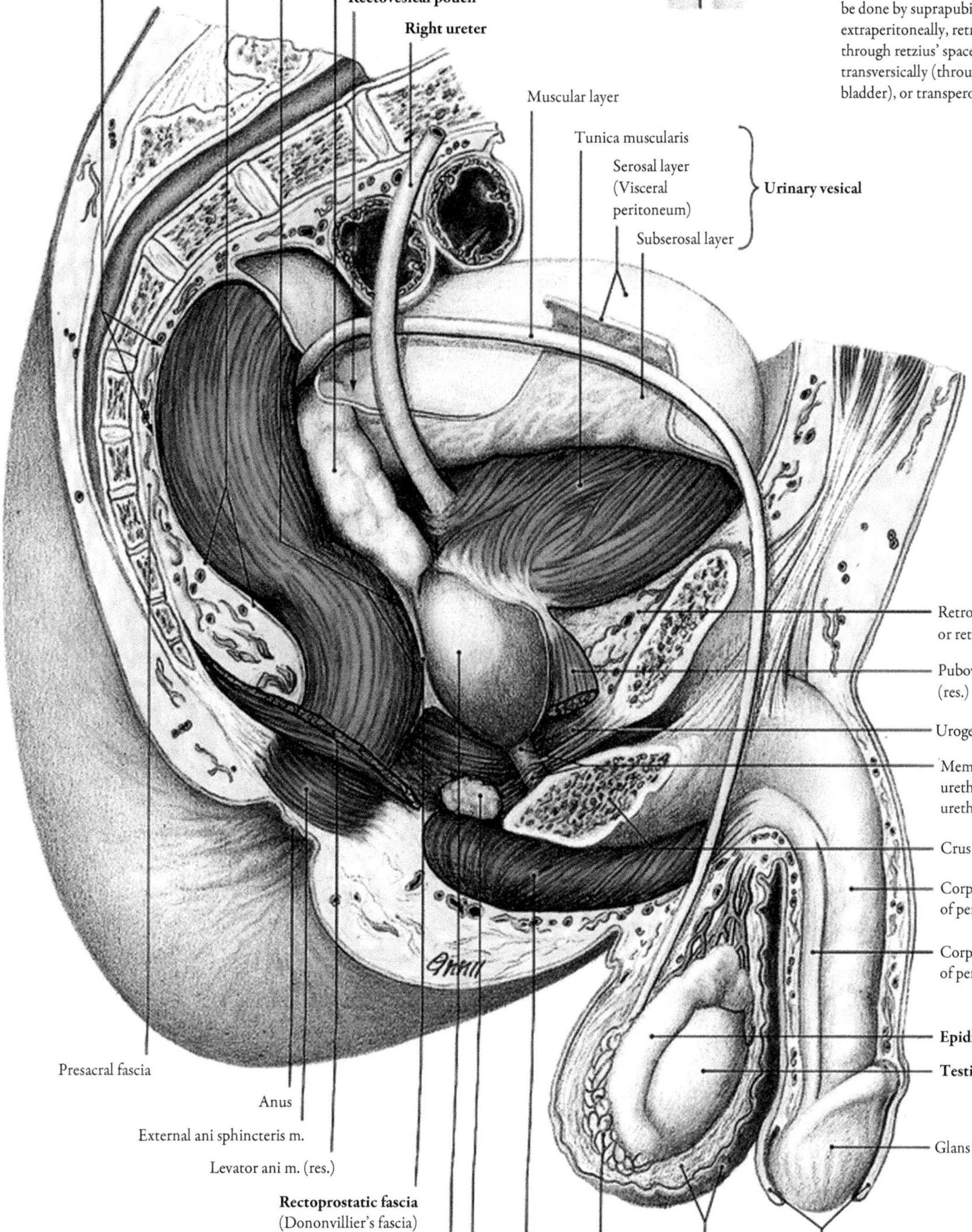

Figure 31.4 Male genitourinary anatomy.

- Urology may choose to stent with a Foley catheter if it is a contusion, stretch injury, or only partial thickness. Healing may then ensue around the catheter. They can consider primary repair, in addition to suprapubic diversion, if it is an isolated penetrating injury to the urethra.
- **Paraphimosis** should be reduced, and **phimosis** with urinary retention is treated with a dorsal slit.
- **High-flow priapism** is treated with aspiration of the corpus cavernosum, irrigation of the corpus, injection of an alpha agonist, and possible shunt construction between the corpus and glans. **Low-flow priapism** in sickle cell disease is treated with exchange transfusion to prevent venular occlusion from sickling.
- **Squamous cell carcinoma** of the penis is treated with resection and 2 cm margins.

GYNECOLOGY

See Figure 31.5.

- The urogenital diaphragm, bulbocavernosus muscle, and the perineal membrane are separated from the external anal sphincter by the perineal body and transverse muscle.
- Cardinal and uterosacral ligaments hold the cervix and vagina in place.
- The infundibulopelvic ligament contains ovarian vessels, and the broad ligament contains uterine vessels.
- The **posterior internal iliac artery division** is primarily the superior gluteal artery that exits above the piriformis muscle
- The **anterior internal iliac artery division** bifurcates into superior vesicular, obturator, middle rectal, and inferior gluteal. The latter traverses posteriorly, below the pyriformis muscle.
- **PREGNANCY**
  - Is associated with a number of physiologic changes including increased blood volume, cardiac output, glomerular filtration rate (GFR), thrombosis, pelvic vein size, and minute ventilation (respiratory rate and tidal volume are both increased). There is a decrease in systemic vascular resistance (SVR), functional residual capacity (FRC), and gastrointestinal (GI) motility.
  - A gestational sac should be visible on ultrasound (US) at a bHCG level of 1500, and the fetal pole at a bHCG level of 6000. If an intrauterine gestational sac is not visualized above these thresholds, one should consider an ectopic location of the pregnancy.
  - **Ectopic pregnancy** can be in the fallopian tube, cornu of the uterus, or intraperitoneal. Treatment is with **methotrexate (MTX)** provided the gestation is not ruptured, is <3 cm, bHCG is low, and there is no cardiac activity on US. Patients who are breastfeeding should not receive MTX. Otherwise the traditional treatment is with **salpingostomy or salpingectomy** if the above criteria are not met.
  - **Missed abortion** has a closed os with a sac on US, but no heart rate. **Threatened abortion** is the same,

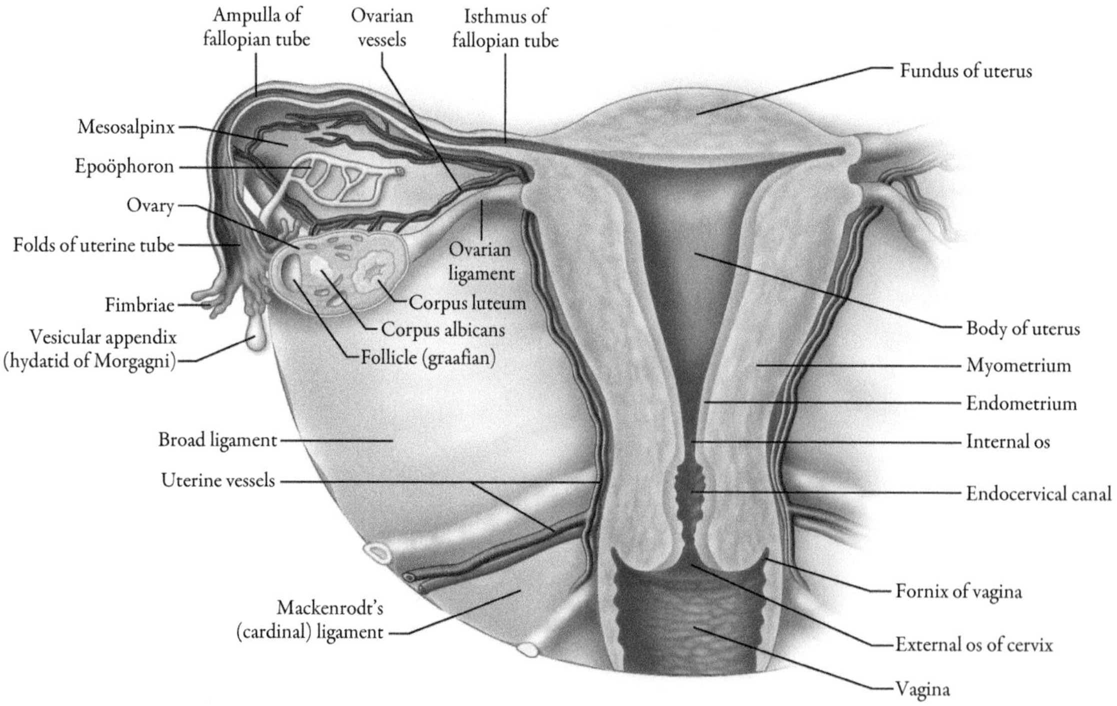

Figure 31.5 Uterine and adnexal anatomy.

but a heartbeat is noted. An incomplete abortion is in progress, and can have tissue protruding through the cervical os.

- **Postpartum hemorrhage** is treated with oxytocin, fundal massage, endometrial curettage, luminal tamponade, B-Lynch or O'Leary sutures, uterine artery ligation or embolization, or total abdominal hysterectomy (TAH).
- **Gestational trophoblastic disease** describes the absence of a fetus and the presence of excessive trophoblastic tissue instead. There may be associated bilateral theca-lutein adnexal cysts. Categories include:
  - A hydatid mole respects the endometrial and myometrial borders.
  - An invasive mole or choriocarcinoma are both more aggressive locally, and the latter may metastasize.
  - Treatment is with dilation and curettage, evacuation, or extraction. Methotrexate is also an option.

- PELVIC CAVITY PATHOLOGY
  - **Endometriosis** is ectopic, benign endometrial tissue seen in 10%–20% of the population. Most often, it is incidentally noted. The laparoscopic appearance does not necessarily correlate with symptom severity. Ectopic endometrial tissue may also be seen at the umbilicus or in areas of old scar. Symptoms include chronic pain, hematochezia, painful defecation, hematuria, and infertility. Treatment is with oral contraceptives (OCPs), NSAIDs, progestin, gonadotropin-releasing hormone (GNRH), or laparoscopic ablation or resection of lesions.
  - **Pelvic inflammatory disease (PID)** most commonly starts post-menstruation, during the early half of the cycle when the cervical mucosal barrier is weakest. Chlamydial infection is more common than gonorrhea. Symptoms include abdominal pain, discharge, and cervical motion tenderness. Treatment is with ceftriaxone and doxycycline.
  - **Tubo-ovarian abscess (TOA)** has focal tenderness and may persist despite oral antibiotics. Treatment is with IV antibiotics, percutaneous drainage, laparoscopic or open drainage, or excision of the tube and ovary if needed.
  - **Gonadal or pelvic vein thrombosis** is secondary to malignancy, infection, trauma, or the postpartum state. Treatment is with anticoagulation. Antibiotics should be added if the thrombus is infected (septic pelvic thrombophlebitis).
  - **Pelvic floor dysfunction** is diagnosed by cystoscopy, urodynamics, fluoroscopy, defecography, manometry, endorectal US, or dynamic MRI. Treatment is with anterior or posterior colporrhaphy, fixation of the vagina to the sacrospinous ligament via a perineal approach, or to the uterosacral ligament or sacrum with mesh via a transperitoneal approach.
  - **Stress urinary incontinence** is treated with anterior vaginal suspension to Cooper's ligament, transvaginal tape sling with polypropylene mesh through the obturator foramen, or urethral bulking agent injections.
- VULVA
  - **Squamous cell carcinoma (SCC)** occurs in nulliparous, obese, smokers with HTN, DM, or chronic glucocorticoid use. Lesions can be secondary to lichen sclerosis, lichen simplex chronicus, Paget's disease, or vulvar intraepithelial neoplasia (VIN). Patients present with a unilateral mass, ulcer, or pruritus. Younger patients with multifocal lesions may have underlying HPV 16 or 18. Treatment is **wide local excision (WLE)** with 1–2 cm margins if the lesion is under 2 cm in size, <1 mm depth invasion, and there is no lymphadenopathy. An **ipsilateral lymph node dissection** can be added for lesions that are >1 mm depth or those with clinically positive nodes. **Bilateral vulvectomy, lymphadenectomy, and possible radiation (XRT)** is considered for lesions greater than 2 cm.
  - **Paget's disease and VIN** are pre-malignant lesions diagnosed by exam and acetowhite stain. Treatment is with topical 5FU, laser ablation, or WLE.
  - **Bartholin cyst** is due to an obstructed gland that normally produces mucous for moisture. If infected, treatment is with incision and drainage, Word catheter, or marsupialization. Excision and biopsy are for older patients with concern for malignancy.
  - **Vulvitis** may be contact (irritant or allergic) or leukoplakia from lichen simplex chronicus or lichen planus. Treatment is with topical steroids.
- VAGINA AND CERVIX
  - **Human papilloma virus (HPV)** is the most common sexually transmitted disease (STD). Serotypes 6–11 cause condyloma accuminata. Perineal skin, vulvar, vaginal, and cervical biopsies can exclude dysplasia, neoplasia, or carcinoma in situ. The latter is associated with HPV strains 16 and 18. Treatment of HPV without dysplasia or neoplasia is **topical imiquimod, 5FU, ablation, or excision.** The HPV vaccine works against serotypes 6, 11, 16, and 18. It is used as primary prophylaxis for females 9–26 years of age and males 11–12 years. See cervical section below for treatment of CIN, carcinoma in situ, and invasive carcinoma.
  - **Vaginal squamous cell carcinoma (vSCC)** in the upper one-third of the vaginal canal is managed like cervical cancer, and if in the lower one-third it is approached like vulvar cancer. Symptoms include bleeding or discharge. Lesions are Stage 3 if the pelvic sidewall is involved, Stage 4 if there is spread throughout the pelvis or distant metastases.

Chemotherapy, radiation, or brace therapy are used for advanced disease.

- **Vesicovaginal fistulae** are diagnosed by instilling methylene blue into the bladder and noting any staining of a vaginal tampon. **Ureterovaginal** fistulae are diagnosed by giving phenazopyridine orally and noting any staining of a vaginal tampon.
- **Rectovaginal fistulae** are diagnosed by instilling rectal methylene blue and noting any staining of a vaginal tampon. Rigid proctoscopy can provide an exact location and also evaluate the health of the rectal mucosa overall.
- **Cervical squamous cell carcinoma (cSCC)** occurs in active smokers with multiple sexual partners.
  - Patients can be screened with **pap smears every 3 years** after the initiation of sexual activity. They should also be screened for HPV. **Colposcopy and biopsy** of acetowhite lesions is performed to follow up positive smears. This helps determine if there is cervical intraepithelial neoplasia, carcinoma in situ, or actual invasive SCC.
  - If **cervical intraepithelial neoplasia (CIN)** is found, treatment depends on the grade. CIN1 can be observed and undergo serial colposcopies and biopsies. CIN 2 or 3 may need a **loop electrodiathermy excision procedure (LEEP) or a cold knife cone resection (CKC).** Carcinoma in situ should receive CKC.
    - If invasive SCC is found, it should be staged by clinical exam and US, CT, or MRI imaging. Treatment is **total abdominal hysterectomy (TAH)** for micro-invasive disease. **Radical hysterectomy** is for disease limited to the cervix or upper two-thirds of the vagina. **Chemoradiation** is for Stage 3b with involvement of the lower one-third of the vagina and pelvic sidewall. **Pelvic exenteration** with urinary and colon diversion is an option for locally recurrent disease after an initial resection.
- **Benign cervical lesions** include endocervical polyps or nabothian cysts from gland obstruction.

- UTERUS
  - **Abnormal endometrial bleeding** is treated with hormonal therapy (clomiphene) if the patient is under 35 years old. Otherwise, biopsy should be done to exclude hyperplasia, atypia, or carcinoma if the patient is older.
  - **Endometrial hyperplasia** is treated with progestins, serial biopsies, and a hysterectomy for any atypia noted.
  - **Endometrial adenocarcinoma** is the most common gynecologic malignancy. The tumor spreads locally. Type 1 is estrogen positive, and type 2 is estrogen receptor negative and more aggressive. Risk factors include nulliparity, a late first pregnancy, obesity, estrogen use, diabetes mellitus, and Lynch II syndrome. Smoking is not a risk factor.
    - Patients have abnormal uterine bleeding. **Endometrial biopsy** via suction catheter is confirmatory.
    - Staging and treatment is with a total abdominal hysterectomy and bilateral salpingo-oophorectomy **(TAH-BSO) plus pelvic and aortic lymphadenectomy, and pelvic washings** with or without radiation. There is no omentectomy. A **radical hysterectomy** takes parametrial tissue lateral to the ureters, uterine arteries at their origin from the anterior division of the internal iliac artery, uterosacral ligaments near the rectum, and the upper one-third of the vagina. Risk reduction TAH-BSO is an option for postmenopausal women with Lynch II.
  - **Adenomyosis** describes endometrial tissue that is embedded in myometrium causing dysmenorrhea, or meno- or metrorrhagia. Treatment is with a hysterectomy.
  - **Leiomyomas** are benign and generally present as pain or abnormal vaginal bleeding. They can occur with increased frequency in pregnancy, during which time they can cause a miscarriage, intrauterine growth retardation (IUGR), or preterm labor. Treatment is with oral contraceptive pills (OCP), progesterone, gonadotropin-releasing hormone (GNRH), myomectomy, or a hysterectomy with or without preoperative embolization. **Leiomyosarcoma** are treated with TAH-BSO without lymphadenectomy.
  - **Endometrial polyps** are most often benign. Diagnosis is by a hysterosonography hysterosalpingogram, or hysteroscopy.
- ADNEXA
  - **Ovarian cancer** is the leading cause of gynecologic death, often due to bowel obstruction.
    - Risk for ovarian malignancy increases with the number of **menstrual cycles** (nulliparity, early menarche, late menopause), Lynch syndrome, or BRCA 1 mutations.
    - Oral contraceptives (OCP), pregnancy, and breastfeeding are protective.
    - **Epithelial tumors** are mostly benign. Carcinoma can be serous, mucinous, or papillary. CA-125 may be elevated.
    - **Stromal** tumor or sarcoma includes granulosa-theca (serum marker inhibin), fibroma-thecoma with or without ascites, or hydrothorax (Meige's syndrome), or Sertoli-Leydig cell tumors. Most are benign, and there is no nodal spread.
    - **Germ cell** tumors include teratoma, dysgerminoma (serum marker LDH), yolk sac (serum marker AFP), embryonal, and choriocarcinoma (serum marker HCG).
    - **Krukenberg tumors** are signet ring cell gastric adenocarcinoma metastatic to the ovary.
    - Diagnosis is by pelvic ultrasound and CT imaging. Screening for high-risk patients with a familial or genetic syndrome is via yearly CA-125 and pelvic US

with or without breast exams and breast imaging via mammogram or MRI.

  - Stage 1 ovary cancers are limited to one or both ovaries. Stage 2 is limited to the pelvis. Stage 3 is throughout the abdomen (the majority of epithelial tumors are in this category at diagnosis). Stage 4 has distant metastases.
- Treatment is **TAH-BSO if Stage 1, and may include pelvic and para-aortic lymphadenectomy, pelvic washings, and omentectomy.**
  - If Stage 2 or greater, **neoadjuvant or adjuvant chemotherapy** may be added.
  - Patients who can be debulked to yield deposits <1 cm may be candidates for **intraperitoneal chemotherapy.**
  - Risk reduction oophorectomy can be considered for a genetic syndrome, BRCA 1 and Lynch II.
- An **incidental ovarian mass** seen at the time of laparoscopy or laparotomy may undergo biopsy as well as omental biopsy and pelvic washings.
- Ovarian masses that have resulted in **acute torsion** should be de-torsed and removed while sparing the ovary. If the ovary or tube are not viable, or the mass is very concerning for malignancy, the tube and ovary can be excised as well.
- **Ovarian cysts** should be removed if postmenopausal and complex in appearance, or premenopausal and a high suspicion for malignancy exists. Simple cysts may be observed.